Advances in Intelligent Systems and Computing

Volume 1247

The series "Advances in Intelligent Systems and Computing" contains publications on theory, applications, and design methods of Intelligent Systems and Intelligent Computing. Virtually all disciplines such as engineering, natural sciences, computer and information science, ICT, economics, business, e-commerce, environment, healthcare, life science are covered. The list of topics spans all the areas of modern intelligent systems and computing such as: computational intelligence, soft computing including neural networks, fuzzy systems, evolutionary computing and the fusion of these paradigms, social intelligence, ambient intelligence, computational neuroscience, artificial life, virtual worlds and society, cognitive science and systems, Perception and Vision, DNA and immune based systems, self-organizing and adaptive systems, e-Learning and teaching, human-centered and human-centric computing, recommender systems, intelligent control, robotics and mechatronics including human-machine teaming, knowledge-based paradigms, learning paradigms, machine ethics, intelligent data analysis, knowledge management, intelligent agents, intelligent decision making and support, intelligent network security, trust management, interactive entertainment, Web intelligence and multimedia.

The publications within "Advances in Intelligent Systems and Computing" are primarily proceedings of important conferences, symposia and congresses. They cover significant recent developments in the field, both of a foundational and applicable character. An important characteristic feature of the series is the short publication time and world-wide distribution. This permits a rapid and broad dissemination of research results.

**** Indexing: The books of this series are submitted to ISI Proceedings, EI-Compendex, DBLP, SCOPUS, Google Scholar and Springerlink ****

More information about this series at http://www.springer.com/series/11156

Zhengbing Hu · Sergey Petoukhov ·
Ivan Dychka · Matthew He
Editors

Advances in Computer Science for Engineering and Education III

Editors
Zhengbing Hu
School of Educational Information Technology
Central China Normal University
Wuhan, Hubei, China

Ivan Dychka
Faculty of Applied Mathematics
National Technical University of Ukraine
Kiev, Ukraine

Sergey Petoukhov
Mechanical Engineering Research Institute
Russian Academy of Sciences
Moscow, Russia

Matthew He
Halmos College of Natural Sciences and Oceanography
Nova Southeastern University
Davie, FL, USA

ISSN 2194-5357 ISSN 2194-5365 (electronic)
Advances in Intelligent Systems and Computing
ISBN 978-3-030-55505-4 ISBN 978-3-030-55506-1 (eBook)
https://doi.org/10.1007/978-3-030-55506-1

This Springer imprint is published by the registered company Springer Nature Switzerland AG
The registered company address is: Gewerbestrasse 11, 6330 Cham, Switzerland

Contents

Computer Science for Manage of Natural and Engineering Processes

Interaction of Group of Bridge Piers on Scour

Andrey Voskoboinick[1], Vladimir Voskoboinick[1(✉)], Vladimir Turick[2], Oleksandr Voskoboinyk[1], Dmytro Cherny[3], and Lidia Tereshchenko[1]

[1] Institute of Hydromechanics of NAS of Ukraine, Kyiv, Ukraine
vlad.vsk@gmail.com
[2] NTUU Igor Sikorsky Kyiv Politecnical Institute, Kyiv, Ukraine
[3] Taras Shevchenko National University of Kyiv, Kyiv, Ukraine

Abstract. Pile groups and complex piers have become more popular in the construction of bridge crossings due to economical and geotechnical reasons. The interaction of bridge piers of various structural solutions, which are in the wake one after another, leads to significant discontinuity and non-linearity of the flow between the piers, and also significantly complicates the process of bed sediment scour. This requires complex scientific research to determine the permissible bed sediment scour near bridge pier groups, since normative calculations do not always give a positive result. Results of experimental researches of formation and development of local and global scours near to bridge piers are submitted. Influence of an arrangement of two bridge transitions which are in a wake one after another, on physics of the formation process scour is shown. The scour before prismatic pier is caused by interaction of the horseshoe vortex structures with the sediment, down flow along the front surface of the pier and vortex systems that arise when the incoming flow is separated from the front faces of the prismatic pier. The scour before the three-row cylindrical pier in the form of the grillage is associated with the action of a horseshoe vortex structure that envelopes the grillage as a whole, horseshoe vortices that arise near each of the cylindrical piers of the grillage and the jet flow that occurs between the first piers of the grillage. With the mutual arrangement of two bridge crossings that are in the wake one after another, the local scour in front of the prismatic pier of the old bridge increases at supercritical flow velocities in shallow water and decreases at subcritical velocities in deep water.

Keywords: Bridge pier · Grillage · Scour · Horseshoe vortex · Jet flow

1 Introduction

The scour of bridge piers more than the permissible norms in most cases leads to the destruction of bridges. Therefore, designers and builders of bridge crossings should take into account the features of the formation of scours and control them in order to minimize the risk of bridge destruction. The main factors that influence on the scour are [1, 2]: a) the hydrodynamic characteristics of the flow (intensity and depth of the flow, average and shear velocity, velocity profile and bed roughness), b) pier parameters (size, geometry, shape, separation between piers, orientation (angle of attack) of the

Z. Hu et al. (Eds.): ICCSEEA 2020, AISC 1247, pp. 3–17, 2021.
https://doi.org/10.1007/978-3-030-55506-1_1

pier relative to the direction of flow) and c) bed sediment parameters (granularity, mass density, particle shape, cohesiveness of the soil).

In many countries of the world, scientific and applied studies of the formation features of the scours, their sources and measures that prevent critical scours are conducted. However, a detailed study of the physics of the flow and dynamics of coherent structures that are responsible for the process of scour at its various stages is still not sufficiently studied, although many experimental measurements and numerical simulations are focused on studying the evolution of the scour near the bridge piers. Bridge flows in most cases of practical application have a wide range of turbulent scales around the piers and they control the process of transfer of surrounding sediment. According to the papers [3–5], the main mechanism that controls the formation and evolution of the scour is the system of horseshoe vortex structures at the base of the pier, which does not depend on the shape of the bridge pier. Vortex structures are formed near the base of the pier when the flow, and in particular the boundary layer approaches the pier and overcomes the significant pressure gradient that forms during the braking of the incoming flow in front of a bluff body. This leads to separation of the boundary layer. In the separation region, a system of vortex structures is formed, which bends around the cylinder in the form of a necklace or horseshoe vortex structures. The horseshoe vortex greatly modernizes the structure of the flow turbulence. The main effect is an increase in local shear stresses on the ground and levels of pressure fluctuations near the bottom in front of the streamlined body.

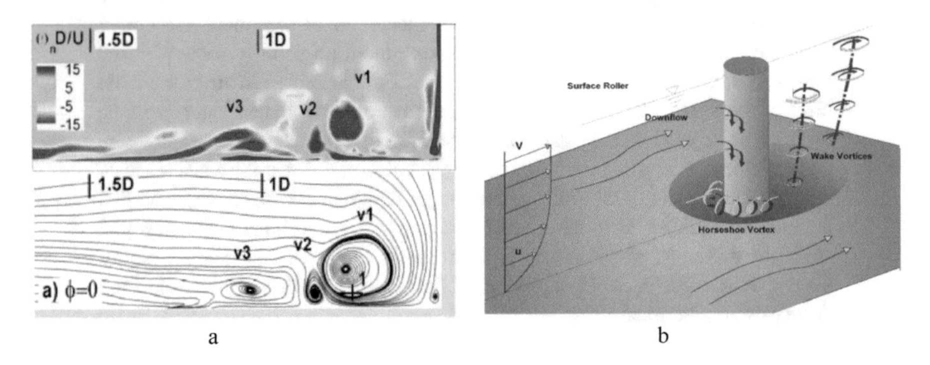

Fig. 1. Generation (a) of the horseshoe vortex structures (from [6]) and formation (b) of the local scour about cylindrical pier (from [1]).

The main coherent structure in the frontal plane of axial symmetry of the bridge pier, which does not depend on the level of incoming flow turbulence, is a system of horseshoe vortices (Fig. 1). These vortex structures interact with each other and the streamlined surface and generate unstable fields of fluctuations of velocity, pressure, vorticity, temperature, and other hydrodynamic and thermal parameters. Flow regimes have a definite effect on the shape and size of the vortex structures. Low-frequency oscillations of the main vortex structures (in particular, a horseshoe vortex) are less energy-intensive in the turbulent flow regime [6]. In the case of laminar flow, low-frequency periodic or quasiperiodic oscillations of the system of horseshoe vortex structures are only a form of possible oscillations. This is consistent with experimental observations that confirm that at very high Reynolds numbers, the horseshoe vortex structure becomes more stable in the sense that the low-frequency oscillations of the core of these vortices are very small. At that time, eddies of smaller scales and oscillations of higher frequencies are observed inside these large-scale structures [1, 6].

Pile groups have become more popular in the construction of bridge crossings due to economical and geotechnical reasons. The shape, depth and size of the scour of the group of bridge piers differ from the scour of a single pier due to the interaction of the vortex and jet flow between the piers and the processes of bed sediment transfer. It should be noted that when flowing around a single pier, a local scour is formed, and when flowing around a pier group, a global scour is additionally formed [7]. The placement, the distance between the piers and the angular arrangement of the piers play a key role in the scour process around the group structure of the bridge piers. If the piers or piles are located at a short distance from each other, then the horseshoe vortices and wake vortices of each pile and their local scours interact. In this case, the flow between the piers is accelerated due to the contraction of the flow by the neighboring piers. Therefore, the calculation of the scour on a single pier, which has a diameter equal to the equivalent diameter of the pile group, which often happens in practice, leads to large errors. This leads to the numerical and physical modeling of the flow around the designed group of bridge piers, taking into account the topography and bed sediment transfer processes. Particular attention should be paid to such modeling, when a group of piers consists of structures of various geometric shapes and sizes. In studies [8, 9] observed that for separation between piers smaller than 1.15 times the diameter of the pier, the pier group becomes like a single body. When the separation between the piers exceeds 5 diameters of the pier, the grouping of the piers does not change the scour depth [10].

Complex piers, consisting of a column, a pile cap and a pile group are usually as foundations for hydraulic or marine building structures. Columns and pile caps have different shapes: rectangular, rectangular with rounded edges, square and circular. The pile shapes are mainly circular, but occasionally square. Unfortunately, the number of studies of complex piers is limited and insufficient, although the use of such piers in bridge construction is growing every year. Among the works done over the past ten years, can mark, example, studies [11–18]. These studies focus on studying the influence of the geometry of complex piers, the configuration of the pile groups, the thickness of the pile-cap, the width of the column on the depth and shape of the scour. Studies shown that the diameter of the piles has the greatest effect on the scour depth. In addition, the flow velocity and the critical velocity of transfer of sediment particles have a significant effect on the scour depth. In general, the scour depth positively correlates with the depth and velocity of the flow, the pile diameter, number of the piles and separation between the piles, which are in line with the flow direction. However, the scour depth is inversely proportional to the average diameter of the sediment particles, the critical flow velocity, the number of the piles and the separation between the piles, which are normal to the flow direction.

The interaction of bridge piers of various structural solutions, which are in the wake one after another, leads to significant discontinuity and non-linearity of the flow between the piers, and also significantly complicates the process of bed sediment scour. Moreover, the characteristic features of the generation of vortex flow and scour in front of a single pier, a pile group and a complex pier correlate with each other when such bridge piers are combined into a complex bridge structure. The study of the features of the generation of vortex and jet flows and the formation of scour holes in the vicinity of such structures requires considerable effort for numerical and physical modeling. The interaction of the group of bridge piers, which have an unequal shape, as well as the wake location of the piers of parallel and closely spaced bridge crossings, has not been studied much and literature data for such structural solutions almost have not been given. This necessitates the conduct of numerical and experimental studies to solve this class of problems.

The purpose of the work is an experimental study of the influence of a group of bridge piers of various constructions that are in-line one after another on the formation of the horseshoe and wake vortex structures also bed sediment scour and accumulation of the sediment about the piers.

2 Experimental Setup, Methodology and Measurement Program

The road-building of bridge crossings in large cities, where the number of vehicles has grown significantly in recent years, requires the placement of bridges close to each other, due to restrictions on transport interchanges and dense urban building. The piers of new bridges are often built in the footprint of the piers of old bridges. In this case, the constructions of bridge piers in most cases are differed from each other. Therefore, there is a need to determine changes in the scour depths in front of the piers of old bridges and to determine the scour near the piers of new bridges. This experimental work, which was carried out in laboratory conditions, is devoted to the solution of such a problem.

Experimental studies were carried out in a hydrodynamic channel, which had a work section length of about 14 m, a width of 1.5 m and a depth of 0.65 m. Water was supplied to the channel by a pump, which was located more than 30 m from the working section. Water from the pump entered inside the channel through an intermediate hydraulic flume about 20 m long. In this flume, the flow disturbances that the pump generated were damped. Then the water entered the water collector, which had the form of a confusor, and then the water flow passed through the honeycomb and mesh devices, which formed a good quality flow in a straight working section. Behind the working section there was a drainage collector in the form of a diffuser, where there were valves that controlled the water level and flow velocity in the working section. Water was discharged into the tank, which was under the laboratory. The pump supplied water into the experimental setup from this tank. Thus, in the experiments, water was used, which had a constant temperature, and, consequently, kinematic viscosity.

Three pairs of bridge pier models were located at a distance of about 8 m from the beginning of the working section and were filled with sifted quartz sand to a height of about 0.2 m from the bottom of the channel. The sand was formed in the form of a flat bottom and in the form of a relief near the piers of the existing bridge on the

Dnieper River. An example of a photograph of the scour near bridge piers for an initially flat bottom is shown in Fig. 2a. The bridge piers in the form of rectangular prisms of the old bridge were located upstream of the bridge piers in the form of three-row cylindrical grillages of the new bridge (Fig. 2b). The pier model of the old bridge had a length of L = 0.095 m and a width of B = 0.04 mm (small scale model) or L = 0.26 m and B = 0.11 m (large scale model). The length of the grillage model was l = 0.2 m, and the width was b = 0.04 m (small model) or l = 0.54 m and b = 0.11 m (big model). The three-row grillage consisted of 31 cylinders with a diameter of d = 0.010 m or d = 0.027 m. The distance between the axis of the bridge pier models was x = 0.31 m or x = 0.84 m and y = 0.35 m or y = 0.95 m, where x is the longitudinal coordinate in the direction of the flow velocity U, and y and z are the transverse coordinates along the width and depth, respectively. The average grain size of quartz sand was d_{50} = 0.00035 m. During the measurements, the flow depth (H) is varied from 0.17 m to 0.34 m, and the average flow velocity (U) is changed from 0.15 m/s to 0.59 m/s.

a

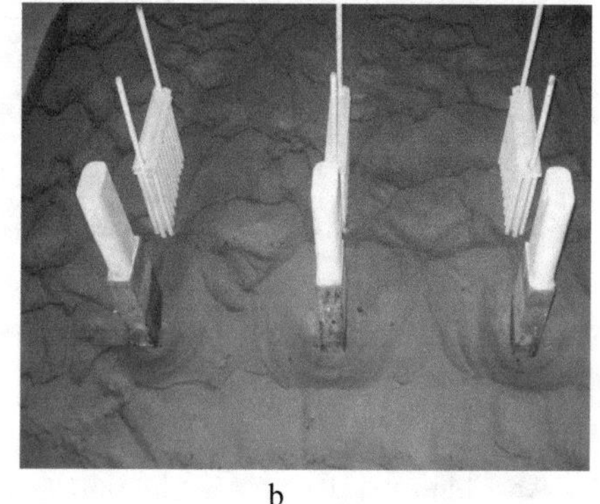

b

Fig. 2. Scour (a) of the prismatic piers and scour (b) of the prismatic piers and the tree-row cylindrical piers.

Special scour depth sensors have been designed and manufactured for research. The operation of the sensors was based on the principle of measuring the depth at which the sensitive element of the full pressure sensor is located. The pressure of the water column was transmitted to the sensitive piezoresistive elements of the sensors through miniature thin-walled tubes. Two types of the scour depth sensors were used in the studies. In the first type of sensors, hemispherical tips were installed at the end of the tubes, which ensured minimal load of the sensors on the sandy bottom of the channel. Measurements of the scour depth of the sediment were carried out continuously and made it possible to record the form and depth of the scour in the research process. In the measurements, an ensemble of four scour depth sensors was used, which were spaced a known distance from each other, as illustrated in Fig. 3a. The design of the measuring unit allowed the scour depth sensors to freely move under their own weight in the guide holes of a holder. The scour depth sensors alternately stopped in their movement at the depth where the measurements are taken, in the process of moving the measuring unit towards the bottom of the channel. When the last sensor stopped, the movement of the measuring structure stopped. Sensor movement data (changes in hydraulic pressure) in the form of electric signals were transmitted to a personal computer via a 16-channel analog-to-digital converter. The profile of the scour hole in various sections near the bridge piers was built according to the measured depths of movement of the sensors and their coordinates. These measurements were carried out periodically as experimental studies were carried out, which made it possible to determine the scour depth and its relief depending on the measurement time.

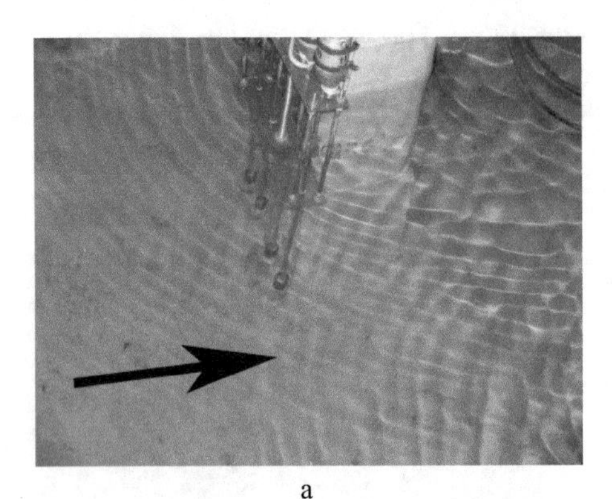

a

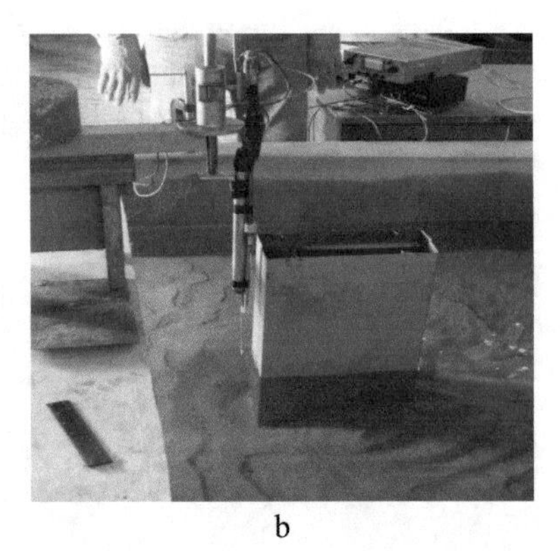

b

Fig. 3. Scour depth sensors: (a) sensors of the full pressure and (b) sensor of the pressure fluctuations and full pressure.

An additional pressure fluctuation sensor included in the design of the second type of sensors. A miniature piezoceramic pressure fluctuation sensor [19–21] was located at a fixed distance from the end of the thin-walled tube of the piezoresistive pressure sensor (Fig. 3b). The sensitive surface of the pressure fluctuation sensor was directed to the bottom of the channel. When the pressure fluctuation sensor touched the ground, its acoustic impedance instantly was changed [19, 22]. The moment of change of the impedance of this sensor was recorded on a computer together with the change in hydraulic pressure, which was measured by the full pressure sensor, and the scour depth of sediment was determined.

After the completion of experimental studies and drying of the relief of the bottom of the channel, photographing and measurements of the characteristic features of the bottom were carried out (profiles of dunes, scour holes and accumulation of the quartz sand). The bottom topography was measured using needle sensors that moved in the coordinate system with an accuracy of 0.1 mm.

The freestream velocity was measured by micro-rotators and thermistor velocity sensors (Fig. 4). Micro-rotators and thermistors were installed by the appropriate holders and brackets to the characteristic places of measurement of the velocity field. Micro-rotators (Fig. 4a) had a diameter of about 0.010 m and measured averaged freestream velocities. This made it possible to construct velocity profiles and determine the mean freestream velocity. Miniature thermistor sensors (diameter of the sensitive surface is about 0.0008 m) were combined into a correlation block (Fig. 4b). These hot-wire anemometers measured the averaged and pulsating velocity components. The small size of these sensors made it possible to introduce minimal distortion into the vortex motion, which was inherent in the flow around the bridge supports. The space-time correlations of the velocity fluctuations, which were measured by an ensemble of thermistor sensors, made it possible to estimate the kinematic characteristics of the vortex structures that were formed and developed inside the scour holes and near the bridge piers.

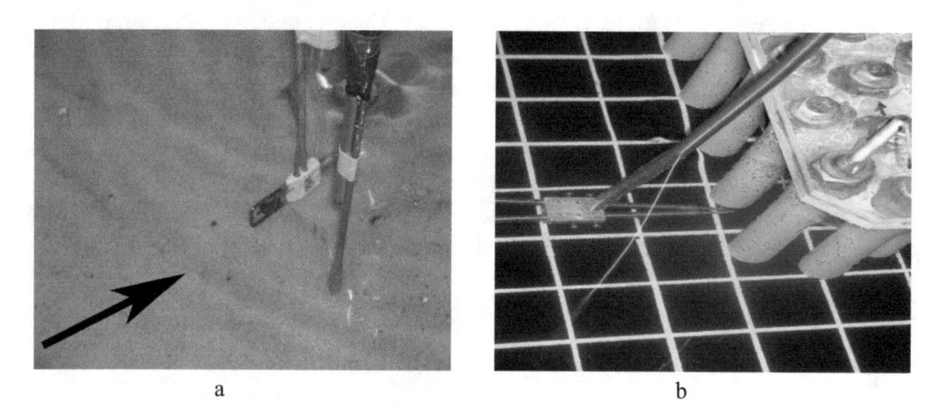

Fig. 4. Sensors of the freestream velocity and the scour depth: (a) the micro-rotator and the pressure fluctuations and (b) the thermistor hot-wire anemometers.

The three-dimensional nature of the velocity field in the vicinity of the bridge piers was studied by visualizing the flow. Visualization was carried out using contrast agents such as colored water-soluble paints and inks, as well as aluminum powder. Dyes were introduced into the stream and recorded by digital video and cameras. Registration and processing of video frames and photographs was carried out on personal computers using the appropriate programs and algorithms. Also, contrasting particles of neutral buoyancy were introduced into the flow, and a technique was used to register the movement of silk streamers (colored silk threads), which took the form of trajectories of motion of vortex structures. Video registration of the movement of the labeled particles made it possible to determine the intensity and direction of motion of these particles, which displayed the velocity field in the volume of the vortex flow.

The largest measurement error was observed when determining the velocity of movement of labeled particles (about 15%) in a confidence interval of 0.95. The error in measuring the average velocity and velocity fluctuations with thermistor sensors and micro-rotators did not exceed 5%. The depth of the local and global scour of the sediments was measured by scour depth sensors with a measurement error of the order of 5%, and the bottom topography was recorded with needle sensors with an error of the order of 1%.

3 Measurement Results

The ability to measure the depth of local scour of the sediment during the course of experimental studies allowed determining the degree of growth of the greatest scour depth of the sediment in time, as illustrated in Fig. 5a. This figure shows the dependence of the depth of local scour in front of the prismatic pier of the old bridge (z/b), on the dimensionless time of research (V_ct/H), where V_c is the critical velocity of the start of movement of sediment particles, which is determined by the dependence $6.19H^{1/6}(d_{50})^{1/3}$ [23]. Here t is the time of the research; H is the depth of the flow above the initially flat bottom. Curves 1 and 2 were obtained during the study for the Froude number $Fr = U/(gH)^{1/2} = 0.38$; $b/d_{50} = 114$ and $Fr = 0.15$; $b/d_{50} = 114$, respectively. Curve 3—data from [24] for $Fr = 0.44$; $b/d_{50} = 14$. Curve 4 and 5—data from [25] and [26] for $Fr = 0.11$; $b/d_{50} = 518$ and $Fr = 0.18$; $b/d_{50} = 181$. The results of the study show that the growth rate of the scour depth is decreased with an exponential law with an increase in the time of research. With an increase in the Froude number, the scour depth is increased. Upon reaching the value of V_ct/H of the order of 40,000, it can be considered with an acceptable degree of accuracy that the scour takes the form of a steady process. Although for a larger fraction of quartz sand grains ($b/d_{50} = 14$), the established scour occurs much later, which can be seen from curve 3 in Fig. 5a.

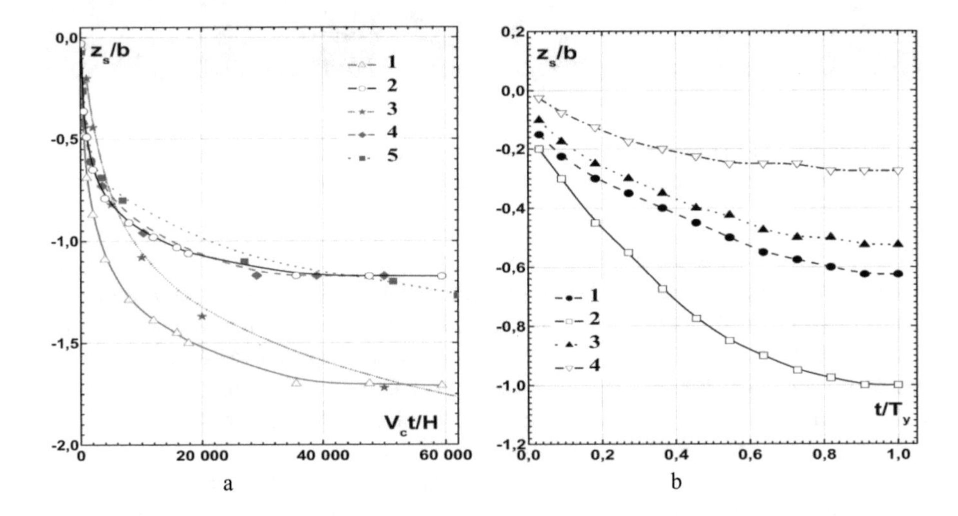

Fig. 5. Scour depth of the sediment: (a) in time and (b) the thermistor hot-wire anemometers.

An increase in the scour depth of sandy soil near the prismatic pier over the period of time that was necessary for the onset of steady scour can be estimated from Fig. 5b. Here, the scour time was normalized by the time of the steady-state process (t/T_{st}) [27, 28]. The measurements were carried out in a section that was parallel to the side wall of the pier, at a distance y = 0.75b from the pier at various distances along the longitudinal axis. Curve 1 is obtained for x = 0.63b; curve 2 - x = 0; curve 3 - x = −b and curve 4 - x = −1.88b. From the above results it follows that the greatest scour depth was observed near the front wall of the scour (front face of the scour). The down flow along the front surface of the scour, the separated flow on the side walls of the pier (originates on the front face) and the horseshoe vortex structures that are formed in front of the front wall of the pier in the lower part of the scour cavity had a significant effect on the sediment at this location.

Local scour of the sediment in the vicinity of the prismatic pier of the model of the old railway bridge passage is axonometric shown in Fig. 6a. In this three-dimensional figure, the coordinates are normalized along the width of the prismatic pier. The data were obtained for a flow velocity of 0.25 m/s and a Froude number of Fr = 0.15. The greatest scour depth was observed in front of the frontal plane of the pier and near the side faces. The upper slope of the frontal depression, which is located upstream, had a smaller angle of inclination relative to the horizontal axis than the angle of the slope of the lower part of the scour hole of the sediment (Fig. 6b).

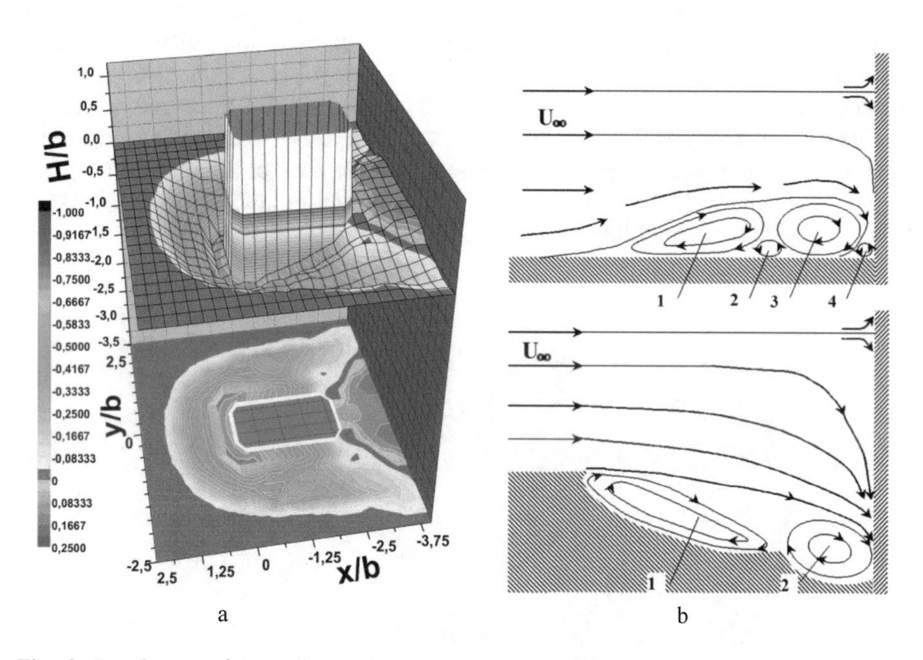

Fig. 6. Local scour of the sediment: (a) axonometry and (b) formation of the horseshoe vortex structures in front of the prismatic bridge pier.

A horseshoe vortex structure (1) was generated in the region of the upper slope, which was formed in the zone of separation of the boundary layer above the sandy bottom of the channel. In the lower part of the scour, another quasistable horseshoe vortex structure (2) was formed. It was generated by a down flow along the frontal surface of the pier, which arose when the incident flow interacted with the front wall of the prismatic pier. This interaction, as well as the separated flow, which is formed vertically located vortex structures on the side walls of the pier and acted as a vacuum cleaner, formed the deepest scour depth that swept the front of the pier in the form of a necklace or horseshoe vortex. The presence of lateral separation vortices led to the appearance of scour, first, near the front lateral faces of the prismatic pier, The horseshoe vortex structures that were developed in the scour carried sediment out of the hole and deposited it behind the pier and along its sides in the form of dunes or accumulations (Fig. 6a).

The placement of the three-row cylindrical pier in the form of a grillage behind the prismatic pier did not change the physics of generation and formation of the scour in front of the prismatic support, but acted on the aft wake part of the vortex flow behind the pier, which led to a change in the relief of the ground near such a complex of piers (Fig. 7). The accumulation of sand behind the prismatic pier practically led to minimal scour in front of the grillage. Near the piers of the new bridge, mainly the accumulation of the sediment was observed, which was deposited in the form of dunes and scour along the length of the grillage. The greatest scour of the sediment near the piers of the new bridge was observed between the first and second lateral cylindrical piers [19, 29, 30]. Here, large-scale horseshoe vortex structures enveloped the entire grillage as a whole (global scour) and small-scale horseshoe vortices enveloped each cylindrical pier (local scour). The jet flow was generated between the cylinders of a three-row grillage. It should be noted that the most intense jet flow was observed between the first and second side piers, as well as between the penultimate and last piers of the grillage [19]. The largest accumulation of the sediment was found in the area of the fifth - sixth cylindrical piers (the middle of the grillage).

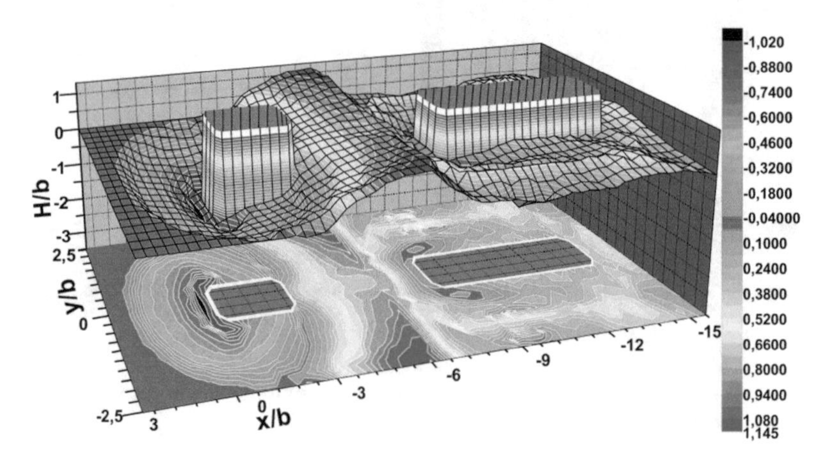

Fig. 7. Local and global scours of the sediment about two bridge piers.

The local scour profiles in the plane of axial symmetry in front of the front wall of the prismatic pier for different depths and flow velocities are shown in Fig. 8a. Curve 1 was measured for U = 0.27 m/s and the Froude number Fr = 0.19; curve 2 - U = 0.59 m/s and Fr = 0.42; curve 3 - U = 0.23 m/s and Fr = 0.13 and curve 4 - U = 0.32 m/s and Fr = 0.17. These data were obtained in the study of the scour only in front of the pier of the old bridge model. It should be noted that the scour profiles had two characteristic sections with different angles of inclination of the depression, which are caused by the formation of two systems of quasi-stable horseshoe vortex structures. In this case, the deeper vortex structure, which is located closer to the pier, had a smaller scale, and, consequently, the scour below it had smaller dimensions. Before the pier, at the junction between the surface of the pier and the bottom, the accumulation of sediment was recorded on all profiles. Therefore, from our point of view, the angular vortex, which had the opposite direction of rotation relative to the horseshoe vortices (see, Fig. 6b), did not exist under conditions of a stable-state process of the scour, or its intensity was small and it could not erode the sediment. This angular vortex was generated during the formation of the scour hole and gradually was degenerated. In addition, between two coherent horseshoe vortex structures, an oppositely rotating vortex was absent in the steady state of the scour. In the scour profile, sand was accumulated at the supposed location of the formation of this vortex. This accumulation sand separated the two slopes of the scour and, upon visual observation, had rather clear outlines. At the same time, this area of sandy accumulation was not stable, but had a spatial oscillating character, which corresponded to the vibration frequencies of large-scale horseshoe vortex structures. So, with an increase in the flow velocity and a decrease in the depth of flow, the scour depth was increased. The width of the scour hole was almost (2–2.5) times greater than the depth of scour and the maximum scour depth was observed in front of the pier at a distance of about 0.25b.

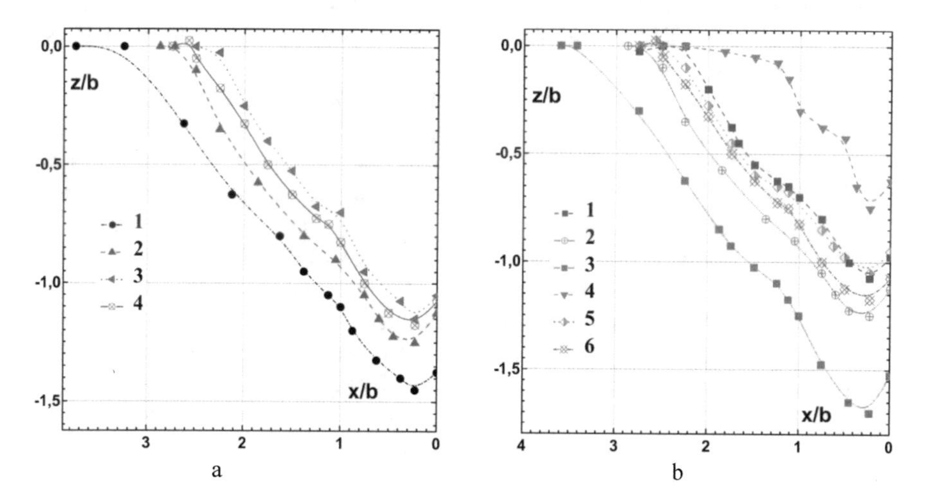

Fig. 8. Scour depth of the sediment: (a) in front of only prismatic pier and (b) in front of tandem location of two models of bridge piers.

The tandem location of the grillage model of the new bridge behind the prismatic pier of the old bridge led to changes in the profiles, which are shown in Fig. 8b for different depths and flow velocities. Curve 1 was obtained for U = 0.17 m/s and Fr = 0.13; curve 2 - U = 0.30 m/s and Fr = 0.22; curve 3 - U = 0.57 m/s and Fr = 0.43; curve 4 - U = 0.16 m/s and Fr = 0.09; curve 5 - U = 0.20 m/s and Fr = 0.11 and curve 6 - U = 0.31 m/s and Fr = 0.17. As noted earlier, the interaction of the two structures of the bridge piers did not lead to significant changes in the process of scour formation in front of the prismatic pier. The mutual influence of bridge pier models particularly led to an increase of the scour hole in front of the prismatic pier of the old bridge by almost 15% for supercritical flow velocities ($U/V_c \approx 1.8$) and shallow water (curve 2 in Fig. 8a and curve 3 in Fig. 8b). This flow regime corresponded to the case of flow with sediment displacement. For the flow regime of a pure water ($U/V_c < 1$), the scour depth decreased by almost 20% (curve 4 in Fig. 8a and curve 6 in Fig. 8b) in a joint study of two bridge crossings.

4 Conclusions

Based on experimental studies of the characteristic features of the generation and formation of local and global scour near models of bridge piers of various designs that were in the wake one after another, the following conclusions can be declared:

1. The local scour of the sediment, formed in the frontal region of the flow around the prismatic pier of the old bridge during the subcritical flow regime, is caused by the interaction of the horseshoe vortex structures enveloping the pier, the down flow along the front surface of the pier, and also the lateral vertical vortex structures that are formed when the flow is separated from front faces of the pier of the old bridge.
2. It was established that for the flow regimes under study, the geometry and configuration of the piers, as well as the nature of the bottom sediment, the steady-state process of formation of local scour occurs when the dimensionless measurement time exceeds 40000. The greatest scour depth is observed near the front wall of the prismatic pier, and when the average flow velocity reaches a value of the critical velocity of the beginning of sediment movement, the scour depth is close to one and a half widths of the prismatic pier.
3. It was found that under the steady-state flow regime, in the axial symmetry plane of the pier, a local scour hole is formed in the form of two slopes of sandy sediment accumulation. The upper slope of the scour, located upstream, has a smaller inclination angle relative to the horizontal axis than the inclination angle of the lower part of the scour adjacent to the front surface of the pier. It was recorded that the width of the scour is almost 2.5 times greater than the scour depth and the maximum scour depth is observed in front of the prismatic pier at a distance of about a quarter of its width.
4. It was established that two quasi-stable coherent horseshoe vortex structures are formed inside the local scour. The first of these occurs when the boundary layer is separated from the leading edge of the scour and forms the upper slope of the scour. The second horseshoe vortex structure of a smaller size is formed during the interaction of the shear layer above the scour hole and the down flow along the front surface of the prismatic pier and forms the lower slope of the scour. Between the horseshoe vortex structures and in the angular junction region of the bottom of the hole and the front surface of the pier, no oppositely rotating vortex structures were detected under the steady-state scour development mode. In these zones of local scour, accumulation of the sand in the form of local sediments was recorded.

5. It was found that scour of the sediment under the conditions of the steady-state process near the construction of the three-row cylindrical pier in the form of the grillage is caused by: a) the formation of the horseshoe vortex structure that surrounds the grillage as a whole (global scour), b) the development of the horseshoe vortices that are generated around each cylindrical pier of the grillage (local scour) and c) the jet flow that is arose between the cylinders of the grillage. The greatest scour was recorded near the first and second lateral piers of the three-row grillage, and the maximum accumulation of sand in the form of dunes was recorded near the mid-cross section of the grillage.
6. It has been established that the placement of the three-row cylindrical grillage of the new bridge behind the prismatic pier of the old bridge does not change the physics of the generation and formation process of local scour in front of the prismatic pier, but acts on the aft wake part of the vortex flow behind the first pier, which leads to a change in the bottom topography near such the bridge group. The accumulation of sand behind the prismatic pier leads to minimal scour of the sediment in front of the grillage. Local scour in front of the prismatic pier of the old bridge is increased by almost 15% at supercritical flow rates ($U > V_c$) in shallow water and is decreased by almost 20% at subcritical velocity in deep water with mutual the location of two bridge crossings in the wake of each other.
7. The research results and recommendations were used during the design and construction of a new bridge over the Dnieper River in Kiev.

References

1. Akhlaghi, E., Babarsad, M.S., Derikvand, E., Abedini, M.: Assessment the effects of different parameters to rate scour around single piers and pile groups: a review. Arch. Computat. Methods Eng. **26**(1), 1–15 (2019)
2. Al-Shukur, A.-H.K., Ali, M.H.: Optimum design for controlling the scouring on bridge piers. Civ. Eng. J. **5**(9), 1904–1916 (2019)
3. Ettema, R., Constantinescu, G., Melville, B.W.: Flow-field complexity and design estimation of pier-scour depth Sixty years since Laursen and Toch. J. Hydraul. Eng. **143**(9), 03117006 (2017)
4. Dargahi, B.: Flow field and local scouring around a cylinder. Royal Institute Technology, Stockholm, Sweden (1987)
5. Sreedhara, B.M., Kuntoji, G., Manu, Mandal, S.: Application of particle swarm based neural network to predict scour depth around the bridge pier. Int. J. Intell. Syst. Appl. **11**(11), 38–47 (2019)
6. Kirkil G., Constantinescu G., Ettema R.: The horseshoe vortex system around a circular bridge pier on a flat bed. In: Proceedings XXXI-st International Association Hydraulic Research Congress, Seoul, Korea, pp. 1–10 (2005)
7. Sumer B.M., Bundgaard K., Fredsoe J.: Global and local scour at pile group. In: Proceedings 15th International Offshore and Polar Engineering Conference, Seoul, Korea, pp. 577–583 (2005)
8. Ataie-Ashtiani, B., Beheshti, A.A.: Experimental investigation of clear-water local scour at pile groups. J. Hydraul. Eng. **132**(10), 1100–1104 (2006)

9. Amini, A., Melville, B.W., Ali, T.M., Ghazali, A.H.: Clear-water local scour around pile groups in shallow-water flow. J. Hydraul. Engng. **138**(2), 177–185 (2012)
10. Hosseini, R., Amini, A.: Scour depth estimation methods around pile groups. KSCE J. Civ. Eng. **19**(7), 2144–2156 (2015)
11. Bateni, S.M., Vosoughifar, H.R., Truce, B., Jeng, D.S.: Estimation of clear-water local scour at pile groups using genetic expression programming and multivariate adaptive regression splines. J. Waterw. Port. C-ASCE. **145**(1), 04018029 (2019)
12. Amini, A., Solaimani, N.: The effects of uniform and nonuniform pile spacing variations on local scour at pile groups. Marine Georesour. Geotechnol. **36**(7), 861–866 (2018)
13. Ferraro, D., Tafarojnoruz, A., Gaudio, R., Cardoso, A.H.: Effects of pile cap thickness on the maximum scour depth at a complex pier. J. Hydraul. Eng. **139**(5), 482–491 (2013)
14. Amini, A., Melville, B.W., Ali, T.M.: Local scour at piled bridge piers including an examination of the superposition method. Can. J. Civ. Eng. **41**(5), 461–471 (2014)
15. Beheshti, A.A., Ataie-Ashtiani, B.: Scour hole influence on turbulent flow field around complex bridge piers. Flow Turbul. Combust. **97**(2), 451–474 (2016)
16. Baghbadorani, D.A., Ataie-Ashtiani, B., Beheshti, A., Hadjzaman, M., Jamali, M.: Prediction of current-induced local scour around complex piers: review, revisit, and integration. Coast. Eng. **133**(3), 43–58 (2018)
17. Moreno, M., Maia, R., Couto, L.: Prediction of equilibrium local scour depth at complex bridge piers. J. Hydraul. Eng. **142**(11), 04016045 (2016)
18. Ramos, P.X., Bento, A.M., Maia, R., Pego, J.P.: Characterization of the scour cavity evolution around a complex bridge pier. J. Appl. Water Eng. Res. **4**(2), 128–137 (2016)
19. Voskobijnyk A.V., Voskoboinick V.A., Voskoboinyk O.A., Tereshchenko L.M., Khizha I. A.: Feature of the vortex and the jet flows around and inside the three-row pile group. In: Proceedings 8th International Conference on Scour and Erosion (ICSE 2016), 12–15 September 2016, Oxford, UK, pp. 897–903 (2016)
20. Voskoboinick, V.A., Makarenkov, A.P.: Spectral characteristics of the hydrodynamical noise in a longitudinal flow around a flexible cylinder. Int. J. Fluid Mech. Res. **31**(1), 87–100 (2004)
21. Voskoboinick, V., Kornev, N., Turnow, J.: Study of near wall coherent flow structures on dimpled surfaces using unsteady pressure measurements. Flow Turbul. Combust. **90**(4), 709–722 (2013)
22. Voskoboinick V.A., Voskoboinick A.V., Areshkovych O.O., Voskoboinyk O.A.: Pressure fluctuations on the scour surface before prismatic pier. In: Proceedings 8th International Conference on Scour and Erosion (ICSE 2016), 12–15 September 2016, Oxford, UK, pp. 905–910 (2016)
23. Laursen, E.M.: An analysis of relief bridge scour. J. Hydraul. Div. **89**, 93–118 (1963)
24. Sturm, T., Sotiropoulos, F., Landers, M., Gotvald, T., Lee, S.-O., Ge, L., Navarro, R., Escauriaza, C.: Laboratory and 3D numerical modelling with field monitoring of regional bridge scour in Georgia. GDOT Res. Project No 2002, Atlanta (2004)
25. Sheppard, D.M.: Large scale and live bed local pier scour experiments: phase 1. Large scale, clearwater scour experiments. FDOT Centr. No BB-473, Final Rep. University of Florida (2003)
26. Miller Jr., W.: Model for the time rate of local sediment scour at a cylindrical structure. Ph. D. Thesis. University of Florida (2003)
27. Bernacki, J., Kolaczek, G.: Anomaly detection in network traffic using selected methods of time series analysis. Int. J. Comput. Net. Inf. Sec. **9**, 10–18 (2015)
28. Nag, A., Karforma, S.: An efficient clustering algorithm for spatial datasets with noise. Int. J. Inf. Tech. Decis. **7**, 29–36 (2018)

29. Dovgiy, S.O., Lyashko, S.I., Cherniy, D.I.: Algorithms of discrete singularities method of computational technologies. Cybernet. Syst. **53**(6), 950–962 (2017)
30. Voskoboinick, V.A., Voskoboinyk, O.A., Cherniy, D.I.: The modeling of different scale hydrologic processes in aquatories. J. Environ. Sci. Nat. Resour. **29**(1), 87–97 (2019)

Mathematical Modeling Dynamics of the Process Dehydration and Granulation in the Fluidized Bed

Bogdan Korniyenko(✉) and Lesya Ladieva

National Technical University of Ukraine "Igor Sikorsky Kyiv Polytechnic Institute", Kyiv 03056, Ukraine
bogdanko@i.ua

Abstract. We consider the study of the dynamics of mathematical models of the processes of dehydration and granulation in a fluidized bed by means of two-phase Euler-Euler. We obtain the pressure transient response of granules, granules temperature and axial velocity of the coolant in the central and peripheral points of the apparatus with a fluidized bed. The offered mathematical model allows to define the change of technological parameters that determine an intensity of the transfer processes in dehydration and granulation of the ammonium sulfate aqueous solutions. Besides, the essential influence of the wall is established. At the same time, the significant difference between changes of pressure, temperature, and speed near the wall surface area of a gas heat carrier is defined. The use of the research results of dynamic modes of the dehydration and granulation processes in the fluidized bed by this mathematical model will allow to take into account all the features of the processes in the creation of a control system.

Keywords: Mathematical modeling · Fluidized bed · Dehydration · Granulation

1 Introduction

The basic requirement to agriculture on the modern stage of development is the introduction of principles of rational land tenure and effective energy saving technologies. Production of new generation mineral fertilizers has an essential value. An applying of fluidization equipment for obtaining of solid composites with the defined properties at the presence of phase transitions allows to combine a number of technological stages. The use of the directed circulation character of motion of granular layer and temperature diffusion, which is complicated by interphase transition turbulence, is the basis of the fluidized bed apparatuses functioning, that are used for the realization of dehydration and granulation processes. Creation and researches of a mathematical model of the dehydration and granulation processes in the fluidized bed considering their stochastic nature are an actual task.

There are plenty of mathematical models of transfer processes in the dispersible systems with the different level of specification. Mechanistic models for predicting of heat transfer coefficients are created in the designing of systems of heat transfer in the

Z. Hu et al. (Eds.): ICCSEEA 2020, AISC 1247, pp. 18–30, 2021.
https://doi.org/10.1007/978-3-030-55506-1_2

fluidized bed. Particularly, there was used an approach that was based on the fleeting conductivity between particles and the surface. There can be distinguished the separate particles models and emulsive phases models [1].

The application of biphasic stream model has advantages over the other models because this model does not require input empiric parameters, such as a middle particle or the package founding time. Hydrodynamics of layer develops freely from the solution of the equation of mass and moment maintenance by the biphasic approach. The biphasic approach was used for the model of heat transfer [2], which describes the model of mass maintenance, the equation of moments and thermal energy in vector form. This model does not take into account the components of turbulence.

Since the biphasic directed character of motion of the bed, which is provided by the organization of process and particularly the construction of gas-distributing grid, is mainly typical for the dehydration and granulation processes in the fluidized bed, then exactly the biphasic approach of Euler-Euler should be used for a mathematical modeling. In the approach of Euler the continuum concept is accepted for the description of phase with the use of Navier–Stokes equation [3–29]. The study of the dynamics of mathematical model of dehydration and granulation processes in fluidized bed using two-phase Euler-Euler model is considered. The transient characteristics of the pressure of granules, temperature of granules and axial velocity of the coolant in the central and peripheral points of the apparatus with fluidized bed are obtained.

2 Review of Mathematical Models of Processes of Dehydration and Granulation in the Fluidised Bed

2.1 Classification of Mathematical Models by the Types of Interfacial Interactions

The mathematical models of apparatus for dehydration and granulation in fluidized bed with the riser flow of coolant were extensively studied [1–3]. According to [2] the models can be divided into (I) simple models based on empirical correlations and (II) models based on the dynamics of bubbles considering physics and hydrodynamics of fluidized bed. According to [1, 2] it was proposed to classify mathematical models in three levels:

Level 1 - the parameters of the model are not connected with a constant size of bubbles in the fluidized bed;

Level 2 - the parameters of the model are connected with a constant size of bubbles in the fluidized bed;

Level 3 - the parameters of the model are connected with a size of bubbles in the fluidized bed, which varies depending on the height of the layer.

According to [4] the models with a constant size of bubbles in the fluidized bed can expand by simple stepwise integration and achieve the same level of complexity as the models of the level 2 and 3.

But the most complete mathematical model can be characterized by the number of phases. The term “phase” is understood to be an area, which contains solid or gas. They can vary in volume fraction of solid, in appearance and hydrodynamic characteristics.

The first approaches to the mathematical modeling of apparatus for dehydration and granulation in fluidized bed were based on a single-phase models, ignoring the segregation of gas and solid particles with the presence of cavities. Thus an attempt was carried out to predict the productivity of a fluidized bed apparatus exclusively by the time distribution of residence, namely an axial mixing of gases. This attempt failed because the transformation in the fluidized bed was described even worse than in the case of ideal mixing [1]. The productivity of a fluidized bed apparatus is got with the contact between the gas and solid particles, taking into account the presence of bubbles [3].

It is essential to consider the features of hydrodynamics. Using the models of Davidson and Harrison for heterogeneous reactions, that are related to differences between the transformation of one phase of the apparatus of ideal mixing and two phase model, the conditionally completely mixed solid phase can be received [1]. Considering staying processes in the diffusion region, when the speed of the process is determined by the rate of mass transfer, the differences between the two models are negligible. However, the advantages of two phase models are significant. That lack of regard of the gas bubbles presence in single-phase approach has led to the emergence of two phase models. These models contain all the important elements that were in all later models - the existence of different phases, the separation of gas flow, mass exchange between the phases.

2.2 Mathematical Model of Fluidized Bed Apparatus with the Use of Stochastic and Chaotic Hydrodynamics

However, these deterministic mathematical models could be improved by using a stochastic approach to the processes occurring in the fluidized bed apparatus. Stochastic mathematical models applied to systems with random processes or complex systems where the randomness is used for necessary simplification. Stochastic approach allows fluctuations in the local hydrodynamics and interfacial exchange while modeling fluidized bed. When operating the fluidized bed apparatus most values that measured, exhibit random fluctuations of relatively high amplitude [3].

For the mathematical modeling of fluidized bed apparatus used also chaotic hydrodynamics. Chaos is a special case of the complex behavior of nonlinear systems [5]. Fluidized bed apparatus are chaotic systems. Experimental research confirmed that local areas of pressure, cavities and concentrations have random fluctuations associated with nonlinear dynamics. That's why descriptions the dynamics of fluidized bed and examine processes in the apparatus for different hydrodynamic regimes can by described by using chaos. Deterministic chaos can occur in a fluidized bed as a result of nonlinear interaction of two bubbles. So you can use the chaotic behavior of fluidized bed for classical the Van-Deemters mathematical model.

2.3 Hydrodynamic Models of Fluidized Bed Apparatus and Their Classification

Very effective are trying to explore the hydrodynamics of multiphase processes in a fluidized bed apparatus by using microbalance models [6]. These mathematical models solve bind conservation equation taking into account the interfacial interaction. In turn, hydrodynamic models are divided into two classes: the continuum model and the model of discrete particles.

Continuum model is also known as the two-liquid model. In this model, the particle and gas phases are considered as two continuums that are completely interosculated. The equation of equilibrium for the mass, momentum and energy must be solved in the coordinate system of Euler. However, solving differential equations of such models requires significant computational resources and makes use of such models for the implementation of control systems with fluidized bed apparatus in real time. For solving this problem can but used discrete bubble Lagrange model that calculates motion for each bubble in a fluidized bed.

Models with discrete particles describe the motion of discrete spherical particles with the influence on them forces of interaction: particle-particle, particle-wall and particle-gas. Golf course in the gas phase can be determined by using equations preserve volume or Navier-Stokes equations. In models with discrete particles may be considered particles of different sizes, whereas in the two-fluid model particles are treated as separate phases. Models with discrete particles are divided into models with soft and hard particles.

Models with soft particles describe interaction particle-particle and particle-wall and got its name because of the assumption that the particles after the collision become deformed. Models with solid particles describe the interaction of quasi-hard spheres in quasi-instantaneous collisions taking into account the force of friction.

Thus, we can classify mathematical models of fluidized bed apparatus considering the dynamics of the solid and gas phases separately. The dynamics of each phase can be described:

1. Account the phase of discrete particles subject to Newton's law - the type of Lagrange models.
2. Account the continuum phase, characterized by the Navier-Stokes equations - Euler approach.

Euler-Euler model, also known as the continuum model or two-liquid model describes the evolution of interactions of solid and gas phases. The interaction between the two phases depends on the hydraulic resistance between phases, on the local relative velocity of phases and local objects intensive fates phases. Also, for the solid phase must be considered the pressure and viscosity.

So, Euler-Euler model gives us the most legal describing of fluidized bed apparatus process.

3 The Two-Phase Model of Euler-Euler

The research of mathematical model dynamics of dehydration and granulation processes in the fluidized bed by using the two-phase model of Euler-Euler is examined in this article. For the process of dehydration and granulation in the fluidized bed, basically the two-phase directional nature of the bed movement is ensured by the organization of the process and in particular by the design of the gas distribution lattice. But when using the pulse equation should be taken into account in addition to the pressure loss on hydraulic resistance, changes of rheological properties due to changes in temperature and increase of water vapor concentration, change of impulse between phases, external mass force, which takes into account dynamic pressure, change of gas density at the different height fluidized bed and increase the torque due to adherence to the solid particles of liquid phase drops.

The multivariate process of dehydration and granulation in the fluidized bed is accompanied by a phase transition, complicated by the formation of a liquid phase on the granules surface with subsequent removal of the liquid phase and the formation by the mass crystallization of the layer of microcrystals of the soluble phase.

The two-phase model of Euler-Euler was used in research. The volume fraction of every phase was calculated by using the continuity equation. The momentum equation for every phase took into account mass transfer between the phases, presence of lifting force and forces that depend on cooperation between the liquefied agent and solid particles on an interphase limit:

$$\begin{cases} \frac{1}{\rho_{ri}}\left[\frac{\partial}{\partial\tau}(\alpha_i\rho_i)+\nabla\cdot(\alpha_i\rho_i\mathbf{V}_i)=\sum\limits_{j=1}^{2}\left(\dot{m}_{ji}-\dot{m}_{ij}\right)\right]; \\ \frac{\partial}{\partial\tau}(\alpha_i\rho_i\mathbf{V}_i)+(\alpha_i\rho_i\mathbf{V}_i\cdot\nabla)\mathbf{V}_i=-\alpha_i\nabla p_i+\nabla\cdot\overline{\overline{\overline{\tau}}}_i+ \\ \qquad +\alpha_i\rho_i\mathbf{g}+\sum\limits_{j=1}^{2}\left[\mathbf{R}_{ji}+\dot{m}_{ji}\mathbf{V}_{ji}-\dot{m}_{ij}\mathbf{V}_{ij}\right]+\mathbf{F}_i+\mathbf{F}_{\mathrm{lift},i}+\mathbf{F}_{\mathrm{vm},i}; \qquad i=\overline{1,2} \\ \frac{\partial}{\partial\tau}(\alpha_i\rho_i h_i)+\nabla\cdot(\alpha_i\rho_i\mathbf{V}_i h_i)=-\alpha_i\frac{\partial p}{\partial\tau}+\overline{\overline{\overline{\tau}}}_i:\nabla\mathbf{V}_i+ \\ \qquad +\nabla\cdot(\lambda_i\nabla T_i)+S_i+\sum\limits_{j=1}^{2}\left[Q_{ji}+\dot{m}_{ji}h_{ji}-\dot{m}_{ij}h_{ij}\right], \end{cases} \tag{1}$$

where at $i=1$ or g – a gaseous phase that is examined - is the air, and at $i=2$ or s – a hard phase that consists of separate granules; α_i – a volume part of i's phase in the compound $\left(\sum\limits_{i=1}^{2}\alpha_i=1\right)$; ρ_i – an actual density of i's phase, kg/m^3; ρ_{ri} – an average density of i's phase, kg/m^3; τ – time, s; $\mathbf{V}_i=\left(V_x,V_y,V_z\right)^{\mathrm{T}}$ – a speed vector of i's phase, m/s; $\dot{m}_{ji}$ – a description of transfer rate of mass from phase the j to the phase i (where $\dot{m}_{ii}=0$), kg/(s·m^3); p – pressure Pa; $p_s=p_2$ – pressure in a friable phase, Pa; $\overline{\overline{\overline{\tau}}}_i=\alpha_i\mu_i\left(\nabla\cdot\mathbf{V}_i+\nabla\cdot\mathbf{V}_i^{\mathrm{T}}\right)+\alpha_i\left(\eta_i-\frac{2}{3}\mu_i\right)\nabla\cdot\mathbf{V}_i I$ – tensor of the stretch-compression stress, Pa; μ_i, η_i – tangential and volume viscosity, respectively, Pa·s; I – single matrix; $\mathbf{g}$ – vector of acceleration, related to the gravitation, m/s^2; $\mathbf{F}_i$ – external mass force, Pa/m; $\mathbf{F}_{lift,i}$ – lifting force, Pa/m; $\mathbf{F}_{vm,i}$ – attached mass force, Pa/m; $\mathbf{R}_{ji}=-\mathbf{R}_{ij}$ (where $\mathbf{R}_{ii}=0$), $\sum\limits_{j=1}^{2}\mathbf{R}_{ji}=\sum\limits_{j=1}^{2}K_{ji}\left(\mathbf{V}_j-\mathbf{V}_i\right)$ – forces, that depend on the friction, pressure, clutch of particles and other factors on the interphase border, N/m^3 (Pa/m); $\mathbf{V}_{ji}=\begin{cases}\mathbf{V}_j \text{ with } \dot{m}_{ji}>0\\ \mathbf{V}_i \text{ with } \dot{m}_{ji}<0\end{cases}$ – speed on the interphase surface, m/s; $h_i=\int\limits_{T_{\mathrm{ref}}}^{T}c_{\mathrm{p}i}dT$ – enthalpy,

J/kg; T – temperature, K; λ_i – thermal conductivity of i's phase, W/(m·K); S_i – a source related to the chemical reactions and thermal radiation, W/m^3; $Q_{ji} = -Q_{ij}$ (where $Q_{ii} = 0$) – the heat exchange intensity between the phases, W/m^3; h_{ji} – enthalpy on the interphase border during evaporation or condensation (crystallization), $h_{ji} = \begin{cases} h_j \text{ with } \dot{m}_{ji} > 0 \\ h_i \text{ with } \dot{m}_{ji} < 0 \end{cases}$, J/kg.

The temperature of granules is considered as the measurement of random motion in particles, which is proportional to the value of the particles random motion area. Random motion of particles rises up due to the mechanical energy that is passed by the particles of granules. This motion creates the internal energy of particles. A transport equation of granules temperature takes into account the convective heat exchange, the hard phase voltage, the flow of energy oscillations, the dispersion of energy of collisions and the exchange of energy between the phases. The temperature of granules is connected to the hard phase voltage:

$$\frac{3}{2}\left[\frac{\partial}{\partial \tau}(\alpha_s \rho_s \Theta_s) + \nabla \cdot (\alpha_s \rho_s \mathbf{V}_s \Theta_s)\right] = \left(-p_s \bar{\bar{I}} + \bar{\bar{\tau}}_s\right) : \nabla \mathbf{V}_s + \nabla \cdot (\lambda_{\Theta_s} \nabla \Theta_s) - \gamma(\Theta_s), \tag{2}$$

where $\left(-p_s \bar{\bar{I}} + \bar{\bar{\tau}}_s\right) : \nabla \mathbf{V}_s$ – an energy release due to the stress in the particles of the hard environment, W/m^3; $\bar{\bar{\tau}}_s = -\frac{\pi}{6}\sqrt{3}\psi \frac{\alpha_s}{\alpha_{s,\max}} \rho_s g_{ss} \mathbf{U}_{s,||} \sqrt{\Theta_s}$ – tensor of the shear stress of granules, Pa; ψ - a coefficient of the mirror interaction between granules and the wall; $\mathbf{U}_{s,||}$ –sliding speed of the granules along the wall, m/s; λ_{Θ_s}– a coefficient of the granules diffusion, kg/(m·s); $\lambda_{\Theta_s} \nabla \Theta_s$ – energy that is released due to the diffusion, W/m^2; $\gamma(\Theta_s)$ – dissipation of the energy due to the collisions of the granules, W/m^3.

4 Experiment

The models of turbulence in the Fluent 6.3 application package are used in this modeling.

The solution of equations of mathematical model allows to consider multivariate process of dehydration and granulation of ammonium sulfate solution as a consequence of the process of interaction of solid particles and to determine the patterns of the movement of continuous and dispersed phase Availability of heat-transfer. An example of numerical solution was conducted by the parameters specified in the table and the following boundary conditions.

Boundary Conditions for Hydrodynamics

The left and right wall of the apparatus are modeled as impermefree sliding hard walls for both phases. On the lower wall is set inflow of hot coolant-air and for the solid phase is taken free of rigid wall.

The upper part of the layer for the liquid and solid phase is taken accordingly: unfinished breaks wall with leakage and impermeproof loose rigid wall, respectively.

Options for the Control Example

№	Name	Conditional Notation	Value	Dimension
1	Initial layer Temperature	T_0	293	K
2	Heat transfer coefficient	α	5.5	1/s
3	The intensity of heat removal from the air layer	β^*	0.032	1/s
4	Concentration field relaxation time	τ^*	8.75	s
5	Coefficient of vertical temperature conduction	a_{22}	from $2{,}7 \cdot 10^{-4}$ to $12{,}5 \cdot 10^{-4}$	m^2/s
6	Output solution costs	G_s	0.0017	m^3/s
7	Concentration of output solution	x_s	0.4	
8	Heat-capacity granules	C_s	1630	J/kg·K
9	Density granules	ρ	1230	kg/m^3
10	Thermal conductivity granules	λ	0,55	W/m·K

Boundary Conditions for Thermal Energy

The left and right wall of the machine is examined with a temperature of 373 K fixed for both phases. Continuous wall leakage is adopted at the top of the liquid phase.

Dynamics of main characteristics of the heat carrier and granules of ammonium sulfate in the two points of the fluidized bed - central and peripheral – is calculated by using the mathematical model (Fig. 1).

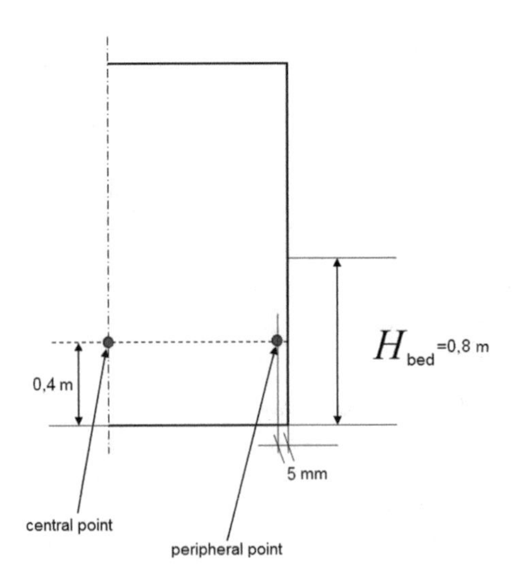

Fig. 1. The location scheme of the central and peripheral points in calculation of the dynamic characteristics in the apparatus with fluidized bed.

The graphs show transient characteristics of granules pressure (Figs. 2 and 3), the temperature of granules (Figs. 4 and 5) and axial speed of a heat carrier (Figs. 6 and 7) in the central and peripheral points of the apparatus with the fluidized bed.

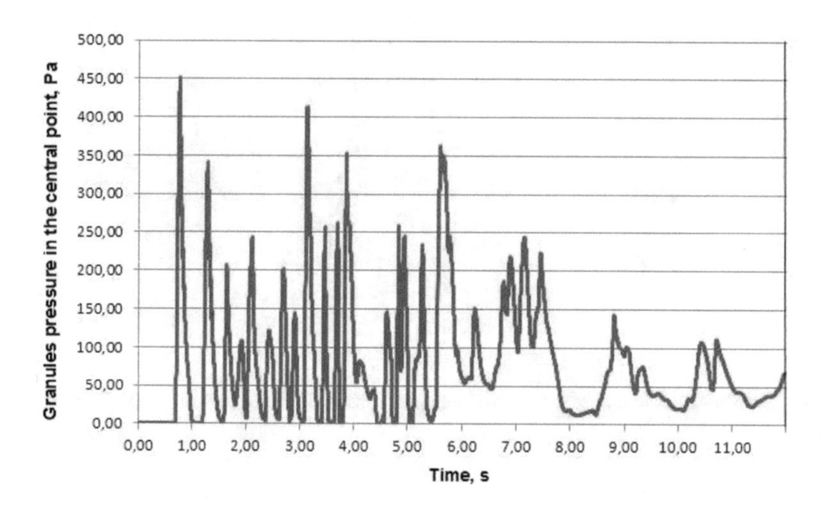

Fig. 2. The transient characteristic of granules pressure in fluidized bed in the central point of apparatus.

As seen from the graphs of transient processes in the fluidized bed, the nonlinear and random systems behavior is typical for them.

The beginning of interaction of liquefied continuous environment is watched at t = 0,8 s and is characterized by the maximal pressure value of 450,0 Pa (Fig. 2), with the next sharp decrease of pressure to zero values. Moreover, the frequency increases to $t \leq 4,0$ s and amplitude decreases to 200,0 Pa at the presence of pulsations. It is explained by the different character of gas environment flow at this point and unstable structure of the granular material in the layer. After $t \geq 6,0$ s, it is observed the sharp decrease of frequency and the stable reduction of amplitude to 50,0–30,0 Pa.

Dynamics of the change of pressure in the fluidized bed in the point, which is located at the distance of 5 mm from the wall of apparatus (Fig. 3), differs significantly by frequency and amplitude from the dynamics of the change of pressure in the central point. Yes, oscillation amplitude in the first 4 s reaches 400,0 Pa at the relatively stable frequency. At $3,8 \leq t \leq 4,5$ s, the oscillation frequency increases significantly at a simultaneous reduction of the amplitude to 200,0–100,0 Pa. At $5,0 \leq t \leq 5,8$ s, the maximal hydraulic difference of 520,0 Pa is reached, that exceeds initial values by 27%. Further, at $t \geq 7,0$ s the oscillation frequency increases, but the oscillation amplitude is in the range of 380,0–150,0 Pa. So, the presence of apparatus wall influences significantly on the fluctuation of pressure difference due to the friction of particles with the wall and expressed pulsation motion of gas environment in this area.

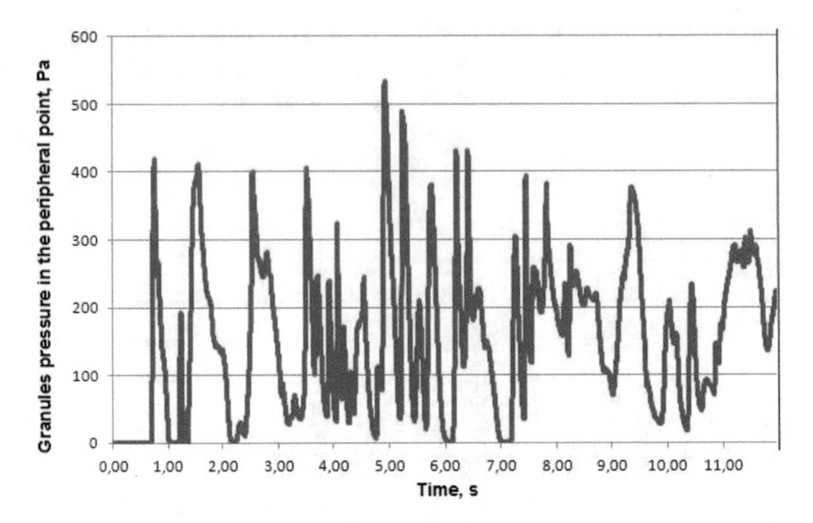

Fig. 3. The transient characteristic of granules pressure in fluidized bed in the peripheral point of apparatus.

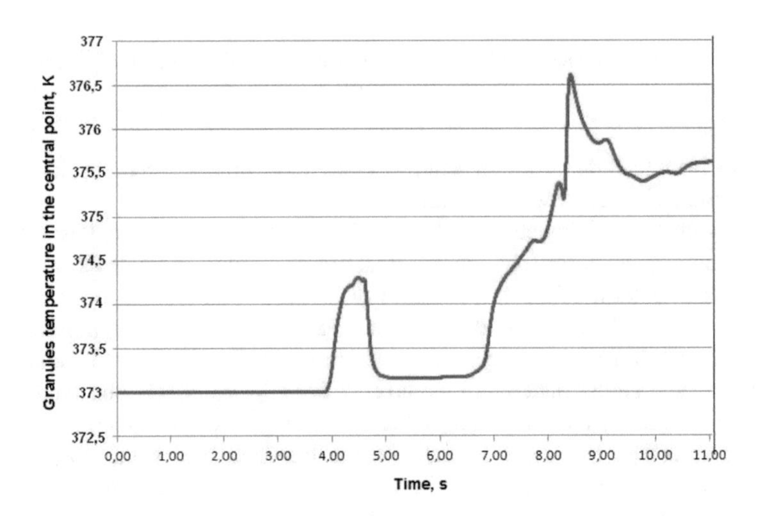

Fig. 4. The transient characteristic of granules temperature in fluidized bed in the central point of apparatus.

The character of the change of granules temperatures in the central point of apparatus is shown on the Fig. 4. Yes, at $t \leq 4{,}0$ s, while gas streams did not reach the central point, the temperature remains unchanging T = 373 K. Accordingly, in the period $4{,}0 \leq t \leq 5{,}0$ s the breakthrough of a gas bubble and insignificant increase of the temperature to 373,2 K are observed. And only after $t \geq 7{,}0$ s, when the relative stabilization of pressure is happening (Fig. 2), there is an increase of temperature to 376,5 K with the next stabilization at the level of 375,5 K. Essentially, beginning from t = 11 s, the stabilization of dispersible environment temperature is happening.

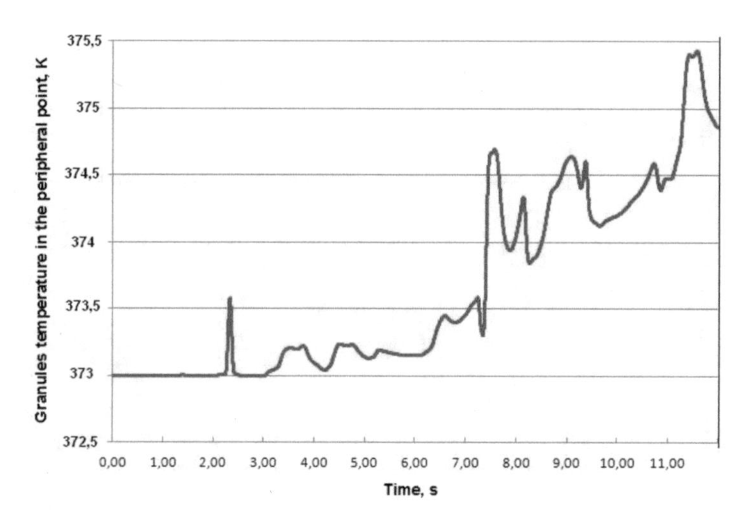

Fig. 5. The transient characteristic of granules temperature in fluidized bed in the peripheral point of apparatus.

In the peripheral point, the temperature change to t $\leq$ 7,0 s is analogical to the previous case, but with some differences. After t $\leq$ 7,0 s, as a result of major pressure pulsations (Fig. 3), the change of the temperature in the point, that is located near the wall of apparatus, has the expressed pulsation character and does not reach the defined level.

Dynamics of the speed change to t $\leq$ 3,0 s carries a pulsation character and is related to distribution of gas environment in the field of granular material layer. Herewith, as a result of fluctuations of granular material, the cases, in which the velocity vector has oppositely directed growth 3,0 $\leq$ t $\leq$ 4,5 s, 6,2 $\leq$ t $\leq$ 6,8 s,

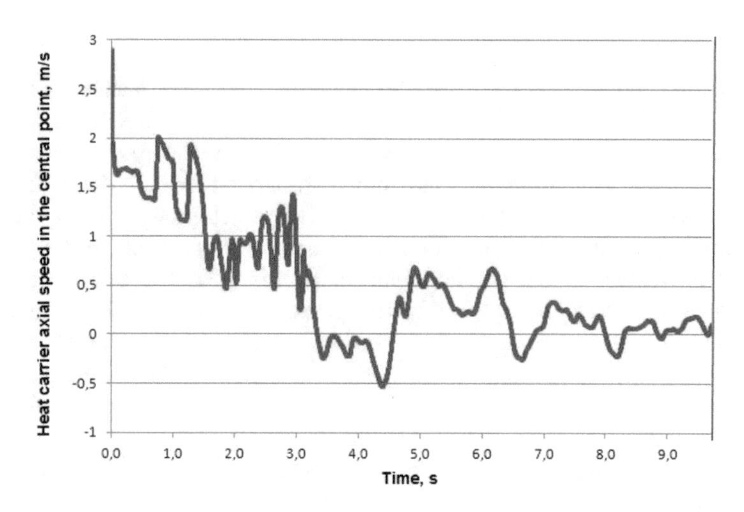

Fig. 6. The transient characteristic of the heat carrier axial speed in the central point of apparatus.

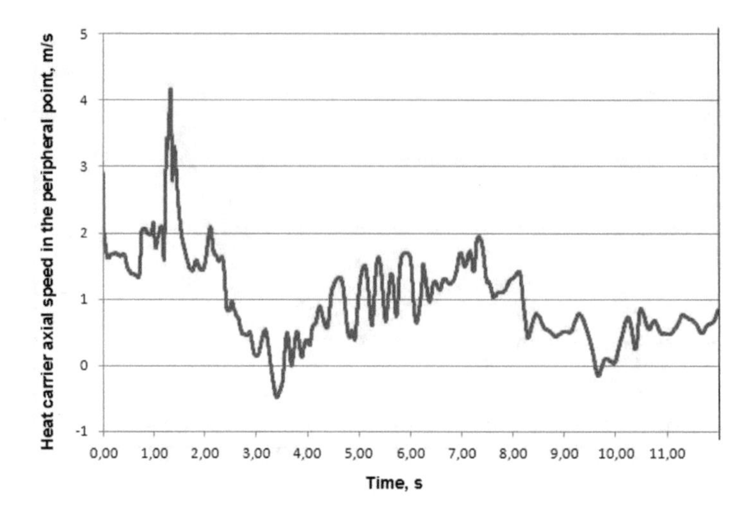

Fig. 7. The transient characteristic of the heat carrier axial speed in the peripheral point of apparatus.

7,8 $\leq$ t $\leq$ 8,0 s, are formed (Fig. 6). Moreover, the value of pulsations decreases significantly after t = 9,0 s.

In the layer, that is located near the wall of apparatus, due to the braking action of the wall on the motion of particles, the pulsations slows significantly, which is represented on the dynamics of the speed change in the peripheral point (Fig. 7). The formation of the twirls is happening near the wall surface area at t = 3,8 s. And only after t $\geq$ 9,0 s, the speed is stabilized in the range of 0,5–0,9 m/s.

5 Conclusions

The offered mathematical model allows to define the change of technological parameters that determine an intensity of the transfer processes in dehydration and granulation of the ammonium sulfate aqueous solutions. As can be seen from the graphs of transient processes in fluidized bed they are characterized by behaviour of nonlinear and random systems. The presence of the wall of the machine significantly affects the fluctuations in pressure, due to the friction of particles with a wall and a pronounced pulse-gate movement of the gas environment in this area. The temperature of the granules goes to a steady regime, but the peripheral point has significant fluctuations. The dynamics of change of the coolant speed has a pulse jet nature and is associated with the distribution of the gas environment in the granular bed material. In a bed located near the wall of the device, due to the braking wall effect on the particle movement significantly slows down the pulsation, displayed on the dynamics of speed changes in the peripheral point. Besides, the essential influence of the wall is established. At the same time, the significant difference between changes of pressure, temperature, and speed near the wall surface area of a gas heat carrier is defined. The use of the research results of dynamic modes of the dehydration and granulation processes in the fluidized bed by this mathematical model will allow to take into account all the features of the processes in the creation of a control system.

References

1. Davidson, J., Harrison, D.: Fluidization. Chemistry, Moscow, 725 p (1973)
2. Grace, J.R. (ed.): Fluidized Beds as Chemical Reactors. In: Gas Fluidization Technology, 428 p. Wiley, Chichester, New York, Brisbane, Toronto, Singapore (1986)
3. Kuniy, D., Lewenshpil, O.: Industrial fluidization. transl. with English, Chemistry, Moscow, 448 p (1976)
4. Kuipers, J.A.M., Hoomans, B.P.B., van Swaaij, W.P.M.: Hydrodynamic models of gas-fluidized beds and their role for design and operation of fluidized bed chemical reactors. Fluidization IX. In: Fan, L.-S., Knowlton, T.M. (eds.) Engineering Foundation, New York, pp. 15–30 (1998)
5. Botterill, J.S.M., Williams, J.R.: The mechanism of heat transfer to gas fluidized beds. Trans. Instn. Chem. Engrs. **41**, 217 (1963)
6. May, W.G.: Fluidized-bed reactor studies. Chem. Eng. Prog. **55**(12), 49–56 (1959)
7. van Deemter, J.J., Drinkenburg, A.A.H.: In fluidization, pp. 334–347. Netherlands University Press, Amsterdam, Netherlands (1967)
8. Orcutt, J.C., Davidson, J.F., Pigford, R.L.: Reaction time distribution in fluidized catalytic reactors. In: Chemical Engineering Progress Symposium Series, vol. 58, no. 38, pp. 1–15 (1962)
9. Partridge, B.A., Rowe, P.N.: Chemical reaction in a bubbling gas-fluidised bed. Trans. Inst. Chem. Eng. **44**, 335–348 (1966)
10. Werther, J.: Mathematical modeling of fluidized bed reactors. Int. Chem. Eng. **20**, 529–541 (1980)
11. Kobayashi, H., Arai, F., Sunawaga, T.: Fluidization models. Chem. Eng. Tokyo **31**, 239–247 (1967)
12. Kato, K., Wen, C.Y.: Bubble assemblage model for fluidized bed catalytic reactors. Chem. Eng. Sci. **24**, 1351–1369 (1969)
13. Zhulynskyi, A.A., Ladieva, L.R., Korniyenko, B.Y.: Parametric identification of the process of contact membrane distillation. ARPN J. Eng. Appl. Sci. **14**(17), 3108–3112 (2019)
14. Galata, L., Korniyenko, B.: Research of the training ground for the protection of critical information resources by iRisk method. Mech. Mach. Sci. **70**, 227–237 (2020). https://doi.org/10.1007/978-3-030-13321-4_21
15. Korniyenko, B.Y., Borzenkova, S.V., Ladieva, L.R.: Research of three-phase mathematical model of dehydration and granulation process in the fluidized bed. ARPN J. Eng. Appl. Sci. **14**(12), 2329–2332 (2019)
16. Kornienko, Y.M., Liubeka, A.M., Sachok, R.V., Korniyenko, B.Y.: Modeling of heat exchangement in fluidized bed with mechanical liquid distribution. ARPN J. Eng. Appl. Sci. **14**(12), 2203–2210 (2019)
17. Bieliatynskyi, A., Osipa, L., Kornienko, B.: Water-saving processes control of an airport. In: MATEC Web of Conferences, 239 (2018). https://doi.org/10.1051/matecconf/201823905003

18. Korniyenko, B., Galata, L., Ladieva, L. Security estimation of the simulation polygon for the protection of critical information resources. Paper presented at the CEUR Workshop Proceedings, vol. 2318, pp. 176–187 (2018)
19. Galata, L.P., Korniyenko, B.Y., Yudin, A.K.: Research of the simulation polygon for the protection of critical information resources. Paper presented at the CEUR Workshop Proceedings, vol. 2067, pp. 23–31 (2017)
20. Korniyenko, B.: Informacijni tehnologii' optymal'nogo upravlinnja vyrobnyctvom mineral'nyh dobryv. K.: Vyd-vo Agrar Media Grup (2014)
21. Korniyenko, B., Osipa, L.: Identification of the granulation process in the fluidized bed. ARPN J. Eng. Appl. Sci. **13**(14), 4365–4370 (2018)
22. Babak, V., Shchepetov, V., Nedaiborshch, S.: Wear resistance of nanocomposite coatings with dry lubricant under vacuum. Sci. Bull. Natl. Min. Univ. Issue **1**, 47–52 (2016)
23. Kravets, P., Shymkovych, V.: Hardware implementation neural network controller on FPGA for stability ball on the platform. In: 2nd International Conference on Computer Science, Engineering and Education Applications, ICCSEEA 2019, Kiev; Ukraine, 26 January 2019–27 January 2019 (Conference Paper), vol. 938, pp. 247–256 (2019)
24. Kandra, D.: Modeling of air temperature using ANFIS by wavelet refined parameters. Int. J. Intell. Syst. Appl. **8**(1), 25–34 (2016). https://doi.org/10.5815/ijisa.2016.01.04
25. Ghiasi-Freez, J., Hatampour, A., Parvasi, P.: Application of optimized neural network models for prediction of nuclear magnetic resonance parameters in carbonate reservoir rocks. Int. J. Intell. Syst. Appl. **7**(6), 21–32 (2015). https://doi.org/10.5815/ijisa.2015.06.02
26. Malekzadeh, M., Khosravi, A., Noei, A.R., Ghaderi, R.: Application of adaptive neural network observer in chaotic systems. Int. J. Intell. Syst. Appl. **6**(2), 37–43 (2014). https://doi.org/10.5815/ijisa.2014.02.05
27. Bhagawati, K., Bhagawati, R., Jini, D.: Intelligence and its application in agriculture: techniques to deal with variations and uncertainties. Int. J. Intell. Syst. Appl. **8**(9), 56–61 (2016). https://doi.org/10.5815/ijisa.2016.09.07
28. Wang, W., Cui, L., Li, Z.: Theoretical design and computational fluid dynamic analysis of projectile intake. Int. J. Intell. Syst. Appl. **3**(5), 56–63 (2011). https://doi.org/10.5815/ijisa.2011.05.08
29. Patnaik, P., Das, D.P., Mishra, S.K.: Adaptive inverse model of nonlinear systems. Int. J. Intell. Syst. Appl. **7**(5), 40–47 (2015). https://doi.org/10.5815/ijisa.2015.05.06

Optimal Control of Point Source Intensity in a Porous Medium

Dmitriy Klyushin and Andrii Tymoshenko(✉)

Faculty of Computer Science and Cybernetics,
Taras Shevchenko National University of Kyiv, Kyiv, Ukraine
inna-andry@ukr.net

Abstract. In this paper, a new combined method for computing the optimal intensity of point sources regulating humidity in a porous medium is proposed. The process is described by Richards-Klute equation. The initial problem is transformed into the linear dimensionless optimal control problem on non-stationary moisture transport in an unsaturated porous medium using Kirchhoff transformation. A variation algorithm is applied to the resulting problem formulation. For this algorithm, the finite difference method is used for both direct and conjugate problems, followed by numerical method application to solve a system of linear algebraic equations. The optimization is achieved by minimizing the deviation of the last state from the target values of this state.

The current paper provides theoretical background and solution for a two-dimensional drip irrigation problem. Theorems on existence and uniqueness are listed for the target functional. All necessary transitions and calculations are shown and demonstrate high accuracy of the method. The proposed method allows to solve the problem of optimal parameter choice for a drip irrigation system, and to improve its effectiveness.

Keywords: Control · Optimization · Richards-Klute equation

1 Introduction

Water transfer in an unsaturated porous medium with point and linear sources remains topical for multiple contemporary researches. These problems are mostly solved by either computer modeling or analytical approach.

The aim of the article is to construct a variation algorithm to identify the optimal source power for moisture transport problems in porous medium. Mathematical model is described by a two-dimensional Richards-Klute equation describing fluid flow in porous medium. Kirchhoff transformation leads to a linear system; then dimensionless equation is stated using scale multipliers. The quality functional is introduced to mark the optimal state, which should be achieved using control over point source intensity. Then necessary theorems proving existence and uniqueness of the solution are shown to guarantee correctness. Finally, variation algorithm and its numerical analog are stated leading to solution.

Z. Hu et al. (Eds.): ICCSEEA 2020, AISC 1247, pp. 31–39, 2021.
https://doi.org/10.1007/978-3-030-55506-1_3

By combining theoretical and numerical methods with control and optimization, the current article offers a complete algorithm for a two-dimensional humidification problem. It can be improved by changing some parts and so has a nice potential for the future research.

2 Literature Review

The popularity of methods depends on their speed, applicability and complexity [1, 2]. Computer modelling allows taking into account real attributes of porous medium based on Richards-Klute equation [3]. Despite the amount of works, the process's controllability and stability are not completely studied, because the problem is quasilinear [4].

According to the applied methods, either finite differences [5], or finite elements [6, 7] are used. To decrease the complexity of solution, some authors [8, 9] offered to use Kirchhoff transformation, which allow solving a linear problem instead. The optimization problem is still discussed [10], mainly according to ground and moisture parameters.

For the problem of moisture transport through the ground, a variation method is used [11–13]. So, linearization fits this approach well with further application of numerical methods for the resulting linear system. In our research, the finite difference method is used to construct the system. As for correctness of the mathematical model, it is discussed in [14–18]. The Kirchhoff transformation for the general problem is introduced in [19].

Alternative approaches to the humidification problem include works [20, 21], based on the Shan–Chen multiphase model of nonideal fluids that allow coexistence of two phases of a single substance. Moreover, new type field irrigation intelligent controller that based on AT89LV52 single chip is introduced in [22].

So, most of contemporary works can be used as an alternative part for the current research. But most of them are focused on modelling and simulation or general theoretical research, while this article is focused on combining the transmission to linear equation, proof of existence and uniqueness with modelling results and control.

3 Mathematical Model

Consider a two-dimensional humidity distribution problem for a limited area with zero initial humidity and a given target humidity at the last time step. The source is assumed having an ideal form, the atmosphere pressure and soil temperature are assumed constant. Mathematical model can be described by the following equation:

$$\frac{\partial \omega}{\partial t} = \frac{\partial}{\partial x}\left[K_x(\omega)\frac{\partial H}{\partial x}\right] + \frac{\partial}{\partial y}\left[K_y(\omega)\frac{\partial H}{\partial y}\right] + \sum_{j=1}^{N} Q_j(t)\delta(x - x_j) \times \delta(y - y_j), \quad (1)$$

$$(x, y, t) \in \Omega_0 \times (0, T],$$

$$\omega|_{x=0} = 0; \omega|_{x=L_1} = 0; \tag{2}$$

$$\omega|_{y=0} = 0; \omega|_{y=L_2} = 0;$$

$$\omega(x, y, 0) = 0, \ (x, y) \in \overline{\Omega}_0.$$

Here, $H = \psi(\omega) - y$ is velocity head, $D_y(\omega) = K_y(\omega)\frac{d\psi}{d\omega}$ is diffusion along y, $\Omega_0 = [(x, y) : \ 0<x<L_1, \ 0<y<L_2]$ stands for the research area, $y = y_0$ is the plane at the ground surface level (Oy is a vertical axe taken downward). We assume that $K_x(\omega) = k_1 k(\omega), \ K_y(\omega) = k_2 k(\omega)$, where $k_1, \ k_2$ are filtration coefficients along $Ox, \ Oy, \ k(\omega)$ stands for humidity function of the ground. To simplify the evaluations, let $k_1 = k_2, \ L_1 = L_2 = 1$. For a transition to a dimensionless equation, following [19], more variables are added:

$$\beta_2 = 0,5\ell, \ \beta_1 = \sqrt{\frac{k_2}{k_1}}\beta_2, \ \alpha = \frac{\langle D_y \rangle \beta_2^2}{T},$$

$$\xi = \frac{\beta_1}{L_1}x, \ \zeta = \frac{\beta_2}{L_2}y, \ \tau = \alpha t.$$

where $\langle D_y \rangle$ is an average value of D_y.

Then, the Kirchhoff transformation is applied [19]:

$$\Theta = \frac{4\pi k_1}{Q^* k_2 \beta_2} \int\limits_{\omega_0}^{\omega} D_y(\omega) d\omega$$

where Q^* is a scale multiplier in case the following conditions are satisfied:

- The relation between $\Theta(\omega)$ and $K_y(\omega)$ is linear: $D_y^{-1}(\omega)\frac{dK_y(\omega)}{d\omega} = \ell = const$
- $\frac{\partial \omega}{\partial t} = \frac{k_2 \beta_2 Q^*}{4\pi k_1}\frac{1}{D_y(\omega)}\frac{\partial \Theta}{\partial t} \simeq \frac{k_2 \beta_2^3 Q^*}{4\pi k_1}\frac{\partial \Theta}{\partial \tau}$.

Some more variables are necessary for the dimensionless problem: $q_j = \frac{Q_j}{Q^*}$ stands for a scaled source power, Ω, Γ are dimensionless equivalents to Ω_0, Γ_0, where Γ_0 is a border of Ω_0.

In this case, it is possible to rewrite the problem (1), (2) as

$$\begin{aligned} \frac{\partial \Theta}{\partial \tau} &= \frac{\partial^2 \Theta}{\partial \xi^2} + \frac{\partial^2 \Theta}{\partial \zeta^2} - 2\frac{\partial \Theta}{\partial \zeta} \\ &+ 4\pi \sum_{j=1}^{N} q_j(\tau)\delta(\xi - \xi_j) \times \delta(\zeta - \zeta_j), (\xi, \zeta, \tau) \in \Omega \times (0, 1], \end{aligned} \tag{3}$$

$$\Theta|_{\xi=0} = 0; \ \Theta|_{\xi=1} = 0;$$

$$\Theta|_{\zeta=0} = 0;\ \Theta|_{\zeta=1} = 0; (\xi, \zeta, \tau) \in \Gamma \times [0, 1]. \tag{4}$$

$$\Theta(\xi, \zeta, 0) = 0, (\xi, \zeta) \in \Omega.$$

The points r_j, $j = \overline{1, N}$, define the location of a source with power $q_j(\tau)$. The target humidity values $\varphi_m(\tau)$ are interpreted as averaging $\Theta(\xi, \zeta, \tau)$ in a small area ω_m around the given points $(\xi_m, \zeta_m) \in \Omega$, $m = \overline{1, M}$. The task is to find $q_j(\tau)$, $j = \overline{1, N}$, minimizing the square deviation of $\Theta(\xi_m, \zeta_m, \tau)$ from $\varphi_m(\tau)$ in norm of $L_2(0, 1)$.

Let the optimal control belong to Hilbert space $(L_2(0, 1))^N$ with the following inner product

$$\langle X, Y \rangle = \sum_{j=1}^{N} \int_0^1 x_j(\tau) y_j(\tau) d\tau.$$

Then the quality functional is

$$J_\alpha(\overline{Q}) = \sum_{m=1}^{M} \int_0^1 \left(\varphi_m(\tau) - \int_\Omega g_m(\xi, \zeta) \Theta(\xi, \zeta, \tau) d\Omega \right)^2 d\tau + \alpha \left\| \overline{Q} \right\|^2, \tag{5}$$

where $\overline{Q}(\tau) = (q_1(\tau), \ldots, q_N(\tau))^T$ is the control vector, $g_m(\xi, \zeta) = \frac{\chi_{\omega m}}{diam \omega_m}$ is the averaging core in ω_m, χ_{ω_m} is an indicator function, $\alpha > 0$ is the regularization constant. Optimal control solves the minimization problem

$$J_\alpha(\overline{Q}^*) = \min_{q \in (L_2(0,1))^N} J_\alpha(\overline{Q}). \tag{6}$$

4 Correctness

By H we denote the space of smooth functions, satisfying (2) by norm:

$$\|u\|_H = \left(\int_Q \left(u_\tau^2 + u_\xi^2 + u_\varsigma^2 + u_{\xi\varsigma}^2 \right) dQ \right)^{1/2}.$$

By H_+ we denote a similar space, satisfying initial and border conditions of the conjugate problem (4). Widening the operator L on H by continuity with respect to initial and border conditions, the following equation is received:

$$L\Theta \equiv \frac{\partial \Theta}{\partial \tau} - \Delta\Theta + 2\frac{\partial \Theta}{\partial \zeta} = f. \tag{7}$$

By applying the approaches according to [12], the following theorems take place.

Theorem 1. *For any function* $f \in (H_+)^*$ *exists only one weak solution of* (7) *in meaning.* $\langle \Theta,\ L^*\Psi \rangle_{L^2(Q)} = \langle f,\ \Psi \rangle_+ \quad \forall \Psi \in H_+,\ \Theta \in L_2(Q)$.

This theorem follows directly from a theorem, proven in [12] as a particular case.

Theorem 2. *If the state of a system is defined as a weak solution of* (7)*, then there is only one control minimizing the functional* (6).

5 Algorithm

To solve the problem (3), (4) iterative algorithm is used [12], consisting of three stages.

1) Solve the direct problem.

$$\begin{aligned} L\Theta^{(k)} &\equiv \frac{\partial \Theta^k}{\partial \tau} - \frac{\partial^2 \Theta^k}{\partial \xi^2} - \frac{\partial^2 \Theta^k}{\partial \zeta^2} + 2\frac{\partial \Theta^k}{\partial \zeta} \\ &= 4\pi \sum_{j=1}^{N} q_j(\tau)\delta(\xi - \xi_j) \times \delta(\zeta - \zeta_j); \end{aligned} \tag{8}$$

$$0 < \tau \le 1; \Theta^k(0) = 0;$$

2) Solve the conjugate problem:

$$\begin{aligned} L^*\Psi^{(k)} &\equiv -\frac{\partial \Psi^k}{\partial \tau} - \frac{\partial^2 \Psi^k}{\partial \xi^2} - \frac{\partial^2 \Psi^k}{\partial \zeta^2} - 2\frac{\partial \Psi^k}{\partial \zeta} \\ &= 2\left(\Theta^k - \phi(\tau)\right);\ 0 \le \tau < 1,\ \Psi^{(k)}(1) = 0; \end{aligned} \tag{9}$$

3) Evaluate the new source power approximation:

$$\frac{Q^{(k+1)} - Q^{(k)}}{\tau_{k+1}} + \Psi^{(k)} + \alpha Q^{(k)} = 0, k = 0, 1, \ldots$$

To solve the direct problem, implicit numerical scheme for the Eq. (8), dividing $0 \le \xi, \zeta \le 1$ with a step $h = \frac{1}{30}$ and time steps $\tilde{\tau} = \frac{1}{100}$ for $0 \le \tau \le 1$.

Using integral-interpolation method, the system for direct problem can be written as

$$\begin{aligned} \Lambda(\xi,\zeta) &= \Lambda_1(\xi) + \Lambda_2(\zeta) \\ &= \frac{\partial \Theta}{\partial \tau} - \left(\varphi(\xi,\zeta) + \frac{1}{h}\varphi_1(\xi) + \frac{1}{h}\varphi_2(\zeta)\right) \end{aligned}$$

Taking into account border conditions, we receive

$$\Lambda_1(\xi) = \begin{cases} 0, & \xi = 0 \\ (\Theta_{\bar{\xi}})_{\xi}, & 0<\xi<1 \\ 0, & \xi = 1 \end{cases}, \Lambda_2(\zeta) = \begin{cases} 0, & \zeta = 0 \\ \left(\Theta_{\bar{\zeta}}\right)_{\zeta} - 2\Theta_{\hat{\zeta}}, & 0<\zeta<1 \\ 0, & \zeta = 1 \end{cases}. \tag{10}$$

By $\hat{y}$ we denote the central difference derivative. Due to border conditions, $\varphi_1(\xi) = \varphi_2(\zeta) = 0$. Jacobi method was used to solve the system of linear equations.

At the third stage the accuracy depends on regularization parameter, which was chosen according to calculation errors. The error of this approach is $O(h^2)$ according to space steps and $O(\tau)$ for time steps.

Three alternatives were chosen to stop the work of the algorithm and provide its finiteness:

- The average difference module between the current and previous values of Θ is less or equal 10^{-7} (accuracy termination);
- The number of iterations exceeded a constant (for example 1000, 2000) providing finiteness of execution.
- Limitation of source power: the algorithm stops when the current approximation of source power is accurate enough. This limitation was used to stop after achieving the 2% accuracy to compare the required number of iterations.

6 Results

As in previous research for one-dimensional problem [13], the accuracy criterion is based on providing the humidity level as near to target function as possible, but the result of calculations is the optimal source power. Based on theorems from the previous section and [12], the convergence is proven, so it is enough to stop the process in a proper moment, for example by limiting the number of iterations or 98% accuracy level termination.

The aim of the experiment is to define, does the algorithm lead to an optimal source power for the target functional and how many iterations should be done. To test the convergence, several approaches calculating the next source power approximation were applied. Although standard method converges, getting a faster approximation was tested.

The target function was taken as a result of modelling with initially chosen optimal source power 10 and this value is equal to 100% result accuracy. Iterations start from zero source power approximation. Several different positions for the point source: near the top left corner, near the middle of the top border, near the middle of the left border and in the center are considered. The right side of equation respectively is:

$$\varphi(\xi,\zeta) = \begin{cases} 4\pi q, & \xi = \frac{7}{30},\ \zeta = \frac{7}{30}; \\ 0, & else, \end{cases} \quad \varphi(\xi,\zeta) = \begin{cases} 4\pi q, & \xi = 0,5,\ \zeta = \frac{7}{30}; \\ 0, & else, \end{cases}$$

$$\varphi(\xi,\zeta) = \begin{cases} 4\pi q, & \xi = \frac{7}{30},\ \zeta = 0,5; \\ 0, & else \end{cases} \quad \varphi(\xi,\zeta) = \begin{cases} 4\pi q, & \xi = 0,5,\ \zeta = 0,5; \\ 0, & else. \end{cases}$$

The deviation of received source power from the optimal was less than 2% when the regularization parameter was chosen equal to 10^{-7}. Another test with the initial source power more than optimal was held. In that case the proposed algorithm also demonstrated the right direction of approximations leading to solution.

The following table shows results of some experiments over the third stage defining new approximations of Q. By "experimental deviation" we mean taking τ_{k+1} equal to scaled current deviation of the achieved result for the current source power approximation from the target function (Table 1).

Table 1. Number of iterations required to get 98% accuracy.

Method description	Iterations
Two-step with τ_{k+1} = 100, all time period optimization	254
Two-step method with experimental deviation, all time period	43
Two-step method with experimental deviation, last moment	40
Three-step method with τ_{k+1} = 100, last time moment	219
Three-step method, experimental deviation, last moment	37

Although the number of iterations was reduced, the wrong scale parameter for experimental deviation-based iterative method may cause problems with convergence, while classic two-step and three-step methods converged in all cases.

To compare the convergence using standard and experimental approaches, see Fig. 1 below.

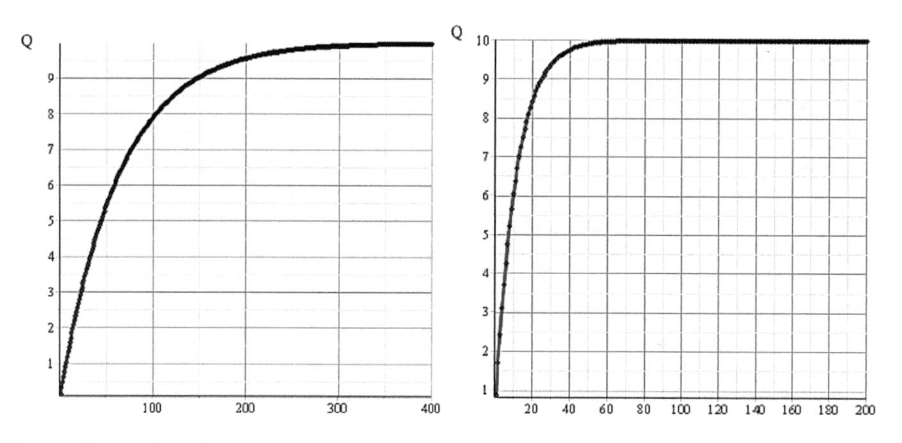

Fig. 1. Standard (left panel) and experimental algorithms (right panel) with 400 and 200 iterations limit (respectively).

So, the offered method demonstrated high accuracy in defining the optimal source power for several options of source placement. Regularization parameter was chosen with respect to calculation errors and received values of theta.

7 Discussion

Among the results mentioned above, the same approach was applied to a three-dimensional model with zero initial humidity and cubic area. The target function was chosen according to the modelling results with source power 10. Although the method converged, it still requires some more proof.

The accuracy depends on the target function (possibility to get results near it), number of iterations (it should be enough for the accuracy) and the choice of time and space discretization steps. Time required for the algorithm is proportional to these steps and there should be a balance between accuracy and time.

To calculate the humidity values in the area, several approaches can be applied, including finite differences like in this article or finite elements [6], but the main focus should be on optimization of the process – that's why controllability problem for a three-dimensional model is still discussed.

8 Conclusion

A variation algorithm was developed to identify the optimal source power for a quasilinear water transfer problem in an unsaturated porous medium. The offered algorithm allows to solve the problem of choosing the proper parameters for the irrigation system and improve its effectiveness.

To improve the applicability, theorems demonstrating existence and uniqueness of solution for a three-dimensional model should be proven. Our future research will be dedicated to either these theorems or a more complicated two-dimensional process.

References

1. Friedman, S.P., Gamliel, A.: Wetting patterns and relative water-uptake rates from a ring-shaped water source. Soil Sci. Soc. Am. J. **83**(1), 48–57 (2019)
2. Hayek, M.: An exact explicit solution for one-dimensional, transient, nonlinear Richards' equation for modeling infiltration with special hydraulic functions. J. Hydrol. **535**, 662–670 (2016)
3. Farthing, M.W., Ogden, F.L.: Numerical solution of Richards' equation: a review of advances and challenges. Soil Sci. Soc. Am. J. **81**(6), 1257–1269 (2017)
4. Zha, Y., et al.: A modified Picard iteration scheme for overcoming numerical difficulties of simulating infiltration into dry soil. J. Hydrol. **551**, 56–69 (2017)
5. List, F., Radu, F.A.: A study on iterative methods for solving Richards' equation. Comput. Geosci. **20**(2), 341–353 (2016)
6. Klyushin, D.A., Onotskyi, V.V.: Numerical modeling of three-dimensional moisture transfer in micro irrigation. J. Numer. Appl. Math. **1**, 54–64 (2016). (in Ukrainian)

7. Zhang, Z.-Y., et al.: Finite analytic method based on mixed-form Richards' equation for simulating water flow in vadose zone. J. Hydrol. **537**, 146–156 (2016)
8. Berninger, H., Kornhuber, R., Sander, O.: Multidomain discretization of the Richards equation in layered soil. Comput. Geosci. **19**(1), 213–232 (2015)
9. Pop, I.S., Schweizer, B.: Regularization schemes for degenerate Richards equations and outflow conditions. Math. Models Methods Appl. Sci. **21**(8), 1685–1712 (2011)
10. Cockett, R., Heagy, L.J., Haber, E.: Efficient 3D inversions using the Richards equation. Comput. Geosci. **116**, 91–102 (2018)
11. Vabishchevich, P.N.: Numerical solution of the problem of the identification of the right-hand side of a parabolic equation. Russian Math. (Iz. VUZ), **47**(1), 29–37 (2003). (in Russian)
12. Lyashko, S., Klyushin, D., Semenov, V., Shevchenko, K.: Eulerian and Lagrangian approach to solving the inverse convection–diffusion problem. Dopov. Nac akad. Nauk Ukr. **10**, 38–43 (2007). (in Ukrainian)
13. Tymoshenko, A., Klyushin, D., Lyashko, S.: Optimal control of point sources in Richards-Klute equation. Adv. Intell. Syst. Comput. **754**, 194–203 (2019)
14. Lyashko, S.I., Klyushin, D.A., Onotskyi, V.V., Lyashko, N.I.: Optimal control of drug delivery from microneedle systems. Cybernetics Syst. Anal. **54**(3), 1–9 (2018)
15. Lyashko, S.I., Klyushin, D.A., Nomirovsky, D.A., Semenov, V.V.: Identification of age — structured contamination sources in ground water. In: Boucekkline, R., et al. (eds.) Optimal Control of Age — Structured Populations in Economy, Demography, and the Environment, pp. 277–292. Routledge, London and New York (2013)
16. Lyashko, S.I., Klyushin, D.A., Palienko, L.I.: Simulation and generalized optimization in pseudohyperbolical systems. J. Autom. Inf. Sci. **32**(5), 108–117 (2000)
17. Lyashko, S.I.: Numerical solution of pseudoparabolic equations. Cybern. Syst. Anal. **31**(5), 718–722 (1995)
18. Lyashko, S.I.: Approximate solution of equations of pseudoparabolic type. Comput. Math. Math. Phys. **31**(12), 107–111 (1991)
19. Shulgin, D.F., Novoselskiy, S.N.: Mathematical models and methods of calculation of humidity transfer during subsurface irrigation. In: Sbornik nauchnyh trudov: Matematika i problemy vodnogo xozyajstva, pp. 73–89. Naukova Dumka, Kiev (1986). (in Russian)
20. Zhang, X.: Lattice Boltzmann implementation for fluids flow simulation in porous media. Int. J. Image Graph. Signal Process. (IJIGSP) **3**(4), 39–45 (2011)
21. Zhang, X.: Single component, multiphase fluids flow simulation in porous media with lattice Boltzmann method. Int. J. Eng. Manuf. (IJEM) **2**(2), 44–49 (2012)
22. Zhang, L., Jiang, J., Liu, L.: Design of field irrigation multi-purpose control device based on idle work compensation. Int. J. Image Graph. Signal Process. (IJIGSP) **3**(1), 38–44 (2011)

Integrated Water Management and Environmental Rehabilitation of River Basins Using a System of Non-linear Criteria

Pavlo Kovalchuk[1], Roman Kovalenko[1], Volodymyr Kovalchuk[1](✉), Olena Demchuk[2], and Hanna Balykhina[3]

[1] Institute of Water Problems and Land Reclamation, National Academy of Agrarian Sciences of Ukraine, Kyiv, Ukraine
kovalchuk.pavlo.ivanovich@gmail.com, romchik89@ukr.net, volokovalchuk@gmail.com
[2] National University of Water and Environmental Engineering, Rivne, Ukraine
ldem1997@ukr.net
[3] National Academy of Agrarian Sciences of Ukraine, Kyiv, Ukraine
maslova-anna@ukr.net

Abstract. Integrated water management by the basin principle using technological, economic and environmental non-linear criteria is proposed. The methodological basis of the approach is the sustainable development paradigm. In line with the conditions of sustainable development, the EU Directive provides development of river basin management plans, according to which a transition from spontaneity to controllability of the water use development and ecological improvement of state of water bodies is carried out. Technological criteria for supporting of water levels in reservoirs have been developed, economic criteria for the dynamics of water intake for various needs have been formalized, the ecological needs of water quality assessment during the washing of rivers from reservoirs are substantiated. Balance equations are given for controlling of water masses dynamics and spread of pollution. A pulse method is proposed within a combined control system for river washing for example of the Ingulets river basin. The scenario analysis of the optimal and existing flushing technologies for the river from the reservoir showed that the proposed technology has economic advantages. Necessary restrictions for environmental criteria are done. At the same time, the existing technology requires significantly larger volumes of water for flushing and does not meet environmental criteria during certain periods of flushing.

Keywords: Integrated water management · River basins · Economic and environmental non-linear criteria

1 Introduction

The methodological basis of water use and environmental rehabilitation of river basins is the improvement of integrated water resources management using a system of technological, economic, environmental and social criteria [1, 2]. It implies a

Z. Hu et al. (Eds.): ICCSEEA 2020, AISC 1247, pp. 40–51, 2021.
https://doi.org/10.1007/978-3-030-55506-1_4

sustainable balanced development of society, aimed at meeting the socio-economic needs of the modern generation and achieving natural and environmental sustainability, ensuring a good ecological state of the environment [3, 4]. Sustainable development conditions in the river basin provide for the transition from spontaneity to controllability of the process of water use development. An integrated approach is proposed for achievement of management objectives. Integration is carried out according to a system of nonlinear criteria. The criteria reflect the effectiveness of decision-making on a basin principle. In addition to the system of criteria, integrated management sets the task of balance models developing for the transfer of water masses from other basins, their dynamics in reservoirs, modeling the processes of pollutant mixing and the water use for economic purposes.

An analysis of recent studies showed that an important direction of river rehabiliation is the washing of rivers from reservoirs [1, 5, 6] and water exchange in reservoirs [7]. In the process of industrial development, a number of man-made reservoirs have been created in the river basins used to store liquid waste and waste mine water. There is a direct discharge of pollution into the river, as well as water pollution from diffuse and point sources [1].

The riverbeds are washed by removing sediment [8, 9]. The riverbeds are washed to protect against malaria during the implementation of recreational projects [10], as well as to protect river deltas from the penetration of ocean water [11, 12]. Separate washing controls are presented in [13–15]. However, these works do not present control models for riverbeds washing, in particular, integrated control using a system of nonlinear criteria.

The purpose of the study is creating of an integrated water resources management system in the river basin, which will ensure the maintenance of the water level in the reservoirs, the saving of water resources during river flushing, the proper water quality when supplying water to consumers, as well as good or excellent ecological status of the rivers.

The research methodology is to develop a system of technological, economic and environmental non-linear criteria, a system of difference equations for the dynamics of water movement and the processes of mixing and transfer of pollution in the control system. A scenario analysis is used [13–15], which provides for simulation modeling of washing options and their multi-criteria optimization according to a set of criteria. The optimal flushing option is compared with the existing one, its advantages are revealed.

2 Management Criteria

Technological criterion $F_1(x,t)$ determines the maintenance of the water level in the reservoir within the specified limits:

$$H_2(t) \leq F_1(x,t) \leq H_1(t), \tag{1}$$

where $H_1(t), H_2(t)$ are the lower and upper water levels during its using.

Economic criteria $F_{2i}(T)$ are mathematically defined as the total water consumption for period T for suppling of water consumers or for flushing of the river:

$$F_{2i}(T) = \int_{0}^{T} Q_i(x_i(t), U_i(t)dt, \; i = 1, \ldots, n, \tag{2}$$

where Q_i is the nonlinear function of water flow for the object $x_i(t) \in X$ and control $U_i(t) \in U$ at period $t \in (0; T)$.

Environmental criterion $F_{31}(x(t), Q(t))$ expresses the degree of risk of pollution export to the river floodplain at some point *x(t)* at flow rates *Q(t)* at time *t*:

$$F_{31}(x(t), Q(t)) = R(x(t), Q(t)) \tag{3}$$

where *R* is a certain function of the risk measure (Fig. 1), which depends on the flow of water from the reservoir *Q(t)*. Environmental releases $Q(t) \geq C(x, t)$ are also formalized, which exceed a certain amount of river flow. For river restoration the water discharge during washing is determined, which ensures the non-return of the saline prism at the water intake. At the same time, environmental goals are also taken into account, set as restrictions on water quality or as criteria for achieving a good ecological state of the river:

$$F_{4j}(x, t) \leq C_j, \; j = 1, \ldots, k, \tag{4}$$

where $F_{4j}(x, t)$ are environmental criteria, C_j are restrictions on water quality indicators according to current standards at the moments $t \in (0; T)$ at measurement points *x*.

It is necessary to improve monitoring research within integrated management systems, especially when water is delivered for irrigation. Water quality for irrigation is characterized according to regulatory documents on environmental and agronomic criteria, and the classes is determined according to these criteria using restrictions in the form of inequalities, the logical-mathematical model of decision-making for single concentrations of the indicator can be represented as:

$$A(S_{ij}) = \begin{cases} I\ class\text{–}"Fit", if\ P_{ij}^1 \leq S_{ij} \leq P_{ij}^2, \\ II\ class\text{–}"Limited\ Fit", if\ P_{ij}^2 \leq S_{ij} \leq P_{ij}^3, \\ III\ class\text{–}"UnFit", if\ P_{ij}^3 \leq S_{ij} \leq P_{ij}^4, \end{cases} \tag{5}$$

where S_{ij} are single concentrations of the indicator, *i* = *1,2* are indices that meet environmental and agronomic criteria; $j \in [1;n_1]U[1;n_2]$ is the ordinal number of environmental or agronomic criteria, n_1, n_2 are numbers of environmental or agronomic criteria; $P_{ij}^1, \ldots, P_{ij}^4$ are restrictions for the classification of indicators; $A(S_{ij})$ is the function that determines the class of the indicator; R are decisive functions in the form of vectors.

The implementation of the ecosystem approach is carried out on the basis of a neural network (Fig. 1) of measurements in space and time of the values of indicators, their sequential graphical analysis along the river bed [13, 15].

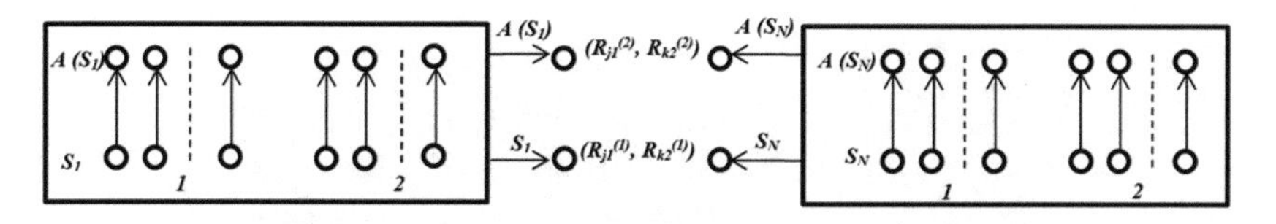

Fig. 1. Assessment of water quality for irrigation by the neuron ensemble.

3 Balance Models of Water and Pollutants

Balance models based on the equation of dynamics and conservation of water masses and pollutants are proposed for simulation. They are constructed in such a way that the flow of water is calculated firstly and then the concentration of chemical indicators is determined as a result of mixing and water exchange. Water exchange in the reservoir occurs so that part of the contaminated water is displaced by the flow of water from the channel:

$$W_i^{n+1} = W_i^n + q_i^n - p_i^n, n = 1, \ldots, N, \tag{6}$$

where W_i^{n+1}, W_i^n are volumes of water resources in the reservoir at the (n + 1) and n-th period in the i-th cell; q_i^n is the volume of water resources coming from the channel and the inflow into the reservoir; p_i^n is the amount of water taken from the reservoir; i is the number of cells.

The concentration of the substance at the n-th period in the i-th cell as a result of mixing is calculated by the formula:

$$U_i^{n+1} = \frac{W_i^n U_i^n + q_i^n C_i^n - p_i^n S_i^n}{W_i^{n+1}}, n = 1, \ldots, N, \tag{7}$$

Where $W_i^n U_i^n$ is the volume of water resources W_i^n with a concentration of U_i^n in the i-th cell at the n-th moment of time; $q_i^n C_i^n$ is the volume of water resources q_i^n coming from the channel to the reservoir with a concentration of C_i^n in the i-th cell at the n-th moment of time; $p_i^n S_i^n$ is the volume of water resources p_i^n taken from the reservoir with the concentration S_i^n in the i-th cell at the n-th moment of time.

As a result of the washing processes, displacing of contaminated water and the mixing processes of the upper and lower layers interact in the rivers (Fig. 2).

To adequately reflect the slowly changing unsteady motion of the water flow in open channels (taking into account the velocities of various layers of water and the lifting force of the flow in the upper layer), we consider a two-layer balance difference model [13], which describes the movement of water in the upper layer of the stream (where significant speed) and the dynamics of the flow in the bottom layer (where the movement of a slowly changing flow occurs) (Fig. 2).

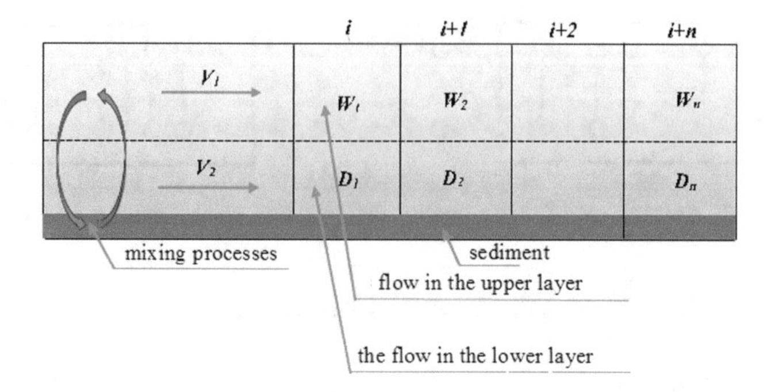

Fig. 2. Scheme of mixing the upper and lower layers of water.

4 Pulse Method in a Combined Control System

As practice shows, washing is carried out by displacing the prism of highly saline waters without significantly mixing them. If the prism of mineralized water is displaced from the sampling, after one or two days the water becomes within the acceptable standards. In such cases, the impulse with high water consumption can be reduced by a certain number of days, which leads to savings in water resources. By saving and increasing the amount of water, an open control line is detected. This positional component (Fig. 3) is determined by planning the sequence of pulses in time with their discharge from the reservoir.

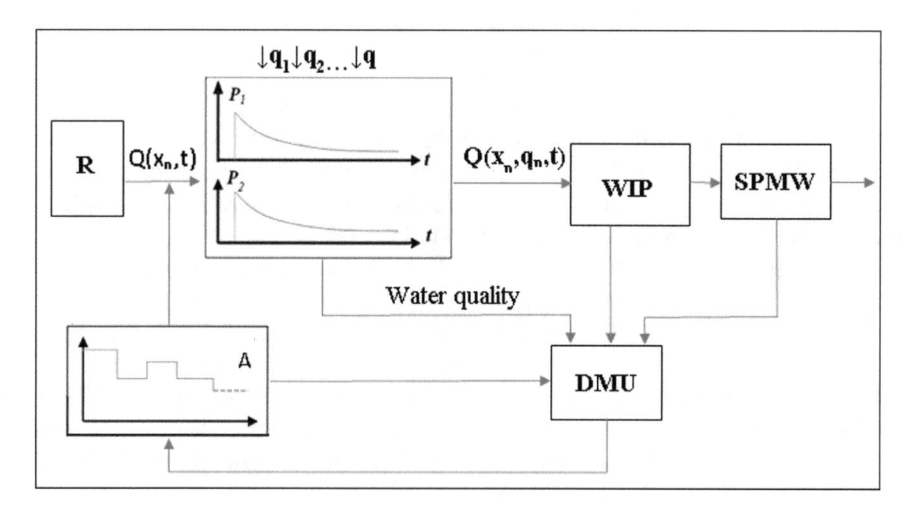

Fig. 3. The scheme of the combined control system for the Ingulets river by pulse method.

The combined control system [16] has both a positional component and a feedback loop (Fig. 3). Feedback controls of water flow from the reservoir according to the water quality in front of the water intake, at the water intake point and the water needs for water intake.

Making decisions about the next impulse is taken according to the function that evaluates the control actions based on the input and output values at the same time, that is, feedback is used:

$$Q(x, t_{n+1}) = F(V_i(t_n), S_i(t_n), V_j(t_n), S_j(t_n), Q(x, t_n)) \quad (8)$$

where $Q(x, t_{n+1})$ is the expense (impulse) at the next (n + 1) time moment; V_j, S_j are chlorine concentrations in the upper and lower layers at the point in front of the water intake; V_i, S_i are concentrations in the upper and lower layers at the water intake point; $Q(x, t_n)$ is the discharge at the previous time from the reservoir; F is a function that determines the decision-making algorithm; R is a reservoir; q_1, q_2,…,q_n are uncontrolled water flows from diffuse and point sources; P_1, P_2 are concentration values at sampling points in front of the water intake and at the main pumping station; WIP is water intake point for irrigation system or for industrial needs; SPMW is the state of the prism of mineralized waters; DMU is decision making unit; A is positional component of the combined control system.

In the presence of uncontrolled interference q_i, models [18] cannot be used to predict water quality at the Andreevka observation point and at the intake to the Ingulets irrigation system.

To substantiate the washing options, a scenario analysis [13] is methodically proposed, based on an ecological and economic approach to the selection of technological solutions.

5 Management Modeling and Analysis Results

For the implementation of integrated water management the Ingulets River Basin was selected (Fig. 4). The water level in the Karachunivske reservoir is maintained, according to criterion (1), due to the replenishment of water from the Dnipro-Ingulets canal (Fig. 4) and the natural flow of the river. The water quality in the reservoir is satisfactory; water is delivered for drinking needs of Kryvyi Rih city and flushing of the river. Flushing is necessary to displace the salt water prism into the Dnipro River, which is formed as a result of pollution by industrial enterprises between the dam of the reservoir and Andreevka village (Fig. 4). The result of washing should be the achievement of a good ecological condition of the river and the supply of water of satisfactory quality to the Ingulets irrigation system.

Fig. 4. The Ingulets river basin.

Figure 4 shows the Inhulets river basin scheme with its tributaries, indicating the main tailings facilities within the Kryvyi Rih and sampling points according to the monitoring data of the State Agency of Water Resources of Ukraine: 5 is the point of water outlet from the Karachunivske reservoir; 12 is Andriyivka water quality measurement point; 13 is the point of water intake for the Ingulets irrigation system.

However, the existing control method without feedback does not provide satisfactory irrigation water quality in certain periods (Fig. 5), it is costly by economic criteria.

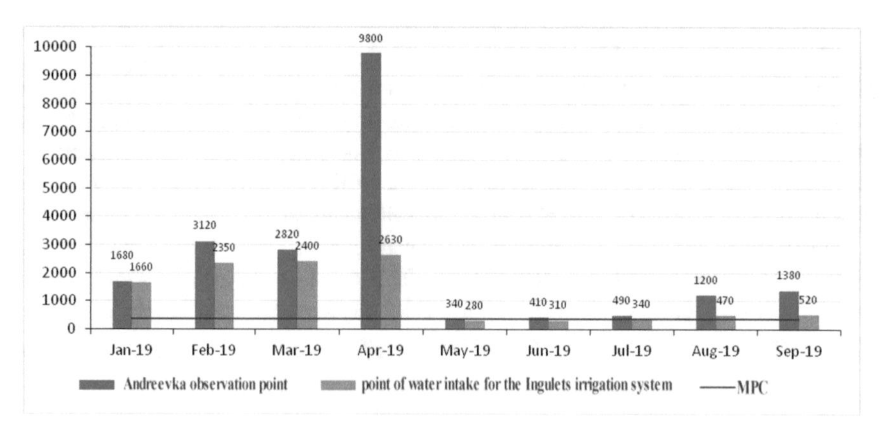

Fig. 5. Water quality according to the anion-chlorine indicator before water intake (Andreevka observation point) and when taken to the Ingulets irrigation system in 2019.

A pulse method is proposed in a combined control system with inputs, outputs and feedback when making decisions. To justify the pulsed flushing method, water samples were taken from different depths (1 m and 3 m from the ground at the main pumping station of the Ingulets irrigation system) before flushing and during flushing. It was found that according to agronomic criteria, as a result of washing, a decrease in the content of the anion-chlorine (Cl^-) index is observed, and the water quality changes from "unsuitable" to "limited suitable" (Fig. 6). As a result of the analysis of the washing process, a rapid improvement (in 1–2 days) of water quality indicators for irrigation was recorded, which is important for water intake to the Ingulets irrigation system.

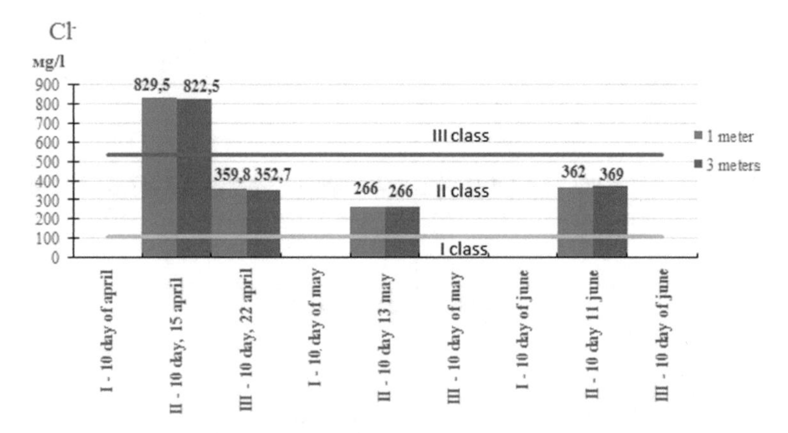

Fig. 6. Dynamics of water quality for irrigation at the water intake point of the Ingulets irrigation system in terms of anion-chlorine (Cl^-) in the process of flushing the Ingulets river from the Karachunivske reservoir

But after some time, the concentration begins to increase again, which means that flushing acts temporarily, and not the entire growing season. So at a certain time moment one more impulse is needed, which will reduce the value of anion-chlorine and other indicators to standard values.

According to the methodology [13, 14], a scenario analysis of integrated control options was carried out using a system of criteria and their comparison with the existing flushing method. The optimal scenario was selected, providing for the operational management of the river flushing according to the content of toxic anion-chlorine ions to prevent the negative process of soil salinization under irrigation conditions. In this scenario the environmental component of the flushing is improved according to criterion (3) so that at the first stage, a small impulse is supplied for 7 days with a flow rate of 8 m^3/s. It ensures that the channel is flushed without water exporting to the river floodplain.

For operational control by criterion (4), the pulse method is used. At the same time, a washing pulse plan is set, which is adjusted by the feedback loop in the combined control system based on the results of water quality measurements at Andreevka observation point and at the point of water intake for the Ingulets irrigation system (Fig. 5). The management scenario takes into account the average ten-day volumes of water intakes for irrigation (Fig. 7).

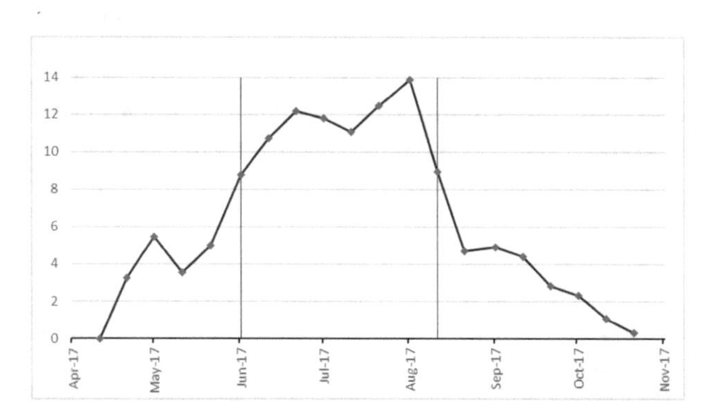

Fig. 7. Water intake volumes for the Ingulets irrigation system in 2017 (in billion m^3).

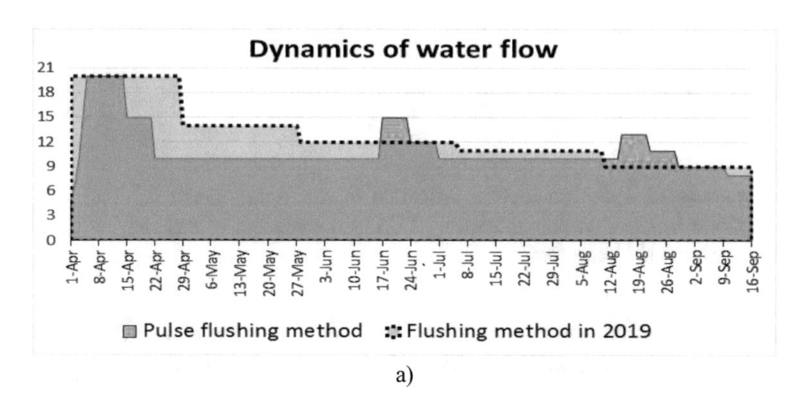

a)

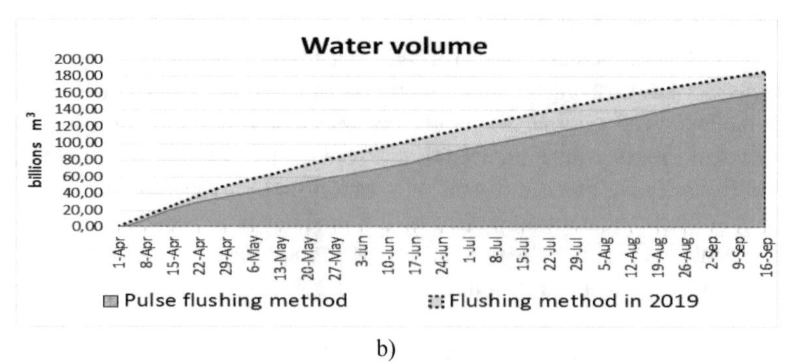

b)

Fig. 8. Dynamics of water flow (a) and water volume (b) in different methods of flushing control.

A comparative analysis of the two flushing methods showed the great efficiency of the combined control system in terms of the amount of water withdrawals for flushing (Fig. 8) and environmental criteria. Comparing with existing method of flushing without feedback in the combined control system, the water quality remains within the MPC [17]. According to the economic criterion (2), the proposed combined management system uses only 160 million cubic meters water for flushing the river, and the existing flushing system is much larger - 190 million cubic meters (Fig. 8) [5]. It is recommended to increase the washing period from April 1 to September 15.

6 Discussion

The most effective model for water use control and ecological rehabilitation of rivers at the present stage is integrated management according to the basin principle using non-linear criteria. A formalized system of criteria for making decisions allows to control water balance in reservoirs according to technological indicators; by environmental criteria it regulates the water flow for flushing the river, conduct environmental releases and maintain satisfactory water quality at the points of water intake; by economic criteria it evaluates and selects the best management scenarios. The pulse method within combined control system is a toolbox for implementing the optimal scenario according to economic criteria, ensuring a good ecological condition and river rehabilitation.

We also analyzed other ways of improving the Ingulets River state, proposed by public organizations. However, dilution of water to standard values before it is discharged into the river requires 5–6 times more water (up to 700 million m^3), which leads to unacceptably significant economic costs, social and technical problems associated with laying a pipeline for discharging contaminated water into the Dnipro river or the Black Sea. At the same time, environmental problems are not solved, since the polluted waters of the quarries and tailings will be filtered into the Ingulets River.

7 Conclusions

For water resources manage within the river basin it is recommended to use an integrated approach according to a system of non-linear technological, economic and environmental criteria. The river flushing management system is selected based on a scenario-based ecosystem analysis of options. The best option is determined by the criteria for saving water resources and the criteria for environmental recovery.

It is recommended as a toolbox for implementation of integrated control to flush the Ingulets river using the pulse method in a combined system that provides both positional and loop control with feedback. An open control line is determined by the planning of pulse sequence in time with water discharge in the Karachunivske reservoir. A feedback loop allows to select or to correct of water discharge based on water quality measurements.

To increase the efficiency of river flushing management, it is recommended to improve the State water monitoring with providing feedback in the integrated management system. Monitoring is carried out by inputs (value of water supply pulse) and outputs (water quality for irrigation intake). For management measurement of water quality is done at the Andreevka observation point and at the Ingulets irrigation system during intakes, the displacement of the prism to the river Dnipro is monitored.

In the future, it is recommended to create a control center in the board of the Ingulets irrigation system for the automated collection and transmission of information, to develop an intelligent system for analysis and management decisions.

References

1. Kovalchuk, P., Kovalenko, R., Balykhina, H.: Methodological features of the concept of water use system management using basin principle. Land Reclam. Water Manag. **107**(1), 17–23 (2018). https://doi.org/10.31073/mivg201801-115
2. Dukhovny, V., Sokolov, V., Manthrithilake, H.: Integrated Water Resources Management: Putting Good Theory into Real Practice. Central Asian Experience, SIC ICWC, Tashkent (2009)
3. UE Water Framework Directive 2000/60/EC. Definition of Main Terms Google Scholar
4. National paradigm of sustainable development for Ukraine. Public Institution «Institute of Environmental Economics and Sustainable Development of the National Academy of Sciences of Ukraine», Kyiv, 72p. (2016). (in Ukrainian)
5. Burlaka, B.: The flushing Inhulets river in 2011. Water Manag. Ukraine **5**, 17–18 (2011). (in Ukrainian)
6. Babiy, P.O., Lisyuk, O.G.: Man-made flood on the river Ros. Water Manag. Ukraine (5), 4–6 (2010). (in Ukrainian)
7. Kovalchuk, P.I., Rozhko, V.I., Balikhina, G.A., Demchuk, O.S.: Mathematical modeling of water exchange scenarios in the Dnipro-Donbass channel system. Math. Comput. Model. Ser.: Tech. Sci. (17), 71–80 (2018). (in Ukrainian)
8. Seng Mah, D., Putuhena, F., Bt Rosli, N.A.: Modelling of river flushing and water quality in a tributary constrained by barrages. Irrigat. Drain. Syst. **25**(4), 427–434 (2011)
9. Bledsoe, B., Beeby, J., Hardie, K..: Evaluation of flushing flows in the fraser river and its tributaries, Technical report, p. 182 (2013)
10. "Environmental assessment accelerated Mahaweli development program" in US Agency for International Development, TAMS, New York, p. 389 (1980)
11. Banas, N.S., Hickey, B.M., Maccready, P.: Dynamics of Willapa Bay Washington: a highly unsteady partially mixed estuary. J. Phys. Oceanogr. **34**, 2413–2427 (2004)
12. Investigation of options to increase the flood mitigation performance of Wivenhoe Dam. Final report. GHD, Brisben, p. 146 (2011)
13. Kovalchuk, P., Balykhina, H., Kovalenko, R., Demchuk, O., Rozhon, V.: Information technology of the system control of water use within river basins. In: Hu, Z., Petoukhov, S., Dychka, I., He, M. (eds.) Advances in Computer Science for Engineering and Education. ICCSEEA 2018. Advances in Intelligent Systems and Computing, vol. 754. Springer, Cham (2019). https://doi.org/10.1007/978-3-319-91008-6_13
14. Kovalchuk, P., Balykhina, H., Demchuk, O., Kovalchuk, V.: Modeling of water use and river basin environmental rehabilitation. In: IEEE XIIth International Scientific and Technical Conference Computer Science and Information Technologies (CSIT 2017), 5–8 September 2017, Lviv, Ukraine, pp. 468–472 (2017)

15. Kovalenko, R.Yu., Kovalchuk, P.I.: Analysis of water quality management methods for irrigation at washes of Ingulets river by Dnipro water. In: Inductive Modeling of Complex Systems, Kiev, no. 6, pp. 90–96 (2014). (in Ukrainian)
16. Ivakhnenko, A., Peka, Y., Vostrov, N.: The Combined Method of Modeling Water and Oil Fields. Naukova Dumka, Kyiv (1984). (in Russian)
17. DSTU 2730:2015. Quality of the environment. The quality of natural water for irrigation. Agronomic criteria. Minekonomrozvytku Ukrainy, Kiev (2015). (in Ukrainian)
18. Kalburgi, P.B., Shareefa, R.N., Deshannavar, U.B.: Development and evaluation of BOD–DO model for river Ghataprabha near Mudhol (India), using QUAL2K. Int. J. Eng. Manuf. (IJEM), 15–25, 8 March 2015 https://doi.org/10.5815/ijem.2015.01.02

Research on Logistics Center Location-Allocation Problem Based on Two-Stage K-Means Algorithms

Meng Wang(✉) and Xuejiang Wei

School of Logistics, Wuhan Technology and Business University, Wuhan 430065, China
535150201@qq.com, xuejiangw@foxmail.com

Abstract. Economy and timeliness are two key issues to be considered in the location of logistics center. This paper makes a logistics center location-allocation model with economy (i.e. total transportation turnover level) as the decision goal, and timeliness (i.e. the maximum transportation distance tolerated) as the constraint condition. Then a two-stage algorithm based on K-means clustering is proposed to solve the model. Firstly, it uses K-means algorithm to calculate the initial location and service area division of logistics center from the perspective of economy. Secondly, in each service area, location scheme is optimized and adjusted to with the lowest total transportation turnover. The example shows that the algorithm can effectively solve the location-allocation problem of logistics center with time constraint.

Keywords: Logistics center · Location-allocation model · Two-stage K-means algorithm · Time constraint

1 Introduction

Logistics center is an important node of the current social and enterprise logistics system, which undertakes the tasks of transportation, safekeeping, carry, wrapping, distribution processed in the region. In the construction of logistics center, large-scale manufacturing enterprises or trading companies usually put economy and timeliness in an important position. Economy is the core problem of logistics center location. We should consider the demand and distribution of customers comprehensively, and strive to make the lowest transportation cost from logistics center to users. The timeliness reflects service level and customer satisfaction directly. In modern enterprise operation, it is closely related to the market competitiveness and plays a more and more prominent role. It is also an important issue that enterprises should consider when selecting the location of logistics center. It is of practical significance to strengthen the research on the location of logistics center with time constraint.

The location of logistics center is generally a multi facility location problem, that is, to select the optimal location of two or more new facilities with several demand points known. Multi facility location, also known as location-allocation problem, not only

Z. Hu et al. (Eds.): ICCSEEA 2020, AISC 1247, pp. 52–62, 2021.
https://doi.org/10.1007/978-3-030-55506-1_5

selects the location, but also determines scale capacity and service allocation scheme of each logistics center.

This research focuses on logistics center location-allocation problem with time constraint, builds a location-allocation model and proposes optimization algorithm, in order to meet the target of both economic and efficiency. Clustering algorithm is an effective way to solve the location-allocation problem, but it needs to be improved to solve the problem of timeliness. In this paper, based on existing literature research and applications, we are engaged in put forward improved algorithm, and verify the validity of algorithm through numerical experiments.

2 Literature Review

In the existing literature on location allocation, Cristiana L. Lara [1] brought out global optimization algorithm for capacitated multi-facility continuous location-allocation problems. Ehsan Mirzaei [2] used exact algorithms for solving a bi-level location–allocation problem considering customer preferences. Mohammad Allahbakhsh [3] focused on crowdsourcing planar facility location allocation problems. Maryam Abareshi [4] made a bi-level model to evaluate the capacitated p-median facility location problem with the most likely allocation solution. Hossein Mojaddadi Rizeei [5] combined an optimised PF probability model with ideal location allocation methods on a geographic information system platform to construct the proposed model for achieving accurate ERC spatial planning. Dong Kaifan [6], Xie Ting [7], Feng jianrui [8], Tian Shubing [9] constructed the location optimization model for the logistics center location problem with time constraint, and used the sequential quadratic programming algorithm (SQP), the shuffled frog leaping algorithm (SFLA) based on global collaborative search, the multi-objective optimization search algorithm, and the adaptive differential evolution algorithm.

Clustering algorithm is convenient and mature, and is often used in facility location planning. Bhupesh Rawat [10] performed a comparative analysis of various clustering algorithms. Sharfuddin Mahmood [11] proposes a modified version of k-means algorithm with an improved technique to divide the data set into specific numbers of clusters with the help of several check point values. Gbadamosi Babatunde [12] used K-means classification algorithm and Multiple Linear Regression to solve practical problem. E. Sandeep Kumar [13] presented decision trees and principal component analysis (PCA) with K-means clustering to compute optimal locations. Yunqian Wang [14] used K-means algorithm to screen potential fire station location. David S. Zamar [15] presented a constrained k-means algorithm and nearest neighbor approach to the BCP, which minimizes travel time and hence fuel consumption. Lu Linglan [16] uses the K-means clustering algorithm based on the density idea, and proposes to determine the optimal clustering number by the difference mean within the cluster. Yu Xiaohan [17] proposed a constraint clustering algorithm based on obstacle distance considering geographical obstacles such as rivers and highways. Li Jiecheng [18] provided solutions for the location of logistics distribution facilities, based on the BIRCH clustering algorithm and gravity method of Dijkstra.

In the existing literature, cluster analysis has a good application in solving the problem of location economy, but it is not enough in time constraint. This paper focuses on the use of K-means algorithm principle for location-allocation problem, while considering time constraint for adjustment and optimization.

3 Location-Allocation Model of Logistics Center with Time Constraint

Besides site selection, there are two important factors in location-allocation problem, which are scale capacity and service allocation scheme. Scale capacity, that is the total amount of distribution of logistics center, is equal to the total amount of demand of each demand point or subordinate node responsible for supply. Service allocation scheme refers to which demand points or subordinate nodes are supplied by each logistics center.

Demand factor is the first premise of logistics center layout planning. All planning schemes must be optimized and adjusted on the basis of meeting customer demand. Otherwise, it will be no research value. The demand factors mainly consider the requirements of service level setting and demand scale distribution. In the optimization analysis of logistics center location, time constraint can be set according to service level, and economic decision-making objectives can be established according to demand data, so as to establish the location allocation model.

Therefore, this paper constructs a logistics center location allocation model with economy (i.e. the total transportation turnover level) as the decision goal, and timeliness (i.e. the maximum transportation distance tolerated) as the constraint condition. The objective function is:

$$\min Z = \sum_{i=1}^{n} \sum_{j=1}^{m} l_{ij} w_j d_{ij} \tag{1}$$

Z is the total transportation turnover. n is the quantity of logistics centers. i is the number of logistics centers (i = 1, 2, 3,…n). m is the quantity of demand points. j is the number of demand points (j = 1, 2, 3,… m). w_j is the demand of No.j demand point. l_{ij} indicates whether No.i logistics center provides service for No.j demand point.

$$l_{ij} = \begin{cases} 1 & \textit{when No.i logistics center serves No.j demand point} \\ 0 & \textit{when No.i logistics center not serves No.j demand point} \end{cases} \tag{2}$$

d_{ij} is the distance from No.i logistics center to No.j demand point. For the convenience of calculation, the longitude and latitude of each node can be converted into the Cartesian coordinate, and the distance between the points can be converted into the actual distance according to a certain conversion coefficient.

$$d_{ij} = K\sqrt{\left(X_i - x_j\right)^2 + \left(Y_i - y_j\right)^2} \tag{3}$$

k is the distance conversion factor, X_i, Y_i are the abscissa and ordinate of No.i logistics center, and x_j, y_j are the abscissa and ordinate of No.j demand point.

According to Formula (1) (3), the objective function of logistics center location is as follows:

$$\min Z = \sum_{i=1}^{m}\sum_{j=1}^{n} l_{ij}w_jK\sqrt{\left(X_i - x_j\right)^2 + \left(Y_i - y_j\right)^2} \tag{4}$$

s.t.

$$l_{ij}K\sqrt{\left(X_i - x_j\right)^2 + \left(Y_i - y_j\right)^2} \leq D_j, (i = 1, 2, \ldots, m;\ j = 1, 2, \ldots, n) \tag{5}$$

$$\sum_{i=1}^{m} l_{ij} = 1, (j = 1, 2, \ldots, n) \tag{6}$$

D_j is the maximum allowable distance to No.j demand point.

Formula (5) indicates that the distance between each proposed logistics center and its service demand point must be less than the maximum allowable distance, to ensure the timeliness of distribution. Formula (6) indicates that the materials required of each demand point are supplied by only one logistics center.

4 Two-Stage Solution Based on K-Means Clustering Algorithm

K-means clustering algorithm is a kind of partition clustering algorithm, which has a good application value in solving resource matching. K-means clustering algorithm, combined with gravity method to determine the location, can effectively solve the problem of facility location-allocation. The core idea of the algorithm is to find out K clustering centers (C_1, C_2, … C_K), so that the quadratic sum of distance from each data point 'X_i' and its nearest clustering center 'C_v' is minimized (it is also called 'Deviation'). The basic steps are as follows:

1) Initialize K sub cluster centers randomly or based on some prior knowledge;
2) Each object in the dataset is assigned to the nearest sub cluster (based on the distance function);
3) According to the current division, recalculate the center of sub cluster (generally the barycenter);
4) Repeat steps 2) and 3) until each sub cluster does not change.

On the basis of K-means clustering algorithm, because the location problem also includes time constraint, that is Formula (5), the model solution can be implemented in two stages. As shown in Fig. 1.

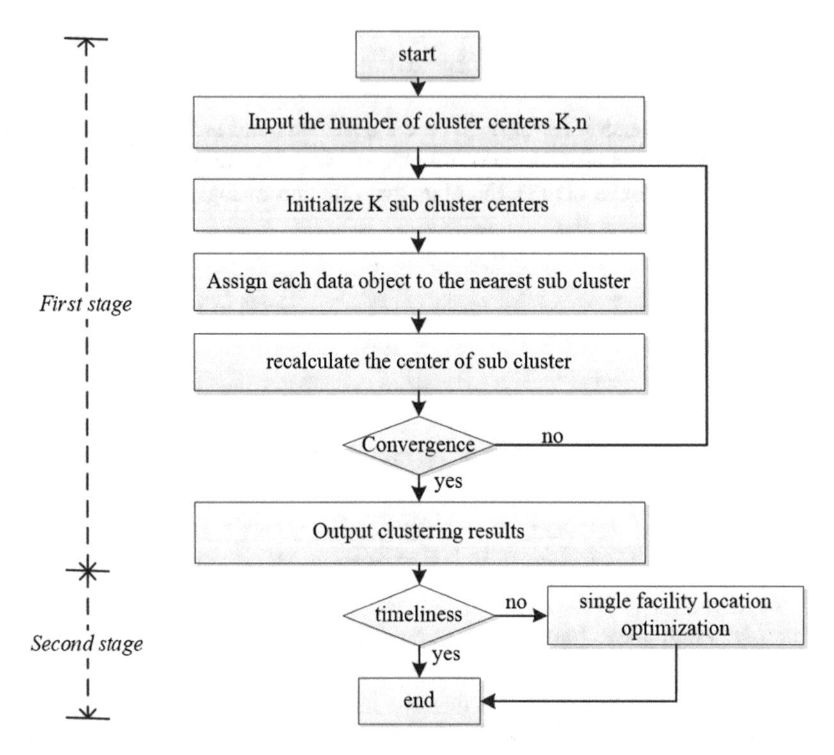

Fig. 1. Two-stage solution based on K-means clustering algorithm

The first stage, without considering the time constraint, only from the perspective of economy, according to Formula (4) and (6), we can preliminary calculation of the location of logistics center and service area division (i.e. the service allocation scheme of logistics center). It is mainly use k-means algorithm.

The second stage, in each service area preliminarily divided, considering time constraint, we use Formula (4) and (5) to optimize and adjust the location scheme (this problem is transformed into single facility location optimization problem), and finally determine the location scheme with the lowest total transportation turnover.

4.1 Location-Allocation Without Considering Timeliness

K-means algorithm can be used to solve the problem without considering the timeliness.

1) Preliminary allocation plan. N support objects are randomly divided into k groups, and the allocation matrix S is initially formed.

$$S = (l_{ij})_{k \times n} \tag{7}$$

2) Location iteration operation. In each group of allocation, gravity-differential method can be used to solve the single facility location problem. The method of least

squares is mainly used here. After several iterations, the location scheme of logistics center is obtained and the total transportation turnover is calculated.

3) Adjust allocation plan. Calculate the transportation turnover from No.j demand point to No.i logistics center.

$$Z_{ij} = w_j K \sqrt{(X_i - x_j)^2 + (Y_i - y_j)^2},\ i = 1, 2, \ldots, m; j = 1, 2, \ldots, n \tag{8}$$

We should sort the calculation results. If the original logistics center allocation scheme of No.j demand point is not the one with the lowest transportation turnover, select the lowest one as the supply point serviced to No.j demand point. We can make a new assignment matrix S'.

4) Repeat step 2) and 3) until the allocation scheme is no longer adjusted, that is, all demand points are supplied by the logistics center with the lowest transportation turnover. At this time, the supply allocation plan is the optimal plan, and the logistics center location is the optimal location considering the economic factors.

4.2 Location Optimization Considering Timeliness

Through the above process, the optimal service allocation scheme and location scheme are obtained, but this scheme does not consider the timeliness requirements. When considering the Formula (5), the above scheme should be optimized and adjusted.

On the basis of the determined service scheme, Formula (4) and (5) can be transformed into:

$$\min Z = \sum_{j=1}^{n} w_j K \sqrt{(X - x_j)^2 + (Y - y_j)^2} \tag{9}$$

s.t.

$$K \sqrt{(X - x_j)^2 + (Y - y_j)^2} \le D_j, (i = 1, 2, \ldots, m; j = 1, 2, \ldots, n) \tag{10}$$

X and Y represent the abscissa and ordinate of the proposed logistics center. In fact, it transfers into a single facility optimal location problem with time constraint.

It is a constrained nonlinear programming problem. To solve this problem, we can use the fmincon() optimization function of MATLAB optimization toolbox to solve it.

Using fmincon () standard model to express the Formula (9) (10) logistics center location model, we can get:

$$f(x) = \sum_{j=1}^{n} w_j K \sqrt{(X - x_j)^2 + (Y - y_j)^2} \tag{11}$$

s.t.

$$C(x) = K \sqrt{(X - x_j)^2 + (Y - y_j)^2} - D_j, (j = 1, 2, \ldots, n) \tag{12}$$

Address coordinates upper and lower:

$$ub = \left[\max\{x_j\}, \max\{y_j\}\right]^T, j = 1, 2, \ldots, n \tag{13}$$

$$lb = \left[\min\{x_j\}, \min\{y_j\}\right]^T, j = 1, 2, \ldots, n \tag{14}$$

There is no linear constraint, so:

$$A = [\,], b = [\,], Aeq = [\,], beq = [\,] \tag{15}$$

fmincon() format is:

$$[x, fval] = fmincon(fun, x_0, A, b, Aeq, beq, lb, ub, nonlcon) \tag{16}$$

x output the result of facility location. *fval* output the value of transportation turnover. *fun* and *nonlcon* are the M-files of objective function and timeliness constraint respectively. A, b, Aeq, beq are null, and lb, ub are the lower and upper limits of location coordinate respectively. The initial feasible solution is the location scheme without considering the timeliness.

By substituting the above M-files and parameters into the fmincon(), the optimal location result (X, Y) can be obtained.

5 Example Analyses

According to the above model and algorithm, this paper takes the location of a regional logistics center as an example for simulation analysis. It is known that the geographical coordinates and demand scale of the subordinate demand nodes in the region are shown in Table 1 (Data source: author's information collection according to enterprise website). It is proposed to establish two regional logistics centers to meet the needs of each node (k = 2). At the same time, the logistics center needs to meet the timeliness requirements of 4-h, that is the maximum distribution distance between the demand point and the logistics center is set as 280 km.

Table 1. Distribution table of demand points of regional logistics center (geographic coordinates)

Demand point	Geographic coordinate	Demand scale	Demand point	Geographic coordinate	Demand scale
1 Huzhou	120.08 e, 30.90 n	3	8 Danyang	119.57 e, 32.00 n	1
2 Xuzhou	117.18 e, 34.27 n	3	9 Ningbo	121.55 e, 29.88 n	1
3 Xiamen	118.08e, 24.48 n	4	10 Yiwu	120.07 e, 29.30 n	1
4 Changle	119.52 e, 25.97n	1	11 Yixing	119.82 e, 31.35 n	1
5 Jinjiang	118.58 e, 24.82 n	1	12 Putian	119.00 e, 25.43 n	1
6 Shanghai	121.47 e, 31.23 n	1	13 Wuxi	120.30 e, 31.57 n	1
7 Zhoushan	122.20 e, 30.00 n	1			

5.1 Location Allocation of Regional Logistics Center Without Considering Timeliness

The geographical coordinates of each demand point in Table 1 are converted into Cartesian coordinates. It can be realized by 'Coord' coordinate transformation software. The results after conversion are shown in Table 2. Considering the road curvature, the distance conversion factor 'K' is set as 1.2.

Table 2. Distribution table of demand points of regional logistics center (Cartesian coordinates)

Demand point	Cartesian coordinate (X, Y)	Demand scale	Demand point	Cartesian coordinate (X, Y)	Demand scale
1 Huzhou	(3435838.060667, 1081799.796966)	3	8 Danyang	(3555508.483514, 1026701.570331)	1
2 Xuzhou	(3798258.349210, 792911.215361)	3	9 Ningbo	(3330933.397048, 1230428.693936)	1
3 Xiamen	(2714615.014886, 913831.083282)	4	10 Yiwu	(3257918.041269, 1090327.045407)	1
4 Changle	(2885258.090502, 1053330.553499)	1	11 Yixing	(3484517.872663, 1054243.443761)	1
5 Jinjiang	(2753949.131823, 963352.244769)	1	12 Putian	(2823185.788924, 1003394.050708)	1
6 Shanghai	(3480707.738588, 1212648.894062)	1	13 Wuxi	(3511508.227895, 1098626.450760)	1
7 Zhoushan	(3348652.521966, 1292560.092180)	1			

1) Determine the initial allocation matrix. Divide the above 13 demand points into 2 groups. The first group is {Huzhou, Xuzhou, Xiamen, Changle, Jinjiang, Shanghai, Zhoushan}. The second group is {Danyang, Ningbo, Yiwu, Yixing, Putian, Wuxi}. The initial allocation matrix is

$$S_0 = (l_{ij})_{2\times 13} = \begin{bmatrix} 1111111000000 \\ 0000000111111 \end{bmatrix}$$

2) Iterative operation. Take $\varepsilon = 0.1$, after the 1st iteration and turnover calculation (can be completed by using the iterative operation function of Excel), Table 3 can be obtained.

   After adjustment according to the rules, the first group is {Xiamen, Changle, Jinjiang, Zhoushan, Ningbo, Yiwu, Putian}, the second group is {Huzhou, Xuzhou, Shanghai, Danyang, Yixing, Wuxi}, and a new allocation matrix is obtained.

$$S_1 = (l_{ij})_{2\times 13} = \begin{bmatrix} 0011101011010 \\ 1100010100101 \end{bmatrix}$$

At this time, the total transportation turnover Z_1 is 8615480.458.

Table 3. 1st iteration result

Preliminary location	Demand point	Distance to center 1	Distance to center 2	Whether to adjust
Logistics center 1 (3372599.886599, 1062817.014239)	1 Huzhou	79231.01847	22766.68937	Y
	2 Xuzhou	604821.7461	535862.7344	Y
	3 Xiamen	809569.5737	909264.481	N
	4 Changle	584920.9413	683580.18	N
	5 Jinjiang	751914.6729	851665.4494	N
	6 Shanghai	221714.0857	164879.842	Y
	7 Zhoushan	277185.3454	287256.8937	N
Logistics center 2 (3454382.572325, 1077794.515976)	8 Danyang	223728.021	135960.1083	N
	9 Ningbo	207255.5693	235569.9162	Y
	10 Yiwu	141522.3217	236236.7674	Y
	11 Yixing	134695.0781	45895.56925	N
	12 Putian	663141.9099	762679.9651	Y
	13 Wuxi	172139.7512	72966.49764	N

Table 4. 3rd iteration result

Preliminary location	Demand point	Distance to center 1	Distance to center 2	Whether to adjust
Logistics center 1 (2714615.014886, 913831.083282)	3 Xiamen	0	909932.5475	N
	4 Changle	264488.1708	683520.4922	N
	5 Jinjiang	75890.09286	852100.9980	N
	12 Putian	168893.9279	762930.4016	N
Logistics center 2 (3454172.372983, 1081279.908875)	1 Huzhou	888629.1046	22010.01818	N
	2 Xuzhou	1308442.7640	538734.0660	N
	6 Shanghai	986769.3800	160826.5764	N
	7 Zhoushan	886246.2866	283397.6128	N
	8 Danyang	1018121.7340	138118.9175	N
	9 Ningbo	831455.7410	232171.9526	N
	10 Yiwu	685502.5635	235755.3038	N
	11 Yixing	939122.5158	48771.1032	N
	13 Wuxi	981647.0312	71882.93257	N

According to the new allocation scheme S_1, the site is relocated. After the 2nd and 3rd iteration, Table 4 is obtained.

It can be seen from Table 4 that the allocation scheme at this time is in line with principle of proximity, and does not need to be adjusted to be the best service allocation scheme. The coordinates of Logistics Center 1 (2714615.014886, 913831.083282) are converted into (118.08 e, 24.48 n) (located in Xiamen, Fujian), with design capacity of 7 units and transportation turnover of 509272.1915. The coordinates of Logistics Center 2 (3454172.372983, 1081279.908875) are converted into (120.08 e, 31.06 n) (located in Wuzhong District, Suzhou, Jiangsu), the design capacity is 13 units, and the transportation turnover is 2853156.652; at this time, the total transportation turnover is 3362428.8435.

5.2 Location Optimization Considering Timeliness

For logistics center 1, the distance from center to each demand point is all less than 280 km, so there is no need to adjust. The address is (118.08 e, 24.48 n) (located in Xiamen, Fujian), and the transportation turnover is 509272.1915.

For logistics center 2, through the distance from Center to each demand point, it can be seen that the distance from logistics center 2 to '2 Xuzhou' and '7 Zhoushan' is more than 280 km, which does not meet the timeliness constraint and needs to be adjusted. Using the fmincon() function of MATLAB, the logistics center address 'x' and total transportation turnover '*fval*' can be outputted.

The output result is 'x = [3591917.69498821; 1059348.79861631]', and the transportation turnover of logistics center 2 is '*fval* = 3484600.085'. The site is located in Gaogang District, Taizhou, Jiangsu. Although the transportation turnover of logistics center 2 increased from 2853156.652 to 3484600.085, the timeliness requirements of 4 h were guaranteed, and the economic and timeliness service objectives were taken into account.

6 Conclusions

To solve the location-allocation problem of logistics centers with time constraint, this paper proposes a two-stage algorithm based on K-means clustering. Firstly, the k-means algorithm is used to calculate the initial location and service area division of logistics center from the perspective of economy. Secondly, in each service area, considering time constraint, the location scheme is optimized and adjusted to with the lowest total transportation turnover.

Calculation example shows that the algorithm can overcome the shortcomings of traditional K-means algorithm in solving the problem with timeliness constraint. So it has good applicability with good balance in economy and timeliness. Through this research, the application of K-means algorithm in facility location can be further expanded.

The limitations of this research are as follows: 1) the effectiveness of the model is not enough to solve the location problem when the number of sub cluster centers changes randomly, so it needs to further increase constraint of facility capacity in location model; 2) the value of distance transformation factor 'K' in the calculation process is arbitrary, which will affect the accuracy of the calculation results, so it needs to combine the regional characteristics in evaluating the value of factor 'K'.

Acknowledgment. This project is supported by: Doctoral research fund subsidized project of WTBU (D2018007); Natural Science Foundation of Hubei Province(2019CFC930); Distinguished Young and Middle-aged Team Program for Scientific and Technological Innovation in Higher Education of Hubei (T201938).

References

1. Lara, C.L., Trespalacios, F., Grossmann, I.E.: Global optimization algorithm for capacitated multi-facility continuous location-allocation problems. J. Global Optim. **71**(4), 871–889 (2018)
2. Mirzaei, E., Bashiri, M., Shemirani, H.S.: Exact algorithms for solving a bi-level location–allocation problem considering customer preferences. J. Ind. Eng. Int. **15**(3), 423–433 (2019)
3. Allahbakhsh, M., Arbabi, S., Galavii, M., et al.: Crowdsourcing planar facility location allocation problems. Computing **101**(3), 237–261 (2019)
4. Abareshi, M., Zaferanieh, M.: A bi-level capacitated P-median facility location problem with the most likely allocation solution. Transp. Res. Part B: Methodol. **123**, 1–20 (2019)
5. Rizeei, H.M., Pradhan, B., Saharkhiz, M.A.: Allocation of emergency response centres in response to pluvial flooding-prone demand points using integrated multiple layer perceptron and maximum coverage location problem models. Int. J. Disaster Risk Reduct. **8**(3) (2019)
6. Dong, K.-F., Gan, H.-C., Zhang, H.-Z.: Distribution center location model based on economics and timeliness. J. Univ. Shanghai Sci. Technol. **35**(04), 336–339+344 (2013). (in Chinese)
7. Xie, T., Zhou, Q., Shi, U.-F.: Nodes location of the shaft radial logistics network of agricultural products based on the analysis of time reliability. J. Changsha Univ. Sci. Technol. (Nat. Sci.) **12**(02), 15–20 (2015). (in Chinese)
8. Feng, J., Gai, W.: Research on multi-objective optimization model and algorithm for reserve site selection of emergency materials. J. Saf. Sci. Technol. **14**(06), 64–69 (2018)
9. Tian, S., Yang, Y., Tian, Y., et al.: On the optimized construction site choice for the oil spilling emergency point in the middle and lower reaches of Yangtze river. J. Saf. Environ. **17**(1), 90–93 (2017). (in Chinese)
10. Rawat, B., Dwivedi, S.K.: Analyzing the performance of various clustering algorithms. Int. J. Mod. Educ. Comput. Sci. (IJMECS) **11**(1), 45–53 (2019)
11. Mahmood, S., Rahaman, M.S., Nandi, D., Rahman, M.: A proposed modification of K-means algorithm. Int. J. Mod. Educ. Comput. Sci. (IJMECS) **7**(6), 37–42 (2015)
12. Babatunde, G., Emmanuel, A.A., Oluwaseun, O.R., Bunmi, O.B., Precious, A.E.: Impact of climatic change on agricultural product yield using K-means and multiple linear regressions. Int. J. Educ. Manag. Eng. (IJEME) **9**(3), 16–26 (2019)
13. Kumar, E.S., Talasila, V., Rishe, N., Kumar, T.V.S., Iyengar, S.S.: Location identification for real estate investment using data analytics. Int. J. Data Sci. Anal. **8**(3), 299–323 (2019)
14. Wang, Y.: Optimization on fire station location selection for fire emergency vehicles using K-means algorithm. In: Proceedings of the 2018 3rd International Conference on Advances in Materials, Mechatronics and Civil Engineering (ICAMMCE 2018), pp. 332–342 (2018)
15. Zamar, D.S., Gopaluni, B., Sokhansanj, S.: A constrained K-means and nearest neighbor approach for route optimization in the Bale collection problem. IFAC-PapersOnLine **50**(1), 12125–12130 (2017)
16. Lu, L.-L., Qin, J.-T.: Multi-regional logistics distribution center location method based on improved K-means algorithm. Comput. Syst. Appl. **28**(08), 251–255 (2019). (in Chinese)
17. Yu, X.-H., Wang, D.: Study on city express delivery area division based on constrained K-means clustering algorithm. J. Harbin Univ. Commer. (Nat. Sci. Ed.) **32**(5), 631–634 (2016). (in Chinese)
18. Li, J.-C., Tao, Y.-D., Sun, Y., Gao, C.: Location algorithm of logistics distribution facilities based on BIRCH clustering. Comput. Syst. Appl. **27**(9), 215–219 (2018). (in Chinese)

Mathematical Modeling of Dynamic Heat-Mass Exchange Processes for a Spray-Type Humidifier

Igor Golinko and Iryna Galytska(✉)

National Technical University of Ukraine "Igor Sikorsky Kyiv Polytechnic Institute", Peremogy Str. 37, 03056 Kiev, Ukraine
conis@ukr.net, irinagalicka@gmail.com

Abstract. A dynamic model of processes heat-mass exchange for a spray-type humidifier, which is used in artificial microclimate systems, is obtained. Based on the analysis of the equations of material and thermal balances, a dynamic model is obtained, which is presented in the form of a system of differential equations, in the state space, in the form of transfer functions. The dynamic model makes it possible to analyze the properties of transients for the spray-type humidifier along the main channels of influence in the coordinates of temperature and air moisture content. Using the example of an industrial air conditioner VEZA KCKP-80, a numerical simulation of transient processes for a spray-type humidifier was carried out. The obtained dynamic model is an integral part for a complex mathematical description of the industrial air conditioner. An essential advantage of the model in the state space is the possibility of using a vector-matrix mathematical apparatus for the synthesis, research, and optimization of a multidimensional control system.

Keywords: Dynamical model · Industrial air conditioner · Spray-type humidifier · Heat- mass exchange processes

1 Introduction

Modern technological processes dictate increased requirements for the microclimate of production, on which the quality of the product and the reduction of its cost largely depend. Industrial air conditioners are a real lever of profit for technologies: light, food, agricultural, pharmaceutical, printing, engineering, electronic and a number of other industries. The performance of an industrial air conditioner is largely dependent on the quality of the control system. A significant contribution to the development of the theory of automatic control for air conditioning systems was made in the works of V. Arkhipov, B.V. Barkalov, Yu.S. Cooper, V.N. Bogoslovsky, A. Sotnikov, A. A. Rimkevich, A.Ya. Kreslin, S.V. Nefelov, M.B. Halamizer, R.W. Haines, J.I. Levenhagen, Z. Falko, H. Zhang, F. Ziller and others.

The automatic control system (ACS) of the air conditioner must be considered within the framework of a single computer-integrated complex, which takes into account the technological relationships between the air conditioning equipment [1].

Z. Hu et al. (Eds.): ICCSEEA 2020, AISC 1247, pp. 63–74, 2021.
https://doi.org/10.1007/978-3-030-55506-1_6

The climatic equipment of the artificial microclimate system for the implementation of control algorithms consists of heat exchangers [2–4] and humidifiers. For humidification of air are used: steam generators, spray-type humidification chambers, ultrasonic, film, capillary-porous and other types of humidifiers [11]. The most common in industrial air conditioners are spray-type humidifiers, in which undergo complex heat and mass transfer processes of air humidification.

For the design and production of humidification chambers, researchers are developing static models of humidifiers. However, to develop effective ACS, it is necessary to take into account the dynamic properties of humidification chambers. The purpose of this research is to develop a mathematical model of the spray-type humidifier, the requirement for which is to study the dynamics of heat and mass transfer of climate equipment in the coordinates of temperature and moisture content. This will allow us to synthesize a complex dynamic model for the development of a control system air conditioner. Dynamic models of heat exchangers (heaters and coolers) are considered in [2–4].

2 Using Dynamical Models

Dynamic models of air conditioners are used for the synthesis and analysis of control systems. When designing and optimizing any control system, the structure and parameters of the controller are primarily dependent on the dynamic behavior of a controlled plant. The choice of whether a different law of control is determined by the dynamic properties of a controlled plant. The choice of a law of control is determined by the dynamic properties of a controlled plant. In automation systems, developers use: classical PI controllers, PID controllers; adaptive controllers [5, 6]; controllers with fuzzy logic [7, 8]; controllers with fractional derivatives [9]; intelligent controllers [10]; controllers based on neural networks and many other types of controllers. In their research, experts compare the operation of control systems with different types of controllers and draw conclusions about the quality of management. In the study of control systems with different types of controllers unchanged in the system there is a controlled plant, which characterizes the dynamic behavior of the technological equipment and formalized in the form of differential equations controlled plant. Thus, an adequate mathematical model of a controlled plant is the basis for the synthesis and analysis of a qualitative control system.

3 Dynamical Model for a Spray-Type Humidifier

The following simplifications were made during the development of an analytical model of the dynamics of heat and mass transfer for a spray-type humidifier: heat exchange with the environment is absent, since the thermal losses of modern heaters do not exceed 5%; the model contains two main dynamic elements with lumped parameters (humidification chamber airspace and water pallet); coefficients of heat and mass exchange averaged and their values are reduced to the corresponding experimentally obtained parameters: E - the coefficient of efficiency of adiabatic humidification and

μ - the coefficient of air humidification [4–6]; the physical properties of the material flows are brought to the averaged values of the working range. The calculation scheme of the water heater is shown in Fig. 1.

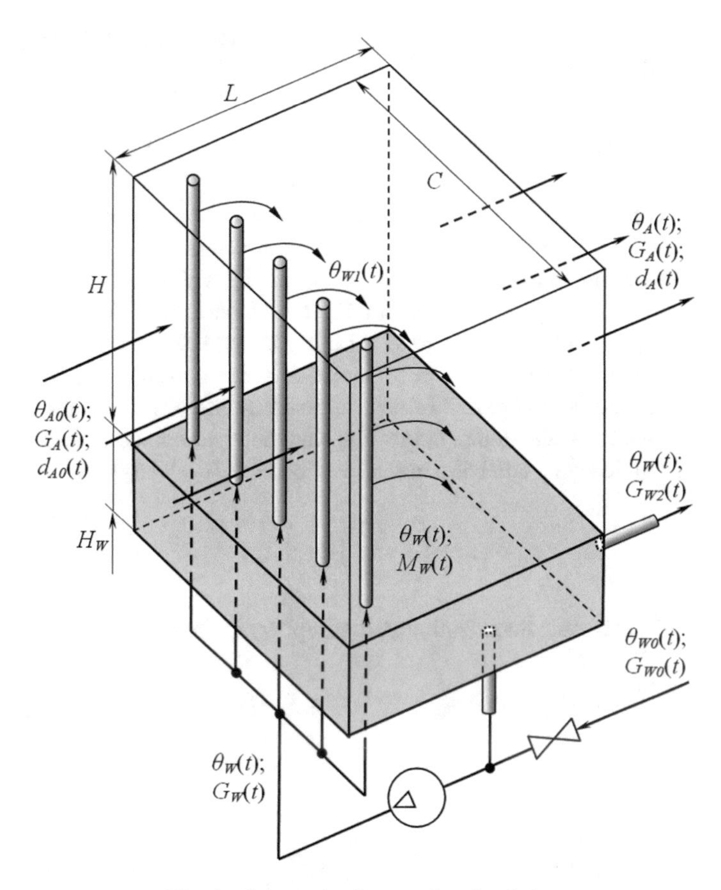

Fig. 1. Schematic diagram for simulation

The following notation is adopted on the diagram. Air with temperature $\theta_{A0}(t)$ and moisture content $d_{A0}(t)$ enters into the camera airspace volume $V_A = L \times C \times H$, air flow $G_A(t)$. Water is sprayed through the jets with water consumption $G_W(t)$ and input temperature $\theta_{W0}(t)$. Water droplets are heated to temperature $\theta_{W1}(t)$, cooling and humidifying the air to parameters $\theta_A(t)$, $d_A(t)$. Part of the water droplets evaporates and is carried out by air as steam. The mass flow rate of steam is less than 1.5% of the mass flow rate of water $G_W(t)$ pumped by the spray gun. To simplify the simulation, suppose that the mass flow rate of the sprayed water and the flow rate of the droplets returning to the pallet of the humidification chamber are equal. There is medium temperature water $\theta_W(t)$ in the volume $V_W = L \times C \times H_W$ of pallet, the mass of water in the pallet $M_W = const$, and an overflow fuse is in the pallet. If the humidification chamber is controlled by the flow rate $G_{W0}(t)$ or temperature of the supply water $\theta_{W0}(t)$, the waste water is drained through the overflow fuse into the sewer. Neglecting the consumption of water for steam generation, we have $G_{W2}(t) \approx G_{W0}(t)$. Consider the thermal and material balances for an air mixture of a humidification chamber and the pallet with water.

The thermal balance for the humid air in humidification chamber with the above assumptions will take the form:

$$G_A\left[c_A(\theta_{A0}-\theta_A)+\frac{r}{1000}(d_{A0}-d_A)\right]-G_W\,c_W(\theta_{W1}-\theta_W)=c_A\,M_A\frac{d\theta_A}{dt}. \quad (1)$$

where c_A, c_W is heat capacity of the air mixture and water for humidification; r is heat of vaporization; d_{W1} is moisture content of air with saturated steam ($\phi = 100\%$) at saturation temperature θ_{W1}; M_A is mass of air in the volume $V_A = L \times C \times H$ of humidification chamber. The right part of Eq. (1) characterizes the accumulation of heat in the air space of the spray-type humidifier. In the left part, the first term characterizes the amount of heat given off by moist air when cooled in the humidification chamber; the second is the amount of heat released by the water into the air.

The recirculation water temperature θ_{W1} can be approximated by knowing the coefficient of efficiency of adiabatic humidification $E \approx (\theta_{A0}-\theta_A)/(\theta_{A0}-\theta_{W1})$ [11]:

$$\theta_{W1}=\left(1-\frac{1}{E}\right)\theta_{A0}+\frac{1}{E}\theta_A; \quad (2)$$

Given (2) in (1), after linearization and grouping, we have such:

$$T_A\frac{d\,\Delta\theta_A}{dt}+\Delta\theta_A=k_0\,\Delta\theta_{A0}+k_1\,\Delta G_A+k_2\,\Delta\theta_W+k_3\,\Delta d_{A0}+k_4\,\Delta d_A, \quad (3)$$

where

$$K_A=G_A\,c_A+\tfrac{G_W\,c_W}{E};\ T_A=\tfrac{c_A M_A}{K_A};\ k_0=\tfrac{1}{K_A}\left[c_A\,G_A-\left(1-\tfrac{1}{E}\right)c_W\,G_W\right];$$
$$k_1=\tfrac{1}{K_A}\left[c_A\,(\theta_{A0}-\theta_A)+\tfrac{r}{1000}(d_{A0}-d_A)\right];\ k_2=\tfrac{c_W\,G_W}{K_A};\ k_3=\tfrac{r\,G_A}{1000\,K_A};$$
$$k_4=-\tfrac{r\,G_A}{1000\,K_A}.$$

Consider the material balance for the humidification chamber airspace. The mass of moisture accumulated in the humidifier airspace is defined as the difference between the mass flow rates of the input/output steam and the vapor generated during adiabatic humidification:

$$\frac{G_A}{1000}(d_{A0}-d_A)+\frac{G_A}{1000}(d_{W1}-d_{A0})=V_A\frac{d\rho_A}{dt}. \quad (4)$$

The density of moist air is defined as [11]

$$\rho_A = \omega\left(1 + \frac{d_A}{1000}\right), \tag{5}$$

where ω is density of dry air under normal conditions.

The moisture content of the air d_{W1} at the saturated vapor line ($\phi = 100\%$) at temperature θ_{W1} can be determined by the *h-d* diagram [4, 11], or approximately by dependence

$$d_{W1} = A\theta_{W1} + B, \tag{6}$$

where A i B are conversion coefficients. With sufficient engineering accuracy [12] for the temperature range 6…12 °C in (7), you can accept the values of the coefficients A = 0.58 g/(°C kg) and B = 2.2 g/kg. Taking into account (5) and (6) in (4), after linearization and grouping, we obtain:

$$T_d \frac{d\,\Delta d_A}{dt} + \Delta d_A = k_5\,\Delta G_A + k_6\,\Delta\theta_{A0} + k_7\,\Delta\theta_A; \tag{7}$$

where

$$T_d = \frac{\omega\,V_A}{G_A};\ k_5 = \frac{1}{G_A}\left[\left(1 - \frac{1}{E}\right)A\,\theta_{A0} + \frac{A}{E}\theta_A + B - d_A\right];\ k_6 = A\left(1 - \frac{1}{E}\right);\ k_7 = \frac{A}{E}.$$

Neglecting water consumption for steam generation, we have a material balance for a pallet with water:

$$G_{W0}(t) - G_{W2}(t) \approx 0. \tag{8}$$

Consider the thermal balance of the pallet with water for the humidification chamber, given (8):

$$G_{W0}\,c_W(\theta_{W0} - \theta_W) + G_W\,c_W(\theta_{W1} - \theta_W) = c_W\,M_W\frac{d\theta_W}{dt} \tag{9}$$

where c_W is heat capacity of the air mixture and the water for humidification; M_W is mass of water in the pallet in the volume $V_W = L \times C \times H_W$ of the humidifying chamber. After linearization and grouping of similarities in (9) with respect to (2), we obtain:

$$T_W\frac{d\,\Delta\theta_W}{dt} + \Delta\theta_W = k_8\,\Delta\theta_{W0} + k_9\,\Delta G_{W0} + k_{10}\,\Delta\theta_{A0} + k_{11}\,\Delta\theta_A, \tag{10}$$

where

$$K_W = G_{W0} + G_W;\ T_W = \frac{M_W}{K_W};\ k_8 = \frac{G_{W0}}{K_W};\ k_9 = \frac{\theta_{W0} - \theta_W}{K_W};\ k_{10} = \left(1 - \frac{1}{E}\right)\frac{G_W}{K_W};$$
$$k_{11} = \frac{G_W}{E K_W}.$$

Differential Eqs. (3), (7) and (10) represent a dynamical model of heat and mass transfer for the spray-type humidifier chamber:

$$\begin{cases} T_A \frac{d\,\Delta\theta_A}{dt} + \Delta\theta_A = k_0\,\Delta\theta_{A0} + k_1\,\Delta G_A + k_2\,\Delta\theta_W + k_3\,\Delta d_{A0} + k_4\,\Delta d_A; \\ T_d \frac{d\,\Delta d_A}{dt} + \Delta d_A = k_5\,\Delta G_A + k_6\,\Delta\theta_{A0} + k_7\,\Delta\theta_A; \\ T_W \frac{d\,\Delta\theta_W}{dt} + \Delta\theta_W = k_8\,\Delta\theta_{W0} + k_9\,\Delta G_{W0} + k_{10}\,\Delta\theta_{A0} + k_{11}\,\Delta\theta_A\,. \end{cases} \quad (11)$$

We present a system of differential Eqs. (11) in state space:

$$\mathbf{X} = \mathbf{AX} + \mathbf{BU}; \quad (12)$$

$$\mathbf{X}' = \begin{bmatrix} \Delta\theta'_A \\ \Delta d'_A \\ \Delta\theta'_W \end{bmatrix};\ \mathbf{A} = \begin{bmatrix} -1/T_A & k_4/T_A & k_2/T_A \\ k_7/T_d & -1/T_d & 0 \\ k_{11}/T_W & 0 & -1/T_W \end{bmatrix};\ \mathbf{X} = \begin{bmatrix} \Delta\theta_A \\ \Delta d_A \\ \Delta\theta_W \end{bmatrix};$$

$$\mathbf{B} = \begin{bmatrix} k_0/T_A & k_3/T_A & k_1/T_A & 0 & 0 \\ k_6/T_d & 0 & k_5/T_d & 0 & 0 \\ k_{10}/T_W & 0 & 0 & k_8/T_W & k_9/T_W \end{bmatrix};\ \mathbf{U} = \begin{bmatrix} \Delta\theta_{A0} \\ \Delta d_{A0} \\ \Delta G_A \\ \Delta\theta_{W0} \\ \Delta G_{W0} \end{bmatrix}.$$

The analytic solution (11) with respect to the original values $\Delta\theta_A$ and Δd_A can be obtained using the Laplace transform. After the corresponding mathematical transformations, the dynamical model of the spray-type humidifier (11) will be represented by a multidimensional model in the Laplace area:

$$\mathbf{Y} = \mathbf{WZ}, \quad (13)$$

Where

$$\mathbf{Y} = \begin{bmatrix} \Delta\theta_A \\ \Delta d_A \end{bmatrix},\ \mathbf{W} = \begin{bmatrix} W_{11} & W_{12} & W_{13} & W_{14} & W_{15} \\ W_{21} & 0 & W_{23} & W_{24} & W_{25} \end{bmatrix},$$

$$\mathbf{Z}^{\mathbf{T}} = [\Delta\theta_{A0}\ \Delta d_{A0}\ \Delta G_A\ \Delta\theta_{W0}\ \Delta G_{W0}];$$

$A(p) = a_3p^3 + a_2p^2 + a_1p + 1$, $W_{11} = (b_2p^2 + b_1p + b_0)/A(p)$,
$W_{12} = (b_5p^2 + b_4p + b_3)/A(p)$, $W_{13} = (b_8p^2 + b_7p + b_6)/A(p)$,
$W_{14} = (b_{10}p + b_9)/A(p)$,
$W_{15} = (b_{12}p + b_{11})/A(p)$, $W_{21} = (c_2p^2 + c_1p + c_0)/A(p)$, $W_{22} = (c_4p + c_3)/A(p)$,

$W_{23} = (c_7p^2 + c_6p + c_5)/A(p),\ W_{24} = c_8/A(p),\ W_{25} = c_9/A(p);$
$k = 1 - k_2k_{11} - k_4k_7,$
$a_1 = [T_A + (1 - k_4k_7)\,T_W + (1 - k_2k_{11})\,T_d]/k,\ a_2 = (T_A\,T_d + T_A\,T_W + T_W\,T_d)/k,$
$a_3 = T_A\,T_d\,T_W/k,\ b_0 = (k_0 + k_2k_{10} + k_4k_6)/k,$
$b_1 = [(k_0 + k_2\,k_{10})\,T_d + (k_0 + k_4k_6)\,T_W]/k,\ b_2 = k_0\,T_d\,T_W/k,\ b_3 = k_3/k,$
$b_4 = k_3(T_W + T_d)/k,\ b_5 = k_3\,T_W\,T_d/k,\ b_6 = (k_1 + k_4\,k_5)/k,$
$b_7 = [(k_1 + k_4\,k_5)\,T_W + k_1\,T_d]/k,\ b_8 = k_1\,T_d\,T_W/k,\ b_9 = k_2\,k_8/k,\ b_{10} = k_2\,k_8\,T_d/k,$
$b_{11} = k_2\,k_9/k,\ b_{12} = k_2\,k_9\,T_d/k,\ c_0 = [k_6(1 - k_2k_{11}) + k_7(k_0 + k_2k_{10})]/k,$
$c_1 = [k_6\,T_A + (k_6 + k_0k_7)\,T_W]/k,\ c_2 = k_6\,T_A\,T_W/k,\ c_3 = k_3\,k_7/k,\ c_4 = k_3\,k_7\,T_W/k,$
$c_5 = [k_1\,k_7 + k_5(k_1 - k_2k_4)]/k,\ c_6 = [k_5\,T_A + (k_5 + k_1k_7)\,T_W]/k,\ c_7 = k_6\,T_A\,T_W/k,$
$c_8 = k_2\,k_7\,k_8/k,\ c_9 = k_2\,k_7\,k_9/k.$
$A(p) = a_3p^3 + a_2p^2 + a_1p + 1,\ W_{11} = (b_2p^2 + b_1p + b_0)/A(p),$
$W_{12} = (b_5p^2 + b_4p + b_3)/A(p),\ W_{13} = (b_8p^2 + b_7p + b_6)/A(p),$
$W_{14} = (b_{10}p + b_9)/A(p),$
$W_{15} = (b_{12}p + b_{11})/A(p),\ W_{21} = (c_2p^2 + c_1p + c_0)/A(p),\ W_{22} = (c_4p + c_3)/A(p),$
$W_{23} = (c_7p^2 + c_6p + c_5)/A(p),\ W_{24} = c_8/A(p),\ W_{25} = c_9/A(p);$
$k = 1 - k_2k_{11} - k_4k_7,$
$a_1 = [T_A + (1 - k_4k_7)\,T_W + (1 - k_2k_{11})\,T_d]/k,\ a_2 = (T_A\,T_d + T_A\,T_W + T_W\,T_d)/k,$
$a_3 = T_A\,T_d\,T_W/k,\ b_0 = (k_0 + k_2k_{10} + k_4k_6)/k,$
$b_1 = [(k_0 + k_2\,k_{10})\,T_d + (k_0 + k_4k_6)\,T_W]/k,\ b_2 = k_0\,T_d\,T_W/k,\ b_3 = k_3/k,$
$b_4 = k_3(T_W + T_d)/k,\ b_5 = k_3\,T_W\,T_d/k,\ b_6 = (k_1 + k_4\,k_5)/k,$
$b_7 = [(k_1 + k_4\,k_5)\,T_W + k_1\,T_d]/k,\ b_8 = k_1\,T_d\,T_W/k,\ b_9 = k_2\,k_8/k,\ b_{10} = k_2\,k_8\,T_d/k,$
$b_{11} = k_2\,k_9/k,\ b_{12} = k_2\,k_9\,T_d/k,\ c_0 = [k_6(1 - k_2k_{11}) + k_7(k_0 + k_2k_{10})]/k,$
$c_1 = [k_6\,T_A + (k_6 + k_0k_7)\,T_W]/k,\ c_2 = k_6\,T_A\,T_W/k,\ c_3 = k_3\,k_7/k,\ c_4 = k_3\,k_7\,T_W/k,$
$c_5 = [k_1\,k_7 + k_5(k_1 - k_2k_4)]/k,\ c_6 = [k_5\,T_A + (k_5 + k_1k_7)\,T_W]/k,\ c_7 = k_6\,T_A\,T_W/k,$
$c_8 = k_2\,k_7\,k_8/k,\ c_9 = k_2\,k_7\,k_9/k.$

4 Simulation of the Dynamic Mode for the Spray-Type Humidifier Chamber

According to the dynamical model (12), (13), let's carry out the simulation of the transition processes of the spray-type humidifier chamber for industrial air conditioner VEZA KCKP-80 [13]. Table 1 shows the thermophysical parameters for the spray-type humidifier chamber.

The calculation of matrices for the model (12), (13) of the spray-type humidifier chamber for industrial air conditioner VEZA KCKP-80 was carried out in the MatLAB environment. We have the following numerical values:

Table 1. Thermophysical parameters of the spray-type humidifier chamber for industrial air conditioner VEZA KCKP-80

Parameter name	Marking	Numerical value	Dimension
Dimensions of the chamber	$H \times L \times C$	2.6 × 3.2 × 2	*m*
The coefficient of adiabatic efficiency	E	0.85	
Air consumption	G_A	25	*kg/sec*
Air density	ω	1.2	*kg/m*^{3}
Air heat capacity	c_A	1010	J/(*kg* °C)
Input air temperature	θ_{A0}	19	°*C*
Output air temperature	θ_A	9.8	°*C*
Moisture content at the entrance of the humidifier	d_{A0}	3.8	*g/kg*
Moisture content at the outlet of the humidifier	d_A	7.5	*g/kg*
Heat of vaporization	r	2256000	J/(*kg* °C)
Conversion coefficient	A	0.58	*g*/(°*C kg*)
Conversion coefficient	B	2.2	*g/kg*
Mass of air in camera	M_A	19.2	*kg*
Volume of humidification chamber	V_A	16	*m*^{3}
Water consumption	G_W	37	*kg/sec*
Water consumption	G_{W0}	1.1	*kg/sec*
Water density	ρ_W	986	*kg/m*^{3}
Water heat capacity	c_W	4185	J/(*kg* °C)
Mass of water in camera	M_W	1200	*kg*
Input water temperature	θ_{W0}	6	°*C*
Output water temperature	θ_W	9	°*C*

$$\mathbf{A} = \begin{bmatrix} -1.7 & -2.91 & 7.99 \\ 0.89 & -1.3 & 0 \\ 0.036 & 0 & -0.032 \end{bmatrix}, \ \mathbf{B} = \begin{bmatrix} 2.71 & 2.91 & 0.049 & 0 & 0 \\ -0.13 & 0 & -0.029 & 0 & 0 \\ -0.005 & 0 & 0 & 0.001 & -0.0025 \end{bmatrix};$$

$$A(p) = 6.8p^3 + 81.8p^2 + 112.9p + 1, \ W_{11} = \left(18.4p^2 + 26.9p + 0.46\right)/A(p),$$

$$W_{12} = \left(19.8p^2 + 26.4p + 0.82\right)/A(p), \ W_{13} = \left(0.33p^2 + 1.02p + 0.03\right)/A(p),$$

$$W_{14} = (0.05p + 0.065)/A(p), \ W_{15} = (-0.14p - 0.18)/A(p);$$

$$W_{21} = \left(-0.91p^2 + 6.66p + 0.21\right)/A(p), \ W_{22} = (17.57p + 0.56)/A(p),$$

$$W_{23} = \left(-0.91p^2 - 1.82p - 0.005\right)/A(p), \ W_{24} = 0.044/A(p), \ W_{25} = -0.121/A(p).$$

The results of transients' simulation by perturbation channels are shown in Fig. 2. From the graphs we can conclude that the inertia of the control channel (Fig. 2, d) is two orders of magnitude higher than the perturbation channel (Fig. 2, a–c), which is explained by the considerable thermal accumulation of water in the pallet compared to the thermal accumulation of the air space in humidifier. For these reasons, the automation of a spray-type humidifier, they refused to regulate the changes in temperature and water flow rate for humidification. It should be added that in Fig. 2, a–c shows the high-speed part of transients in airspace, which affects the control system settings. Considering the heat accumulation of water in the pallet (which is two orders of magnitude higher than the heat accumulation of the air space), the transient processes of the humidifier air space continue to pass at the rate of heat exchange in the pallet.

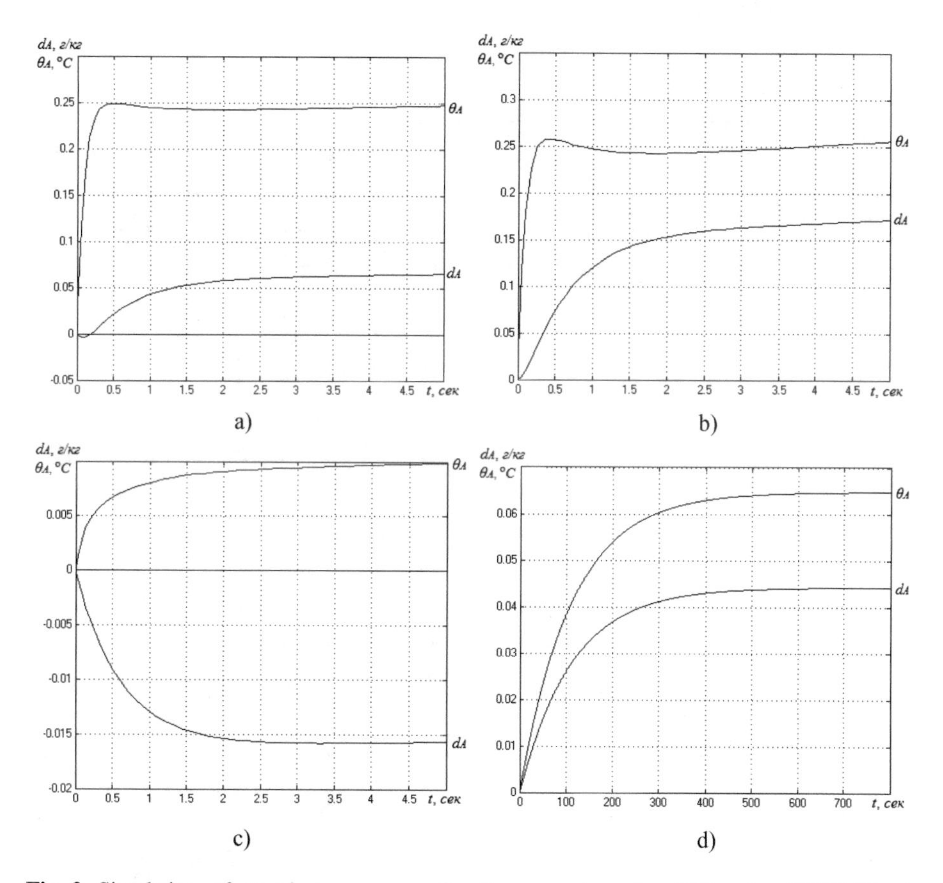

Fig. 2 Simulation of transient processes for the spray-type humidifier the KCKP-80 air conditioner: a) $\Delta\theta_{A0} \to \Delta\theta_A$, $\Delta\theta_{A0} \to \Delta d_A$, $\Delta\theta_{A0} = 1$ °C; b) $\Delta d_{A0} \to \Delta\theta_A$, $\Delta d_{A0} \to \Delta d_A$, $\Delta d_{A0} = 1$ g/kg; c) $\Delta G_{A0} \to \Delta\theta_A$, $\Delta G_{A0} \to \Delta d_A$, $\Delta G_{A0} = 1$ kg/sec; d) $\Delta\theta_{W0} \to \Delta\theta_A$, $\Delta\theta_{W0} \to \Delta d_A$, $\Delta\theta_{W0} = 1$ °C

In Fig. 3, a–c, show complete transitions for the considered channels of influence. The graphs are presented in Fig. 2, d and Fig. 3, d confirm the ineffectiveness of regulatory influences.

According to the simulation results, which are presented as transients in Fig. 2 and Fig. 3, shows that the feature of this controlled plant is the low inertia of the perturbation channels compared to the control channels. The resulting transient processes for the channels of perturbation and regulation have an aperiodic nature without delay. By the nature of the transients, the PI-controller should be recommended for controlling the spray-type humidifier. Given the low inertia of transients and the absence of delay, the proportional part of the controller will quickly transfer the plant to a new operating mode, and the integral part will provide accurate monitoring of the controller's task. According to practical recommendations, the use of PID controllers does not justify itself for controlling high-speed plant.

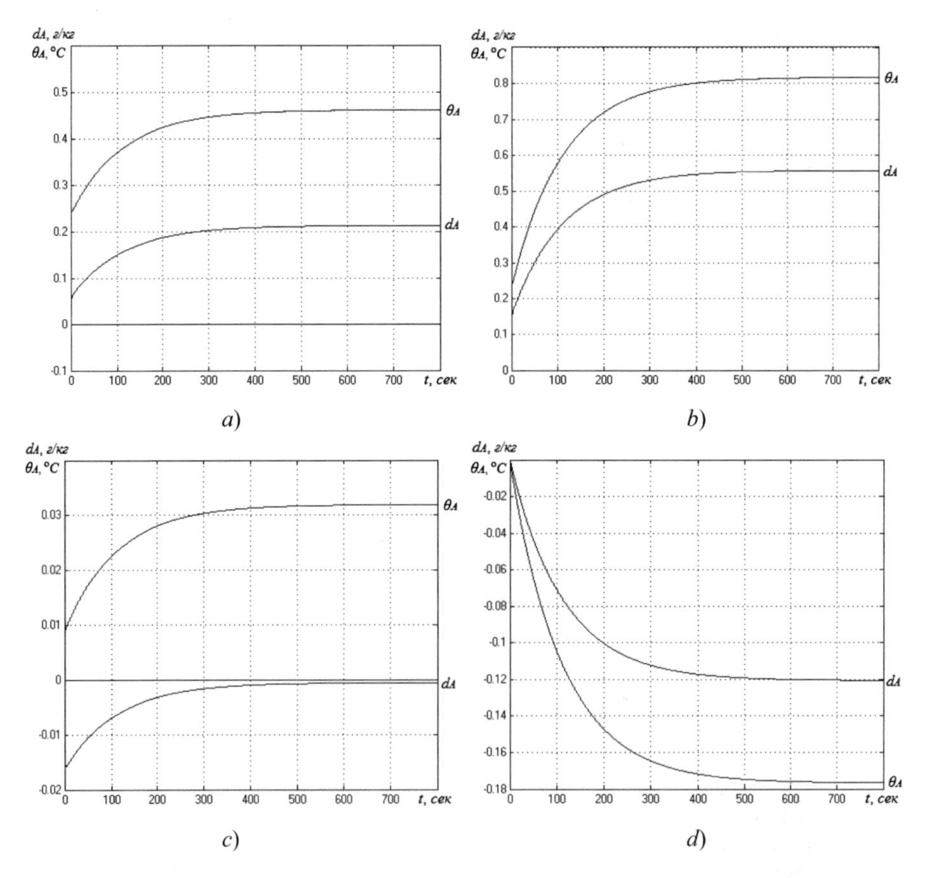

Fig. 3 Simulation of transient processes for the spray-type humidifier the KCKP-80 air conditioner: *a*) $\Delta\theta_{A0} \rightarrow \Delta\theta_A$, $\Delta\theta_{A0} \rightarrow \Delta d_A$, $\Delta\theta_{A0} = 1$ °C; *b*) $\Delta d_{A0} \rightarrow \Delta\theta_A$, $\Delta d_{A0} \rightarrow \Delta d_A$, $\Delta d_{A0} = 1$ g/kg; *c*) $\Delta G_{A0} \rightarrow \Delta\theta_A$, $\Delta G_{A0} \rightarrow \Delta d_A$, $\Delta G_{A0} = 1$ kg/s; *d*) $\Delta G_{W0} \rightarrow \Delta\theta_A$, $\Delta G_{W0} \rightarrow \Delta d_A$, $\Delta G_{W0} = 1$ kg/s

5 Conclusions

Based on the results of this research, the following conclusions can be drawn.

1. A dynamic model of the spray-type humidifier of industrial air conditioner is obtained, which is represented by equivalent dependencies in the form of a system of differential Eqs. (11), in the state space (12), in the form of transfer functions (13). The choice between dynamic models (11)–(13) is determined by the synthesis methods of the control system.
2. In contrast to existing static models, the proposed model describes the dynamics of heat and mass transfer processes in the coordinates of temperature and moisture content of air.
3. It is convenient to use the dynamic model of the spray-type humidifier in the state space (12) as a component of the complex model for industrial air conditioner when designing the ACS of a computer-integrated technological complex of air conditioning.
4. The resulting mathematical model is part of the research and is an integral part of an industrial air conditioner. In the future, it is planned to obtain a complex dynamic model of an industrial air conditioner, on the basis of which a synthesis of a multidimensional regulator microclimate in a room can be made. Comparison of existing and synthesized control systems will allow to evaluate the effectiveness of the created microclimate system. This will take the air-conditioning control system to a new level and ensure efficient use of energy for artificial climate systems [1].

References

1. Golinko, I.M.: Principles of automatic control systems synthesis for industrial air conditioners. Autom. Technol. Bus.-Process. **8**, 33–42 (2016)
2. Golinko, I., Galytska, I.: Mathematical model of heat exchange for non-stationary mode of water heater. In: ICCSEEA 2019: Advances in Computer Science for Engineering and Education II, pp. 58–67 (2019)
3. Golinko, I.M.: The dynamic model of heat-mass exchange for water cooling of industrial air conditioning. Sci. Technol. NTUU "KPI" **6**, 27–34 (2014)
4. Nikolaeva, K.A., Golinko, I.M.: Dynamic model of a heater in the state space. In: 14th International Scientific and Practical Conference "Modern Problems of Scientific Support of Power Engineering", Kyiv, p. 28 (2016). (in Ukrainian)
5. Poonam, S., Agarwal, S.K., Kumar, N.: Advanced adaptive particle swarm optimization based SVC controller for power system stability. Int. J. Intell. Syst. Appl. (IJISA) **7**(1), 101–110 (2014)
6. Puangdownreong, D.: Multiobjective multipath adaptive Tabu search for optimal PID controller design. Int. J. Intell. Syst. Appl. (IJISA) **7**(8), 51–58 (2015)
7. Misra, Y., Kamath, H.R.: Design algorithm and performance analysis of conventional and fuzzy controller for maintaining the cane level during sugar making process. Int. J. Intell. Syst. Appl. (IJISA) **7**(1), 80–93 (2014)

8. Yazdanpanah, A., Piltan, F., Roshanzamir, A., Mirshekari, M., Mozafari, N.: Design PID baseline fuzzy tuning proportional derivative coefficient nonlinear controller with application to continuum robot. Int. J. Intell. Syst. Appl. (IJISA) **6**(5), 90–100 (2014)
9. Soukkou, A., Belhour, M.C., Leulmi, S.: Review, design, optimization and stability analysis of fractional-order PID controller. Int. J. Intell. Syst. Appl. (IJISA) **8**(7), 73–96 (2016)
10. Lakshmi, K.V., Srinivas, P., Ramesh, C.: Comparative analysis of ANN based intelligent controllers for three tank system. Int. J. Intell. Syst. Appl. (IJISA) **8**(3), 34–41 (2016)
11. Belova, E.M.: Central Air Conditioning Systems in Buildings. Euroclimate, Moscow (2006)
12. Kokhansky, A.I., Kolpakchi, E.M.: Identification of the transfer functions of the charge air cooler. Autom. Ship Tech. Equip. **12**, 68–77 (2007)
13. Air conditioning systems. VEZA. Catalog 2012, 166 p. (2012)

Packing Optimization Problems and Their Application in 3D Printing

A. M. Chugay[1(✉)] and A. V. Zhuravka[2]

[1] Institute for Mechanical Engineering Problems, National Academy of Sciences of Ukraine, Kiev, Ukraine
chugay.andrey80@gmail.com
[2] Kharkiv National University of Radio Electronics, Kharkiv, Ukraine
http://www.ipmach.kharkov.ua

Abstract. Packing optimization problems have a wide spectrum of real-word applications. One of the applications of the problems is 3D printing problem. For additive production technology it is very important to implement various optimization methods which will help to save time and material. In this work, we propose approach (model and optimization methods) that allows us to get dense packing of objects to be printed in working chamber of a 3D printer and simultaneously minimize the height of working chamber. This reduces print costs by minimizing the number of layers of 3D-printing and reducing the number of printer's starts. For solving 3D packing problem phi-functions technique is proposed as a constructive means of the mathematical modeling of a given problem. On the basis of the phi-function a mathematical model of the problem is constructed and solution approach is proposed. A number of examples presented in paper demonstrate the efficiency of our methodology.

Keywords: Packing optimization problems · 3D printing technology · Working chamber · Mathematical modeling · Phi-function · Non-linear optimization

1 Introduction

Packing optimization problems have a wide spectrum of real-word applications, including transportation, logistics, chemical and aerospace engineering, shipbuilding, robotics, additive manufacturing, materials science. In this paper optimization methods of process 3D-printing technology for SLS [1] additive manufacturing is developed. This technology uses high power laser sintering for small particles of plastic, ceramic, glass or metal flour in three-dimensional structure.

Selective laser sintering (SLS) empowers the fast, flexible, cost-efficient, and easy manufacture of prototypes for various application of required shape and size by using powder based material. The physical prototype is important for design confirmation and operational examination by creating the prototype unswervingly from CAD data.

The main feature of this technology is the use of powder, consisting of particles of metal coated polymer. After the sintering process piece is placed in a high temperature

Z. Hu et al. (Eds.): ICCSEEA 2020, AISC 1247, pp. 75–85, 2021.
https://doi.org/10.1007/978-3-030-55506-1_7

kiln to burn plastic and fusible took the bronze. The advantages of the technology include no need for material support. Parts immersed into a powder, which works on as a support (Fig. 1) [2].

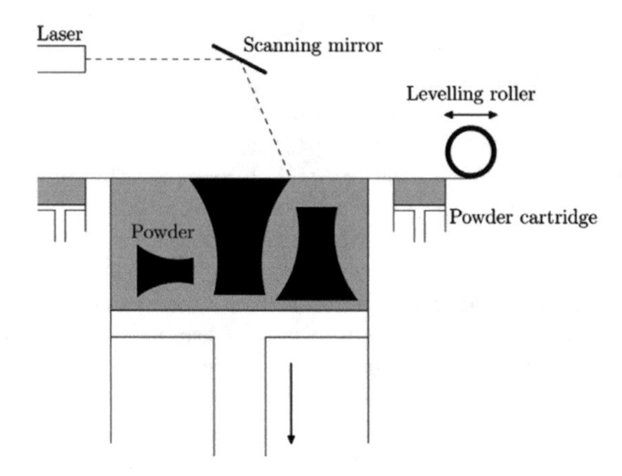

Fig. 1. Scheme of 3D-printing using SLS-technology [2]

In the past 10 years, 3D-prototyping technologies are evolving rapidly: new, improved existing, new uses of existing technology. The purpose of research is the development of modern information technologies that will improve 3D-printing process using advanced additive technology. It is proposed to develop methods for accelerating printing cycle due to the simultaneous printing of several models of providing dense filling the entire volume of the working chamber 3D printer using SLS technology. For this purpose, layout optimization problems solved three-dimensional objects on the basis of constructive means of mathematical and computer modeling, mathematical models.

The process of creating new prototypes can be roughly divided into two main components: the creation of virtual manufacturing models and prototypes. This study is intended for developing technologies improve these stages of prototyping.

The first step is to create virtual 3D models of objects (computer models of prototypes).

One of the objectives of the second phase is to reduce the time and cost of making prototypes of products. For each starts of SLS printer requires time and energy for heating and maintaining temperature. Thus, in paper [3], the authors provide data on what savings can be achieved by optimizing the packing of objects to be created.

So in [3] what savings can be achieved by optimizing the placement of objects to be created.

Model and methods proposed in this work allow us to optimize the process of 3D printing for the following indicators:

- simultaneous printing of several prototypes to providing dense filling the entire volume of the 3D printer working chamber;
- minimize the time and cost of prototyping products by significantly reducing printing cycle.

To implement the described tasks in this paper the packing optimization problem of non-oriented convex 3D objects is solved. Our approach is based on the mathematical modelling of relations between geometric objects by means of phi-function technique. That allowed us to reduce the problem solving to nonlinear programming.

2 Literature Review

Analysis of publications devoted to solving the problem of optimal placement of objects, has shown that the problem of packing arbitrary 3D objects (allowing continues rotation), taking into account the minimum allowable distance is the least studied in class placement problems. Among the objects are usually considered cylinders or parallelepipeds and arbitrary spatial shape approximated by sets δ-parallelepipeds. To solve the problems layout using heuristic and meta-heuristic algorithms, resulting in the loss of optimal solutions.

Thus, there is a need to build adequate mathematical models for problems using real packaging (without first approximation) spatial forms of accommodation and containers, as well as the specified minimum acceptable distance and other technological limitations. An important issue is also the development of efficient algorithms for solving problems using modern layout of local and global NLP-solvers.

It is well known that the 3D object packing problem is NP-complete. Because of its NP complexity, the problem is difficult to solve satisfactorily. So, in order to find its approximate solution a lot of researchers use a very wide variety of techniques [4–6], including heuristics (heuristics based on different approximation rules [7], genetic algorithms [8], simulated annealing algorithm [9], artificial bee colony algorithm [10]), extended pattern search [8], traditional optimization methods [12, 13] and different mixed approaches utilizing heuristics and methods of non-linear mathematical programming [14].

As mentioned in [7] solution processes consist of the following loop procedures: 1) ordering a sequence of objects; 2) applying geometric procedures to the objects according to their position in the sequence; 3) calculating an objective function.

The geometric procedures can be implemented by methods which differ by: the path of object movement, complexity of rotation modeling and whether an intersection object is allowed during solution phases.

In the majority of papers, either orientation changes of 3D objects are not allowed or only discrete changes in the orientation for given angles (45 or 90°) are allowed. For example, in [2] only the parallel translation algorithm is used for packing convex polytopes. In [14] the authors propose the HAPE3D algorithm which can be applied to an arbitrarily shaped polyhedral which, in its turn, can rotate around each coordinate axis at eight different angles only.

In [15] the authors point out that for 3D packing problems making calculations of 0 to 360° orientations of objects with respect to each axis is impossible.

At present, due to the difficulty in constructing adequate mathematical models, there are few works that solve 3D packing problems provided that continuous rotations of geometric objects are allowed. The solutions to the problems are considered in [16–19].

3 Problem Statement

In order to minimize the time used for prototyping using SLS-technology facilities should be located as tightly as possible and the number of layers should be minimized. Therefore, the problem of minimizing layers can be formulated as a problem of dense packing items in minimum container height (Fig. 2).

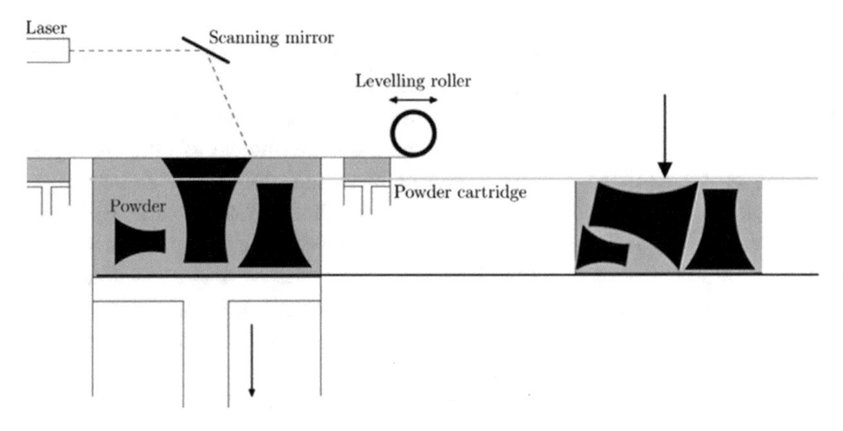

Fig. 2. Minimizing of the height of the occupied part of the 3D printer working chamber

Let set of convex 3D objects P_i, $i \in I = \{1, 2, \ldots, n\}$, and direct rectangular parallelepiped $C = \{X \in R^3 : w_2 \leq x_1 \leq w_1, l_2 \leq x_2 \leq l_1, \eta_2 \leq x_3 \leq \eta_1\}$ are given. The vector $\Upsilon = (\eta_1, \eta_2)$ determines the height container C. Objects P_i can take an arbitrary spatial shape [17].

Placing P_i in R^3 vector defined by translation vector $v_1 = (x_i, y_i, z_i)$ and rotation angles $\theta_i = (\phi_i, \psi_i, \omega_i)$ $i \in I$. Thus, vector $u_i = (v_i, \theta_i) = (x_i, y_i, z_i, \phi_i, \psi_i, \omega_i)$ determines location P_i in R^3, and vector $u = (u_1, u_2, \ldots, u_n) \in R^m$ where $m = 6n$, determines location P_i, $i \in I$, in R^3.

Further object P_i translated by vector v_i and rotated on angles θ_i denote as $P_i(u_i)$ and cuboid C with variable size Υ by $C(\Upsilon)$.

Problem. It is needed to define the vector $u \in R^m$ which provides placement of $P_i(u_i)$, $i \in I$, into the cuboid $C(\Upsilon)$ so that height $H(\Upsilon) = (h_2 - h_1)$ reached the minimum value (Fig. 3).

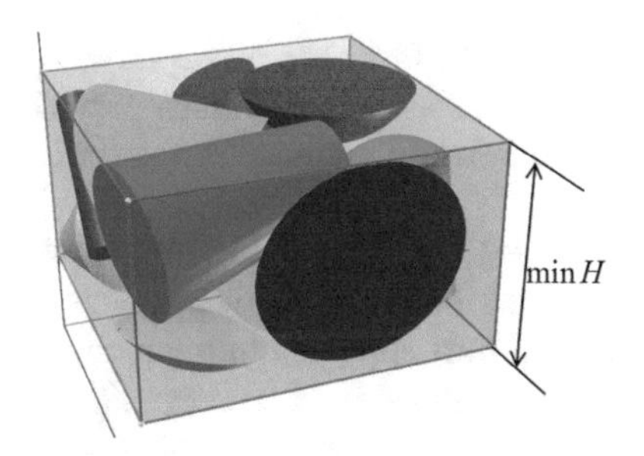

Fig. 3. Packing optimization problem statement

4 Mathematical Model and Solution Method

On the basis of $\Phi-$ functions [16–19] mathematical model of the problem can be written as a classical problem of nonlinear programming

$$H(\Upsilon^{*}) = \min_{(u,\Upsilon)\in\Lambda} H(\Upsilon), \tag{1}$$

where

$$\begin{aligned}\Lambda = \{(u,\Upsilon) \in R^{m+2} :\\ \Phi_{ij}(u_i, u_j) \geq 0, 0 < i < j \in I,\\ \Phi_i(u_i, \Upsilon) \geq 0, i \in I,\\ \eta_2 - \eta_1 \geq 0\}.\end{aligned} \tag{2}$$

The system (2) $\Phi_{ij}(u_i, u_j)$ – quasi-phi-function or phi-function (conditions non-overlapping of objects P_i and P_j), $\Phi_i(u_i, \Upsilon)$ – phi-function for P_i and object $cl(R^3 \backslash C^*)$ (specifies conditions of placement P_i in container C).

Proved that $\Lambda = \bigcup_{q=0}^{\zeta} \Lambda_q$ where each of the subregions Λ_q determined by the system of inequalities which left side is infinitely differentiated functions. Thus, to find the global minimum point of problem (1)–(2) should solve the problem

$$H(\Upsilon^{*}) = \min\{H(\Upsilon^{*q}), q \in Q\},$$

where

$$H(\Upsilon^{*,q}) = \min_{(u,\Upsilon)\in\Lambda_q} H(\Upsilon), q \in Q.$$

Because $\zeta > n!$, then cannot find a reasonable solution to this problem.

It is well known that three-dimensional objects packing problem is NP-complex, so to find their approximate solution uses a very wide range of techniques, ranging from heuristics (heuristics based on different rules, genetic algorithms, simulation annealing) to traditional optimization methods. But none of the known authors does not solve the problem of three-dimensional packaging capabilities provided continuous rotation of objects due to the difficulty of building adequate mathematical models and calculations of large volume.

Therefore, to find the approach to global a minimum of (1)–(2) an approach that consists of the following stages: building the three points, search for points of local minima and their aims for part-bust approach to the global minimum.

For initial points cover P_i clusters S_i minimum radii ρ_i^{∇}, $i \in I$. We assume that $\Upsilon = \Upsilon^0$ and provides accommodation in C. The radii ρ_i spheres S_i, $i \in I$, are variable and form a vector $\rho = (\rho_1, \rho_2, \ldots, \rho_n) \in R^n$. This makes it possible to formulate the following problem

$$\Pi(\rho^0) = \max \Pi(\rho) = \max_{(v,\rho)\in\Omega} \sum_{i=1}^{n} \rho_i \tag{3}$$

where

$$\begin{aligned} &\Omega = \{(v,\rho) \in R^{4n}, \Phi_{ij}^{SS}(v_i, v_j, \rho_i, \rho_j) \geq 0, \\ &0 < i < j \in I, \Phi_i^S(v_i, \rho_i) \geq 0, i \in I, \\ &s_i(\rho_i) = \rho_i - \rho_i^{\nabla} \geq 0, i \in I\}; \end{aligned} \tag{4}$$

$$\begin{aligned} \Phi_{ij}^{SS}(v_i, v_j, \rho_i, \rho_j) &= (x_i - x_j)^2 + (y_i - y_j)^2 \\ &+ (z_i - z_j)^2 - (\rho_i + \rho_j)^2; \end{aligned}$$

$$\begin{aligned} &\Phi_i^S(v_i, \rho) = \min\{x_i - \rho_i - w_2^0, y_i - \rho_i - l_2^0, \\ &z_i - \rho_i - \eta_2^0, w_1^0 - x_i - \rho_i, \\ &l_1^0 - y_i - \rho_i, \eta_1^0 - z_i + \rho_i\}. \end{aligned}$$

Asking starting point (v^0, ρ^0) where $v_i^0 \in C(\Upsilon^0)$, $\rho^0 = 0$ and to this point we calculate global maximum point (v^*, ρ^*) problem (3)–(4).

Then take polyhedrons $P_i(v_i^*)$ instead of clusters $S_i(v_i^*)$, generate random angles $\theta_i^s = (\phi_i^s, \psi_i^s, \omega_i^s) \in [0, \pi]$, $i \in I$, fix them and solve the problem

$$H(\Upsilon^*) = \min_{(v,\Upsilon)\in\Gamma\subset R^{3n+2}} H(\Upsilon), \tag{5}$$

where

$$\begin{aligned} &\Gamma = \{(v,\Upsilon) \in R^{3n+2} : \Phi_{ij}(v_i, v_j) \geq 0, 0 < i < j \in I, \\ &\Phi_i(v_i,\Upsilon) \geq 0, i \in I, \end{aligned} \tag{6}$$

$$w_1 \geq 0, l_1 \geq 0, \eta_1 \geq 0, w_2 - w_1 \geq 0,$$
$$l_2 - l_1 \geq 0, \eta_2 - \eta_1 \geq 0\}.$$

Obviously, $\Gamma = \bigcup_{q=0}^{\zeta} \Gamma_q$ where Γ_q determined linear system of inequalities. This means that the search for points of local minima is reduced to solving a sequence of linear programming problems.

Let the point (v^*, Υ^*) is a point of local minimum of (5)–(6). Then build the starting point $(u^s, \Upsilon^s) = (v^*, \theta^s, \Upsilon^*)$. To this point we find the starting point (u^{0*}, Υ^{0*}) local minimum problem (1)–(2).

Directed partial search of local minima problem (1)–(2) reduces to solving these auxiliary problems.

We assume that the coefficients homothetic transformation $h = (h_1, h_2, \ldots, h_n)$ are variable. This makes it possible to formulate the problem

$$H(\Upsilon) = \min_{Y = (u, \Upsilon, h) \in \Delta} H(\Upsilon), \tag{7}$$

where

$$\Delta = \{Y \in R^{7n+2}, \Phi_{ij}(u_i, u_j, h_i, h_j) \geq 0,$$
$$0 < i < j \in I, \Phi_i(u_i, \Upsilon, h_i) \geq 0, \tag{8}$$

$$h_i \geq 0, i \in I, H(\Upsilon^{0*}) - H(\Upsilon) \geq \varepsilon, w_1 \geq 0, l_1 \geq 0,$$
$$\eta_1 \geq 0, w_2 - w_1 \geq 0,$$
$$l_2 - l_1 \geq 0, \eta_2 - \eta_1 \geq 0\};$$

$$\varepsilon = (\frac{1}{2})^t 0.1 H(\Upsilon^{0*}), t = 0,1,2.\ldots. \tag{9}$$

For the starting point $Y^0 = (u^{0*}, \Upsilon^{0*}, h^\nabla)$ we find a local minimum point $Y^{*1} = (u^{*1}, \Upsilon^{*1}, h^{*1})$. Next, we construct the sequence $h^1_{i_1} \geq h^1_{i_2} \geq \ldots . \geq h^1_{i_n}$. Based on this sequence and sequences $1 = h^\nabla_1 \geq h^\nabla_2 \geq \ldots . \geq h^\nabla_n$ form a point $(\tilde{u}, \tilde{h})$, where $\tilde{u}_j = u^{1*}_{i_j}$, $\tilde{h}_j - \min\{h^1_{i_j}, h^\nabla_j\}$, $j \in I$.

Next, to the starting point $(\tilde{u}, \tilde{h})$ we calculate point (u^0, h^0) local maximum problem

$$F(h^*) = \max_{(u,h) \in D} F(h) = \max \sum_{i=1}^{n} h_i, \tag{10}$$

where

$$
\begin{aligned}
D = \{ & (u,h) \in R^{7n}, \Phi_{ij}(u_i, u_j, h_i, h_j) \geq 0, \\
& 0 < i < j \in I, \\
& \Phi_i(u_i, h_i) \geq 0, \\
& h_i^{\nabla} - h_i \geq 0, h_i \geq 0, i \in I \}.
\end{aligned} \tag{11}
$$

If $F(h^0) = \sum_{i=1}^{n} h_i^0 = \sum_{i=1}^{n} h_i^{\nabla} = b$, all P_i, $i \in I$, packed in $C(\Upsilon^1)$. In this case the starting point (u^0, Υ^{0*}) re-solve problem (1)–(2) and so on. If $j \neq i_j, j \in I$, at least one pair of indexes, $F(h^0) < b$ then increase t in (9) 1 and solve the problem (7)–(8) and so on until you get $F(h^0) = b$ or $j = i_j, j \in I$. In the case of $F(h^0) = b$ go to solving the problem (1)–(2). If $j = i_j$, $j \in I$, the previous local minimum point of problem (1)–(2) was adopted as an approximation to global minimum point of the problem.

Note that the search for local extremes of formulated optimization problems developed a special method of decomposition, which can significantly reduce computing costs by significantly reducing the number of irregularities in the process of finding local extremes. The key idea optimization procedure allow us to generate subsets of feasible region at every step as follows. Based on the analysis of the starting point in the constrains system of problems we include additional restrictions on the parameters of each object placement. This allows moving objects within a individual rectangular container. Then remove phi-inequality for all pairs of objects, individual containers that do not overlap. Thus, we reduce the number of restrictions and, in the case quasi-phi-features of extra variables. Then a search for the local minimum point of subproblem is performed. The resulting local extremum of subtask used as a starting point for the next iteration.

To solve optimization problems a modern solver IPOPT was used [20].

5 Computation Experiments

We present a number of examples to demonstrate the efficiency of our methodology. We have run all experiments on an Intel I5 2320 computer, programming language C#, Windows 10 OS. Figure 4 demonstrate examples of 3D objects packing obtained by the proposed methods.

In order to test effectiveness of the developed approach a number of benchmarks given in [14] were solved. The results are shown in the Table 1.

Fig. 4. Examples of 3D objects packing

Table 1. Result of comparison with benchmarks presented in [14]

Approach	*HAPE3D*	*Our approach*
	The result of packing of 36 objects	
Volume	12480	10720
Runtime (sec)	9637	4789
Result illustration		
	The result of packing of 40 objects	
Volume	61950	56012
Runtime (sec)	99952	24543
Result illustration		

6 Conclusions

3D-printing procedure using SLS-technology is fairly long time (printing can take many hours or even days) and require large financial costs associated with running a printer, camera heating and temperature stabilization. In 3D printing technology it is very important to implement various optimization approaches which will help to save time and material.

It should be noted that the time required for sintering the powder is much less than the time required preparing each layer of powder. The paper proposes to reduce the time and cost by providing simultaneous printing maximum number of products providing optimum packaging facilities. We propose approach that allows us to get dense packing of objects to be printed in working chamber of a 3D printer and simultaneously minimize the height of working chamber. This reduces print costs by minimizing the number of layers of 3D-printing and reducing the number of printer's starts. For this end the optimization packing problem of a given set of convex 3D objects without their overlapping into cuboid of minimal volume is solved. Phi-functions are proposed to be used as a constructive means of the mathematical modeling of a given problem. On the basis of the phi-function a mathematical model of the problem is constructed, and its main properties which influence the choice of the strategy for solving the problem are examined. The obtained mathematical model presents the problem in the form of a classical problem of nonlinear programming, which makes it possible to use modern solvers for searching for a solution. Effective methods for construction starting points and searching for locally optimal solutions based on homothetic transformations are proposed. To search for local extrema of the formulated optimization problems, a special method of decomposition has been developed, which allows us to significantly reduce computational costs due to a considerable reduction in the number of inequalities. The key idea of the optimization procedure allows us to generate subsets of the domain of admissible solutions at each stage of searching for a local extremum. Parallel computations were used to search for local extrema, which made it possible to reduce time expenditures.

A number of examples presented in paper demonstrate the efficiency of our methodology.

The methods proposed in the work can be used for solving the problem of packing non-convex objects.

References

1. Ramya, A., Vanapalli, S.: 3D printing technologies in various applications. Int. J. Mech. Eng. Technol. **7**(3), 396–409 (2016)
2. Egeblad, J., Nielse, B.K., Brazil, M.: Translational packing of arbitrary polytopes. Computat. Geometry **42**, 269–288 (2009)
3. Baumers, M., Dickens, P., Tuck, C., Hague, R.M.: The cost of additive manufacturing: machine productivity, economies of scale and technology-push. Technol. Forecast. Soc. Chang. **102**, 193–201 (2016)

4. Akash, N.: A fast heuristic algorithm for solving high-density subset-sum problems. Int. J. Math. Sci. Comput. **2**, 55–61 (2017)
5. Chakroborty, S., Hasan, M.B.: A proposed technique for solving scenario based multi-period stochastic optimization problems with computer application. Int. J. Math. Sci. Comput. (IJMSC) **4**, 12–23 (2016)
6. Aliyeva, A.G., Shahverdiyeva, R.O.: Application of mathematical methods and models in product - service manufacturing processes in scientific innovative technoparks. Int. J. Math. Sci. Comput. (IJMSC) **3**, 1–12 (2018)
7. Verkhoturov, M., Petunin, A., Verkhoturova, G., Danilov, K., Kurennov, D.: The 3D object packing problem into a parallelepiped container based on discrete-logical representation. IFAC-PapersOnLine **49**(12), 1–5 (2016)
8. Karabulut, K., İnceoğlu, M.: A Hybrid genetic algorithm for packing in 3D with deepest bottom left with fill method. In: Advances in Information Systems, vol. 3261, pp. 441–450 (2004)
9. Cao, P., Fan, Z., Gao, R., Tang, J.: Complex housing: modeling and optimization using an improved multi-objective simulated annealing algorithm. In: Proceedings of the ASME, vol. 60563, p. V02BT03A034 (2016)
10. Li, G., Zhao, F., Zhang, R., Du, J., Guo, C., Zhou, Y.: A parallel particle bee colony algorithm approach to layout optimization. J. Comput. Theor. Nanosci. **13**(7), 4151–4157 (2016)
11. Torczon, V., Trosset, M.: From evolutionary operation to parallel direct search: Pattern search algorithms for numerical optimization. Comput. Sci. Stat. **29**, 396–401 (1998)
12. Birgin, E.G., Lobato, R.D., Martinez, J.M.: Packing ellipsoids by nonlinear optimization. J. Glob. Optim. **65**, 709–743 (2016)
13. Fasano, G.: A global optimization point of view to handle non-standard object packing problems. J. Glob. Optim. **55**(2), 279–299 (2013)
14. Liu, X., Liu, J., Cao, A., Yao, Z.: HAPE3D - a new constructive algorithm for the 3D irregular packing problem. Front. Inf. Technol. Electron. Eng. **16**(5), 380–390 (2015)
15. Joung, Y.-K., Noh, S.D.: Intelligent 3D packing using a grouping algorithm for automotive container engineering. J. Comput. Design Eng. **1**(2), 140–151 (2014)
16. Stoyan, Y.G., Chugay, A.M.: Packing different cuboids with rotations and spheres into a cuboid. Adv. Decis. Sci. (2014). https://www.hindawi.com/journals/ads/2014/571743
17. Stoyan, Y.G., Semkin, V.V., Chugay, A.M.: Modeling close packing of 3D objects. Cybern. Syst. Anal. **52**(2), 296–304 (2016)
18. Stoyan, Y.G., Chugay, A.M.: Mathematical modeling of the interaction of non-oriented convex polytopes. Cybern. Syst. Anal. **48**, 837–845 (2012)
19. Grebennik, I.V., Pankratov, A.V., Chugay, A.M., Baranov, A.V.: Packing n-dimensional parallelepipeds with the feasibility of changing their orthogonal orientation in an n-dimensional parallelepiped. Cybern. Syst. Anal. **46**(5), 793–802 (2010)
20. Wachter, A., Biegler, L.T.: On the implementation of an interior-point filter line-search algorithm for large-scale nonlinear programming. Math. Program. **106**(1), 25–57 (2006)

Development of Low-Cost Internet-of-Things (IoT) Networks for Field Air and Soil Monitoring Within the Irrigation Control System

Volodymyr Kovalchuk[1(✉)], Oleksandr Voitovich[1], Dmytro Demchuk[2], and Olena Demchuk[3]

[1] Institute of Water Problems and Land Reclamation, NAAS of Ukraine, Kiev, Ukraine
volokovalchuk@gmail.com, aleks@krakow.in
[2] National Technical University of Ukraine "Ihor Sikorsky Kyiv Polytechnic Institute", Kiev, Ukraine
ddima199703@gmail.com
[3] National University of Water and Environmental Engineering, Rivne, Ukraine
ldem1997@ukr.net

Abstract. The article presents the development and testing of low-cost Internet-of-Things (IoT) networks for air and soil monitoring in agricultural fields. The sensor ensemble for completing an air and soil monitoring point (MP) was founded. The Watermark sensors and automatic tensiometers for soil moisture monitoring at 2–3 depths, temperature and relative humidity sensors for air monitoring, and rain sensors for the validation of irrigation rate and precipitation recording were proposed. The structure of sensors and modules of the base station (BS), which collected the monitoring data in the field and sent it to the server, was proposed. The Internet-of-Things (IoT) network architecture and the data transmission methods were also proposed. To transmit monitoring data to a remote server and to make irrigation decisions, it was suggested to use LoRa radio and mobile Internet. The practical testing of networks was carried out in the farm of the Kherson region (Ukraine) in large area fields under sprinkling. The effect of crop height and the location of forest strips in the area between the BS and the monitoring points, and the necessary number of MP for one BS were determined. For stand-alone remote from BS monitoring points, it was recommended to use mobile Internet for data transmission. Unlike similar studies, the article analyzes the entire monitoring chain: from the formation of an air and soil sensor ensemble at the MP to the construction of a network of monitoring points in the field and equipment of BS. The complexity of the work is that it is recommended to combine the use of both LoRa and GSM for the transmission of the collected monitoring data. The monitoring data obtained through these networks is used as a feedback loop in the decision support system for sprinkling irrigation control. The areas of further studies of this problem were analyzed and proposed as well.

Keywords: Internet-of-Things network · Wireless sensors · Field air and soil monitoring · The Irrigation Control System · Arduino Pro Mini

Z. Hu et al. (Eds.): ICCSEEA 2020, AISC 1247, pp. 86–96, 2021.
https://doi.org/10.1007/978-3-030-55506-1_8

1 Introduction

Internet-of-Things (IoT) is one of the fastest growing industries and is being actively used by society. It is used in industry, smart homes, medicine, and agriculture to monitor and manage crop conditions. Its use allows to receive high yields and to save irrigation water. In agriculture, a deterrent for the creation of monitoring networks is expensive equipment from large companies such as Davis Instruments, Pessl Instruments, Tevatronic [1–3].

Recently, low-cost Internet-of-Things (IoT) networks with various automatic sensors have been dynamically evolving to monitor soil moisture and microclimate parameters in the field. The importance of such research is confirmed by the fact that the European Union has funded several projects in this area of research. For example, two projects from the European Union's Horizon 2020 Research and Innovation Program are ENORASIS (January 2012–December 2014) [4] and WAZIUP (February 2016–January 2019) [5].

Analysis of the sources helps us to understand the place of each sensor and to select the appropriate ways for transmitting monitoring data. For monitoring parameters such as soil moisture and microclimate in the field, both separately operating sensors [6–10] and a field radio network with a large number of sensors are used [11–16]. The availability low-cost of controllers and sensors, the open of libraries for their use, contribute to the development of a variety of new devices and wider application of this equipment. It accelerates the development of the Internet-of-Things (IoT) in general. Various radio technologies available on the market are used for wireless communication: LoRa modules [11, 16], ZigBee [13], Wi-Fi [15], rarely direct Internet access through the use of GSM modules [12, 15]. It is also problematic to select the sensor that is optimal in these conditions for field monitoring due to the wide variety of their operating principles and their designs.

The developments in the field monitoring of soil moisture and microclimate are often used in greenhouses [6], garden plots [10] with drip irrigation [1, 14–16], for irrigation of small areas as a whole. In the conditions of a sprinkling irrigation for watering of large areas there is the problem of the implementation active control for actuators such as starting pumping equipment, switching on/off sprinklers. The implementation still encounters "an insurmountable obstacle" – the technique is overwhelmingly non-automated. Watering is started by a person.

Another aspect of using low cost monitoring networks for a sprinkler watering is the lack of a comprehensive Irrigation Control System – DSS, which integrates sensor networks into a single system. We have developed the DSS [17] that generates recommendations on the server based on field measurements of low-cost Internet-of-Things (IoT) nets sensors, as well as using forecasting weather factors [18] and modeling of moisture transfer with formalized water-physical properties of soils.

Consequently, the purpose of the publication is to select the optimal sensor ensemble and to determine the optimum sensor structure at one point of the field using a single controller; to develop the optimal architecture of the wireless Low-Cost Internet-of-Things (IoT) sensor networks within the Irrigation Control System.

The methodology for achieving goals is to analyze and to select the most appropriate sensors and a controller, to design networks based on low-cost components, and finally to carry out a field experiment to test the operation of monitoring networks. All of the above is intended to use the monitoring data by the Irrigation Control System.

2 Materials and Methods

2.1 The Sensors

The field monitoring of air and soil conditions is based on measuring atmospheric pressure, temperature and air humidity and soil moisture, collect data of rainwater amount and wind speed. A DHT22 digital sensor was used to measure the temperature and relative humidity of the air. A BMP280 digital sensor was used to measure the atmospheric pressure.

In general, soil moisture is measured today using dielectric conductivity sensors (Fig. 1c) or electric resistance sensors for examining the amount of soil moisture [8]. Water potential is measured a Watermark (Fig. 1b) [7, 19, 21] based on electric resistance of gypsum or automatic tensiometers [1, 10], including tensiometer our own design (Fig. 1a) [20]. The water potential is converted into soil moisture and vice versa using formalized water-physical properties of soils [22].

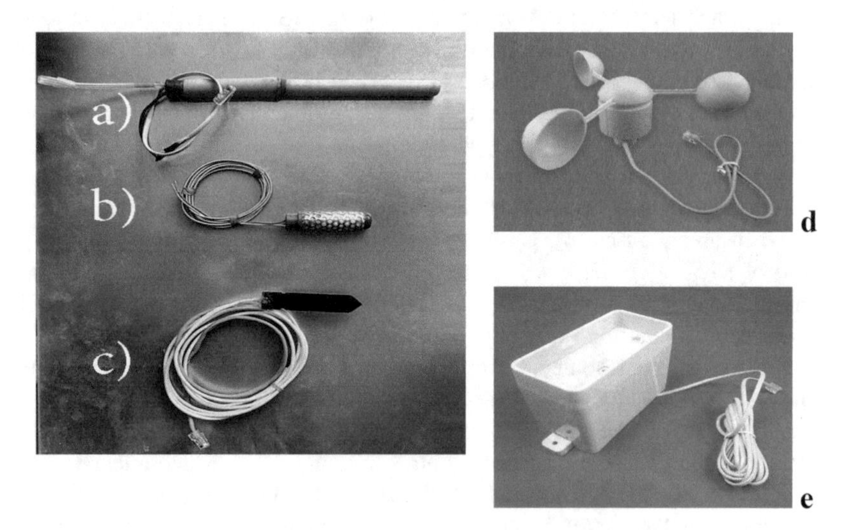

Fig. 1. Sensors for monitoring soil moisture or water potential: (a) automatic tensiometer; (b) watermark; (c) dielectric conductivity sensor. The anemometer for wind speed measurement (d) and a rainwater amount (rain-gauge) sensor (e)

Automatic measurements of soil water potential were carried out by using automatic tensiometers [20] or Watermark (Irrometer) sensors [21]. The choice criterion was the need to use in Irrigation Control System data based on the measurement of a soil water potential [17]. Data of amount rain water is a sum of precipitation amount and irrigation rate. For this aim China-made sensors from the home weather station are used (Fig. 1d and e).

2.2 The Microcontrollers

Today, there are many variants of popular controllers from different manufacturers operating on a similar principle to the Open-Hardware Platform. In our development we use ESP2866, ESP32, Atmega328P, Arduino Pro Mini. In our case, the criterion for the choice was: cost, required performance and compatibility with the Arduino IDE programming environment.

2.3 Wireless Sensor Networks and Base Stations (BS)

Main idea for building our Internet-of-Things (IoT) networks is maximal use of the wireless data transfer principle. The sensor nodes located at the field monitoring points periodically transmit the measurement data to the server using the BS. For this purpose, a radio network is organized on the field by which the nodes transmit data to the base station. The radio network uses LoRa technology and combines sensors with BS. The base station has a connection to the mobile Internet and through the GPRS it sends the accumulated data from the sensors to the server. Remotely located and separated fields transmit data to the server directly via GPRS.

2.4 Software

The server side consists of a Web server and a MySQL database. Data processing and visualization is done in PHP and JavaScript. The user can directly access this information through the Internet browser on a computer or a smartphone. Mentioned DSS processes the received soil and air monitoring data, uses it as feedback in irrigation management to generate recommendations and to display information in the form of tables and charts, but its detailed structure and operation is the subject of another issue [17].

3 Research Results

3.1 An Ensemble of Sensors at the Point of Field Monitoring

The developed irrigation control system [17] uses a soil moisture transfer model for predicting soil moisture. The model requires the monitoring data in the form of water potential. So we use sensors the Watermark and the tensiometer that can measure the water potential. The soil and air monitoring point (or sensor node) in the field MP (Fig. 2) is usually located near the start of the sprinkler to monitor the start of watering.

Let us consider the basic layout of the components of a monitoring point using the Arduino Pro Mini. Atmega328 microcontroller, based on the board, contains a minimal set of interfaces for connecting all the necessary sensors. The board has 8 pins with a digital-to-analog converter and 14 digital pins. Among the interfaces that we use, we can distinguish the following:

- one serial UART interface for programming the controller and the possibility of diagnosing its operation;
- 2 pins of external interrupts that respond to contact closure and give a possibility to connect reed switch-based sensors such as a rain gauge or an anemometer;
- serial peripheral interface (SPI) for connecting the microcontroller to one of the peripheral modules, such as LoRa radio module;
- I2C serial interface that allows you to connect a pair of BMP280 pressure sensor.

The DS18B20 and DHT22 temperature sensors connect to any two free digital pins. To connect a single Watermark sensor, a current generator circuit that is implemented reads the sensor resistance and uses 2 digital and 2 analog pins. The presence of free pins gives a possibility to place two of these resistance readers. In the general circuit diagram of the device, the following structure can be distinguished:

- digital pins 0 (RX), 1 (TX) - used for UART;
- digital pins 2, 3 are reserved for rain and wind sensors;
- digital pins 4, 5 are reserved for the possible connection of a GSM module;
- digital pins 10, 11, 12, 13 represent the SPI interface and are used to connect the LoRa module;
- analog pins A4, A5 (I2C) are used to connect two tensiometers with pressure sensors BMP280;
- analog pins A2, A3 are reserved for the possibility of connecting temperature sensors DS18B20 and DHT22;
- digital pins 6, 7, 8, 9 & analog pins A0, A1, A6, A7 are used in the resistance meter circuit for two Watermark sensors.

There are two versions of the Pro Mini: one running at 3.3 V at 8 MHz, the other at 5 V at 16 MHz. It is advisable to use the 3.3 V version for field power supply of the field monitoring point.

To form an ensemble of sensors at the point of field monitoring (Fig. 2) 2 Watermark (Irrometer) sensors [21] or 2 tensiometers of our own design [20], or 1 Watermark and 1 tensiometer were used. Most crops during irrigation form 0–50 cm root system in the soil layer. Therefore, 2 sensors were located at depths of 0–20 cm and 20–40 cm, respectively. Soil temperature, which is needed to calculate corrections when measuring Watermark, was measured in a soil layer 0–20 cm with a DS18B20 sensor. An automatic tensiometer measures vacuum using a BMP280 digital sensor located inside the tensiometer body in an airtight flexible shell.

The DHT22 digital sensor measures microclimate parameters: temperature and relative air humidity at the field monitoring point. The data obtained are used on the server to calculate evapotranspiration according to the Shtoiko formulas [18, 23].

An analogue rain-gauge sensor (see Fig. 1e and 2), which validates watering rates and measures precipitation, is important in sprinkler irrigation. Therefore, each monitoring point is completed by using them. Each MP is equipped with Arduino Pro Mini microcontroller, LoRa module, a battery, a solar panel and a charging module for measuring and data transferring.

Fig. 2. Complete set and general view of the monitoring point (MP)

3.2 Wireless Monitoring Data Transmission and a Base Station Development

The sensor nodes on fields are often installed at large distances from one another. The development of a fully functional field radio network with a data being transmitted and with synchronization of all elements of the network may take an unreasonably long time. Therefore, we have proposed a simplified structure of the network of "star"-type, with a central host server (BS) surrounded by client stations (MPs). The issue of the reliability of monitoring data transmission was solved by the excessive number of data packets sent. At the required frequency of measurements once an hour, the client stations sent information from the sensors 3–4 times an hour.

Ra-02 is a wireless transmission module based on SEMTECH's SX1278 wireless transceiver. It may particularly suitable for the reading domestic meters, smart home applications and security alarm systems. Using a long-range transmission, we have established a connection between the sensor nodes on approximately 2 km distance.

The base station performs the functions of LoRa Internet Gate (Gateway) and the weather station. The LoRa gateway receives data from the MP via radio and sends it to the server via GPRS internet. In order to perform the LoRa (Gateway) functions, the following list of modules is required: Arduino Pro Mini microcontroller, GSM module, LoRa module, battery, solar panel and charging module. To improve stability and reliability, two batteries are used: one to supply the GSM module and another one for the rest of the components (Fig. 3).

In order to serve as a weather station, the base station is additionally equipped with the respective sensors: temperature and air humidity sensor (on DHT22), atmospheric pressure sensor (on BMP280), a rain-gauge, and a wind speed sensor (Fig. 3a).

3.3 Separated Monitoring Point (SMP)

While monitoring the fields located on a distance, it is reasonable to use the separated point of monitoring, SMP (Fig. 3c), which combines the properties of the base station with the MP station but differs from both by the absence of LoRa modules. There is no radio network established in the field, and all measured values SMP sends to the Internet server via its own GSM module.

3.4 Low-Cost Internet-of-Things (IoT) Network Practical Test

The purpose of the field test was to verify the completeness of the MP sensor ensemble, the reliability of low-cost MP and BS equipment, and to test the architecture of LoRa's networks and real data transmission distance. Practical tests of monitoring networks were conducted in Kherson region, Ukraine (Fig. 4). Sprinkler-irrigated farmland was vast. Each field was about 60–65 ha in area. At these circumstances, it was advisable to use low-power radio and mobile Internet to transmit the monitoring information [24]. Given the small volume of each package transmitted data, it was possible to use a LoRA modulated radio transmission. This increased the data transmission range by two times compared to conventional radio transmissions [1]. The mentioned above distances (up to 2 km) allowed us to build a radio network around the base station at a distance of 1–2 fields (Fig. 4).

Fig. 3. The BS general view (a), a board prototype (b); a SMP prototype (c)

As mentioned above, the classic "star"-type topology network was used for the low-cost Internet-of-Things (IoT) soil and air monitoring network. A base station was installed equidistantly from other field monitoring points (Fig. 4). All monitoring stations were periodically connected to BS through the LoRa radio communication sending measured data from sensors. The base station periodically sent the received data to the server via GSM connection. It was advisable to install SMP with direct access to GSM internet to cover the far away fields.

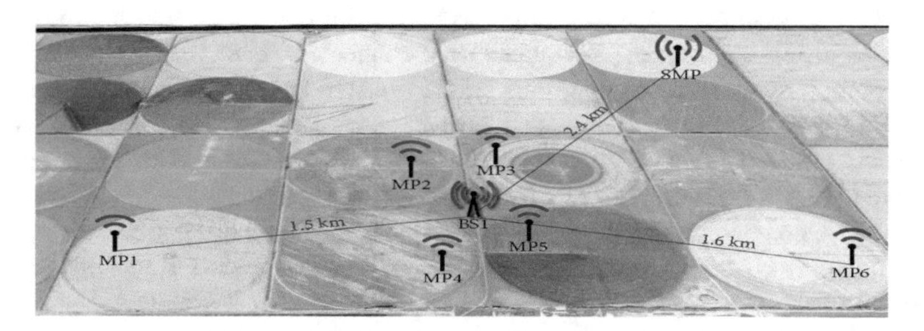

Fig. 4. Soil and air monitoring network architecture with one BS and field with SMP

The presented architecture of the monitoring networks was tested in plain terrain. The flat terrain and undersized crops (rapeseed, winter wheat, soybeans and alfalfa) made it possible to reach a maximum distance of 2 km (unlike [24], where distance was up to 5 km). Tall corn and sunflower plants reduced the distance to 1–1.5 km.

The base station may cover an area of up to 16 (Fig. 4) fields. But when PM4^{-th} corn grew, monitoring of the MP1^{-th} field became problematic. At the same time, MP5^{-th} wheat had absolutely no effect on MP6^{-th} soybean monitoring (Fig. 4). Forest strips also reduced the distance between MP and BS by one and a half or two times. Therefore, according to test results, the smooth functioning of the entire low-cost Internet-of-Things network is ensured by the connection of 4–6 MP per 1 BS (Fig. 5).

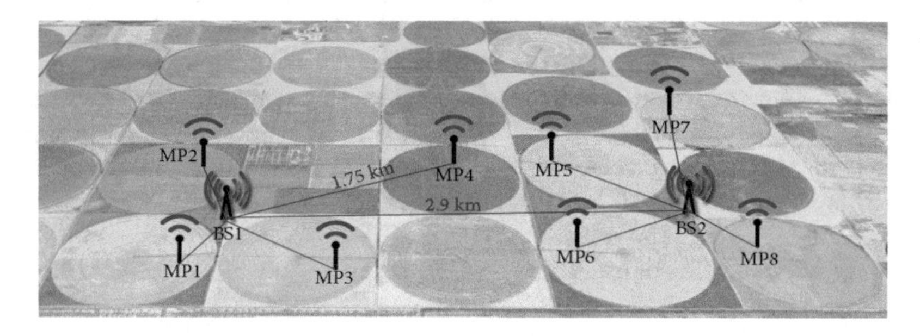

Fig. 5. The recommended architecture of a field soil and air monitoring with several BS

The developed networks can be deployed in any geographical location, provided that there is a GSM data connection allowing the data from the sensors to be transmitted to a database hosted on a remote Internet server. There is an option to place the server, which collects the monitoring data, right on the farm [5].

4 Research Perspectives and Discussion

1. Expensive monitoring systems [2, 3] use powerful controllers to which up to 6–8 soil humidity sensors are simultaneously connected. This is good for scientific research. But in real farm conditions, as the research has shown, it would be enough to cover 2–3 soil depths. Expensive systems also use a high-cost block of weather sensors. The irrigation control system [17, 25, 27] based on low-cost monitoring of soil and air uses a minimum of weather sensors to calculate evapotranspiration [23, 25, 26]. Our low-cost monitoring network is characterized by the fact that soil and air monitoring sensors, rainwater amount sensors are used simultaneously in the field.
2. Possible technologies should be applied to determine the phases of plant grows, since watering in different phases of development requires different watering rates and a threshold for humidity to start watering [17]. The solution is to analyze the images directly from the field as in [5].
3. Requires additional research on the number of monitoring points per a field. More reliable results can be obtained by setting the second monitoring point closer to the completion of irrigation, i.e. set the second feedback loop.
4. Additional studies are needed to determine the maximum and optimal number of MP that can be connected to a single BS, although we may recommend 4–6.
5. Need research of MP antenna position for increasing data transmission distance.

5 Conclusions

The possibility of use the low-cost Internet-of-Things air and soil monitoring networks is scientifically and practically confirmed when using irrigation control system. On the basis of the principle of measuring the soil water potential, such optimal sensors were selected for soil monitoring: Watermark and automatic tensiometers of our own design. Based on the previous studies on evaluation of evapotranspiration [23] and weather Data mining [18] allowed to identify and recommend low-cost sensors for air monitoring in a field (DHT22, BMP280 as more accurate) and rain-gauge for rainfall and watering rates amount. The board is based on the Atmega328 microcontroller. The variants of a complete set of measuring devices, depending on their purpose in the field, are offered.

Some practical schemes based on a «star» topology of low-cost Internet-of-Things (IoT) network and wireless data transmission were developed and tested. Based on the obtained results, to build a LoRa radio network around the base station at a distance up

1–2 km or of 1–2 fields with performance 4–6 MP/1 BS was recommended. More distance requires using SPM with GSM data transmisions.

The scientific contribution of the work is to improve irrigation control systems by developing low-cost Internet-of-Things networks for air and soil monitoring. The networks provide a feedback to the Irrigation Control System as well as reliable accurate operation of the system computing modules, hosted on the server. That enables to integrate the sensor networks and software into a single system. Applying this integrated system provides the saving of irrigation water and obtaining higher yields due to improving the irrigation management.

References

1. Tevatronic LTD. http://tevatronic.net/. Accessed 1 May 2019
2. Pessl Instruments GmbH. http://metos.at/fieldclimate/. Accessed 11 Nov 2018
3. Davis Instruments. https://www.davisinstruments.com/. Accessed 16 Jan 2019
4. ENORASIS. https://cordis.europa.eu/project/id/282949. Accessed 3 Sept 2018
5. WAZIUP. https://www.waziup.eu/. Accessed 11 Sept 2019
6. Ferrarezi, R.S., Dove, S.K., van Iersel, M.W.: An automated system for monitoring soil moisture and controlling irrigation using low-cost open-source microcontrollers. HortTechnology **25**(1), 110–118 (2015)
7. Fisher, D.K.: Automated collection of soil-moisture data with a low-cost microcontroller circuit. Appl. Eng. Agric. **23**(4), 493–500 (2007)
8. Maddah, M., Olfati, J.A., Maddah, M.: Perfect irrigation scheduling system based on soil electrical resistivity. Int. J. Veg. Sci. **20**(3), 235–239 (2014). https://doi.org/10.1080/19315260.2013.798755
9. Fisher, D.K., Gould, P.J.: Open-source hardware is a low-cost alternative for scientific instrumentation and research. Mod. Instrum. **1**(02), 8 (2012). https://doi.org/10.4236/mi.2012.12002
10. Thalheimer, M.: A low-cost electronic tensiometer system for continuous monitoring of soil water potential. J. Agric. Eng. **44**(3), XLIV (2013). https://doi.org/10.4081/jae.2013.e16
11. Payero, J.O., Mirzakhani-Nafchi, A., Khalilian, A., Qiao, X., Davis, R.: Development of a low-cost Internet-of-Things (IoT) system for monitoring soil water potential using Watermark 200SS sensors. Adv. IoT **7**(03), 71 (2017). https://doi.org/10.4236/ait.2017.73005
12. Bitella, G., Rossi, R., Bochicchio, R., Perniola, M., Amato, M.: A novel low-cost open-hardware platform for monitoring soil water content and multiple soil-air-vegetation parameters. Sensors **14**(10), 19639–19659 (2014). https://doi.org/10.3390/s141019639
13. Song, J.J., Zhu, Y.L.: Environment monitoring system for precise agriculture based on wireless sensor network. In: Applied Mechanics and Materials, vol. 475, pp. 127–131 (2014). Trans Tech Publications. https://doi.org/10.4028/www.scientific.net/AMM.475-476.127
14. Khan, F., Shabbir, F., Tahir, Z.: A fuzzy approach for water security in irrigation system using wireless sensor network. Sci. Int. **26**(3) (2014)
15. Villarrubia, G., Paz, J.F.D., Iglesia, D.H., Bajo, J.: Combining multi-agent systems and wireless sensor networks for monitoring crop irrigation. Sensors **17**(8), 1775 (2017). https://doi.org/10.3390/s17081775

16. Dupont, C., Vecchio, M., Pham, C., Diop, B., Dupont, C., Koffi, S.: An open IoT platform to promote eco-sustainable innovation in Western Africa: real urban and rural testbeds. Wirel. Commun. Mob. Comput. **2018**, 1–17 (2018). https://doi.org/10.1155/2018/1028578
17. Gadzalo, Ya., Romashchenko, M., Kovalchuk, V., Matiash, T., Voitovich O.: Using smart technologies in irrigation management. In: International Commission on Irrigation and Drainage, 3rd World Irrigation Forum (WIF3), pp. 1–6 (2019). (ICID Id: W.1.3.02)
18. Kovalchuk, V., Demchuk, O., Demchuk, D., Voitovich, O.: Data mining for a model of irrigation control using weather web-services. In: Hu, Z., Petoukhov, S., Dychka, I., He, M. (eds.) Advances in Computer Science for Engineering and Education. ICCSEEA 2018. Advances in Intelligent Systems and Computing, vol. 754. Springer, Cham (2019). https://doi.org/10.1007/978-3-319-91008-6_14
19. Radman, V., Radonjić, M.: Arduino-based system for soil moisture measurement. In: Proceedings of the 22nd Conference on Information Technologies (IT), vol. 17, pp. 289–292 (2017)
20. Kovalchuk, V.P., Voitovich, O.P., Demchuk, D.O.: Ukrainian Patent for Utility Model UA 132271, 25 February 2019. https://sis.ukrpatent.org/en/search/detail/1223767/
21. Irrometer: WATERMARK Sensor Model 200SS. Irrometer Co., Inc., Riverside. Source: https://www.irrometer.com/pdf/sensors/403%20WATERMARK%20Sensor-WEB.pdf
22. Van Genuchten, M.T.: A closed-form equation for predicting the hydraulic conductivity of unsaturated soils. Soil Sci. Soc. Am. J. **44**(5), 892–898 (1980)
23. Romashchenko, M.I., Bohaienko, V.O., Matiash, T.V., Kovalchuk, V.P., Danylenko, Iu.Iu.: Influence of evapotranspiration assessment on the accuracy of moisture transport modeling under the conditions of sprinkling irrigation in the south of Ukraine. Arch. Agron. Soil Sci. (2019). http://doi.org/10.1080/03650340.2019.1674445
24. Jawad, H.M., Nordin, R., Gharghan, S.K., Jawad, A.M., Ismail, M.: Energy-efficient wireless sensor networks for precision agriculture: a review. Sensors **17**, 1781 (2017). https://doi.org/10.3390/s17081781
25. Goap, A., Sharma, D., Shukla, A.K., Krishna, C.R.: An IoT based smart irrigation management system using machine learning and open source technologies. Comput. Electron. Agric. **155**, 41–49 (2018). https://doi.org/10.1016/j.compag.2018.09.040
26. Vinduino R3 sensor station. https://www.vinduino.com/portfolio-view/lora-sensor-station. Accessed 29 Nov 2018
27. Anusha, K., Mahadevaswamy, U.B.: Automatic IoT based plant monitoring and watering system using Raspberry Pi. Int. J. Eng. Manuf. (IJEM) **8**(6), 55–67 (2018). https://doi.org/10.5815/ijem.2018.06.05

Evaluation Research on Technology Innovation Efficiency of High-End Equipment Manufacturing Based on DEA Malmquist Index Method—Taking Guangdong Province as an Example

Shan Wu[1,2], Haotai Li[1(✉)], Jinlong Zhang[1], and Bingchan Fan[2]

[1] School of Management, Huazhong University of Science and Technology, Wuhan, China
hariny@163.com
[2] Research Center for Mathematical Modeling, Wuhan Technology and Business University, Wuhan, China

Abstract. Innovation is the core driving force of high-end equipment manufacturing industry, and the evaluation of its efficiency has important theoretical and practical value. In this paper, taking Guangdong Province as an example, which is a strong province of high-end equipment manufacturing industry in China. DEA Malmquist index method is utilized. Different from other DEA methods, this method is used to study the variation of efficiency, instead of just giving the efficiency value. In order to reduce the repeatability of the existing index system and ensure the scientificity and accuracy of the evaluation results, two new input index parameters, namely, full-time equivalent input of R&D personnel and internal expenditure of R&D funds, were constructed innovatively. The results show that in order to improve the technological innovation efficiency of high-end intelligent manufacturing in Guangdong province, it is not enough to focus on innovation itself, but also to focus on the transformation of innovation achievements.

Keywords: High-end equipment manufacturing industry · Technological innovation efficiency · Data envelopment analysis

1 Introduction

High-end equipment manufacturing industry (HEMI), also known as advanced equipment manufacturing industry, is a strategic emerging industry led by high and new technology, lying at the high end of the value chain and the core of the industrial chain and determining the overall competitiveness of the entire industrial chain. The development of HEMI is of great strategic significance for China's transformation from a big manufacturing country to a powerful country. "Made in China 2025" lists high-end equipment innovation project as one of the five major projects.

The evaluation of innovative capability of high-end manufacturing industry has attracted the attention of many countries, and different evaluation methods and index

Z. Hu et al. (Eds.): ICCSEEA 2020, AISC 1247, pp. 97–107, 2021.
https://doi.org/10.1007/978-3-030-55506-1_9

systems have been put forward. Lee and Smith [1] made an empirical study on the relationship between R&D, long-term performance and market share of high-tech enterprises in the United States and Japan, and found that there was a significant positive correlation between R&D and the other two. Neelankevil and Alaganar [2] used Granger causality test to evaluate the innovation efficiency of high-tech industry. Romijn and Mike [3] used multiple regression analysis to study the performance and efficiency of high-tech industries.

The technological innovation efficiency (TIE) in high-end equipment manufacturing and related industries had also been studied in China. Some scholars used cross-sectional data to study individual industries and sub-sectors, such as Zhang Qinghui and Wang Jianping [4] used BCC model to measure the comprehensive efficiency of independent innovation, pure technical efficiency (PTE) and scale efficiency (SE), and returns to scale (RTS) of 17 high-tech industries in China. At the same time, through projection analysis, the adjustment suggestions of inefficient units were porjected. Wei Feng and Jiang Yonghong [5] constructed input-output index system. Super-DEA model and grey relational analysis method were used to study the regional and industrial technological innovation efficiency of small and medium-sized enterprises in Anhui Province. In addition, some scholars used panel data to study the TIE of related industries with time span. For example, Feng Miao and Teng Jiajia [6] used Data Envelopment Analysis (DEA) to evaluate the overall TIE of Jiangsu high-tech industry from both horizontal and vertical perspectives. Huang Haixia and Zhang Zhihe [7] used panel data from 2005 to 2012 and Malmquist index method to measure TFP of the strategic emerging industries' technological innovation in 28 provincial administrative regions and three Regions (East, Central and West).

To sum up, it can be seen that the current research on the TIE is mainly in the broad sense of high-tech industry. However, the literature on high-tech evaluation of high-end manufacturing industry is almost void.

From the perspective of research methods, most scholars regard the process of technological innovation as a whole. Based on various perspectives, they study the overall efficiency of technological innovation, discuss the changing trend and regional differences of the overall efficiency, directly analyze the initial input and final output, without considering the intermediate process, and ignore the internal R & D structure of technological innovation, which is easy to cause the "black box" in the process of technological innovation evaluation.

Guangdong Province is the top three provinces of science and technology innovation in China. The "Three-year Plan of Action for the Struggle of Industrial Transformation and Upgrading in Guangdong Province (2015–2017)" states that Guangdong Province will actively adapt to the new normal economic development and promote economic transformation and upgrading through industrial transformation and upgrading [8]. High-end equipment has also been listed as the main attacking direction of manufacturing upgrading in Guangdong Province, and effective evaluation of its innovation efficiency has important theoretical and practical value. However, there are few studies on the efficiency evaluation of technological innovation of high-end manufacturing in Guangdong Province. The research results not only play a positive role in improving the industrial technological innovation system, enhancing the efficiency of industrial technological innovation, and promoting the rapid and healthy development of the whole industry, but also have important reference value for other provinces, especially for the transformation and upgrading of manufacturing industry in the central and western underdeveloped regions.

2 Methods

DEA-Malmquist index method was originally proposed by Malmquist (1953). Caves et al. (1982) first applied the index to the calculation of productivity change. Since then, combining with DEA theory established by Charnes et al. (1978), it has been widely used in productivity calculation. In empirical analysis, researchers generally use the Malmquist index based on DEA constructed by Fare et al. (1994). Based on panel data, this index calculates Total Factor Productivity (TFP).

$$M_{j0}^{t+1}(X_{j0}^{t+1}, Y_{j0}^{t+1}, X_{j0}^{t}, Y_{j0}^{t}) = \left[\frac{D_{j0}^{t}(X_{j0}^{t+1}, Y_{j0}^{t+1})}{D_{j0}^{t}(X_{j0}^{t}, Y_{j0}^{t})} \bullet \frac{D_{j0}^{t+1}(X_{j0}^{t+1}, Y_{j0}^{t+1})}{D_{j0}^{t+1}(X_{j0}^{t}, Y_{j0}^{t})}\right]^{\frac{1}{2}} \tag{1}$$

Where, $(X_{j0}^{t+1}, Y_{j0}^{t+1})$ and $\left(X_{j0}^{t}, Y_{j0}^{t}\right)$ represent the input and output vectors of j_0th decision making units in t and $t+1$ periods respectively. $D_{j0}^{t}(X_{j0}^{t}, Y_{j0}^{t})$ and $D_{j0}^{t+1}(X_{j0}^{t}, Y_{j0}^{t})$ represent the distance functions of production points in t period and $t+1$ period, respectively, based on the production possibility boundary of t period.

In order to decompose the result into Technical Efficiency Change (TEC) and Technology Change (TC) separately, Eq. (1) is deformed, and the result is as follows:

$$M_{j0}^{t+1}(X_{j0}^{t+1}, Y_{j0}^{t+1}, X_{j0}^{t}, Y_{j0}^{t}) = \frac{D_{j0}^{t+1}(X_{j0}^{t+1}, Y_{j0}^{t+1})}{D_{j0}^{t}(X_{j0}^{t}, Y_{j0}^{t})} \left[\frac{D_{j0}^{t}(X_{j0}^{t+1}, Y_{j0}^{t+1})}{D_{j0}^{t+1}(X_{j0}^{t+1}, Y_{j0}^{t+1})} \bullet \frac{D_{j0}^{t}(X_{j0}^{t}, Y_{j0}^{t})}{D_{j0}^{t+1}(X_{j0}^{t}, Y_{j0}^{t})}\right]^{\frac{1}{2}} \tag{2}$$

Where, $\frac{D_{j0}^{t+1}(X_{j0}^{t+1}, Y_{j0}^{t+1})}{D_{j0}^{t}(X_{j0}^{t}, Y_{j0}^{t})}$ represents Technical Efficiency change, which refers to the result of efficiency change caused by system change.

If TEC > 1, it indicates an increase in efficiency; otherwise, efficiency declines.

$\left[\frac{D_{j0}^{t}(X_{j0}^{t+1}, Y_{j0}^{t+1})}{D_{j0}^{t+1}(X_{j0}^{t+1}, Y_{j0}^{t+1})} \bullet \frac{D_{j0}^{t}(X_{j0}^{t}, Y_{j0}^{t})}{D_{j0}^{t+1}(X_{j0}^{t}, Y_{j0}^{t})}\right]^{\frac{1}{2}}$ denotes for Technology Change, which is the result of innovation or the introduction of new technology.

If TC > 1, it indicates technological progress, production boundary promotion; otherwise, it means technological recession. When the constraints of constant returns to scale are removed, efficiency changes based on variable returns to scale, can be further decomposed into pure efficiency change (PTEC) and scale change (SC), namely

$$\text{TEC} = \text{PTEC} * \text{SC}$$

Thus it can be seen

$$TFP = PTEC * SC * TC$$

3 Empirical Analysis

3.1 Construction of Index System

In the process of constructing the evaluation index of TIE of high-end equipment in Guangdong Province, objectivity principle, multidimensional principle and mutual exclusion principle are followed [9, 10].

In order to further study the efficiency of technological innovation, technology innovation is divided into two stages: technology formation stage and technology application stage. The input in the technology application stage not only includes the output in the technology formation stage, but also includes the full time equivalent of R&D personnel, because the technology application stage also needs the input of R&D personnel. Thus, the evaluation index system of TIE of HEMI in Guangdong province is constructed as shown in Table 1.

Table 1. Two-stage evaluation index system for TIE

Index system of technology formation stage		Index system of technology application stage	
Input	R&D personnel full time equivalent (person year) (X_1)	Input	R&D personnel full time equivalent (person year) (Y_1)
	R&D internal expenditure (10 thousand yuan) (X_2)		Patents (pieces) (Y_2)
Output	Patents (pieces) (Y)	Output	New product sales revenue (10 thousand yuan) (Z)

The efficiency of technological innovation is related to many factors. Therefore, the most representative indicators of "people", "money" and "things" are selected to build an indicator system. At the stage of technology formation, the reasons for selecting indicators are as follows:

1) R&D personnel equivalent of full time reflects the number of R&D personnel who are truly and effectively engaged in scientific research, as well as the intensity, quality and efficiency of R&D personnel's scientific research work. It represents the investment of "person".
2) R&D internal expenditure refers to the total expenditure actually used in the unit for carrying out R&D activities [11]. It is the most important financial condition for independent innovation to be realized, reflecting the "money" investment.
3) The number of patent applications refers to the number of patent applications submitted to the state intellectual property administration, which reflects the

intermediate results of R&D activities, and is also the input in the stage of technology application, reflecting the "things" investment.

In the technology application stage, the reasons for selecting indicators are as follows:

1) New product sales revenue refers to the revenue realized by the enterprise in selling new products from the main business revenue and other business revenue, which is the final result of R&D activities. In view of the data unavailable, the sales revenue of new products is replaced by the output value of new products.
2) In the stage of technology application, patents have been formed, and the formation process of new product output value mainly depends on R & D personnel, which does not need more R&D funds, Therefore, only full time R&D personnel equivalent and the number of patent applications are taken as input indicators.

3.2 Sample Selection and Data Source

As far as possible to fit the key development direction of HEMI in Guangdong Province, the following directions are selected referring to the industry classification standards issued by the National Bureau of Statistics: (A) general equipment manufacturing industry, (B) special equipment manufacturing industry, (C) transportation equipment manufacturing industry, (D) electrical machinery & equipment manufacturing industry, (E) computers & communications & other electronic equipment manufacturing industry and (F) Instrument manufacturing industry.

In 2009, the State Council held three symposiums on the development of emerging industries. In the same year, the general office of the people's Government of Guangdong Province issued the "opinions on the implementation of the adjustment and revitalization plan for the equipment manufacturing industry of Guangdong Province". Therefore, 2009 is selected as the starting point to study the technological innovation efficiency of HEMI in Guangdong. At present, the R&D data of HEMI in Guangdong Province has been published to 2013. From the data of 2009–2013, the development trend of TIE of HEMI in Guangdong province can be basically analyzed. The empirical data adopted are all from the "Guangdong Science and Technology Statistical Yearbook" and "Guangdong High and New Technology Industry Development Research Report" in 2009–2013.

3.3 Dynamic Analysis of Technology Formation Stage

Learning from the general approach, it is assumed that the lag period of technological innovation of HEMI in Guangdong province is one year, that is, the data from 2009 to 2012 for input index and the data from 2010 to 2013 for output index are adopted.

According to the constructed index system of technology formation stage, and using DEA-Malmquist index method, Malmquist index and decomposition results of technology formation stage are obtained, based on the panel data of six subdivisions of HEMI in Guangdong Province from 2009 to 2013 (see Table 2).

Table 2. Malmquist index and its decomposition of technology formation stage

Time period	TEC	TC	PTEC	SC	TFP
2009–2010	0.8681	0.9291	0.9355	0.9280	0.8066
2010–2011	1.2815	0.6781	1.2137	1.0559	0.8690
2011–2012	0.8549	1.3637	0.8239	1.0377	1.1659
2012–2013	0.7583	1.6804	1.0267	0.7385	1.2742
Average value	0.9215	1.0962	0.9900	0.9309	1.0102

Note: If the index value is greater than 1, it means that the corresponding index is increasing; If the index value is less than 1, it means that the corresponding index is decreasing; The difference between the index value and 1 is the extent of increase or decrease.

As can be seen from Table 2, the growth rate of TFP in the whole period is only 1.02%, which indicates that the utilization efficiency of technological innovation resources of HEMI in Guangdong province is relatively low during the early technology formation period. Since TFP is the product of TEC and TC, the growth rate of technical efficiency change is −7.85%, and the average annual growth rate of technical change is 9.62%, so the stagnation of TFP is mainly the decline of technical efficiency. At the same time, it should be noted that TEC only increased in 2010–2011, and all the remaining years showed negative growth, and during 2012–2013, the decrease was 24.17% at the most, which indicated that the industrial structure of Guangdong HEMI needed to be adjusted urgently. PTEC and SC together determine TEC, while PTEC has hardly changed in 2009–2013. Therefore, the main reason for low technical efficiency is the negative growth of SC, which indicates that the development scale of manufacturing industry should be moderate.

Table 3. TFP, TEC and TC of technology formation stage of sub industries

Time period	2009–2010			2010–2011		
Industry	TFP	TEC	TC	TFP	TEC	TC
A	0.8024	0.8763	0.9157	0.8591	1.3432	0.6396
B	0.6694	0.7311	0.9156	0.9795	1.5189	0.6449
C	0.7216	0.7652	0.9429	0.9814	1.3530	0.7254
D	1.0432	1.1065	0.9428	0.8943	1.2329	0.7254
E	0.7438	0.7889	0.9429	1.2642	1.7662	0.7158
F	0.9156	1.0000	0.9156	0.4611	0.7370	0.6257
Time period	2011–2012			2012–2013		
Industry	TFP	TEC	TC	TFP	TEC	TC
A	1.1058	0.8259	1.3389	1.2282	0.7437	1.6514
B	1.1844	0.8811	1.3443	1.4351	0.8523	1.6838
C	0.8768	0.6373	1.3758	1.0219	0.5870	1.7408
D	1.3486	1.0000	1.3486	1.7152	1.0000	1.7152
E	0.8536	0.6206	1.3755	1.1713	0.6677	1.7542
F	1.8996	1.3569	1.4000	1.1828	0.7651	1.5459

The dynamic changes and decomposition results of total factor productivity in the technology formation stage of six sub-industries of HEMI in Guangdong province from 2009 to 2013 are shown in Table 3.

According to Table 3, TFP of 6 sub industries of HEMI in Guangdong Province has ups and downs in 4 periods: From 2009 to 2010, TFP in industry D increased slightly, with an increase of 4.32%. TFP in the other five industries declined significantly, with the largest decline being industry B, with a decrease of 33.06%. From 2010 to 2011, except for the 26.42% increase in TFP of industry E, the other five industries all saw a decline, with the largest decline being F, which reached 53.89%. From 2011 to 2012, the situation began to improve. Only the TFP of C and E decreased by about 15%, while the rest all increased rapidly. Industry F increased the most, up to 89.96%. From 2012 to 2013, TFP of six sub industries of HEMI in Guangdong Province achieved overall growth, with the largest increase of D, 71.52%, and the smallest increase of C, 2.19%.

TFP is obtained by TEC multiplying TC. In order to find out the reason of TFP change, we need to further decompose it. From Table 3, it can be seen that only TEC of D showed a slight growth trend from 2009 to 2010. Only F showed a decreasing trend in TEC, while the rest showed an obvious increasing trend during 2010–2011. The highest one was E, which reached 76.62%. From 2011 to 2012, only TEC of F showed an increasing trend. During 2012–2013, except for TEC of D, the other five industries all fell sharply, the most obvious being C, which was 41.30%. It can also be seen from Table 3 that during the two periods from 2009 to 2010 and from 2010 to 2011, six segments of HEMI in Guangdong Province experienced technological regression. In the two periods of 2011–2012 and 2010–2011, the situation reversed and the technological progress was obvious, with an increase of over 30% and 50% respectively.

In the stage of technology formation, the annual TFP and its decomposition results of six subdivisions of HEMI in Guangdong Province are shown in Table 4, during 2009–2013.

Table 4. Mean value of Malmquist index and its decomposition in the technological formation stage of six subdivisions

Industry	A	B	C	D	E	F
TEC	0.9221	0.9556	0.7889	1.0807	0.8717	0.9353
TC	1.0667	1.0752	1.1313	1.1215	1.1296	1.0552
PTEC	0.9526	1.0617	0.9307	1.0000	1.0000	1.0000
SC	0.9526	1.0617	0.9307	1.0000	1.0000	1.0000
TFP	0.9837	1.0275	0.8925	1.2121	0.9847	0.9869

It can be seen from Table 4 that the six subdivisions of HEMI in Guangdong Province are in the state of technological progress, and the technological efficiency is in the state of recession except for Industry D. However, technological progress did not promote the overall improvement of TFP, only TFP of B and D increased, the other four industries TFP decreased. This shows that in the stage of technology formation, the technological innovation efficiency of HEMI in Guangdong Province largely depends on TEC. Technological progress promotes the technological innovation of HEMI in Guangdong Province, but it does not play a decisive role.

3.4 Dynamic Analysis of Technology Application Stage

In this stage, the panel data of six subdivisions of HEMI in Guangdong Province from 2009 to 2013 are still used. According to the constructed index system of technology application stage and the DEA Malmquist index method, the Malmquist index and its decomposition results of the whole sample industry technology application stage of HEMI in Guangdong Province from 2009 to 2013 are obtained (see Table 5). In the stage of technology application, the TFP of HEMI in Guangdong province fluctuates, and the efficiency of technology innovation is sometimes low and sometimes high. The efficiency of technological progress declined year by year, with the largest decline of 41.24% from 2011 to 2012, indicating that the technology formed in the first stage did not play a good role. The technical efficiency change index decreased only in 2010–2011, but increased in other years, increasing by nearly 60% in 2011–2012.

Table 5. Malmquist index and its decomposition of technology application stage

Time period	TEC	TC	PTEC	SC	TFP
2009–2010	1.2947	0.9428	1.1035	1.1732	1.2206
2010–2011	0.9080	0.9339	0.8715	1.0419	0.8480
2011–2012	1.5983	0.5876	0.9995	1.5991	0.9392
2012–2013	1.0517	0.9644	1.0681	0.9847	1.0143
Average value	1.1857	0.8405	1.0066	1.1779	0.9965

As in the technology formation stage, the reasons for TFP change can be analyzed by analyzing efficiency change and technology change. It can be seen from Table 6 that: except for the technological progress of industry C in 2009–2010, which increased by about 8%, all other industries and all periods are in the state of technological retrogression. Except for three industries (B, D, F) in 2010–2011 and two industries (A, F) in 2012–2013, the rest are in the state of technical efficiency improvement.

Table 6. TFP/TEC/TC of technology application stage of sub industries

Time period	2009–2010			2010–2011		
Industry	TFP	TEC	TC	TFP	TEC	TC
A	0.9403	1.0248	0.9176	1.0339	1.1090	0.9323
B	1.5264	1.6636	0.9175	0.7841	0.8408	0.9326
C	1.0801	1.0000	1.0801	0.9413	1.0000	0.9413
D	1.0036	1.0937	0.9175	0.8592	0.9215	0.9324
E	1.3293	1.4489	0.9174	1.1329	1.2149	0.9325
F	1.5992	1.7430	0.9175	0.5007	0.5369	0.9326
Time period	2011–2012			2012–2013		
Industry	TFP	TEC	TC	TFP	TEC	TC
A	1.2402	2.1336	0.5813	0.9000	0.9324	0.9653
B	1.1336	1.9503	0.5812	1.0370	1.0742	0.9654
C	0.6207	1.0000	0.6207	0.9600	1.0000	0.9600
D	0.8754	1.5062	0.5812	1.2655	1.3110	0.9653
E	0.9231	1.5882	0.5812	1.2240	1.2680	0.9653
F	0.9736	1.6749	0.5813	0.7847	0.8129	0.9652

In the stage of technology application, the annual TFP and its decomposition results of six subdivisions of HEMI in Guangdong Province are shown in Fig. 1, during 2009–2013.

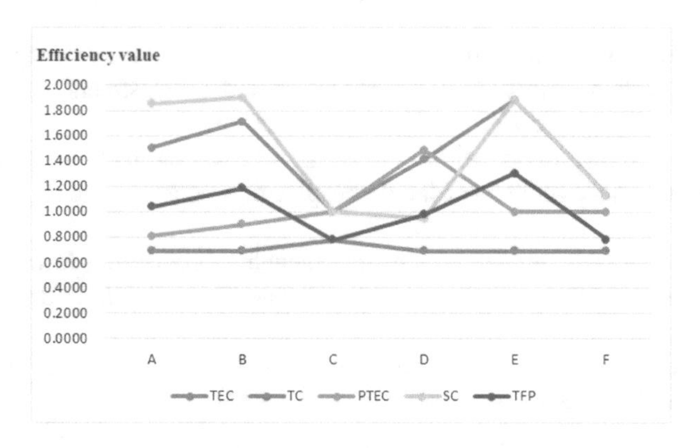

Fig. 1. Mean value of Malmquist index and its decomposition in the technological application stage of six subdivisions of HEMI

As can be seen from Fig. 1, all the six subsectors show a state of technological regression in terms of technological change, while in terms of technical efficiency change index, they all show an upward trend. Although both industry B and industry E have a significant increase in the change index of technical efficiency, the TFP of technological innovation is reduced due to the significant technological regression. It can be seen that, in the stage of technology application, the efficiency of technology innovation is affected by the change of technology efficiency and the change of technology.

4 Conclusions

Based on the TIE evaluation of HEMI in Guangdong Province from 2009–2013, it can be seen that:

1) The technological innovation efficiency of HEMI in Guangdong Province, in the stage of technological formation, largely depends on the improvement of technological efficiency; However, in the stage of technological application, and the change of technological efficiency and technological change together affect the total factor productivity.
2) In these six industries, only industry E in the two stages of technology formation and application has been improved simultaneously, which shows that this industry not only attaches importance to the formation of technology, but also attaches importance to the timely transformation of the new technology into economic benefits. This is closely related to its industry characteristics of fast update and high agility. Once the technology is formed, it will be put into application quickly.

In order to reasonably determine the investment scale of each factor of technological innovation and improve the use efficiency of the factor, the following Suggestions are proposed based on the specific situation of technological innovation of high-end equipment in Guangdong province:

1) The internal investment of R&D funds can promote the improvement of technological innovation ability to some extent, but the investment scale should be appropriately controlled to improve the use efficiency of funds;
2) Strengthen technical cooperation with scientific research institutes and institutions of higher learning, and increase the development, production and sales of high-tech products.
3) Pay attention to the transformation of achievements after the formation of technology, form a virtuous circle, and promote the efficiency of technological innovation;
4) Try to shorten the period from technology formation to technology application to ensure the market competitiveness of new products.

R&D personnel and R&D funds, one part of the investment in the technology formation stage, the other part of the investment in the technology application stage, how to determine the appropriate investment ratio, is the next stage of the work to improve the content.

Acknowledgement. This paper is supported by major projects of the National Social Science Foundation (NO. 16ZDA013).

References

1. Lee, J., Shim, E.: Moderating effects of R&D on corporate growth in US and Japanese hi-tech industries: an empirical study. J. High Technol. Manag. Res. (6), 179–191 (1995)
2. Romijn, H., Mike, A.: Innovation networking and proximity: lessons from small high technology firms in the UK. Reg. Stud. **36**(1), 81–88 (2002)
3. Neelankevil, J.P., Alaganar, V.T.: Strategic resource commitment of high technology firms an international comparison. J. Bus. Res. (6), 493–502 (2003)
4. Zhang, Q., Wang, J.: Evaluation of independent innovation efficiency of China's high-tech industry based on DEA. Sci. Technol. Manag. Res. **31**(10), 9–13+17 (2011). (in Chinese)
5. Wei, F., Jiang, Y.: Evaluation of technological innovation efficiency of small and medium-sized enterprises in Anhui Province and analysis of influencing factors. China Sci. Technol. Forum (08), 100–106 (2012). (in Chinese)
6. Feng, Y., Teng, J.: Evaluation of technological innovation efficiency of high-tech industry in Jiangsu Province. Sci. Sci. Manag. Sci. Technol. **31**(08), 107–112 (2010). (in Chinese)
7. Huang, H., Zhang, Z.: Technological innovation efficiency of China's strategic emerging industries based on DEA Malmquist index model. Technol. Econ. **34**(01), 21–27+68 (2015). (in Chinese)
8. People's Government of Guangdong Province: Three year action plan for industrial transformation and upgrading of Guangdong Province (2015–2017), 21 Mar 2013. http://www.gd.gov.cn/gkmlpt/content/0/143/post_143801.html. (in Chinese)
9. Chen, W., Jing, R., Zhang, H., et al.: Evaluation of technological innovation efficiency of large and medium-sized industrial enterprises in northeast China – based on DEA-Malmquist index method. East China Econ. Manag. **31**(02), 66–71 (2017)
10. Hou, R.: Research on the evolution and driving difference of regional technology innovation efficiency improvement in China based on Malmquist index. Ind. Technol. Econ. **35**(01), 122–129 (2016)
11. Yuan, Q., Wu, L., Zhang, P.: Research on the path of improving innovation efficiency of large manufacturing enterprises in China under green growth. Sci. Technol. Progr. Countermeas. **34**(22), 85–92 (2017)

Investigation of Load-Balancing Fast ReRouting Model with Providing Fair Priority-Based Traffic Policing

Oleksandr Lemeshko, Maryna Yevdokymenko(✉), Oleksandra Yeremenko, and Anastasiia Shapovalova

Kharkiv National University of Radio Electronics, 14 Nauky Ave., Kharkiv, Ukraine
{oleksandr.lemeshko.ua, maryna.yevdokymenko, oleksandra.yeremenko.ua}@ieee.org, anastasiia.shapovalova@nure.ua

Abstract. The paper is devoted to the investigation of the load-balancing Fast ReRouting model with providing fair priority-based Traffic Policing. A primary place in the model is distinguished for the conditions of network elements protection, as well as network bandwidth together with the provision of load-balancing. The novelty of the studied model is in the fact that the flows, which are the source of network overload, were limited first. In addition, as the results of the study have shown, the use of the linear-quadratic optimality criterion of routing solutions gave the process of Traffic Policing a fair and balanced character, when, under overload conditions, the flows with higher priority were limited but to a lesser extent than flows with low priority. In order to achieve this result, the Fast ReRouting problem has been stated in an optimization form with the introduction of a linear-quadratic objective function. The numerical research results of the application of the proposed model confirmed the effectiveness and adequacy of the obtained routing solutions.

Keywords: Fast ReRouting · Load-balancing · Traffic Policing · Fair priority

1 Introduction

Nowadays, one of the most important aspects of configuring existing and designing new infocommunication networks (ICNs) is to ensure their high resilience [1–5]. This is conditioned by the fact that network traffic growth caused by an increase in the number of different services and facilities provided to end-users. Moreover, existing infocommunication networks often cannot cope with the load and, as a result, structural failures (failures of nodes, links, etc.) and functional (a sharp decrease in the Quality of Service due to network overload) occur [6–15]. In this regard, it is recommended to use the technological solutions of the Network Layer [16–20] and Fast ReRouting (FRR) protocols, which are one of the most effective means of increasing network resilience [21–24]. FRR is based on the fact that in addition to the primary route, a set of backup paths is also pre-calculated, onto which packet flows are switched almost

Z. Hu et al. (Eds.): ICCSEEA 2020, AISC 1247, pp. 108–119, 2021.
https://doi.org/10.1007/978-3-030-55506-1_10

instantly in case of a failure of a network element. However, despite the obvious advantages of using Fast ReRouting, there are a number of drawbacks associated with the redundancy in the use (reservation) of the network resource, which often reduces the network performance and the quality of service (QoS) in general. Therefore, one of the main requirements for Fast ReRouting protocols is also to ensure the efficient use of available network resources by applying the Traffic Engineering (TE) technology [25–30]. However, under the conditions of network overload associated with the lack of network resources, the use of TE cannot guarantee a given level of QoS. Therefore, together with TE-FRR, it is also necessary to use the traffic management functionality at the network edge by implementing the Traffic Policing (TP) functionality [31–33]. This can be achieved only by having appropriate mathematical models and methods that would ensure a consistent solution of the TE-FRR and TP problems in ICNs. In this regard, a flow-based model of Fast ReRouting with load-balancing and fair priority-based Traffic Policing is proposed in this work.

2 Analysis of Existing Solutions on Fast ReRouting in Networks

As the analysis [12, 15, 21–30] has shown, the vast majority of mathematical methods for Fast ReRouting in ICNs are based on heuristic models. The main advantage of such solutions is their acceptable computational complexity of the subsequent protocol implementation. However, the calculation of the primary and backup routes with the implementation of the specified protection schemes for network elements, but without consideration of characteristics of the transmitted packet flows, does not allow to properly protect its bandwidth, especially in the conditions of the ICN overload.

Therefore, more and more often scientists develop and use flow-based models and methods of Fast ReRouting in research [5, 6, 15, 22, 27, 31]. This class of models and methods allows not only to calculate the desired paths (primary and backup), but also to solve the problem of load-balancing along these routes. A special place among these models is taken by solutions based on the implementation of the principles of Traffic Engineering (TE) when balancing the load along the primary and backup routes [22, 31]. With their help, a sufficiently effective and balanced utilization of the ICN link resource primarily is provided, which positively affects the overall level of QoS in the network in general.

In [31], an approach to solving the Fast ReRouting problem with the protection of ICN bandwidth, which develops a model [15] for the case of multipath routing, was proposed. Its main advantage is ensuring consistency in solving FRR problems based on ensuring balanced utilization of its link resource according to TE principles, as well as implementing Priority-Based Traffic Policing (PB-TP) on the network edge when it is overloaded. Moreover, the calculation of control variables that determine the order of TE-FRR and PB-TP was carried out in the course of solving the optimization problem of the linear programming class, which positively affected the computational complexity of the obtained solutions. In addition, the specifics of the considered solution is that in the case of ICN overload, a differentiated TP was provided on the basis of absolute priorities, when under the network overload low priority flows were initially limited (up to their full blocking), and only then the limitation concerned higher priority flows. Thus, higher priority flows start being limited only when the low priority packet flows are completely blocked, which in practice is not always acceptable.

Hence, in this paper we propose a compromise solution presented by the Load-Balancing Fast ReRouting Model, which provides a Fair Priority-Based Traffic Policing, when all flows are subject to limitation in case when the network is overloaded, but low priority flows are limited to a higher extent and high priority ones are limited to a lesser extent.

3 Load-Balancing Fast ReRouting Model with Providing Fair Priority-Based Traffic Policing

To describe the network structure, we use the graph $G=(R,E)$, in which the set of vertices $R=\{R_i;\ i=\overline{1,m}\}$ models the set of routers, and the set of arcs $E=\{E_{i,j};\ i,j=\overline{1,m};\ i\neq j\}$ describes the set of communication links connecting network routers. The number of links is determined as $n=|E|$. The set of routers adjacent to R_i is denoted by $R_i^*=\{R_j:\ \exists E_{j,i}\in E;\ j=\overline{1,m};\ i\neq j\}$.

Then to describe a number of functional characteristics of ICNs within the framework of the proposed mathematical model, let K be the set of packet flows that are transmitted over the network; s_k is the source node of the k th packet flow ($k\in K$); d_k is the destination node of the k th packet flow ($k\in K$); λ^k is the average packet intensity of the k th flow (packets per second), and $\varphi_{i,j}$ is the link bandwidth between the i th and j th nodes under $i,j=\overline{1,m};\ i\neq j$.

In this case, the result of solving the fast rerouting problem is the determination of the routing variables $x_{i,j}^k$ and $\bar{x}_{i,j}^k$, which characterize the fraction of the k th flow intensity in the communication links belonging to the primary and backup routes, respectively. When using the multipath routing, the following restrictions are imposed on these variables:

$$0\le x_{i,j}^k\le 1 \text{ and } 0\le \bar{x}_{i,j}^k\le 1. \tag{1}$$

To ensure the connectivity of routes in the network as a whole, conditions are introduced for maintaining the flow for routing variables of the primary and backup routes, which differ from the previously known ones [6, 22, 32] by allowing to describe the processes of traffic policing at the network edge under conditions of its overload:

$$\sum_{j:E_{i,j}\in E} x_{i,j}^k-\sum_{j:E_{j,i}\in E} x_{j,i}^k=\begin{cases}0; k\in K, R_i\neq s_k, d_k;\\ 1-\beta^k; k\in K, R_i=s_k;\\ \beta^k-1; k\in K, R_i=d_k;\end{cases} \tag{2}$$

$$\sum_{j:E_{i,j}\in E} \bar{x}_{i,j}^k-\sum_{j:E_{j,i}\in E} \bar{x}_{j,i}^k=\begin{cases}0; k\in K, R_i\neq s_k, d_k;\\ 1-\bar{\beta}^k; k\in K, R_i=s_k;\\ \bar{\beta}^k-1; k\in K, R_i=d_k,\end{cases} \tag{3}$$

where β^k and $\bar{\beta}^k$ are the fractions of the intensity of the k th flow that will receive a denial of service using the primary or backup multipaths in the ICN, respectively.

As shown in [22, 31], the condition for protecting communication link $E_{i,j} \in E$, which guarantees that this link will not be used by the backup route for multipath routing, has the form:

$$0 \leq \bar{x}_{i,j}^k \leq \delta_{i,j}^k \tag{4}$$

at

$$\delta_{i,j}^k = \begin{cases} 0, & \text{when protecting the link } E_{i,j}; \\ 1, & \text{otherwise.} \end{cases} \tag{5}$$

The protection conditions for the node $R_i \in R$ are as follows:

$$0 \leq \bar{x}_{i,j}^k \leq \delta_{i,j}^k \text{ at } R_j \in R_i^*,\ j = \overline{1,m}. \tag{6}$$

The fulfillment of the above conditions (5) and (6) guarantee that when protecting the node $R_i \in R$, it is forbidden to use the backup route of all communication links that come out of this node.

The linear conditions for bandwidth protection and load-balancing in the network presented in [22] have the following form:

$$\sum_{k \in K} \lambda^k \cdot u_{i,j}^k \leq \alpha \cdot \varphi_{i,j},\ E_{i,j} \in E, \tag{7}$$

at

$$x_{i,j}^k \leq u_{i,j}^k \text{ and } \bar{x}_{i,j}^k \leq u_{i,j}^k, \tag{8}$$

where $u_{i,j}^k$ and α are also control variables. In addition, $u_{i,j}^k$ characterize the upper bound of the routing variables values used in the calculation of the primary and backup routes, and the variable α quantitatively determines the upper bound of the utilization on the network communication links.

The following restrictions are imposed on the given control variables [31]:

$$0 \leq u_{i,j}^k \leq 1, \tag{9}$$

$$0 \leq \alpha \leq \alpha_{TH}, \tag{10}$$

where α_{TH} is the maximum allowed value of the upper bound (threshold) of the links utilization, which is determined by the requirements for the level of Quality of Service in the network.

In contrast to the results presented in [31], in this paper it is proposed to use the linear-quadratic form of the objective function of the control variables α, β^k and $\bar{\beta}^k$, which is to be minimized

$$J = \sum_{k \in K} w_k \cdot (\beta^k)^2 + \sum_{k \in K} \bar{w}_k \cdot (\bar{\beta}^k)^2 + c \cdot \alpha \to \mathbf{min}, \tag{11}$$

where weighting coefficients must satisfy the following condition

$$w_k > \overline{w}_k > w_p > \overline{w}_p > \ldots > c, \tag{12}$$

despite the fact that the priority of packets of the k th flow (PR^k) exceeds the priority of packets of the p th flow (PR^p).

Quantitatively, the objective function (11) characterizes the conditional cost (metric) of the coordinated solution for such network problems as FRR, TE, and TP. A similar relationship between the weighting coefficients in the objective function (12) is dictated by the fact that traffic policing should be carried out only when the network is overloaded. As it will be shown below, the introduction of the quadratic form (11) of the variables β^k and $\bar{\beta}^k$ makes it possible to implement fair Traffic Policing based on relative priorities. This will be manifested in the fact that when the network is overloaded, traffic restriction will be balanced, i.e. high-priority flows will be limited to a lesser extent than low-priority packet flows. This eliminates the situation when full blocking of low-priority flows is allowed, which has taken place when using the linear analogue of the objective function (11) [31].

4 Results of Investigating the Proposed Solution of Fast Rerouting with Load-Balancing and Fair Priority-Based Traffic Policing

The investigation of the proposed fast rerouting model with fair priority-based traffic policing has been carried out on a set of network topologies and protection schemes for network elements. Features of the calculations will be demonstrated on the source data when using the network structure that consists of sixteen nodes and twenty-four communication links. The gaps in the links indicate their bandwidths (Fig. 1).

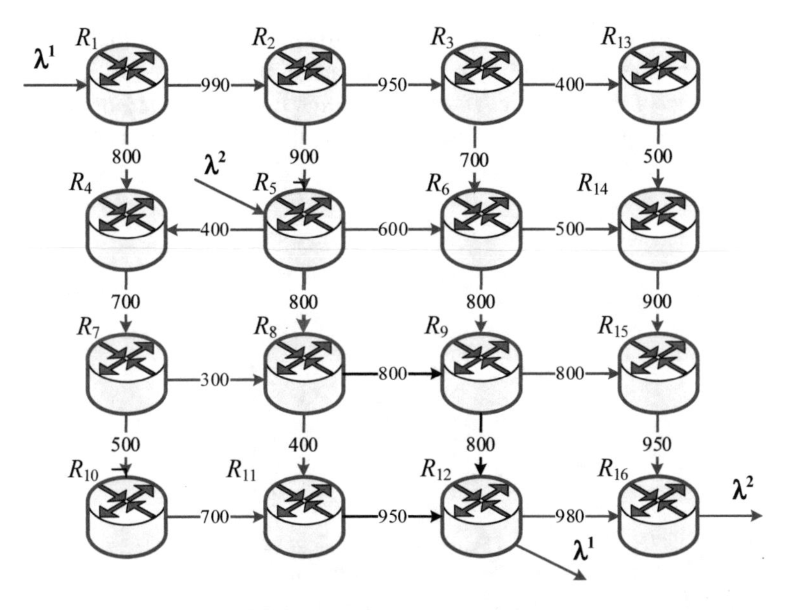

Fig. 1. Network structure for numerical example.

To solve the problem of fault-tolerant routing with load-balancing based on Traffic Engineering and supporting fair priority-based Traffic Policing, the $E_{5,8}$ communication link has been selected as an example of the protected network element.

In order to consider the influence of the priorities of the transmitted packets during the investigation, the case of two flows transmitted in the network has been simulated, which had the following characteristics:

- 1st flow: from R_1 to R_{12} with intensity $\lambda^1 = 1\ldots1200$ 1/s and priority of flows $PR^1 = 4$;
- 2nd flow: from R_5 to R_{16} with intensity $\lambda^2 = 1\ldots1200$ 1/s and priority of flows $PR^2 = 1$.

Assume that the threshold of the upper bound of the network links utilization in the network equals to α_{TH} = 0.75, while the upper bound $\alpha_{i,j}$ for an arbitrary communication link $E_{i,j}$ has been determined according to the expression [31]:

$$\alpha_{i,j} = \frac{\sum\limits_{k \in K} \lambda^k \max(x_{i,j}^k, \bar{x}_{i,j}^k)}{\varphi_{i,j}}, \tag{13}$$

where $\alpha_{i,j}$ is determined for the worst case from the point of view of link $E_{i,j}$ utilization by flows transmitting along either primary or backup routes.

Figure 2 shows the dynamics of changes in the upper threshold of congestion of network communication links (α) depending on the intensity of two packet flows arriving the network.

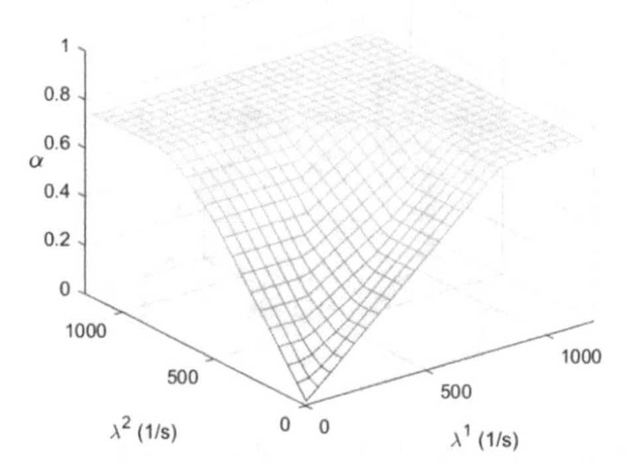

Fig. 2. The dependence of the upper bound of the utilization on the network communication links on the intensities of the first (λ^1) and second (λ^2) flows of packets that are transmitted in the network at $\alpha_{TH} = 0.75$.

Based on the research results presented in Fig. 2, we can conclude that with the increase in the load on the network (the intensities of the first and second packet flows), the upper bound of the utilization on the network links also has increased. At the same time, it should be noted that with a low network load, there was no restriction of flow intensities at the network edge. However, under the network overload, when $\alpha \rightarrow \alpha_{TH}$, traffic restriction began. As shown in Fig. 3 and Fig. 4, traffic policing functions were implemented based on the consideration of flow priorities. Compared with the solution proposed in [31] and based on the implementation of the linear analogue of the objective function (11), the results did not lead to a complete blocking of the packet flow with the lowest priority.

When considering the results shown in Fig. 3 and Fig. 4, it should be noted that the fair traffic policing mechanism for two flows of different priority in the network is carried out in accordance with condition (12) and starts when the maximum allowed value of the threshold of the links utilization is reached, i.e. $\alpha = \alpha_{TH}$. It is important to understand that according to (12) under congestion conditions, the prior and most intense restriction will be experienced by the flow, the packets of which have the lowest IP priority and are transmitted along the backup path.

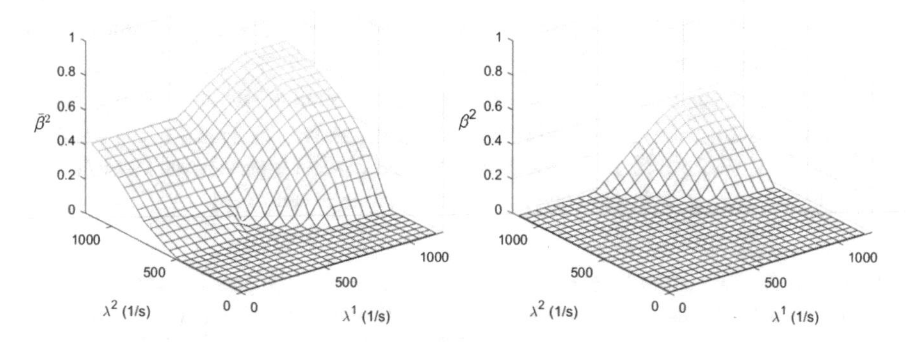

Fig. 3. The results of solving the fair priority-based traffic policing problem for the second flow of packets transmitted along the backup (a) and primary (b) paths.

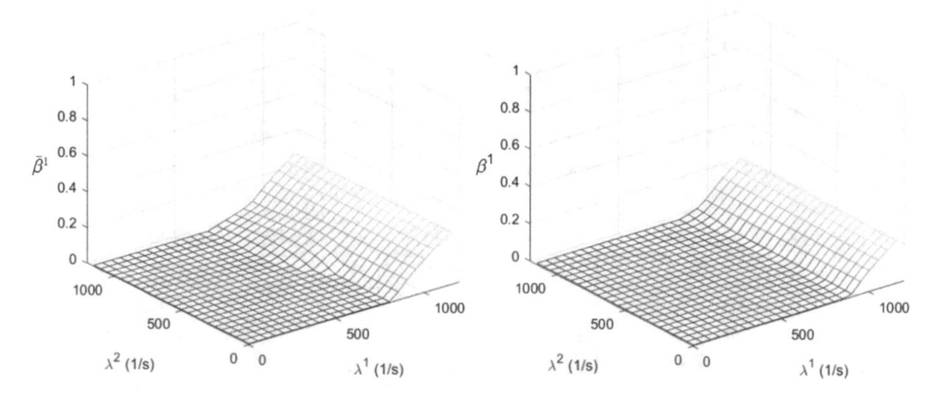

Fig. 4. The results of solving the fair priority-based traffic policing problem for the first packet flow transmitted along the backup (a) and primary (b) paths.

This is clearly seen in Fig. 3a, when the restriction of the second flow, which had the priority $PR^2 = 1$ and transmitted along the backup path, began at about $\lambda^2 = 388$ 1/s. Moreover, with low intensity of the first (more prioritized) flow ($\lambda^1 \leq 500$ 1/s), the restriction of the second flow transmitted along the backup path was less intense than with the high intensity of the first flow ($\lambda^1 > 500$ 1/s).

As shown in Fig. 3b, in accordance with the fair Traffic Policing model, the second flow, if it was transmitted along the primary path, began to be limited much later, at $\lambda^2 = 737$ 1/s. At the same time, the same flow was not completely blocked when using the backup path (Fig. 3a). The results obtained indicated that the second flow was less restricted when using the primary path than when using the backup route, which confirms the adequacy of the proposed solution based on condition (12).

In the framework of the proposed TE-FRR with fair traffic policing model, a similar situation is also specific to the first flow with the higher IP priority, i.e. $PR^1 = 4$

(Fig. 4). The first flow, when using the backup route, will be limited (Fig. 4a) starting from $\lambda^1 = 853$ 1/s; and when using the primary route it will be limited starting from $\lambda^1 = 931$ 1/s (Fig. 4b).

For clarity, let us look at the more detailed results obtained when $\lambda^1 = 900$ 1/s and $\lambda^2 = 900$ 1/s with priorities of flows as mentioned above. In Table 1, the results of solving the TE-FRR problem for the two described flows are presented. The utilization $\alpha_{i,j}$ for each network link has been found in accordance with (13).

Table 1. Order of multipath routing of two flows using the proposed load-balancing fast rerouting model with providing fair priority-based traffic policing (link $E_{5,8}$ protection).

Link	1st flow		2nd flow		$\alpha_{i,j}$
	Primary path	*Backup path*	*Primary path*	*Backup path*	
$E_{1,2}$	660.01	587.32	0	0	0.67
$E_{2,3}$	560.00	560.00	0	0	0.70
$E_{1,4}$	239.99	239.99	0	0	0.25
$E_{2,5}$	100.01	27.32	0	0	0.11
$E_{3,6}$	560.00	560.00	0	0	0.80
$E_{5,4}$	0	0	0	0	0
$E_{5,6}$	27.32	27.32	452.68	452.68	0.80
$E_{4,7}$	239.99	239.99	0	0	0.34
$E_{5,8}$	72.69	0	367.31	0	0.55
$E_{6,9}$	587.32	587.32	52.68	52.68	0.80
$E_{7,8}$	239.99	239.99	0	0	0.80
$E_{8,9}$	11.79	5.34	348.21	0	0.45
$E_{7,10}$	0	0	0	0	0
$E_{8,11}$	300.89	234.66	19.11	0	0.80
$E_{9,12}$	599.11	592.66	40.89	40.89	0.80
$E_{10,11}$	0	0	0	0	0
$E_{11,12}$	300.89	234.66	19.11	0	0.34
$E_{3,13}$	0	0	0	0	0
$E_{13,14}$	0	0	0	0	0
$E_{6,14}$	0	0	400.00	400.00	0.80
$E_{14,15}$	0	0	400.00	400.00	0.44
$E_{9,15}$	0	0	360.00	11.79	0.45
$E_{15,16}$	0	0	760.00	411.79	0.80
$E_{12,16}$	0	0	60.00	40.90	0.06

Due to the network overload to ensure that condition (10) is met, it has been found that:

- when using the backup path, the first flow with high priority receives a denial of service at the network edge with the intensity 72.68 1/s;

- when using the primary path, the second flow with low priority receives a denial of service at the network edge with the intensity 80 1/s;
- when using the backup path, the second flow with low priority receives a denial of service at the network edge with the intensity 447.31 1/s.

5 Conclusions

The model of Fast ReRouting with load-balancing and fair priority-based Traffic Policing has been developed and studied in the article. The model is represented by the conditions for implementing the multipath routing (1), flow conservation along the primary (2) and backup (3) routes, which allow describing the processes of traffic policing at the network edge under its overload. A primary place in the model is taken by the protection (reservation) conditions of ICN elements (4)–(6), as well as network bandwidth in accordance with load-balancing implemented in the network (7)–(10). The compliance of the obtained routing solutions with the Traffic Engineering principles has been confirmed by the introduction and fulfillment of condition (7), when the upper bound of each communication link utilization during the fast rerouting of a set of flows in the network did not exceed the established threshold α_{TH}, the value of which is set based on QoS requirements.

The advantage of the proposed model is that it provides a consistent solution to the problems of TE, FRR and fair priority-based Traffic Policing in ICNs. The novelty of the presented model is in the fact that the flows, which are the source of network overload, were limited first. In addition, as the numerical research results of the study have shown, which have been obtained using the MatLab package, the use of the linear-quadratic optimality criterion (12) of routing solutions gave the process of traffic restriction a fair and balanced character, when, under overload conditions, the flows with higher priority were limited but to a lesser extent than flows with low priority. This made it possible to implement the load-balancing fast rerouting with providing fair priority-based traffic policing strategy based on relative priorities.

References

1. White, R., Banks, E.: Computer Networking Problems and Solutions: An Innovative Approach to Building Resilient, Modern Networks, 1st edn. Addison-Wesley Professional (2018)
2. Stallings, W.: Foundations of Modern Networking: SDN, NFV, QoE, IoT, and Cloud. Addison-Wesley Professional (2015)
3. Rak, J., Papadimitriou, D., Niedermayer, H., Romero, P.: Information-driven network resilience: research challenges and perspectives. Opt. Switch. Netw. **23**(2), 156–178 (2017)
4. Tipper, D.: Resilient network design: challenges and future directions. Telecommun. Syst. **56**(1), 5–16 (2013)
5. Rak, J.: Resilient Routing in Communication Networks, 1st edn. Springer (2015)

6. Lemeshko, O., Arous, K., Tariki, N.: Effective solution for scalability and productivity improvement in fault-tolerant routing. In: International Scientific-Practical Conference Problems of Infocommunications Science and Technology (PIC S&T) Proceedings, pp. 76–78. IEEE (2015). https://doi.org/10.1109/INFOCOMMST.2015.7357274
7. Ruban, I.V., Churyumov, G.I., Tokarev, V.V., Tkachov, V.M.: Provision of survivability of reconfigurable mobile system on exposure to high-power electromagnetic radiation. In: Selected Papers of the XVII International Scientific and Practical Conference on Information Technologies and Security (ITS 2017), pp. 105–111. CEUR Workshop Processing (2017)
8. Lemeshko, O., Yeremenko, O., Tariki, N.: Solution for the default gateway protection within fault-tolerant routing in an IP network. Int. J. Electr. Comput. Eng. Syst. **8**(1), 19–26 (2017). https://doi.org/10.32985/ijeces.8.1.3
9. Smelyakov, K., Sandrkin, D., Ruban, I., Vitalii, M., Romanenkov, Y.: Search by image. New search engine service model. In: 2018 International Scientific-Practical Conference Problems of Infocommunications. Science and Technology (PIC S&T) Proceedings, pp. 181–186. IEEE (2018). https://doi.org/10.1109/INFOCOMMST.2018.8632117
10. Lemeshko, O., Nevzorova, O., Ilyashenko, A., Yevdokymenko, M.: Hierarchical coordination method of inter-area routing in backboneless network. In: Advances in Intelligent Systems and Computing, vol. 938, pp. 90–102. Springer (2020). https://doi.org/10.1007/978-3-030-16621-2_9
11. Lemeshko, O.V., Yeremenko, O.S.: Dynamics analysis of multipath QoS-routing tensor model with support of different flows classes. In: 2016 International Conference on Smart Systems and Technologies (SST) Proceedings, pp. 225–230. IEEE (2016). https://doi.org/10.1109/SST.2016.7765664
12. Koryachko, V.P., Perepelkin, D.A., Byshov, V.S.: Development and research of improved model of multipath adaptive routing in computer networks with load balancing. Autom. Control Comput. Sci. **51**(1), 63–73 (2017). https://doi.org/10.3103/S0146411617010047
13. Smelyakov, K., Dmitry, P., Vitalii, M., Anastasiya, C.: Investigation of network infrastructure control parameters for effective intellectual analysis. In: 2018 14th International Conference on Advanced Trends in Radioelecrtronics, Telecommunications and Computer Engineering (TCSET) Proceedings, pp. 983–986. IEEE (2018). https://doi.org/10.1109/tcset.2018.8336359
14. Lemeshko, O., Yevdokymenko, M., Anad Alsaleem, N.: Development of the tensor model of multipath QoE-routing in an infocommunication network with providing the required quality rating. Eastern-Eur. J. Enterp. Technol. **5**(2(95)), 40–46 (2018). https://doi.org/10.15587/1729-4061.2018.141989
15. Cruz, P., Gomes, T., Medhi, D.: A heuristic for widest edge-disjoint path pair lexicographic optimization. In: 6th International Workshop on Reliable Networks Design and Modeling Proceedings (RNDM), pp. 9–15. IEEE (2014)
16. Lakshman, N.L., Khan, R.U., Mishra, R.B.: MANETs: QoS and investigations on optimized link state routing protocol. Int. J. Comput. Netw. Inf. Secur. (IJCNIS) **10**(10), 26–37 (2018). https://doi.org/10.5815/ijcnis.2018.10.04
17. Najafi, G., Gudakahriz, S.J.: A stable routing protocol based on DSR protocol for mobile ad hoc networks. Int. J. Wirel. Microwave Technol. (IJWMT) **8**(3), 14–22 (2018). https://doi.org/10.5815/ijwmt.2018.03.02
18. Mallapur, S.V., Patil, S.R., Agarkhed, J.V.: A stable backbone-based on demand multipath routing protocol for wireless mobile ad hoc networks. Int. J. Comput. Netw. Inf. Secur. (IJCNIS) **8**(3), 41–51 (2016). https://doi.org/10.5815/ijcnis.2016.03.06
19. Moza, M., Kumar, S.: Analyzing multiple routing configuration. Int. J. Comput. Netw. Inf. Secur. (IJCNIS) **8**(5), 48–54 (2016). https://doi.org/10.5815/ijcnis.2016.05.07

20. Krishna, S.R.M., Ramanath, S., Prasad, K.: Optimal reliable routing path selection in MANET through novel approach in GA. Int. J. Intell. Syst. Appl. (IJISA) **9**(2), 35–41 (2017). https://doi.org/10.5815/ijisa.2017.02.05
21. Papan, J., Segec, P., Paluch, P., Uramova, J., Moravcik, M.: The new multicast repair (M-REP) IP fast reroute mechanism. Concurr. Comput.: Pract. Exp., e5105 (2018). https://doi.org/10.1002/cpe.5105
22. Lemeshko, O., Yeremenko, O., Yevdokymenko, M.: MPLS traffic engineering solution of multipath fast reroute with local and bandwidth protection. In: Advances in Intelligent Systems and Computing, vol. 938, pp. 113–125. Springer (2020). https://doi.org/10.1007/978-3-030-16621-2_11
23. Hasan, H., Cosmas, J., Zaharis, Z., Lazaridis, P., Khwandah, S.: Development of FRR mechanism by adopting SDN notion. In: 24th International Conference on Software, Telecommunications and Computer Networks (SoftCOM) Proceedings, pp. 1–7 (2016)
24. Zhang, X., Cheng, Z., Lin, R., He, L., Yu, S., Luo, H.: Local fast reroute with flow aggregation in software defined networks. IEEE Commun. Lett. **21**(4), 785–788 (2017). https://doi.org/10.1109/LCOMM.2016.2638430
25. Tomovic, S., Radusinovic, I.: A new traffic engineering approach for QoS provisioning and failure recovery in SDN-based ISP networks. In: 23rd International Scientific-Professional Conference on Information Technology (IT) Proceedings, pp. 1–4. IEEE (2018)
26. Golani, K., Goswami, K., Bhatt, K., Park, Y.: Fault tolerant traffic engineering in software-defined WAN. In: Symposium on Computers and Communications (ISCC) Proceedings, pp. 01205–01210. IEEE (2018)
27. Lemeshko, O., Yeremenko, O., Yevdokymenko, M.: Tensor model of fault-tolerant QoS routing with support of bandwidth and delay protection. In: 2018 XIIIth International Scientific and Technical Conference Computer Sciences and Information Technologies (CSIT) Proceedings, pp. 135–138. IEEE (2018). https://doi.org/10.1109/stc-csit.2018.8526707
28. Lin, S.C., Wang, P., Luo, M.: Control traffic balancing in software defined networks. Comput. Netw. **106**, 260–271 (2016)
29. Mendiola, A., Astorga, J., Jacob, E., Higuero, M.: A survey on the contributions of software-defined networking to traffic engineering. IEEE Commun. Surv. Tutor. **19**(2), 918–953 (2017). https://doi.org/10.1109/COMST.2016.2633579
30. Prabhavat, S., Nishiyama, H., Ansari, N., Kato, N.: On load distribution over multipath networks. IEEE Commun. Surv. Tutor. **14**(3), 662–680 (2012). https://doi.org/10.1109/SURV.2011.082511.00013
31. Lemeshko, O., Yeremenko, O., Yevdokymenko, M., Shapovalova, A., Ilyashenko, A., Sleiman, B.: Traffic engineering fast reroute model with support of policing. In: 2nd Ukraine Conference on Electrical and Computer Engineering (UKRCON) Proceedings, pp. 842–845. IEEE (2019). https://doi.org/10.1109/UKRCON.2019.8880006
32. Lemeshko, O.V., Garkusha, S.V., Yeremenko, O.S., Hailan, A.M.: Policy-based QoS management model for multiservice networks. In: International Siberian Conference on Control and Communications (SIBCON) Proceedings, pp. 1–4. IEEE (2015). https://doi.org/10.1109/SIBCON.2015.7147124
33. Lemeshko, A.V., Evseeva, O.Y., Garkusha, S.V.: Research on tensor model of multipath routing in telecommunication network with support of service quality by greate number of indices. Telecommun. Radio Eng. **73**(15), 1339–1360 (2014). https://doi.org/10.1615/TelecomRadEng.v73.i15.30

Tensor Multiflow Routing Model to Ensure the Guaranteed Quality of Service Based on Load Balancing in Network

Oleksandr Lemeshko[1], Oleksandra Yeremenko[1(✉)], Maryna Yevdokymenko[1], and Ahmad M. Hailan[2]

[1] Kharkiv National University of Radio Electronics, 14 Nauky Ave., Kharkiv, Ukraine
{oleksandr.lemeshko.ua, oleksandra.yeremenko.ua, maryna.yevdokymenko}@ieee.org

[2] Thi-Qar University, Nasiriya, Iraq
ahmad.m.hailan@utq.edu.iq

Abstract. The article proposes a tensor multiflow routing model with the aim of ensuring the guaranteed Quality of Service based on load balancing in a network. The novelty of this model is that it is focused on providing specified numerical values of Quality of Service indicators such as flow rate and average packet delay. This was achieved by introducing into the multiflow routing model structure the corresponding Quality of Service conditions, which were formulated in an analytical form using the tensor research methodology. At the same time, the space metric was determined by both the service discipline of packets on the routers interfaces and the statistical model of the served flows. In addition, the conditions are formulated in such a way that for each flow transmitted in the network, the resulting average end-to-end packet delays along the set of calculated paths were the same and did not exceed the permissible values. The numerical example has demonstrated the adequacy of the proposed model in terms of the correctness of the obtained calculation results with regard to providing specified values of quality indicators that may differ for each of the packet flows transmitted in the network.

Keywords: QoS · Multipath routing · Flow-based model · Tensor model · Load balancing · Packet rate · End-to-end delay

1 Introduction

Providing a specified level of Quality of Service (QoS) is a rather complicated scientific and practical task, requiring the coordinated work of many network protocols, mechanisms, and applications of all seven OSI Layers [1–6]. An important place among technological means of providing end-to-end QoS is given to routing protocols [6, 7]. At the same time, the QoS level directly depends both on the characteristics of the routes used (their length (hops), throughput, reliability, etc.) [8–19], and on the implemented principles of load balancing among them [20–30].

Z. Hu et al. (Eds.): ICCSEEA 2020, AISC 1247, pp. 120–131, 2021.
https://doi.org/10.1007/978-3-030-55506-1_11

As the analysis [6, 7] showed, in a number of existing routing protocols, for example, in the RIP protocol, load balancing is supported only along paths with the same metric. A rather effective solution from the point of view of using a network resource is load balancing along paths with different metrics, such as in the EIGRP protocol, when the share of the transmitted flow along one path or another is proportional to its metric.

The most promising of the known is the approach based on the implementation of the principles of Traffic Engineering (TE) [20–30] when the minimum of the upper bound of the load on the network links acts as a criterion for the optimality of load balancing. However, even its implementation in practice only indirectly improves the level of Quality of Service but does not guarantee the set values of certain QoS indicators – performance, average delay, packet loss probability.

One of the perspective approaches seems to be the use of tensor models of QoS routing [31–33]. In this case, the key role in the structure of such a mathematical routing model is occupied by the conditions for ensuring the guaranteed Quality of Service. Then the main goal of balancing the load on the network is to ensure the specified numerical values of specific (selected) QoS indicators.

The aim of the research is obtaining the solution for tensor multiflow routing based on load balancing, directed towards providing Quality of Service over a set of parameters (flow rate and average packet delay).

To achieve the set aim, the following research problems have to be solved:

- using the flow-based model based on multipath routing strategy;
- network geometrization with the introduction of geometric space and coordinate systems formed by the distinct links and a set of routes in the network;
- analytical formulation of QoS ensuring conditions over a set of parameters;
- stating the optimization problem of tensor load balancing multiflow routing with QoS ensuring and selection of optimality criterion;
- numerical research of the tensor load balancing model and obtaining results, which prove its adequacy.

The article is structured in the following way. In Sect. 2, a work analysis related to Traffic Engineering in modern networks is given. In Sect. 3, constraints and conditions for the tensor multiflow routing model are presented. In Sect. 4, end-to-end QoS constraints are formulated. In Sect. 5, numerical research results are described and analyzed. In Sect. 6, results and their applicability for load balancing with guaranteed QoS are discussed. Section 7 concludes the article.

2 Work Analysis Related to Traffic Engineering

Currently, as shown by the analysis, the most promising and effective is the flow-based approach, which is described in many scientific papers [8, 9, 18–21, 31–33]. It is important to note that mathematical models and routing methods should be as adapted as possible to the requirements of the network concepts of Traffic Engineering, Load Balancing Routing, and QoS-based Routing, namely:

- implementation of a multipath routing strategy;
- ensuring load balancing along paths with the same and different metrics;
- maintaining the Quality of Service concurrently in many different types of indicators (time, rate and reliability).

Ensuring load balancing is an integral part of technological solutions to traffic management that support the Quality of Service in modern infocommunication networks, aimed at optimizing network resource utilization and delivering traffic in a network as a whole, for example, within the concept of Traffic Engineering [20]. For example, [25] presents a Mixed Integer Programming (MIP) solution for network optimization based on TE tunneling schemes, which explicitly defines routes for each flow. Thus, [25] presents a method for calculating the optimal solution in terms of the number of tunnels, which helps to prevent network overload and effectively balance the load in communication links at a given network topology. In doing so, the formulated MIP optimization problem was solved using a high-speed solver, and the resulting optimal solutions for several variants of the network topology, including GEANT, proved the workability of the proposed overload prevention solution.

In [26], the solution of the problem of coordinated route selection and flow distribution in SDN networks was considered. To account for the sensitivity of the serviced flows to packet delay, this task was formulated as optimization while minimizing end-to-end delay, subject to data flow constraints and Quality of Service requirements. Since the formulated optimization problem is NP-complete, a modified minimum cost algorithm was proposed. By choosing iteratively minimum cost routes to transfer flows in an updated auxiliary network, optimal solutions for routing and balancing flows across the network can be obtained.

In turn, [27] proposed a new TE approach when using virtual channels on SDN networks that provide QoS guaranteed services when transmitting data between two points in an Internet Service Provider (ISP) network. The approach is based on the OpenFlow-based SDN architecture, which allows for the high precision of distributing the transmitted flow across multiple paths. The logic of traffic management of the SDN controller is divided into offline and online components. The online component manages the dynamic receipt of virtual channel requests as required by the Service Level Agreements. While the standalone component is responsible for periodically optimizing traffic distribution across the network. In this way, the virtual channel request acceptance rate is increased and the degradation of Best Effort traffic in a scalable manner is minimized.

3 Tensor Multiflow Routing Model

Consider the tensor routing model, which will be used for load balancing under self-similar traffic properties with guaranteed QoS, where structural and functional parameters are described as follows [18, 19]:

$S = (U, V)$ graph representing the network structure;
U set of vertices simulating network routers ($U = \{u_i, i = \overline{1,m}\}$);
V

set of edges representing links where the edge $v_z = (i,j) \in V$ models the zth link connecting the ith and jth routers $(V = \{v_z = (i,j); z = \overline{1,n}; i,j = \overline{1,m}; i \neq j\})$;

s_k source node;

d_k destination node;

K set of flows circulating in the network ($k \in K$);

λ^k_{req} average intensity (packet rate) of the kth flow;

$x^k_{i,j}$ routing variables, which determine the fraction of intensity of each distinct kth flow from the ith to jth node via a destined jth interface under condition of multipath routing strategy $0 \leq x^k_{i,j} \leq 1$;

$\varphi_{i,j}$ link capacity measured in packets per second (1/s)

In the tensor load balancing model the routing variables $x^k_{i,j}$ have to be determined. In addition, from the aspect of the network nodes overload prevention, the flow conservation conditions must be fulfilled. Therefore, these conditions for the multiple flows transmission are defined for the source (s_k), destination (d_k), and transit nodes for every distinct kth flow in the following way [31]:

$$\begin{cases} \sum\limits_{j:(i,j)\in V} x^k_{i,j} - \sum\limits_{j:(j,i)\in V} x^k_{j,i} = 1,\ k \in K,\ i = s_k; \\ \sum\limits_{j:(i,j)\in V} x^k_{i,j} - \sum\limits_{j:(j,i)\in V} x^k_{j,i} = 0,\ k \in K,\ i \neq s_k, d_k; \\ \sum\limits_{j:(i,j)\in V} x^k_{i,j} - \sum\limits_{j:(j,i)\in V} x^k_{j,i} = -1,\ k \in K,\ i = d_k. \end{cases} \tag{1}$$

Furthermore, the capacity constraints related to the links utilization have to be met:

$$\sum_{k\in K} \lambda^k_{req} x^k_{i,j} \leq \varphi_{i,j},\quad (i,j) \in V. \tag{2}$$

Next, in order to find the routing variables the following objective function is subject to minimization:

$$J = \sum_{(i,j)\in V} \sum_{k\in K} h^x_{i,j} \cdot \lambda^k_{req} \cdot x^k_{i,j}. \tag{3}$$

Here in the objective function (3) $h^x_{i,j}$ is the link $(i,j) \in V$ routing metric.

4 End-to-End QoS Constraints Formulation

With the aim of the end-to-end QoS guarantee, the additional conditions should be introduced into the model in analogy to [31–33] to meet QoS requirements for every separate flow transmitting in the network. Then it is needed to be done the tensor presentation of the network where its structure is defined by the discrete n-dimensional space (n is the number of links). Additionally, the two coordinate systems (CSs) that are orthogonal in relation to each other are considered in the network: CS of network edges $\{v_k, k = \overline{1,n}\}$, as well as CS of linearly independent circuits $\{\pi_i, i = \overline{1,\mu}\}$ and node pairs $\{\eta_j, j = \overline{1,\phi}\}$. Then, for example, in the case of modeling the operation of a network router interface by the queuing system M/M/1, the tensor E provides the space metric. Moreover, according to the following expression, the coordinates of tensor projection in the coordinate system of edges E_v will be calculated that expressed by diagonal matrix elements, the size of which is $n \times n$, correspondingly to each separate flow in the network [31, 33]:

$$e_{zz}^{v} = \frac{1}{\lambda_v^z(\phi_z - \lambda_v^z)} \quad (4)$$

where λ_v^z defines the intensity of the observed flow in the z th link with sequentially numbered links, and φ_z is the reserved z th link bandwidth for considered flow.

Afterward for appropriate presentation, the twice contravariant metric tensor $G_v = [E_v]^{-1}$ will be used. Its projection has been transformed due to the changing of the coordinate system:

$$G_{\pi\eta} = A^t G_v A \quad (5)$$

where $G_{\pi\eta}$ defines the projection of the tensor G in the coordinate system of circuits and node pairs; A is the covariant coordinate transformation matrix of the $n \times n$ size due to transition from CS of circuits and node pairs to CS of edges; finally, $[\cdot]^t$ is the transpose operator.

As it has been explained in [31–33]

$$G_{\pi\eta} = \left\| \begin{array}{c|c} G_{\pi\eta}^{\langle 1\rangle} & G_{\pi\eta}^{\langle 2\rangle} \\ \hline G_{\pi\eta}^{\langle 3\rangle} & G_{\pi\eta}^{\langle 4\rangle} \end{array} \right\|, \quad G_{\pi\eta}^{\langle 4\rangle} = \left\| \begin{array}{c|c} G_{\pi\eta}^{\langle 4,1\rangle} & G_{\pi\eta}^{\langle 4,2\rangle} \\ \hline G_{\pi\eta}^{\langle 4,3\rangle} & G_{\pi\eta}^{\langle 4,4\rangle} \end{array} \right\|$$

where $G_{\pi\eta}^{\langle 1\rangle}$ and $G_{\pi\eta}^{\langle 4\rangle}$ are $\mu \times \mu$ and $\phi \times \phi$ submatrices respectively, $G_{\pi\eta}^{\langle 2\rangle}$ is a $\mu \times \phi$ submatrix, $G_{\pi\eta}^{\langle 3\rangle}$ is a $\phi \times \mu$ submatrix; $G_{\pi\eta}^{\langle 4,1\rangle}$ is the first element of the matrix $G_{\pi\eta}^{\langle 4\rangle}$; $[\cdot]^{-1}$ is an inversion operation of a matrix.

In accordance with results obtained in [31–33], the QoS guarantee conditions for each kth flow in correspondence to the requirements of delay τ_{req}^k and the packet flow rate λ_{req}^k has the form:

$$\lambda_{req}^k \left(G_{\pi\eta}^{\langle 4,1\rangle} - G_{\pi\eta}^{\langle 4,2\rangle} \left[G_{\pi\eta}^{\langle 4,4\rangle} \right]^{-1} G_{\pi\eta}^{\langle 4,3\rangle} \right)^{-1} \leq \tau_{req}^k. \quad (6)$$

It should be noted that expression from the left side of the inequality (6) is devoted for calculation of the kth packet flow average end-to-end delay – τ_{MP}^k.

5 Numerical Research

The abilities of the tensor multiflow routing model with ensuring the guaranteed Quality of Service based on load balancing (1)–(6) will be demonstrated with the network structure illustrated in Fig. 1. In this case, the network contained the following elements: 5 routers and 7 links. In Fig. 1 in the breaks of links, the corresponding bandwidth $\varphi_{i,j}$ (1/s) is indicated.

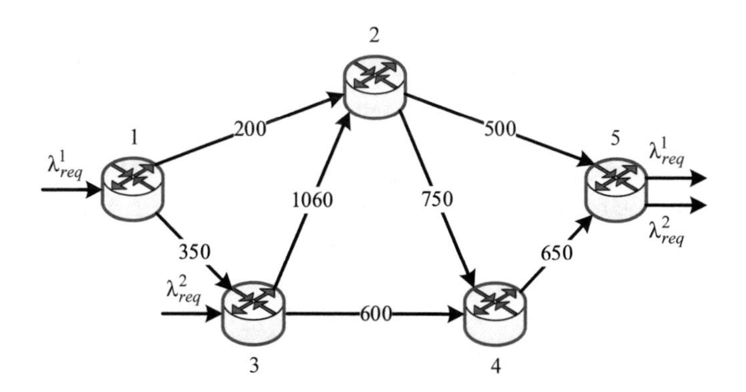

Fig. 1. Network structure under investigation.

Assume that two flows with the following characteristics are transmitted in the network:

- first flow with the intensity $\lambda^1_{req} = 400$ 1/s at the required average end-to-end packet delay $\tau^1_{req} = 85$ ms, source and destination are first and fifth nodes respectively;
- second flow with the intensity $\lambda^2_{req} = 720$ 1/s at the required average end-to-end packet delay 75 ms, source and destination are third and fifth nodes respectively.

Table 1 presents the calculation results obtained for each flow separately, obtained by a centralized calculation. The developed model was simulated in the MATLAB environment using the fmincon subroutine (Optimization Toolbox).

During the transmission of the first flow for the structure S_1, the metric tensor G_v has the form:

$$G_v = \left\| \begin{matrix} 9017.9 & 0 & 0 & 0 & 0 & 0 & 0 \\ 0 & 21852 & 0 & 0 & 0 & 0 & 0 \\ 0 & 0 & 66655 & 0 & 0 & 0 & 0 \\ 0 & 0 & 0 & 27084 & 0 & 0 & 0 \\ 0 & 0 & 0 & 0 & 42495 & 0 & 0 \\ 0 & 0 & 0 & 0 & 0 & 2878.4 & 0 \\ 0 & 0 & 0 & 0 & 0 & 0 & 3123.9 \end{matrix} \right\|.$$

Table 1. Results of modeling

Link	Link capacity, 1/s	Rate of the 1st flow, 1/s	Rate of the 2d flow, 1/s	Average delay, ms
(1,2)	200	131.32	–	14.6
(1,3)	350	268.68	–	12.3
(3,2)	1060	151.28	484.144	2.3
(3,4)	600	117.36	251.856	4.3
(2,4)	750	87.72	177.768	2.1
(2,5)	500	194.92	290.304	67.7
(4,5)	650	205.08	429.696	65.7

According to expression (4), and knowing the structure of the matrix A of the covariant coordinate transformation during the transition from CS of circuits and node pairs to CS edges, we obtain the projection $G_{\pi\eta}$ of the tensor G in CS of circuits and node pairs:

$$G_{\pi\eta} = \left\| \begin{array}{c|c} G_{\pi\eta}^{\langle 1 \rangle} & G_{\pi\eta}^{\langle 2 \rangle} \\ \hline G_{\pi\eta}^{\langle 3 \rangle} & G_{\pi\eta}^{\langle 4 \rangle} \end{array} \right\| =$$

$$= \left\| \begin{array}{ccc|cccc} 9017.9 & 0 & 0 & 0 & 9017.9 & 0 & 0 \\ 0 & 27084 & 0 & 0 & 0 & 27084 & -27084 \\ 0 & 0 & 2878.4 & 2878.4 & -2878.4 & 0 & 0 \\ \hline 0 & 0 & 2878.4 & 6002.4 & -2878.4 & 0 & -3123.9 \\ 9017.9 & 0 & -2878.4 & -2878.4 & 121050 & -66655 & -42495 \\ 0 & 27084 & 0 & 0 & -66655 & 115590 & -27084 \\ 0 & -27084 & 0 & -3123.9 & -42495 & -27084 & 72703 \end{array} \right\|,$$

$$G_{\pi\eta}^{\langle 4 \rangle} = \left\| \begin{array}{c|c} G_{\pi\eta}^{\langle 4,1 \rangle} & G_{\pi\eta}^{\langle 4,2 \rangle} \\ \hline G_{\pi\eta}^{\langle 4,3 \rangle} & G_{\pi\eta}^{\langle 4,3 \rangle} \end{array} \right\| = \left\| \begin{array}{c|ccc} 6002.4 & -2878.4 & 0 & -3123.9 \\ \hline -2878.4 & 121050 & -66655 & -42495 \\ 0 & -66655 & 115590 & -27084 \\ -3123.9 & -42495 & -27084 & 72703 \end{array} \right\|.$$

Which implies that

$$\tau_{MP}^{1} = \lambda_{req}^{1} \left(G_{\pi\eta}^{\langle 4,1 \rangle} - G_{\pi\eta}^{\langle 4,2 \rangle} \left[G_{\pi\eta}^{\langle 4,4 \rangle} \right]^{-1} G_{\pi\eta}^{\langle 4,3 \rangle} \right)^{-1} = 400 \cdot 2.057 \cdot 10^{-4} = 0.0823 \text{ (s)}$$

so that the QoS ensuring conditions (6) are satisfied for the first flow according to the corresponding requirements: 85 ms for the average delay and 400 1/s for the packet flow rate.

When transmitting the second flow for the corresponding structure S_2, which does not include links between the first and second, as well as the first and third nodes, the metric tensor G_v takes the form:

$$G_v = \begin{Vmatrix} 206250 & 0 & 0 & 0 & 0 \\ 0 & 58120 & 0 & 0 & 0 \\ 0 & 0 & 86143 & 0 & 0 \\ 0 & 0 & 0 & 4287.4 & 0 \\ 0 & 0 & 0 & 0 & 6544.9 \end{Vmatrix}$$

Thus, the obtained projection of the tensor G in the CS of the circuits and node pairs $G_{\pi\eta}$ is as follows:

$$G_{\pi\eta} = \begin{Vmatrix} G_{\pi\eta}^{\langle 1 \rangle} & G_{\pi\eta}^{\langle 2 \rangle} \\ G_{\pi\eta}^{\langle 3 \rangle} & G_{\pi\eta}^{\langle 4 \rangle} \end{Vmatrix} =$$

$$= \begin{Vmatrix} 58120 & 0 & 0 & 0 & -58120 \\ 0 & 4287.4 & 4287.4 & -4287.4 & 0 \\ 0 & 4287.4 & 10832 & -4287.4 & -6544.9 \\ 0 & -4287.4 & -4287.4 & 296680 & -86143 \\ -58120 & 0 & -6544.9 & -86143 & 150810 \end{Vmatrix},$$

$$G_{\pi\eta}^{\langle 4 \rangle} = \begin{Vmatrix} G_{\pi\eta}^{\langle 4,1 \rangle} & G_{\pi\eta}^{\langle 4,2 \rangle} \\ G_{\pi\eta}^{\langle 4,3 \rangle} & G_{\pi\eta}^{\langle 4,3 \rangle} \end{Vmatrix} = \begin{Vmatrix} 10832 & -4287.4 & -6544.9 \\ -4287.4 & 296680 & -86143 \\ -6544.9 & -86143 & 150810 \end{Vmatrix}.$$

Therefore

$$\tau_{MP}^2 = \lambda_{req}^2 \left(G_{\pi\eta}^{\langle 4,1 \rangle} - G_{\pi\eta}^{\langle 4,2 \rangle} \left[G_{\pi\eta}^{\langle 4,4 \rangle} \right]^{-1} G_{\pi\eta}^{\langle 4,3 \rangle} \right)^{-1} = 720 \cdot 9.72 \cdot 10^{-5} = 0.070\,(\mathrm{s})$$

which means that the Quality of Service conditions (6) are met for the second flow according to the required values of the average delay of 75 ms and the packet flow rate of 720 1/s.

6 Discussion of Results and Their Applicability for Load Balancing with Guaranteed QoS

The results of the study of the proposed tensor multiflow routing model based on load balancing on a number of numerical network examples confirmed its effectiveness and adequacy. The novelty and main advantage of the presented approach are as follows. Previously known solutions that are based on the implementation of TE principles directly contribute to a more balanced network load and only indirectly improve the QoS level. The presented model is a further development and improvement of previously known approaches related to individual solutions for TE and QoS routing [18–20, 31–33]. Using the proposed model initially focuses on providing for each packet flow the required (specified) QoS level by multipath routing without involving additional means of managing network resources, for example, reservation protocols. This was achieved by obtaining (due to the use of the tensor approach) and introducing into the proposed model the conditions for providing QoS in terms of transmission rate and average end-to-end delay.

The practical implementation of the proposed multiflow routing model will increase the performance requirements of network routers, since there is a need for real-time multidimensional nonlinear optimization problems solving. On the other hand, the most relevant application of the proposed tensor approach to load balancing relates to SDN technologies (SD-WAN, Hybrid SD-WAN, T-SDN), since network controllers in the general case have sufficient computing power to solve the formulated optimization problems.

7 Conclusion

The article proposes tensor multiflow routing model based on load balancing, represented by expressions (1)–(6). The novelty of the proposed model is that, in contrast to the well-known load balancing solutions aimed at improving the QoS level in the network as a whole, it is focused on providing specified numerical values of the selected indicators of Quality of Service – flow rate and average packet delay. This was achieved by introducing the appropriate QoS conditions into the multiflow routing model structure (6). It was possible to formulate conditions (6) in an analytical form due to the use of the tensor research methodology when the network structure was described by a one-dimensional simplicial complex, and the functional description was presented by a bivalent mixed tensor in a discrete metric space. In this case, the space metric (4) was determined by both the service discipline of packets on the routers interfaces and the statistical model of the served flows. Thus, the conditions for ensuring the Quality of Service are invariant in their form with respect to the metric of the space in question.

In addition, conditions (6) are formulated in such a way that for each flow, the resulting average end-to-end packet delays along the set of calculated paths were the same and did not exceed permissible values. The research results confirmed the adequacy of the obtained solutions (Table 1) from the point of view of providing specified values of quality indicators, which may differ for each of the packet flows transmitted over the network.

Further development of the load balancing tensor multiflow routing model is seen in the enhancement of the set of QoS indicators included in the model, for example, packet loss probability, by introducing the corresponding conditions of QoS ensuring. Taking all into account, the proposed solution is focused on application in Software-Defined Networks, in which the corresponding controller is responsible for calculating the routes.

References

1. Dodd, A.Z.: The Essential Guide to Telecommunications (Essential Guide Series), 6 edn. Prentice Hall (2019)
2. Parker, P.M.: The 2021–2026 World Outlook for Software-Defined Wide Area Network (SD-WAN) Appliances. ICON Group International, Inc. (2019)
3. Blokdyk, G.: Managed Hybrid WAN SD-WAN The Ultimate Step-By-Step Guide. 5STARCooks (2018)
4. Goransson, P., Black, C., Culver, T.: Software Defined Networks: A Comprehensive Approach. Morgan Kaufmann (2016)
5. Janevski, T.: QoS for Fixed and Mobile Ultra-Broadband. Wiley IEEE Press (2019)
6. Carthern, C., Wilson, W., Rivera, N., Bedwell, R.: Cisco Networks: Engineers' Handbook of Routing, Switching, and Security with IOS, NX-OS, and ASA. Apress (2015)
7. Szigeti, T., Zacks, D., Falkner, M., Arena, S.: Cisco Digital Network Architecture: Intent-Based Networking for the Enterprise. Cisco Press (2018)
8. Nevzorova, O., Arous, K., Hailan, A.: Flow-based model of hierarchical multicast routing. In: 2015 Second International Scientific-Practical Conference Problems of Infocommunications Science and Technology (PIC S&T) Proceedings, pp. 50–53. IEEE (2015). https://doi.org/10.1109/INFOCOMMST.2015.7357266
9. Lemeshko, O., Nevzorova, O., Vavenko, T.: Hierarchical coordination method of inter-area routing in telecommunication network. In: 2016 International Conference Radio Electronics & Info Communications (UkrMiCo) Proceedings, pp. 1–4. IEEE (2016). https://doi.org/10.1109/UkrMiCo.2016.7739626
10. Papan, J., Segec, P., Paluch, P., Uramova, J., Moravcik, M.: The new multicast repair (M-REP) IP fast reroute mechanism. Concurr. Comput.: Pract. Exp., e5105 (2018). https://doi.org/10.1002/cpe.5105
11. Lakshman, N.L., Khan, R.U., Mishra, R.B.: MANETs: QoS and investigations on optimized link state routing protocol. Int. J. Comput. Netw. Inf. Secur. (IJCNIS) **10**(10), 26–37 (2018). https://doi.org/10.5815/ijcnis.2018.10.04
12. Najafi, G., Gudakahriz, S.J.: A stable routing protocol based on DSR protocol for mobile ad hoc networks. Int. J. Wirel. Microwave Technol. (IJWMT) **8**(3), 14–22 (2018). https://doi.org/10.5815/ijwmt.2018.03.02
13. Smelyakov, K., Sandrkin, D., Ruban, I., Vitalii, M., Romanenkov, Y.: Search by image. New search engine service model. In: 2018 International Scientific-Practical Conference Problems of Infocommunications. Science and Technology (PIC S&T) Proceedings, pp. 181–186. IEEE (2018). https://doi.org/10.1109/INFOCOMMST.2018.8632117
14. Churyumov, G.I., Tkachov, V., Tokariev, V., Diachenko, V.: Method for ensuring survivability of flying ad-hoc network based on structural and functional reconfiguration. In: Selected Papers of the XVIII International Scientific and Practical Conference on Information Technologies and Security (ITS 2018). CEUR Workshop Processing, pp. 64–76 (2018)
15. Mallapur, S.V., Patil, S.R., Agarkhed, J.V.: A stable backbone-based on demand multipath routing protocol for wireless mobile ad hoc networks. Int. J. Comput. Netw. Inf. Secur. (IJCNIS) **8**(3), 41–51 (2016). https://doi.org/10.5815/ijcnis.2016.03.06
16. Moza, M., Kumar, S.: Analyzing multiple routing configuration. Int. J. Comput. Netw. Inf. Secur. (IJCNIS) **8**(5), 48–54 (2016). https://doi.org/10.5815/ijcnis.2016.05.07
17. Krishna, S.R.M., Ramanath, S., Prasad, K.: Optimal reliable routing path selection in MANET through novel approach in GA. Int. J. Intell. Syst. Appl. (IJISA) **9**(2), 35–41 (2017). https://doi.org/10.5815/ijisa.2017.02.05

18. Yeremenko, A.S.: A two-level method of hierarchical-coordination QoS-routing on the basis of resource reservation. Telecommun. Radio Eng. **77**(14), 1231–1247 (2018). https://doi.org/10.1615/TelecomRadEng.v77.i14.20
19. Yeremenko, O.S., Lemeshko, O.V., Nevzorova, O.S., Hailan A.M.: Method of hierarchical QoS routing based on the network resource reservation. In: 2017 IEEE First Ukraine Conference on Electrical and Computer Engineering (UKRCON) Proceedings, pp. 971–976. IEEE (2017). https://doi.org/10.1109/UKRCON.2017.8100393
20. Wang, N., Ho, K., Pavlou, G., Howarth, M.: An overview of routing optimization for internet traffic engineering. IEEE Commun. Surv. Tutor. **10**(1), 36–56 (2008). https://doi.org/10.1109/COMST.2008.4483669
21. Lemeshko, O., Yeremenko, O.: Enhanced method of fast re-routing with load balancing in software-defined networks. J. Electr. Eng. **68**(6), 444–454 (2017). https://doi.org/10.1515/jee-2017-0079
22. Lin, S.C., Wang, P., Luo, M.: Control traffic balancing in software defined networks. Comput. Netw. **106**, 260–271 (2016). https://doi.org/10.1016/j.comnet.2015.08.004
23. Koryachko, V.P., Perepelkin, D.A., Byshov, V.S.: Development and research of improved model of multipath adaptive routing in computer networks with load balancing. Autom. Control Comput. Sci. **51**(1), 63–73 (2017). https://doi.org/10.3103/S0146411617010047
24. Koryachko, V.P., Perepelkin, D.A., Byshov, V.S.: Enhanced dynamic load balancing algorithm in computer networks with quality of services. Autom. Control Comput. Sci. **52**(4), 268–282 (2018). https://doi.org/10.3103/S0146411618040077
25. Munemitsu, T., Kotani, D., Okabe, Y.: A mixed integer programming solution for network optimization under tunneling-based traffic engineering schemes. In: 2018 IEEE 42nd Annual Computer Software and Applications Conference (COMPSAC) Proceedings, vol. 2, pp. 769–776. IEEE (2018). https://doi.org/10.1109/COMPSAC.2018.10335
26. Meng, F., Chai, R., Zhang, C.: Delay minimization based joint routing and flow allocation for software defined networking. In: 2017 9th International Conference on Wireless Communications and Signal Processing (WCSP) Proceedings, pp. 1–6. IEEE (2017). https://doi.org/10.1109/WCSP.2017.8171146
27. Tomovic, S., Radusinovic, I.: Traffic engineering approach to virtual-link provisioning in software-defined ISP networks. In: 2017 25th Telecommunication Forum (TELFOR) Proceedings, pp. 1–4. IEEE (2017). https://doi.org/10.1109/TELFOR.2017.8249296
28. Prabhavat, S., Nishiyama, H., Ansari, N., Kato, N.: On load distribution over multipath networks. IEEE Commun. Surv. Tutor. **14**(3), 662–680 (2012). https://doi.org/10.1109/SURV.2011.082511.00013
29. Koubàa, M., Amdouni, N., Aguili, T.: Efficient traffic engineering strategies for optimizing network throughput in WDM all-optical networks. Int. J. Comput. Netw. Inf. Secur. (IJCNIS) **7**(6), 39–49 (2015). https://doi.org/10.5815/ijcnis.2015.06.05
30. Ageyev, D., Kirichenko, L., Radivilova, T., Tawalbeh, M., Baranovskyi, O.: Method of self-similar load balancing in network intrusion detection system. In: 2018 28th International Conference Radioelektronika (RADIOELEKTRONIKA) Proceedings. pp. 1–4. IEEE (2018). https://doi.org/10.1109/radioelek.2018.8376406
31. Lemeshko, A.V., Evseeva, O.Y., Garkusha, S.V.: Research on tensor model of multipath routing in telecommunication network with support of service quality by greate number of indices. Telecommun. Radio Eng. **73**(15), 1339–1360 (2014). https://doi.org/10.1615/TelecomRadEng.v73.i15.30

32. Lemeshko, O.V., Yeremenko, O.S.: Dynamics analysis of multipath QoS-routing tensor model with support of different flows classes. In: 2016 International Conference on Smart Systems and Technologies (SST) Proceedings, pp. 225–230. IEEE (2016). https://doi.org/10.1109/SST.2016.7765664
33. Lemeshko, O., Yeremenko, O., Yevdokymenko, M.: Tensor model of fault-tolerant QoS routing with support of bandwidth and delay protection. In: 2018 XIIIth International Scientific and Technical Conference Computer Sciences and Information Technologies (CSIT) Proceedings, pp. 135–138. IEEE (2018). https://doi.org/10.1109/stc-csit.2018.8526707

The Method of Correction Functions in Problems of Optimization of Corroding Structures

D. G. Zelentsov(✉), L. I. Korotka, and O. R. Denysiuk

Ukrainian State University of Chemical Technology,
Gagarina Av. 8, Dnipro 49005, Ukraine
dgzelentsov@gmail.com

Abstract. The paper proposes and substantiates the use of the method of correction functions in modeling the processes of corrosion deformation in structures operating in aggressive environments. This approach is especially relevant for optimal design problems, when the calculation of the constraint functions involves modeling this process. In this method a solution to the durability problem corresponding to a specific value of the vector of variable parameters is sought as a product of two functions: an approximate solution of the problem, and a correction function (error of the approximate solution). An approximate solution is sought for a simplified model of corrosion deformation of a structure, i.e. some of data on which a solution depends is ignored. This data is used to construct the correction function. On the other hand, the correction function doesn't depend on data determining design parameters. Therefore, it is invariant with respect to the constructions of the considered class. The paper proposes to consider the correction function as a neural network with a single output neuron - the error of an approximate solution. The calculation of the constraint functions, thus, implies an approximate solution to the problem, which can be obtained with minimal computational costs after refinement of a result. It is shown that the joint use of the computational methods of solid mechanics and computational intelligence methods (including artificial neural networks and genetic methods) will significantly improve the quality of solving complex applied problems in the mechanics of corroding structures.

Keywords: Corroding structures · Durability problem · Artificial neural networks · Polynomial approximation

1 Introduction

In the process of optimal design and calculation of bearing capacity of mechanical structures which function, normally, in aggressive external media (AM), approaches and methods that are used to solve the problem are very important (Dey and Roy 2015). Special attention should be paid to mutually correlating issues related to computational costs and error estimates of obtained numerical results.

Z. Hu et al. (Eds.): ICCSEEA 2020, AISC 1247, pp. 132–142, 2021.
https://doi.org/10.1007/978-3-030-55506-1_12

At an initial stage of modeling the behavior of corroding structures (CS), the accuracy of obtained results did not receive sufficient attention in most studies. Thus, with this approach, the material consumption of structures was too high.

Two related groups of equations are used to construct a model of corrosion deformation, namely the system of solid mechanics equations and the system of differential equations (SDE) describing an accumulation of geometric damage in structural elements. The equations of the first group are used to calculate stress functions in right parts of the SDE. In general, the finite element method (FEM) is used to solve it, so only a numerical solution of the SDE is possible. Parameters of the numerical solution are input parameters and do not change in the process of solving the problem. Therefore, an error of obtained result in this case is not predictable, since for each specific problem the corresponding SDE is solved.

The number of iterations in solving the optimization problem can reach tens or hundreds of thousands, at each iteration the constraint functions (CF) are calculated. When solving problems of discrete optimization of corroding trusses, the use of optimization methods based on derivatives is not possible due to the fact that a solution of the optimization problem belongs to a non-metric set (a set of indices). To solve such problems, evolutionary modeling methods, in particular, genetic algorithms (GA) can be used (Assimi and Jamali 2017; Webb et al. 2017). Their disadvantages include, first of all, low efficiency compared to the methods based on derivatives, since GAs operate not with a single point of the solution space, but with a set of these points (population). In this case, the only way to improve efficiency is to reduce the cost of calculating the CF. At the same time, it is required to provide the specified accuracy, and it is not possible to reduce computational costs by reducing a number of time grid nodes, when solving the SDE describing corrosion wear in structural elements, without the loss of accuracy.

The objective of the paper is to propose an approach that makes it possible to improve accuracy and efficiency when solving optimization problems for CS with changing geometric parameters.

The method is based on the joint use of solid mechanics methods, numerical methods for solving differential equations and methods of computational intelligence, in particular, neural networks and genetic methods.

2 Literature Review

With previously existing algorithms (Zelentsov and Korotkaya 2018) it was not possible to control a calculation error for constraint functions when searching for a solution of an optimization problem. The error of solution of the SDE, which defines the error in constraint functions, depends on a numerical solution parameter (in this paper a distance between nodes of a time grid), and, consequently, a number of calls to the FEM procedure. Results of numerical experiments showed that when changing the numerical solution parameter, the durability of a design, which was previously considered optimal, may be less than it was required. Therefore, a resulting solution in this case does not satisfy the system of constraints. Oh the other hand, use of such solution parameters that, with acceptable probability, will make it possible to define the durability of a design with an error that is not exceeding an admissible one for any point of solution space, will lead to very high computational costs.

Different authors propose the use of neural networks in problems of corrosion assessment (Oladipo et al. 2017; Zelentsov and Denysiuk 2019). In particular, in paper (Denysiuk 2016) it was proposed to use neural networks to control the error of the numerical solution in solving the optimization problem. However, the neural network accuracy control module did not always provide the required accuracy, since the factor of change of internal forces in elements of statically indeterminate trusses was ignored when obtaining the training sample.

3 Problem Statement

As an object of research in this paper trusses intended for operation in aggressive media are considered. The weight optimization problem can be formulated as follows: it is required to determine types and sizes of truss elements specified on the set of indices of dimension n, so that the volume of a structure would be minimal, and throughout the specified service life it would retain its bearing capacity, that is, satisfy the conditions of strength and stability. In the form of an optimization problem, this statement has the form:

$$F(\bar{x}) = \sum_{i=1}^{N} A_i(\bar{x}) \cdot L_i \to \min;\ \bar{x} \in X_D;$$

$$X_D:\ \left\{\bar{x} \in I^n \left| g_1(\bar{x}) = [\sigma] - \sigma_i(\bar{x}, t^*) \geq 0;\ g_2(\bar{x}) = \sigma_i * (\bar{x}, t^*) - \sigma_i(\bar{x}, t^*) \geq 0;\ i = \overline{1, N}\right.\right\}. \tag{1}$$

Here A_i and L_i are cross-sectional area and length of i-th truss element; N is a number of rods in the model; $\bar{x}$ is a vector of varied parameters (types and sizes of cross-sections); σ_i and σ_i^* are current stress and critical buckling stress for an i-th element; $[\sigma]$ is the value of yield stress; t* is the specified durability.

The specificity of optimization problems of this class is as follows:

- the constraint functions include time;
- a solution is sought on a discrete non-metric set (set of indices I^n).

The calculation of constraint functions involves calculation of the stress state in a structure at a given time, taking into account the corrosion process occurring in it.

Modeling the process of corrosive deformation involves solving a SDE describing the process of geometric damage accumulation in truss elements:

$$\frac{d\delta_i}{dt} = v_0 \cdot \Phi\left(\sigma_i(\bar{\delta})\right) \quad \delta_i\big|_{t=0} = 0; \quad i = \overline{1, N}, \tag{2}$$

where δ_i is a depth of corrosion damage (damage parameter); v_0 is a corrosion rate in the absence of stress; Φ is a known stress function; t is time.

The question of how a structure will exhaust its bearing capacity is purely theoretical. In this case, the constraint system of the optimization problem (1) can be reduced to a single constraint:

$$t(\bar{x}, \bar{c}, \bar{\alpha}) - t^* \geq 0, \tag{3}$$

where $\bar{c}$ is a vector of parameters of an aggressive medium; $\bar{\alpha}$ is a vector containing all other parameters that affect the SSS of a structure (boundary conditions, loading conditions, etc.).

Thus, the constraint (3) check assumes the joint solution of two groups of equations.

To calculate stress in the right parts of (2), the solid mechanics equations are used: the system of equations of equilibrium and compatibility of deformations, Cauchy relations and Hooke's law. As a system of equations of the FEM, they have the form:

$$\begin{cases} \bar{R} = K^{-1} \cdot \bar{u} \\ \bar{\varepsilon} = D \cdot \bar{u} \\ \bar{\sigma} = E \cdot \bar{\varepsilon} \end{cases}, \tag{4}$$

Here K, D, E are matrices of stiffness, differentiation and elasticity; $\bar{R}$, $\bar{u}$, $\bar{\varepsilon}$ and $\bar{\sigma}$ are vectors of nodal loads, displacements, strains and stresses.

Thus, calculation of the CF in the optimization problem is reduced to solving numerically the Cauchy problem for the SDE (2) together with the FEM problem (4).

The aim of the paper is to develop a method that will minimize computational costs and at the same time ensure sufficient accuracy in calculation of the constraint functions.

4 The Method of Correction Functions

The scheme of solving the optimization problem is shown in Fig. 1. It follows from it that obtaining a solution to the problem of optimal design of CS with varying geometric characteristics requires sufficiently large computational costs. At the same time, the error of an obtained solution for the reasons shown above is unknown.

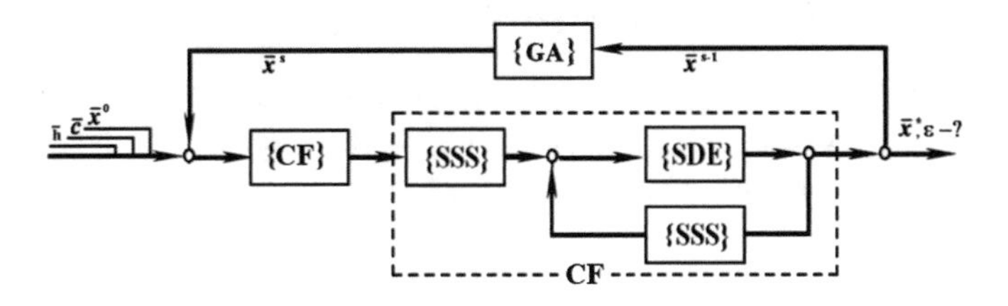

Fig. 1. The scheme of solving the optimization problem

Here, modules {SSS} and {SDE} are respectively modules for calculating stress-strain state and solving the SDE describing the corrosion wear; {RF} is a module for calculating the CF; {GA} is a module for implementing the genetic algorithm.

In the context of the given problem, the main question is how an error of calculating the constraint functions depends on errors of individual components of the corrosion deformation model, and how this error changes in the process of solving the optimization problem. The errors of a design model, models of material and limit state obviously do not depend on a specific value of the vector of varied parameters. An error of a corroding cross-section model will be determined by an accuracy of calculating cross-sectional geometric characteristics at any time and belongs to the category of removable errors. This error also does not depend on a specific value of the variable parameter vector.

It is obvious that the following variants are possible with respect to the error in solution of the optimization problem.

A subset of the solution space points for which an error of the CF calculation will exceed a maximum allowable one. If the solution of the optimization problem is one of these points, then this solution is unacceptable, since the durability of the "optimal" design may be less than planned.

A subset of points for which an error is less than the allowable one. When calculating the CF at such points, the computational costs are excessive.

A third subset, the cardinality of which is incomparably less than for the previous two, for which an error of the CF calculation is within a permissible value. In this case, the conditions of accuracy and efficiency are met.

The main strategy to increase the accuracy and efficiency of the algorithm for solving the problem of optimal design of CS involves a maximum increase in the cardinality of the third subset. Thus, it is necessary to abandon the thesis about the constant parameters of the numerical method for solving the SDE. These parameters were determined in the process of solving the optimization problem on the basis of information about the vector of variable parameters, the parameters of an AM and the permissible error in the problem solution. The implementation of such a strategy is possible using methods of computational intelligence.

It is not possible to use an analytical solution of the SDE (2) because it is unacceptable to ignore time-varying forces:

$$\sigma_i = \frac{Q_i}{A_i}: \quad \begin{cases} A_i = A_0 - P_0 \cdot \delta_i + s \cdot \delta_i^2 = A(\delta_i) \\ Q_i = Q(\bar{\delta}(t)) \end{cases} \tag{5}$$

where A_0 and P_0 are initial area and perimeter of a cross-section.

The paper proposes a new approach to solving the problem (3). It is proposed to look for the solution of the problem (3) in the form:

$$t(\bar{x}, \bar{c}, \bar{\alpha}) = \tilde{t}(\bar{x}, \bar{c}) \circ \varphi(\bar{c}, \bar{\alpha}), \tag{6}$$

where $\tilde{t}(\bar{x}, \bar{c})$ is an approximate solution, which in the ideal case ignores the factor of change in a force vector; $\varphi(\bar{c}, \bar{\alpha})$ is a function that makes it possible to take this factor into account. This approach will be justified if the process of determining $\tilde{t}(\bar{x}, \bar{c})$ and $\varphi(\bar{c}, \bar{\alpha})$ is known and not time-consuming.

The solution $\tilde{t}(\bar{x}, \bar{c})$ provided that Q_i = const $\left(i = \overline{1, N}\right)$ can be easily obtained using an analytical formula that determines the durability of a rod under axial loading:

$$t^* = \min_i\{t_i\}. \tag{7}$$

In the paper, it's assumed that a structure loses its load-bearing capacity in case of failure of any single element. Therefore, in this case, instead of determining the durability of a structure as a whole, it is sufficient to determine the durability of its individual finite element. It's possible if the effect of remaining ($N - 1$) structural elements on the corrosion deformation process in this element is formalized, so a law establishing the dependence between internal forces in elements and depth of their corrosion damage Q(t) is necessary. This approach will obviously reduce computational costs.

Ideally, the function $\varphi(\bar{c}, \bar{\alpha})$ will depend on the nature of the force change in such an element. Let the change of forces in an element be described by a polynomial of degree n:

$$Q(t) = P_n(t) = \alpha_0 + \alpha_1 t + \alpha_2 t^2 + \ldots + \alpha_n t^n, \tag{8}$$

where α_i $\left(i = \overline{0, n}\right)$ are unknown coefficients of the polynomial.

It should be noted that it is impossible to predict the durability of a structure in advance.

With this approach, the function $\varphi(\bar{c}, \bar{\alpha})$ is essentially an error $\tilde{t}(\bar{x}, \bar{c})$ with respect to the exact solution $t(\bar{x}, \bar{c}, \bar{\alpha})$. Here vector $\bar{\alpha}$ is the coefficient vector of the polynomial (8). The actual durability of a rod will depend not only on the vector $\bar{c} = \{v_0, k, \sigma_0, A_0, P_0\}$, but also on the coefficients of the polynomial (8). Here k is a coefficient of influence of stress on the rate of corrosion process, σ_0 is initial stress. Then:

$$\varphi = \frac{|t * - \tilde{t}|}{\tilde{t}}. \tag{9}$$

Here $\tilde{t}$ is an approximate value of durability, which is a solution to the equation:

$$\frac{d\delta}{dt} = v_0\left(1 + k \cdot \frac{Q}{A_0 - P_0 \cdot \delta + s\delta^2}\right), \tag{10}$$

the exact durability value will be determined from the solution of the equation:

$$\frac{d\delta}{dt} = v_0\left(1 + k \cdot \frac{P_n(t)}{A_0 - P_0 \cdot \delta + s\delta^2}\right). \tag{11}$$

When finding an approximate solution $\tilde{t}$, it is unacceptable to completely ignore the change of forces in an element, therefore, the formula (6) must be changed. Otherwise, finding the coefficients of the polynomial (8) is not possible.

Based on the facts above, it can be argued that it is necessary to choose the method for determining an approximate solution $\tilde{t}$, which will allow to obtain: 1) an approximate solution of the durability problem; 2) a number of an element that determines the durability of a system as a whole; 3) coefficients of the polynomial $P_n(t)$ describing the force changes in this element.

At the same time, the computational cost of solving the problem $\tilde{t}$ should be significantly less than that of solving the problem of obtaining an accurate (reference) or asymptotically accurate solution t^*.

The numerical-analytical algorithm for solving the SDE describing the process of damage accumulation takes into account the effect of the depth of corrosion damage on the stress change, but ignores the effect of the axial force change. For statically definable corroding trusses (when the internal forces in rods are constant), it is possible to obtain an exact solution to the durability problem. The error in solution for statically indeterminate CS is determined only by degree of change in internal forces.

As a model of damage accumulation, the following equation will be used:

$$\frac{d\delta}{dt} = v_0(1 + k\sigma). \tag{12}$$

Its solution t* for Q = const has the form:

$$t^* = \frac{\delta^*}{v_0} - \frac{2kQ}{v_0 d}\ln\left\{\frac{(P_0 + d - 2a\cdot\delta^*)(P_0 - d)}{(P_0 - d - 2a\cdot\delta^*)(P_0 + d)}\right\}. \tag{13}$$

Here δ^* is the limit value of the depth of corrosion damage; $c = A_0 + kQ$; $d = \sqrt{P_0^2 - 4ac}$. The limit value δ^* is determined from the conditions of strength, stability and cross-sectional continuity.

In the case where the axial force in an element is not constant (Q $\neq$ const), the SDE (12) turns into a system of differential equations of the form:

$$\frac{d\delta_i}{dt} = v_0\left(1 + k\cdot\sigma\left(\delta_i; Q_i\left(\overline{\delta}\right)\right)\right). \tag{14}$$

In the numerical-analytical algorithm for solving the SDE (2), the increment of the corrosion depth $h_\delta^s = \delta^s - \delta^{s-1} = const$ is given, and the corresponding increment of time h_t^s is determined by the formula:

$$h_t^s = \frac{h_\delta^s}{v_0} - \frac{2kQ^{s-1}}{v_0 d}\ln\left\{\frac{\left(P^{s-1} + d - 2ah_\delta^s\right)\left(P^{s-1} - d\right)}{\left(P^{s-1} - d - 2ah_\delta^s\right)\left(P^{s-1} + d\right)}\right\}. \tag{15}$$

The system of mechanical Eqs. (4) is solved in each node of a finite-difference grid (FDG). So, for the time interval h_t^s, the change in the cross-sectional area is taken into account, but the change in the axial force is ignored. The latter determines the error of a solution.

Let us suppose that a law of internal force change in the rod element, the durability of which determines the durability of the structure as a whole, is known, and so is the approximate solution $\tilde{t}$ obtained using the numerical-analytical algorithm for a fixed number of nodes. In this case, it is possible to construct a function approximating the error of the solution on the basis of previously unaccounted factors $\varphi(\bar{c}, \bar{\alpha})$.

It's necessary to note that the numerical-analytical method will give a very approximate solution of the problem, since the increments $h_\delta \rightarrow h_t$ are large enough. It is obvious that the number of nodes on the axis δ and t correspond to the degree of the polynomial (9), thus: $\lim_{n\rightarrow\infty} \tilde{t}(n) = t^*$, $\lim_{n\rightarrow\infty} \bar{\alpha}(n) = \bar{\alpha}^*$.

As noted earlier, the dependencies Q = Q(t) can be obtained based on results of the approximate solution. We will approximate it with a polynomial of degree 3. To determine the coefficients of the polynomial, it's possible to use values of the internal forces in truss elements in four nodes of the FDG (including t = 0).

Figure 2 shows the artificial neural network that is used to approximate the error of the approximate solution under active strength constraints. The input parameters of the network are initial area A_0 and perimeter P_0 of a section, initial stress value σ_0, parameter of an aggressive medium v_0 and coefficients of the polynomial α_1, α_2, α_3. The output parameter of the network is the error of the approximate solution.

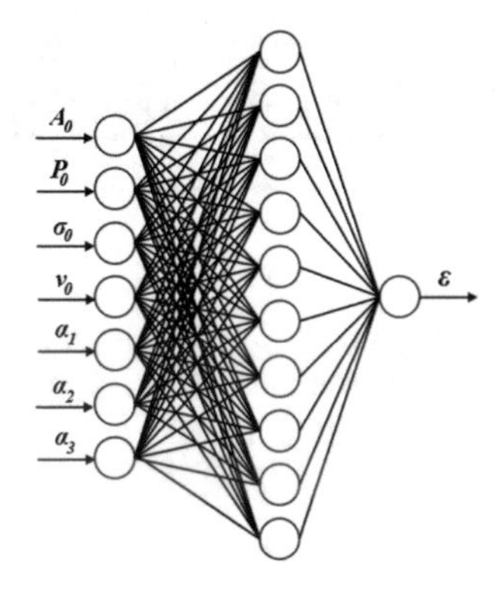

Fig. 2. The architecture of neural network.

A general algorithm of the proposed method can be represented as follows.

1. For a given vector of variable parameters, the problem of stress-strain state is solved and initial values of forces and stresses in rods are found. Based on obtained values, durability of all elements is calculated using (13) and the element to which its minimum value corresponds (hereinafter – the leading element) is determined.
2. A SDE of the form (2) is solved using the numerical-analytical method; a number of nodes of the FDG is determined by a degree of the polynomial describing the change of forces in the leading element. The result of this step is an approximate solution of the durability problem and a value of the coefficients of the polynomial.
3. On the basis of data on cross-sectional parameters of the leading element, initial stress in it and coefficients of the approximating polynomial, an error of the approximate solution is calculated using artificial neural networks (ANNs).

4. Based on the approximate solution of the durability problem and its error, a more accurate solution is obtained.

5 Results and Discussion

For numerical illustration of the proposed method, a number of problems of weight optimization of corroding trusses with rods made of shaped profiles were solved. A type (I-beam, channel beam, equal and unequal angles) and a profile size were taken as variable parameters. An integer genetic algorithm was used to solve the optimization problem (Ashlock 2006).

Neural networks approximating the error of the numerical-analytical solution were trained for each type of cross-section using the backpropagation algorithm. The optimal structure of neural networks was determined using the methods given in (Korotkaya 2011; Zelentsov and Korotkaya 2018; Du and Swamy 2014).

Table 1 shows the results of ANNs testing for a stretched I-beam rod.

Table 1. The results of testing the neural network.

<0.2%	0.2–0.4%	0.4–0.6%	0.6–0.8%	0.8–1.0%	>1.0%
332	282	162	106	72	46

Test data consisted of 1000 samples. The maximum network error was 3.26%.

The following table shows the results of calculating the durability of the five-rod truss, made from different profiles (Table 2).

Table 2. Test results for the method

Problem	Types and sizes of truss elements	t_{et}, years	t_{pr}, years	ε_{nn}, %	t_{ac}, years	ε, %
1	Equal angle 125 × 129 × 9	2.3217	2.2084	5.71	2.3345	0.55
2	Unequal angle 140 × 90 × 10	2.5914	2.7345	6.01	2.5702	0.81
3	I-beam 160 × 81	1.5727	1.5023	4.95	1.5767	0.25
4	Channel beam 180 × 70	1.6659	1.6925	2.03	1.6608	0.31
5	Channel beam 200 × 76	1.8382	1.8732	1.75	1.8404	0.11

The table shows: a number of solved problem, type and sizes of truss elements, a reference solution, an approximate solution obtained on four grid nodes, an error of the approximate solution (the output parameter of the neural network), a more accurate solution based on an error value and an its error relative to a reference solution.

As follows from the results, the accuracy of a solution does not depend on geometric characteristics of a section, a type of constraints and values of initial stresses.

Some results confirming the high efficiency of the proposed method are given in Table 3. Here the computational costs of different methods for calculating constraint functions with the maximum permissible error $\varepsilon^* = 1\%$ are compared. The optimization problem was solved for five-rod and fifteen-rod trusses using a genetic algorithm.

Table 3. Analysis of effectiveness for the methods.

Method	Number of calls to the FEM procedure	
	5-rod truss	15-rod truss
GA	3 375 141	10 287 508
NN + GA	1 522 317	3 618 484
STM + NN + GA	565 944	1 528 043
Method of correction functions	46 714	94 396

A number of calls to the FEM procedure was taken as an efficiency measure:

- when using a fixed time step to calculate the CF, at which a calculation error does not exceed ε^* on the whole set of variable parameters;
- when using a neural network precision control module;
- when using a neural network precision control module and the sliding tolerance method (STM);
- when using the method of correction functions proposed in current paper.

Data for the first three methods is given in (Zelentsov and Korotkaya 2018; Zelentsov and Denysiuk 2019).

The results of numerical experiments show that the proposed method can reduce computational costs by an order of magnitude comparing to known methods.

It is shown that the accuracy of the results of solving durability problem is significantly affected by the factor of changes in the internal forces in structural elements due to the corrosion process. The negative influence of this factor on the accuracy of calculations is almost completely eliminated using the correction function.

6 Conclusions

The method based on the use of neural networks approximating the error of an approximate solution to the problem of the durability of corroding structures and its subsequent refinement has significantly reduced computational costs while ensuring a required accuracy. Application of the proposed method is most effective for problems of optimal design of structures of this class, since the durability problem is solved at each iteration of the optimization algorithm to calculate the constraint functions.

The method proposed by the authors made it possible for the first time to use a neural network as a correction function, invariant with respect to the constructions of the class in question, when solving problems of optimal design of corroding structures.

The proposed method can be generalized to other classes of structures (beams, plates, shells) in further research.

The results obtained during the numerical experiments confirm the correctness of the chosen approach.

References

Ashlock, D.: Evolutionary Computation for Modeling and Optimization. Springer, New York (2006)

Assimi, H., Jamali, A.: Sizing and topology optimization of truss structures using genetic programming. Swarm Evol. Comput, **37**, 90–103 (2017)

Denysiuk, O.R.: Determination of rational parameters for the numerical solution of systems of differential equations. Bull. Kherson Nat. Tech. Univ. **3**(58), 208–212 (2016). (in Russian)

Du, K.-L., Swamy, M.N.S.: Neural networks and statistical learning. Enjoyor Labs, Enjoyor Inc., China (2014). http://dx.doi.org/10.1007/978-1-4471-5571-3

Dey, S., Roy, T.K.: Multi-objective structural optimization using fuzzy and intuitionistic fuzzy optimization technique. Int. J. Intell. Syst. Appl. (IJISA) (5), 57–65 (2015)

Korotkaya, L.I.: Application of neural networks in numerical solution of some systems of differential equations. Eastern-Eur. J. Enterp. Technol. **51**(3/4), 24–27 (2011)

Oladipo, B.A., Ajide, O.O., Monye, C.G.: Corrosion assessment of some buried metal pipes using neural network algorithm. Int. J. Eng. Manuf. (IJEM) **7**(6), 27–42 (2017)

Webb, D., Alobaidi, W., Sandgern, E.: Structural design via genetic optimization. Mod. Mechan. Eng. **7**(3), 73–90 (2017)

Zelentsov, D., Denysiuk, O.: Neural network algorithm for accuracy control in modelling of structures with changing characteristics. In: Hu, Z., Petoukhov, S., Dychka, I., He, M. (eds.) Advances in Computer Science for Engineering and Education. ICCSEEA 2018. Advances in Intelligent Systems and Computing, vol. 754. Springer, Cham (2019)

Zelentsov, D.G., Korotkaya, L.I.: Technologies of Computational Intelligence in Tasks of Dynamic Systems Modeling: Monograph. Balans-Klub, Dnepr. (2018). http://dx.doi.org/10.32434/mono-1-ZDG-KLI. (in Russian)

Improving the Reliability of Simulating the Operation of an Induction Motor in Solving the Technical and Economic Problem

Mykola Tryputen[1], Vitaliy Kuznetsov[2(✉)], Alisa Kuznetsova[3], Maksym Tryputen[3], Yevheniia Kuznetsova[2], and Tetiana Serdiuk[4]

[1] Dnipro University of Technology, Av. Dmytra Yavornytskoho, 19, Dnipro, Ukraine
nikolay.triputen@gmail.com
[2] National Metallurgical Academy of Ukraine, Gagarina Avenue, 4, Dnipro, Ukraine
{wit1975,wit_jane2000}@i.ua
[3] Oles Honchar Dnipro National University, 35, D. Yavornitsky Avenue, 4 Building of DNU, Dnipro, Ukraine
{karamel75,triputen2014}@i.ua
[4] Dnipro National University of Railway Transport Named after Academician V. Lazaryan, Dnipro, Ukraine
serducheck-t@rambler.ru

Abstract. The article is devoted to the control of the statistical laws of linear voltages reproduced on a computer during computational studies on an energy-economic model when solving the problem of choosing protection tools for induction motors operating in workshop electric networks of industrial enterprises with low-quality electricity. Simultaneous and continuous estimation of average values, variances, autocorrelation and cross-correlation functions of harmonics of linear voltages is based on the adaptive approach.

The article presents mathematical expressions for correcting the values of statistical quantities with the accumulation of information and proposes structural schemes for a comprehensive analysis of the results of reproduction of random variables and functions on a computer. The results of checking the values of the average values and dispersions of six harmonics of simulated linear voltages supplying an induction motor with a power of 7.5 kW are shown for the conditions of the rolling shop No. 1 of Dneprospetsstal LLC.

Keywords: Induction motor · Electric networks · Linear voltages · Protection tools · Economic model · Electric energy

1 Introduction

Noisy electricity within workshop networks of industrial enterprises affects unfavorably the performance of electromechanical transducers, i.e. asynchronous alternating-current motors. Unsymmetry and anharmonicity of three-phase networks as well as

Z. Hu et al. (Eds.): ICCSEEA 2020, AISC 1247, pp. 143–152, 2021.
https://doi.org/10.1007/978-3-030-55506-1_13

their harmonic components result in the increased temperature of the motor windings; the decreased power coefficient and efficiency; the increased amount of reactive power being consumed; and the reduced service life. The above-mentioned worsens operational capability along with the electric equipment reliability while reducing the performance efficiency on the whole.

Minimization of the noisy electricity effect on the performance efficiency can be achieved at the expense of the use of "individual" LC-filters, and "cluster" devices to compensate effect of such noisy electricity at a workshop level; as well as at the expense of control of supply voltage distortion within the areas of its onset. Certain measures may also be untaken if they are not expedient economically.

Each of the listed alternatives is characterized by a specific implementation cost, and economic potential. Selection of one of them is rather complex problem being solved in terms of computer-based experiments.

2 Literature Review and Problem Statement

It is a well-known fact that there is certain negative effect of poor-quality power supply upon operational characteristics of induction motors (IM) [1–3]. Reduced quality of power voltage results in pulsation of the moment generated by the motor, drop of starting and critical IM moment, increase in vibration, early wear of bearing and gear components, increased steel losses due to higher harmonic field constituents in a gap, reduction in such power indices of induction motor operation as efficiency coefficient and power coefficient. Moreover, availability of noisy electric energy within workshop grids of industrial enterprises results in the accelerated physical ageing; in the decreased power efficiency of equipment in use; and in the increased risk of industrial emergency situations. Paper [4] has shown that the problem solution should be sought at technical-and-economic level involving methods of mathematical modeling. Papers [4] and [5] have proposed a technique to make optimum decision as for electric equipment operation under the conditions of noisy power. The technique relies upon economic evaluation of various alternatives to recover supply voltage up to the preset quality indices. Moreover, its suitability has been demonstrated in terms of induction motor operation. According to the technique, power indices of electromechanical transducer are calculated involving the current quality power indices within the enterprise power grid [6], [7], and [8], and basing upon electric model, and thermal model [9]. If indices, calculated in such a way, differ substantially from preset ones, various alternatives of engineering solutions, intended to recover electric power supplying the motor, are considered. Cost of each of the alternatives is estimated and final decision, concerning its further operation, is made.

Method relies upon the use of power and economic model of certain electric equipment; taken as a whole, it helps optimize selection of technical means aimed at electric energy quality recovery according to cost criterion involving restrictions to power indices of the electrical consumer. However, calculation of different variants is based upon the knowledge of statistic regularities of linear voltage change under specific operation conditions of the equipment. That supposes carrying out of a number of expensive and long-term experiments using real object. To reduce both cost of the experiments as well as their period, it has been proposed to substitute industrial experiments for computational ones. For that purpose, power and economic model is supplemented by a unit to form linear voltages and to control them. Probability model of linear voltages to be applied in workshops of industrial enterprises is represented in [10].

3 Power and Economic Model of Electric Equipment

Figure 1 demonstrates one of the variations of power and economic model making it possible to perform computational studies of IM operation. In this context, making a correct decision is possible, if only linear voltages are simulated in accordance with their statistic regularities. Basing upon specific features of linear voltage simulation [10] it is required to control average values, dispersion, autocorrelation, and cross-correlation functions of harmonics of linear voltages. Moreover, the listed values and functions should be evaluated simultaneously and continuously during the modeling process. Such an evaluation can be performed relying upon adaptive approach.

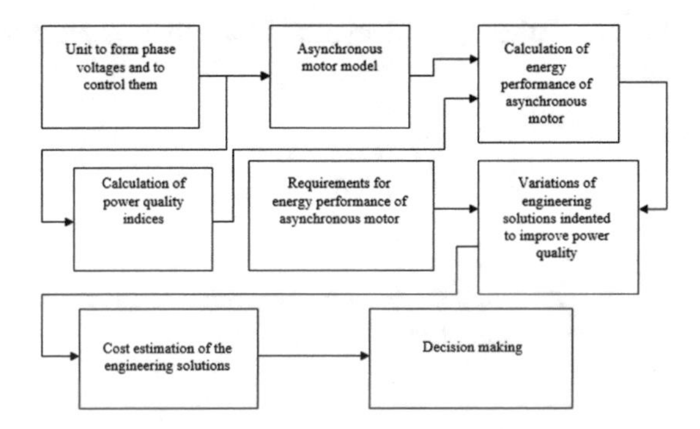

Fig. 1. Schematic diagram of power and economic model of electric equipment

Average value of continuous stationary random process at t time moment is determined using the formula:

$$\overline{x}(t) = \frac{1}{t}\int_{0}^{t} x(t)dt \tag{1}$$

where $x(t)$ is continuous stationary random process.

Differentiate left side and right side of expression (1) with respect to t:

$$\frac{d\overline{x}(t)}{dt} = -\frac{1}{t^2}\int_{0}^{t} x(t)dt + \frac{1}{t}x(t)$$

or:

$$\frac{d\overline{x}(t)}{dt} = -\frac{1}{t}\overline{x}(t) + \frac{1}{t}x(t)\frac{1}{t}(x(t) - \overline{x}(t)). \tag{2}$$

If the random process is represented by a discrete (impulse) function, then expression (2) is:

$$\overline{x}[iT] - \overline{x}[(i-1)T] = \frac{1}{i}(x[iT] - \overline{x}[(i-1)T]) \tag{3}$$

where T is time discretization of $x(t)$ function; $i = \overline{1,n}$ is discretization interval number; and n is a total of discretization intervals.

It is more convenient to demonstrate expression (3) as follows:

$$\overline{x}[iT] = \overline{x}[(i-1)T] + \frac{1}{i}(x[iT] - \overline{x}[(i-1)T]) \tag{4}$$

Dispersion of continuous stationary random process at t time moment is determined by means of the formula:

$$D_x(t) = \frac{1}{t}\int_0^t (x(t) - \overline{x}(t))^2 dt \tag{5}$$

and values of autocorrelation function and cross-correlation function for different time shifts τ are determined by means of the formulas:

$$R_{x,x}(t,\tau) = \frac{1}{t}\int_0^t ((x(t) - \overline{x}(t))(x(t-\tau) - \overline{x}(t)))dt \tag{6}$$

$$R_{x,y}(t,\tau) = \frac{1}{t}\int_0^t ((x(t) - \overline{x}(t))(y(t-\tau) - \overline{y}(t)))dt \tag{7}$$

where $y(t)$ is continuous stationary random process, and $\overline{y}(t)$ average $y(t)$ value at t time moment.

After performing transformation of (5), (6), and (7) expressions, being analogous to the above mentioned ones, we obtain the following for continuous random functions:

$$\frac{dD_x(t)}{dt} = \frac{1}{t}((x(t) - \overline{x}(t))^2 - D_x(t)) \tag{8}$$

$$\frac{dR_{x,x}(t,\tau)}{dt} = \frac{1}{t}((x(t) - \overline{x}(t))(x(t-\tau) - \overline{x}(t)) - R_{x,x}(t,\tau)) \tag{9}$$

$$\frac{dR_{x,y}(t,\tau)}{dt} = \frac{1}{t}((x(t) - \bar{x}(t))(y(t-\tau) - \bar{y}(t)) - R_{x,y}(t,\tau)) \tag{10}$$

In a digital form, (8), (9), and (10) expressions are:

$$D_x[iT] = D_x[(i-1)T] + \frac{1}{i}((x[(i-1)T] - \bar{x}[(i-1)T])^2 - D_x[(i-1)T]) \tag{11}$$

$$\begin{aligned} R_{x,x}[iT,\tau] = R_{x,x}[(i-1)T,\tau] + \frac{1}{i}((x[(i-1)T] - \bar{x}[(i-1)T]) \\ \times ((x[(i-1)T] - \bar{x}[(i-1)T])) - R_{x,x}[(i-1)T,\tau]) \end{aligned} \tag{12}$$

$$\begin{aligned} R_{x,y}[iT,\tau] = R_{x,y}[(i-1)T,\tau] + \frac{1}{i}((x[(i-1)T] - \bar{x}[(i-1)T]) \\ \times ((y[(i-1)T] - \bar{y}[(i-1)T])) - R_{x,y}[(i-1)T,\tau]) \end{aligned} \tag{13}$$

Figure 2 represents structural scheme of a control system implementing (2), (8), (9), and (10) algorithms to evaluate statistic characteristics of continuous implementations of random functions of first harmonics of amplitudes (phases) of linear voltages *AB* and *BC* $U_{mAB1}(t)$ and $U_{mBC1}(t)$ ($\psi_{AB1}(t)$ and $\psi_{BC1}(t)$).

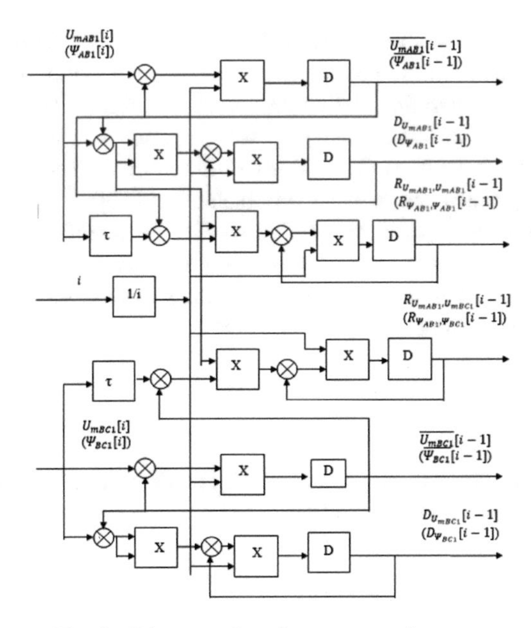

Fig. 2. Diagram of analogous control system

Figure 3 represents structural scheme of a control system implementing (4), (11), (12), and (13) algorithms to evaluate statistic characteristics of discrete implementations of the same random functions according to [10].

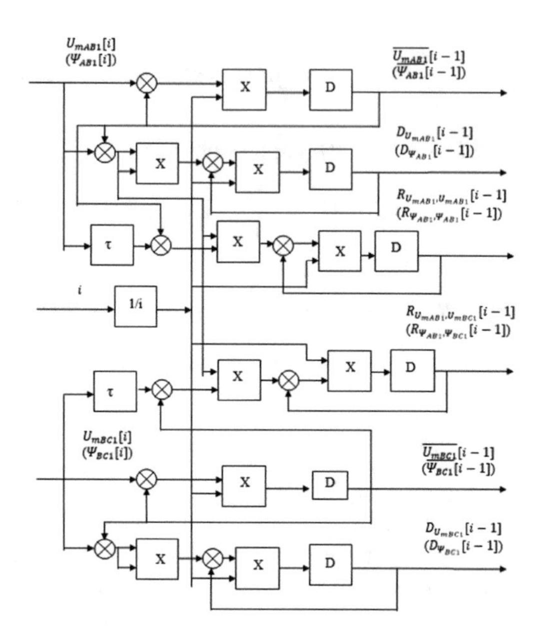

Fig. 3. Diagram of discrete control system

The control system scheme, shown in Fig. 2, can be used in the process of analogous modeling of linear phase voltages; the scheme, shown in Fig. 3, is applicable in the context of digital modeling. Letter D specifies discrete integrator, i.e. digrator. Values of the averages and dispersions of the generated random functions, obtained during the modeling, have been checked for significance of their variation from those hypothetic average values and dispersions obtained in [10].

Zero hypothesis checking in terms of α H_0: $\overline{x} = x_0$ significance level concerning equality between an overall average $\overline{x}$ of normal population with the known dispersion D_0 and hypothetic value x_0 in terms of a competing hypothesis $H_1 : \overline{x} \neq x_0$ has been performed basing upon the criterion value [12]:

$$U_{observ} = \frac{(\overline{x} - x_0)\sqrt{n}}{D_0}$$

and critical point u_{cr} of two-sided critical region determined according to Laplace function table relying upon the equation:

$$\phi(u_{cr}) = \frac{(1-\alpha)}{2}.$$

In this context, n is the number of observations.

If $|U_{observ}| < u_{cr}$, then there is no necessity to reject zero hypothesis.

Determine $\phi(u_{cr}) = 0,475$ for $\alpha = 0,05$ where $u_{cr} = 1,96$.

Table 1 demonstrates checking results of the average random sequences of harmonics of linear voltages generated according to [11, 12] and evaluated with the help of a control system (Fig. 3) if $n = 30$.

Table 1. Checking results of average harmonics of linear voltages

Linear voltage U_{AB}						
Harmonic	Amplitude, V			Phase, degrees		
	Average x_0	Average $\bar{x}$	$\lvert U_{observ} \rvert$	Average x_0	Average $\bar{x}$	$\lvert U_{observ} \rvert$
1	529.82	531.26	$\lvert 1.8 \rvert$	–	–	–
2	4.23	4.38	$\lvert 0.7 \rvert$	63	59.95	$\lvert -1.58 \rvert$
3	17.60	16.85	$\lvert -1.35 \rvert$	206	208.01	$\lvert 1.34 \rvert$
4	1.51	1.58	$\lvert 1.63 \rvert$	92	94.88	$\lvert 1.71 \rvert$
5	18.54	17.75	$\lvert -1.51 \rvert$	130	134.99	$\lvert 1.87 \rvert$
6	3.05	2.95	$\lvert -1.02 \rvert$	290	286.24	$\lvert -1.67 \rvert$
Linear voltage U_{BC}						
Harmonic	Amplitude, V			Phase, degrees		
	Average x_0	Average $\bar{x}$	$\lvert U_{observ} \rvert$	Average x_0	Average $\bar{x}$	$\lvert U_{observ} \rvert$
1	532.09	533.38	$\lvert 1.70 \rvert$	–	–	–
2	3.98	4.41	$\lvert 1.88 \rvert$	78	74.63	$\lvert -1.83 \rvert$
3	19.13	18.38	$\lvert -1.44 \rvert$	235	236.7	$\lvert 1.33 \rvert$
4	1.55	1.64	$\lvert 1.92 \rvert$	111	114.27	$\lvert 1.74 \rvert$
5	16.77	17.35	$\lvert 1.25 \rvert$	114	109.11	$\lvert -1.85 \rvert$
6	4.15	4.33	$\lvert 0.94 \rvert$	325	327.08	$\lvert 0.97 \rvert$
Linear voltage U_{CA}						
Harmonic	Amplitude, V			Phase, degrees		
	Average a_0	Average $\bar{y}$	$\lvert U_{observ} \rvert$	Average a_0	Average $\bar{y}$	$\lvert U_{observ} \rvert$
1	530.41	531.85	$\lvert 1.90 \rvert$	–	–	–
2	3.71	4.07	$\lvert 1.78 \rvert$	94	91.23	$\lvert -1.55 \rvert$
3	18.27	19.09	$\lvert 1.69 \rvert$	182	182.31	$\lvert 0.19 \rvert$
4	1.50	1.57	$\lvert 1.63 \rvert$	83	85.42	$\lvert 1.77 \rvert$
5	16.01	16.56	$\lvert 1.08 \rvert$	165	169.22	$\lvert 1.71 \rvert$
6	3.82	3.57	$\lvert -1.88 \rvert$	310	315.34	$\lvert 1.89 \rvert$

Zero hypothesis checking in terms of α $H_0 : D_x = D_0$ significance level concerning equality between unknown overall dispersion D_x with the known dispersion D_0 and hypothetic value D_0 in terms of a competing hypothesis $H_1 : D_x \neq D_0$, has been performed basing upon the criterion value [11, 12].

$$\chi^2_{observ} = \frac{(n-1)D_x}{D_0}$$

Zero hypothesis is accepted if $\chi^2_{l.cr.(1-\alpha/2;k)} < \chi^2_{observ} < \chi_{r.cr.(\alpha/2;k)}$ inequality is met. In this context, $k = n - 1$ is the number of degrees of freedom; $\chi^2_{l.cr.(1-\alpha/2;k)}$ and $\chi^2_{r.cr.(\alpha/2;k)}$ are left and right critical points determined according to Laplace function table. In the context of $n = 30$ and $\alpha = 0,05$, $\chi^2_{l.cr.(1-\alpha/2;k)} = 16$ and $\chi^2_{r.cr.(\alpha/2;k)} = 42.6$

Table 2 demonstrates checking results of dispersions of random consequences of harmonics of linear voltages generated with the help of digital generators.

Table 2. Control results of dispersions of harmonics of linear voltages

Linear voltage U_{AB}						
Harmonic	Amplitude, V			Phase, degrees		
	Dispersion D_0	Dispersion D_x	χ^2_{observ}	Dispersion D_0	Dispersion D_x	χ^2_{observ}
1	19.11	21.19	32.16	–	–	–
2	1.42	1.24	25.34	112	70.06	18.14
3	9.35	6.41	19.87	68	62.16	26.51
4	0.06	0.04	21.19	85	68.06	23.22
5	8.29	11.35	39.72	214	276.06	37.41
6	0.27	0.31	33.07	152	180.51	34.44
Linear voltage U_{BC}						
Harmonic	Amplitude, V			Phase, degrees		
	Dispersion D_0	Dispersion D_x	χ^2_{observ}	Dispersion D_0	Dispersion D_x	χ^2_{observ}
1	17.36	11.87	19.83	–	–	–
2	1.56	1.02	18.92	102	96.20	27.35
3	8.19	8.83	31.27	49	40.82	24.16
4	0.06	0.05	22.88	106	137.99	37.75
5	6.44	5.59	25.17	210	191.97	26.51
6	1.11	0.82	21.45	138	144.00	30.47
Linear voltage U_{CA}						
Harmonic	Amplitude, V			Phase, degrees		
	Dispersion D_0	Dispersion D_x	χ^2_{observ}	Dispersion D_0	Dispersion D_x	χ^2_{observ}
1	17.28	1.59	21.13	–	–	–
2	1.25	1.75	40.52	96	87.66	26.48
3	7.14	9.38	38.10	78	97.23	36.15
4	0.06	0.04	17.06	56	66.86	34.62
5	7.66	6.59	24.93	183	250.90	39.76
6	0.53	0.36	19.67	240	338.90	40.95

Experimental validation is the most reliable method to confirm adequacy of any mathematical model. The rolling shop No. 1 of Dneprospetsstal LLC was selected as the experimental one; the rolling shop contains powerful semiconductor converter which operation is accompanied by distortions in the workshop power grid (asymmetry and nonsinusoidality). During the experiment, oscillograms of currents used by IM of 7.5 kW power have been obtained. In the process of the experiment, there was an access to a zero point of the motor; thus, oscillograms of phase currents and voltages were taken. Measuring of active resistances of windings has shown their symmetry and correspondence to the certified values. IM shaft load was of random character changing within a wide range from 2.3 up to 12.8 kW.

Figure 4 demonstrates a window of CED Expert software in the process of oscillographic testing of signals during operation of tested electric motor under loading.

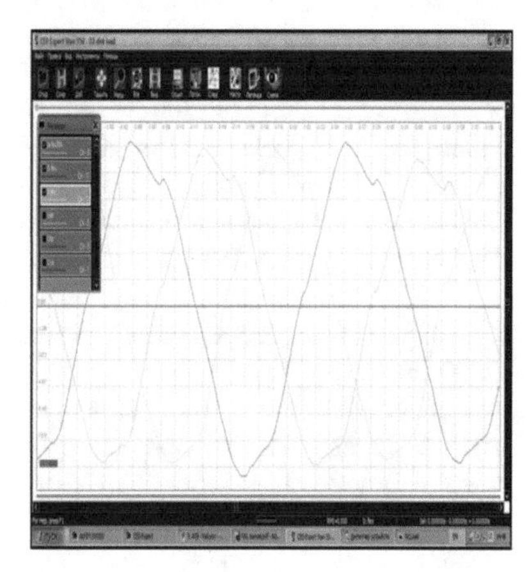

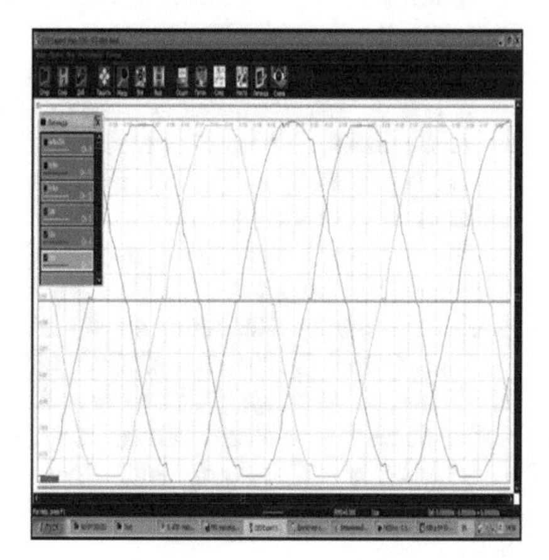

Fig. 4. Oscillograms of currents (at the left) and voltages (at the right) within the considered electric motor while operating under loading

4 Discussion and Results

Analysis of complex processes in terms of computer-based experiments involves errors resulting from the fact that a discrete function is represented as a set of its values in the context of different groups of arguments; if numbers, obtaining from the calculations, are rounded; and if decimal numbers as well as binary numbers are converted into floating-point numbers. Such errors may give rise to absolutely unexpected results.

Complementing of power-economic model of asynchronous motor with the control system of static characteristics of linear voltages helps regulate correctness of the random sequences being generated during computational experiments to select cost-effective alternative for recovering quality of electric power being supplied to the electric energy motor. The control systems have been synthesized with the use of adaptation concept based upon the mathematical expressions obtained in the process of the analysis. The paper represents estimation results as for the control of averages and amplitude dispersions, and phases of six harmonics of linear voltages obtained during the computer-based modeling. Estimations of the averages and dispersions of the generated random sequences have been verified as for the value of their differences from the corresponding hypothetic averages and dispersions.

References

1. Sayenko, D.: Analytical methods for determination of the factual contributions impact of the objects connected to power system on the distortion of symmetry and sinusoidal waveform of voltages. Przegląd Elektrotechniczny **91**, 81–85 (2015)
2. Zhezhelenko, I.V.: Selected problems of nonsinusoidal modes within power grids of enterprises. Energoatomizdat, Zhezhelenko, I.V., Saenko, Yu.L., Baranenko, T.K., Gorpinich, A.V., Nesterovich, V.V, (eds.) p. 294. Energoatomizdat (2007)
3. Pedra, J.: Estimation of typical squirrel-cage induction motor parameters for dynamic performance simulation. IEEE Proc. Gener. Transm. Distrib. **153**(2), 197 (2006). https://doi.org/10.1049/ip-gtd:20045209
4. Seritan, G.C., Cepişcã, C., Guerin, P.: The analysis of separating harmonics from supplier and consumer. Electrotehnicã Electronicã Automaticã (EEA) **55**(1), 14–18 (2007)
5. Yevheniia, K., Vitaliy, K., Mykola, T., Alisa, K., Maksym, T., Mykola, B.: Development and verification of dynamic electromagnetic model of asynchronous motor operating in terms of poor-quality electric power. In: 2019 IEEE International Conference on Modern Electrical and Energy Systems (MEES), Kremenchuk, Ukraine, pp. 350–353 (2019). https://doi.org/10.1109/mees.2019.8896598
6. Electric energy. Requirements for electric energy quality within general-purpose power grids: GOST 13109–97. – IPK. Publication House of Standards, p. 15 (1998)
7. Vitaliy, K., Nikolay, T., Yevheniia, K.: Evaluating the effect of electric power quality upon the efficiency of electric power consumption. In: 2019 IEEE 2nd Ukraine Conference on Electrical and Computer Engineering, Lviv, Ukraine, 2–6 July 2019, pp. 556–561 (2019). https://doi.org/10.1109/ukrcon.2019.8879841
8. Romashykhin, I., Rudenko, N., Kuznetsov, V.: The possibilities of the energy method for identifying the parameters of induction motor. In: 2017 International Conference on Modern Electrical and Energy Systems (MEES), Kremenchuk, pp. 128–131 (2017). https://doi.org/10.1109/mees.2017.8248869
9. Kachan, Yu.G., Nikolenko, A.V., Kuznetsov, V.V.: Identifying Parameters and Testing Adequacy of Thermal Model of Asynchronous Motor Operating Under the Conditions of Noisy Electric Energy, Kremenchuk, pp. 56–59 (2011). (Messenger of Kremenchuk State Polytechnical University: #1)
10. Kachan, Y., Nikolenko, A.V.: On the voltage modeling within power grids of industrial enterprises. Electrotech. Power Ind. **1**, 72–75 (2012)
11. Gmurman, V.: Theory of Probability and Mathematical Statistics. Vysshaya shkola (2003)
12. Kommaev, V., Kalinina, V.: Theory of Probability and Mathematical Statistics. INFA-M (2001)

Research and Application of Storage Location Optimization in Warehousing Center of TPL Enterprises

Yong Wang[1,2,4], Pei-lin Zhang[1], Qian Lu[3(✉)], Daniel Tesfamariam Semere[2], and Gang Li[4]

[1] School of Transportation, Wuhan University of Technology, Wuhan, Hubei, China
[2] Department of Production Engineering, KTH Royal Institute of Technology, Stockholm, Sweden
[3] Department of Planning, Puren Hospital, Wuhan, Hubei, China
myqian0412@163.com
[4] School of Logistics, Wuhan Technology and Business University, Wuhan, Hubei, China

Abstract. As the demand for logistics industry becomes more and more strict, various TPL companies have made corresponding changes to their own infrastructure, management style and future plans. For the warehousing center, the change is inevitable. The unchanging management mode and the stagnant management thought cannot meet the requirements of modern logistics. In the distribution industry, the optimization of the position of the goods gives us a chance to change, which can make us improve the deficiencies and bring benefits to the company while seeking new development direction. This paper is to question the existence of the warehousing center of Huangpi District, and optimize it. Showing the basic situation, and then to the shortage of the goods warehousing center put forward. Suggestions for improvement, the optimization of goods make corresponding adjustment, to make optimization to achieve the desired effect, finally make a summary of warehousing center position optimization.

Keywords: Storage location optimization · Location type · Warehousing center · SKU

1 Introduction

The Third-party Logistics (TPL) refers to the concentration of risks in production and operation companies to do good jobs in main business. The production and operation enterprise entrusts the logistics activities originally handled by itself to a professional logistics service company in the form of contract. This approach achieves the entire logistics management and contract logistics.

With the rapid development of modern logistics industry, warehousing logistics is booming with the development of logistics industry. Warehousing is a series of behaviors that store items in a specific place such as a warehouse and wait for use. These specific sites include houses, certain large containers, and large buildings. In the

Z. Hu et al. (Eds.): ICCSEEA 2020, AISC 1247, pp. 153–166, 2021.
https://doi.org/10.1007/978-3-030-55506-1_14

past, storage is mainly used to store products that are not used in time. These products need static storage and are stagnant in social production. The dynamic storage is to sort out the stored products, do a good job of classification, regular inspection, processing and circulation, successfully complete the production, and put the products into the market. Both static and dynamic storage are required Good use of storage skills to manage. In the warehouse management system, how to use the locator to store the goods reasonably and skillfully is the important content that we need to further study and discuss. And the storage location optimization can help us solve the problem.

The significance of storage location optimization mainly includes the following aspects:

(1) It solves the problem of excessive resource waste. Picking times are frequent, and the number of picking lots is set in the center area. It effectively improves the efficiency of picking.
(2) It allocates the inventory capacity reasonably, and distributes the more numerous items evenly, thus avoiding crowding during picking.
(3) It is convenient for high fork drivers to put on the shelves. It optimizes the path, develops standard drag regulations, and stores the area. It doesn't require too many complicated operations, in one step.
(4) Decrease in the proportion of picking errors. Separation of similar goods or confusing goods increases inventory accuracy.
(5) It specializes in warehouse standards. Through the storage location optimization, various operations in the warehouse proceeded smoothly, which improved the overall operation level.

The paper summarized the shortcomings of the company B warehousing center in terms of storage location management, and proposed corresponding improvement measures. The optimization case in the paper is the author's daily order production and arrival volume through the warehouse. Analyze the daily activities of warehouse goods, and develop appropriate storage locations for goods to achieve the purpose of location optimization.

2 Literature Review

Analysis of the development of China's warehousing logistics in recent years [1], we will find that the development of warehousing logistics is becoming more and more rapid. According to the development trend of modern logistics [2], warehousing logistics occupies a high position. The development of warehousing logistics has attracted people's attention to warehousing management, and the requirements for warehousing management have also increased. In order to strengthen the management of warehousing, it is necessary to optimize the storage location. Storage location optimization is an extremely important part of warehousing management [3].

Manufacturing companies are different from other industries [4]. Relatively speaking, location optimization is the core link of supply chain management. The optimization of the storage location [5] is conducive to the further development of the warehousing industry. Science is developing, and the use of high-tech equipment in the warehousing industry has made the storage location optimization management more and more modern [6].

Different from the past, the new location management storage combination [7], the storage utilization is greatly improved. Some large-scale production enterprises and some large-scale circulation enterprises in special industries have increased their investment in storage facilities.

In recent years, a large number of three-dimensional warehouses have been built [8]. Even the three-dimensional library with a high degree of automation [9], in the organization system, has set up an independent warehousing or logistics company to serve the society while undertaking its own warehousing [10] and transportation business. Automated warehouse storage location optimization not only increases warehouse utilization, staff efficiency, production order efficiency, etc. [11].Small and medium-sized manufacturing companies [12] often choose self-operated warehouses rather than public warehouses.

At present, most of the self-operated warehouses of small and medium-sized manufacturing enterprises use a layout design that mixes the sorting and stocking and storage places. That is, the existing storage area is utilized, and the stacking height is optimized relative to the size of the outbound station when necessary [13] to improve efficiency.

3 Methodology and Proposed Criteria

Storage location optimization is to facilitate the smooth progress of various activities within the warehouse to ensure the normal production of daily orders.

So we are going to pay attention to the following points when optimizing the location [14]:

3.1 Ensuring Safe Selection of Commodities

① Products that are afraid of moisture, mildew, and rust, such as cloth shoes, cotton cloth, tea, cigarettes, hardware goods, etc., should choose dry or sealed goods.
② Products that are afraid of light, heat, and soluble, such as rubber products, colored paper, grease, ink, candy, etc., should choose a low-temperature cargo space.
③ Products that are afraid of freezing, such as bottled ink, western medicine preparations, certain cosmetics and other juice products, should choose a position of no less than 0°.
④ Dangerous goods such as flammable, explosive, toxic, corrosive, radioactive, such as alcohol, benzene, resin glue, sulfuric acid, hair-printing paper, camphor, and matches, should be stored in suburban warehouses.
⑤ Commodities that are prone to physical or chemical reactions cannot be stored in the same area. Such as daily soap and paper, hardware and knitwear, can not be stored together due to performance. Tea, cigarettes, bakelite products, grease cosmetics and other commodities have different degrees of volatility and odor, and must be reserved for special storage.

⑥ Different items of firefighting methods should be stored separately.
⑦ Among the commodities stored in the same area, the possibility of infestation should be considered.

3.2 Convenient Storage and Delivery Principles

① Delivery method. Goods that adopt the delivery method shall be close to the site of tally and loading because of the need for the operation of the tally, the delivery of goods by truck, the delivery of goods, etc. For goods that adopt the pick-up method, the storage location should be close to the warehouse exit, which is convenient for vehicles coming in and out.
② Operating conditions. All kinds of goods need different tools to operate in and out of the warehouse. Therefore, it is necessary to consider the condition of loading and unloading equipment in storage area and the operation method of warehousing goods.
③ The speed of goods entering and leaving the warehouse. For fast-forward and fast-moving goods, choose a convenient location that is conducive to the entry and exit of the vehicle. For goods that are slow to store for a long time, the cargo space should not be close to storage door. For the whole and zero out of the goods, consider the conditions for sporadic delivery. For goods that are zero-in and out, consider the ability to ship in a centralized manner.

3.3 Striving to Save the Choice of Warehouse Capacity

The optimization of the storage location is also in line with the choice of saving. Store the maximum amount of merchandise in the smallest storage capacity [15].

When the storage load and height are basically fixed, the volume and weight of the goods should be separated from each other, so that the weight and volume of the storage location and goods are closely combined. The goods are placed in heavy storage location and lightly placed in small storage locations.

In addition, in the optimization of the storage location and the specific use, it can also be based on the fact that the warehouse goods have different speeds of throughput, and the characteristics of the operation are difficult to match, and the products with hot sale and long storage, difficult operation and labor saving are matched, which are stored in the same storage location.

In doing so, not only can the functions of the warehouse use be fully utilized, but also the phenomenon of unevenness between the various storage areas can be maintained.

4 Numerical Experiments

4.1 Current Situation Analysis of Warehousing Center of Company B

Huangpu Warehousing Center of Company B is located at the intersection of Industrial Park 2 Road and Tengfei East Road, Huangpi District, Wuhan. The warehouse center covers an area of 300 mu, with a total construction area of 200,000 m^2. It has 30 loading and unloading ports, and the inventory amount can reach 30 million Yuan. In addition to meeting the daily distribution of more than 10 supermarket stores of Company B, it can meet the distribution of 1000 varieties for the society. The warehousing center is mainly engaged in small household appliances, and is also engaged in the wholesale and retail of other commodities. The company plans to build a large-scale modern logistics center integrating industrial products wholesale, commodity distribution, and warehousing center.

The daily production orders of Company B's warehouse center are about 30,000 orders, and the average daily arrival volume is 40,000 pieces. Among them, warehouse products include small appliances and beautiful makeup. The daily business involves the allocation, distribution, and robotic arm. The transfer business has developed rapidly, involving Nanchang, Xinxiang, Changsha and Zhengzhou. The robot arm orders 3,000 pieces daily, with a maximum record of 50,000 pieces per day. It is a new development business in the warehouse center. The whole warehouse is divided into five warehouses, each of which covers an area of about 20,000 square meters, and is produced in a single warehouse, each of which completes daily orders.

4.2 Solution to the Problem of Storage Position Management

(1) Develop uniform pallet specifications

The specifications of the storage trays of the goods are inconsistent, and various pallet specifications appear in the same product.

In response to this situation, the first step in the receipt of goods should be strictly controlled. For example, the standard of pallet specifications is uniform, and the receiving personnel are obliged to require the loading and unloading workers to place them according to the standard supporting regulations when placing the goods.

For different batches of goods, if the volume is different, take the volume as a reference, and formulate a pallet specification standard. When entering the cargo space, separate the storage and make a warning sign.

(2) Mark the internal and external sales barcode consistent goods

At the time of receipt, the warehouse will label the goods with the same barcode inside and outside the box, and set up a special partition to store such goods. The storehouse specially arranges personnel to register the goods with the same barcode inside and outside the box, and make the mark The warehouse and the staff are organized to unpack. The storehouses set up a special area for such goods, and strengthen the promotion work for employees, with special emphasis.

(3) Set the storage location according to the actual sales volume of the product.

The warehouse management personnel should regularly analyze the outbound quantity of the goods on a weekly or monthly basis, and compare the changes in the outbound quantity of each time period to formulate the storage plan for the goods.

For example, the location of the best-selling product is set closer to the production area, and the location of the unsalable goods is slightly farther from the production area.

(4) Place a commodity in a storage position

The storehouse arranges special personnel to check the goods in the warehouse every day. Once a phenomenon of multiple goods is found, it must be dealt with in time. The professional placement of a product in one location can not only reduce the workload of the pickers, but also improve the overall operational level of the warehouse.

(5) Main product and gift are stored in the same area.

The main products and gifts belong to the same business. When the same merchant arrives, they can choose to store the goods in the same area as the merchant type, and arrange the main products and gifts as far as possible in the adjacent positions. This storage avoids excessive waste of labor resources and reduces the workload for the picking staff. When the commodities are on the shelves, they don't have to think too much about how to arrange the goods.

(6) Put the product code of the goods in accordance with the safety regulations

Every employee should pay more attention to the illegal operation in the warehouse. Employees should be brave enough to suggest improvements to their superiors. The head of the warehouse organizes the super high cargo pile. Any person should wear a helmet when picking the goods in the store area to prevent unnecessary danger [16].

4.3 Research on the Optimization Method of Storage Location in Company B

Optimization ideas and goals

1. Optimization ideas

   (1) Collect the basic information needed
   Product information, product demand seasons, product requirements, and product documentation can all be accessed through WMS (Warehouse Management System) or ERP (Enterprise Resource Planning). The liquidity of goods is expressed in: sales volume, sales forecast, and inventory of each category of goods. When the sales volume and inventory of goods change, we have to make corresponding adjustments. Analysis of the various parts of the entire warehouse, such as historical data, turnover rate, etc., allows us to explore the law and arrange the goods reasonably.
   (2) Optimal determination
   We analyze the collected data, analyze the characteristics of the goods in the warehouse, and consider the influence of various factors. Regularly adjust the cargo space to test the impact of the cargo position on the overall production operation. We finally determine an optimal storage location principle.

(3) Software implementation
Through the computer software to collect information related to the location optimization, and learn from the experience of the warehouse to optimize the successful warehouse. Use the computer to analyze the daily order volume in the warehouse and use data to convey the optimization effect.

2. Optimization goals

We collect data, analyze data, and adjust the location. We optimize the location, and ultimately we want to reduce the production cost and bring benefits to the company. The optimization of storage location is an ongoing process. We need to constantly adjust the storage location, and regularly analyze the data of picking efficiency, quantity of stocks, and quantity of shelves. We minimize the amount of labor in each part and come up with a relatively optimal strategy. The goods on the goods will also change according to the season or some uncertain factors. We need to make appropriate adjustments in time to optimize the location to the best effect. The optimization of storage location should proceed from the whole. The optimized storage location should improve the efficiency of picking, reduce costs and reduce labor [17].

4.4 Optimization Analysis of Storage Location

1. Optimization of storage location based on sales volume

Analyze the flow of goods into and out of the warehouse within one week, and register the best sellers (Take some commodities as an example) (Tables 1 and 2).

Table 1. Commodities flow into and out of the warehouse table

Name of owner	BL	DL	FL	JY	HY	SB	WY	QY	Total amount March 11–17
Total sales	38000	150000	2000	3600	200	47	5890	140000	204737
Total adjustment	300	4000	580	0	7600	50	0	12	12542
Total racks	0	0	40	600	0	0	0	100	740
Total amount	38300	154000	2620	4200	7800	97	5890	140112	353019

Sort the total amount of goods and analyze the trend of the products within one week (The total amount is from high to low)

Table 2. Sorting table of total commodities

Name of owner	DL	BL	BL	HY	WY	JY	FL	SB
Total amount	140112	154000	38300	7800	5890	4200	2620	97
Ranking	1	2	3	4	5	6	7	8

Set the position of the merchandise placed according to the above total amount (Use rank instead of owner) (Fig. 1).

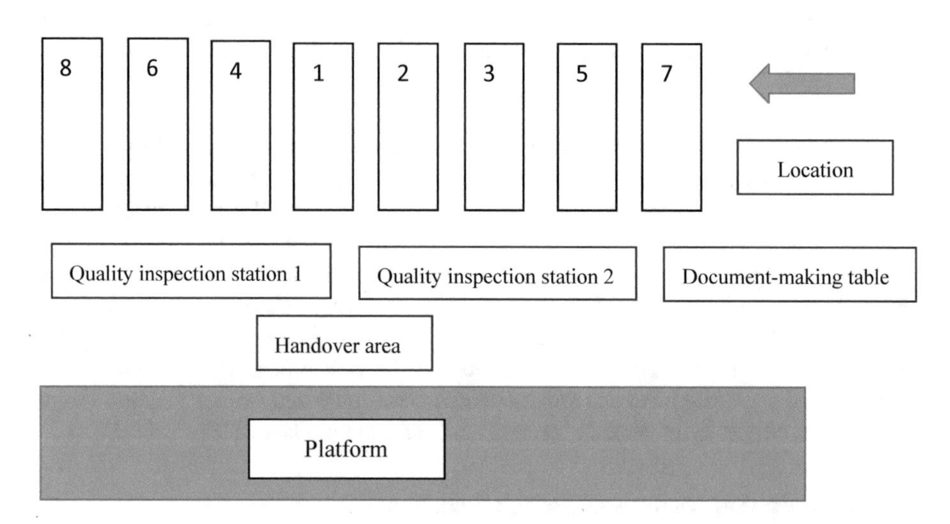

Fig. 1. Ranking of commodity locations

2. Optimized storage location based on the volume of goods

The goods inside the warehouse are both inbound and outbound. When the sales volume increases, the arrival volume will definitely increase year on year. The increase in arrival volume makes the optimization of the cargo space more important. If the arrival of the goods in the early period does not allocate the goods, it will lead to a series of problems in the later stage. Such as: unreasonable distribution of goods, so that large quantities of goods can be placed without goods; Slow-moving products occupy the golden area and increase the burden of picking [18].

Analyze the arrival quantity of goods within one week, register and rank the goods with large quantity of goods (Take some owners as an example) (Table 3).

Table 3. Commodities arrival ranking table

Name of owners	AL	RD	BT	RS	MX	WS	NU	AM
Arrival volume	13012	8400	5300	4600	3890	3600	620	10
Ranking	1	2	3	4	5	6	7	8

(2) Analysis of the arrival quantity of goods within one week, register and rank the goods with large quantities. (volume is measured in terms of the number of pallets) (Table 4).

Table 4. Commodities volume ranking table

Name of owners	BT	RD	MX	WS	RS	NU	AL	AM
Total amount	360	250	237	50	44	32	13	1
Ranking	1	2	3	4	5	6	7	8

Set the position where the product is placed according to the above total volume (Use rank instead of owner) (Figs. 2 and 3).

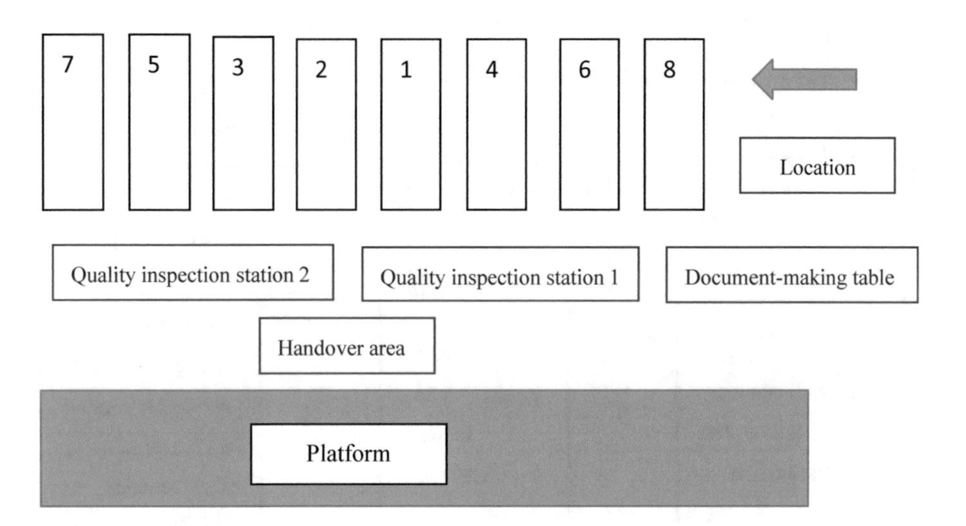

Fig. 2. Ranking of commodity locations

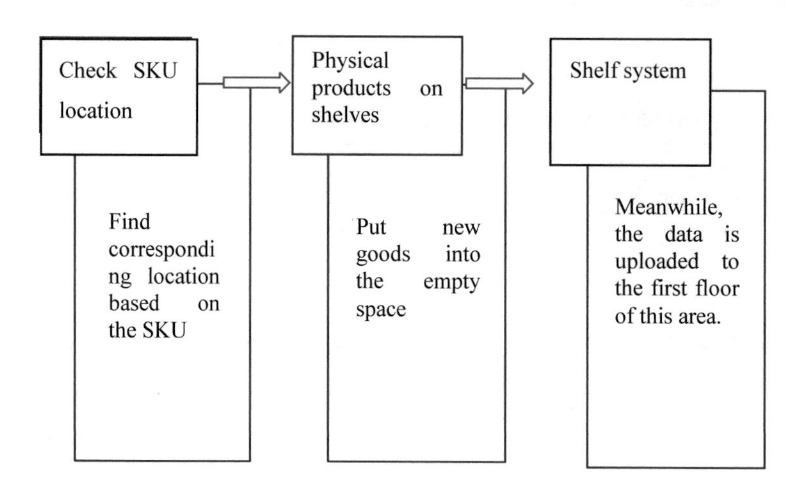

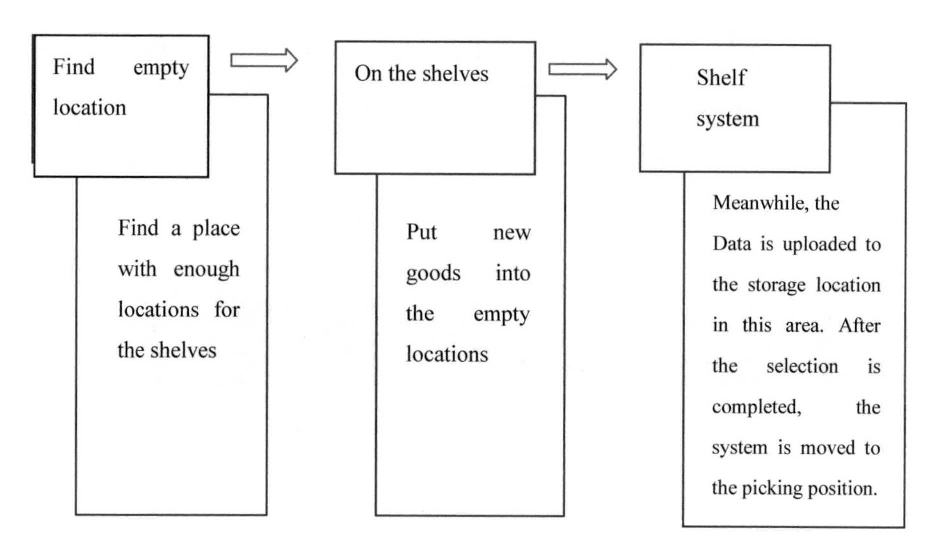

Fig. 3. Product shelf flow chart

When the number of pallets is more than 100: When the goods are removed from the shelves, the goods of the aisles are given priority. After the passage is emptied, the overhead goods are removed in turn (Fig. 4).

Fig. 4. Optimized position goods placement map (3 in total, more than 100)

When the number of pallets is 10–100: from the inside to the outside, from top to bottom, the products are out of stock by a vertical bar and a vertical bar. Ensure that the remaining items are concentrated in one area. Other items can be placed on the vacated whole area (Fig. 5).

Fig. 5. Optimized position goods placement map (10–100)

When the number of pallets is no more than 10: on the same side of the aisle, the same SKU, the products are in the stock by vertical bar and a vertical bar. And then the goods are adjusted according to the actual situation, and the logo is made (Fig. 6).

Fig. 6. Optimized position goods placement map (10–100)

5 Management Insights and Main Conclusion

With the rapid development of society, warehousing logistics is used in all walks of life. People are increasingly demanding the logistics industry. Only warehouse warehouses that respond quickly to market demands and meet customer requirements can be eliminated. Therefore, this requires modern warehousing logistics to keep up with the times and constantly improve its management level, professional skills and knowledge.

After the optimization of the storage locations, it can not only effectively reduce the production cost of the company, but also bring convenience to daily work. If need three or four people to complete the daily commodity inventory work, it is much easier and easier after the location optimization [20].

Optimization effect

1. Developed a standard shelf-up strategy, making it easier and more convenient to put on shelves and replenishing goods, greatly improving the space utilization of the storage location.
2. Store the goods with higher picking ratio in the vicinity of the production area, shorten the picking time, and improve the efficiency of employee picking.

3. According to the daily order production and sales volume in the warehouse, to reasonably allocate the storage locations and balance the cargo capacity of each district.
4. Reduce the burden on the staff when each large quantity arrives, and directly put the goods on the shivers in accordance with the optimization principle, saving time and effort.
5. Avoid unnecessary mis-sending commodity incidents and reduce losses for the company [19].

The location optimization scheme written in the paper is currently only suitable for warehouse use in the small electricity industry. Other types of warehouses may not be able to achieve cost savings for the company if they are set according to the optimization scheme of this paper.

We need to practice in different types of warehouses, and finally get a location optimization solution for all types of warehouses.

According to different requirements of warehousing center, the functions and applications of warehousing center are also different. However, the objective of the location optimization is not the same, and it can be divided into the following:

(1) Overall, the optimization of storage location can facilitate the smooth production of orders in the warehouse, greatly improving production efficiency and reducing production costs.
(2) Reasonable storage location optimization enables a series of "storage" and "take" actions in the warehouse operation to be carried out efficiently and simply.
(3) From the perspective of employees, storage location optimization not only brings benefits to the company, but also reduces the burden on the daily production of employees.

Acknowledgment. This project is supported by Key Projects of CAST (China Association of Science and Technology) Project (2018CASTQNJL33); Fundamental Research Funds for the Central Universities (2019-JL-008); MOE Project of Humanities and Social Sciences 14YJCZH154); WTBU Academic Team (XSTD2015004).

References

1. Li, Y.: Optimization of storage management strategy for automated warehouses. SME Manag. Technol. (late issue) (04), 150–151 (2017)
2. Xin, F., Chenglin, W., Patil, A.: Study on the optimization of storage location in integrated storage and sorting system. Logist. Sci. Technol. **40**(02), 142–148 (2017)
3. He, C.: Research on cargo space optimization under the mode of goods-to-person operation. Zhejiang University of Science and Technology, pp. 22–25 (2017)
4. Anitha, P., Malini, M., Patil, A.: A review on data analytics for supply chain management: a case study, Int. J. Inf. Eng. Electron. Bus. (IJIEEB) **10**(5), 30–39 (2018)
5. Fu, X., Zhang, B., Wang, W.: Study on storage location allocation algorithm of multi-mode automatic access system. Ind. Instr. Autom. (03) 56–59 (2016)

6. Yang, J.: Research on cargo space allocation and picking path optimization in shelf warehousing center. Beijing Jiaotong University, pp. 25–26 (2016)
7. Teng, X., Teng, X.J.: Research on warehouse location optimization in logistics center. Mod. Bus. **32**, 14–15 (2016)
8. Junjun, H.: Research on optimization method of warehouse location. Jiangsu Bus. Theory (04), 31–32 (2016)
9. Ting, P.-H.: An efficient and guaranteed cold-chain logistics for temperature-sensitive foods: applications of RFID and sensor networks. IJIEEB **5**(6), 1–5 (2013)
10. Xueying, J.: Research on the optimization of spare parts storage position of DA company. Harbin University of Science and Technology, pp. 14–15 (2015)
11. Xie, J.: Research on optimization of warehousing center of D company. Shenzhen University, pp. 23–28 (2016)
12. Benotmane, Z., Belalem, G., Neki, A.: A cost measurement system of logistics process. Int. J. Inf. Eng. Electron. Bus. (IJIEEB) **10**(5), 23–29 (2018)
13. Yong, W., Deng, X.: Empirical study on performance evaluation of agricultural product supply chain based on factor analysis. China Bus. Mark. **3**, 10–16 (2015)
14. Dogan, I.: Analysis of facility location model using Bayesian networks. Expert Syst. Appl. Int. J. **39**, 1092–1104 (2012)
15. Wang, Y.: Study on location model of cold chain logistics distribution center, pp. 11–25. Changsha University of Science and Technology, Changsha (2008)
16. Zhao, P., Liu, B., Xu, L., Wan, D.: Location optimization of multidistribution centers based on low-carbon constraints. Discrete Dyn. Nat. Soc. **2013**, 1–6 (2013). Article ID 427691
17. Jing, W., Zhongqin, M.: Selection of multi-distribution center location based on low carbon. Rev. Fac. Ingeniería U.C.V. **31**(7), 11–22 (2016)
18. Li, Ye, Liu, X., Chen, Y.: Selection of logistics center location using axiomatic fuzzy set and TOPSIS methodology in logistics management. Expert Syst. Appl. **38**(6), 7901–7908 (2011)
19. Wang, Y., Zhang, P., Lu, Q., Semere, D.T., Du, W.: Supplier measurement of fresh supply chain in sustainable environment. EKOLOJI **28**(107), 1995–2004 (2019)
20. Badri, M.A., Davis, D.L., Davis, D.: Decision support models for the location of firms in industrial sites. Int. J. Oper. Prod. Manag. **15**(1), 50–62 (1995)

Long-Term Operational Planning of a Small-Series Production Under Uncertainty (Theory and Practice)

Alexander Anatolievich Pavlov(✉)

National Technical University of Ukraine "Igor Sikorsky Kyiv Polytechnic Institute", 37, Prospekt Peremohy, 03056 Kyiv, Ukraine
pavlov.fiot@gmail.com

Abstract. We propose an efficient original solution to the problem of long-term operational planning under uncertainty conditions for arbitrary objects with a network representation of technological processes. We suppose that the coefficients of the functional that determine the plan's quality may change before its implementation. Therefore we should take into account this instability factor during planning. We show the significance of the general theory for a wide class of problems by the example of small-series production. The solution presented can be used in the framework of enterprise resource planning systems, logistics systems, supply chain management systems, etc. Three factors determine the efficiency of obtained solutions: (a) using the original optimization theory developed by the author for one class of combinatorial problems under uncertainty; (b) the efficiency of previously developed algorithm for the problem solving in a deterministic formulation; (c) using efficient optimization models to find stable values of expert weights by the empirical matrix of pairwise comparisons which can contain zero elements.

Keywords: Uncertainty · Combinatorial optimization · Multi-stage scheduling problem · Small-Series production · Compromise solution

1 Introduction

The author in papers [1, 2] outlined the basics of original general theory of solving for a whole class of combinatorial optimization problems under uncertainty. The class of problems is wide enough and was distinguished for the first time. The uncertainty considered relates to the weight coefficients of the functional which is a linear convolution of weight coefficients and arbitrary numerical characteristics of a feasible solution. There must exist an individual efficient algorithm to solve each of the problems in a deterministic formulation. The class of problems under consideration includes, in particular, NP-hard combinatorial optimization problems by criteria of the total weighted tardiness of jobs minimization on one machine, the total weighted completion time of jobs minimization on one machine, transportation problem, network flow problems, etc.

The purpose of this paper is to improve methods of optimization under uncertainty, to substantiate the efficiency and practical significance of the created theory by the

Z. Hu et al. (Eds.): ICCSEEA 2020, AISC 1247, pp. 167–180, 2021.
https://doi.org/10.1007/978-3-030-55506-1_15

example of long-term planning in a small-series production with a network representation of technological process in uncertainty conditions. Suppose that we have an operational long-term plan and a functional that determines its quality. Under the uncertainty we mean the possibility of the weight coefficients of the functional to change their values before the plan's implementation. The problem is to find a compromise plan that takes such uncertainty into account in the best way according to one of the introduced criteria [1, 2].

The problem is solved for the first time in such formulation. In this paper, the author uses his theory to originally modify the four-level model of planning (including operational planning) and decision making introduced in [3, 4]. Thus, the previously obtained results are combined, and for the first time we propose an original solution to the operational planning problem under uncertainty by the example of small-series production. This research contributes to intractable combinatorial problems solving under the conditions of uncertainty of the weight coefficients of the functional. Also, the study serves for efficient finding of optimal compromise solutions for various sets of weight coefficients. This helps to improve the quality of operational planning and, accordingly, affects positively the output quality indicators of the operational plan.

Optimal plans building for discrete objects of various nature has been the subject of many recent publications. In particular, the papers related to scheduling theory problems [5–13], modeling of planning and control objects [14–16], and planning under uncertainty [17, 18]. One of the conclusions formulated in [18] is that a large portion of the existing research studies the deterministic state of production planning and ignores its inherent uncertain nature. This may lead to considerable errors and imprecise decisions in practice. The problem of planning formalization may be associated with managing huge financial flows. Therefore, ignoring uncertainty may lead to large financial losses. We performed a comprehensive literature analysis by keywords "linear convolution" and "combinatorial optimization under uncertainty." We have found many publications, e.g., [19–22], with the use of linear convolution but all of them were devoted to multi-criteria problems solving in different areas of economics and technology by reducing them to single-criterion problems. We did not find a use of linear convolution of weight coefficients and arbitrary numerical characteristics of a feasible solution for effective solution under uncertainty of problems from the class that we distinguished. For example, the method of a compromise solution finding to the two-criterion linear programming problem [23] uses the LP-metrics model. However, our approach is fundamentally different already in that we need to solve only one, not three optimization problems to find a compromise solution. The problem we formulate below is also considered for the first time. In the works we considered, e.g., [17, 24, 25], there was no definition of a sufficiently wide class of combinatorial optimization problems that we have singled out below. We have not found papers dealing with this type of uncertainty. Complex and time-consuming methods of stochastic and robust optimization were developed to resolve various forms of uncertainty. However, the critical importance of taking into account dynamism and uncertainty (in particular, during planning and scheduling in practice) was noted in the majority of publications we reviewed. The need to take this uncertainty into consideration in decision-making models is undoubted. Thus, the proposed approach to solving the problem formulated below, which is often encountered in practice, has not been considered in the literature and is being implemented for the first time. The results stated below are based on the original properties of linear convolution obtained by the author. The properties are different from everything we saw in other papers.

The rest of this paper is organized as follows. Section 2 formulates the class of combinatorial optimization problems we study, the optimality criteria, the fundamentals of the general theory and methodology for a solution obtaining. In Sect. 3, we formulate the operational planning problem which is a multi-stage scheduling problem (MSSP) of a general form. We provide the basics of a suboptimal algorithm for its solving. Section 4 formulates and solves the problem of finding a compromise operational plan in the conditions of uncertainty. We give the methodology of applying the general theory to solve specific problems on the example of small-series production. Section 5 discusses the results obtained.

2 General Theoretical Provisions

The main provisions of the theory described in [1, 2] are as follows. We study the class of combinatorial optimization problems of the following form [1]:

$$\min_{\sigma\in\Omega}\sum_{i=1}^{s}\omega_i k_i(\sigma), \tag{1}$$

where ω_i are numbers, $k_i(\sigma)$ is i-th arbitrary numerical characteristic of a feasible solution σ. Ω is the set of all feasible solutions.

There is an efficient solving algorithm for the problem (1) in such (deterministic) formulation. Any structural change in Ω (for example, adding a linear constraint) makes application of the algorithm impossible. Examples of the problems that conform the above class are the transportation problem, network flow problems, a range of NP-hard combinatorial problems [3, 4].

There are L sets of weights $\left\{\omega_i^l,\ i=\overline{1,s}\right\}$, $l=\overline{1,L}$. Each one may be a set of coefficients $\omega_1,\ldots,\omega_s$ of the problem (1) at the stage of fulfillment of its solution [1]. We can specify probabilities $P_l>0$, $l=\overline{1,L}$, $\sum_l P_l=1$, for each of the possible sets of weights (such probabilities do not exist if the uncertainty is not described by a probabilistic model). We need to find a feasible solution $\sigma\in\Omega$ that satisfies one of the following *conditions* [1]:

1. Let us denote:

$$f_{opt}^l=\min_{\sigma\in\Omega}\sum_{i=1}^{s}\omega_i^l k_i(\sigma),\ \{\sigma_l\}=\arg\min_{\sigma\in\Omega}\sum_{i=1}^{s}\omega_i^l k_i(\sigma),\ L_l=\sum_{\substack{m=1\\m\neq l}}^{L}\left(\sum_{i=1}^{s}\omega_i^m k_i(\sigma_l)-f_{opt}^m\right)$$

Remark 1. If $\{\sigma_l\}$ consists of more than one solution, we keep the one on which we have $\min_{\{\sigma_l\}} L_l$ and denote this solution by σ_l (we show in [1] how to obtain σ_l for the case when Ω is finite).

Suppose that $L_p = \min_l L_l$ (L_p corresponds to a solution σ_p). We need to find σ that reaches

$$\min_{\sigma\in\Omega} \sum_{l=1}^{L} \left(\sum_{i=1}^{s} \omega_i^l k_i(\sigma) - f_{opt}^l \right). \tag{2}$$

2. Find a feasible solution that satisfies the condition

$$\min_{\sigma\in\Omega} \sum_{l=1}^{L} a_l \left(\sum_{i=1}^{s} \omega_i^l k_i(\sigma) - f_{opt}^l \right) \tag{3}$$

where $\forall a_l > 0$ are the coefficients set by experts.

3. Let us introduce a random variable $F = \sum_{i=1}^{s} \bar{\omega}_i k_i(\sigma) - \bar{f}_{opt}$ where s + 1-dimensional discrete random variable $\bar{\omega}_1 \ldots, \bar{\omega}_s, \bar{f}_{opt}$ is specified by the table:

$$\left\{ \begin{array}{c} \omega_1^l, \ldots, \omega_s^l, f_{opt}^l \\ P_l > 0,\ l = \overline{1, L} \end{array} \right\}.$$

We need to find a solution to this problem:

$$\min_{\sigma\in\Omega} MF = \min_{\sigma\in\Omega} \sum_{l=1}^{L} P_l \left(\sum_{i=1}^{s} \omega_i^l k_i(\sigma) - f_{opt}^l \right). \tag{4}$$

4. Find a feasible solution $\sigma(\Delta_1, \ldots, \Delta_L)$ for which

$$\Delta_i \le l_i,\ l_i > 0,\ i = \overline{1, L}. \tag{5}$$

Here, $\sigma(\Delta_1, \ldots, \Delta_L)$ is a feasible solution $\sigma \in \Omega$ that has the specified in brackets deviations from the optimums for each set of weights: $\Delta_l = \sum_{i=1}^{s} \omega_i^l k_i(\sigma) - f_{opt}^l$, $l = \overline{1, L}$.

Finding a solution that satisfies one of the four given conditions follows from the following statement and its corollaries [1].

Statement 1 [1]. The following is true for arbitrary $a_l > 0$, $l = \overline{1, L}$:

$$\arg\min_{\sigma\in\Omega}\sum_{l=1}^{L}a_l\left[\sum_{i=1}^{s}\omega_i^l k_i(\sigma)-f_{opt}^l\right]=\arg\min_{\sigma\in\Omega}\sum_{i=1}^{s}\left(\sum_{l=1}^{L}a_l\omega_i^l\right)k_i(\sigma). \quad (6)$$

Corollary 1. Solving of the problem $\min\limits_{\sigma\in\Omega}\sum\limits_{l=1}^{L}a_l\left[\sum\limits_{i=1}^{s}\omega_i^l k_i(\sigma)-f_{opt}^l\right]$ reduces to solving of one problem of the form (7):

$$\min_{\sigma\in\Omega}\sum_{i=1}^{s}\left(\sum_{l=1}^{L}a_l\omega_i^l\right)k_i(\sigma). \quad (7)$$

Let $\sigma^2(a_1,\ldots,a_L)$ denote the set of solutions of the problem (7) at coefficients $a_1,\ldots,a_L$; $\sigma(a_1,\ldots,a_L,\Delta_1,\ldots,\Delta_L)$ a feasible solution $\sigma\in\sigma^2(a_1,\ldots,a_L)$ that for given coefficients $a_l>0,\ l=\overline{1,L}$ has the specified $\Delta_l=\sum\limits_{i=1}^{s}\omega_i^l k_i(\sigma)-f_{opt}^l,\ l=\overline{1,L}$.

Corollary 2. Suppose that a solution $\sigma(a_1,\ldots,a_L,\Delta_1,\ldots,\Delta_L)\neq\sigma_1\vee\ldots\vee\sigma_L$ belongs to the set $\sigma^2(a_1,\ldots,a_L)$. Then, $\nexists\sigma(\Delta_1',\ldots,\Delta_L')\in\Omega$ for which $\Delta_i'\leq\Delta_i,\ i=\overline{1,L},\ \Delta^{\mathrm{T}}=(\Delta_1,\ldots,\Delta_L)\neq\Delta'^{\mathrm{T}}=(\Delta_1',\ldots,\Delta_L')$.

Corollary 3. To obtain a solution according to condition 1 it is necessary to set $a_l=1,\ l=\overline{1,L}$ in the problem (1). In this case, the solution to problem (15) does not coincide with the solution σ_p corresponding to $\min\limits_{l}L_l$ if there is such a feasible solution $\sigma(\Delta_1,\ldots,\Delta_L)$, $\Delta_l=\sum\limits_{i=1}^{s}\omega_i^l k_i(\sigma)-f_{opt}^l$ for which

$$\sum_{l=1}^{L}\Delta_l<L_p. \quad (8)$$

Corollary 4. Suppose that $\sigma(a_1,\ldots,a_L,\Delta_1,\ldots,\Delta_L)\neq\sigma_1\vee\ldots\vee\sigma_L$ and $\exists\sigma(a_1,a_2,\ldots,a_{i-1},a_i',a_{i+1},\ldots,a_L,\Delta_1',\ldots,\Delta_L')\neq\sigma_1\vee\ldots\vee\sigma_L,\ \ a_i'\neq a_i$. Then it is true that

$$\begin{aligned}(a_i'-a_i)(\Delta_i'-\Delta_i)<0,\ (\Delta_i'-\Delta_i)\left[(a_1\Delta_1'+\ldots+a_{i-1}\Delta_{i-1}'+a_{i+1}\Delta_{i+1}'+a_L\Delta_L')\right.\\ \left.-\ (a_1\Delta_1+\ldots+a_{i-1}\Delta_{i-1}+a_{i+1}\Delta_{i+1}+a_L\Delta_L)\right]<0.\end{aligned} \quad (9)$$

Corollary 5. Suppose that $\sigma(1,\ldots,\ 1,\Delta_1,\ldots,\Delta_L)\neq\sigma_p$ and does not satisfy the condition 4. Then, by logic of the inequalities (9), we can organize a sequential procedure for the problem (7) solving by increasing $\forall a_i$ at each iteration if $\Delta_i>l_i$ and decreasing $\forall a_j$ if $\Delta_j<l_j$. As a result, we either find a solution σ satisfying the condition 4 or obtain a set of solutions $\{\sigma\}^1$ each of which violates the condition 4. It is true that $\sigma(1,\ldots,\ 1,\Delta_1,\ldots,\Delta_L)\in\{\sigma\}^1$. Denote $\{\sigma\}^2=\{\sigma_1,\ldots,\sigma_L\}$. Then a compromise solution for the condition 4 is a solution $\bar{\sigma}\in\{\sigma\}^1\cup\{\sigma\}^2$ that reaches

$$\min \sum_{t} C_{j_t} \left(\Delta_{j_t} - l_{j_t} \right), \ \forall t \ \Delta_{j_t} > l_{j_t}$$

where $C_j > 0$, $j = \overline{1, s}$, are expert coefficients.

Corollary 6. We can find a feasible solution σ that satisfies the condition 2 or 3 by solving one problem of the form (7).

3 Operational Plan Building for Small-Series Production in a Deterministic Formulation

Let us consider one of the planning problems in the most general formulation. Let the technological process be set in the network form, an example of the network is shown in Fig. 1.

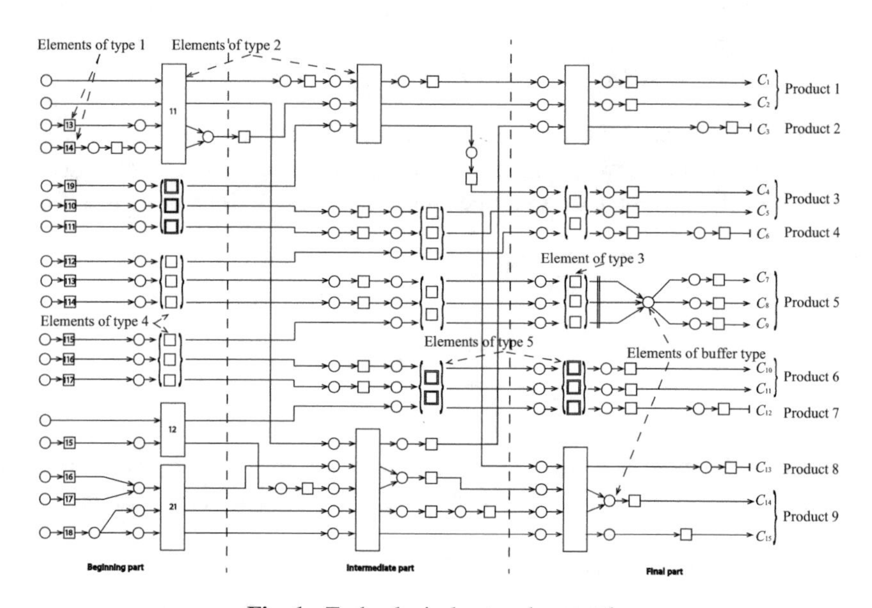

Fig. 1. Technological network example

A network is a combination of elements of type 1–5 [4, 26, 27]. It is an oriented acyclic graph similar to flow network with several sources and several sinks. Element of type 1 corresponds to a machine which processes only one job. Elements of type 2–5 correspond in the network to machines (machine sets) which must process a set of jobs in random order without interruption:

- element of type 2 is a single machine that processes a set of jobs sequentially;

- element of type 3 are identical parallel machines that process a set of jobs with a common due date in parallel;
- element of type 4 are identical parallel machines that process a set of jobs with different due dates in parallel;
- element of type 5 are unrelated parallel machines of various productivities that process a set of jobs with different due dates in parallel;
- element of buffer type is used to designate the (dis)assembly of several jobs.

At any time point, a single machine may process only one job. The network is an oriented graph with two types of vertices. A vertex of type 1 is represented by a circle ○, a vertex of type 2 is a square □ or a different machine notation. An arrow (arc) *entering* a circle indicates a completed job. An arrow *coming out* of a circle is a new uncompleted job ready for processing. A vertex of type 1 must be immediately followed by a vertex of type 2 which is a machine the job is processed on.

A detailed description of the network structure and algorithms for its construction and functioning are given in [4, 27].

The result of the operational plan's implementation is the fulfillment of n products (or n products series). Let C_i be the completion time of i-th product (i-th product series). We need to build an operational plan optimizing one of the following criteria:

The total profit maximization:

$$\max \sum_{i=1}^{n} \omega_i(T) \cdot (T - C_i) = \min \sum_{i=1}^{n} \omega_i(T) \cdot C_i \tag{10}$$

where T is the plan period, $\omega_i(T) = \text{const} > 0$ are the expert coefficients;

$$\max \left\{ \sum_{i=1}^{n} \omega_i U_i \right\},\ U_i = \begin{cases} 1, C_i = d_i \\ 0, C_i \neq d_i \end{cases}, \tag{11}$$

where d_i is the due date of i-th product (i-th product series);

$$\min \left\{ \sum_{i=1}^{n} \omega_i \max(0,\ C_i - d_i) \right\}; \tag{12}$$

$$\max \left\{ \sum_{i=1}^{n} \omega_i U_i \right\},\ U_i = \begin{cases} 1, C_i \leq d_i \\ 0, C_i > d \end{cases}; \tag{13}$$

$$\min \sum_{i=1}^{n} \omega_i |C_i - d_i|. \tag{14}$$

The formulated problem is a MSSP of a general form. An efficient approximation algorithm for its solving was described in [4, 27]. The methodology of this algorithm's construction is as follows.

Based on the formalized network representation of a technological process [4, 27], we implement procedures of its two-level aggregation: combining jobs into aggregated units (the first level of aggregation) and combining critical paths of products with intersection on their common vertices of the precedence graph representing the set of resources as a single machine (the second level of aggregation). The formal model of the second level of aggregation is a single-stage scheduling problem: one machine, n jobs, the partial order of their processing is given in the form of a directed acyclic graph, the optimality criterion is one of (10)–(14). Considering criterion (10), we have the NP-hard problem of minimizing the total weighted completion time of aggregated jobs with the partial order specified by a directed acyclic graph (TWCT). We proposed an efficient PSC-algorithm to solve TWCT problem in [28].

We reasonably replaced problems with criteria (11)–(14) by an approximating TWCT problem with the criterion of $\min \sum_{i=1}^{n} \omega_i C_i$ (here, ω_i, $i = \overline{1, n}$, equal to the weights of criteria (11)–(14)). A simplified justification for such approximation is the following. In the optimal solution of TWCT problem, taking into account the restrictions imposed on the processing order by a directed acyclic graph constructed on the critical paths of products, the optimal solution to the approximating TWCT problem is a priority-ordered schedule in which the product sequence is split into maximum priority sets G_i, $i = \overline{1, k}$, $p(G_1) \geq p(G_2) \geq \ldots \geq p(G_k)$. $p(G_1) = \sum_{i \in G_1} \omega_i \Big/ \sum_{i \in G_1} l_i$ where ω_i is the profit from the sale of the product i (product series i) in criteria (10), (11), (13); ω_i is the cost of (penalty for) the tardiness or earliness/tardiness of product i (product series i) per time unit in criteria (12), (14); l_i is the processing time of product i (product series i). Thus, if we at first assign to processing jobs of the optimal schedule corresponding to the set G_1, then the products first assigned to processing will reach the maximum averaged profit per time unit of the production cycle with the length $\sum l_i$, $i \in G_1$ (in criteria (10), (11), (13)) or the maximum averaged penalty per time unit of the production cycle (in criteria (12), (14)). After the set G_1 we assign to processing products from the set G_2, etc.

We formalized the coordinated planning procedure carried out in the aggregated network of the first level of aggregation. Within this procedure, the solution obtained at the second level of aggregation determines the order of assignment to processing for aggregated jobs of products (or product groups), and the order meets the specified criterion (one of (10)–(14)). For example, when planning by criterion (11), we first assign aggregated jobs of the first product in TWCT optimal sequence obtained at the second level of aggregation, so that they complete just in time. Then we assign to processing aggregated jobs for the second product, etc. In this way, violations of the due dates will be allowed only for products with a small averaged profit per time unit of the production cycle.

Coordinated planning results in finding of suboptimal by the criterion (one of (10)–(14)) completion times of all products (product series). The completion times become the due dates for the final jobs in MSSP (see Fig. 1). We solve the MSSP sequentially from the assignment of final jobs, then moving on to their predecessors, etc., according to the criterion: the earliest time for a job's assignment to processing is as late as possible. The suboptimal algorithm for this criterion is based on sequential solving of a number of NP-hard single-stage scheduling problems. An efficient solution of the problems was proposed in [4, 29, 30] based on the theory of PSC-algorithms [31].

4 Finding a Compromise Operational Plan Under Uncertainty

The problem formulated in Sect. 3 and the algorithm for its solving (i.e., an operational plan building for MSSP (see Fig. 1) by one of the criteria (10)–(14)) belongs to the class of combinatorial optimization problems introduced in Sect. 2 in the deterministic formulation (1). Indeed, in our case, the i-th arbitrary numerical characteristic of a feasible solution $k_i(\sigma)$ is:

1. the completion time of i-th product C_i in criterion (10);
2. $U_i = \begin{cases} 1, C_i = d_i \\ 0, C_i \neq d_i \end{cases}$ in criterion (11);
3. $\max(0,\ C_i - d_i)$ in criterion (12);
4. $U_i = \begin{cases} 1, C_i \leq d_i \\ 0, C_i > d \end{cases}$ in criterion (13);
5. $|C_i - d_i|$ in criterion (14).

The suboptimal algorithm for MSSP solving is efficient only when used in the formulation given in Sect. 3.

In accordance with Sect. 2, we formulate the problem of finding a compromise operational plan under uncertainty as follows. At the time of the operational plan's implementation, the set of coefficients $\omega_i,\ i = \overline{1,n}$ (the set of weights according to one of the criteria (10)–(14)) may take one value from the set $\left\{\omega_i^l,\ i = \overline{1,n}\right\},\ l = \overline{1,L}$. Find a compromise operational plan that minimizes one of the criteria (2)–(5).

We solve the problem as the following.

Criterion (2). The algorithm presented in aggregated form in Sect. 3 solves the problem of finding the operation plan in a deterministic formulation by one of the criteria (10)–(14) in which the weights $\omega_i,\ i = \overline{1,n}$, are equal to

$$\omega_i = \sum_{l=1}^{L} \omega_i^l,\ i = \overline{1,n}. \tag{15}$$

Criterion (3). The algorithm presented in aggregated form in Sect. 3 solves the problem of finding the operation plan in a deterministic formulation by one of the criteria (10)–(14) in which the weights ω_i, $i = \overline{1, n}$, are equal to

$$\omega_i = \sum_{l=1}^{L} a_l \omega_i^l, \ i = \overline{1, n}, \tag{16}$$

where $a_l > 0$, $l = \overline{1, L}$, are expert coefficients determining the importance of deviation from the l-th optimal solution in a deterministic formulation.

Criterion (4). In this case,

$$\omega_i = \sum_{l=1}^{L} P_l \omega_i^l, \ i = \overline{1, n}. \tag{17}$$

where P_l is the event probability that at the implementation stage of the operational plan weights in the functional (one of (10)–(14)) are $\omega_i = \omega_i^l$, $l = \overline{1, L}$, $i = \overline{1, n}$.

Criterion (5). In accordance with Statement 1 and its corollaries (Sect. 2), using the algorithm for an operational plan building in a deterministic formulation with a set of weight coefficients (one of the sets (15)–(17)), we apply the iterative procedure given in Corollary 5 of Statement 1 (Section 2). The procedure either finds a compromise solution that fulfills the Criterion (5), or minimizes the functional

$$\min \sum_{t} C_{j_t} \left(\Delta_{j_t} - l_{j_t} \right), \quad \forall_t \ \Delta_{j_t} > l_{j_t} \tag{18}$$

(Corollary 5 of Statement 1, Section 2).

Remark 2. In [2] we present the procedure for finding expert weights $a_l > 0$, $C_l > 0$, $l = \overline{1, L}$, by empirical matrices of pairwise comparisons (EMPC). It guarantees the stability of the weights' values under small perturbations in the values of nonzero elements of the EMPC.

Remark 3. An EMPC may contain zero elements.

5 Discussion of the Results

We have given the solution to a class of multi-stage scheduling problems. The class is a formal scheduling model of a small-series production operating in conditions of uncertainty. An efficient solution to this problem is based on:

- previously obtained results on the efficient solving of this problem in a deterministic formulation using the theory of PSC-algorithms [3, 4];
- the original theory developed by the author to solve one class of combinatorial optimization problems under uncertainty [1, 2].

The results obtained are constructive and can be used in long-term planning problems under uncertainty for arbitrary objects with a network representation of technological processes. The strictly proved theoretical results described in this paper (see Statement 1 and its corollaries) do not require a confirmation by statistical studies since the efficiency of the problem solving under uncertainty is determined by the efficiency of its solving in a deterministic formulation. Studies of the efficiency of combinatorial optimization problems solving in deterministic formulations were presented in [3, 4, 7–9]. [3, 4] also contain an example of the planning problem solving which we cannot give here because of the paper's limitations. We could publish an article based on this paper with a voluminous (up to 25 pages) example of the problem solving in conditions of uncertainty.

We can estimate the efficiency of the presented methodology for small-series production from the data of statistical research of the small-series production planning system based on the above described four-level model. Table 1 shows the time of an operational plan construction and the deviation of its functional value obtained by other APS-systems (SAP R/3, IT-Enterprise) which use fast heuristic algorithms, in comparison with our planning system that uses PSC-algorithms. The deviation of the functional value is given in percentage from the value obtained by our system.

Table 1. Comparison of systems for the planning problem solving

Param.	Solving time by our system in seconds			Solving time by other systems in seconds			The deviation of the functional value from ours, %		
n	Max	Min	Avg	Max	Min	Avg	Max	Min	Avg
100	0.31	0.11	0.16	0.28	0.08	0.11	12.48	8.43	10.56
200	0.66	0.20	0.33	0.57	0.18	0.24	13.04	8.82	9.80
500	2.61	0.92	1.32	2.21	0.69	0.89	12.13	9.54	10.52
1000	7.31	2.55	3.51	5.81	1.77	2.29	13.67	11.42	12.23
5000	131.24	45.51	64.12	97.54	29.21	37.42	15.11	8.93	12.84
10000	511.51	175.32	225.81	335.15	99.25	127.12	14.89	11.78	12.61

The research was performed on a PC with 3 GHz Intel processor. We generated 125 examples for each dimension n (the number of jobs in network technology). Then we solved each of the 750 examples by various systems taking into account the criterion of minimizing the weighted completion time of jobs.

The results show that, as a rule, other systems operate from 1.5 to 2 times faster but their efficiency by the functional value is from 8.5 to 15% worse. This confirms the practical significance of our research.

References

1. Pavlov, A.A.: Optimization for one class of combinatorial problems under uncertainty. Adapt. Syst. Autom. Control **1**(34), 81–89 (2019). https://doi.org/10.20535/1560-8956.1.2019.178233
2. Pavlov, A.A.: Combinatorial optimization under uncertainty and formal models of expert estimation. Bull. Nat. Tech. Univ. KhPI Ser. Syst. Anal. Control Inf. Technol. **1**, 3–7 (2019). https://doi.org/10.20998/2079-0023.2019.01.01
3. Zgurovsky, M.Z., Pavlov, A.A.: Combinatorial Optimization Problems in Planning and Decision Making: Theory and Applications, 1st edn. Springer, Cham (2019). https://doi.org/10.1007/978-3-319-98977-8
4. Zgurovsky, M.Z., Pavlov, A.A.: Trudnoreshaemye zadachi kombinatornoy optimizacii v planirovanii i priniatii resheniy. Naukova dumka, Kyiv (2016). (in Russian)
5. Pavlov, A.A., Misura, E.B., Melnikov, O.V., Mukha, I.P.: NP-hard scheduling problems in planning process automation in discrete systems of certain classes. In: Hu, Z., Petoukhov, S., Dychka, I., He, M. (eds.) Advances in Computer Science for Engineering and Education, ICCSEEA 2018. Advances in Intelligent Systems and Computing, vol. 754, pp. 429–436. Springer, Cham (2019). https://doi.org/10.1007/978-3-319-91008-6_43
6. Pavlov, A.A., Khalus, E.A., Borysenko, I.V.: Planning automation in discrete systems with a given structure of technological processes. In: Hu, Z., Petoukhov, S., Dychka, I., He, M. (eds.) Advances in Computer Science for Engineering and Education, ICCSEEA 2018. Advances in Intelligent Systems and Computing, vol. 754, pp. 177–185. Springer, Cham (2019). https://doi.org/10.1007/978-3-319-91008-6_18
7. Pavlov, A.A., Misura, E.B., Melnikov, O.V., Mukha, I.P., Lishchuk, K.I.: Study of theoretical properties of PSC-algorithm for the total weighted tardiness minimization for planning processes automation. In: Hu, Z., Petoukhov, S., Dychka, I., He, M. (eds.) Advances in Computer Science for Engineering and Education II, ICCSEEA 2019. Advances in Intelligent Systems and Computing, vol. 938, pp. 152–161. Springer, Cham (2020). https://doi.org/10.1007/978-3-030-16621-2_14
8. Pavlov, A.A., Misura, E.B., Melnikov, O.V., Mukha, I.P., Lishchuk, K.I.: Statistical research of efficiency of approximation algorithms for planning processes automation problems. In: Hu, Z., Petoukhov, S., Dychka, I., He, M. (eds.) Advances in Computer Science for Engineering and Education II, ICCSEEA 2019. Advances in Intelligent Systems and Computing, vol. 938, pp. 398–408. Springer, Cham (2020). https://doi.org/10.1007/978-3-030-16621-2_37
9. Pavlov, A.A., Misura, E.B., Melnikov, O.V., Mukha, I.P., Lishchuk, K.I.: Approximation algorithm for parallel machines total tardiness minimization problem for planning processes automation. In: Hu, Z., Petoukhov, S., Dychka, I., He, M. (eds.) Advances in Computer Science for Engineering and Education II, ICCSEEA 2019. Advances in Intelligent Systems and Computing, vol. 938, pp. 459–467. Springer, Cham (2020). https://doi.org/10.1007/978-3-030-16621-2_43
10. Alhumrani, S.A., Qureshi, R.J.: Novel approach to solve resource constrained project scheduling problem (RCPSP). Int. J. Mod. Educ. Comput. Sci. (IJMECS) **8**(9), 60–68 (2016). https://doi.org/10.5815/ijmecs.2016.09.08
11. Aggarwal, A., Verma, R., Singh, A.: An efficient approach for resource allocations using hybrid scheduling and optimization in distributed system. Int. J. Educ. Manag. Eng. (IJEME) **8**(3), 33–42 (2018). https://doi.org/10.5815/ijeme.2018.03.04

12. Bala, M.I., Chishti, M.A.: Load balancing in cloud computing using Hungarian algorithm. Int. J. Wirel. Microw. Technol. (IJWMT) **9**(6), 1–10 (2019). https://doi.org/10.5815/ijwmt.2019.06.01
13. Sajedi, H., Rabiee, M.: A metaheuristic algorithm for job scheduling in grid computing. Int. J. Mod. Educ. Comput. Sci. (IJMECS) **6**(5), 52–59 (2014). https://doi.org/10.5815/ijmecs.2014.05.07
14. Knyshov, G.V., Nastenko, I.A., Kondrashova, N.V., Nosovets, O.K., Pavlov, V.A.: Combinatorial algorithm for constructing a parametric feature space for the classification of multidimensional models. Cybern. Syst. Anal. **50**(4), 627–633 (2014). https://doi.org/10.1007/s10559-014-9651-3
15. Nastenko, I., Pavlov, V., Nosovets, O., Zelensky, K., Davidko, O., Pavlov, O.: Solving the individual control strategy tasks using the optimal complexity models built on the class of similar objects. In: Shakhovska, N., Medykovskyy, M.O. (eds.) CCSIT 2019. Advances in Intelligent Systems and Computing, vol. 1080, pp. 535–546. Springer, Cham (2020). https://doi.org/10.1007/978-3-030-33695-0_36
16. Nastenko, I.A., Pavlov, V.A., Nosovets, O.K., Davydko, O., Pavlov, O.: Optimization models for calculation of personalized strategies. In: Idemudia, E. (ed.) Handbook of Research on Social and Organizational Dynamics in the Digital Era, pp. 305–323. IGI Global, Hershey (2020). https://doi.org/10.4018/978-1-5225-8933-4.ch015
17. Titov, V.V., Bezmelnitsyn, D.A., Napreeva, S.K.: Planirovanie funktsionirovaniya predpriyatiya v usloviyakh riska i neopredelennosti vo vneshney i vnutrenney srede. Nauchno-tekhnicheskie vedomosti SPbGPU. Ekonomicheskie nauki **10**(5), 172–183 (2017). https://doi.org/10.18721/JE.10516. (in Russian)
18. Jamalnia, A., Yang, J.-B., Feili, A., Xu, D.-L., Jamali, G.: Aggregate production planning under uncertainty: a comprehensive literature survey and future research directions. Int. J. Adv. Manuf. Technol. **102**, 159–181 (2019). https://doi.org/10.1007/s00170-018-3151-y
19. Noghin, V.D.: Linejnaja svertka kriteriev v mnogokriterial'noj optimizacii. Artif. Intell. Decis. Making **2014**(4), 73–82 (2014). (in Russian)
20. Gorelik, V.A., Zolotova, T.V.: Problem of selecting an optimal portfolio with a probabilistic risk function. J. Math. Sci. **216**(5), 603–611 (2016). https://doi.org/10.1007/s10958-016-2921-z
21. Kharchenko, A., Halay, I., Zagorodna, N., Bodnarchuk, I.: Trade-off optimal decision of the problem of software system architecture choice. In: 2015 Xth International scientific and technical conference “Computer Sciences and Information Technologies” (CSIT), pp. 198–205 (2015). https://doi.org/10.1109/STC-CSIT.2015.7325465
22. Bindima, T., Elias, E.: A novel design and implementation technique for low complexity variable digital filters using multi-objective artificial bee colony optimization and a minimal spanning tree approach. Eng. Appl. Artif. Intellig. **59**, 133–147 (2017). https://doi.org/10.1016/j.engappai.2016.12.011
23. Mirzapour, S.M.J., Hashem, A., Malekly, H., Aryanezhad, M.B.: A multi-objective robust optimization model for multi-product multi-site aggregate production planning in a supply chain under uncertainty. Int. J. Prod. Econ. **134**, 28–42 (2011). https://doi.org/10.1016/j.ijpe.2011.01.027
24. Iemets, O.O., Barbolina, T.M.: Liniyni optymizatsiyni zadachi na razmishchennyakh z imovirnisnoyu nevyznachenistyu: vlastyvosti i rozv’yazannya. Syst. Res. Inf. Technol. **2016** (1), 107–119 (2016). https://doi.org/10.20535/SRIT.2308-8893.2016.1.11. (in Ukrainian)
25. Shang, C., You, F.: Distributionally robust optimization for planning and scheduling under un-certainty. Comput. Chem. Eng. **110**, 53–68 (2018). https://doi.org/10.1016/j.compchemeng.2017.12.002

26. Zgurovsky, M.Z., Pavlov, A.A., Misura, E.B., Melnikov, O.V., Lisetsky, T.N.: Metodologiya postroeniya chetyrekhurovnevoy modeli planirovaniya, prinyatiya resheniy i operativnogo planirovaniya v setevykh sistemakh s ogranichennymi resursami. Visnyk NTUU KPI Inform. Oper. Comput. Sci. **61**, 60–84 (2014). (in Russian)
27. Zgurovsky, M.Z., Pavlov, A.A.: Algorithms and software of the four-level model of planning and decision making. In: Combinatorial Optimization Problems in Planning and Decision Making: Theory and Applications, 1st edn. Studies in Systems, Decision and Control, vol. 173, pp. 407–518. Springer, Cham (2019). https://doi.org/10.1007/978-3-319-98977-8_9
28. Zgurovsky, M.Z., Pavlov, A.A.: The total weighted completion time of tasks minimization with precedence relations on a single machine. In: Combinatorial Optimization Problems in Planning and Decision Making: Theory and Applications, 1st edn. Studies in Systems, Decision and Control, vol. 173, pp. 291–344. Springer, Cham (2019). https://doi.org/10.1007/978-3-319-98977-8_7
29. Zgurovsky, M.Z., Pavlov, A.A.: Optimal scheduling for two criteria for a single machine with arbitrary due dates of tasks. In: Combinatorial Optimization Problems in Planning and Decision Making: Theory and Applications, 1st edn. Studies in Systems, Decision and Control, vol. 173, pp. 17–38. Springer, Cham (2019). https://doi.org/10.1007/978-3-319-98977-8_2
30. Zgurovsky, M.Z., Pavlov, A.A.: Optimal scheduling for vector or scalar criterion on parallel machines with arbitrary due dates of tasks. In: Combinatorial Optimization Problems in Planning and Decision Making: Theory and Applications, 1st edn. Studies in Systems, Decision and Control, vol. 173, pp. 39–105. Springer, Cham (2019). https://doi.org/10.1007/978-3-319-98977-8_3
31. Zgurovsky M.Z., Pavlov A.A.: Introduction. In: Combinatorial Optimization Problems in Planning and Decision Making: Theory and Applications, 1st edn. Studies in Systems, Decision and Control, vol. 173, pp. 1–14. Springer, Cham (2019). https://doi.org/10.1007/978-3-319-98977-8_1

Estimation of Hurst Parameter for Self-similar Traffic

Anatolii Pashko[1(✉)], Tetiana Oleshko[2], and Olga Syniavska[3]

[1] Taras Shevchenko National University of Kyiv, Kyiv, Ukraine
aap2011@ukr.net
[2] National Aviation University, Kyiv, Ukraine
lllota@ukr.net
[3] Uzhgorod National University, Uzhhorod, Ukraine
olja_sunjavska@ukr.net

Abstract. Recent studies have shown that the network traffic of modern networks has the properties of self-similarity. And this requires finding adequate traffic simulation methods and download processes in modern telecommunications networks. Models of self-similar traffic and the process of loading telecommunication networks are based on the methods of fractional Brownian motion (FBM) modeling. Traffic characteristics of modern telecommunications networks vary widely and depend on a large number of network settings and settings, protocol characteristics and user experience. The self-similarity of the fractional Brownian motion is characterized by the Hurst index. The article explores methods for estimating the Hurst index. Realizations of fractional Brownian motion with known Hurst index are considered. For these realizations the obtained Hurst's estimates.

Keywords: Fractional Brownian motion · Hurst index · Self-similar traffic · Gaussian process

1 Introduction

Fractional Brownian motion models are used in many industries. Thus, in addition to the problems of modeling and estimating traffic in the theory of telecommunication networks [1–5], fractional Brownian motion models are used in financial mathematics [6], queuing systems [7], in solving problems of computational mathematics [8], in medical studies [9, 10], to evaluate the [11].

The choice of the fractional Brownian motion is due to the fact that FBM has self-similarity properties. This FBM property is characterized by the Hurst index.

The traffic model was investigated on the basis of FBM. Delay time and probability of data loss were estimated in [12].

Processing traffic data is one of the urgent problems of infocommunications. Any signal can be considered as an overlay of a useful signal and background noise, which is a combination of stochastic and fractal components.

The numerical indicators of these properties are, respectively, the Hurst exponent, stability index, increment correlation coefficients, which generalize the autocorrelation

Z. Hu et al. (Eds.): ICCSEEA 2020, AISC 1247, pp. 181–191, 2021.
https://doi.org/10.1007/978-3-030-55506-1_16

function. Obviously, the estimation of the Hurst exponent is a priority in the analysis of self-similar processes [13].

Currently, there are many methods for estimating the Hurst parameter, but all of them are focused on such special cases of the processes, when the self-similarity property is combined with either a long-term dependence (fractional Brownian motion) [13] or heavy tails [14].

When evaluating the Hurst parameter, RS analysis, time analysis of variance and deviation analysis (DFA) are most commonly used. A common property of these methods is that they are all based on the use of the statistical properties of second-order samples (variance, standard deviation, correlation coefficients). In [13–15], the fractional moment method was developed, which is equally operable for both Gaussian and heavy tails. This method can be applied to estimate the Hurst parameter.

All methods considered are approximate. The properties of non-bias and consistency are not satisfied for them.

The paper studies methods for estimating the Hurst parameter based on the use of Baxter sums and Levy-Baxter limit theorems [16]. The obtained estimates are unbiased and consistent.

2 Evaluation of the Hurst Index

Suppose $B = \{B_t,\ t \geq 0\}$ is fractional Brownian motion with Hurst parameter α – Gaussian random process with zero mean and covariance function $r(s,t) = \frac{1}{2}\left(|s|^{2\alpha} + |t|^{2\alpha} - |s-t|^{2\alpha}\right),\quad s,t \geq 0$.

Suppose that the Hurst parameter α is unknown, such that $\alpha \in (0, \alpha^*]$, where $\alpha^* \in (0,1)$ is fixed. Also, assume that the random process $B = \{B_t,\ t \geq 0\}$ is observed at the points $\left\{\frac{k}{a_n},\ 0 \leq k \leq a_n\right\}$, where $a_n \in \mathrm{N},\ n \geq 1;\ a_n \to \infty,\ n \to \infty$. Suppose that for an arbitrary $\beta > 0$ the series $\sum_{n=1}^{\infty} a_n^{-\beta}$ is convergent.

Let for natural number p, $\Delta B_{k,n}^{(p)}$ be increment of order p of fractional Brownian motion B, $k = 0, \ldots, a_n - 1$. In particular,

$$\Delta B_{k,n}^{(1)} = B_{\frac{k+1}{a_n}} - B_{\frac{k}{a_n}},$$

$$\Delta B_{k,n}^{(2)} = B_{\frac{k+2}{a_n}} - 2B_{\frac{k+1}{a_n}} + B_{\frac{k}{a_n}},$$

$$\Delta B_{k,n}^{(3)} = B_{\frac{k+3}{a_n}} - 3B_{\frac{k+2}{a_n}} + 3B_{\frac{k+1}{a_n}} - B_{\frac{k}{a_n}},$$

$$\ldots\ldots\ldots\ldots\ldots\ldots\ldots\ldots\ldots\ldots\ldots\ldots,$$

$$\Delta B_{k,n}^{(p)} = \sum_{i=0}^{p} (-1)^i C_p^i B_{\frac{k}{n} + \frac{p-i}{n}}.$$

Consider the sequences of Baxter sums for fractional Brownian motion B_H:

$$\widehat{S}_n^{(p)} = a_n^{2\alpha-1} \sum_{k=0}^{a_n-1} \left(\Delta B_{k,n}^{(p)}\right)^2, \; S_n^{(p)} = a_n^{2\alpha-1} \widehat{S}_n^{(p)}, \; n \geq 1. \tag{1}$$

Suppose

$$V_p(k, \alpha) = \frac{1}{2} \sum_{i,j=0}^{p} (-1)^{i+j+1} C_p^i C_p^j |k + (i-j)|^{2\alpha}, \; k \geq 0, p \geq 1.$$

In particular, $V_1(0, \alpha) = 1; \quad V_2(0, \alpha) = 4 - 4^{\alpha}$.

The direct calculation makes it possible to obtain the following formulas for mathematical expectation and variance of a random variable $\widehat{S}_n^p$, $p \geq 1$:

$$E\widehat{S}_n^{(p)} = V_p(0, \alpha);$$

$$\operatorname{Var} \widehat{S}_n^{(p)} = \frac{1}{n} \left(2V_p^2(0, \alpha) + 4 \sum_{k=1}^{n-1} \left(1 - \frac{k}{n}\right) V_p^2(k, \alpha) \right).$$

Theorem 1. [16] Statistics

$$\widehat{\alpha}_n^{(1)} = \frac{1}{2} \left(1 - \frac{\ln S_n^{(1)}}{\ln a_n}\right), \; n \geq 1$$

is a strongly consistent estimate of the Hurst parameter α.

To find the variance estimate $\mathrm{D}\, \widehat{S}_n^{(1)}$, we use the following lemma.

Lemma 1. [16] Suppose $B = \{B_t, \; t \geq 0\}$ is fractional Brownian motion with Hurst parameter $\alpha \in (0, \alpha^*]$. Then at $\alpha^* \in (0, 1)$ the following inequality holds:

$$\sup_{\alpha \in (0, \alpha^*]} \mathrm{D}\, \widehat{S}_n^{(1)} \leq \frac{D_1}{a_n},$$

Where

$$D_1 = \begin{cases} 2(3 + 2\zeta(4 - 4\alpha^*)), \; \alpha^* \in \left(0, \frac{3}{4}\right), \\ 2(3 + 2(1 + \ln a_n)), \; \alpha^* = \frac{3}{4}, \\ 2\left(3 + 2\frac{a_n^{4\alpha^*-3}}{4\alpha^*-3}\right), \; \alpha^* \in \left(\frac{3}{4}, 1\right), \end{cases} \tag{2}$$

$$\zeta(s) = \sum_{n=1}^{\infty} \frac{1}{n^s}, \; s > 1.$$

Similarly, it can be shown that statistics $\widehat{\alpha}_n^{(p)} = \frac{1}{2}\left(1 - \frac{\ln S_n^{(p)}}{\ln a_n}\right)$, $p \geq 2$ are strongly consistent estimates of the Hurst parameter α.

Lemma 2. Suppose $B = \{B_t,\ t \geq 0\}$ is fractional Brownian motion with Hurst parameter α. Then the inequality holds:

$$\sup_{\alpha \in (0,1)} D\widehat{S}_n^{(p)} \leq \frac{D_p}{a_n},$$

where

$$D_p = 2\left(K_p + \frac{1}{2}L_p + \frac{1}{2}M_p^2 \zeta(4p-4)\right), p \geq 2, \tag{3}$$

$$V_p(m, \alpha) = \frac{1}{2}\sum_{i,j=0}^{p} (-1)^{i+j+1} C_p^i C_p^j |m + (i-j)|^{2\alpha},\ k \geq 0,$$

$$K_p = \sup_{\alpha \in (0,1)} V_p^2(0, \alpha),$$

$$L_p = \sup_{\alpha \in (0,1)} V_p^2(1, \alpha),$$

$$M_p = \sup_{\alpha \in (0,1)} |2\alpha(2\alpha-1)(2\alpha-2)\cdot \ldots \cdot (2\alpha-(2p-1))|, p \geq 2,$$

$\zeta(\cdot)$ is Riemann zeta function.

From Lemma 2 for $p = 2$, we obtain: $V_2(0, \alpha) = 4 - 4^{\alpha}$; and $V_2(1, \alpha) = 7 - 4 \cdot 2^{2\alpha} + 3^{2\alpha}$.

Estimates were found for p = 2

$$K_2 = \sup_{\alpha \in (0,1)} V_2^2(0, \alpha) = \sup_{\alpha \in (0,1)} (4 - 4^{\alpha})^2 = 9,$$

$$L_2 = \sup_{\alpha \in (0,1)} V_2^2(1, \alpha) = \sup_{\alpha \in (0,1)} \left(7 - 4 \cdot 2^{2\alpha} + 3^{2\alpha}\right)^2 = 16,$$

$$M_2 = \sup_{\alpha \in (0,1)} |2\alpha(2\alpha-1)(2\alpha-2)(2\alpha-3)| = 0,563.$$

Further, with the inequality in Lemma 2, previously obtained values K_p, L_p, M_p and considering that $\zeta(8-4\alpha) \leq \zeta(4) = \frac{\pi^4}{90}$ for $\alpha \in (0,1)$, we obtain the following inequality for p = 2:

$$\sup_{\alpha\in(0,1)} D\widehat{S}_n^{(2)} \le \frac{2}{a_n}\left(9+\frac{16}{2}+\frac{1}{2}\cdot(0.563)^2\cdot\frac{\pi^4}{90}\right)\approx\frac{34.344}{a_n}.$$

Theorem 2. Suppose $B=\{B_t,\ t\geq 0\}$ is fractional Brownian motion with Hurst parameter α. Then $\widehat{S}_n^{(p)} \to V_p(0,\alpha), p\geq 1$ with probability one as $n\to\infty$.

3 Confidence Intervals

Theorem 3. The interval $\left(\widehat{\alpha}_n^{(p)} - l_\varepsilon(n),\ \widehat{\alpha}_n^{(p)} + r_\varepsilon(n)\right)\cap(0,\ 1)$, where

$$\widehat{\alpha}_n^{(p)} = \frac{1}{2}\left(1-\frac{\ln\, S_n^{(p)}}{\ln\, a_n}\right),$$

$$l_\varepsilon(n)\geq -\frac{1}{2}\frac{\ln\left(1-\sqrt{\frac{2D_p}{a_n\varepsilon}}\right)}{\ln a_n}\ \text{provided that}\ \sqrt{\frac{2D_p}{\varepsilon}}<1,$$

$$r_\varepsilon(n)\geq\frac{1}{2}\frac{\ln\left(\sqrt{\frac{2D_p}{a_n\varepsilon}}+1\right)}{\ln a_n},$$

$S_n^{(p)}$ is defined by the equality (1), D_p is determined for $p=1$ by the equality (2), and for $p\geq 2$ is defined by the equality (3), is the confidence interval for the Hurst parameter α with confidence level $(1-\varepsilon)\in(0,1)$.

Another estimate of the Hurst parameter α is obtained from the likelihood convergence of the unit of Baxter sums in the Theorem 2 with $p=1,2$. Indeed, for arbitrary $\alpha\in(0,\ 1)$ there are convergence

$$\frac{\widehat{S}_n^{(2)}}{\widehat{S}_n^{(1)}}\to\frac{V_2(0,\alpha)}{V_1(0,\ \alpha)}=4-4^\alpha.$$

with probability 1 as $n\to\infty$.

Let's consider function $\theta_{2,1}(\alpha)=\frac{V_2(0,\alpha)}{V_1(0,\ \alpha)}=4-4^\alpha,\ \alpha\in(0,\ 1)$, which is continuous and decreasing at interval $(0,\ 1)$, $\theta_{2,1}(0+)=3$, $\theta_{2,1}(1-)=0$. Suppose $\alpha_{2,1}(\theta),\ \theta\in(0,\ 3)$ is function, inverse to the function $\theta_{2,1}(\alpha),\ \alpha\in(0,\ 1)$. Suppose

$$\widehat{\theta}_n=\frac{S_n^{(2)}}{S_n^{(1)}}=\frac{\widehat{S}_n^{(2)}}{\widehat{S}_n^{(1)}},\ n\geq 1. \tag{4}$$

Then $\widehat{\theta}_n \to \theta = \theta(\alpha)$ with probability 1 as $n \to \infty$. From the following considerations it follows

Theorem 4. Statistics

$$\widehat{\alpha}_n^{(2,1)} = 1 - \frac{1}{2}\log_2\left(\widehat{\theta}_n + 1\right), \ n \geq 1$$

is a strongly consistent estimate of the Hurst parameter α.

4 Constructing a Confidence Intervals

Lemma 3. [16] Suppose $\{X_k | 0 \leq k \leq a_n, \ a_n \in \mathrm{N}\}$, $\{Y_k | 0 \leq k \leq a_n, \ a_n \in \mathrm{N}\}$ are sets of random variables with finite moments of 4th order such that $EX_k = EY_k = 0$, $EX_k^2 = EX_0^2, \ EY_k^2 = EY_0^2, \ 0 \leq k \leq a_n$;

$$S_1 = \sum_{k=0}^{a_n} X_k^2, \ S_2 = \sum_{k=0}^{a_n} Y_k^2, \ \delta = \frac{EX_0^2}{EY_0^2}.$$

Then for arbitrary $\varepsilon > 0$ there is inequality:

$$P\left\{ \left| \frac{S_1}{S_2} - \delta \right| > \varepsilon \right\} \leq \frac{\mathrm{Var}\ Q_1}{(EQ_1)^2} + \frac{\mathrm{Var}\ Q_2}{(EQ_2)^2},$$

where $Q_1 = (\delta - \varepsilon)S_2 - S_1$, $Q_2 = S_1 - (\delta + \varepsilon)S_2$.

Statistics $\widehat{\theta}_n, \ n \geq 1$, defined in equality (4), because Theorem 3, is a strongly consistent estimate of the parameter $\theta = \theta(\alpha)$. Using Lemma 3, we construct a confidence interval for the parameter θ, and then get the confidence interval for the Hurst parameter α. Suppose $(1 - \varepsilon) \in (0, 1)$ is confidence level. We define positive number $m_\varepsilon(n)$ so that

$$P\left\{ \left| \widehat{\theta}_n - \theta \right| \geq m_\varepsilon(n) \right\} \leq \varepsilon.$$

For statistics $\widehat{\theta}_n = \frac{\widehat{S}_n^{(2)}}{\widehat{S}_n^{(1)}}$ we have:

$Q_1 = (\theta - m_\varepsilon(n))\widehat{S}_n^{(1)} - \widehat{S}_n^{(2)} \ Q_2 = \widehat{S}_n^{(2)} - (\theta + m_\varepsilon(n))\widehat{S}_n^{(1)}$.

For mathematical expectations of random variables Q_1, Q_2 have:

$$EQ_1 = EQ_2 = -m_\varepsilon(n)V_1(0, H) = -m_\varepsilon(n).$$

To find upper estimate of variances of these random variables we apply inequality $(a + b)^2 \leq 2(a^2 + b^2), \ a, \ b \in R$ and Lemma 1:

$$\mathrm{D}Q_1 \leq 2(\theta - m_\varepsilon(n))^2 \mathrm{D}\, S_n^{(1)} + 2\mathrm{D}\, S_n^{(2)} \leq \frac{2}{a_n}(\theta - m_\varepsilon(n))^2 D_1 + \frac{2}{a_n} D_2.$$

Similarly,

$$DQ_2 \leq \frac{2}{a_n}(\theta + m_\varepsilon(n))^2 D_1 + \frac{2}{a_n} D_2.$$

We will select $\gamma_n(p)$ so that the following inequalities hold:

$$\frac{D\,Q_1}{(EQ_1)^2} \leq \frac{2(\theta - m_\varepsilon(n))^2 D_1 + 2D_2}{m_\varepsilon^2(n) a_n} \leq \frac{\varepsilon}{2},$$

$$\frac{\mathrm{D}\,Q_2}{(EQ_2)^2} \leq \frac{2(\theta + m_\varepsilon(n))^2 D_1 + 2D_2}{m_\varepsilon^2(n) a_n} \leq \frac{\varepsilon}{2}. \tag{5}$$

For $\alpha^* \in (0,\ 1)$ we have:

$$D_1 = \begin{cases} 2(3 + 2\zeta(4 - 4\alpha^*)), \ \alpha^* \in \left(0,\ \frac{3}{4}\right), \\ 2(3 + 2(1 + \ln a_n)), \ \alpha^* = \frac{3}{4}, \\ 2\left(3 + 2\frac{a_n^{4\alpha^* - 3}}{4\alpha^* - 3}\right), \ \alpha^* \in \left(\frac{3}{4},\ 1\right), \end{cases}$$

where $\zeta(\cdot)$ is Riemann zeta function and

$$D_2 \approx \frac{34.344}{a_n}.$$

For $\alpha \in (0,\ \alpha^*]$, $\theta \in (0,\ 3)$ we have: $\frac{2(3 + m_\varepsilon(n))^2 D_1 + 2D_2}{m_\varepsilon^2(n) a_n} \leq \frac{\varepsilon}{2}$.

Solve this inequality with respect to $m_\varepsilon(n)$ provided $\frac{\varepsilon}{4} a_n - D_1 > 0$, that is true for sufficiently large $n \in \mathrm{N}$:

$$2(3 + m_\varepsilon(n))^2 D_1 + 2D_2 \leq \frac{\varepsilon}{2} a_n m_\varepsilon^2(n),$$

$$\left(\frac{\varepsilon}{4} a_n - D_1\right) m_\varepsilon^2(n) - 6D_1 m_\varepsilon(n) - (9D_1 + D_2) \geq 0,$$

$$m_\varepsilon(n) \geq \frac{6D_1 + \sqrt{D}}{2\left(\frac{\varepsilon}{4} a_n - D_1\right)},$$

provided that the inequality $D = \varepsilon a_n (9D_1 + D_2) - 4D_1 D_2 \geq 0$ is true for sufficiently large n. Therefore, the following theorem holds.

Theorem 5. Suppose $\alpha \in (0,\ \alpha^*]$, where $\alpha^* \in (0,\ 1)$ is fixed, $\frac{\varepsilon}{4}a_n - D_1 > 0$, $(1-\varepsilon)$, $\varepsilon \in (0,\ 1)$ is confidence level, D_1 is calculated by the formula (2), D_2 is calculated by the formula (3). Then the following inequality holds:

$$P\left\{ H \in \left(H_{n,l}, H_{n,r}\right)\right\} \geq 1-\varepsilon,$$

where

$$H_{n,l} = \varphi\left(\min\left(\widehat{\theta}_n + m_\varepsilon(n),\ 3\right)\right),$$

$$H_{n,r} = \varphi\left(\max\left(\theta(H^*),\ \widehat{\theta}_n - m_\varepsilon(n)\right)\right),$$

$$\widehat{\theta}_n = \frac{\widehat{S}_n^{(2)}}{\widehat{S}_n^{(1)}},\ m_\varepsilon(n) \geq \frac{6D_1 + \sqrt{D}}{2\left(\frac{\varepsilon}{4}a_n - D_1\right)},$$

$$D = \varepsilon a_n(9D_1 + D_2) - 4D_1D_2 \geq 0,$$

$$\varphi(\theta) = 1 - \frac{1}{2}\log_2(\theta+1),\ \theta \in (0,\ 3).$$

In [17, 18] for the traffic simulation model of fractional Brownian motion was used

$$\breve{S}_\alpha(t,M) = \sum_{k=1}^{M}\left(\breve{a}_k \sin\left(\breve{x}_k t\right)X_k + \breve{b}_k\left(1-\cos\left(\breve{y}_k t\right)\right)Y_k\right). \tag{6}$$

Representation (6) was obtained from the results by Dzhaparidze and Zanten [19]

$$W_\alpha(t) = \sum_{k=1}^{\infty}\left(b_k(1-\cos(y_k t))\xi_k + a_k\sin(x_k t)\eta_k\right),$$

where $\{\xi_k, \eta_k\}$ are independent standard Gaussian random variables, $\{x_k\}$ are zeros of Bessel function $J_{-\alpha}(x)$, $\{y_k\}$ are zeros of the Bessel function $J_{1-\alpha}(x)$, $a_k = \frac{\pi^\alpha\sqrt{2C}}{x_k^{\alpha+1}J_{1-\alpha}(x_k)}$, $b_k = \frac{\pi^\alpha\sqrt{2C}}{y_k^{\alpha+1}J_{-\alpha}(y_k)}$, $C = \frac{\Gamma(2\alpha+1)\sin(\pi\alpha)}{\pi^{2\alpha+1}}$.

The Fig. 1 shows the implementation of fractional Brownian motion for $\alpha = 0.6$.

The Hurst index estimate for (4) is $\widehat{\alpha}_n^{(2,1)} = 0.637$. The confidence interval is $[0.554; 0.673]$.

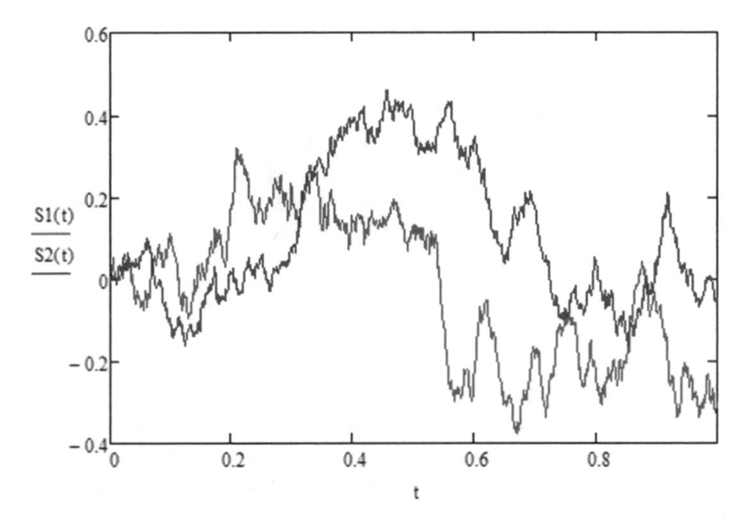

Fig. 1. Realization of fractional Brownian motion with $\alpha = 0.6$.

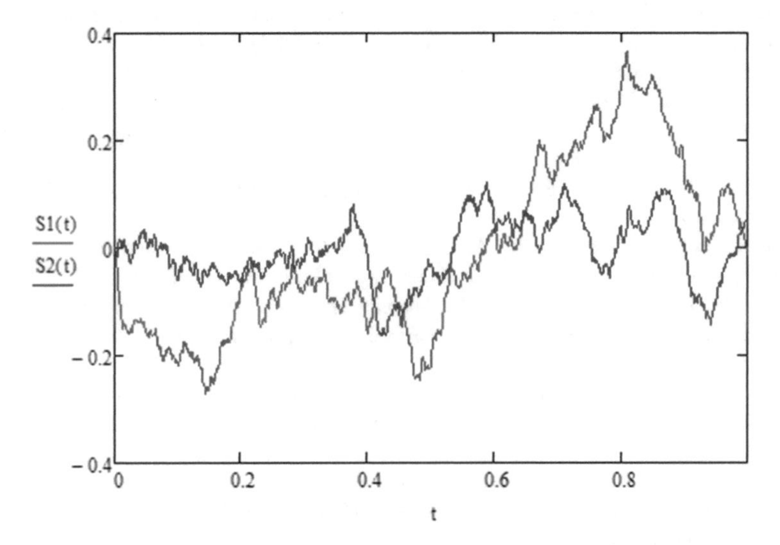

Fig. 2. Realization of fractional Brownian motion with $\alpha = 0.7$.

The Fig. 2 shows the implementation of fractional Brownian motion for $\alpha = 0.7$. The Hurst index estimate for (4) is $\widehat{\alpha}_n^{(2,1)} = 0.721$. The confidence interval is [0.638; 0.757].

It is the confidence interval for the Hurst parameter α with confidence level $\varepsilon = 0.1$.

In this case, the estimate of the Hurst index can be used to assess the quality of modeling.

5 Conclusions

We have investigated methods for estimating the Hurst parameter based on the use of Baxter sums and Levy-Baxter limit theorems. The estimates obtained are unbiased and consistent. Confidence intervals for these estimates are obtained. Particular cases of the second and third order are considered. The example of calculation for realization of fractional Brownian motion with known Hurst index is represented.

References

1. Norros, I.: A storage model with self-similar input. Queueing Syst. **16**, 387–396 (1994)
2. Kilpi, J., Norros, I.: Testing the Gaussian approximation of aggregate traffic. In: Proceedings of the second ACM SIGCOMM Workshop, Marseille, France, pp. 49–61 (2002)
3. Sheluhin, O.I., Smolskiy, S.M., Osin, A.V.: Similar Processes in Telecommunication. Wiley, Hoboken (2007)
4. Chabaa, S., Zeroual, A., Antari, J.: Identification and prediction of internet traffic using artificial neural networks. Intell. Learn. Syst. Appl. **2**, 147–155 (2010)
5. Gowrishankar, S., Satyanarayana, P.S.: A time series modeling and prediction of wireless network traffic. Int. J. Interact. Mob. Technol. (iJIM) **4**(1), 53–62 (2009)
6. Mishura, Yu.: Stochastic Calculus for Fractional Brownian Motion and Related Processes. Springer, Berlin (2008)
7. Kozachenko, Yu., Yamnenko, R., Vasylyk, O.: φ-sub-Gaussian random process. Vydavnycho-Poligrafichnyi Tsentr “Kyivskyi Universytet”, Kyiv (2008). (in Ukrainian)
8. Sabelfeld, K.K.: Monte Carlo Methods in Boundary Problems. Nauka, Novosibirsk (1989). (in Russian)
9. Goshvarpour, A., Goshvarpour, A.: Chaotic behavior of heart rate signals during Chi and Kundalini meditation. Int. J. Image Graph. Signal Process. (IJIGSP), **2**, 23–29 (2012)
10. Hosseini, S.A., Akbarzadeh, T.M.-R., Naghibi-Sistani, M.-B.: Qualitative and quantitative evaluation of EEG signals in epileptic seizure recognition. Int. J. Intell. Syst. Appl. **06**, 41–46 (2013)
11. Prigarin, S., Hahn, K., Winkler, G.: Comparative analysis of two numerical methods to measure Hausdorff dimension of the fractional Brownian motion. Siberian J. Num. Math. **11** (2), 201–218 (2008)
12. Ageev, D.V.: Parametric synthesis of multiservice telecommunication systems in the transmission of group traffic with the effect of self-similarity. Electron. Sci. Spec. Edn. Prob. Telecommun. **1**(10), 46–65 (2013). (in Russian)
13. Gajda, J., Wylomanska, A., Kumar, A.: Fractional Lévy stable motion time-changed by gamma subordinator. In: Communications in Statistics - Theory and Methods (2018). https://doi.org/10.1080/03610926.2018.1523430
14. Kirichenko, L., Shergin, V.: Analysis of the properties of ordinary Levy motion based on the estimation of stability index. Int. J. Inf. Content Process. **1**(2), 170–181 (2014)
15. Shergin, V.L.: Estimation of the stability factor of alpha-stable laws using fractional moments method. Eastern Eur. J. Enterpr. Technol. **6**, 25–30 (2013)
16. Kozachenko, Yu., Kurchenko, O., Syniavska, O. : Levy-Baxter theorems for random fields and their applications. Shark, Uzhgorod (2018). (in Ukrainian)

17. Pashko, A.: Simulation of telecommunication traffic using statistical models of fractional Brownian motion. In: 4th International Scientific-Practical Conference Problems of Infocommunications. Science and Technology. Conference Proceedings, 10–13 October, pp. 414–418 (2017)
18. Pashko, A.: Accuracy of simulation for the network traffic in the form of Fractional Brownian Motion. In: Pashko, A., Rozora, I. (eds.) 14th International Conference on Advanced Trends in Radioelectronics, Telecommunications and Computer Engineering. Proceedings, pp. 840–845 (2018)
19. Dzhaparidze, K., Zanten, J.: A series expansion of fractional Brownian motion. CWI. Probability, Networks and Algorithms[PNA] R0216 (2002)

Subduction in the Earth's Lithosphere, Modeled as a Compressed Flow of Cubes Using the Discrete Element Method

Dmitri Vengrovich(✉) and Vasyl Kulich(✉)

Institute of Geophysics, NAS of Ukraine,
Bohdan Khmelnytskyi str., 63 G, Kiev, Ukraine
vengrovich@gmail.com, kulichvas@gmail.com

Abstract. The Earth's crust as an array of plates consists of blocks with a hierarchical size distribution. We present results of numerical simulation of the stress localization in Earth's crust in such tectonic processes as plates collision and subduction taking into account the discrete structure of the lithosphere. The motion simulation of an array consisting of cubes, in conditions of compression was carried out by the new algorithm Discrete Element Method (DEM) that has been tested in experiments with two cubes collision. During the movement, the three-dimensional hook periodically appears in an array of blocks. This effect is manifested in the increase of pair contact quantity, mechanical energy and in the growth of interaction forces between cubes array and border. Energy dissipation is rising especially. In blocks array, the force chains arise towards the external load. The originality of the study consists of the geophysical application of the discrete element method to consolidated media.

Keywords: Massif of cubes · Discrete element method · External loads

1 Introduction

Space heterogeneous block structure with a hierarchical size distribution is indicative of a geophysical media conception. Sizes ratio of the blocks of adjacent levels k ranges within fairly narrow limits k $\sim$ 3 [1, 2]. The interaction of blocks defines many processes in the geophysical media. The properties of earthquakes are due to the friction regularities between blocks [3]. Structures of blocks can be modeled by granular media with various forms and sizes. The behavior of granular media is significantly different from the properties of solids, liquids, and gases. Most of the commonly used techniques to study and simulate numerically geophysical processes assume the Earth or lithosphere to be a continuum with averaged physical properties and are, accordingly, founded upon the laws of continuum thermodynamics and mechanics. We have focused on studying such an aspect of these problems as the presence of a structure in the material constituting the Earth's crust. The numerical modeling algorithm was developed [4] and tested [5, 6]. The successful results of using this theory indicate that many geophysical processes in the Earth's crust can be investigated from a unified point of view on the crust as a structured block media. This work is a continuation of the experimental seismic wave

Z. Hu et al. (Eds.): ICCSEEA 2020, AISC 1247, pp. 192–202, 2021.
https://doi.org/10.1007/978-3-030-55506-1_17

propagation study caused by the irregularity of lithospheric stress as a result of the plates block structure [7]. The whole field of research on the granular media dynamics predicts the formation of localized structures during compression in the form of prestressed granules chains. They study the dynamics of such environments, and recently the passage of waves through them. In contrast to solid materials, waves in granular materials are more complex due to the heterogeneous nature of these systems. Fluctuations in the stresses and velocities in such medium are sometimes several times higher than average values. External loads on the granular medium lead to the formation of force chains in it in the direction of loading [8, 9]. Cubes of various sizes are taken as the granular medium elements in our case. We are used DEM for computer simulation the processes in a medium of cubic granules. DEM was introduced to simulate the large deformations in jointed rock masses [10]. The selection of sharp-edged elements allows modeling of the important characteristics of such systems [11]. It is the most effective means of studying the granular materials properties and granular flow phenomena from a microscopic point of view [12, 23]. The method was tested in experiments with a collision of 2 cubes.

2 Numerical Experiments with Cubes Massif in Limited Conditions Under External Load and Algorithm for DEM

2.1 Theory

Results of numerical simulation of the shear motion of a cubes massif by DEM using the "Cubluck" program was presented in [13]. Now, this program is used for modeling the motion of a block massif, consisting of cubes of near sizes, in limited conditions. The position of the i-th cube is defined by the center mass coordinates r_{ci} and three parameters defining its rotational motion. Thus, the granular medium composed of N cubes has $6\ N$ degrees of freedom and satisfies the following equations of motion:

$$m_i d^2 \boldsymbol{r}_{ci}/dt^2 = \sum\nolimits_j \boldsymbol{F}_{ij} \tag{1}$$

$$d\boldsymbol{K}_{ci}/dt = \sum\nolimits_j \boldsymbol{M}_c(\boldsymbol{F}_{ij}). \tag{2}$$

Here m_i, $\boldsymbol{r}_{ci}$ are the mass and vector of a center for the i-th cube; $\boldsymbol{F}_{ij}$ is the contact force appearing between i-th and j-th cubes; $\boldsymbol{K}_{ci}$ is the kinetic moment of i-th cube with respect of its center; $\boldsymbol{M}_c(\boldsymbol{F}_{ij})$ is torque with respect to the center of i-th cube.

Due to the cube symmetry, the rotational motion of cube and sphere around their centers has many common features and, in particular, $\boldsymbol{K}_{ci} = I_{ci}\boldsymbol{\omega}_{ci}$, where $\boldsymbol{\omega}_{ci}$ is the vector of angular velocity for the cube rotating about its center, I_{ci} is the corresponding moment of inertia for the cube, $I_{ci} = \ const$. Therefore (2) are simpler than the Euler equations for rotation of arbitrary shapes figures.

The moment of contact force acting on cube i during the contact with cube j is $\boldsymbol{M}_c(\boldsymbol{F}_{ij}) = \boldsymbol{r}_{ij}^c \times \boldsymbol{F}_{ij}$, where $\boldsymbol{r}_{ij}^c$ is the vector pointing from cube center to contact point of adjacent cubes.

The description of the translational motion of cube center is relatively simple task, whereas the consideration of its rotational motion demands more effort. To overcome well-known obstacles, we apply the quaternion technique [14]. In the global coordinates system, the kinematic equation for rotation quaternion has following form $d\Lambda/dt = \omega_{ci} \circ \Lambda/2$, where $\Lambda = (\lambda_0, \lambda_1, \lambda_2, \lambda_3) = \lambda_0 + \lambda_1 \boldsymbol{i}_1 + \lambda_2 \boldsymbol{i}_2 + \lambda_3 \boldsymbol{i}_3 = \lambda_0 + \boldsymbol{\lambda}$ is quaternion. During calculations, we also perform the quaternion normalization adjusting the norm following the modified equation

$$d\Lambda/dt = \boldsymbol{\omega}_{ci} \circ \Lambda/2 - q\Lambda(\|\Lambda\| - 1). \tag{3}$$

The parameter q varies from 0 to 1. We put $q = 0.5$.

Cubes vertex coordinates during rotation are found from the formula $\boldsymbol{r}' = \Lambda \bigcirc \boldsymbol{r} \bigcirc \Lambda^{-1}$.

Thus, relations (1,2,3) form the closed system of equations for the description interaction of pair cubes. To solve Eqs. (1) and (2), Beeman's algorithm is used [15]. Equation (3) is solved by the Runge-Kutta 4th order method.

Cubes interact when they intersect. It is assumed that the cubes are not deformed. The search for contacts between them is divided into 2 stages. Far contact occurs when crossing the spheres described around the cubes. When this condition is met, we can find near contact.

To describe near contact independent from the type of contact, we use algorithm [11] which considers the particle as a domain bounded by halfspaces $(\boldsymbol{\xi} - \boldsymbol{c} - \boldsymbol{a})\boldsymbol{a} \leq 0$, where $\boldsymbol{c}$ is the center of mass for particle, $\boldsymbol{\xi}$ is the coordinate vector of an arbitrary point in halfspace, $\boldsymbol{a}$ is a vector pointing from particle center of mass perpendicular to edge plane of halfspace.

The resulting figure of 2 cubes intersection is a convex polyhedron. The algorithm for finding its vertices is to evaluate all intersection points of all combinations of two planes of faces of one cube and the face plane of another cube.

Interaction between bodies is defined by a single force that is broken into normal force and tangential friction force. Point coordinates, where the contact force is applied, coincide with the mean values of vertices coordinates of intersection figure $\boldsymbol{x}^O = \sum_j \boldsymbol{x}_j/m$, where m is their number.

The direction of force is defined by the unit normal in contact. To define this vector $\boldsymbol{n}^f$, we use the contact line which is the broken line of intersection for surfaces of two contacting particles [12]. On the base of all segments of the contact line, we construct the fitting plane (FP). To do this, the singular value decomposition (SVD) [16–18] is used. Normal to this plane coincides with contact normal and define the direction for the normal force of interaction between particles.

We used the model for normal force, which has been proposed in [19]. It does not depend on the type of contacts, is suitable for the particles with various shapes and defines the interaction between bodies. Using this model, contact interaction between bodies is described by the single force which depends on volume V of intersection domain of interacted bodies and penetration depth d,

$$F^n = kE^*(Vd)^{0.5},$$

where coefficient $k = 4/3 \cdot (\pi)^{-0.5}$, $V = 1/6 \cdot d \cdot d_{max} \cdot d_p$, and d_{max} is the maximal distance between vertices of intersection figure (direction $\boldsymbol{l}_m$), d_p is maximal distance between vertices in the direction $\boldsymbol{n}_p = \boldsymbol{n}^f \, x \, \boldsymbol{l}_m$.

For identical materials, the effective Young's moduli E^* satisfies the relation $E^* = 0.5E/(1-\nu^2)$ where E and ν are Young's moduli and Poisson ratios for contacting bodies. The penetration depth is defined as a distance of penetration of the intersection domain in normal force direction and is derived by the following formula: $d = \max(\boldsymbol{n}^f \cdot \boldsymbol{p}^c) - \min(\boldsymbol{n}^f \cdot \boldsymbol{p}^c)$.

Besides normal forces, during contact, the tangential forces arise. These forces are evaluated via the Coulomb model of friction. This model distinguishes static friction (coupling) $F^s < \mu^s F^n$ and kinetic friction (slip) $F^k = \mu^k F^n$, where μ^s and μ^k are the empirical coefficients. At the application of this model some difficulty arises, since during contact, we don't know the type of contact and all forces applied to the particle. Another problem concerns the jumps of friction forces. To avoid this problem, authors of the article [19] offer the model for friction in which the discontinuities of friction forces are smoothed. The static friction is approximated via the slip model with very low velocities and then

$$\begin{aligned} F^f &= ((2\mu^{s*} - \mu^k)x^2/(x^4 + 1) + \mu^k - \mu^k/(x^2 + 1))F^n, \\ \mu^{s*} &= \mu^s(1 - 0.09\left(\mu^k/\mu^s\right)^4), \; x = v^t/v^s. \end{aligned}$$

Here v^t is the tangential velocity, v^s is the velocity of transition from static to kinetic friction. The direction of friction force F^f is parallel to relative tangential velocity of cubes at contact point. Code "Cubluck" has been written on the base of described model.

2.2 Verification

The algorithm and program have been verified. The collision of two polycarbonate cubes was modeled by the DEM method. The symmetrical face-edge collision was determined in the begin conditions. The initial arrangement of the cubes is depicted in Fig. 1(1).

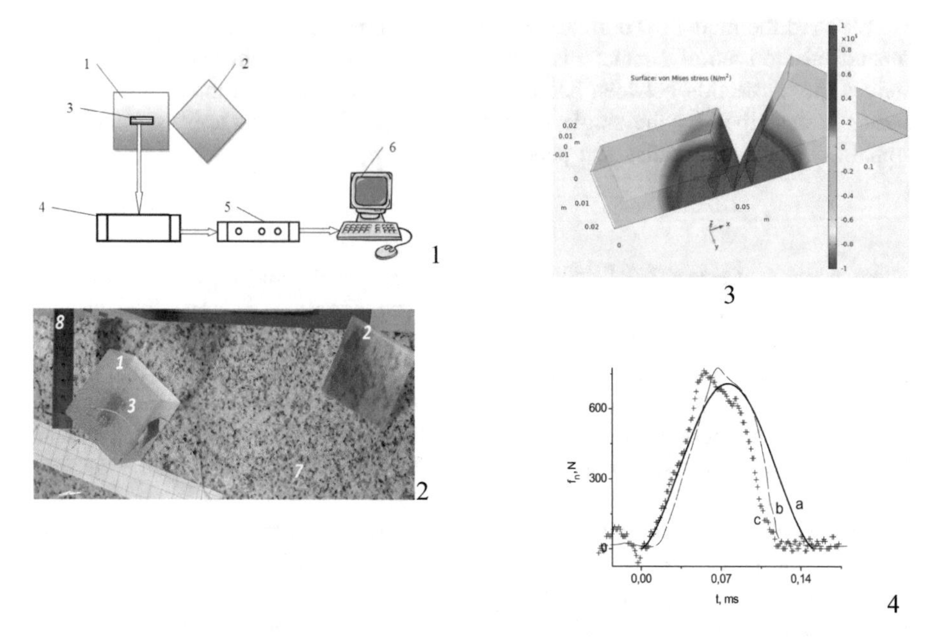

Fig. 1. Comparison of DEM, FEM calculations of two cubes collision and experiments. 1. Scheme of the experiment; 2. Photo of cubes after interaction; 3. Distribution of stresses in the cubes during interaction (here is a quarter of the area due to symmetry); 4. Evolution of the normal interaction force of cubes during contact, obtained by 3 methods: a – DEM calculations; b – FEM calculations; c – results of experiments.

In the global coordinate system, one cube is rotated by 45° around axis Oz, other is not rotated. Material properties: edges of cubes are 48.5 mm in size; density ρ = 1.777 g/sm^3; Young's module E = .2 10^9 Pa; Poisson ratios ν = 0.35. Calculations by the DEM method were compared with experiments and calculations by the FEM method. The experimental scheme is depicted in Fig. 1(1): 1-hitting cube; 2-cube for which hits; 3-strain gauge; 4-strain station; 5-oscilloscope RS-500; 6-computer. The location of the cubes after interaction in the experiment is shown in Fig. 1(2): 7-polished granite slab; the hitting cube was accelerated by a steel strip 8 to initial speed v_{x0} = 0.3 m/s.

Distribution of stresses in the cubes during contact received in the calculations by FEM method depicted in Fig. 1(3). The evolution of the normal interaction force between cubes during contact obtained by 3 methods is shown in Fig. 1(4): a - DEM calculations; b - FEM calculations; c - results of experiments. The figure shows that the normal force amplitude is slightly smaller, and the contact time is larger in DEM calculations compared to FEM calculations and experiments. Similar results of calculations are observed.

2.3 Simulation

Computer simulation of the motion of a block massif was carried out. The cubes had mechanical properties characteristic of acrylic plastic: density $\rho = 1.2$ g/cm^3; Young's modulus $E = 3. * 10^9$ N/m^2; Poisson's ratio $\nu = 0.3$. Friction force parameters: $\mu_s = 0.7$; $\mu_k = 0.4$; $v_s = 0.01$ m/s.

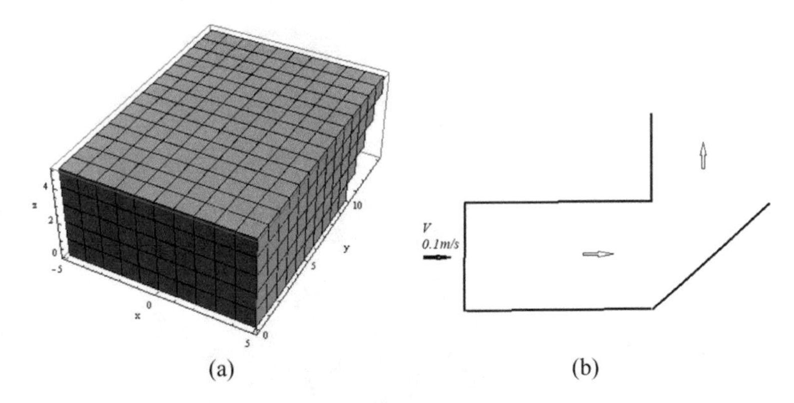

Fig. 2. The initial location of a three-dimensional massif of 600 cubes and schematic representation of the cubes massif motion.

The cubes sizes were considered as variables with base size 1 cm. The dimensions of the edges are reduced to 1 cm × $(1-p)$, where p is a random value $0 < p < 0.01$. The initial arrangement of the cubes is shown in Fig. 2(a). Initially, cubes are unmoved and their edges parallel to the axes of a rectangular Cartesian coordinate system. 10 cubes are located in wide and 5 cubes - in height. In length, the number of cubes increases by 1 from 10 to 14 with increasing height. Their centers have the coordinates as for a massif of identical cubes with size 1 cm.

For ordered movement, 6 boundary cubes of 50 cm size are installed, which do not interact among ourselves. The edges of 5 of them are parallel to axes of a rectangular Cartesian coordinate system. Cube № 4 was rotated 45° relative to Ox axis. Cube № 2 moves along axis Oy at a constant velocity $V_y = 0.1$ m/s, the other 5 are stationary.

Coordinates of centers of boundary cubes (in cm): № 1 (0., 10., −25.); № 2 (0, −25. (initial), 5.); № 3 (30., 10., 5.) № 4 (0., 10. + 25. $\sqrt{2}$., 0.); № 5 (−30., 10., 5.); № 6 (0., −15., 30.).

Boundary lines in the plane $x = 0$ and the cubes movement direction under the action of external loading are shown schematically in Fig. 2(b). The system has a free area where cubes fly under the influence of external loading.

Cubes array simulates the behavior of a moving tectonic plate of the earth's crust in subduction zone. Boundary cubes 1 and 4 model an immovable tectonic plate. For better visualization, the model is inverted with respect to the natural tectonic plates arrangement in subduction zone.

The calculations were performed in dimensionless variables. The characteristic values are x = 0.01 m; m = 0.001 kg; t = 0.001 c. From them the characteristic force F = 10 N, velocity v = 10 m/s and others are deduced.

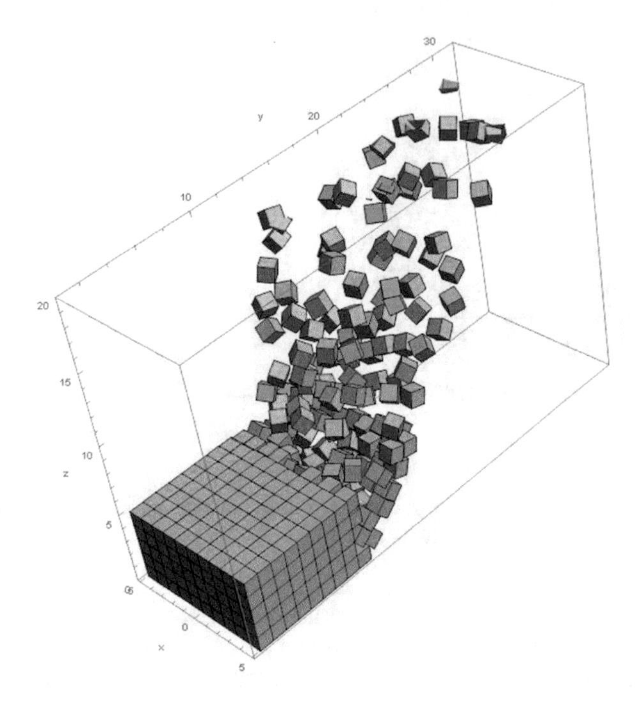

Fig. 3. Cubes massif position with a uniform shift of the boundary cube on 1 cm per 0.1 s.

Cubes array location with a uniform shift of the boundary in 1 cm per 0.1 s is shown in Fig. 3. Under influence of external loading, the cubes fly up to a height to 60 cm.

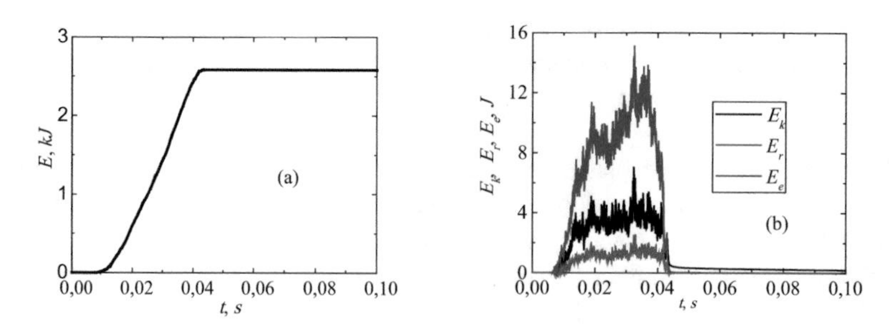

Fig. 4. Changing the total E (a), kinetic E_k, rotational E_r and elastic E_e energies (b) of cubes system within 0.1 s.

Mechanical energy changes unevenly over time. There is a sharp increase in the total mechanical energy of the system during 10 ms–43 ms, caused mainly by the contribution of dissipative energy E_d (Fig. 4(a). There is a sharp increase in the kinetic, rotational, and elastic energies of the cubic system at this time interval (Fig. 4(b)). In further time, these types of energy decrease, the contribution of dissipation (friction) is negligible.

The number of paired contacts N_c of cubes in the allocated time range increases sharply too, as shown in Fig. 5.

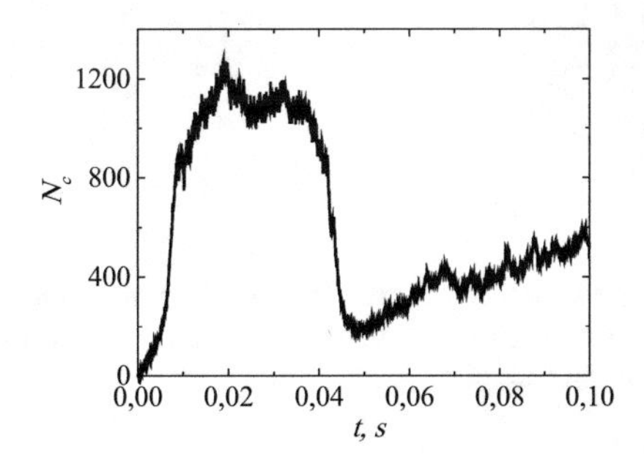

Fig. 5. Evolution of the number of paired contacts N_c in cubes system within 0.1 s

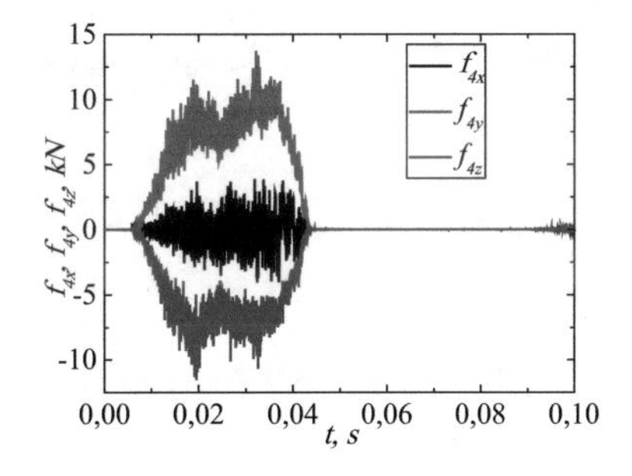

Fig. 6. Components of sum force acting on the boundary cube No. 4 within 0.1 s

The total interaction forces of boundary cubes with the massif of cubes in the allocated time range are increasing dramatically. For example, the total force components acting on the boundary cube № 4 are shown in Fig. 6. Distribution of paired contacts in a system of 600 cubes at the moment of 20 ms, when the number of contacts between them reaches 1000 pieces, is shown in Fig. 7a. The whole system of cubes is covered by interaction. Cylinders connect the centers of cubes of contacting pairs.

The distribution of interaction forces between 600 cubes at the moment of 20 ms of external loading is shown in Fig. 7(b). Maximum forces are formed in the direction of external load acting, and considerable forces in the direction of free region. The larger diameter of the cylinder corresponds to the bigger force. The maximum interaction force is 2.2 kN.

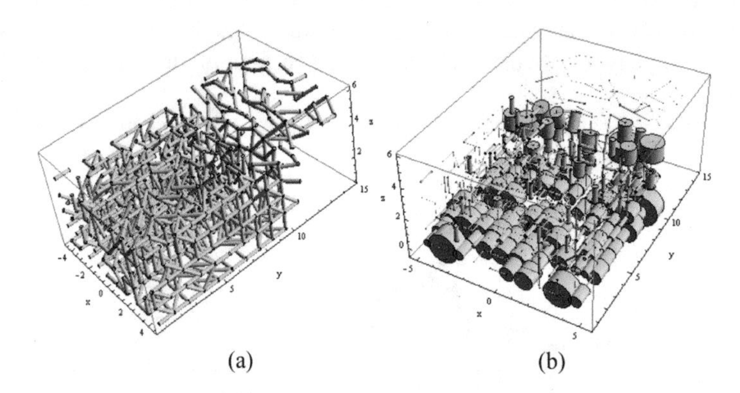

(a) (b)

Fig. 7. Distribution of paired contacts (a) and forces (b) between cubes at time 20 ms of external load action, the maximum interaction force is 2.2 kN. Larger diameter of the cylinder corresponds to bigger force in Fig. 7(b).

The described increase of cubes interaction by time interval 10–43 ms in the massif, obtained as a result of computer simulation the motion of cubes under influence of external loading by DEM, can be considered as a collective volumetric hook in the system. A similar phenomenon is observed during earthquakes [20–22].

Continued calculations up to 0.22 s led to the appearance of a hook 2 times else in the cubic system as shown in Fig. 8.

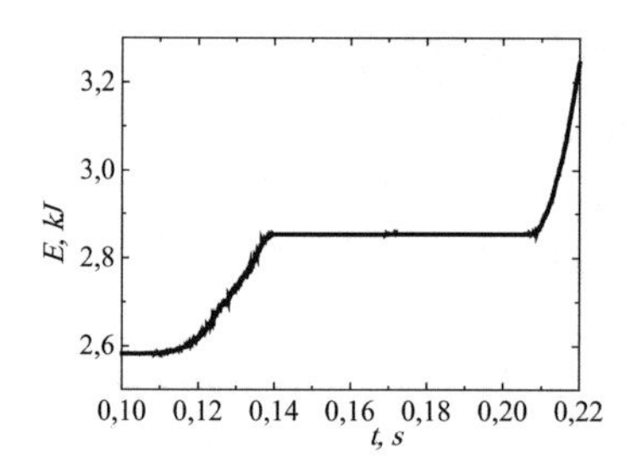

Fig. 8. Change in the total energy E of cubes system in time from 0.1 to 0.22 s

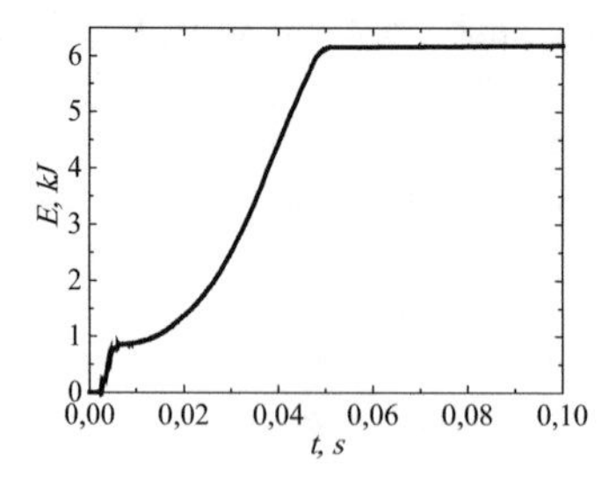

Fig. 9. Changing the total energy of a cubes system with constant dimensions of 1 cm over a time 0.1 s

The above calculations were repeated for a cubes massif with the same size of edges 1 cm. The evolution of the total energy of the system within 0.1 s is shown in Fig. 9. A hook appears 2 times in 0.1 s of a process in a massif with the same cubes. As can be seen from the figure, in the system the total energy increases more than 2 times for 0.1 s compared to the calculations for a massif with different cubes sizes (Fig. 4a).

3 Concluding Remarks

Computer simulation of a three-dimensional cubes massif motion in limited conditions under the influence of external loading was carried out. In the process of moving a collective volumetric hook periodically arises, that comprises most cubes. Force chains are formed toward the direction of external load acting. It shows that the simulated effects of the motion of a block massif consisting of cubes of close sizes in comprehensive compression conditions can correspond to natural Earth's crust processes. As a remarkable achievement of this approach should be considered a high speed of calculation by program "Cubluck", due to the use of quaternions theory and the simplicity of the cubes rotations calculation. On the other hand, realistic forms of blocks in an array can be created by the cubes bonding algorithms.

References

1. Sadovsky, M.A.: Selected Works: Geophysics and Explosion Physics. Sadovsky, M.A.; Exec. Adushkin, V.V. (ed.), 440 p. Nauka, Moscow (2004). (in Russian)
2. Kocharyan, G.G.: Fault Geomechanics. GEOS, Moscow, 424 p. (2016). (in Russian)
3. Scholz, C.H.: Earthquakes and friction laws. Nature **391**, 37–42 (1998)
4. Starostenko, V.I., Danilenko, V.A., Vengrovich, D.B., Kutas, R.I., Stovba, S.M., Stephenson, R.: Modeling of the evolution of sedimentary basins including the structure of the natural medium and self-organization processes. Izvestiya, Phys. Solid Earth. **37**(12), 1004–1014 (2001)
5. Vengrovich, D.B., Sheremet, G.P.: Irregularity of lithospheric stress as a result of plates structure. In: 18th International Conference on Geoinformatics - Theoretical and Applied Aspects (2019). https://doi.org/10.3997/2214-4609.201902152
6. Vengrovich, D.B.: Simulation of salt diapirism in Dnieper-Donets Basin for hydrocarbon exploration. In: Geoinformatics 9th EAGE International Conference on Geoinformatics - Theoretical and Applied Aspects (2010). https://doi.org/10.3997/2214-4609.201402780
7. Vengrovich, D.B.: Tectonic and seismological settings of subduction. In: 16th International Conference Geoinformatics -Theoretical and Applied Aspects (2017). https://doi.org/10.3997/2214-4609.201701858
8. Majmudar, T.S., Behringer, R.P.: Contact force measurements and stress-induced anisotropy in granular materials. Nature **435**, 1079–1082 (2005). https://doi.org/10.1038/nature03805
9. Grinchuk, P.S., Danilova-Tretiak, S.M., Stetukevich, N.I.: Force chains influence on the conductivity of granular media. Reports of the NAS of Belarus, vol. 57, no. 6, pp.105–112 (2013). (in Russian)
10. Cundall, P.A.: Formulation of a three-dimensional distinct element model. Part I: a scheme to detect and represent contacts in a system composed of many polyhedral blocks. Int. J. Rock Mech. Min. Sci. Geomech. Abstr. **25**(3), 107–116 (1988)

11. Nassauer, B., Liedke, T., Kuna, M.: Polyhedral particles for the discrete element method. Granular Matter **15**(1), 85–93 (2013). https://doi.org/10.1007/s10035-012-0381-9
12. Zhao, S., Zhou, X., Liu, W.: Discrete element simulations of direct shear tests with particle angularity effect. Granular Matter **17**(6), 793–806 (2015). https://doi.org/10.1007/s10035-015-0593-x
13. Mykulyak, S., Kulich, V., Skurativskyi, S.: Simulation of shear motion of angular grains massif via the discrete element method. In: Hu, Z., Petoukhov, S., Dychka, I., He, M. (eds.) Advances in Computer Science for Engineering and Education. AISC, vol. 754, pp. 74–81. Springer, Cham (2019). https://doi.org/10.1007/978-3-319-91008-6_8
14. Branets, V.N., Shmyglevsky, I.P.: Application of Quaternions in Problems of Orientation of a Rigid Body, 320 p. Nauka, Moscow (1973). (in Russian)
15. Beeman, D.: Some multistep methods for use in molecular dynamics calculations. J. Comput. Phys. **20**(2), 130–139 (1976)
16. Shakarji, C.V.: Least-squares fitting algorithms of the NIST algorithm testing system. J. Res. Natl. Inst. Stand. Technol. **103**(6), 633–641 (1998)
17. Forsythe, G.E., Malcolm, M.A., Moler, C.B.: Computer Methods for Mathematical Computations. Prentice Hall Inc., Englewood Cliffs (1977)
18. Awati, A.S., Patil, M.R.: Inpainting of structural reconstruction of monuments using singular value decomposition refinement of patches. Int. J. Image Graph. Sig. Process. (IJIGSP) **11**(5), 44–53 (2019). https://doi.org/10.5815/ijigsp.2019.05.05
19. Nassauer, B., Kuna, M.: Contact forces of polyhedral particles in discrete element method. Granular Matter **15**(3), 349–355 (2013). https://doi.org/10.1007/s10035-013-0417-9
20. Rebetsky, Yu.L.: Signs of metastability of the crust in preparation for a catastrophic earthquake - features of the tectonic stress field. Trigger effects in geosystems. In: IV All-Russian Conference, Moscow, pp. 19–29 (2017). (in Russian)
21. Dobrovolsky, I.P.: The Theory of Tectonic Earthquake Preparation, 218 p. Nauka, Moscow (1991). (in Russian)
22. Sherman, S.I.: Tectonophysical signs of the formation of strong earthquake foci in seismic zones of Central Asia. Geodyn. Tectonophysics. **7**(4), 495–512 (2016). https://doi.org/10.5800/GT-2016-7-4-0219
23. El Islam, A.A.N., Abdelghani, A., Houari, B.: Electromagnet separator of different particles. Int. J. Eng. Manuf. **5**, 22–31 (2018). https://doi.org/10.5815/ijem.2018.05.03. Published Online September 2018 in MECS

The Spatial Characteristics of High-Speed Railway Noise Emission at Different Speeds

LiLi Li[1(✉)], Qing Xie[2], and Sha Wu[2]

[1] Wuhan Railway Vocational and Technical College, Wuhan 430205, China
13638693558@163.com

[2] Jingzhou Environmental Protection Science and Technology Co., Ltd., Jingzhou 434000, China

Abstract. It is the rapid development of high-speed railway in China that has brought great convenience and progress to the social economy. The overall operation of high-speed rail has a limited impact on the external environment, which is mainly caused by operation noise. The main sources of noise in high-speed railway operation are the mechanical noise of the train itself, the wheel-rail collision noise and the frictional noise generated by the train and the air. By monitoring the environmental noise of high-speed railway in different running states, the results show that the running speed of high-speed railway is positively correlated with noise emission intensity. When high-speed railway runs at low speed, the main noise source mainly comes from wheel, rail and machinery itself. While above a certain speed, the total noise source is dominated by the frictional noise generated by the vehicle and the air. At the same speed, the noise generated by high-speed railway is different for different project locations which the tunnel entrance is the highest while the bridge and straight road section are basically not affected by the external structure. At the same time, the sound barrier has an effective protection to the external environment, which can greatly reduce the noise emission of high-speed rail.

Keywords: High speed railway · Noise · Monitoring

1 Instruction

As one of the important symbols of urbanization development, high-speed railway has greatly improved the life and efficiency of urban residents, reduced the travel time between regions, and it is an important modern means of transportation [1]. As an important urban transportation facility, it is one of the important research hot-spots that the impact of noise emission characteristics on the environment. In addition, China is leading in high-speed railway construction, also is forward-looking about the study on high-speed rail environmental issues [2, 3].

It is one of the key points of environmental research on traffic noise. As early as in the era of common railway development, railway noise was the object of continuous attention. Subsequently, different types of research were carried out for different train types, such as electrified locomotives and magnetic levitation trains [4, 5]. According to the research results of various countries, railway noise mainly comes from the collision

Z. Hu et al. (Eds.): ICCSEEA 2020, AISC 1247, pp. 203–209, 2021.
https://doi.org/10.1007/978-3-030-55506-1_18

between railway tracks and wheels, the mechanical operation of the train itself, and the friction between the train and air [6, 7], and the intensity of noise source is different in different velocity intervals. This is closely related to the type of phased noise. According to the study, the relationship between the occurrence speed interval of high-speed rail is shown in Fig. 1 [8, 9]. The main noise source in the long journey stage is aerodynamic noise, that is, noise generated by train operation and air friction. The theoretical analysis is shown in Fig. 2 [10]. Due to the increasing maturity of the theory and technology of modern high-speed railway at this stage, which greatly improved the vehicle body condition and operation facility, so that the focus of noise emission source intensity has also changed [11], which the original short-circuit rail is replaced by long-distance seamless rail [12], the interior of the vehicle body is completely closed, the mechanical structure is refined and the speed is greatly increased [13, 14]. On this basis, the main factor of noise emission is the friction between high-speed train operation and air. Speed-up safely is the important goal of high-speed railway operation. On the basis of speed-up safely, it is the key point of further research how to control noise [7]. This paper analyzes the characteristics of noise emission by monitoring the operation noise of high-speed railway and provides the direction of external noise control.

The significance of this paper is to understand the noise emission characteristics of high-speed railway at different speeds and the effectiveness of the main noise emission control measures. Basing these characteristics, improvement of protective measures can reduce the external impact of high-speed railway.

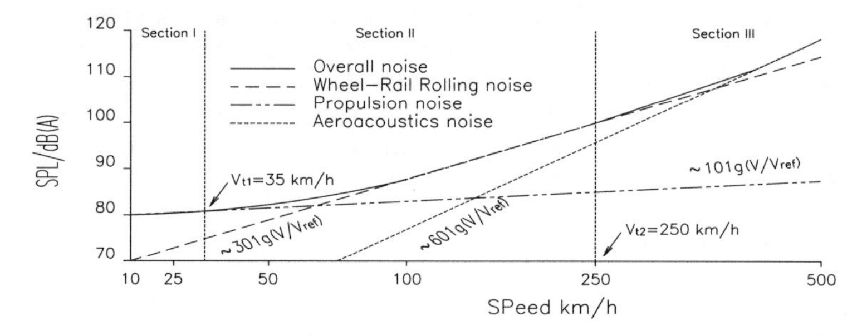

Fig. 1. Train noise source and speed distribution characteristics (Drawing by the author for showing the noise source from high-speed railway from the monitoring results)

Based on the above characteristics, we calculated and determined the monitoring noise reference index as the equivalent continuous noise level. The principle is as follows:

$$\mathrm{L_{eq}} = 10\lg\left[\frac{1}{t_2 - t_1}\int_{t_1}^{t_2} 10^{\frac{L}{10}} dt\right]$$

Where:

L_{eq} – equivalent continuous A-weighted sound pressure level

L – Instantaneous value of noise level

As a rapidly growing infrastructure, high-speed railway greatly convenient for residents, but it is inevitably that part of the high-speed railway runs through urban buildings or residential areas [15]. The research on noise [16, 17] emission characteristics of high-speed railway can provide the basis for noise protection and noise reduction of high-speed railway, and reduce the negative impact of high-speed railway.

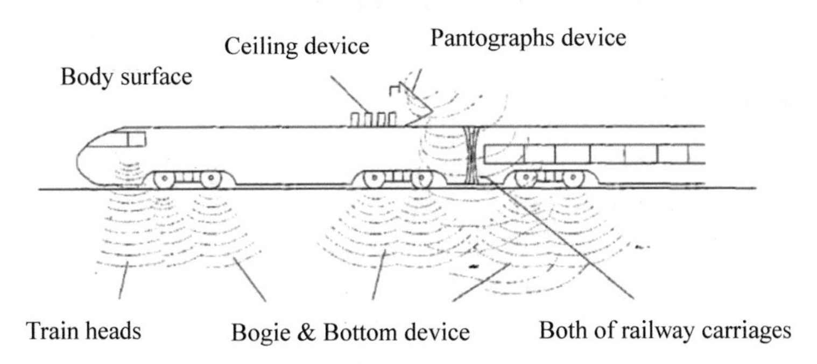

Fig. 2. Schematic diagram of aerodynamic noise (Drawing by the author for showing the noise source from high-speed railway)

The operation principle of the monitoring instrument is to collect the parameters involved in the formula by who derives, and then calculate the results.

2 Background and Purpose of Monitoring

2.1 Monitoring Plan and Methods

In this study, the high-speed railway line between Wuhan and Jiujiang was selected as the monitoring object, and continuous monitoring was conducted at 8:00–18:00 on June 20 to June 25, 2018. According to "railway boundary noise limit and measurement method" (GB12525-90), we tested one hour equivalent sound level under the condition of average density of train operation. It was selected as the monitoring point 30 m outside the boundary of the railway corridor, and 1.2 m above the ground. Due to the impact of noise protection measures on the residential area, monitoring point A of the neighboring residents is located 30 m outside the corridor, and monitoring point B is located 80 m outside the corridor.

Monitoring daytime vehicle operation, we select the station in and out of the direction (monitoring points L1 and L2), D train operation section (L3), tunnel entrances (L4), the bridge gateway (monitoring points L5), near the residential area A (about 5 units, monitoring points L6), near the residential area B (about 6 units, monitoring points about L7). The high-speed railway of this line is expected to have 20–24 trains every day of full load, and the designed speed is 200–300 km/h. According to the actual operation survey, due to the new line, the passenger flow is not full, currently 10–12 trains every day. At the same time, parallel trains in the same direction and reverse direction are avoided when high-speed railway passes.

2.2 Monitoring Purposes

The purpose of monitoring is to determine the noise emission difference of high-speed railway at different speeds and the noise emission difference of high-speed railway in different engineering positions.

The research team chose days for monitoring noise emission by actual railway running with the noise meter. The monitoring methods are following the plan above.

3 Results and Discussion

3.1 The Relationship Between the Characteristics of High-Speed Railway Running Speed and Noise Emission

The monitoring results show that the variation trend of noise emission under different operating conditions is shown in Table 1.

Table 1. Summary of noise monitoring results at different positions (Field measurement record)

Monitoring location	Speed per hour (km/h)	Noise value dB (A)	Location (m)	Measures
Into the station (L1)	80	55.5	30	–
Outbound (L2)	100	56.7	30	–
Operation (L3)	260	58.5	30	–
Tunnel (L4)	200	59.0	30	–
Bridge (L5)	200	57.5	30	–
Residential areas A (L6)	200	58.5	30	Sound barrier
		56.5	45	
Residential areas B (L7)	200	56.5	80	–
		56.0	100	

It can be seen from Table 1 that during the monitoring period, the noise emission range of high-speed railway operation of this line is within the range of 55.5–59.0 dB, in which the noise emission monitoring result is obviously positively correlated with the vehicle running speed. Once the high-speed railway enters the urban built-up area, it will rapidly decrease its running speed and control its speed at 80 km/h. When it runs out of the urban built-up area, its speed will gradually increase to100 km/h. In addition to the friction noise caused by the speed, it is mainly the noise changes generated by important engineering facilities.

From Table 1, the noise is higher in high-speed operation than that in low-speed operation, which conforms to the results of literature research and the current design characteristics of high-speed railway. It is the main cause of noise that the friction between high-speed trains and air, which visual hearing is similar to the sharp wind in the original distance. Similarly, under the condition of running speed, the external engineering characteristics and engineering structure also have a direct impact on the noise. The noise at the entrance and exit of the tunnel is slightly higher than that of the train passing through the bridge or the straight section. The tunnel structure will affect the direction and intensity of noise radiation. Without acoustic environmental protection measures, that is, without acoustic barriers, the residential areas outside the corridor should not reach the standard of noise emission. The monitoring result of residential area B from 80 m away from the corridor reached the standard, which was similar to the 30 m monitoring result of the sound barrier. It was also shown from another perspective that, in the case of residential areas near the corridor, the sound barrier effectively reduced the noise value. In the case of lack of protection in the near distance, the noise value could not reach the standard of the acoustic functional area in the non-traffic environment. In addition, on the premise that the straight-line distance between A and B in residential areas is 55 m different, the difference in noise monitoring results is relatively small, except the sound barrier function, the high speed rail noise energy is strong, the transmission distance is long, not easy to attenuation.

3.2 Noise Emission from High-Speed Railway Operation Affects the Life of Surrounding Cities

High-speed railway is mainly used for passenger transport. As an important tool for urban operation, high-speed railway stations should not be too far away from the city center, nor be placed in areas with too dense buildings in the city center. The layout of high-speed railway station is generally placed in the sub-center of the city and the main traffic passage will pass through some urban edge buildings and a small number of living areas. It is inevitably that noise emission of high-speed railway has an impact on the external environment, especially on residential areas and office environment. In view of this problem, designers, research experts and management departments of high-speed railway have taken a variety of feasible measures. Among them, the effect of sound barrier is remarkable that also can be seen from the above monitoring results that sound barrier plays a significant role in reducing the noise of high-speed railway in a short distance. According to the characteristics of high-speed railway operation, residential areas A and B were selected for further monitoring of specific noise effects.

Table 2 shows that when the train runs through residential areas A and B, away from the corridor 30 m that residential areas B without sound barrier which noise is 70.2 dB, that belongs to a high noise and interferes with life. As mentioned above, it lacks sound barrier protection. In order to run smoothly, high-speed railways usually have high roadbed sections, and residential areas have elevation differences from high-speed railway corridors. Through supplementary monitoring of noise emission level in the vertical direction, corresponding monitoring is carried out for different heights of the same horizontal distance The results show that the position with the shortest linear distance from the corridor of high-speed railway has the highest noise intensity, and the acoustic barrier can ensure the effective control of noise emission at different heights in the near distance. The noise monitoring results of residential area B without sound barrier at 30 m position show that it is about 20 dB higher than residential area A with sound barrier. Monitoring results 80 m away from the corridor show that noise of residential area A has certain attenuation but less than 10 dB, it may be that the emission source of high-speed railway is strong, the noise energy is strong, and not easy to attenuate, also a part of the reason is the local acoustic environment level. In residential areas without barriers, B attenuation results are less than 15 dB. It indicates that the local level of area A is relatively high, and normal distance attenuation can greatly reduce the noise propagation. However, for the high-altitude environment without shielding, the noise propagation distance is still very far.

Table 2. Summary of noise monitoring results at different positions (Field measurement record)

Monitoring points	From the gallery	Ground 1.2 m monitoring results	Ground 3 m monitoring results	Ground 7 m monitoring results
Residential areas A (L6)	30 m	58.5	59.0	58.8
	80 m	50.2	51.0	51.0
Residential areas B (L7)	30 m	70.2	72.5	71.0
	80 m	56.5	59.0	62.0

4 Summary and Conclusion

1) The monitoring results show that the noise of high-speed railway running mainly comes from the friction between high-speed train body and airflow. The noise monitoring results in high-speed railway operation stage are obviously larger than the noise of vehicle acceleration and deceleration, The results show that the noise level is positively correlated with the vehicle speed, and running speed from low to high corresponds to noise results from low to high.
2) The impact of the train passing through a special section of road is related to the external structure of the train passing through the section, such as the tunnel and the bridge. Besides the influence of noise speed, the external structure also has a certain influence on the noise propagation. In view of the fact that the selection of high-speed railway will generally avoid the impact on the life of urban residents, as long as the nearby residential areas that cannot completely avoid the impact can be reduced as long as they keep a certain distance.
3) It is the key point of further study on the condition of train return in the same direction or reverse direction and multi-vehicle parallel. The influence of noise emission on urban residents' life and special protected areas under complicated vehicle conditions is urgently needed to be studied.

Acknowledgment. This project is supported by Environmental protection of the new section of Wuhan-Jiujiang passenger dedicated line from Daye north station to Yangxin station has been completed and accepted (2018-2-0049).

References

1. Khanaposhtani, M.G., Gasc, A., Francomano, D., et al.: Effects of highways on bird distribution and soundscape diversity around Aldo Leopold's shack in Baraboo, Wisconsin, USA. Landscape Urban Plann. **192**, 103–136 (2019)
2. Paraskevi, B., Pavlos, K., Apostolos, K.: Effects of road traffic noise on the prevalence of cardiovascular diseases: the case of Thessaloniki. Greece. Sci. Total Environ. **703**, 134–177 (2019)
3. Janello, C., Donavan, P.R.: Mapping heavy vehicle noise source heights for highway noise. Transp. Res. Rec. **2672**(24), 134–143 (2018)
4. Kamineni, A., Duda, S.K., Chowdary, V., et al.: Modelling of noise pollution due to heterogeneous highway traffic in India. Transp. Telecommun. J. **20**(1), 22–39 (2019)
5. Elmenhorst, E.M., Griefahn, B., Rolny, V., et al.: Comparing the effects of road, railway, and aircraft noise on sleep: exposure-response relationships from pooled data of three laboratory studies. Int. J. Environ. Res. Public Health **16**(6), 1073 (2019)
6. Alexander, A., El-Aassar, A., MacDonald, J., et al.: Technical approaches to developing a highway noise programmatic agreement. Transp. Res. Rec. **2673**(1), 102–109 (2019)
7. Pathak, S.S., Lokhande, S.K., Kokate, P.A., et al.: Assessment and Prediction of Environmental Noise Generated by Road Traffic in Nagpur City, India. In: Singh, V., Yadav, S., Yadava, R. (eds.) Environmental Pollution. Water Science and Technology Library, pp. 167–180. Springer, Singapore (2018)
8. Lylloff, O.A.: Efficient nonnegative least squares solvers for beamforming deconvolution. Lyngby Technical University of Denmark (2014)
9. Zhu, Z., Cheng, Z.: Numerical simulation research and experimental verification of aerodynamic force of high-speed railway sound barrier. Railway Stand. Des. **11**, 77–80 (2011). (in Chinese)
10. Zhou, X., Xiao, X., He, B., et al.: Prediction and verification of noise reduction effect of high-speed railway sound barrier. J. Mech. Eng. **49**(10), 14–19 (2013). (in Chinese)
11. Sirin, O., Ohiduzzaman, M., Kassem, E.: Evaluation of noise performance of multi-lane highways in the State of Qatar. In: IOP Conference Series: Materials Science and Engineering, vol. 517, no. 1, pp. 012–106. IOP Publishing (2019)
12. Kurita, T., Mizushima, F.: Environmental measures along Shinkansen lines with FASTECH360 high -speed test trains. JR EAST Tech. Rev. **16**, 47–55 (2010)
13. Verbeek, T.: The relation between objective and subjective exposure to traffic noise around two suburban highway viaducts in Ghent: lessons for urban environmental policy. Local Environ. **23**(4), 448–467 (2018)
14. Bazaras, J.: Internal noise modeling problems of transport power equipment. Transport **XXI** (1), 19–25 (2006)
15. Sassa, T., Sato, T., Yatsui, S.: Numberical analysis of aerodynmic noise radiation from a high-speed train surface. J. Sound Vibr. **247**(3), 407–416 (2001)
16. Gourav, T.S.: An efficient adaptive based median technique to de-noise colour and greyscale images. Int. J. Modern Educ. Comput. Sci. **11**(2), 48–53 (2018)
17. Sunitha, P., Satya Prasad, K.: Speech enhancement based on wavelet thresholding the multitaper spectrum combined with noise estimation algorithm. Int. J. Modern Educ. Comput. Sci. **11**(9), 48–55 (2019)

Research of Dynamic Stress on Assembly Process of Interference Fit Between Axle and Hole of Planetary Gear

Mengya Zhang[1], Yanhui Cai[1], Kun Chen[1(✉)], Jie Mei[1], Xiaofei Yin[1], and Jin Liu[2]

[1] School of Logistics Engineering, Wuhan University of Technology, Wuhan 430063, China
kmno40311@163.com
[2] China Merchants Hoi Tung Trading Company Limited, Hong Kong 999077, China

Abstract. In planetary gear reducer, axle and hole of planetary gear are assembled by interference fit. The value and distribution of contact pressure in interference fit mainly depend on factors such as interference fit value and coefficient of friction of mating surface. These factors directly affect the working state and service life of the planet carrier. In order to study assembly stress of interference fit between axle and hole of planetary gear under different conditions, the suitable interference fit value and surface friction coefficient of the interference fit between axle and hole of planetary are discussed. ABAQUS software is used to simulate assembly process of interference fit under different conditions. The dynamic pressing force, the optimal interference and the optimal friction coefficient of the planetary axle are obtained. The comparison between simulation results and calculation results shows that the assembly process simulation of ABAQUS has high accuracy.

Keywords: Planetary gear reducer · Interference fit value · Friction coefficient · ABAQUS simulation

1 Introduction

The axle of planet wheel and the hole of planet carrier are assembled with interference fit, which is very important for the transmission torque of planet gear. If interference fit value is too small, there will not be enough extrusion pressure to transmit torque. Small vibration and even sliding may occur between axle of planetary gear and planetary carrier, which will eventually cause planetary gear reducer to fail in advance. If interference fit value is too large, extrusion pressure will become larger. If the extrusion pressure becomes larger, it will directly affect the strength of the planet carrier, and may also cause stress concentration. As time goes on, the impact will gradually accumulate more and more, and eventually may cause fatigue failure [1]. At present, wired sensors are widely used in stress-strain detection, and wireless sensors have been developed rapidly. Hussein h. Shia et al. studied the application of particle swarm optimization (PSO) and other algorithms in wireless sensor detection [2]. In addition, Shuangxue

Z. Hu et al. (Eds.): ICCSEEA 2020, AISC 1247, pp. 210–219, 2021.
https://doi.org/10.1007/978-3-030-55506-1_19

Fu et al. conduct contact analysis of gear transmission system under dynamic and static conditions [3]. By comparing analysis results with experimental data, it proves the authenticity and feasibility of gear contact analysis. They found that contact analysis results under dynamic environment are closer to real working conditions of gear measured in the experimental data than the static analysis results, which are more accurate. At the same time, the maximum stress of contact surface between axle of planetary gear and hole of planetary gear in the assembly process is affected seriously by friction coefficient. Therefore, in order to study appropriate interference fit value and friction coefficient between gear of planetary and axle of planetary, the model of planetary gear and planetary axle is established by using UG software. Then, the assembly process between axle and hole of planetary gear is simulated and analysed with ABAQUS finite element analysis software. And stress cloud chart and curve between pressing force of planetary axle and pressing displacement of planetary axle are generated after the assembly. Finally, by comparing the calculated results with the experimental results, the optimal interference fit value and friction coefficient are obtained.

At present, some progress in study of interference coordination has been made by China. Based on ABAQUS finite element simulation software, 3D model of interference fit of bushing parts is established. By changing interference fit value, stress distribution law of contact surface under different interference fit value is studied. They analyze difference between finite element simulation results and theoretical results with reference to their theoretical calculations. And relationship between contact pressure and parameters of interference fit under condition of finite element simulation is also studied [4]. By superimposing the Lapp displacement function and the displacement potential function, the interference fit displacement function is obtained. It is verified that this function satisfies the differential equations and boundary conditions of the spatial axisymmetric problem under the interference fit condition, and solution of stress and displacement in the problem is obtained by this equation. Finally, they also analyze why interference fit causes stress concentration and propose corresponding measures based on stress concentration [5].

Adana Özel et al. obtain stress calculation formula for interference fit, and use finite element analysis software to simulate different kinds of interference fits [6]. Hamid jahed et al. optimize parameters (including cylinder thickness, contact pressure and self-reinforcement ratio) of each layer of cylinder in multi-layer interference connection [7], and obtain the optimal size of each layer of cylinder and the optimal stress distribution in the initial state. The stress distribution of three-layer cylinder interference fit model established by finite element analysis software is analyzed [8].

There are some limitations in traditional theoretical calculation method. Under some conditions, calculation results are not accurate. In order to prove this conclusion, Boutoutaou H has established two interference fit models of defective contact surface and ideal contact surface. By using finite element software, it is found that defective contact surface will have a great impact on interference fit [9]. Shanshan Li et al. carry out finite element analysis and calculation on driving gear of single pair gear during meshing process of planetary reducer gear. By analyzing contact stress distribution of object under dynamic and static conditions, the point with weakest bending strength is obtained. The results show that the root of the driving gear is most likely to break [10].

In conclusion, there is little research and analysis on interference fit of axle and hole of planetary gear. In addition, these theories and simplified models sometimes fail to accurately calculate parameters under interference fit actually. So stress situation between containing part and contained part cannot be truly reflected by the calculation, which makes design and analysis of parts with interference fit have certain limitations. Through finite element analysis of contact state of three-dimensional model, the stress between mating parts which makes research of this paper has a certain practical significance can be simulated more reliably.

2 Structure and Working Principle of Planetary Reducer

The main structure of planetary gear reducer includes sun wheel (a), the inner ring gear (b), the planet wheel (c), the planet carrier (x), as shown in Fig. 1.

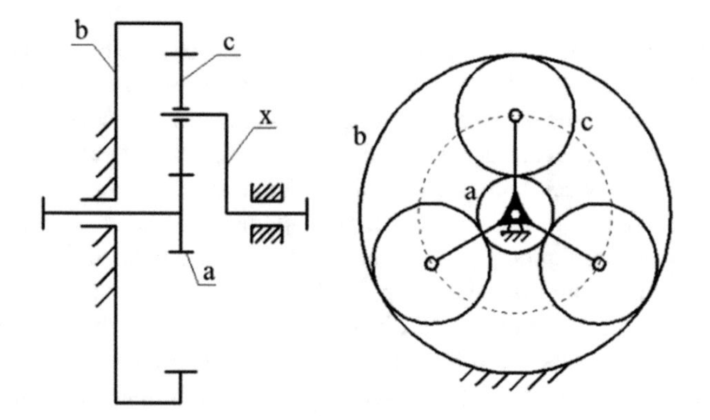

Fig. 1. Structure diagram of planetary gear reducer.

Yaxiang Zhou compares all kinds of gear transmission according to the parameter requirements, and finally get the scheme of mixed gear train, and draw the mechanism diagram [11]. As shown in Fig. 1, the planetary gear reducer includes three parts: an input, a transmission, and an output. The input part includes a driving motor, an input axle, and a sun gear. This part is mainly connected to the input shaft through the drive motor, and the input shaft drives the sun gear to rotate. After sun gear rotating, planetary gear rotates together with sun gear by meshing. Because of existence of inner ring gear, planetary gear rotates around axle of planetary gear while revolving around sun gear. This is transmission part. Through the orbital motion of the planet wheels, the planetary axles transmit torque to the planet carrier for rotation. Through continuous transmission of torque, torque is finally output through output axle connected to back of planet carrier. The planet carrier and output axle are referred to as output portion of planetary gear reducer.

3 Calculation of Contact Surface Stress of Interference Fit

In study of interference fit, it has often been case that components are simplified into thick-walled cylinders for research. For coordination problem of cylinders, it is generally calculated using empirical formulas. And sometimes it is calculated using Lamé equation in elastoplastic state. Generally, there are two factors to destroy interference fit: the stress on contact surface of interference fit and press in force during assembly process.

3.1 Stress on Contact Surface of Interference Fit

When using theory of elasticity and thick wall cylinder to calculate stress on contact surface of interference fit, firstly, it is assumed that the pressure on the contact surface is uniformly distributed after assembly. Then parts with interference fit is simplified to thick wall cylinder with interference fit for analysis. Based on knowledge of elasticity, calculation formula of contact surface stress of interference fit can be derived as follows [12]:

$$\mathrm{P} = \frac{\delta}{\mathrm{d}\left(\frac{C_1}{E_1} + \frac{C_2}{E_2}\right)} \quad (1)$$

$$C_1 = \frac{d^2 + d_1^2}{d^2 - d_1^2} - \mu_1 \quad (2)$$

$$C_2 = \frac{d_2^2 + d^2}{d_1^2 - d^2} - \mu_2 \quad (3)$$

δ——interference;
d——mating diameter;
d1——inner diameter of the contained part (solid contained part $d_1 = 0$);
d2——outer diameter of the containing part;
C1, C2——rigidity coefficient of the contained part and the contained part;
E1, E2——elastic modulus of the contained part and the contained part;
μ1, μ2——Poisson's ratio of the contained part and the contained part.
If the contained part is a solid axis, the formula can be reduced to:

$$p = \frac{\delta}{d\left(\frac{1-\mu_1}{E_1} + \frac{C_2}{E_2}\right)} \quad (4)$$

If the same material is used between two parts, the formula can be simplified as follows:

$$p = \frac{E\delta}{2d}\left[1 - \left(\frac{d}{d_2}\right)^2\right] \tag{5}$$

E——elastic modulus of material;

3.2 Pressing Force in the Assembly Process

When using the press in method for the interference assembly, the pressing speed is slow, and the press in process shall be stable to ensure that the whole process has no impact, which means that the press in force and the friction force of components are approximately equal. So the calculation formula of pressing force in the assembly process can be calculated according to following formula:

$$F = \pi dlpf \tag{6}$$

d——mating diameter;
l——mating length;
p——contact stress;
f——friction coefficient;

4 Simulation Analysis

Liquan Wang et al. use nonlinear finite element theory to analyze the contact force of the retaining ring connection structure [13]. Using intelligent optimization control method which are Particle swarm optimization, random occurring distributed time delay particle swarm optimization, each PID controller is tuned [14]. The chattering problem of SMC has been handled by using pseudo sliding function. Further results have been analyzed by comparing them with the basic conventional controllers [15]. Output voltage can be stabilized in constant amplitude and frequency by designing the appropriate controller for the considered system [16]. With development of finite element method and enrichment of nonlinear basic knowledge, nonlinear analysis in finite element method is widely used. This method can be used to simulate interference fit more accurately. Through simulation, the stress and strain of contact surface and deformation displacement can be obtained, which makes analysis of interference fit more real and effective and design work easier.

4.1 Establishment of Experimental Model of Planetary Gear Reducer

As shown in Fig. 2, each part model of planetary gear reducer is assembled, which is built by using UG software. In this paper, the double-sided integrated planetary carrier is selected. Planetary gear parameters are: teeth number 25, modulus 1.5, tooth width 4 mm, pressure angle 20°. The parameters of sun gear are: teeth number 18, modulus 1.5, tooth width 4 mm, pressure angle 20°. The input shaft and the sun gear are made into an integrated structure. The parameters of inner ring gear are: teeth number 18, modulus of 1.5, tooth width of 4 mm and pressure angle of 20°. Input axle and sun gear are integrated into one structure. Axle of planetary is chamfered at one end for assembly purposes.

Parameters of inner gear ring are: module 1.5, teeth number 69, tooth width 5 mm, pressure angle 20°, external diameter of internal gear 115 mm, teeth number of gear shaper 100.

Fig. 2. Assembled model using UG software

4.2 Simulation and Analysis of Result Based on ABAQUS

In simulation analysis of this assembly process, the settings of ABAQUS software mainly include several processes of dividing the part grid, defining material properties, setting analysis steps, setting interactions, and defining boundary conditions. With completing the software setup, a new job can be created for simulation. And finally stress cloud chart of each stage of assembly process can be obtained.

In this paper, the base hole system is used to fit the axial holes. H7/n6, H7/p5, H7/r6 and H7/s6 are selected as four kinds of interference fit. After selecting the maximum size of corresponding axle, assembly process of interference fit is analyzed.

4.2.1 Simulation and Analysis of Result Based on ABAQUS

The stress cloud diagram of four kinds of interference fit after assembly is obtained by simulation, and the relationship between force and displacement during assembly can also be derived. The stress cloud diagram after assembly is shown in Fig. 3.

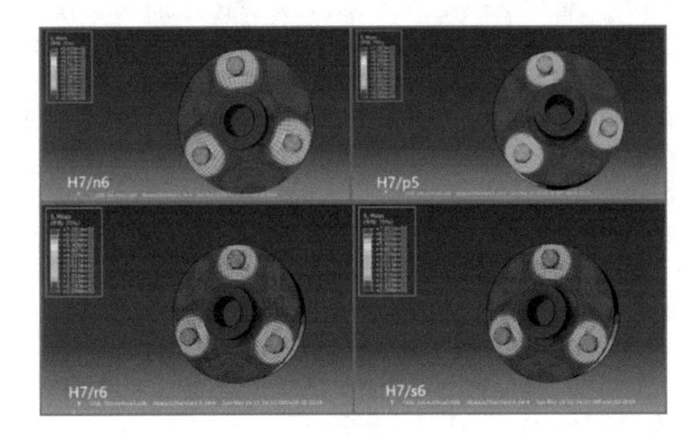

Fig. 3. Stress cloud chart after assembly process using ABAQUS software.

The curve between pressing force of axle and pressing displacement of axle during assembly process is shown in Fig. 4 below.

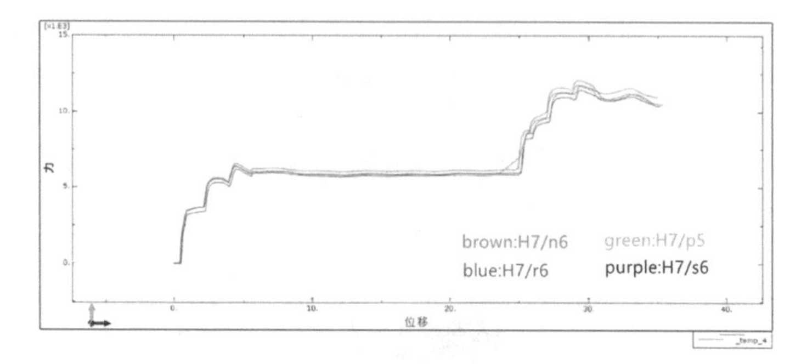

Fig. 4. The curve between pressing force of axle and pressing displacement of axle during assembly process using ABAQUS software.

The maximum stress of four interference fits in the assembly process is shown in Table 1 below.

Table 1. Maximum stress of four interference fits during assembly process using UG software.

Interference fit	H7/n6	H7/p5	H7/r6	H7/s6
Axle diameter (mm)	10.019	10.021	10.028	10.032
σ_{max} (MPa)	802.2	802.5	807.6	807.7

According to results of finite element simulation, for four kinds of interference fit, with the increase of interference, the stress on the contact surface increases linearly. At the same time, the maximum stress appears at corner of end surface of planetary hole in the initial stage of assembly. Therefore, in actual assembly conditions, chamfering is required on axle of planetary and hole of planetary carrier. In addition, in the case of neglecting the stress concentration, the stress of the most contact surface between axle of planetary and hole of planet carrier is close to calculating value of simplified hole of planetary and axle of planet model. Through theoretical calculation, pressing force of planetary axle passing through the first part of planetary axle is about 5.28kN, and pressing force of planetary axle passing through the second part of planetary axle is about 10.46 kN. It can be seen from functional relationship diagram between pressing force of planetary axle and pressing displacement of planetary axle that pressing force is similar to theoretical calculating value, indicating that the simulation process has certain reliability. The stress cloud chart shows that the smaller interference fit value can reduce stress concentration and failure probability of parts in interference fit. The maximum stress of four sizes of planetary axle in the assembly process is 802.2 MPa, 802.5 MPa, 807.6 MPa and 807.7 MPa respectively. The material selected in this paper is 42CrMo, and the yield strength of this material is 930 MPa.

Thus four sizes of planetary axle assembly process will not damage the part itself.

4.2.2 Simulation and Analysis of Result of Different Friction Coefficient Based on ABAQUS

The stress cloud diagram with changing friction coefficient after assembly process is shown in Fig. 5.

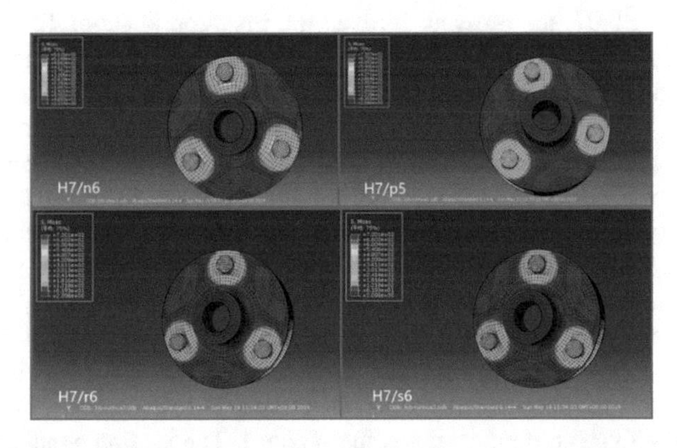

Fig. 5. Stress cloud chart after assembly process using ABAQUS software.

After changing coefficient of friction, the curve between pressing force of axle and pressing displacement of axle during the assembly process is shown in Fig. 6 below.

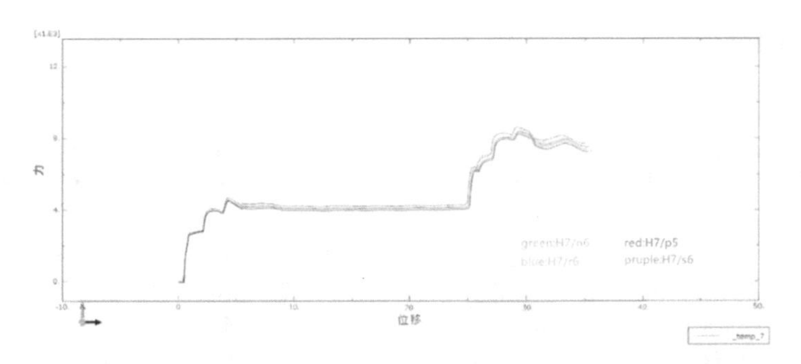

Fig. 6. The curve between pressing force of axle and pressing displacement of axle during assembly process using ABAQUS software.

The maximum stress of four interference fits in the assembly process is shown in Table 2.

Table 2. Maximum stress of four interference fits during assembly process using UG software.

Interference	H7/n6	H7/p5	H7/r6	H7/s6
Axle diameter (mm)	10.019	10.021	10.028	10.032
σ_{max} (MPa)	826.1	829	833	834.1

By comparing the curves between pressing force and pressing displacement of same planetary axle with two friction coefficients, it can be found that the two curves are similar in shape, but pressing force is different with stable pressing force. When the friction coefficient equals 0.1, the pressing force at the first stop is about 6 kN, and the peak value of pressing force at the installation stage is about 12 kN. When the friction coefficient equals 0.07, the pressing force at the first stop is about 4kN, and the peak value of pressing force at the installation stage is about 10 kN. This phenomenon is in line with the engineering practice. The friction force of the contact surfaces of interference fit and required pressing force is enlarged by increase of friction coefficient. When the same interference fit value is used for planetary axles, the maximum stress generated by interference fit process with a friction coefficient of 0.07 is greater than the maximum stress generated with a friction coefficient of 0.1. Through the analysis, it can be known that for interference fit, increasing friction coefficient will lead to an increase in roughness of the contact surface, which will eventually reduce actual effective interference fit value of mating parts. After the actual effective interference is reduced, the stress on the contact surface is reduced. Therefore, for four interference fits studied in this paper, when the friction coefficient is 0.1, the stress of contact surface produced by the four interference fits is the smallest. The friction coefficient of contact surface can seriously affect the stress of contact surface during assembly. Therefore, the surface roughness, lubrication and other factors should be considered in the follow-up study, and the appropriate surface friction coefficient and interference fit value should be selected.

5 Conclusion

The extent to which friction and interference affect the contact surface stress during assembly is discussed in this paper. Using ABAQUS software, the assembly process of planetary gear model established by UG is simulated. The experimental results show that interference fit value is basically linear with the stress on contact surface between axle of planetary gear hole of planetary gear, and the measured stress is close to theoretical calculation result. The friction coefficient has great influence on the contact surface stress during the assembly process. For the eight interference fits studied in this paper, the stress of contact surface produced by the H7/N6 fits friction and coefficient equalling 0.1 is the smallest. The stress cloud chart shows that the maximum stress point on contact surface is at the end of the hole of planet carrier after assembly process. So the stress concentration should be reduced through structural design. In addition, the relationship between friction coefficient and stress of contact surface is not linear. Therefore, in the follow-up study, the surface roughness, lubrication and other factors should be considered to select the appropriate surface friction coefficient and interference.

Acknowledgements. This project is supported by the National Key Research and Development Program of China (No. 2017YFC0805703).

References

1. Zhao, J., Lin, T., Zong, S., et al.: Fatigue life and its influencing factors of planetary wheel-bearing interference fit joints. J. Dalian Univ. Tech. **56**(04), 355–361 (2016). (in Chinese)
2. Shia, H.H., Tawfeeq, M.A., Mahmoud, S.M.: High rate outlier detection in wireless sensor networks: a comparative study. Int. J. Modern Educ. Comput. Sci. **11**(4), 13–22 (2019)
3. Fu, S., Zhou, C., Han, X.: Analysis of the effective load distribution and gear strength between meshing teeth under typical working conditions. Mech. Transm. **40**(08), 102–106 (2016). (in Chinese)
4. Wang, K., Danhong, T., Wang, Y.: Study on contact pressure of bushing interference fit surface. Des. Manuf. Diesel Engine **24**(04), 30–33 (2018). (in Chinese)
5. Ozturk, F.W.T.: Simulations of interference and interfacial pressure for three disk shrink fit assembly. Gazi Univ. J. Sci. **23**(2), 233–236 (2010)
6. Ozel, A., Yazici, B., Akpinar, S., Aydin, M.D., Temiz, Ş.: A study on the strength of adhesively bonded joints with different adherends. Compos. B Eng. **62**, 167–174 (2014)
7. Jahed, H., Farshi, B., Hosseini, M.: Fatigue life prediction of autorettage tubes using actual material behavior. Int. J. Press. Vessels Pip. **83**(10), 749–755 (2006)
8. Fan, X., Sun, L., Wang, Y., et al.: Analytical solution of axial hole interference fit. Machinery **38**(09), 26–30+43 (2011). (in Chinese)
9. Boutoutaou, H., Bouaziz, M., Fontaine, J.F.: Modeling of interference fits taking form defects of the surfaces in contact into account. Mater. Des. **32**(7), 3692–3701 (2011)
10. Li, S., Han, L., Liang, Y.: Finite element analysis of contact stress of helical gear based on ANSYS. Mech. Eng. Autom. **04**, 23–24 (2009)
11. Zhou, Y.: Design and kinematic analysis of planetary gear reducer. Sci. Tech. Innov. Prod. **06**, 80–82 (2017). (in Chinese)
12. Mi, H.: Elasticity, 1st edn, pp. 88–95. Tsinghua University Press, Beijing (2013). (in Chinese)
13. Wang, L., Sun, Z., Meng, Q.-x.: Finite element analysis for the contact problem of check ring joint structure in underwater skirt pile gripper. Equip. Manuf. Tech. **4**, 9–11 (2006)
14. Marie, M.J., Mahdi, S.S., Tarkan, E.Y.: Intelligent control for a swarm of two wheel mobile robot with presence of external. Int. J. Modern Educ. Comput. Sci. **11**(11), 7–12 (2019)
15. Kapoor, N., Ohri, J.: Improved PSO tuned classical controllers (PID and SMC) for robotic manipulator. Int. J. Modern Educ. Comput. Sci. (IJMECS) **7**(1), 47–54 (2015)
16. Khani Maghanaki, P., Tahani, A.: Designing of fuzzy controller to stabilize voltage and frequency amplitude in a wind turbine equipped with induction generator. Int. J. Modern Educ. Comput. Sci. **7**(7), 17–27 (2015)

MECE Guided Systematic Analysis and Fussy Assessment of Major Sports Event Broadcasting Center Service Quality—Take the 7th Military World Games as an Example

Ziye Wang[1(✉)], Fei Lu[2], Xinpeng Zhao[1], and Qingying Zhang[3]

[1] College of Journalism and Communication, Wuhan Sports University, Wuhan 430079, China
kathy8899@126.com
[2] Executive Committee of the 7th Military World Games, Wuhan 430056, China
[3] School of Logistics Engineering, Wuhan University of Technology, Wuhan 430063, China

Abstract. The service of broadcasting center is indispensable for sports events. While recording events in various forms, it can timely share the process, results and wonderful moments of the competition to the public. Whether the services provided by the broadcasting center for journalists and media organizations are in place directly affects the quality of event communication. Taking the 7th world military games as an example, this paper analyzes the functions, characteristics and functions of the event broadcasting center by using the MECE principle, probes into the construction requirements of the event broadcasting center of Wuhan military games, confirms what problems need to be analyzed by using the principle of logical thinking, and seeks for the breakthrough point to solve the problems. On this basis, a structural model of the factors influencing the service quality of the event broadcasting center is established. According to the steps of establishing factor set, evaluation set, weight set, single factor fuzzy evaluation and fuzzy comprehensive evaluation of the service quality of the relay center of Wuhan military games, the comprehensive evaluation of the service quality of the relay center of Wuhan military games is completed, which is significant to provide valuable reference for the media service of large-scale events.

Keywords: Military World Games · Large-scale event broadcast center · System analysis · System evaluation

1 Instruction

Sports competitions, especially large-scale sports events, are increasingly entering the public life [1]. The 7th Military World Games held in Wuhan China, in October 2019, abbreviated as “Wuhan Military Games”, or WH-MWG, 7MWG-WH for short, is the largest, most participants and most influential games in the history of the world military games. In the 10 day period, 9,308 soldiers from 109 countries participated in 27 major

Z. Hu et al. (Eds.): ICCSEEA 2020, AISC 1247, pp. 220–230, 2021.
https://doi.org/10.1007/978-3-030-55506-1_20

and 329 minor events, including shooting, swimming, track and field, basketball, etc. It has attracted hundreds of millions of people all over the world [2].

In such a large-scale world-class event, broadcasting is very important and complex, and is capable to be accomplished by setting up a broadcasting center [3–5]. It is of vital significance to improve the service quality of the broadcast center [6]. By using the method of fuzzy evaluation to assess the service quality of the broadcasting center, it is capable to gain useful experience which is worth popularizing.

2 The Role of the Event Broadcasting Center for the Military Games

From the perspective of the system, the construction of the event broadcasting center involves many aspects of work, with many influencing factors and complex system architecture, which requires a comprehensive and detailed analysis using scientific methods [7, 8].

2.1 Simulation System and Estimation Indexes

With the continuous progress of society and the public's attention to health and sports activities, the social role of sports is becoming increasingly more important [9].

The basic conditions and requirements for the success of a large-scale sports event are not only the high-level performance of athletes, the substantial improvement of sports performance, and the orderly management of the event, but also the media communication of the event [10, 11]. Samaranch, the former president of the International Olympic Committee, once said: "there are three elements to successfully host the Olympic Games: first, good organizational work; second, national sports achievements; third, media publicity." It is not hard to imagine that without media coverage, an event is just a sports gathering, with outstanding achievements not announced, outstanding athletes not publicized, and large-scale sports events lack their own most valuable communication significance [12].

Media communication is the bridge between sports events and the public [13]; the media pays attention to the inside and outside of the venue, which is the extension of the audience's eyes and ears; the media reports determine the influence of the host city of the event [14].

With the enhancement of the communication power and influence of modern large-scale sports events, the service quality of sports event broadcasting center has been paid much more attention than ever before [15].

The expressions of this quality of service are as follows: (1) to spread the information of large-scale sports events more quickly, timely and accurately, to balance the reports of media from all walks of life and all kinds of media, so that the media can provide active and positive reports on mega-events; (2) to spread the groups of major sports events in the global scope through the reports of media on big sports events organization concept; (3) the broadcasting center promotes the cultural communication of the host country and city of large-scale sports events by optimizing services, builds a good image of the host country and city, and better realizes the economic and social benefits of mega-events [16, 17].

2.2 Broadcast Characteristics of Military Games

From the standpoint of the service providers, the work that the broadcasting center needs to complete can be summarized as providing various types of services such as news work, transportation, catering, accommodation, etc. for the mass media from all over the world during the competition period [18, 19]. Specifically speaking, it is necessary to focus on the overall goal of the event organization, follow the established competition mode, follow the competition principles formulated by each event, do a good job in all preparatory work, and provide work facilities and news services for registered journalists to report large-scale events, so as to fully meet the needs of the specialization, informatization and service of sports media.

Different from other sports competitions, the athletes come from the military forces of various countries. There are special projects of Navy, land force and air force in the projects, and the venues are relatively scattered. The successful completion of broadcasting activities requires not only professional narrators who are familiar with different projects, but also commentators and announcers, distributed venues and broadcasting centers keep a good connection and cooperation between the various events. This brings some challenges to the work of the broadcasting center.

2.3 Requirements for the Construction of Wuhan Military Games Event Broadcasting Center

Focusing on the theme of the MWG of "Sharing Friendship and Building Peace together", promoting the city image of "Modernization, Internationalization and Ecology" of Wuhan, and meeting the needs of domestic and foreign media organizations reporting on the WH-MWG, the design and construction of the broadcasting center of the Wuhan Military Games should follow the following principles:

(1) Principle of Safety

The content of security principle mainly includes: media oriented security, operation and broadcasting security, as well as personnel and equipment security [20].

(2) Principle of Specialization

The so-called specialization mainly refers to the professionalization of hardware facilities and media services.

(3) Principle of Practicability

Its main meaning is the practicality of function design, operation rules and hardware construction.

(4) The Principle of Frugality

It should been fully considered about the needs of domestic and foreign media, according to the world military sports media service practices, combined with the characteristics of Wuhan, China, and design a main media center that not only follows international practices but also suits the characteristics of WH-MWG [21]. Under the condition of meeting the needs of domestic and foreign TV institutions, ENG shooting is taken to be adopted for some TV relay projects. It is also necessary to fully solicit the software and hardware requirements of the main media center from the domestic and foreign media participating in the coverage of the WH-MWG.

(5) The Principle of Combining International Practice with China's Reality

According to international practice, first-class working facilities and media services should be provided for journalists of writing, photography and radio and television. Under the premise of internationalization, standardization and scientificization, resources should be allocated and utilized reasonably to realize the optimization of media service and make it in line with international management and standards. In strict accordance with the requirements of the Organizing Committee and the military, it is necessary to do a good job in the preparation and operation of the main media center of the WH-MWG. It is also necessary to make functional planning according to the architectural characteristics of the main media center of MWG and the needs of journalists.

3 MECE Analysis of the Function of the Event Broadcasting Center

3.1 Connotation of MECE Principle

MECE principle is a very evidently useful tool in the pyramid principle of management [22]. It emphasizes the important role of thinking infrastructure in mastering the principle of logical thinking. Using the concept of MECE to analyze a major issue is to avoid overlap and omission, which is called "mutual independence and complete exhaustion". This method can effectively grasp the core of the problem and find the way to solve it.

"Mutual independence" means that the classification of problems is in the same dimension, which is easily distinguished clearly and cannot be overlapped, while "complete exhaustion" means comprehensive and thorough in the ways of analysis. MECE analysis has two steps: first, confirm what the problems are; second, find the entry point of MECE.

Generally, when facing practical problems, it is necessary to put forward hypothesis or assumptions and find solutions on the premise of fully mastering the facts.

MECE is a very useful tool to test the correctness of the initial hypothesis. It is to analyze the problem step by step in a thorough way to find a solution. In solving business problems (or any other problems), a clear thinking must be formed. Even when there are many problems that are confusing, the analyst's thinking must be unrestricted, honest and complete.

3.2 Application of MECE Principle

MECE principle can be used to classify and analyze the problems that need to be solved reasonably, so as to find the most important link at present.

(1) Follow the Principle of MECE and Confirm the Problem Items first.

The main difficulties faced by the function analysis work of the broadcasting center are as follows:

- the links involved in broadcasting are very complex;
- there are many events and scattered venues;
- different broadcasting methods may be separate for different events, with different focus and broadcasting methods;
- broadcasting teams from different countries and regions may have different working habits and broadcasting styles, as well as different needs;

Therefore, it is needed to analyze many problems in detail to form a preliminary understanding of the problem.

(2) Classify the Problems According to MECE Principle.

By using the method of logical tree analysis, the main influencing factors of the service quality of the broadcasting center of the Military Games are found out one by one, and their meanings are defined.

Then, using the method of fuzzy analysis, the service quality of the broadcasting center is systematically evaluated.

4 Comprehensive Evaluation of the Service Quality of the Broadcasting Center of the Military Games

To judge whether the service quality of the event broadcasting center meets the requirements, it is necessary to use scientific methods for system analysis and evaluation [23–26].

Combining with MECE principle analysis logic tree, it is capable to get the structure model of influencing factors of service quality of event broadcast center shown in Fig. 1. According to the following steps, the comprehensive evaluation of the service quality of the broadcasting center of the military games is carried out. The evaluation structure model is divided into target level, criterion level and index level, among which, the criterion level includes 6 dimensions: security and reliability, advanced technology, practicability, standardization, openness and expansibility.

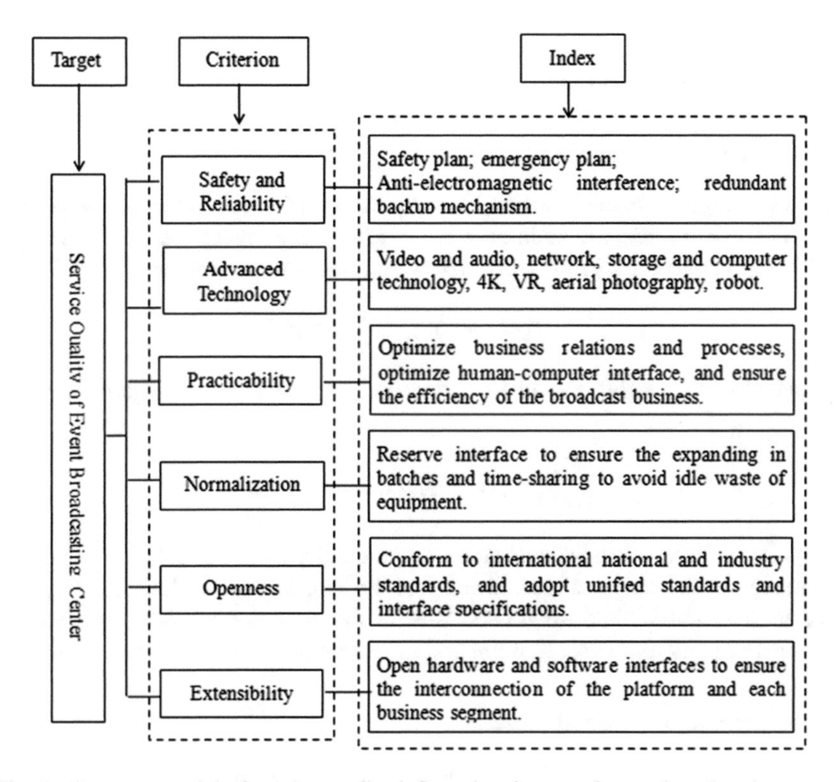

Fig. 1. Structure model of service quality influencing factors of event broadcasting center

4.1 Establish Factor Set

Factor set *U* is a set of factors affecting the evaluation object, expressed as:

$$U = \{u_1, u_2, u_3, \ldots, u_n\}$$

Among them, u_I is the factor that affects the evaluation object. In the fuzzy comprehensive evaluation model of the service quality of the broadcast center, it refers to different evaluation factors.

4.2 Establishment of Evaluation Set

Evaluation set *V* is the set of all possible judgment results that the evaluator makes to the evaluation object. Expressed as:

$$V = \{v_1, v_2, v_3, \ldots, v_n\}$$

The purpose of the comprehensive evaluation of the event broadcasting center service quality is to get a more reasonable evaluation result of the service quality level from the evaluation concentration on the basis of considering all the influencing factors of the service quality. The evaluation set is: $V = \{$very satisfied, relatively satisfied, satisfied, basically satisfied, not satisfied$\}$.

4.3 Building the Weight Set

In the evaluation of various factors, the importance of each evaluation factor is different. In order to represent the importance of each factor, the corresponding weight w_i should be given to each factor U_i, which is composed of weight set W, that is

$$W = \{w_1, w_2, w_3, \ldots, w_n\} \tag{1}$$

Generally, all weights should be normalized and meet the nonnegative condition, that is

$$\sum_{i=1}^{n} W_i = 1 \; W_i \geq 0 \, (i = 1, \ldots, n) \tag{2}$$

The fuzzy comprehensive evaluation of the quality level of broadcast center is the combination of the weight of each factor and the single factor evaluation, so it is very important to determine the weight set. It is a usual way through AHP (analytic hierarchy process) and expert scoring method to determine the weight set.

4.4 Single Factor Fuzzy Evaluation

The single factor fuzzy evaluation of the quality level of broadcast center starts from the single factor in the factor set U to determine the membership degree of the evaluation object to each element in the evaluation set. When the evaluation object is evaluated according to the i-th factor U_i in the factor set, the membership degree of the j-th element V_j in the evaluation set is R_{ij}, and R_i is a single factor evaluation set, which is expressed as follows:

$$R_i = \{R_{i1}, R_{i2}, R_{i3}, \ldots, R_{in}\} \tag{3}$$

The evaluation set of n factors is composed into a single factor evaluation matrix R.

$$R = \begin{bmatrix} R_1 \\ \ldots \\ R_n \end{bmatrix} = \begin{bmatrix} r_{11} & \ldots & r_{1n} \\ \ldots & \ddots & \ldots \\ r_{n1} & \ldots & r_{nn} \end{bmatrix} \tag{4}$$

4.5 Fuzzy Comprehensive Evaluation of the Service Quality of the Event Broadcasting Center

When the weight set W and the single factor evaluation matrix R are known, the fuzzy transformation can be used for comprehensive evaluation.

$$B = W{\cdot}R = (w_1, w_2, w_3, \ldots, w_n)\cdot\begin{bmatrix} r_{11} & \cdots & r_{1n} \\ \cdots & & \cdots \\ r_{n1} & \cdots & r_{nn} \end{bmatrix} = (b_1, b_2, b_3, \ldots, b_n) \quad (5)$$

In the formula, B is called the fuzzy comprehensive evaluation set, and "•" represents some fuzzy synthetic operation, and $b_i(i = 1, 2,\ldots,n)$ is the fuzzy comprehensive evaluation index, which is called the evaluation index for short. The meaning of b_i is the membership degree of the evaluation object to the i-th element in the evaluation set when all factors are considered comprehensively.

According to the calculation results of fuzzy comprehensive evaluation set of quality level, the service quality level is determined according to the principle of maximum membership degree. $V_i(i = 1,2,\ldots,n)$, the corresponding evaluation set element of $b_i(i = 1, 2,\ldots, n)$, the largest fuzzy comprehensive evaluation index in the fuzzy comprehensive evaluation set, is selected as the comprehensive evaluation result of the broadcast service quality level of the military games.

4.6 Comprehensive Evaluation of Service Quality of WH-MWG Broadcasting Center

According to the comprehensive evaluation index system of service quality of broadcasting center, the fuzzy comprehensive evaluation model is used to evaluate.

(1) Evaluation of Service Capacity and Quality Level of Broadcasting Center

The six influencing factors for the evaluation of the service capacity and quality level of the broadcasting center are analyzed and determined as follows: safe and reliable, advanced technology, convenient and practical, expandable, standardized and open.

After determining the influencing factors, each factor is given a certain weight through the expert scoring method. In this case, the weights are *(0.25, 0.15, 0.20, 0.15, 0.15, 0.10)*. The expert's evaluation results are as follows:

$$\mathrm{B}_1 = (0.25\,0.15\,0.20\,0.15\,0.15\,0.10)\cdot\begin{bmatrix} 0.31 & 0.30 & 0.15 & 0.13 & 0.11 \\ 0.30 & 0.35 & 0.15 & 0.15 & 0.05 \\ 0.31 & 0.35 & 0.10 & 0.10 & 0.10 \\ 0.30 & 0.25 & 0.15 & 0.15 & 0.15 \\ 0.25 & 0.25 & 0.15 & 0.15 & 0.15 \\ 0.30 & 0.30 & 0.15 & 0.14 & 0.11 \end{bmatrix}$$
$$= (0.2970\,0.3025\,0.1400\,0.1495\,0.1110) \quad (6)$$

(2) Evaluation of Service Function Quality Level of Broadcasting Center

The influencing factors of the service function quality level evaluation of the broadcasting center include building layout and function, transportation supporting, operation time, public broadcasting system, safety and emergency services, cleaning and maintenance, and protection of sponsors' rights and interests. In this case, the weight values given to these factors are *(0.20, 0.16, 0.14, 0.10, 0.15, 0.10, 0.15)* according to the assessment results of experts. The evaluation results are as follows:

$$\begin{aligned} B_2 &= (0.20\,0.16\,0.14\,0.10\,0.15\,0.10\,0.15) \cdot \begin{bmatrix} 0.30 & 0.20 & 0.20 & 0.15 & 0.15 \\ 0.30 & 0.25 & 0.25 & 0.10 & 0.10 \\ 0.25 & 0.30 & 0.20 & 0.15 & 0.10 \\ 0.30 & 0.30 & 0.20 & 0.15 & 0.05 \\ 0.25 & 0.25 & 0.25 & 0.20 & 0.05 \\ 0.30 & 0.25 & 0.20 & 0.20 & 0.05 \\ 0.25 & 0.25 & 0.25 & 0.15 & 0.10 \end{bmatrix} \\ &= (0.2780\,0.2520\,0.2230\,0.1545\,0.0925) \end{aligned} \tag{7}$$

(3) Comprehensive Evaluation of Service Quality Level of Broadcasting Center

In this case, the weights given are *0.60* and *0.40* respectively. By the evaluation results of service capability quality level and service function quality level of broadcast center, the comprehensive evaluation results of service quality level of broadcast center are as follows:

$$\begin{aligned} A &= (w_1, w_2)\begin{pmatrix} B_1 \\ B_2 \end{pmatrix} \\ &= (0.60.4)\begin{pmatrix} 0.29700.30250.14000.14950.1110 \\ 0.27800.25200.22300.15450.0925 \end{pmatrix} \\ &= (0.28940.28230.17320.15150.1036) \end{aligned} \tag{8}$$

(4) Analysis of Evaluation Results

According to the evaluation results, "extremely satisfied" accounts for *28.94%*, "relatively satisfied" accounts for *28.23%*, "satisfied" accounts for *17.32%*, "basically satisfied" accounts for *15.15%*, and "not satisfied" accounts for *10.36%*. The different evaluation results are in accordance with normal distribution, while the overall service quality level is high, which shows that the operation organization plan can better meet the requirements of the Military Games Broadcast. The whole scheme is feasible and superior.

5 Conclusion

News broadcasting center is an important part of large-scale sports events. It can not only record the process and result of the event, but also let people from all over the world share the content of the event. It can also carry forward the spirit of sports, establish friendship among people of all countries, and promote world peace and development.

The establishment of the broadcast center is definitely considered to be complicated system engineering. Using the concept and principle of MECE to analyze the function and effect of the event broadcasting center one by one is to analyze the structure of system elements and find solutions to problems by using the method of logic tree on the basis of listing the existing problems.

Through the application of fuzzy analysis method, this paper makes a comprehensive evaluation on the service quality of the broadcasting center of the WH-MWG, that is, to determine the influencing factors of the evaluation on the service ability and quality level of the live broadcast center, and to give certain weight to each factor through the method of expert scoring. Then from two aspects of service capability quality level and service function quality level, the comprehensive evaluation results of service quality level of broadcasting center are obtained. The results show that the construction and operation of the event broadcasting center of the 7th MWG is successful, which has played a very good role in promoting the success of the competition and has been amply proved by the actual effect.

Acknowledgment. This paper is supported by—(1) Hubei Teaching Research Project "Research on the Reform of Sports Media Professional Training Model Based on OBE Concept"; (2) Wuhan Sports University Young Teachers' Scientific Research Fund Project "The MECE study under the background of Sports Events media convergence".

References

1. Mcevoy, C.D.: An Investigation of the Relationship between Television Broadcasting and Game Attendance[J]. Int. J. Sport Manage. Market. **2**(3), 222–235 (2014)
2. Kim, J., Kim, Y., Kim, D.: Improving well-being through Hedonic, Eudaimonic, and social needs fulfillment in sport media consumption. Sport Manage. Rev. SMR **20**(3), 309–321 (2017)
3. Hoehn, T.: Broadcasting and sport. Oxford Rev. Econ. Policy **19**(4), 552–568 (2003)
4. Hambrick, M.E.: Sport communication research: a social network analysis. Sport Manage. Rev. **20**, 170–183 (2017)
5. Son, J.W., Lee, A., Kim, S.J.: Knowledge construction for the broadcasting content by using audience oriented data. In: IEEE/WIC/ACM International Conference on Web Intelligence and Intelligent Agent Technology (WI-IAT), 89–92. ACM (2015)
6. Lan, J.: Research on gap model construction and its application method of comprehensive sports center service quality. J. Comput. Theor. Nanosci. **19**(6), 1803–1806 (2013)
7. Wasson: System analysis, design, and development. Insight **11**(4), 499–504 (2015)
8. Easterling, D., Arnold, E.M., Jones, J.A.: Achieving synergy with collaborative problem solving: the value of system analysis. Found. Rev. **5**, 215–221 (2013)
9. Moradi, M., Honari, H., Naghshbandi, S., et al.: The association between informing, social participation, educational, and culture making roles of sport media with development of championship sport. Procedia – Soc. Behav. Sci. **46**(5), 5356–5360 (2012)
10. Xiaofeng, X.: Value development of sports media. Sci. Tech. Inf. **15**, 107 (2014). (in Chinese)
11. Ziye, W., Qingying, Z., Mengya, Z.: Training platform construction of omni-media live broadcast of sport event. In: Advances in Computer Science for Engineering and Education II, pp. 205–216 (2018)

12. Yoshida, M., James, J.D., Cronin, J.J.: Sport event innovativeness: Conceptualization, measurement, and its impact on consumer behavior. Sport Manage. Rev. **16**(1), 68–84 (2013)
13. Nah, S., Yamamoto, M.: The integrated media effect: rethinking the effect of media use on civic participation in the networked digital media environment. Am. Behav. Sci. **62**(8), 1061–1078 (2018)
14. Ribeiro, T.M., Correia, A., Biscaia, R., et al.: Examining service quality and social impact perceptions of the 2016 Rio de Janeiro Olympic games. Int. J. Sports Mark. Spons. **19**(2), 160–177 (2018)
15. Bauer, E., Adams, R.: Application Service Quality. Service Quality of Cloud-Based Applications. Wiley, New York (2014)
16. Wang, J., Zhao, S., Wei, P., et al.: Evaluation indicators system of sports public services performance of local government. J. Shenyang Sport Univ. **15**(1), 1–4 (2011). (in Chinese)
17. Kotevsk, Z., Tasevska, I.: Evaluating the potentials of educational systems to advance implementing multimedia technologies. Int. J. Modern Educ. Comput. Sci. (IJMECS) **9**(1), 26–35 (2017)
18. Sherman, H.D., Zhu, J.: Analyzing performance in service organizations. MIT Sloan Manage. Rev. **54**(4), 37–42 (2013)
19. Fetaji, B., Fetaji, M., Ebibi, M., et al.: Analyses of impacting factors of ICT in education management: case study. Int. J. Modern Educ. Comput. Sci. (IJMECS) **10**(2), 26–34 (2018)
20. Fei, Y., Tao, T., Hongwei, Y.: Research on concept and allocation principle of safety integrity level. J. Beijing Jiaotong Univ. **41**(5), 79–84 (2017). (in Chinese)
21. Roiland, D.: Frugality, a positive principle to promote sustainable development. J. Agric. Environ. Ethics **29**(4), 571–585 (2016)
22. Wang, Z., Feng, S., Zhao, X.: MECE method and its application in sports event interpretation. In: Proceedings of ISMSS2019, pp. 340–343 (2019)
23. Hong, Z.: Research on the comprehensive evaluation of sports management system with analytical hierarchy process. In: International Conference on Intelligent Transportation, pp. 665–668 (2016). IEEE
24. Dauer, M., Jaeger, J., Bopp, T., et al.: Protection security assessment - system evaluation based on fuzzification of protection settings. In: Innovative Smart Grid Technologies Europe (ISGT EUROPE), 2013 4th IEEE/PES, pp. 128–142. IEEE (2013)
25. Liu, S., Chen, P.: Research on fuzzy comprehensive evaluation in practice teaching assessment of computer majors. Int. J. Modern Educ. Comput. Sci. **7**(11), 12–19 (2015)
26. Khan, A.A., Madden, J.: Speed learning: maximizing student learning and engagement in a limited amount of time. Int. J. Modern Educ. Comput. Sci. (IJMECS) **8**(7), 22–30 (2016)

Vehicles' Joint Motion Model Based on Dynamic Soft Rough Set

Volodymyr Sherstjuk[1(✉)], Maryna Zharikova[1], Igor Sokol[2], and Ruslan Levkivskiy[3]

[1] Kherson National Technical University, Kherson, Ukraine
vgsherstyuk@gmail.com, marina.jarikova@gmail.com
[2] Maritime Institute of Postgraduate Education, Kherson, Ukraine
kherson.sokol@gmail.com
[3] Kherson State Maritime Academy, Kherson, Ukraine
levka.ru55555@gmail.com

Abstract. This work presents a novel model of the safe vehicles' joint motion in confined spaces based on the proposed safety-leveled dynamic soft rough topology. The soft topological space is used to build a spatial model based on the ordered set of safety levels while rough approximation is used for its blurring. The proposed model represents joint motion space as a union of blurred dynamic topological subspaces, each of which has a similar level of safety estimation. It is intended to be used in the real-time navigation support systems for large teams of unmanned vehicles to provide spatially-distributed dynamic safety assessment that is critical to the diagnosis of the current navigation situation. The proposed model provides enough accuracy of the safety condition evaluation and acceptable performance, so it can be effectively used to solve real-time trajectory planning tasks during the joint vehicles' motion in confined spaces.

Keywords: Spatial model · Joint motion · The grid of cells · Blurred boundaries · Soft rough set · Dynamic soft rough topology

1 Introduction

Modern technologies often involve a significant number of manned and unmanned vehicles moving together within confined spaces, e.g. smart fishing, smart forest-fire fighting, smart traffic systems, etc. In many cases, vehicles moving in different environments simultaneously can be used here. For example, teams solving the fishing task can involve a multitude of aerial, surface, and underwater vehicles while teams solving the forest fire fighting task can involve ground and aerial vehicles.

The joint motion of many vehicles is limited by certain restrictions such as the space and the reaction of an environment [1]. The latter usually gives rise to various disturbances with respect to different space points. A growth of vehicles' number and size, a significant increase in their speed and density of movement lead to various incidents and accidents. Therefore, the most essential challenge of the joint activity of vehicles is to keep their safety. The joint activity of vehicles can be considered as

Z. Hu et al. (Eds.): ICCSEEA 2020, AISC 1247, pp. 231–242, 2021.
https://doi.org/10.1007/978-3-030-55506-1_21

simultaneous processes arising as a result of vehicles' interactions in limited space, which often give rise to danger and risk for each other. The totality of such interactions forms a dynamic system, which evolves in space and time, and can be simultaneously affected by a significant number of factors due to highly dynamic, unpredictable, nondeterministic, and partially observable environments. Given time limit conditions, the control of such a dynamic system is a complex and non-trivial task. Thus, it requires real-time navigation support [2].

Today, studying the navigation support methods intended for the joint motion control within confined spaces is one of the actively researched areas. Some approaches to path planning for highly mobile vehicles operating in uncertain environments based on genetic algorithms, fuzzy and hybrid systems have been proposed. Systematic reviews of such approaches are represented in [3–6]. Unfortunately, most approaches are local in nature, that is, they are not intended to build the trajectories of a group of vehicles. Besides that, they are based on continuous trajectories and do not consider the features of confined spaces. Since such problems are not currently worked enough, they are of great interest. Thus, the development of a safe joint navigation support model for the group of vehicles within the confined space is a topic of interest. The objective of this paper is to propose a vehicles' joint motion model based on the safety-leveled dynamic soft rough topology for the navigation support system. Thus, we need to develop a topological space divided into dynamic safe and danger areas corresponding to the safety assessments, to blur the boundaries of safe and danger areas using soft rough sets allowing to assess safe motion conditions, and to provide a simulation to examine the efficiency of the proposed model.

The paper is organized in the following way: Sect. 2 introduces the problem of safe joint motion of a group of vehicles, Sect. 3 explains a discretized spatial model of confined space, Sect. 4 describes approach to the estimation of safety conditions, Sect. 5 represents soft rough topological model decomposing the confined space into blurred safety regions, and finely Sect. 6 deals with simulation results.

2 Safe Path Planning Problem

Consider a set of vehicles $U = \{u_0, u_1, \ldots u_n\}$ distributed over a certain confined space C representing AOI. Suppose each $u_i \in U$ moves inside C changing its position over time and avoiding collisions with other vehicles. Currently, the path planning task for each u_i stands for obtaining the path $Pt(u_i)$ that consists of a sequence of necessary positions (waypoints) of vehicles at given time moments (timepoints) up to achieving a certain goal. Due to unpredictable environments, the vehicle u_i is exposed to several disturbances, which should be compensated by changing the originally planned path to avoid unsafe areas. Thus, the real trajectory $Tr(u_i)$ of the vehicle u_i coincides rarely with its planned path $Pt(u_i)$. Clearly, when some vehicles change their trajectories, it changes the circumstances for other ones and forces them to change their own trajectories dynamically. This process is iterative. The narrower the space of vehicles' joint movement is and the denser they move in it, the more intensively they interact creating a spatially bounded dynamical system and the more difficult it is to control them. Intensive interactions of vehicles give rise to a problem of dynamic joint motion re-planning to provide safe motion conditions for all interacting vehicles.

2.1 Safe Motion Conditions Assessment

The most common method of safety conditions assessment is proposed for a pair of rapprochement objects u_i and u_j based on the definition of linear and temporal characteristics of the joint motion process, respectively, distance D_{ij} and time T_{ij} to the closest point of approach [7]. Thus, the assessment is based on the subsequent comparison of D_{ij} and T_{ij} with the given permissible values of distance D_z and time T_z remaining to the closest point of approach. However, in the case of the simultaneous movement of many vehicles, this method is not applicable because pairwise comparisons entail high computational complexity due to the iterative interaction processes. Another approach [8] allows breaking down the circumjacent area into safe and dangerous domains to eliminate the ingress of any other vehicles or unsafe objects into the safe domain around the considered vehicle. Thus, any intrusion into the boundary of the safe domain having the shape of a circle, ellipse, etc. will be qualified as a threat. The shape and size of the domain depend on a set of factors of stochastic nature that make it impossible to determine domain boundaries precisely [9].

Obviously, uncertainty complicates the determining of safe motion conditions. Therefore, "fuzzy boundaries" of the safe domain [10] and "fuzzy safety domain" [11] were proposed to resolve uncertainty, where the size and shape of fuzzy domains depend on the safety level considered as the membership degree of the current situation to a certain fuzzy set of "safe situations". However, the fuzzy sets' membership degrees are expected to be determined with any statistical or expert methods, so the fuzzy approach is weakly applicable under the real-time conditions. Thus, the bottleneck of the safe motion conditions assessment is the difficulty of determining the accepted conditions of the closest approach or the safety domain boundaries.

2.2 Rough Safe Motion Conditions Assessment

To solve the problem of safe motion conditions assessment we use a rough set approach [12]. Consider a certain concept $X \in C$ and an equivalence relation R, a lower approximation consists of all objects that necessarily belong to the concept X while an upper approximation consists of all objects that possibly belong to the concept X. The space between the lower and upper approximations is the boundary area, which consists of all objects that cannot be unambiguously mapped as shown in Fig. 1.

Thus, using the lower and higher approximations we can determine intervals, which define the safe boundaries of the spatial areas $BND_{H1}, \ldots BND_{Hm}$ as $POS_{H(i+1)} = NEG_{Hi} \cup BND_{Hi} \cup POS_{Hi}$ for each $H(i+1)$ (Fig. 2). Clearly, uncertain safe motion conditions can be represented in a "rough" way without a priori information because the boundary area does not require the assignment of any distribution (probability, possibility, etc.) or membership functions.

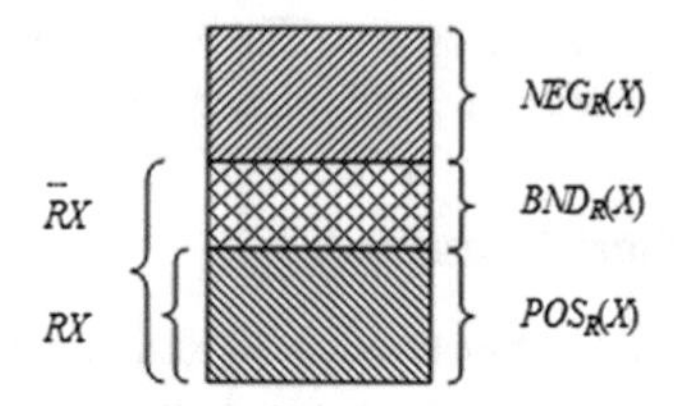

Fig. 1. Determining the areas of the rough set.

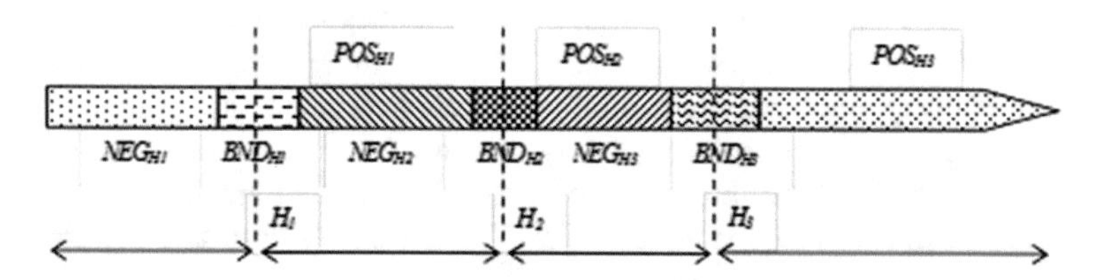

Fig. 2. Determination of the boundaries of the spatial areas.

2.3 The Proposed Approach

Using the rough sets, we can divide the space C into three subspaces – the unambiguously safe area corresponding to $POS_R(X)$, the unambiguously dangerous area corresponding to $NEG_R(X)$, and that area $BND_R(X)$ for which we do not know the degree of safety or danger. The latter can be represented as the rough or fuzzy set, giving the following partition into more dangerous or less dangerous areas. All we need is to define the equivalence relation R correctly. However, estimations of the safe and dangerous areas change subsequently to the changes in vehicles' trajectories, so the partitions of the space C can also change. Thus, we need to build a topological space divided into safe and danger areas, which should be dynamic and correspond to the safety assessments [12].

2.4 Requirements to the Spatial Model

Partial observable and uncertain environments force us to find ways to deal with inaccuracy, incompleteness, and inconsistency of vehicles' positions and motion parameters' observations with respect to the spatial model C. There are several well-known approaches to deal with the uncertainty and vagueness based on the fuzzy [13], rough [14], and soft [15] set theories, each of which has its inherent difficulties as pointed out in [16]. Due to the absence of important a priori information, such as membership functions for fuzzy sets, equivalence relations for rough sets, or parameterizations for soft sets, these approaches cannot ensure the adequacy of the spatial model of joint vehicles' motion. Therefore, many researchers combine some of these approaches proposing to use rough fuzzy, fuzzy rough, rough soft, fuzzy soft sets [17–19], etc., which provide enough advantages to build a blurred spatial model. The soft rough sets approach proposed in [20] is of particular interest because it provides a possibility to build soft topological space of the joint vehicles' motion whereas the rough sets will contribute to outline the safe subspaces within it. Such subspaces can be approximated. Since the aim of this paper is to develop the safe vehicles' joint motion model in confined spaces, we need to build a dynamic rough soft topology.

3 Spatial Model of Confined Space

Let Y be a set of a certain nature and T be a set of time points. Consider a time scale imposed by a partial order $<_T$ over time points from T with initial value t_0.

Consider C is a three-dimensional Euclidean space. Suppose C is a linear uniform space with respect to the norm $\|y\|_c = \min_{t \in [t_0, T)} (y(t))$, where $y \in Y$, $t \in T$, and $\xi_C(y_1, y_2) = \|y_1 - y_2\|$ is a suitable metric. Suppose e_1, e_2, e_3 is a basis in C where the metric ξ_C remains uniform. Thus, each decomposition $v = \alpha_1 e_1 + \alpha_2 e_2 + \alpha_3 e_3$ defines coordinates $v(\alpha_1, \alpha_2, \alpha_3)$ representing the position v of the vehicle u_i within C. The change of vehicle's coordinates within C over time t describes its motion.

Let's impose a metrical grid of coordinate lines within C using the norm ξ_C and a linear map f such that coordinate lines build a set D of cubic cells of a certain size $\delta \times \delta \times \delta$, $f : C \to D$. Thus, we discretize AOI by a grid $D = \{d_{xyz}\}$ of isometric cubic cells d_{xyz}, where x, y, z correspond to e_1, e_2, e_3 respectively, so the vehicle's location can be discrete. Each cell is now the spatial object of the minimal size corresponding to a connected subspace of C bounded to certain coordinates. Each cell can also be associated with a set of attribute values A called the cell state. Our discretization assigns equal values of the attributes to each point within the cell d.

Space C can be also divided into a finite set of disjoint three-dimensional objects represented as geometric shapes, which outline boundaries of homogeneous areas. The homogeneous spatial areas uniform in terms of their attributes' values and approximated by the set of cells are called regions and denoted by h. Each region has the properties of continuity and connectivity and cannot overlap or cover one another. We can also represent area spatially distributed over C containing a non-continuous plurality of separate regions and described by definite homogenous assessments of safe motion conditions. Such a set of regions is called the zone and denoted by H.

4 Estimation of Safety Motion Conditions

Let ξ_B be a metric with the properties similar to ξ_C, and χ be an anisometric surjection such that $\chi : \xi_C \to \xi_B$. A set of domain-dependent limits $\rho_i(t) = \{\rho_0(t), \ldots \rho_l(t)\}$ can be defined for each u_i at a time t based on the metric ξ_B. Suppose $Pos_{u_i}(t)$ describes the position of the vehicle u_i at the time t. Thus, for each couple (u_i, u_k) of vehicles, we can evaluate the limit $\|Pos_{u_i}(t) - Pos_{u_k}(t)\|_{\xi_B} \to \rho_i(t)$. In a similar way, we can use a time norm $\|T\|$ over T and the corresponding metric ξ_T to evaluate time-dependent limits $\varphi_i(t) = \{\varphi_0(t), \ldots \varphi_q(t)\}$. Since in both cases the subset of the motion and environmental parameters $A_S \subseteq A$ of the cell $d \in D$ corresponding to the current position $Pos_{u_i}(t)$ of the vehicle u_i at the time t is a basis for determining such limits, we assume that A_S uniquely identifies the safety estimation of cells at the time t.

Safety motion conditions can be represented as safety areas. Based on $\rho_i(t)$ and/or $\varphi_i(t)$, we can obtain domain- and time-dependent safety areas around u_i represented as regions within C. For example, in order to avoid collisions we can describe forbidden (h_A), dangerous (h_B), restricted (h_C), and unrestricted regions (h_D) delimited by certain boundaries. Generally, such regions can be represented as non-spherical geometric shapes with vague boundaries due to the uncertainty of observations [21].

Suppose a function $\vartheta_d(t, Pos_{u_i}(t), \rho_i(t), \varphi_i(t)) \rightarrow h_{ji}(t)$ matches the cell d with a certain region h_j for the vehicle u_i at the time t with a certain degree ϑ_{dji}. Since the cell d can be simultaneously matched with the different regions of the different vehicles, it is necessary to convolve their estimations: $\vartheta_d(t) = \oplus_{j=1}^{m}\left(\otimes_{i=1}^{n}\vartheta_{dji}(t)\right)$. Taking into account that limits $\rho_0(t), \ldots \rho_l(t)$ are connected through a partial order $\preccurlyeq_\rho$ such that $\rho_0(t) \preccurlyeq_\rho \ldots \preccurlyeq_\rho \rho_l(t)$, we can quantify the values $\vartheta_d(t)$ at each cell within C to some levels $\omega(d, t)$ based on the partial order $\preccurlyeq_\rho$ and a certain scale $\Omega = \{\omega_1, \ldots \omega_m\}$. The number m of scale elements should be a compromise. When it increases, the model accuracy also increases, but the computational complexity grows even more. An example of the 6-levels scale proposed in [21] is shown in Table 1. Thus, the safety level of each cell can be assessed based on the estimations within an interval (0, 1), and the discretized AOI can be partitioned into a multitude of regions, each of which has some safety level of six possible. Regions with the same safety level can be combined into certain zones within C.

Table 1. Safety levels

Safety level, ω	Limit value	Safety degree	Safety assessment
ω_1	$\rho_0(t)$	≈ 1	Safe
ω_2	$\rho_1(t)$	≥ 0.8	Almost safe
ω_3	$\rho_2(t)$	$0.6 \div 0.8$	Undangerous
ω_4	$\rho_3(t)$	$0.4 \div 0.6$	Unsafe
ω_5	$\rho_5(t)$	$0.2 \div 0.4$	Dangerous
ω_6	$\rho_6(t)$	≤ 0.2	Forbidden

5 Topological Model

5.1 Basic Topological Model

Let us consider the discretized confined space D endowed with the topological properties [22]. Since the cell is a homogeneous area D in terms of attribute values A, the set of cells is A-indiscernible: $(\forall d_1, d_2 \in D)(\forall a \in A)[f(d_1, a) = f(d_2, a)]$ where $f(d, a)$ is a function that returns the value of a parameter a for the cell d. Thus, using the equivalence relation $R_D^A = D \times D$ (reflexive, symmetric, and transitive) on the set of all cells within D, we can define a corresponding equivalence class $R_D^A(d)$, approximation space $apr_D = \left(D, R_D^{A_i}\right)$, and topological space $T_D^A = (D, Def(apr_D))$ [23].

However, we are not so much interested in the set A as in the subset $A_S \subseteq A$ of motion and environmental parameters affecting the safety assessments. This subset defines the A_S-indiscernibility relation $R_D^{A_S}$ such that $(\forall a \in A_S) R_D^{A_S} = \{(d_m, d_n) \in D \times D|, f(d_m, a) = f(d_n, a)\}$. Respectively, all cells having the equal values of attributes $a_i, \ldots a_k \in A_S$ belong to a certain equivalence class $R_D^{A_S}(d)$ from the set of equivalence classes generated by A_S-indiscernibility relation: $(\forall a \in A_S)(\forall d_m, d_n \in D)\left[\left(R_D^{A_S}(d_m) = R_D^{A_S}(d_n)\right) \Leftrightarrow f(d_m, a) = f(d_n, a)\right]$. It means that all different points that belong to the different cells d_m, d_n of the equivalence class $R_D^{A_S}(d)$ have equal values of these attributes. That is why, all cells that belong to a certain region h are A_S-indiscernible, and the region h is a connected subset of a certain equivalence class $R_D^{A_S}(d)$. Thus, $apr_D^{A_S} = \left(D, R_D^{A_S}\right)$ is the approximation space based on the family of composite sets $Def\left(apr_D^{A_S}\right)$, and $T_D^{A_S} = \left(D, Def\left(apr_D^{A_S}\right)\right)$ is the topological space generated by A_S-indiscernibility relation within D.

5.2 Defining a Soft Set of Cells

Suppose the grid D is a universe and A is the cell's attribute set. Consider a mapping F that takes A into the set of all subsets of the set D. Thus, a pair (F, A) can be considered as a parameterized family of subsets of the cell set D, which constitutes a soft set of cells over D denoted by Υ_D. Consider $A_S \in A$ is a subset of time-varying safety-critical cell's attributes. Correspondingly, a pair (F, A_S) is considered as a set of A_S-approximate elements of the soft set [24]. Taking into account the set of safety levels represented in Table 1, we can break down the set of cells D into six subsets of A_S-approximate elements (ω_i-elements, $\omega_i \in \Omega$, $i = 1..m$). The confined space can be represented as the soft set $\Upsilon_D(t) = \left\{(\omega_i, \Upsilon_D(\omega_i, t)) : \omega_i \in 2^\Omega, \Upsilon_D(\omega_i, t) \in 2^D\right\}$, where $\Upsilon_D(\omega_i, t)$ is an ω_i-element of the soft set, namely a set of cells which have the safety level $\omega_i \in \Omega$ at the time t. Since $Def(\Upsilon_D(t))$ is a set of all compositional sets of the soft set $\Upsilon_D(t)$, we obtain a dynamic soft topological space $T_D^\Omega(t) = (D, Def(\Upsilon_D(t)))$.

5.3 Defining Rough Elements of the Soft Set

Since determining the cells' safety degrees precisely is quite difficult, the simplest way to blur soft topology is to use rough sets as proposed in Sect. 2.2. Let $R_D^{\omega_i}$ be an indiscernibility relation in the set of cells D such as $R_D^{\omega_i} = \{(d_m, d_n) \in D \times D | f(d_m, \omega_i) = f(d_n, \omega_i), \omega_i \in \Omega\}$ where ω_i is the safety level of the cell $d_j \in D$. Thus, $apr_D^{\omega_i} = (D, R_D^{\omega_i})$ is a Pawlak approximation space [12]. Let $\Upsilon_D(t)$ be a soft set in the set of cells D. Its lower approximation can be defined as $\underline{\Upsilon_D}(\omega, t) = \{\forall \omega_i \in \Omega(R_D^{\omega_i}(d) \subseteq \Upsilon_D(\omega_i, t) | d \in D)\}$ while the upper approximation is $\overline{\Upsilon_D}(\omega, t) = \{\forall \omega_i \in \Omega(R_D^{\omega_i}(d) \cap \Upsilon_D(\omega_i, t) \neq \emptyset | d \in D)\}$. The rough set of cells, the safety level of which is ω_i at the time t, is determined by two approximations and constitute exactly the ω_i-element of the soft set such that $\widehat{\Upsilon}_D(\omega_i, t) = \left\{\underline{\Upsilon_D}(\omega_i, t), \overline{\Upsilon_D}(\omega_i, t)\right\}$. If $\underline{\Upsilon_D}(\omega_i, t) = \overline{\Upsilon_D}(\omega_i, t)$, the corresponding ω_i-element of the soft set $\Upsilon_D(t)$ is a strict set, in the other cases it is the rough set. The negative area of the rough set $\widehat{\Upsilon}_D(\omega_i, t)$ can be defined as $NEG\left(\widehat{\Upsilon}_D(\omega_i, t)\right) = D - \overline{\Upsilon_D}(\omega_i, t)$ while its boundary area $BND\left(\widehat{\Upsilon}_{W_D}(w, t)\right) = \overline{\Upsilon_D}(\omega_i, t) - \underline{\Upsilon_D}(\omega_i, t)$. Clearly, $Def\left(\widehat{\Upsilon}_D(\omega_i, t)\right)$ is a family of all rough sets, which represent the sets of cells having the safety level ω_i, and defines a rough topology in D. Thus, the pair $T_D^{\omega_i}(t) = \left(D, Def\left(\widehat{\Upsilon}_D(\omega_i, t)\right)\right)$ is the rough topological space at the time t. The proposed rough approximation can be essentially improved using the tolerance [25] or ever similarity relation [26] instead of the equivalence relation. The tolerance relation gives a coverage of the set of cells D and enables blurring the boundaries of the safety regions in contrast to the equivalence relation, which gives a partition of the set D into disjoint subsets.

5.4 Building the Soft Rough Topological Model

Following Sects. 5.2 and 5.3, we can define the grid D as the soft rough set of cells $\hat{\Upsilon}_D(t) = \left\{\left(\omega_i, \left\{\underline{\Upsilon_D}(\omega_i, t), \overline{\Upsilon_D}(\omega_i, t)\right\}\right) : \forall \omega_i \in \Omega, \Upsilon_D(\omega_i, t) \in 2^D\right\}$. Accordingly, we obtain a ω_i-leveled dynamic soft-rough topological space $\hat{T}_D(t) = \bigcup_{i=1}^{m} \hat{T}_D^{\omega_i}(t) = \bigcup_{i=1}^{m} \left(D, Def\left(\widehat{\Upsilon}_D(\omega_i, t)\right)\right)$ being a soft union of ω_i-elementary rough topological spaces and representing the AOI at time t. Let us define the ω_i-tolerance relation $\widehat{R}_H^{\omega_i}$ on a set of regions H in the same way as A_S-indiscernibility relation on a set of cells D. Thus, $\widehat{R}_H^{\omega_i}(h)$ is the equivalence class of the set of regions H generated by $\widehat{R}_H^{\omega_i}$, so $apr_H^{\omega_i} = \left(H, \widehat{R}_H^{\omega_i}\right)$ is the rough approximation space and correspondingly $\widehat{T}_H^{\omega_i} = (H, Def(apr_H^{\omega_i}))$ is the rough topological space. Since regions are dynamic, the topological space $\widehat{T}_H^{\omega_i}$ is also dynamic. Clearly, we obtain the ω_i-leveled soft-rough dynamic topological space $\tilde{T}^{A_j}_{\widehat{T}_D\,A_i}(t)$.

6 Results and Discussion

The vehicles' joint motion model based on 6-leveled soft-rough dynamic topological space corresponding to Table 1 has been developed. The proposed model has been implemented using the C++ programming language and integrated into the real-time navigation support system Breeze [21] based on onboard microcontroller STM32F429 (Cortex M4 180 MHz, 2 Mb Flash/256 Kb RAM). To examine the validity and the efficiency of the proposed model, the simulation of 12 vehicles' motion has been run within the discretized spatial model. In order to solve the task of safe trajectory planning, the interior and the closure of the topological space have been defined at each time point. The desired vehicle's trajectory can be represented by slicing the grid D where each slice d_j is taken corresponding to each next time moment t_j. Thus, we can obtain a motion corridor for each vehicle as shown in Fig. 3. The intersection of such corridors represents safety subspaces as the soft-rough topological space based on the permissible safety level (in this case, $\omega_i = 1$). During the simulation, we have compared the total time of evaluation each next slices d_j needed to safe trajectory planning and obtained based on a dynamic soft rough set of cells with results of simulation based on the crisp, fuzzy, and fuzzy soft topological models as shown in Fig. 4.

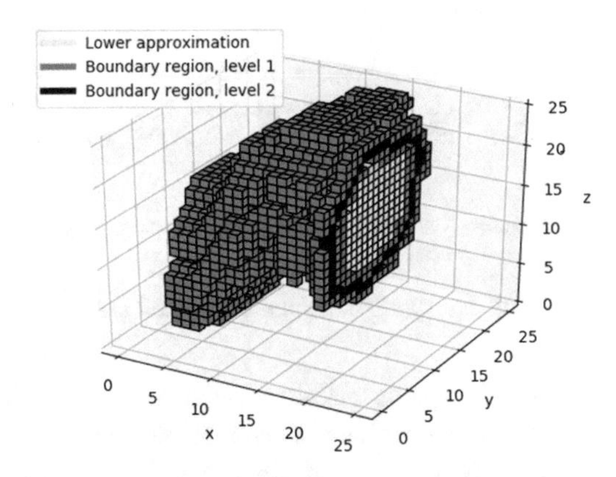

Fig. 3. The motion corridor representation within AOI.

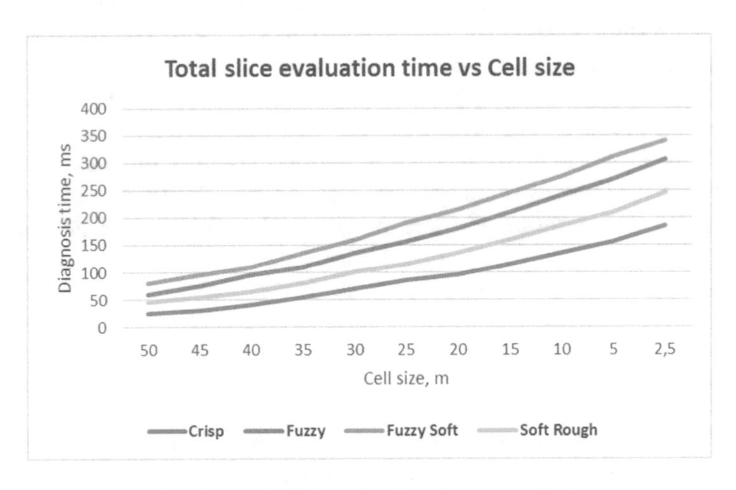

Fig. 4. Total slice evaluation time vs. cell size

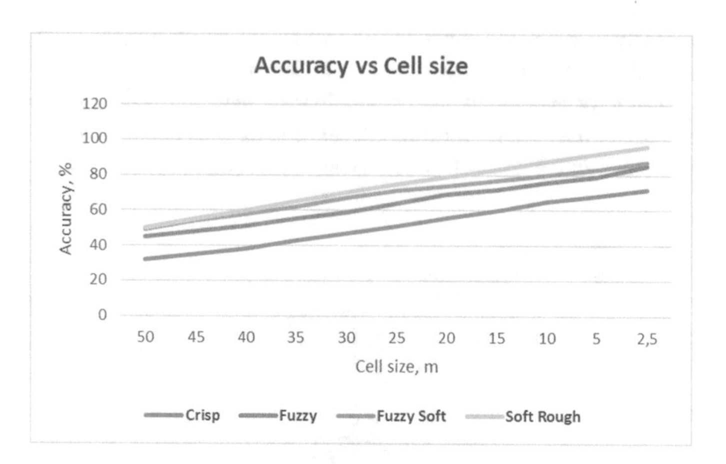

Fig. 5. Approximation error vs. cell size

Obviously, the simulation results mainly depend on the sampling of the spatial model and the kind of the used topology. Therefore, we have varied the cell size to investigate its impact on the total slice evaluation time. Figure 4 depicts the results of the total time evaluating while Fig. 5 depicts the results of evaluating the model accuracy (accuracy has been calculated as a ratio of the evaluated safety area to the real one within each slice). Clearly, the proposed joint motion model based on the 6-leveled soft-rough topological space provides enough accuracy of the safety condition evaluation with respect to the other kinds of topologies while winning in time except for the crisp topology. However, the crisp topology yields in accuracy whereas fuzzy and fuzzy soft topologies are worse in terms of performance. The difference between the soft-rough and fuzzy-based topologies becomes greater, the smaller is the cell size. Thus, the simulations results depict that the proposed model provides the total time of evaluation each safety slice of topological space below 250 ms for any cell sizes up to 2,5 m, which makes the model of practical use in real-time for any reasonable speed.

7 Conclusions

The vehicles' joint motion model proposed in the paper is a novel tool for the safe vehicle trajectory planning in confined spaces based on the safety-leveled dynamic soft rough topology. The soft topology is used to build the spatial model based on the ordered set of safety levels while rough approximation is used for its blurring. The model represents the confined space as a union of blurred topological subspaces, each of which has a similar level of safety estimation and is dynamic due to the dynamic nature of the safety assessments. The proposed model is intended to be used in the real-time navigation support systems for large teams of unmanned vehicles to provide spatially-distributed dynamic safety assessment that is critical to the diagnosis of the current navigation situation. In this paper, we have described in short some implemented elements of the model and have given some results of the simulations.

The simulations have shown that the proposed model provides enough accuracy of the safety condition evaluation and acceptable performance, so it can be effectively used to solve real-time trajectory planning tasks during the joint vehicles' motion in confined spaces. It can be investigated in future works in order to analyze the properties of the dynamic soft rough topology and their further improvement.

References

1. Aenugu, V., Woo, P.-Y.: Mobile robot path planning with randomly moving obstacles and goal. Int. J. Intell. Syst. Appl. (IJISA) **4**(2), 1–15 (2012). https://doi.org/10.5815/ijisa.2012.02.01
2. Sherstjuk, V.: Scenario-case coordinated control of heterogeneous ensembles of unmanned aerial vehicles. In: Proceedings of 2015 IEEE 3rd International Conference on Actual Problems of Unmanned Aerial Vehicles Developments (APUAVD 2015), Kyiv, pp. 275–279 (2015). https://doi.org/10.1109/APUAVD.2015.7346620
3. Miao, H., Huang, X.: A heuristic field navigation approach for autonomous underwater vehicles. Intell. Autom. Soft Comput. (2014). https://doi.org/10.1080/10798587.2013.872326
4. Mahmoudzadeh, S., Powers, D.M.W., Yazdani, A.M., Sammut, K., Atyabi, A.: Efficient AUV path planning in time-variant underwater environment using differential evolution algorithm. J. Mar. Sci. Appl. **17**(4), 585–591 (2018). https://doi.org/10.1007/s11804-018-0034-4
5. Seif, R., Oskoei, M.A.: Mobile robot path planning by RRT* in dynamic environments. Int. J. Intell. Syst. Appl. (IJISA) **7**(5), 24–30 (2015). https://doi.org/10.5815/ijisa.2015.05.04
6. Bhadoria, A., Singh, R.K.: Optimized angular a star algorithm for global path search based on neighbor node evaluation. Int. J. Intell. Syst. Appl. (IJISA) **6**(8), 46–52 (2014). https://doi.org/10.5815/ijisa.2014.08.05
7. Zak, B.: The problems of collision avoidance at sea in the formulation of complex motion principles. Int. J. Appl. Math. Comput. Sci. **14**, 503–514 (2004)
8. Goodwin, E.: A statistical study of ship domain. J. Navig. **28**, 328–344 (1975)
9. Lazarowska, A.: Decision support system for collision avoidance at sea. Polish Marit. Res. **19**, 19–24 (2012). https://doi.org/10.2478/v10012-012-0018-2
10. Zhao, J., Wu, Z., Wang, F.: Comments of ship domains. J. Navig. **46**, 422 (1993). https://doi.org/10.1017/S0373463300011875
11. Pietrzykowski, Z., Uriasz, J.: The ship domain – a criterion of navigational safety assessment in an open sea area. J. Navig. **62**(01), 93–108 (2009). https://doi.org/10.1017/S0373463308005018
12. Sallam, E., Medhat, T., Ghanem, A., Ali, M.E.: Handling numerical missing values via rough sets. Int. J. Math. Sci. Comput. (IJMSC) **3**(2), 22–36 (2017). https://doi.org/10.5815/ijmsc.2017.02.03
13. Nguyen, X.T., Nguyen, V.D.: Support-intuitionistic fuzzy set: a new concept for soft computing. Int. J. Intell. Syst. Appl. (IJISA) **7**(4), 11–16 (2015). https://doi.org/10.5815/ijisa.2015.04.02
14. Nagaraju, M., Tripathy, B.K.: Study of covering based multi granular rough sets and their topological properties. Int. J. Inf. Tech. Comput. Sci. (IJITCS) **7**(8), 61–67 (2015). https://doi.org/10.5815/ijitcs.2015.08.09

15. Molodtsov, D.A.: Soft set theory – first results. Comput. Math Appl. **37**(4–5), 19–31 (1999). https://doi.org/10.1016/S0898-1221(99)00056-5
16. Sherstjuk, V., Zharikova, M., Sokol, I.: Approximate spatial model based on fuzzy-rough topology for real-time decision support systems. In: Proceedings on IEEE First Ukraine Conference on Electrical and Computer Eng. (UKRCON), Kyiv, pp 1037–1042 (2017). https://doi.org/10.1109/UKRCON.2017.8100408
17. Maji, P.K., Roy, A.R., Iswas, R.B.: An application of soft sets in a decision-making problem. Comput. Math Appl. **44**, 1077–1083 (2002). https://doi.org/10.1016/S0898-1221(02)00216-X
18. Broumi, S., Smarandache, F., Dhar, M., Majumdar, P.: New results of intuitionistic fuzzy soft set. Int. J. Inf. Eng. Electron. Bus. (IJIEEB) **6**(2), 47–52 (2014). https://doi.org/10.5815/ijieeb.2014.02.06
19. Meng, D., Zhang, X., Qin, K.: Soft rough fuzzy sets and soft fuzzy rough sets. Comput. Math Appl. **62**, 4635–4645 (2011). https://doi.org/10.1016/j.camwa.2011.10.049
20. Feng, F., Li, Y., Leoreanu-Fotea, V.: Application of level soft sets in decision making based on interval-valued fuzzy soft sets. Comput. Math Appl. **60**, 1756–1767 (2010). https://doi.org/10.1016/j.camwa.2010.07.006
21. Zharikova, M., Sherstjuk, V.: Case-based approach to intelligent safety domains assessment for joint motion of vehicles ensembles. In: Proceedings of the 4th International Conference on Methods and Systems of Navigation and Motion Control (MSNMC), Kyiv, pp. 245–250 (2016). https://doi.org/10.1109/MSNMC.2016.7783153
22. Allam, A.A., Bakeir, M.Y., Abo-Tabl, E.A.: Some methods for generating topologies by relations. Bull. Malays. Math. Soc. Ser. 2 **31**, 35–45 (2008)
23. Abdel-Monsef, M.E., Embaby, O.A., El-Bably, M.K.: New approach to covering rough sets via relations. Int. J. Pure Appl. Math. (IJPAM) **91**(3), 329–347 (2014). https://doi.org/10.12732/ijpam.v91i3.6
24. Neog, T.J., Sut, D.K.: An introduction to the theory of imprecise soft sets. Int. J. Intell. Syst. Appl. (IJISA) **4**(11), 75–83 (2012). https://doi.org/10.5815/ijisa.2012.11.09
25. Skowron, A., Stepaniuk, J.: Tolerance approximation spaces. Fundamenta Informaticae **27**(2–3), 245–253 (1996). https://doi.org/10.5555/2379560.2379571
26. Nguyen, X.T., Nguyen, V.D., Nguyen, D.D.: Rough fuzzy relation on two universal sets. Int. J. Intell. Syst. Appl. (IJISA) **6**(4), 49–55 (2014). https://doi.org/10.5815/ijisa.2014.04.05

Resonance and Rational Numbers Distribution: An Universal Algorithm of Discrete States Occurrence in the Spectra of Various Nature Systems

Victor A. Panchelyuga(✉), Maria S. Panchelyuga, and Olga Yu. Seraya

Institute of Theoretical and Experimental Biophysics of the Russian Academy of Sciences, 142290 Institutskaya Str., 3, Pushchino, Moscow Region, Russia
victor.panchelyuga@gmail.com

Abstract. The existence of the universal fractal spectra was demonstrated in a number of studies. The article considers a general approach demonstrating the occurrence of such spectra for various nature systems. It is based on two extremely general concepts: resonance and roughness of real physical systems. Application of the algorithm developed in the study causes the occurrence of two complementary fractals on the sets of rational and irrational numbers accordingly. Besides, it was shown that the power of equivalence classes of rational numbers is connected with well-known fact that resonance appears more easily for pairs of frequencies, which are small natural numbers. The results are used for the analysis of the spectrum of the periods of various nature systems.

Keywords: Resonance · Rational numbers · Irrational numbers · Algorithm · Fractals · Scaling · Periods · Spectra of periods · Universal spectra of periods · Chain fractions

1 Introduction

Our studies [1, 2] reveal a spectrum of periods in the fluctuations of the various nature processes. Its main features are universality (this spectrum can be found in the time series of fluctuations of the processes of various nature) and fractality (a discrete set of the periods, constituents of the spectrum, can be described by a self-similar exponential function). The further investigations showed that these features can be observed also in studies [6–24] that will be reviewed in Sect. 5.

It is clear that the description of such property as "universality" requires a most general approach that finally leads to such property as "fractality". The article suggests the unique approach based on two primary concepts: resonance and roughness of real physical systems.

Z. Hu et al. (Eds.): ICCSEEA 2020, AISC 1247, pp. 243–252, 2021.
https://doi.org/10.1007/978-3-030-55506-1_22

It is commonly known that resonance is a relation r of frequencies of two oscillations p and q, expressed by a rational number:

$$r = \frac{p}{q}, \quad p, q \in \mathbb{N}, \quad r \in \mathbb{Q}, \tag{1}$$

where $\mathbb{N}$ is a set of natural numbers, and $\mathbb{Q}$ is a set of rational numbers. If $r \in \mathbb{Q}'$, where $\mathbb{Q}'$ is a set of irrational numbers, the resonance is impossible.

The foregoing definition of resonance, in spite of its common nature, presents immediately the following question. It is well know that any physical system is « rough» in the sense that the values of its parameters are always subject to irremovable fluctuations. So in a real physical system, r cannot correspond to a single rational number, since as a result of fluctuations of the parameters p and q, a rational relation r selected initially cannot be held accurately. However, it is known that the distribution of the irrational numbers on the number axis is everywhere compact set and, hence, there is the infinity of irrational numbers in the neighborhood of any rational number. Thus, a slightest fluctuation of physical system parameters will cause the violation of the condition $r \in \mathbb{Q}$ and, as a result, impossibility of the resonance. Nevertheless, in spite of the roughness of the real physical systems, the resonance phenomena exist. And here comes a question: how can this fact be in accordance with the definition of the resonance as $r \in \mathbb{Q}$?

The second question closely related to the first one and logically following from it, as will be shown further, is: why in the real physical systems resonance appears more easily for the r, at which p and q are small numbers?

The solving of the problems is considered in Sects. 2–4. Section 5 provides a literature review considering the experimental results that can be interpreted in the terms of the approach suggested in the study.

2 The Rational Numbers Distribution

Article [3] contains an idea that rational and irrational numbers are distributed along the number axis in a nonuniform manner. To justify the idea, a procedure of construction of the rational numbers set $\mathbb{Q}$ suggested in [4] as a completely ordered set of the completely ordered sets organized as a uniquely defined system and based on the following chain fraction:

$$\{Q_i^{a_i}\} = \cfrac{1}{a_1 \pm \cfrac{1}{a_2 \pm \cfrac{1}{\cdots \atop a_i \pm \frac{1}{\cdots}}}} \tag{2}$$

where $a_1, a_2, \ldots, a_i = \overline{1, \infty}$, $i \to \infty$. Expression (2) gives rational numbers belonging to an interval [0, 1]. It is common knowledge that intervals [0, 1] and [1, ∞) are in a one-to-one correspondence, that is any patterns obtained for [0, 1] basing (2) are true for [1, ∞) also. For $i \to \infty$ expression (2) gives:

$$\{Q_i^{a_i} | i \to \infty\} \to \mathbb{Q}. \tag{3}$$

It is obvious that in case (3) it is impossible to speak about any distribution, since the rational numbers are located along the number axis everywhere dense. The condition $i \to \infty$ means that the parameters p and q must be defined with the infinite accuracy, which is impossible as was mentioned earlier. Therefore, for the real physical systems i must be limited, being the condition of the «roughness» of a physical system. Let us consider a case $i = 3$ as an example:

$$\{Q_1^{a_1}\} = \frac{1}{a_1},\ i = 1,\ a_1 = \overline{1, N}; \tag{4}$$

$$\{Q_i^{a_i}\} = \frac{1}{a_1 \pm \frac{1}{a_2}} = \frac{a_2}{a_1 a_2 \pm 1},\ i = 1, 2,\ a_1, a_2 = \overline{1, N}; \tag{5}$$

$$\{Q_i^{a_i}\} = \frac{1}{a_1 \pm \frac{1}{a_2 \pm \frac{1}{a_3}}} = \frac{a_2 a_3 \pm 1}{a_1(a_2 a_3 \pm 1) \pm a_3},\ i = \overline{1, 3},\ a_1, a_2, a_3 = \overline{1, N}. \tag{6}$$

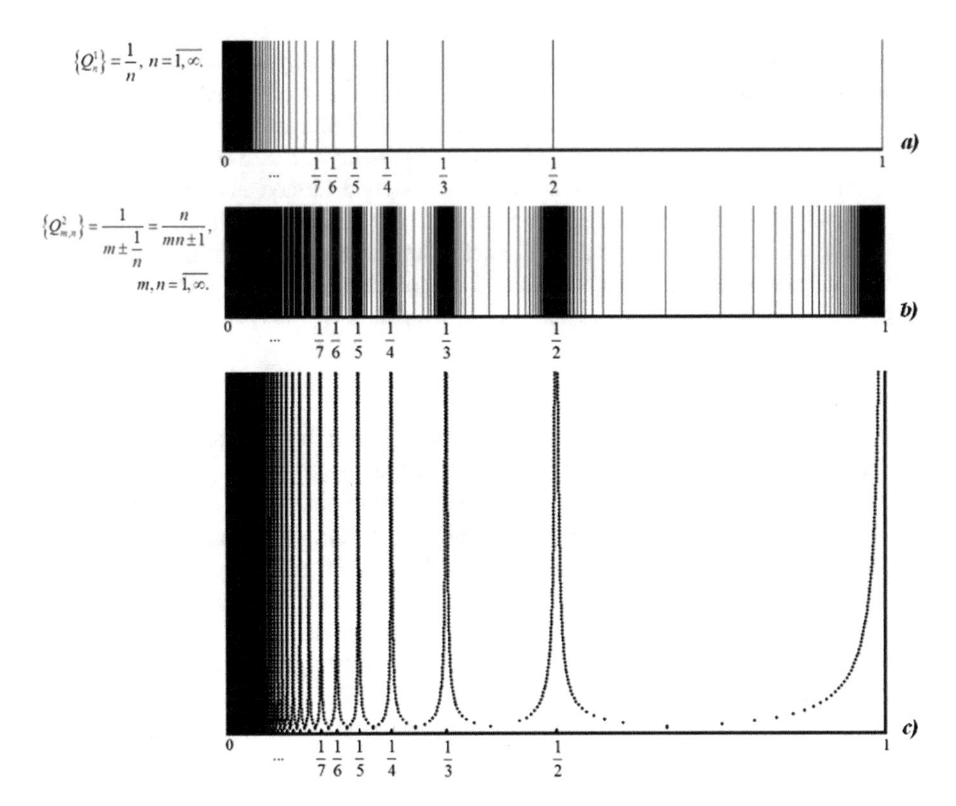

Fig. 1. Rational numbers distribution, a)–c).

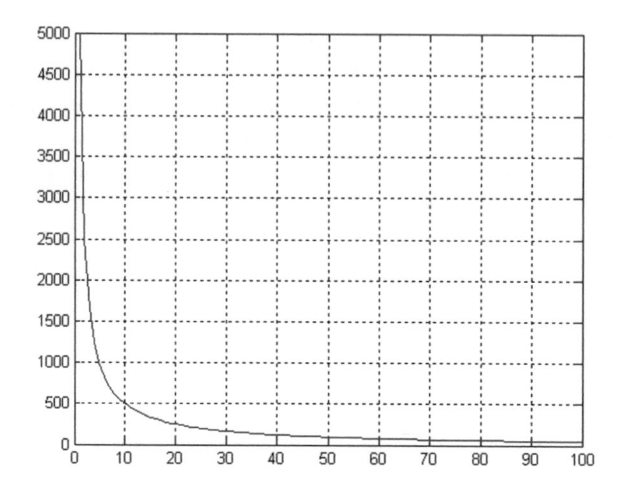

Fig. 2. Power of the equivalence classes for N = 5000. X-axis: value of q; Y-axis: power of the equivalence classes.

Figure 1*a*) and Fig. 1*b*) present the results of the calculations basing on (4) and (5) for $N = 100$. A limitation of the graphic resolution does not let to demonstrate the case described by expression (6). Figure 1*c*) shows the distribution function of rational numbers for [0, 1]. As one can see, consideration of a physical system roughness results in the existence of the sharp rational maximums demonstrated at the Fig. 1*c*).

The condition of resonance (1) cannot have a sense that a resonance is a unique pair of numbers p and q, the relation of which r is expressed as a rational number. In real situation an infinite set of the pairs p and q correspond to the same rational number, r. Supposing that each natural number is a frequency of some oscillator, the power of this set or equivalence class gives a number of oscillators that are in the resonance.

Let us make an estimate the powers of the equivalence classes. For this purpose, let us present the condition of the resonance (1) for $p, q \in [1, \infty) \in \mathbb{N}$ in the form of:

$$p \bmod q = 0 \tag{7}$$

Let $p > q$, then all q, satisfying condition (7) will be the integer divisors of p. Equation (7) considering the physical system roughness: $p, q \in [1, N] \in \mathbb{N}$, where N is rather big but finite natural number, allows for estimation of powers of equivalence classes for $q = \overline{1, N}$. The results of corresponding calculations are presented at Fig. 2.

As one can see from Fig. 2, the power of the equivalence classes decreases very rapidly with the increase of q. It is obvious from Fig. 2 that the power of equivalence classes corresponding to the initial numbers of the natural axis is several orders higher than the power of the equivalence classes for the other numbers. This is just the circumstance explaining, in our opinion, the fact that the resonance the most «readily» occurs for the r, for which p and q are the initial numbers of the natural axis.

3 The Irrational Numbers Distribution

Similarly, to the rational maximums presented at Fig. 1, we may consider a problem of searching for the irrational numbers maximums. Contrary to rational maximums, to which maximal interaction of the parts of some systems corresponds, the parameters corresponding to the irrational maximums are minimal interaction and maximal stability of the system. The authors of [3] suppose that irrational maximums correspond to the minimums in the distribution of rational numbers. The study assumes that the parts of «the most irrational numbers» may be played by the algebraic numbers that are the roots of the equation

$$\alpha^2 + \alpha b + c = 0 \tag{8}$$

Let $c = -1$. Then

$$\alpha = \frac{1}{\alpha + b} = \frac{1}{b + \frac{1}{b + \frac{1}{b + \dots}}} = \frac{\sqrt{b^2 + 4} - b}{2} \tag{9}$$

The infinite chain fraction (9) gives the worse approximation of irrational number α, the less is value of b. Therefore, the worst approximation will occur in the case $b = 1$:

$$\alpha_1 = \frac{1}{1 + \frac{1}{1 + \frac{1}{1 + \dots}}} = \frac{\sqrt{5} - 1}{2} = 0.6180339 \tag{10}$$

Case (10) corresponds to so called «golden ratio». The further calculations on the base of (9) give the value of α_2:

$$\alpha_2 = \frac{1}{2 + \frac{1}{2 + \frac{1}{2 + \dots}}} = \frac{\sqrt{8} - 2}{2} = 0.4142135. \tag{11}$$

Equation (11) gives so called «silver ratio». The following values are provided by Eq. (12):

$$\alpha_3 = \frac{1}{3 + \frac{1}{3 + \frac{1}{3 + \dots}}} = \frac{\sqrt{13} - 3}{2} = 0.3027756 \tag{12}$$

The chain fractions like (10)–(12) are called periodical and are designated as $\alpha_1 = [\bar{1}]$, $\alpha_2 = [\bar{2}]$, $\alpha_3 = [\bar{3}]$. The values of $\alpha_4 - \alpha_7$ are as follows:

$$\begin{aligned} \alpha_4 &= [\bar{4}] = 0.2360679, \\ \alpha_5 &= [\bar{5}] = 0.1925824, \\ \alpha_6 &= [\bar{6}] = 0.1622776, \\ \alpha_7 &= [\bar{7}] = 0.1400549, \\ &\ldots\ldots\ldots\ldots\ldots\ldots\ldots\ldots . \end{aligned} \tag{13}$$

The α1…α7… values presented in (10–13) are named «metal ratios».

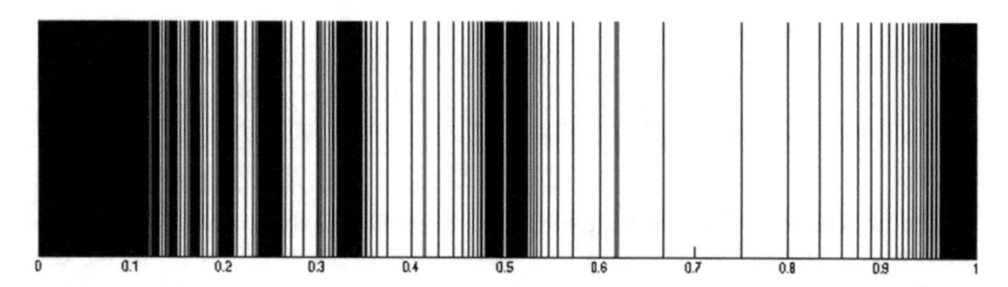

Fig. 3. Irrational maximums distributions according to (10)–(13). Algebraic numbers.

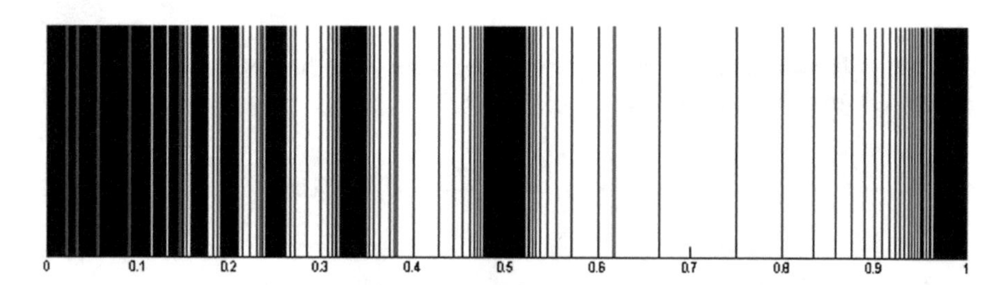

Fig. 4. α_1^n - degrees of the golden ratio.

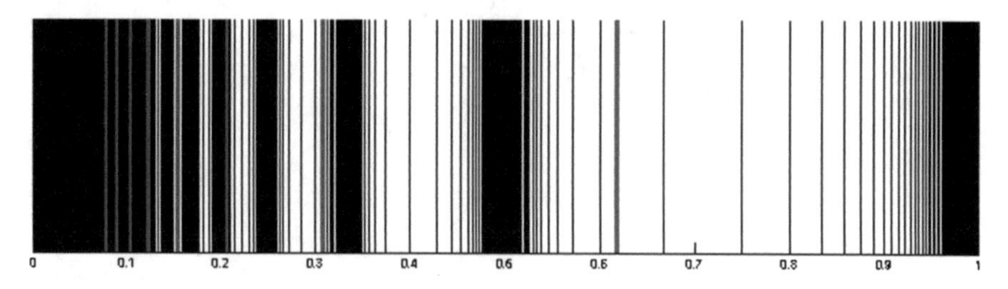

Fig. 5. The numbers of α_1/n type.

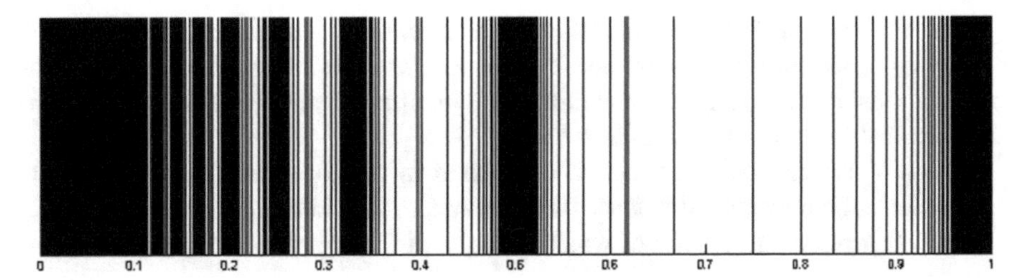

Fig. 6. Division of a segment in the golden ratio.

Figure 3 presents the results of the calculations according to (10)–(13). The red vertical lines correspond to the values of $\alpha 1 \ldots \alpha 7$ (10)–(13). The black lines correspond to the distribution of the rational numbers presented at Fig. 1*b*).

As follows from Fig. 3, the algebraic numbers tend to rational maximums with the increase of *b*. The result tells that the numbers are not the best candidates to play roles of «the most irational numbers» .

Figure 4 presents another version of construction of irrational maximums basing on so called generalized golden ratio [5]. As one can see from the figure, in this case the result is also far not the sought one. It may be noted that both irrational sequences presented at Fig. 3 and Fig. 4 start from the numbers corresponding to the golden ratio. The same is true for other sequences present in abundant literature on the golden ratio. Let us consider some of them. Figure 5 presents a sequence basing of the golden ratio as well, α_1/n. As one can easily see from Fig. 5 the sequence has the same features as the sequences at Figs. 3 and Fig. 4.

The first member of the sequences at Fig. 3, 4 and 5, α_1, corresponds exactly to the rational minimum shown at Fig. 3c). It is not unreasonable to assume that all other members of the irrational sequence the same correspond to the rational minimums shown at Fig. 3c). Such a property presents in the sequence shown at Fig. 6. The property of this sequence is that its members divide the segments between rational maximums in the golden ratio proportion. As one can see from Fig. 6, the members of this sequence coincide exactly with the rational minimums. In other words, the sequences shown at Fig. 3b) and Fig. 6 are complementary and they both have the fractal properties.

4 Rational and Irrational Fractals as an System

All the above-described results are based on the concepts of resonance and roughness of a physical system. These concepts being applied to the set of real numbers lead to two interrelated distributions on the set of rational and the set of irrational numbers. Distribution maximums of rational numbers, Fig. 1*b*) correspond to the maximal sensitivity of a system to the external influence, and to maximal interaction of the system parts. The resonance phenomenon is more stable and occurs easier, when the relations of the resonance frequencies belong to one of the rational maximums and the values of *p* and *q* are small integral numbers that lie in the very beginning of the natural number axis.

The minimums in the rational number distribution correspond to the maximum density of the irrational numbers distribution. Irrational maximums coincide with the minimal sensitivities to any external influences, minimal interaction of the system parts and maximal stability of its structure.

We suppose that both distribution considered in the paper are complementary to each other and being applied to some physical system should be considered together as a single distribution, two parts of which describe different properties of one the same system.

5 Results and Discussion

The ideas like those considered above were used in studies [6–16] that, among other issues, consider a large number of the examples of the use of an approach similar to the presented above for the description of real physical systems. For example, the author of [6], proceeding from the results obtained at the analysis of the chain systems, develops an approach named as Global Scaling with the main conclusion very similar to ours. In studies [7, 15] the concept is used to analyze distribution of masses and orbital periods of the celestial bodies of the Solar System. Study [11] analyzes masses, radii, distances from the Sun, orbital periods and rotation periods of the celestial bodies, and study [16] deals with the masses and orbital periods of Saturn, Jupiter and Uranus satellites. A series of studies [8–10, 14] analyzes masses of elementary particles and chemical elements. The study [12] analyzes the numerical values of life cycles of the excited electronic states of H, He and Li atoms, and of Li^+ ion. The author of [13] examines the ground state and excited states of the hydrogen atom.

Study [17] considers the cycles observed in the systems of various natures (geologic, astronomic, and biologic) with the periods between 57.3 years and 1.64 billion years. Its authors found the synchronism in the behavior of the analyzed cycles suggesting, as they suppose, a general reason of the astronomic origin. In addition to the synchronism, they tell about the existence of a self-similar universal scale caused by the fractal distribution of the matter in the Universe.

Studies [18, 19] analyze the spectra of some nature processes (fluctuations of the temperatures of the Earth surface and of the geomagnetic field, changes in the width of tree-rings, crop yields, etc.), and describe the results of the spectral analysis of the macroscopic fluctuations of some nature processes (speeds of the biochemical reactions, rate of the ^{239}Pu preparations decay, sizes of the trees annular rings, the Sun radio emission (λ = 10.7 sm), annual precipitations and ground temperatures at the predetermined regions of Russia, etc.) base on the different time series produced in ITEB RAS, IRE RAS and some other institutions. The time intervals of the initial time series fell into the range from hundreds and tens thousand years to hours and minutes. It is shown that the typical frequencies of the processes examined in [18, 19] form a large scale, hardly determined self-similar hierarchy described with an equation of the form similar to (5). The discrete self-similar distributions may be found not only in time duration but in space dimensions as well. For example, studies [20–22] describe such distributions for the sizes of the solids both of the nature origin and resulting the processes of rock fragmentation. Study [23, 24] follow the fractal structures at cosmological scales.

6 Conclusions

A model proposed in the study is based on the definition of the resonance (1) and the obvious hypothesis on the roughness of a real physical system. The generality of the concepts predetermines the universality of the results obtained. The resulting pair of the complementary fractal distributions provides a model for the description of the discrete spectra in the system of any nature [6–30]. The development of the specific mathematical algorithms for such descriptions will be an objective of the future investigations.

References

1. Panchelyuga, V.A., Panchelyuga, M.S.: Fractal dimension and histogram method: algorithm and some preliminary results of noise-like time series analysis. Biophysics **58**(2), 283–289 (2013)
2. Panchelyuga, V.A., Panchelyuga, M.S.: Local fractal analysis of noise-like time series by the all-permutations method for 1–115 min periods. Biophysics **60**(2), 317–330 (2015)
3. Dombrowski, K.: Rational numbers distribution and resonance. Prog. Phys. **1**, 65–67 (2005)
4. Khintchine, A.Ya.: Continued fractions. University of Chicago Press, Chicago (1964)
5. Stakhov A.P.: Codes of golden proportion. Moscow (1984)
6. Muller, H.: Fractal scaling models of resonant oscillations in chain systems of harmonic oscillators. Prog. Phys. **2**, 72–76 (2009)
7. Muller, H.: Fractal scaling models of natural oscillations in chain systems and the mass distribution of the celestial bodies in the Solar system. Prog. Phys. **1**, 62–66 (2010)
8. Muller, H.: Fractal scaling models of natural oscillations in chain systems and the mass distribution of particles. Prog. Phys. **3**, 61–66 (2010)
9. Muller, H.: Emergence of particle masses in fractal scaling models of matter. Prog. Phys. **4**, 44–47 (2012)
10. Ries, A., Fook, M.V.L.: Fractal structure of nature's preferred masses: application of the model of oscillations in a chain system. Prog. Phys. **4**, 82–89 (2010)
11. Ries, A., Fook, M.V.L.: Application of the model of oscillations in a chain system to the solar system. Prog. Phys. **1**, 103–111 (2011)
12. Ries, A., Fook, M.V.L.: Excited electronic states of atoms described by the model of oscillations in a chain system. Prog. Phys. **4**, 20–24 (2011)
13. Ries, A.: The radial electron density in the hydrogen atom and the model of oscillations in a chain system. Prog. Phys. **3**, 29–34 (2012)
14. Ries, A.: A bipolar model of oscillations in a chain system for elementary particle masses. Prog. Phys. **4**, 20–28 (2012)
15. Muller, H.: Scaling of body masses and orbital periods in the solar system. Prog. Phys. **11**(2), 133–135 (2015)

16. Muller, H.: Scaling of moon masses and orbital periods in the systems of saturn, jupiter and uranus. Prog. Phys. **11**(2), 165–166 (2015)
17. Puetz, S.J., Prokoph, A., Borchardt, G., Mason, E.W.: Evidence of synchronous, decadal to billion year cycles in geological, genetic, and astronomical events. Chaos Solitons Fractals **62**, 55–75 (2014)
18. Shabel'nikov, A.V.: Impact of cosmophysical factors on the climate and biosphere of the Earth. Biophysics **37**(3), 479–482 (1992)
19. Shabel'nikov, A.V., Kiryanov, K.G.: Secular, annual and diurnal fluctuations of the parameters of some natural processes. Biophysics **43**(5), 874–877 (1998)
20. Sadovsky, M.A.: On distribution of the sizes of solid separates. DAN USSR **269**(1), 69–72 (1983)
21. Yu Yurchenko, L., Berdikov, V.F., Sukhonos, S.I.: Invariance in some physical properties of coarse silicon carbide when the conditions of its crushing and grinding are changed. DAN USSR **293**(3), 610–613 (1987)
22. Bovenko, V.N., Zh Gorobets, L.: About the display of discreteness of solids. DAN USSR **292**(5), 1095–1100 (1987)
23. Baryshev, Yu., Teerikorpi, P.: Discovery of Cosmic Fractals, p. 408. World Scientific, Singapore (2002)
24. Pietronero, L.: The fractal structure of the universe: correlations of galaxies and clusters. Phys. A **144**, 257 (1987)
25. Hu, Z., Bodyanskiy, Y.V., Tyshchenko, O.K., Samitova, V.O.: Fuzzy clustering data given in the ordinal scale. Int. J. Intell. Syst. Appl. (IJISA), **9**(1), 67–74 (2017). https://doi.org/10.5815/ijisa.2017.01.07
26. Hu, Z., Bodyanskiy, Y.V., Tyshchenko, O.K.: Samitova, V.O.: Fuzzy clustering data given on the ordinal scale based on membership and likelihood functions sharing. Int. J. Intell. Syst. Appl. (IJISA), **9**(2), 1–9 (2017). https://doi.org/10.5815/ijisa.2017.02.01
27. Zhou, J., Yang, X., Hu, W.: Nonlinear time series predication of slope displacement based on smoothing filtered data. Int. J. Intell. Syst. Appl. (IJISA), **1**(1), 30–41 (2009). https://doi.org/10.5815/ijisa.2009.01.04
28. Goshvarpour, A., Goshvarpour, A.: Classification of heart rate signals during meditation using Lyapunov exponents and entropy. Int. J. Intell. Syst. Appl. (IJISA), **4**(2) 35–41 (2012). https://doi.org/10.5815/ijisa.2012.02.04
29. Bhatia, P.K., Singh, S., Kumar, V.: On applications of a generalized hyperbolic measure of entropy. Int. J. Intell. Syst. Appl. (IJISA), **7**(7), 36–43 (2015). https://doi.org/10.5815/ijisa.2015.07.05
30. Nazimuddin, A.K.M., Ali, S.: Periodic pattern formation analysis numerically in a chemical reaction-diffusion system, Int. J. Math. Sci. Comput. (IJMSC), **5**(3), 17–26 (2019). https://doi.org/10.5815/ijmsc.2019.03.02

Perfection of Computer Algorithms and Methods

On Unsupervised Categorization in Deep Autoencoder Models

Serge Dolgikh[1,2(✉)]

[1] National Aviation University, Komarova Ave., 1, Kiev 03058, Ukraine
serged.7@gmail.com
[2] Solana Networks, 301 Moodie Dr., Ottawa K2H9C4, Canada

Abstract. In this study the authors investigated categorization parameters of the latent space created by a class of deep autoencoder neural network models in unsupervised observation of the environment, without prior knowledge either in the form of ground truth data, or novelty detection framework based on pre-known categories. Representations created with deep autoencoder models with data representing samples of Internet datagrams demonstrated the emergence of a higher-level concept sensitive structure that can be measured and visualized. The parameters of distributions of higher-level concept samples in the latent space were measured and reported with strong positive correlation observed between the categorization performance of the model and classification accuracy in its latent space. The results provide empirical support for the connection between categorization ability of unsupervised models with self-encoding and classification accuracy of the corresponding representations.

Keywords: Artificial intelligence · Deep neural networks · Unsupervised learning · Density-based clustering

1 Introduction

1.1 Unsupervised Representations

Study of representations with the intent to identify and separate the most informative components in general data has a long history. Unsupervised hierarchical representations created with RBM and DBN proved to be efficient and improved the accuracy of subsequent classification [1–3].

In a different approach to identifying new patterns in general data, applications of clustering methods in novelty detection were developed, such as OLINDDA method by Spinosa et al. [4], applications of concept clustering, self-organizing neural networks and density-based clustering [5–7] aggregation methods [8] and other works, see Pimentel et al. for a comprehensive review of the field [9].

Effects of spontaneous high-level concept sensitivity in unsupervised models were observed by GoogleLab team (Le et al. [10]). By applying a deep sparse autoencoder model to image classification, they observed spontaneous formation of concept sensitive neurons activated by images in certain higher-level category, such as ‘cat’s face’.

Z. Hu et al. (Eds.): ICCSEEA 2020, AISC 1247, pp. 255–265, 2021.
https://doi.org/10.1007/978-3-030-55506-1_23

The effect was observed after training the model the with very large datasets of images in an entirely unsupervised mode without any exposure to ground truth. In [11] a spontaneous formation of grid-like cells, similar to those observed in mammals was detected in a recurrent neural network with deep reinforcement learning. The emergence of concept-sensitive structure has been observed in unsupervised training of deep autoencoder models with data representing Internet traffic in [12]. It was demonstrated that such emergent structures can be used in "landscape-based" approach to learning that offers higher flexibility and have significantly lower ground truth requirements compared to common methods.

These results offer a cue that certain neural networks whether artificial or biological, may naturally structure the input data by similarity in completely unsupervised observation of the environment, perhaps correlated with higher-level concepts.

The hypothesis supported by the above results is that distributions in representations created by some unsupervised models in self-training can be correlated with certain higher-level concepts in the original (i.e. input) data, and that relationship can be used as a foundation to learn new, previously unknown concepts in a way that is intuitive, flexible and requires minimal supervision.

2 Methods

2.1 Model and Data

The model used in the study to produce a transformed representation of input data space is a deep autoencoder neural network of near-symmetrical layout, with significant compression in the central layer (Fig. 1). We used compression factor, that is, the ratio of the dimensionality of the input to the central layer of the model up to 10.

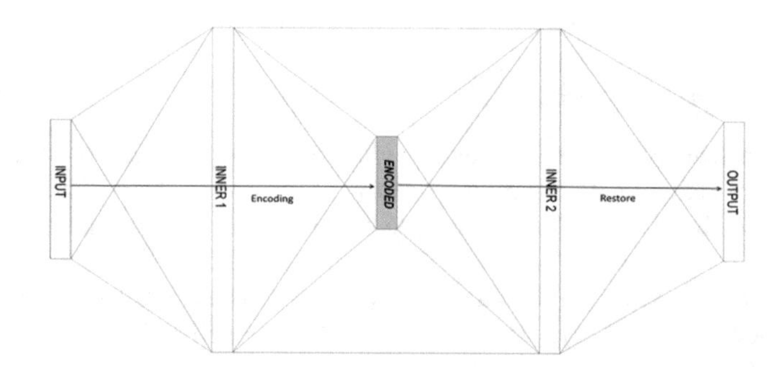

Fig. 1. Deep symmetric autoencoder model with compression.

It was hypothesized in [12] that in a feedforward autoencoder network, such a layout of with strong compression in the central layer, that has to be propagated forward to the output with high accuracy of reproduction, may facilitate spontaneous categorization or clustering of data in the central layer by higher-level concept similarity during unsupervised autoencoder training.

The choice of a deep autoencoder model as the mean for producing structured and purportedly, correlated with certain higher-level concepts encoded space is supported by the following arguments:
Being a universal approximator [13], feedforward neural networks have virtually unlimited versatility and are suitable to represent even complex data types, as images, video and other.

The effect of spontaneous emergence of higher-level concept sensitive structure in deep neural network and autoencoder models was reported in the earlier studies [8–10];

Deep neural networks are widely present in biologic systems that are also highly successful in self-learning with minimal ground truth data [11, 14];

In the authors view these arguments substantiate the choice of deep autoencoder models as a starting point for a study of spontaneous categorization by higher level concepts and learning based on the emergent unsupervised information landscape.

The data in the study is represented by recordings of Internet data packets in a public Internet network (packet dumps). A publicly available dataset captured in New Zealand (WITS, [15]) was processed as a set of sessions associated to the same source and destination with parameters representing size and timing statistics of data packets in conversations [16]. A detailed description of models and the data is given in [12].

Being a live recording in a core Internet network, the data had a wide representation of traffic patterns, with over 4,000 distinct applications represented in the dataset. In a sense, it can be compared to a recording of sound in a busy shopping mall, with the task to classify conversations in by some characteristics of the speaker, such as gender, age, occupation, etc. For that reason, it is believed it is well suited test the validity of the developed approach with data of significant diversity and variation.

2.2 Training and Classification

The models are first trained in an unsupervised autoencoder mode to achieve good reproduction of inputs. Two measures of the quality of reproduction i.e., the average deviation of the output of the model from the input were used:

1) Cost function, Mean Squared Error, had starting value in the range of 0.25 dropping to 0.001–0.002 after 100 epochs of self-supervised training; and

2) Accuracy, measured as the match of *softmax(input)*, *softmax(output)* or, a coincidence of the maximum value position in the input sample to that in the output, thus a measure related to covariance of the input and output samples. Measured in this way, the accuracy has increased after 100 epochs from $\sim 1\%$ to, on average, 95%.

Trained models can perform encoding transformation from input data space to the activations of the central layer of the model (or as referred to throughout, encoded or latent space), as in (1), [12]:

$$\mathrm{Enc}(\mathrm{X}) = \mathrm{encoder.predict}(\mathrm{X}) \tag{1}$$

where X is the input sample, and encoder, a submodel that transforms the input layer to activations of the central layer of the model.

In encoded space of a trained model, the emergent structure can be identified by applying an unsupervised, density-based clustering method, such as Mean Shift [17]. It allows to identify clusters of encoded samples in the latent space without ground truth data. The resulting cluster for a sample X in the input data space can be calculated as ((2), [12]):

$$\text{Cl}(\text{X}) = \text{struct.predict}(\text{Enc}(\text{X})) \tag{2}$$

where struct is density-based unsupervised clustering algorithm pre-trained on a generic unsupervised sample in encoded space.

To perform classification, a classifier such as Nearest Neighbor in this study, can be trained with a subset of labeled samples in encoded space. The resulting classifier can be applied to classify samples in the input space ((4), [12]):

$$\text{Cat}(\text{X}) = \text{classifier.predict}(\text{Enc}(\text{X})) \tag{3}$$

where Cat(X) is the predicted category of X by a pre-trained classifier.

A note on the difference between (2) and (3) above: whereas Cl(X) represents the implicit or "native" cluster associated with X that can be calculated by entirely unsupervised means, without ground truth data, Cat(X) represents the external category to which X is classified and cannot be calculated without ground truth data relevant to the category, and training of a category classifier.

2.3 Measurement and Visualization

The structure in encoded space that emerges as a result of unsupervised training, also referred to as "unsupervised landscape", can be measured and observed by the following methods:

1. By applying an unsupervised clustering in the encoded space and identifying clusters populated by samples of the application category.
2. By measuring parameters of application category samples distribution in encoded space.
3. By applying multi-dimensional histogram methods in the encoded space.
4. By plotting and direct observation and measurement of category samples in the encoded space.

3 Results

3.1 Unsupervised Categorization

By categorization we mean the ability of some models to group data samples in compact structures or "clusters" in encoded space by higher-level category similarity. To measure categorization ability of models, two types of data samples were used:

1) the distribution of application category samples transformed to encoded space defines the "category space", that is, the region in encoded space where application samples are located. The size of application sample has to be significant with respect to the total population of the application in the dataset; a fraction of 0.1–0.3 was used;

2) the "generic sample", a set of non-labeled data points that is used to identify and measure the size and shape of the region in encoded space that is populated by all applications, in other words, the image of the input data set in the latent space of the model. The typical size of the generic sample was 0.1–0.2 of the dataset.

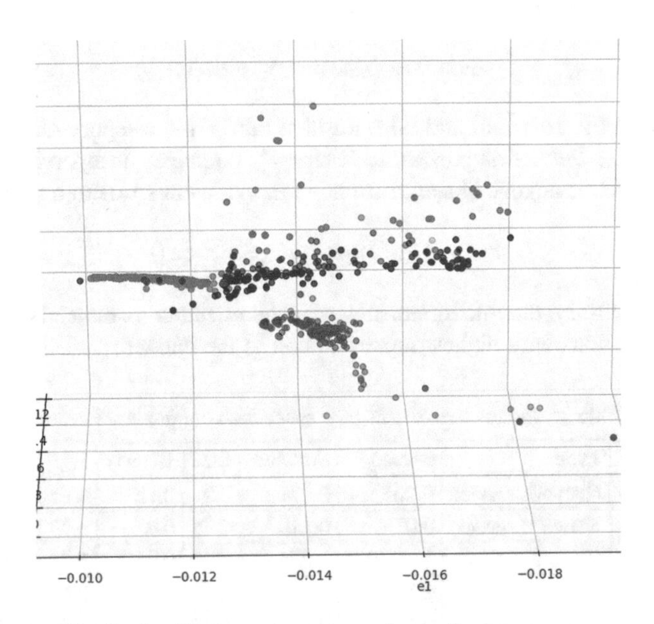

Fig. 2. Application category samples in the latent space.

In Fig. 2 category samples of three common application categories (DNS, network service: green; Escale, online game: magenta; MSN instant messenger: red) were visualized in the latent space of an autoencoder model with three latent dimensions.

For each application category C, that is a distinct Internet application, the following parameters were measured:

Spread, or the relative volume of the category sample *cat(C)* to generic sample *gen*:

$$\mathrm{Spr(C)} = \mathrm{Vol}\ ((\mathrm{cat(C)}))/\mathrm{Vol}\ (\mathrm{gen})$$

Structure, defined as the number of visually identifiable features, or calculated by the unsupervised clustering method clusters in the application category sample:

$$\mathrm{Str(C)} = \mathrm{cluster\_count(cat(C))}\ /\ \mathrm{total\_clusters(gen)}$$

Size, the maximum and mean size of an individual feature in the category sample relative to the mean size of generic sample;

Density, the apparent density of category features identified visually on a scale (0, 1); or a calculated based on the number of points in the category sample and its volume:

$$\mathrm{Den(C) = Count(cat(C))/Vol(cat(C))}$$

Shape, the observed shape of category features, identified visually.
Accuracy: a measure of classification accuracy for the application category classifier trained with a subset of the category sample, defined as the accuracy of classification for in- and out-of-class samples:

$$\mathrm{Acc(C) = (Recall(C),\ FPR(C))} \tag{4}$$

where Recall, FPR are recall and false positive rate of the category classifier.

Finally, *Cumulative Categorization Factor* F_{cc} combines unsupervised categorization parameters of category image distribution in the latent space to a single measure:

$$\mathrm{Fcc = log(Den/(Spr \times Str \times Sz))} \tag{5}$$

Table 1 summarizes the measurements of categorization parameters for common Internet applications with highest representation in the dataset.

Table 1. Parameters of category space for common applications.

Category	Type	Spread	Structure	Size	Density	F_{cc}	Accuracy
DNS	Network service	0.005	0.071	0.2	0.75	4.02	0.915
NTP	Network service	0.01	0.119	0.15	0.7	3.59	0.958
Telnet	Remote session	1×10^{-4}	0.023	0.02	0.96	7.30	0.974
SMTP	Mail	0.1	0.071	0.2	0.6	2.62	0.970
MSN	Messenger	0.015	0.048	0.18	0.7	2.74	0.921
XBox	Online game	0.2	0.143	0.13	0.6	2.21	0.808
BitTorrent	File sharing	0.35	0.238	0.3	0.4	1.23	0.709
Streaming	Media stream	0.1	0.119	0.2	0.65	2.47	0.892
HTTP*	www	0.4	0.332	0.3	0.4	0.94	0.762

* Hypertext Transfer Protocol, used by multiple applications.

As seen from these results, for most measured applications the category space is represented by a small number of well-defined category clusters. Not surprisingly, the measurements show that the categories with the most expressed structure in encoded space, that is, a small number of compact and dense clusters (DNS, NTP and Telnet) produced the highest accuracy of classification.

On the opposite end of the categorization spectrum we observed applications with greater variability of content and behavior, such as streaming, BitTorrent, and Web protocol (HTTP). In fact, the latter should not be considered as a distinct application as it can carry many applications different in content and behavior, so the higher number of associated clusters and relative sparsity of the category space in this case is hardly surprising.

Calculation of correlation coefficient between cumulative categorization factor F_{cc} and classification accuracy of a classifier trained with labeled samples in the latent space produced a strong positive value of *0.756*, a clear indication of a positive correlation between categorization properties of the application distribution in the representation space of the model and its classification performance.

In the next step the authors attempted to evaluate the categorization ability of the model by comparing it with the structure of raw, not encoded data space. To this end, a randomly selected sample of a given application category was transformed to the encoded space and the parameters of its distribution measured. The parameters of the encoded representation of application category space were then compared to those of the same sample in the input data space.

First, we defined "significant category clusters" as those that contain the fraction of the category data points above certain minimum threshold (the value of 5% was used). In the next step, three parameters of the sample were calculated for the same category sample in the input data space, and its image in the encoded space of the model: 1) resolution, defined as the ratio of the number of significant category clusters to the total number of clusters calculated by the clustering method; 2) concentration, the fraction of the category sample contained in significant clusters; and 3) relative density, the ratio of the number of category points in the cluster to its characteristic size, calculated as a weighted average over all significant category clusters. These results presented in Table 2, with parameters of the original (input) data in the left three columns and those of the latent representation – in the right.

Table 2. Categorization performance of deep autoencoder models.

Category	Resolution (Original)	Concentration, %	Density	Resolution (Latent)	Concentration	Density
DNS	0.66	90	273.5	4.8	89	491.5
NTP	2.6	98	150.2	9.6	100	266.9
Telnet	0.6	96	1870	2.4	96	3520
SMTP	1.3	93	181.7	4.8	94	405.3
MSN	0.66	96	535.7	4.8	96	921.9
XBox	1.3	74	133.0	3.6	85	297.0
BitTorrent	1.3	65	110.4	3.6	73	249.0
Streaming	1.3	61	77.8	9.6	80	233.0
HTTP	2.0	79	59.5	3.6	81	215.3

Comparing the parameters of categorization allows to see clearly the effects of encoding transformation on the input data. Whereas the dimensionality of the data space was reduced by an order of magnitude, the resulting category image was more structured, more concentrated and denser than the original sample.

For comparison, similar measurements were performed with Principal Component Analysis whereby category samples were transformed by PCA to the same number of principle components as the dimensionality of encoded space of autoencoder model.

By all parameters autoencoder models outperformed PCA transformation, although by a smaller margin than unprocessed data above.

These results demonstrate that the encoding representation created in self-supervised training can result in a compact, structured, and categorized representation of the input data. The ability of these models to produce such distinct and compact representations of higher-level category spaces in the entirely unsupervised, self-controlled process of observation and processing of input data is intriguing and, as will be demonstrated in the following sections, can be used to successfully learn new concepts, not known to the model previously with minimal supervision.

3.2 Early Learning

In the earlier studies [18] an approach to early learning of new concepts was suggested whereby the emergent unsupervised structure described in the previous section and can be harnessed to achieve effective learning of new concepts with minimal ground truth data, down to a single encounter. The approach is based on certain general assumptions about the data:

1. *Significance*: the parameters in which input data is represented are significant for the category being learned, that is, are sufficient to differentiate it among the data of other categories.
2. *Representation*: the new category being learned is well represented in the unsupervised data so that associated structure can be formed in unsupervised training.
3. *Attention*: a function additional to the learning model that we'll refer to as "focus" or "attention", brings a sample or a small "signal" set of samples of the new category to the learning system as something important it needs to learn.
4. *Empirical test*: the learning model is able to confirm or disprove its prediction via applying it to a test producing confident result.

The "signal-sampling" method of early learning was applied to the categories studied in the previous section to evaluate the correlation between the learning performance that is, the accuracy of the initial "signal" learning and the parameters of the category distribution in the representation space. The results are presented in Table 3, with "Signal accuracy" indicating the classification accuracy of a category classifier trained with a single true sample [16], while "Trained accuracy" refers to that of a classifier pre-trained with a set of labeled samples in the encoded space.

Calculation of the correlation coefficient of the categorization factor of the category distribution and signal accuracy of the classifier again produced a strong positive value of *0.65*. These results support the hypothesis that categorization parameters of the category sample in the latent space and the ability of classifier to learn and predict it are closely correlated.

Table 3. Categorization vs. accuracy, autoencoder models.

Application category	Categorization factor	Signal accuracy	Trained accuracy
DNS (net.service)	4.02	0.861	0.915
NTP (net.service)	3.59	0.842	0.958
Telnet (session)	7.30	0.843	0.974
SMTP (email)	2.62	0.702	0.930
MSN (messenger)	2.74	0.748	0.921
XBox (online game)	2.21	0.792	0.808
BitTorrent	1.23	0.592	0.709
Streaming	2.47	0.798	0.892
HTTP	0.94	0.312	0.762

3.3 Summary

In the study the authors presented the measurements of categorization parameters of category distributions created by a class of deep autoencoder models in unsupervised training.

Demonstrated a clear and strong positive correlation between categorization parameters of a concept in the latent space and classification performance.

The results provide empirical support for the connection between categorization ability of unsupervised models with self-encoding and the improvement in classification accuracy in supervised training with unsupervised representations.

4 Conclusion

Let X be the input data space, $C = \{C_k\}$ a finite set of certain higher-level concepts, and $\{X_k\}$, category subspaces of X. We shall call a transformation $T: X \rightarrow Y$ a categorizing transformation, or categorizing encoder if for all concepts in the concept set C, the encoded image $Y_k = T(X_k)$ is well-expressed, that is, is composed of a finite set of clusters with density above certain minimum threshold D_{min}.

The set of categorizing encoders on C can then be ordered by the measure of categorizing ability as demonstrated in the study. For example, if the categorization measure is based on the parameters of resolution, concentration and density, the results of Sect. 3.1 would indicate that deep autoencoder models rank higher by categorization ability for the studied data than the identity transformation (that is, unprocessed input data) and PCA. Such a scale may lead to development of models with better categorizing performance leading to more efficient learning that is, achieving better classification accuracy with smaller training data.

Why use categorizing encoders? There are already quoted indications in the earlier results [10–12] that some models can produce encoded spaces that improve categorization by higher-level concept in entirely unsupervised, passive observation of the environment, potentially improving the efficiency of learning for such concepts.

Whereas autoencoder models studied here may not be the only mean to produce categorizing transformations, due to versatility and approximation power of deep neural networks their capacity to categorize spontaneously, in passive and unsupervised observation of the environment, merits further in-depth study with potential applications in many promising areas such as visual information analysis [19] among many others, natural language processing and recognition [20] and others.

References

1. Hinton, G., Osindero, S., Teh, Y.W.: A fast learning algorithm for deep belief nets. Neural Comput. **18**(7), 1527–1554 (2006)
2. Bengio, Y.: Learning deep architectures for AI. Foundation and Trends Mach. Learn. **2**(1), 1–127 (2009)
3. Fischer, A., Igel, C.: Training restricted Boltzmann machines: an introduction. Pattern Recogn. **47**, 25–39 (2014)
4. Spinosa E., de Carvalho A.C.P.L.F., Gama J.: OLINDDA: a cluster-based approach for detecting novelty and concept drift in data streams. In: 2007 ACM Symposium on Applied Computing (SAC), pp. 448–452 (2007)
5. Fanizzi, N., d'Amato, C., Esposito, F.: Conceptual clustering and its application to concept drift and novelty detection. In: Bechhofer, S., Hauswirth, M., Hoffmann, J., Koubarakis, M. (eds.) ESWC 2008. LNCS, vol. 5021, pp. 318–332. Springer, Heidelberg (2008)
6. Cassisi, C., Ferro, A., Guigno, R., et al.: Enhancing density-based clustering: parameter reduction and outlier detection. Inf. Syst. **38**(3), 317–330 (2013)
7. Abubaker, M., Ashour, W.: Efficient data clustering algorithms: improvements over Kmeans. Int. J. Intell. Syst. Appl. (IJISA) **5**(3), 37–49 (2013)
8. La Red Martínez, D.L., Acosta, J.C.: Aggregation operators review - mathematical properties and behavioral measures. Int. J. Intell. Syst. Appl. (IJISA) **7**(10), 63–76 (2015)
9. Pimentel, M., Clifton, D., Clifton, L., et al.: A review of novelty detection. Sig. Process. **99**, 215–249 (2014)
10. Le, Q.V., Ransato, M.A., Monga, R., et al.: Building high-level features using large scale unsupervised learning. arXiv:1112.6209 (2012)
11. Banino, A., Barry, C., Kumaran, D.: Vector-based navigation using grid-like representations in artificial agents. Nature **557**, 429–433 (2018)
12. Dolgikh, S.: Spontaneous concept learning with deep autoencoder. Int. J. Comput. Intell. Syst. **12**(1), 1–12 (2018)
13. Hornik, K., Stinchcombe, M., White, H.: Multilayer feedforward neural networks are universal approximators. Neural Netw. **2**(5), 359–366 (1989)
14. Hassabis, D., Kumaran, D., Summerfield, C., et al.: Neuroscience inspired Artificial Intelligence. Neuron **95**(2), 245–258 (2017)
15. WITS passive datasets, Waikato University, New Zealand. https://wand.net.nz/wits. Accessed 11 Feb 2018
16. Wright, C., Monrose, F., Masson, G,M.: HMM profiles for network traffic classification. In: 2004 ACM workshop on visualization and data mining for computer security, pp. 9–15. ACM (2004)
17. Comaniciu, D., Meer, P.: Mean shift: a robust approach toward feature space analysis. IEEE Trans. Pattern Anal. Mach. Intell. **24**(5), 603–619 (2002)

18. Dolgikh, S.: Categorized representations and general learning. In: Aliev, R.A., Kacprzyk, J., Pedrycz, W., Jamshidi, M., Babanli, M.B., Sadikoglu, F.M. (eds.) ICSCCW 2019. AISC, vol. 1095, pp. 93–100. Springer, Cham (2020)
19. Petersson, H., Gustafsson, D., Bergstrom, D.: Hyperspectral image analysis using deep learning. In: 6th International Conference Image Proceedings Theory, Tools and Applications (IPTA), Oulu, U.S.A (2016)
20. Gazeau, V., Varol, C.: Automatic spoken language recognition with neural networks. Int. J. Inf Technol Comput Sci (IJITCS) **10**(8), 10–17 (2018)

Performance of Vectorized GPU-Algorithm for Computing ψ-Caputo Derivative Values

Vsevolod Bohaienko(✉)

V.M. Glushkov Institute of Cybernetics of NAS of Ukraine, Kyiv, Ukraine
sevab@ukr.net

Abstract. Modeling of diffusion in fractal-structured media using multidimensional fractional differential equations has high computational complexity. To lower time spent on simulation, the paper addresses the issues of efficient GPU-implementation of ψ-Caputo derivative values computation while solving three-dimensional time-fractional diffusion equation by locally one-dimensional finite difference scheme. We focus on arithmetic operations vectorization while performing computations on GPU. For the approximation procedure based on integral kernel's expansion into series, we present a vectorized GPU-algorithm and experimentally show that it allows speeding up computations comparing with straightforward GPU implementation. The paper presents the results of computational experiments showing that GPU implementation of non-local fractional derivative's part computation up to 4-times accelerates fractional diffusion equation solution in the case when linear equation systems solution is performed by OpenMP-parallelized implementation on CPU. When considering the computation of fractional derivative values, GPU algorithm gives up to 130-times speed-up. Vectorization of computational procedure gives an additional up to 4 times decrease of non-local fractional derivative's part computation time comparing to the basic GPU algorithm. The best thread group allocation approach for the considered scheme is to have thread group size equal to the number of terms in the truncated series.

Keywords: GPU algorithms · Finite-difference method · Diffusion equation · ψ-Caputo derivative

1 Introduction

Diffusion processes in media with memory effect are efficiently described by differential equations of fractional order [1–5] that include different forms of fractional derivatives with respect to the time variable. As fractional derivatives are non-local integral-differential operators, computation of their values is highly complex. To lower computational complexity, the most commonly used approaches of different nature are parallel computing (e.g. [6, 7]) and methods based on the approximation of integral kernels (e.g. [2, 8, 9]).

Parallel computing techniques, especially for shared memory systems including GPUs, can be efficiently used for splitting finite-difference schemes such as locally

Z. Hu et al. (Eds.): ICCSEEA 2020, AISC 1247, pp. 266–275, 2021.
https://doi.org/10.1007/978-3-030-55506-1_24

one-dimensional ones [10, 11] that reduce multi-dimensional problems to a set of one-dimensional ones.

In the paper, we focus on the issue of efficient GPU-implementation of ψ-Caputo derivative's values computation when solving three-dimensional time-fractional diffusion equation by locally one-dimensional finite-difference scheme. The objectives of the research is to construct GPU algorithms for calculating non-local part of fractional derivative values using scalar and vector arithmetic operation and experimentally study their performance obtaining dependencies of execution time and speed-up on algorithms' and problem's parameter values. In Sect. 2 we give a brief review of works in the field of GPU usage for solving fractional differential equations. In Sect. 3 we present a mathematical statement of the problem and a finite-difference scheme for its solution. Optimized computation scheme for ψ-Caputo fractional derivative is given in Sect. 4 while the proposed parallel algorithms are described in Sect. 5. Further, in Sect. 6 we give the results of numerical experiments on performance and speed-up of the algorithms.

2 Literature Review

The studies on the usage of GPU in simulations on the base of fractional differential equations mostly focus on the equations with classical Caputo derivatives with respect to the time variable [12, 13] or Riesz derivatives with respect to the space variables [14]. In the first case, the main parallelized operations are summation operations and memory access optimizations [15] are deeply studied in this context. As GPUs have fast shared and slow global memory, reuse of data in shared memory was proposed in [12] for solving fractional reaction-diffusion equation by finite-difference scheme. When using semi-analytical methods, the Mittag-Leffler function's calculation on GPU was used in [16]. For the case of space-fractional equations, known GPU implementations of linear algebra algorithms are mostly used. Vectorization optimizations when solving fractional differential equations were studied only for CPU case in [17] and showed up to 2-time speed-up. As the generalized ψ-Caputo derivative with functional parameter [18] is a newly presented mathematical construct, a little research was conducted concerning the usage of GPU while solving differential equations with such derivatives.

3 Problem Statement and Finite-Difference Scheme

We consider the following three-dimensional time-fractional diffusion equation [19]:

$$\begin{aligned} \sigma D_{t,g}^{(\beta)} C(x,y,z,t) &= D\left(\frac{\partial^2 C(x,y,z,t)}{\partial x^2} + \frac{\partial^2 C(x,y,z,t)}{\partial y^2} + \frac{\partial^2 C(x,y,z,t)}{\partial z^2}\right) \\ &+ F(x,y,z,t),\ \ 0 \le x \le 1,\ 0 \le y \le 1,\ 0 \le z \le 1,\ t > 0,\ 0 < \beta \le 1 \end{aligned} \tag{1}$$

where $C(x,y,z,t)$ is the diffusive substance concentration, σ is the porosity, D is the diffusion coefficient, F is the given source-term function, and $D_{t,g}^{(\beta)}$ is the generalized ψ-Caputo fractional derivative with respect to the time variable t and a function $g(t)$ that has the following form [18]:

$$D_{t,g}^{(\beta)}C(x,y,z,t) = \frac{1}{\Gamma(1-\beta)}\int_0^t \frac{\partial C(x,y,z,t)}{\partial t}(g(t)-g(\tau))^{-\beta}d\tau,$$

$$g(t) \in C^1[0,+\infty),\ g'(t) > 0\ (t \geq 0),\ g(0) = 0.$$

Within this paper, it is also required that $g(t)$ has an infinitely differentiated inverse function $f(t)$.

For the Eq. (1) we, following [19], pose first-order initial and boundary conditions.

We solve the Eq. (1) using locally one-dimensional finite-difference scheme [20] on the uniform grid

$$\omega = \{(x_i, y_j, z_k, t_l) : x_i = ih_1\ (i = \overline{0, n_1+1}),\ y_j = jh_2\ (j = \overline{0, n_2+1}),$$
$$z_k = kh_3\ (j = \overline{0, n_3+1}),\ t_l = l\tau\}$$

where h_1, h_2, h_3, τ are the steps with respect to the spatial and the time variables.

The finite-difference scheme, similarly to [11], has the following form:

$$C_{ijk}^{(l+1/3)} - \frac{\tau D}{k_1 h_1^2}(C_{i-1,j,k}^{(l+1/3)} - 2C_{ijk}^{(l+1/3)} + C_{i+1,j,k}^{(l+1/3)}) = C_{ijk}^{(l)} + \frac{\tau}{3k_1}F_{ijk}^{(l)}, \tag{2}$$

$$C_{ijk}^{(l+2/3)} - \frac{\tau D}{k_1 h_2^2}(C_{i,j-1,k}^{(l+2/3)} - 2C_{ijk}^{(l+2/3)} + C_{i,j+1,k}^{(l+2/3)}) = C_{ijk}^{(l+1/3)} + \frac{\tau}{3k_1}F_{ijk}^{(l)}, \tag{3}$$

$$C_{ijk}^{(l+1)} - \frac{\tau D}{k_1 h_3^2}(C_{i,j,k-1}^{(l+1)} - 2C_{ijk}^{(l+1)} + C_{i,j,k+1}^{(l+1)}) = C_{ijk}^{(l+2/3)} + \frac{\tau}{3k_1}F_{ijk}^{(l)},\ k_1 = \frac{b_l^{(l+1)}}{\Gamma(1-\beta)}, \tag{4}$$

$$F_{ijk}^{(l)} = F_{ijk} + \frac{1}{\Gamma(1-\beta)}\sum_{s=0}^{l-2} b_s^{(l)} \frac{C_{ijk}^{(s+1)} - C_{ijk}^{(s)}}{\tau}, \tag{5}$$

$$b_s^{(i)} = \int_{t_s}^{t_{s+1}} (g(t_i) - g(\tau))^{-\beta} d\tau. \tag{6}$$

Discretized form of first-order boundary conditions are further added to the equation systems (2)–(4) resulting in three-diagonal linear equation systems, which are solved by the Thomas algorithm [20]. As the Thomas algorithm has linear computational complexity, total complexity of obtaining a solution of (2)–(4) on the time step l has an order of $O(l \cdot n_1 \cdot n_2 \cdot n_3)$.

Systems (2)–(4) consists of independent linear equation systems giving ability for their efficient implementation in multithreaded mode and on GPU.

4 Scheme for Computing ψ-Caputo Fractional Derivative Values

Computation of the discretized form (5) of non-local part of fractional derivative in straightforward manner has $O(l)$ computational complexity. This significantly influences solution speed when doing simulations upon large time ranges. Thus, we consider the algorithm for performing computations upon (5) based on the series expansion of the integrals (6) [21, 22].

Using the presentation of $b_s^{(i)}$ in the form

$$b_s^{(j)} = \sum_{n=0}^{\infty}\left((-1)^n\binom{-\beta}{n}g(t_j)^{-\beta-n}S_n\right),\ S_n(t_s,t_{s+1}) = \sum_{m=0}^{\infty}\left[B_m\frac{f^{((m+1)}(g(t_{s+1}))}{m!}\right],$$

$$B_m = \int\limits_{g(t_s)}^{g(t_{s+1})} x^n(x-g(t_{s+1}))^m dx,\ B_0 = \int\limits_{g(t_s)}^{g(t_{s+1})} x^n dx = \frac{1}{n+1}\left(g(t_{s+1})^{n+1} - g(t_s)^{n+1}\right),$$

$$B_{i+1} = -\frac{n+i+2}{g(t_{s+1})(i+1)}\left(B_i - \frac{g(t_s)^{n+1}(g(t_s)-g(t_{s+1}))^{i+1}}{g(t_{s+1})(i+1)}\right),$$

computations upon (5) can be organized [22] as

$$\sum_{s=0}^{l-2} b_s^{(l)}\frac{C_{ijk}^{(s+1)} - C_{ijk}^{(s)}}{\tau} \approx \frac{1}{\tau}\sum_{n=0}^{K}\left((-1)^n\binom{-\beta}{n}g(t_l)^{-\beta-n}S_{n,l-1}\right),$$

$$S_{n,l} = S_{n,l-1} + (C^{(l)} - C^{(l-1)})S_n(t_{l-1},t_l),\ S_{n,0} = 0. \tag{7}$$

where K is the given number of terms in series.

When using the scheme (7), total computational complexity of obtaining problem's solution on one time step has an order of $O(K \cdot n_1 \cdot n_2 \cdot n_3)$. Scheme (7) also allow saving of only a solution on the two last time steps comparing with a need to store all previous solutions when doing computations upon (5). This, in turn, allows solving the considered equation on larger grids and time intervals.

5 Parallel Implementation

We consider multithreaded OpenMP implementation of systems' (2)–(4) solution with each thread solving a block of linear systems. Calculation of non-local part of ψ-Caputo fractional derivative upon (7) is performed for every node of the grid using GPU.

The usage of GPU is organized as follows. The values of $S_n(t_{l-1}, t_l)$ are calculated on CPU on the step $l-1$ in parallel with GPU computations upon (7) and uploaded on the step l into GPU memory along with the solution $C^{(l-1)}$. Each GPU thread calculates the value of $F_{ijk}^{(l)}$ for a specific node (i, j, k) and, after GPU performed the computations, values of $F_{ijk}^{(l)}$ are copied back into CPU memory.

To use GPU's local memory, we form thread groups of a controllable size χ with computations organized in $\lceil l/\chi \rceil$ substeps. On the substep s, the values of $S_n(t_{l-1}, t_l), s\lceil l/\chi \rceil \leq n < (s+1)\lceil l/\chi \rceil$ are loaded into local memory and a corresponding part of $S_{n,l}$ is updated by each thread. Further, the resulting K-terms sum in (7) is computed with $(-1)^n \binom{-\alpha}{n} g(t_{l-2})^{-\alpha-n}$ in advance calculated by the threads in parallel and loaded into local memory.

This algorithm was, except for thread blocking influence, studied in [22]. In the context of its further development, we propose to represent summations in (7) in vector form. Summations in (7) are split into 16-terms vector operations, which correspond to float16/double16 OpenCL datatypes, obtaining

$$\sum_{s=0}^{l-2} b_s^{(l)} \frac{C_{ijk}^{(s+1)} - C_{ijk}^{(s)}}{\tau} \approx \frac{1}{\tau} \sum_{n=0}^{\lceil K/16 \rceil} \left(\vec{v}_n^{(1)} \left(S_{n \cdot 16, l-1}, \ldots, S_{(n+1) \cdot 16, l-1} \right) \right),$$

$$\vec{v}_n^{(1)} = \left((-1)^{n \cdot 16} \binom{-\beta}{n \cdot 16} g(t_l)^{-\beta - n \cdot 16}, \ldots, (-1)^{(n+1) \cdot 16} \binom{-\beta}{(n+1) \cdot 16} g(t_l)^{-\beta-(n+1) \cdot 16} \right)$$

$$\begin{aligned} \left(S_{n \cdot 16, l}, \ldots, S_{(n+1) \cdot 16, l} \right) &= \left(S_{n \cdot 16, l-1}, \ldots, S_{(n+1) \cdot 16, l-1} \right) \\ &+ \left(C^{(l)} - C^{(l-1)} \right) \left(S_{n \cdot 16}(t_{l-1}, t_l), \ldots, S_{(n+1) \cdot 16}(t_{l-1}, t_l) \right). \end{aligned} \tag{8}$$

6 Numerical Experiments

We conduct the numerical experiments to test the performance of the proposed parallel algorithms obtaining dependencies of their speed-up on finite-difference grid size and the parameter K of the scheme (7), and finding optimal thread group size. As the main studied indicators are the indicators of performance, we use a model problem with analytical polynomial solution expecting obtaining close-to-linear speed-ups from the developed algorithms. To measure speed-up, we collect data on total execution times of algorithms' implementations while solving a model problem. Data analysis for dependencies acquisition is performed using least squares fitting.

We consider the model problem for the Eq. (1) with $g(t) = t^{\gamma}$. When $F(x,y,z,t) = \frac{\Gamma(1+2/\gamma)}{\Gamma(1-\beta+2/\gamma)} t^{2-\beta\gamma} - 2D(x^2y^2 + y^2z^2 + x^2z^2)$, its solution is $C(x,y,z,t) = C_0(x,y,z,t) = x^2y^2z^2 + t^2$ for the case of $\sigma = 1$ and when initial and boundary conditions are $C(x,y,z,0) = C_0(x,y,z,t)$, $C(0,y,z,t) = C_0(0,y,z,t)$, $C(1,y,z,t) = C_0(1,y,z,t)$, and similarly for other space variables.

The system (2)–(4) was solved for 40 time steps of size $\tau = 0.0025$, $\gamma = 0.9$, $\beta = 0.8$, grid size of $N \times N \times N$, $N = 20,\ldots,70$, and $K = 16,\ldots,64$ on a single node of SCIT-4 cluster of VM Glushkov Institute of cybernetics of NAS of Ukraine. The used node contains two Intel(R) Xeon(R) Bronze 3104 CPUs with total of 12 cores and Nvidia RTX 2080 Ti GPU.

Average times spent by OpenMP and GPU OpenCL implementations for obtaining a solution on one time step are given in Table 1 along with the speed-up of GPU algorithm. For GPU implementation, thread group size was equal to the number of terms K in truncated series (7). The times here were expectedly close to constant with the increase of time step number and depended on grid size parameter N as $O(N^3)$ for both CPU and GPU implementations. Comparing with total summation upon (5) average square error of problem's solution worsens not more than by 6% ($K = 16$). With the increase of K accuracy worsening quadratically decreased.

Overall speed-up from GPU usage increases with the increase of K. When the value of N grows, it increases up to $N = 40$ remaining close to constant for larger values of N. Such behavior can be explained by non-full usage of GPU resources when $N < 40$. As even for mathematically optimized computation scheme (7) calculation of fractional derivative's values greatly contribute to the total algorithm's complexity, its GPU implementation allowed decreasing solution time up to 4 times in the conducted experiments.

Table 1. Algorithms performance on one time step.

	K	$N = 20$	30	40	50	60	70
OpenMP implementation time, ms	16	5,13	18,21	43,59	85,38	163,08	233,85
	32	7,18	26,15	60,26	118,97	225,13	328,46
	64	10,77	37,18	91,03	180,77	330,51	502,82
GPU implementation time, ms	16	4,87	11,79	24,87	46,41	93,33	117,18
	32	5,13	10,51	23,59	46,15	95,38	116,41
	64	4,87	11,54	23,59	48,97	90,77	122,05
GPU speed-up, %	16	105,26%	154,35%	175,26%	183,98%	174,73%	199,56%
	32	140,00%	248,78%	255,43%	257,78%	236,02%	282,16%
	64	221,05%	322,22%	385,87%	369,11%	364,12%	411,97%

Average time spent by OpenCL kernel to perform computations upon (7) and (8) with different K and thread group size measured using internal OpenCL profiling tools is given in Table 2 along with the corresponding time spent by OpenMP implementation on CPU.

While doing GPU computations upon (7) with thread block size equal to K, speed-up comparing to the execution on CPU increased logarithmically (Fig. 1) with the increase of N and reached 130-times maximum for $N = 70, K = 32$. Thread block size equal to 32 was the value at which the highest speed-up was achieved for every tested value of N. For greater values, speed-up lowered to 52-times for 64 threads per group comparing to 62-times for 16 threads per group.

Vectorized GPU algorithm's usage gives an additional performance boost comparing to GPU algorithm that perform calculations upon (7) (Fig. 2). Computations upon (8) were from 8% to 25% faster independently of N for $K = 16, 32, 48$. For higher values of K, effect of vectorized GPU implementation of (8) linearly increase with the increase of N reaching 3-times speed up on $K = 64$, $N = 70$. Comparing to OpenMP implementation, in this best case, vectorized GPU algorithm gave 153-times speed-up.

As the speed-up of GPU algorithms lowers for K and thread group size more than 32, we also tested a variant, when thread group size remained equal to 16 (Fig. 2). In this case, performance of vectorized algorithm significantly worsens while for the basic GPU algorithm slight unstable decrease in execution time was observed only for large grids ($N > 50$).

Table 2. Time spent on computations upon (7) and (8), ms.

K	Group size	$N = 20$	30	40	50	60	70
OpenMP							
16		0,36	6,89	20,23	40,31	70,13	119,12
32		2,41	14,84	36,90	73,90	132,18	213,73
64		6,00	25,87	67,67	135,69	237,56	388,09
GPU upon (7)							
16	16	0,07	0,19	0,40	0,74	1,24	1,91
32	32	0,08	0,18	0,37	0,66	1,08	1,64
32	16	0,11	0,31	0,66	1,24	2,06	3,20
64	64	0,14	0,34	1,33	2,64	4,66	7,42
64	16	0,19	0,56	1,19	2,23	3,70	5,77
Vectorized GPU upon (8)							
16	16	0,06	0,18	0,37	0,68	1,13	1,76
32	32	0,06	0,16	0,34	0,60	0,99	1,51
32	16	0,13	0,37	0,79	1,47	2,46	3,83
64	64	0,10	0,26	0,72	1,19	1,74	2,53
64	16	0,33	1,03	2,22	4,20	6,99	10,91

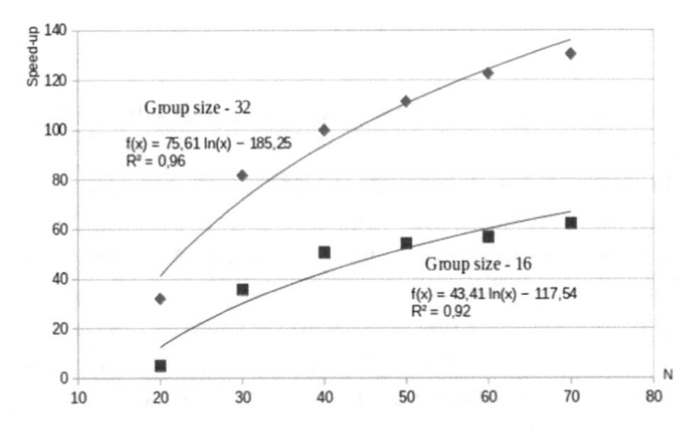

Fig. 1. GPU algorithm's speed-up.

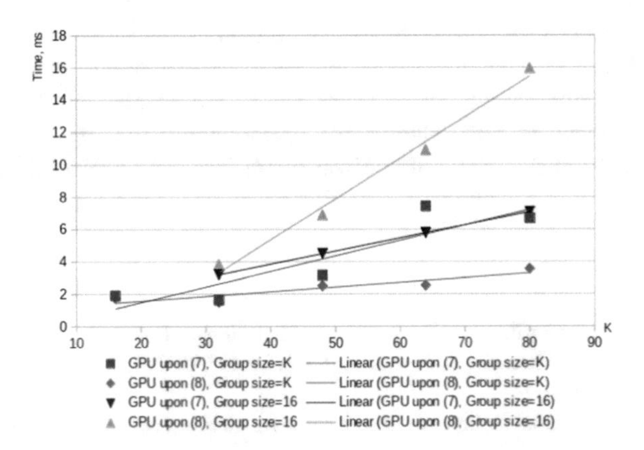

Fig. 2. Different GPU-algorithms' execution times.

Summarizing the obtained results, we can state the following main dependencies of algorithm's performance on the parameters:

- the increase in the parameters N^3 and K that determine the volume of computations, expectedly leads to a linear increase of execution time for both analyzed algorithms. Speed-up in this case has a logarithmic increase in the conducted experiment tending to theoretically expected constant speed-up when all GPU resources are optimally used. The speed-up of the vectorized algorithm comparing to the non-vectorized one increases with the increase of the volume of computations;
- algorithms' speed-up with the increase of thread group size behaves non-linearly. It increases with the increase of thread group size on initial stage and starts to decrease passing the optimal value that was in the conducted experiments equal to 32. The lowest execution time is achieved when thread group size is equal to the number K of the terms in the series (7).

7 Conclusions

Calculation of fractional derivatives' non-local parts significantly influence time needed to solve fractional derivative equations comprising 60% or more of total execution time with this percent increasing with the increase of time step number. In the case of the generalized ψ-Caputo derivative, the described approximating procedure based on integral's kernel expansion into series allows obtaining constant computational complexity and memory consumption with respect to the time step number and lowers time needed to compute the values of non-local parts of fractional derivatives to $\sim 15\%$ of total execution time. However, this is still a significant contribution that can be further lowered using GPU implementations of the procedure. The proposed straightforward and optimized, vectorized GPU algorithms showed up to 130-times speed-up of this operation execution comparing to the execution on CPU.

The presented computational algorithms give an ability to simulate diffusion in fractal structured media on larger grids and for larger time intervals. As we considered a generalized form of Caputo-type fractional derivative, the proposed approach can be further studied for the specific cases of ψ-Caputo derivative and other forms of fractional derivatives, integral kernels of which can be represented similarly to (7). Also, considering efficiency of GPU-algorithm's vectorization, further studies can be performed in the field of the usage of hardware implemented tensor arithmetic.

References

1. Gorenflo, R., Mainardi, F.: Fractional calculus: integral and differential equations of fractional order. In: Fractals and Fractional Calculus in Continuum Mechanics, pp. 223–276, Springer, Wien (1997)
2. Podlubny, I.: Fractional Differential Equations. Academic Press, New York (1999)
3. Bulavatsky, V.M.: Mathematical modeling of dynamics of the process of filtration convection diffusion under the condition of time nonlocality. J. Autom. Inf. Sci. **44**(2), 13–22 (2016)
4. Sokolovskyy, Y., Levkovych, M.: Two-dimensional mathematical models of visco-elastic deformation using a fractional differentiation apparatus. Int. J. Modern Educ. Comput. Sci. (IJMECS) **10**(4), 1–9 (2018)
5. Pakhira, R., Ghosh, U., Sarkar, S.: Study of memory effect in an inventory model with linear demand and shortage. Int. J. Math. Sci. Comput. (IJMSC) **5**(2), 54–70 (2019)
6. Diethelm, K.: An efficient parallel algorithm for the numerical solution of fractional differential equations. Fract. Calc. Appl. Anal. **14**(3), 475–490 (2011)
7. Gong, C., Bao, W., Tang, G.: A parallel algorithm for the Riesz fractional reaction-diffusion equation with explicit finite difference method. Fract. Calc. Appl. Anal. **16**(3), 654–669 (2013)
8. Gong, C., Bao, W., Liu, J.: A piecewise memory principle for fractional derivatives. Fract. Calc. Appl. Anal. **20**(4), 1010–1022 (2017)
9. Ford, N.J., Simpson, A.C.: The numerical solution of fractional differential equations: speed versus accuracy. Numer. Algorithms **26**(4), 333–346 (2001)
10. Chen, A., Li, C.P.: A novel compact ADI scheme for the time-fractional subdiffusion equation in two space dimensions. Int. J. Comput. Math. **93**(6), 889–914 (2016)

11. Bulavatsky, V.M., Bogaenko, V.A.: Numerical simulation of fractional-differential filtration-consolidation dynamics within the framework of models with non-singular kernel. Cybern. Syst. Anal. **54**(2), 193–204 (2018)
12. Liu, J., Gong, C., Bao, W., Tang, G., Jiang, Y.: Solving the Caputo fractional reaction-diffusion equation on GPU discrete. Dyn. Nat. Soc. **2014**, 820162 (2014)
13. Baban, A., Bonchis, C., Fikl, A., Rosu, F.: Parallel simulations for fractional-order systems. In: 18th International Symposium on Symbolic and Numeric Algorithms for Scientific Computing (SYNASC), pp. 141–144 (2016)
14. Wang, Q., Liu, J., Gong, C., Zhang, Y., Xing, Z.: A GPU-based fast solution for Riesz space fractional reaction-diffusion equation. In: 18th International Conference on Network-Based Information Systems, Taipei, pp. 317–323 (2015)
15. Gong, C., Bao, W., Tang, G., Jiang, Y., Liu, J.: Computational challenge of fractional differential equations and the potential solutions: a survey. Math. Prob. Eng. **2015**, 258265 (2015)
16. Golev, A., Penev, A., Stefanova, K., Hristova, S.: Using GPU to speed up calculation of some approximate methods for fractional differential equations. Int. J. Pure Appl. Math. **119** (3), 391–401 (2018)
17. Gong, C., Bao, W., Tang, G., Yang, B., Liu, J.: An efficient parallel solution for Caputo fractional reaction–diffusion equation. J. Supercomput. **68**(3), 1521–1537 (2014)
18. Almeida, R.: A Caputo fractional derivative of a function with respect to another function. Commun. Nonlinear Sci. Numer. Simul. **44**, 460–481 (2017)
19. Bogaenko, V.A., Bulavatsky, V.M.: Computer modeling of the dynamics of migration processes of soluble substances in the case of groundwater filtration with free surface on the base of the fractional derivative approach. Dopov. Nac. akad. nauk Ukr. **12**, 21–29 (2018). (in Russian)
20. Samarskii, A.: The Theory of Difference Schemes. CRC Press, New York (2001)
21. Bohaienko, V.O.: A fast finite-difference algorithm for solving space-fractional filtration equation with a generalised Caputo derivative. Comput. Appl. Math. **38**(3), 105 (2019)
22. Bohaienko, V.O.: Efficient computation schemes for generalized two-dimensional time-fractional diffusion equation. In: Papers of the International scientific and practical conference ITCM – 2019, Ivano-Frankivsk, Ukraine, pp. 238–241 (2019)

On Calculation by the Linearization Method of Mixed Parametric and Self-oscillations at Delay

Alishir A. Alifov(✉)

Mechanical Engineering Research Institute of the Russian Academy of Sciences, Moscow 101990, Russia
a.alifov@yandex.ru

Abstract. The interaction of parametric oscillations and self-oscillations in the presence of a time lag in the elastic force and a non-ideal (limited power) energy source in the system is considered. The method of direct linearization of non-linearities is used to solve the system of nonlinear equations of motion. On its basis, the equations of stationary and non-stationary motions are derived. The stability of stationary oscillations is considered. Calculations are carried out to obtain information on the effect of delay on the characteristics of stationary modes of motion. The calculated data obtained by the direct linearization method and the asymptotic method of nonlinear mechanics, which is widely used to study nonlinear dynamical systems, are compared. Both methods gave the same quantitative and qualitative result, while in the case of the direct linearization method, it took several orders of magnitude less labor and time than with the asymptotic method.

Keywords: Parametric oscillations · Self-oscillations · Time lag · Energy source · Limited power · Elastic force · Method · Direct linearization

1 Introduction

In automatic control systems, regulators, tracking systems, various technical devices, vibration machines, electronic devices, etc. systems with time lag are widespread. The study of oscillations in systems with a time lag was carried out in a large number of works, among which we can specify [1–5] and many others. These works do not take into account the limited power of the energy source that supports the operation of the system. However, the functioning of real physical systems is due to some energy source with limited power. An attempt to study nonlinear oscillatory systems with time lag and limited power of an energy source is made in a small number of papers [6–8 et al.].

The study of the dynamics of devices in the vast majority of cases is associated with the solution of differential equations, which also describe different types of oscillatory processes. Approximate methods of nonlinear mechanics are used to study nonlinear oscillatory systems [9–11], which are characterized by high labor intensity, increasing with increasing degree of nonlinearity. High labor costs are one of the main problems of nonlinear dynamics of various objects [12–16]. Oscillator networks, which play an

Z. Hu et al. (Eds.): ICCSEEA 2020, AISC 1247, pp. 276–284, 2021.
https://doi.org/10.1007/978-3-030-55506-1_25

important role in physics, electronics, neural networks, biology, chemistry, etc., are characterized by this problem.

From the above approximate methods of nonlinear mechanics are significantly different direct linearization methods [17–20 et al.]. The latter are quite simple to use and have several orders of magnitude less labor. The aim of the work is the development of methods for calculating the interaction of types of oscillations in systems whose dynamics are supported by energy sources of limited power. Below, using the direct linearization method, the interaction of parametric oscillations and self-oscillations in the presence of an energy source of limited power and delay is studied. The mechanical model of the frictional self-oscillating system, the equations of its motion, and the stability of oscillations are considered. To obtain information on the influence of the properties of the energy source and the delay on the dynamics of the system, calculations were performed.

2 Equation of Motion

Differential equations of motion of the system (Fig. 1) have the form

$$m\ddot{x} + k_0\dot{x} + (c_0 + b\cos pt)x = T(U) - c_1x_\tau \tag{1}$$

$$J\ddot{\varphi} = M(\dot{\varphi}) - r_0T(U)$$

where m is body mass that comes into contact with a tape driven by an engine with a torque characteristic $M(\dot{\varphi})$ (the difference between the engine torque and the moment of resistance to rotation) and a total moment of inertia of the rotating parts J, k_0 is the damping coefficient, $(c_0 + b\cos pt)x$ is parametric excitation with the amplitude $b = const$ and frequency p, $c_0 = const$, $T(U)$ is nonlinear friction force, which causes self-oscillations, $U = V - \dot{x}$, $V = r_0\dot{\varphi}$, $r_0 = const$ is radius of the point of application of the friction force $T(U)$ at the point of contact of the body and the tape, $\dot{\varphi}$ is the rotational speed of the engine, $x_\tau = x(t-\tau)$, τ is a constant reflecting the elastic force time lag.

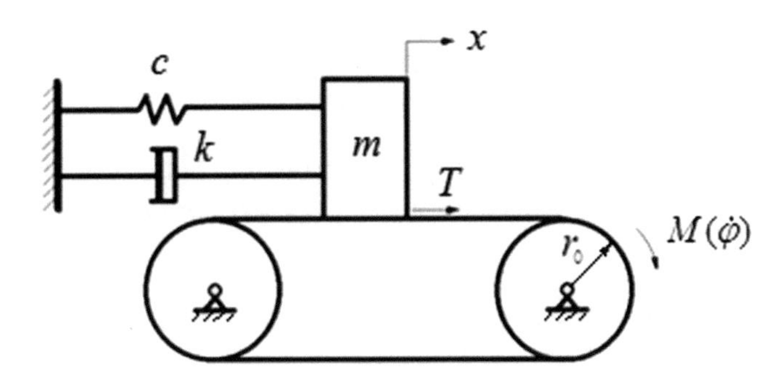

Fig. 1. System model.

Represent the friction force $T(U)$ in the form

$$T(U) = R[\operatorname{sgn} U + F(\dot{x})],\ F(\dot{x}) = \sum_i \delta_i U^i = \sum_{n=0}^{5} \alpha_n \dot{x}^n \tag{2}$$

where $\delta_i = const$, $s = 2, 3, 4, \ldots$, $i = 1, 2, 3, \ldots, R$ is the normal reaction force, sgn $U = 1$ at $U > 0$ and sgn $U = -1$ at $U < 0$,

$$\alpha_0 = \delta_1 V + \delta_2 V^2 + \delta_3 V^3 + \delta_4 V^4 + \delta_5 V^5$$

$$\alpha_1 = -(\delta_1 + 2\delta_2 V + 3\delta_3 V^2 + 4\delta_4 V^3 + 5\delta_5 V^4)$$

$$\alpha_2 = \delta_2 + 3\delta_3 V + 6\delta_4 V^2 + 10\delta_5 V^3$$

$$\alpha_3 = -(\delta_3 + 4\delta_4 V + 10\delta_5 V^2)$$

$$\alpha_4 = \delta_4 + 5\delta_5 V$$

$$\alpha_5 = -\delta_5$$

Using the direct linearization method [17], we replace the nonlinear function $T(\dot{x})$ with a linear function

$$T_*(\dot{x}) = B_T + k_T \dot{x} \tag{3}$$

where B_T, k_T is the linearization coefficients

$$B_T = \sum_n N_n \alpha_n \upsilon^n, \quad n = 0,\ 2,\ 4 \qquad (n \text{ is even number})$$

$$k_T = \sum_n \alpha_n \bar{N}_n \upsilon^{n-1}, \quad n = 1,\ 3,\ 5 \qquad (n \text{ is odd number})$$

$$N_n = (2r+1)/(2r+1+n),\ \bar{N}_n = (2r+3)/(2r+2+n),\ \upsilon = \max|\dot{x}|.$$

The symbol r represents the linearization accuracy parameter [17].

In view of (3), nonlinear equations (1) take the form

$$\begin{aligned} m\ddot{x} + k\dot{x} + (c_0 + b\cos pt)x &= R(\operatorname{sgn} U + B_T) - c_1 x_\tau \\ J\ddot{\varphi} &= M(\dot{\varphi}) - r_0 R(\operatorname{sgn} U + B_T + k_T \dot{x}) \end{aligned} \tag{4}$$

where $k = k_0 - R k_T$.

The solution of system (4) can be constructed by the method of changing variables with averaging [17]. It allows you to study both stationary and non-stationary processes. When averaging is carried out, the $\dot{\varphi}$ variable is replaced by Ω and, accordingly, in the expressions α_n, instead of V, $u = r_0 \Omega$ takes place.

3 Solutions of Equations

In the method of changing variables with averaging [17] for a nonlinear equation with linearized nonlinear functions $\bar{F}(\dot{x})$ and $\bar{f}(x)$

$$\ddot{x}+\bar{F}(\dot{x}) + \bar{f}(x) = H(t, x) \tag{5}$$

using the solution forms $x = a\cos\psi$, $\dot{x} = -ap\sin\psi$, $\psi = pt+\xi$, the following standard forms of the equation for determining non-stationary values of υ and ξ are obtained:

$$\frac{d\upsilon}{dt} = -\frac{k\upsilon}{2} - H_s(\upsilon,\xi), \qquad \frac{d\xi}{dt} = \frac{\omega^2 - p^2}{2p} - \frac{1}{\upsilon}H_c(\upsilon,\xi)$$

where $\upsilon = ap$,

$$H_s(\upsilon,\xi) = \frac{1}{2\pi}\int_0^{2\pi} H(\cdots)\sin\psi\, d\psi, \quad H_c(\upsilon,\xi) = \frac{1}{2\pi}\int_0^{2\pi} H(\cdots)\cos\psi\, d\psi$$

The first equation (4) belongs to type (5), which allows us to determine and calculate the expressions $H_s(\upsilon,\xi)$ and $H_c(\upsilon,\xi)$ for (4). And the solution of the second equation (4) can be constructed using the procedure described in [20]. System solutions are different for $u \geq ap$ ($U > 0$) and $u < ap$ ($U < 0$). We have the following equations for determining the values of the amplitude a, phase ξ and velocity u:

a) $u \geq ap$

$$\begin{aligned}
\frac{da}{dt} &= -\frac{ka}{2m} + \frac{ba}{4pm}\sin 2\xi + \frac{c_1 a}{2pm}\sin p\tau \\
\frac{d\xi}{dt} &= \frac{\omega_0^2 - p^2}{2p} + \frac{b}{4pm}\cos 2\xi + \frac{c_1}{2pm}\cos p\tau \\
\frac{du}{dt} &= \frac{r_0}{J}\left[M(\frac{u}{r_0}) - r_0 R(1+B_T)\right];
\end{aligned} \tag{6, a}$$

b) $u < ap$

$$\begin{aligned}
\frac{da}{dt} &= -\frac{a}{2m}\left[k + \frac{4R}{\pi a^2 p^2}\sqrt{a^2p^2 - u^2}\right] + \frac{ba}{4pm}\sin 2\xi + \frac{c_1 a}{2pm}\sin p\tau \\
\frac{d\xi}{dt} &= \frac{\omega_0^2 - p^2}{2p} + \frac{b}{4pm}\cos 2\xi + \frac{c_1}{2pm}\cos p\tau \\
\frac{du}{dt} &= \frac{r_0}{J}\left[M(\frac{u}{r_0}) - r_0 R(1 - B_T) - \frac{r_0 R}{\pi}(3\pi - 2\psi_*)\right]
\end{aligned} \tag{6, b}$$

where $\psi_* = 2\pi - \arcsin(u/ap)$, $\omega_0^2 = c_0/m$.

From (6,a) stationary values of the amplitude and phase are determined

$$4\left[m\,(\omega_0^2 - p^2) + c_1 \cos p\tau\right]^2 + 4[pk - c_1 \sin p\tau]^2 = b^2$$
$$tg2\xi = \frac{c_1 \sin p\tau - pk}{m\,(\omega_0^2 - p^2) + c_1 \cos p\tau}. \tag{7}$$

For calculations in the case of $u < ap$ the technique described in [7] was used. The amplitude of stationary oscillations at $u < ap$ is determined by the approximate equality $ap \approx u$.

To find stationary velocity values from the condition $\dot{u} = 0$ follows the equation

$$M(u/r_0) - S(u) = 0$$

where

a) $u \geq ap$

$$S(u) = r_0 R(1 + B_T);$$

b) $u < ap$

$$S(u) = r_0 R\left[(1 - B_T) + \pi^{-1}(3\pi - 2\psi_*)\right].$$

Taking into account the approximate equality $ap \approx u$ the expression $S(u)$ at $u < ap$ is simplified.

4 Stability of Stationary Oscillations

To study the stability of stationary oscillations, we compose equations in variations for (6,a and 6,b) and use the Routh-Hurwitz criteria. Stationary oscillations are stable if the conditions are met

$$D_1 > 0,\; D_3 > 0,\; D_1 D_2 - D_3 > 0 \tag{8}$$

where

$$D_1 = -(b_{11} + b_{22} + b_{33})$$

$$D_2 = b_{11}b_{33} + b_{11}b_{22} + b_{22}b_{33} - b_{12}b_{21}$$

$$D_3 = b_{12}b_{21}b_{33} - b_{11}b_{22}b_{33}.$$

The coefficients in (8) with $u \geq ap$ have the form

$$b_{11} = \frac{r_0}{J}(Q - r_0 R\frac{\partial B_T}{\partial u}),\ b_{12} = -\frac{r_0^2 R}{J}\frac{\partial B_T}{\partial a},\ b_{21} = \frac{aR}{2m}\frac{\partial k_T}{\partial u}$$

$$b_{22} = -\frac{1}{2m}(k - aR\frac{\partial k_T}{\partial a}) + \frac{b}{4pm}\sin 2\xi + \frac{c_1}{2pm}\sin p\tau,\ b_{23} = \frac{ab\cos 2\xi}{2pm}$$

$$b_{33} = -\frac{b\sin 2\xi}{2pm}$$

where $Q = \frac{d}{du}M(\frac{u}{r})$.

With $u < ap$ only the coefficients change

$$b_{11} = \frac{r_0}{J}\left[Q - r_0 R\frac{\partial B_T}{\partial u} - \frac{2r_0 R}{\pi\sqrt{a^2p^2 - u^2}}\right],\ b_{12} = -\frac{r_0^2 R}{J}\left[\frac{\partial B_T}{\partial a} + \frac{2u}{\pi a\sqrt{a^2p^2 - u^2}}\right]$$

$$b_{21} = \frac{a}{2m}\left[R\frac{\partial k_T}{\partial u} + \frac{4uR}{\pi a^2p^2\sqrt{a^2p^2 - u^2}}\right]$$

$$b_{22} = -\frac{1}{2m}(k - aR\frac{\partial k_T}{\partial a} + \frac{4Ru^2}{\pi a^2p^2\sqrt{a^2p^2 - u^2}})$$

Taking $u = r_0\Omega$ into account when averaging, for the friction force (2) we have

$$\frac{\partial B_T}{\partial u} = \frac{\partial \alpha_0}{\partial u} + N_2(ap)^2\frac{\partial \alpha_2}{\partial u} + N_4(ap)^4\frac{\partial \alpha_4}{\partial u}$$

$$\frac{\partial k_T}{\partial u} = \bar{N}_1\frac{\partial \alpha_1}{\partial u} + \bar{N}_3(ap)^2\frac{\partial \alpha_3}{\partial u} + \bar{N}_5(ap)^4\frac{\partial \alpha_5}{\partial u}$$

$$\frac{\partial B_T}{\partial a} = 2ap^2(N_2\alpha_2 + 2N_4\alpha_4 a^2p^2),\ \frac{\partial k_T}{\partial a} = 2ap^2(\bar{N}_3\alpha_3 + 2\bar{N}_5\alpha_5 a^2p^2)$$

$$\alpha_0 = \delta_1 u + \delta_2 u^2 + \delta_3 u^3 + \delta_4 u^4 + \delta_5 u^5,\ \alpha_1 = -(\delta_1 + 2\delta_2 u + 3\delta_3 u^2 + 4\delta_4 u^3 + 5\delta_5 u^4)$$

$$\alpha_2 = \delta_2 + 3\delta_3 u + 6\delta_4 u^2 + 10\delta_5 u^3,\ \alpha_3 = -(\delta_3 + 4\delta_4 u + 10\delta_5 u^2)$$

$$\alpha_4 = \delta_4 + 5\delta_5 u,\ \alpha_5 = -\delta_5$$

$$\frac{\partial \alpha_0}{\partial u} = \delta_1 + 2\delta_2 u + 3\delta_3 u^2 + 4\delta_4 u^3 + 5\delta_5 u^4,\ \frac{\partial \alpha_1}{\partial u} = -2(\delta_2 + 3\delta_3 u + 6\delta_4 u^2 + 10\delta_5 u^3)$$

$$\frac{\partial \alpha_2}{\partial u} = 3(\delta_3 + 4\delta_4 u + 10\delta_5 u^2),\ \frac{\partial \alpha_3}{\partial u} = -4(\delta_4 + 5\delta_5 u),\ \frac{\partial \alpha_4}{\partial u} = 5\delta_5,\ \frac{\partial \alpha_5}{\partial u} = 0$$

When calculating $\partial B_T/\partial u$, $\partial B_T/\partial a$ only even degrees of n are taken into account (respectively α_0, α_2, α_4). To calculate $\partial k_T/\partial u$, $\partial k_T/\partial a$ odd powers of n are taken into account (respectively α_1, α_3, α_5).

5 Calculations

Calculations were performed to obtain information on the effect of delay on stationary oscillation modes. The characteristic of the friction force was chosen in the widespread (it was also observed in the space experiment [21]) form $T(U) = R(\operatorname{sgn} U - \delta_1 U + \delta_3 U^3)$ in practice, where δ_1 and δ_3 are positive constants. As the calculated parameters are used: $\omega_0 = 1\ \text{s}^{-1}$, $m = 1\ \text{kgf} \cdot \text{s}^2 \cdot \text{cm}^{-1}$, $k_0 = 0.02\ \text{kgf} \cdot \text{s} \cdot \text{cm}^{-1}$, $b = 0.07\ \text{kgf} \cdot \text{cm}^{-1}$, $r_0 = 1\ \text{cm}$, $J = 1\ \text{kgf} \cdot \text{s} \cdot \text{cm}^2$, $R = 0.5\ \text{kgf}$, $\delta_1 = 0.84\ \text{s} \cdot \text{cm}^{-1}$, $\delta_3 = 0.18\ \text{s}^3 \cdot \text{cm}^{-3}$, $c_1 = 0.05\ \text{kgf} \cdot \text{cm}^{-1}$. The calculated values for the delay are $p\tau = 0;\ \pi/2;\ \pi$.

Figure 2 shows some calculation results in the case of $u = 1.2\ \text{cm} \cdot \text{s}^{-1}$. In the calculations, the linearization accuracy parameter $r = 1.5$ was used. Curve 1 corresponds to the delay value $\tau = 0$, curve 2 to $p\tau = \pi/2$, curve 3 to $p\tau = \pi$, the a_s symbol denotes the amplitudes of self-oscillations at the corresponding delay values. Exactly the same curves are obtained if the widely used asymptotic method of averaging nonlinear mechanics is used to solve (1) [9]. Stable sections are shown by solid lines, unstable by dashed lines. Oscillations with amplitudes are stable if the steepness $Q = \frac{d}{du} M(\frac{u}{r_0})$ of the characteristic of the energy source $M(u/r_0)$ is within the shaded sectors.

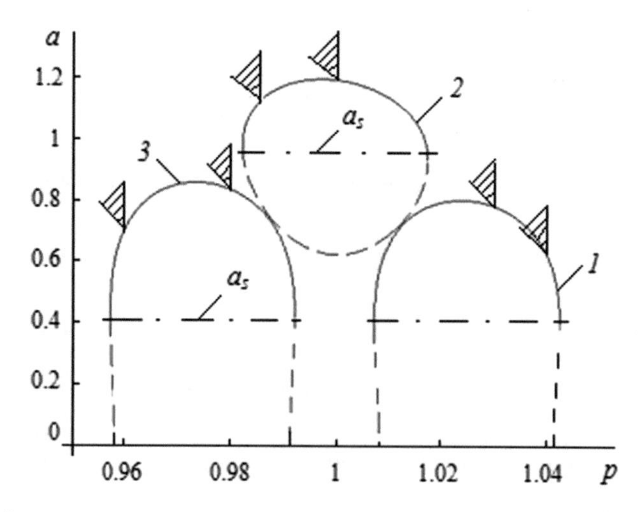

Fig. 2. Amplitude-frequency curves and sectors (shaded) characteristics of the source of energy for sustained oscillations.

6 Discussion and Conclusion

The broad nature of the oscillatory processes play a large role in the functioning of many technical objects for various purposes. All real dynamic systems are nonlinear, and the use of linear models (valid only for small intervals of parameter changes) for their description is associated with mathematical difficulties - finding solutions to nonlinear equations. These difficulties (high labor and time costs) are inherent in all known methods by which approximate solutions of nonlinear equations of motion of dynamical systems are constructed. Not every scientist in the technical field knows how to use these methods, not to mention the engineers and designers who develop devices for various purposes. Therefore, the forefront is the simplicity of the method, low labor costs and time of its application. These positive features have direct linearization methods, which enrich the methods for calculating nonlinear systems. Using these methods, we examined in this article the interaction of self-oscillations and parametric oscillations. A feature of the system is the presence of limited excitation and delay in elasticity. To study the effect of delay on the dynamics of the system, calculations were carried out by two methods: the linearization method and the asymptotic method of nonlinear mechanics. The calculation results by these methods completely coincide, while using the direct linearization method is much simpler. It turned out that the delay has a significant effect on the shape of the amplitude-frequency curves and their location relative to the natural frequency of the system.

References

1. Rubanik, V.P.: Oscillations of Quasilinear Systems with Lag. Nauka, Moscow (1969). (in Russian)
2. Zhirnov, B.M.: Single-frequency resonance oscillations of a frictional self-oscillating system with delay with external disturbance. J. Appl. Mech. **14**(9), 102–109 (1978). (in Russian)
3. Astashev, V.K., Hertz, M.E.: Auto-oscillations of a visco-elastic rod with limiters under the action of a lagging force. J. Mashinovedenie **5**, 3–11 (1973). (in Russian)
4. Butenin, N.V., Neymark, Yu.I., Fufaev, N.A.: Introduction to the Theory of Nonlinear Oscillations. Nauka, Moscow (1976). (in Russian)
5. Kononenko, V.O.: Nonlinear Vibrations of Mechanical Systems. Naukova Dumka, Kiev (1980). (in Russian)
6. Rubanik, V.P., Starik, L.K.: On the stability of self-oscillations of a cutter in the case of an imperfect energy source. Scientific Works of Universities Lit. SSR. Vibrotekhnike, vol. 2 (11), pp. 205–212 (1971). (in Russian)
7. Alifov, A.A., Frolov, K.V.: Interaction of Nonlinear Oscillatory Systems with Energy Sources. Hemisphere Publishing Corporation, Taylor & Francis Group, New York (1990). ISBN 0-89116-695-5
8. Alifov, A.A., Abdiev F.K.: The interaction of forced oscillations and self-oscillations with friction with a delayed argument. VINITI, no. 358-85 dep., Moscow, Russia (1985). (in Russian)
9. Bogolyubov, N.N., Mitropolsky, Yu.A.: Asymptotic Methods in the Theory of Nonlinear Oscillations. Nauka, Moscow (1974). (in Russian)

10. Hayasi, C.: Forced Oscillations in Nonlinear Systems. Nippon Printing and Publishing Co., Osaka (1953)
11. Blekhman, I.I. (ed.): Vibrations in technology: directory, vol. 2. Oscillations of nonlinear mechanical systems. Engineering, Moscow, Russia (1979). (in Russian)
12. Gourary, M.M., Rusakov, S.G.: Analysis of oscillator ensemble with dynamic couplings. In: AIMEE 2018. The Second International Conference of Artificial Intelligence, Medical Engineering, Education, pp. 150–160 (2018)
13. Acebrón, J.A., et al.: The Kuramoto model: a simple paradigm for synchronization phenomena. Rev. Modern Phys. **77**(1), 137–185 (2005)
14. Bhansali, P., Roychowdhury, J.: Injection locking analysis and simulation of weakly coupled oscillator networks. In: Li, P., et al. (eds.) Simulation and Verification of Electronic and Biological Systems, pp. 71–93. Springer (2011)
15. Ashwin, P., Coombes, S., Nicks, R.J.: Mathematical frameworks for oscillatory network dynamics in neuroscience. J. Math. Neurosci. **6**(2), 1–92 (2016)
16. Ziabari, M.T., Sahab, A.R., Fakhari, S.N.S.: Synchronization new 3D chaotic system using brain emotional learning based intelligent controller. Int. J. Inf. Technol. Comput. Sci. (IJITCS) **7**(2), 80–87 (2015). https://doi.org/10.5815/ijitcs.2015.02.10
17. Alifov, A.A.: Methods of Direct Linearization for Calculation of Nonlinear Systems. RCD, Moscow, Russia (2015). (in Russian). ISBN 978-5-93972-993-2
18. Alifov, A.A.: On the calculation by the method of direct linearization of mixed oscillations in a system with limited power-supply. In: Hu, Z., Petoukhov, S., Dychka, I., He, M. (eds.) Advances in Computer Science for Engineering and Education II, ICCSEEA 2019. Advances in Intelligent Systems and Computing, vol 938, pp. 23–31. Springer, Cham (2020). https://doi.org/10.1007/978-3-030-16621-2_3
19. Alifov, A.A.: Method of the direct linearization of mixed nonlinearities. J. Mach. Manuf. Reliab. **46**(2), 128–131 (2017). https://doi.org/10.3103/s1052618817020029
20. Alifov, A.A., Farzaliev, M.G., Jafarov, E.N.: Dynamics of a self-oscillatory system with an energy source. Russ. Eng. Res. **38**(4), 260–262 (2018). https://doi.org/10.3103/s1068798x18040032
21. Bronovec, M.A., Zhuravljov, V.F.: On self-oscillations in systems for measuring friction forces. Izv. RAN, Mekh. Tverd. Tela, no. 3, pp. 3–11 (2012). (in Russian)

Adaptive Strategies in the Multi-agent "Predator-Prey" Models

Petro Kravets[1], Volodymyr Pasichnyk[1], Nataliia Kunanets[1], Nataliia Veretennikova[1](✉), and Olena Husak[2]

[1] Information Systems and Networks Department, Lviv Polytechnic National University, Lviv, Ukraine
Petro.O.Kravets@lpnu.ua, vpasichnyk@gmail.com, nek.lviv@gmail.com, nataver19@gmail.com
[2] Department of Applied Mathematics and Information Technologies, Yuriy Fedkovych Chernivtsi National University, Chernivtsi, Ukraine
gusakolena17@gmail.com

Abstract. The problem of coordination and self-organization of multi-agent strategies based on a predator-prey stochastic game model with local connections between agents is investigated. The game recurrent method and algorithm of formation of the coordinated agent strategies in the course of minimization of average loss functions are developed. Computer modelling of a predator-prey stochastic game is executed. The influence of model parameters on convergence of a game method is studied as well. Modern multi-component network systems, in particular IoT systems, require the development of new approaches to their modeling and formation of appropriate methods. The authors propose to use the stochastic game methods obtained for this purpose.

Keywords: Coordination · Self-organization · Stochastic game · Pursuit task · Predator-prey model

1 Introduction

The modern development of network computer technologies makes it necessary to develop and implement distributed information and software systems, their complexity is that any centralized management of them is ineffective or impossible. The decentralized agent-oriented management methods are used to ensure the viability of such systems under conditions of uncertainty [1].

The use of stochastic game methods to model the behavior of multiagent systems is one of the original modern approaches to creating distributed multi-component network complexes of various purposes. Particularly good results from such an approach in modeling are observed in the reproduction of IoT class network architectures, when thousands of elements operate simultaneously in networks, which by their nature of behavior can be represented as stochastic multiagent aggregates. This feature of the authors' approach is original, useful and effective.

A software agent is an autonomous software system with elements of artificial intelligence that can interact with other agents and people using the resources of the

Z. Hu et al. (Eds.): ICCSEEA 2020, AISC 1247, pp. 285–295, 2021.
https://doi.org/10.1007/978-3-030-55506-1_26

information network. Multi Agent System (MAS) is a system formed by several interacting intelligent agents [2–4]. The interaction of agents is localized in a decentralized system. The decision making by each agent is based on available local information and interaction within the multi-agent system. System functioning, as a rule, is carried out in the conditions of the aprioristic uncertainty caused by external or internal factors.

The success of a distributed solution to a common task depends on the coordination level of agents. Coordination is to ensure coordinated, well-organized work of all MAS units [5–8]. Coordination can be centralized or decentralized. The methods of decentralized coordination of agents will be more effective with the growth of the structural and functional complexity of the distributed system in the conditions of uncertainty in terms of the speed of solving the problem. Coordination is necessary to agree with the individual goals and agents' behavior, when each agent improves or does not worsen the value of its utility function, and the system as a whole improves the quality of solving a common task. Methods for solving the coordination problem are based on the results of the classical management theory, the study of operations, the game theory, planning and the results of other areas of mathematics and cybernetics.

Decentralized coordination is based on the interaction between agents. Such interaction may be implicit when agents act independently and influence each other due to changing states of the common environment, or explicit, communicative, when agents interact further directly with each other. The information exchange can be global, or within the limits of locally determined coalitions. Agents can share knowledge about the explored environment, data on current and projected strategy choices, values of received benefits, etc.

Coordination is a necessary factor in the system of self-organization. Self-organization is a purposeful process of creating, recreating, organizing or improving the organization (structure and functions) of a complex dynamic system at the expense of internal factors without the corresponding external influence.

MAS self-organization is an ability of an agent team with locally-determined links and goals to achieve sustainable, coordinated behavior strategies in conditions of uncertainty at the expense of self-learning, an ability to function as a unit and to ensure the implementation of the global goal of system development. Self-organization and complex forms of MAS behavior can be manifested in the implementation of the simplest actions of agents [9–12].

The stochastic game is one of the means of modeling the MAS [13, 14]. Along with the individual characteristics, the game theory and the MAS theory have common research subjects. We use the stochastic game model considering the factors of competition, interaction, cooperation, training, self-organization that are inherent in the agent team. This model provides an adaptive search for points of equilibrium of win functions on a unit simplex of mixed player strategies. In a stochastic game, players are learned to choose the optimal medium-level strategies (actions), rebuilding their own vectors of dynamic mixed strategies (conditional probabilities of action options).

The game coordination of agent strategies in the process of MAS self-organization is an actual scientific and practical problem, not sufficiently studied now. From the cognitive point of view, it is important to visualize the coordinated strategies as a result of the computer experiment as one of the signs in MAS self-organization.

Studying the processes of MAS coordination and self-organization will be done based on the game model of pursuit of "predator-prey" [15–17]. Nature-based patterns of pursuit with a variety of scenarios (for example, a pursuit by a shark of a school of fish, attacking a swarm of birds, etc.) are used to work out the strategies of agents' behavior in distributed decision-making systems. Predator-agents and prey-agents are distinguished in the pursuit models that have antagonistic goals. The goal of predator agents is to reach one of the preys during the pursuit, otherwise to minimize the distance to the prey. The purpose of the prey-agents is to save their own lives, otherwise to maximize or keep a safe distance to the predator. The choice of this task is due to its simplicity and ease of visualization of coordinated strategies as signs of the game self-organization of the system.

The **purpose** of this work is to determine the conditions and mechanisms for local coordination of agent actions, which leads to the self-organization of the MAS on the example of the "predator-prey" stochastic game. In order to achieve the goal, it is necessary to solve the following tasks such as to construct a model of a stochastic game, to develop a method and an algorithm for solving a stochastic game, to perform computer simulation of a stochastic game to identify the factors of MAS coordination and self-organization.

2 Formulation of a Game Task

Solving the self-organization task of the MAS will be done based on adaptive methods of stochastic games considering the system distribution and the common character of the decision-making process, purpose antagonism (between a predator and a prey) and consistency (within the group of predators or preys).

A stochastic game $(D, \{U^i\}_{i \in D}, \{P^i(\xi^i)\}_{i \in D})$ is given by a set of player agents $D \neq \varnothing$, vectors of pure strategies $U^i = (u^i[1], u^i[2], \ldots, u^i[N])$ and a priori unknown distribution $P^i(\xi^i)$ of random variables ξ^i $\forall i \in D$. So, one of the agents is a predator, and the rest are preys. Pure strategies $u[k]$, $k = 1..N$ determine the directions of possible agent movements.

The repeating game develops at discrete time moments $n = 1, 2, \ldots$. Each agent $i \in D$ carries out an independent random selection from one of $N \geq 2$ their own pure strategies $u_n^i = u^i \in U^i$.

The MAS structure defines a locally-determined mechanism for the formation of random losses $\xi_n^i = \xi_n^i(u_n^{D_i})$, which are functions of common strategies $u^{D_i} \in U^{D_i} = \underset{j \in D_i}{\times} U^j$ from local subsets of agents $D_i \subseteq D$, $D_i \neq \varnothing$ $\forall i \in D$. Random losses $\{\xi_n^i\}$ are independent $\forall u_n \in U$, $\forall i \in D$, $n = 1, 2, \ldots$, have a constant mathematical expectation $M\{\xi_n^i(u^{D_i})\} = v(u^{D_i}) = const$ and a limited second moment $\sup_n M\{[\xi_n^i(u^{D_i})]^2\} = \sigma^2(u^{D_i}) < \infty$. Stochastic characteristics of random losses $\{\xi_n^i\}$ $\forall i \in D$ are not known priori to the agents. After option choosing u_n^i $\forall i \in D$ players receive the current loss, which consists of three components:

$$\zeta_n^i = \lambda \xi_n^i[1] + (1 - \lambda)\xi_n^i[2] + \mu_n, \tag{1}$$

where $\lambda \in [0, 1]$ is a weight coefficient; $\mu_n \sim Normal(0, d)$ is an additive white Gaussian noise, which is a normally distributed random value with zero mathematical expectation and dispersion $d > 0$. The first component (1) defines a fine for breaking the movement coordination of neighboring agents:

$$\xi_n^i[1] = \sum_{j \in D_i} \chi(u_n^i \neq u_n^j),$$

where $\chi() \in \{0, 1\}$ is an indicator function of an event, u_n^i is a current prey strategy of $i \in D$.

The second component (1) defines a fine for violation of a replication condition (repetition) of the left-hand or right-hand actions in the direction of the agent's movement:

$$\xi_n^i[2] = \chi(s \in D_i)\chi(u_n^i \neq u_n^s) + \chi(s \notin D_i)\chi(u_n^i \neq u_n^{Side(i)}),$$

where s is a predator identifier; u_n^s is a current predator strategy; $Side(i)$ is an operator function for identifying the neighbor agent located on the left $Side(i) = Left$ or on the right $Side(i) = Right$ from the prey's direction.

The quality of choosing pure strategies at the time moment n is determined by average losses

$$Z_n^i = \frac{1}{n} \sum_{t=1}^{n} \zeta_t^i \; \forall i \in D. \tag{2}$$

The game purpose is to minimize the functions (2) of average losses:

$$\overline{\lim_{n \to \infty}} Z_n^i \to \min_{u_n^i} \; \forall i \in D. \tag{3}$$

So, interacting with the environment based on the observation of current losses $\{\zeta_n^i\}$, each player $i \in D$ has to learn to choose pure strategies $\{u_n^i\}$ in such time course $n = 1, 2, \ldots$ to ensure that the criteria system is implemented (3). Solutions of the game task will satisfy one of the conditions of collective equilibrium, for example, by Nash, Slater, Pareto, depending on the method of formation of strategy sequences $\{u_n^i\}$ $\forall i \in D$.

3 Method of Solving the Problem

In a competitive environment, there is a need to overcome monopoly power in the support of resource network services, and users have an opportunity to choose providers. Preference is given to providers who have a more loyal pricing policy and more convenient means of service. At the same time, providers are focusing on modern information technologies, including the Internet of Things. Providers are interested in attracting a large number of users to their services while campaigning for the transition to receiving services at their corporations.

In order to choose solutions in the conditions of uncertainty, it is applied Markov adaptive methods, which provide the optimal choice of options on the average at the next time moments based on processing the current information about the system.

Formation of the sequence of solution options $\{u_n^i\}$ will be based on the dynamic vectors of mixed strategies $p_n^i = (p_n^i(1), p_n^i(2), \ldots, p_n^i(N)) \; \forall i \in D$, and their elements $p_n^i(j)$, $j = 1..N$ are probabilities of choosing pure strategies provided the pre-history realization of pure strategy choice $\{u_t^i | \, t = 1, 2, \ldots, n-1\}$ and obtaining the corresponding losses $\{\zeta_t^i | t = 1, 2, \ldots, n-1\}$. Mixed strategies get a meaning on N-dimensional unit simplex:

$$S^N = \left\{ p \left| \sum_{j=1}^{N} p(j) = 1; \, p(j) \geq 0 \, (j = 1..N) \right. \right\}.$$

The values of pure strategies are determined on the condition that:

$$u_n^i = \left\{ u^i(l) | \, l = \arg\min_l \sum_{k=1}^{l} p_n^i[k] > \omega \; (k, l = 1..N) \right\}, \tag{4}$$

where $\omega \in [0, 1]$ is a random variable with a uniform distribution.

It is necessary to determine the method of changing the vectors of mixed strategies $p_n^i \; \forall i \in D$, which will provide generating of random pure strategy $\{u_n^i\}$ according to (4) for performance of criteria system (3). It is constructed the method of solving a stochastic game based on the stochastic approximation of complementary slackness condition of a determined game, valid for mixed strategies at the point of the Nash equilibrium [18, 19].

It is defined the polylinear function of the average losses of the determined game:

$$V^i(p^{D_i}) = \sum_{u^{D_i} \in U^{D_i}} v^i(u^{D_i}) \prod_{j \in D_i; u^j \in u^{D_i}} p^j(u^j),$$

where $v(u^{D_i}) = M\{\xi_n^i(u^{D_i})\}$.

Then the vector condition of complementary slackness (CS) will look like

$$\mathrm{CS}^{\rightarrow} = \nabla_{p^i} V^i(p^{D_i}) - e^{N_i} V^i(p^{D_i}) = 0 \quad \forall i \in D,$$

where $\nabla_{p^i} V^i(p^{D_i})$ is a gradient of the average loss function; $e^N = (1_j | j = 1..N)$ is a vector, whose all components are equal to 1; $p^{D_i} \in S^{M_i}$ is combined mixed strategies of players from local sets D_i, given on a convex unit simplex S^{M_i} $(M_i = \prod_{j \in D_i} |D_i|)$.

It should be performed the weighing of the condition of complementary slackness by the vector elements of mixed strategies to consider solutions to the vertices of a unit simplex:

$$diag(p^i)(\overrightarrow{CS}) = 0 \quad \forall i \in D, \tag{5}$$

where $diag(p^i)$ is a square diagonal matrix of order N_i, built from vector elements p^i.

Taking into account that $diag(p^i)[\nabla_{p^i} V^i - e^{N_i} V^i] = = E\{\zeta_n^i[e(u_n^i) - p_n^i] | p_n^i = p^i\}$, where $E\{\}$ is a function of mathematical expectation, it is obtained a recurrence dependence from (5) on the basis of the stochastic approximation method:

$$p_{n+1}^i = \pi_{\varepsilon_{n+1}}^N \{p_n^i - \gamma_n \zeta_n^i [e(u_n^i) - p_n^i]\}, \tag{6}$$

where $\pi_{\varepsilon_{n+1}}^N$ is a projector for a unit ε-simplex $S_{\varepsilon_{n+1}}^N \subseteq S^N$ [13]; $p_n^i \in S_{\varepsilon_n}^N$ is mixed strategies of i-agent; $\gamma_n > 0$ is a monotonically decreasing sequence of nonnegative quantities, which regulates the step size of the method; $\varepsilon_n > 0$ is a monotonically decreasing sequence of nonnegative quantities, which regulates the expansion rate of ε-simplex; $\zeta_n^i \in R^1$ is a current loss of an agent; $e(u_n^i)$ is a unit vector indicator of the option choice $u_n^i \in U^i$. Design for an extendable ε_n-simplex $S_{\varepsilon_{n+1}}^N$ provides the fulfilment of the condition $p_n^i[j] \geq \varepsilon_n$, $j = 1..N$, necessary for completeness of statistical information about selected pure strategies, and the parameter $\varepsilon_n \to 0$, $n = 1, 2, \ldots$ is used as an additional control element for the convergence of the recurrent method.

A stochastic game begins with uninformed vectors of mixed strategies with element values $p_0^i(j) = 1/N$, where $j = 1..N$. At the next time moments, the vector dynamics of mixed strategies is determined by Markov recursive method (6). At the time moment n, each player $i \in D$ selects a pure strategy u_n^i based on the mixed strategy p_n^i, and by the time $n+1$ receives the current loss ζ_n^i, after that it is calculated the mixed strategy p_{n+1}^i according to (6).

Due to the dynamic rebuilding of mixed strategies based on the processing of current losses, the method (6) provides an adaptive choice of pure strategies in time.

Parameters γ_n and ε_n determine the convergence conditions for the stochastic game and can be given as follows:

$$\gamma_n = \gamma n^{-\alpha}, \ \varepsilon_n = \varepsilon n^{-\beta}, \tag{7}$$

where $\gamma > 0$; $\alpha > 0$; $\varepsilon > 0$; $\beta > 0$.

The strategy convergence (6) to the optimal values with probability 1 and in the quadratic mean is determined by the ratios of parameters γ_n and ε_n (7), which must satisfy the basic conditions of stochastic approximation [13, 20]. The effectiveness of taken decisions is estimated by:

1) the function of by Nash, Pareto:

$$Z_n = \frac{1}{L}\sum_{i=1}^{L} Z_n^i, \tag{8}$$

where $L = |D|$ is a power of a player set;

2) the coordination coefficient of player strategies:

$$\mathrm{K}_n = \frac{1}{nL}\sum_{t=1}^{n}\sum_{i=1}^{L}\chi\left(\xi_n^i[1] + \xi_n^i[2] = 0\right), \tag{9}$$

where $\chi() \in \{0, 1\}$ is an indicator event function.

4 Computer Simulation Results

The solution of the stochastic game of a "predator-prey" pursuit will be done using the game method (6) with the parameters such as $D = \{(x, y)\}$, $x = 1..m$, $y = 1..m$, $m = 7$, $s = (4, 4)$, $N = 4$, $U = (\mathit{front} = 0, \mathit{right} = 1, \mathit{behind} = 2, \mathit{left} = 3)$, $\lambda = 0.5$, $\gamma = 1$, $\varepsilon = 0.999/N$, $\alpha = 0.01$, $\beta = 2$, $n_{\max} = 10^5$.

Method (6) provides solving a stochastic game in pure strategies (on the vertex of a simple simplex). Initial agent strategies are uncoordinated, which is determined by the choice of an arbitrary direction of movement $u \in U$. Agents are trained to coordinate their actions during the game. The obtained variants of coordinated strategies of the trained game are presented in Fig. 1.

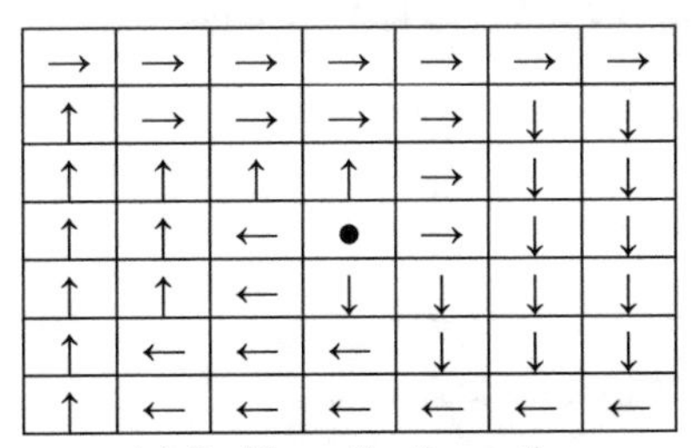

a) left-side replication action

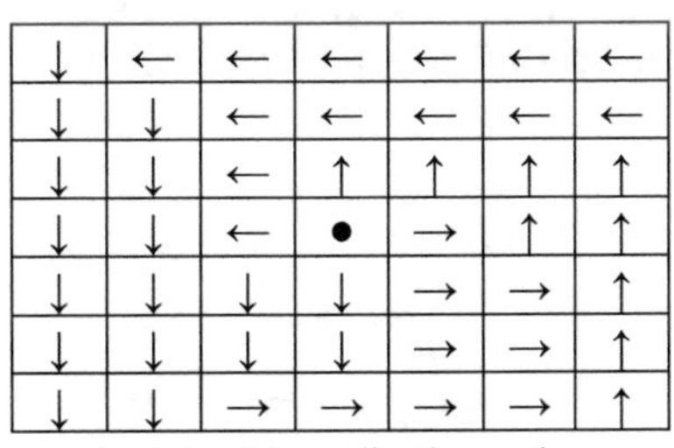

b) right-side replication action

Fig. 1. Self-organization of a stochastic pursuit game

The position of a predator is represented by a point in the central part of the figure. The arrows indicate the movement direction of the prey-agents for the trained stochastic game. By tracking the direction of arrows, it is seen that the right-side self-organization is at the left-hand side ($\mathit{Side}(i) = \mathit{Left}$) of the neighboring agents, and, conversely, the left-side system self-organization is in the right-side orientation ($\mathit{Side}(i) = \mathit{Right}$).

In Fig. 2 the function charts of average losses of players Z_n (8) and the coefficient of coordinated strategies K_n (9) are shown in a logarithmic scale. These functions characterize the effectiveness of the self-organization of a stochastic game.

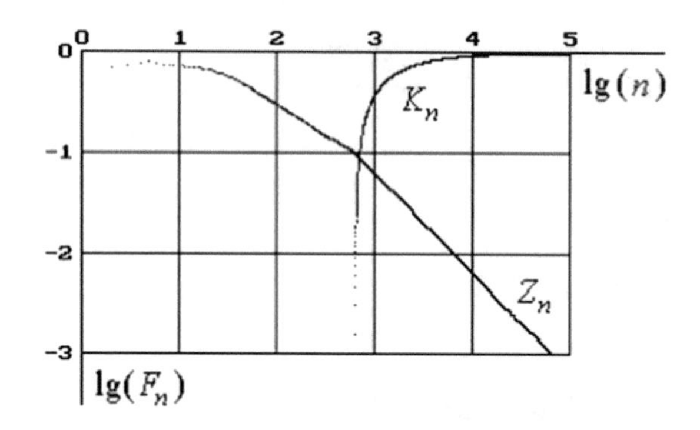

Fig. 2. Characteristics of system self-organization

Reducing the function of average losses in time indicates the convergence of the game method. The coefficient change of coordinated strategies illustrates the progress achieved in the learning process of increasing the coordination of players' actions, starting with $\sim 10^3$ steps of the stochastic game.

The stability of a stochastic game coordination under the influence of disturbance in the form of white noise is investigated. Influence of noise dispersion d on the efficiency of the game method (6) is shown in Fig. 3. The coordination coefficient K_n is calculated on the sample of n = 100 realization of the stochastic game for $\alpha = 0.01$ and $\beta = 2$. The coordination of player strategies exceeds the level of 80% for values $d \in [0;\, 0,36]$. The increase in the noise intensity ($d > 0,36$) leads to a decrease in the coefficient of strategy coordination.

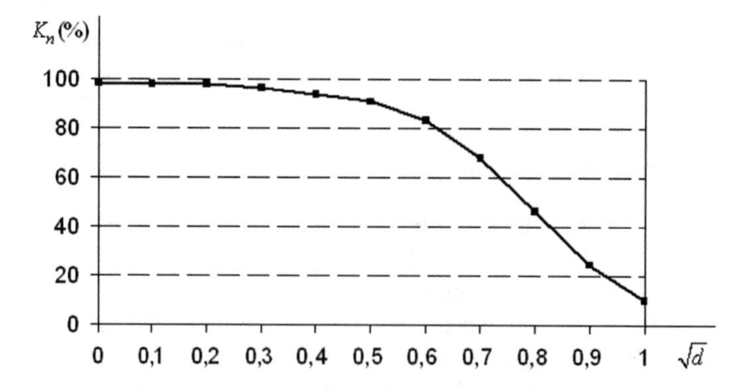

Fig. 3. Dispersion influence on game coordination

In addition to the parameters of the recurrent method, the dimension of the stochastic game has a significant influence on coordination. Dependence of the coordination coefficient K_n on the player number $L = m \times m$ is shown in Fig. 4. Acceptable (more than 80%) coordination of strategies has a place for a game with the agent number $L \leq 121$. The value of the coordination coefficient is calculated during the $n_{\max} = 10^5$ steps of a stochastic game.

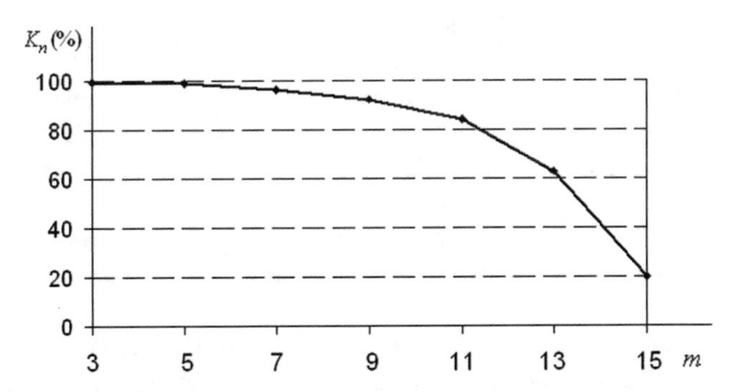

Fig. 4. The player number on the game coordination effect

The growth of the dimensionality of the stochastic game leads to an increase in the average number of steps required to achieve an appropriate level of agent strategy coordination.

The proposed method (6) for solving the stochastic game of agents belongs to the class of reaction methods (based on the processing the reaction of the environment on the agent actions) and has a relatively low degree of convergence, which is associated with a priori system uncertainty. The information gathering is carried out in the learning process by adaptive reorganization of mixed strategy vectors in proportion to the values of current losses. This drawback can be overcome by the high speed of modern computer facilities and the possibility of parallelizing the task or using the networked distributed computing.

5 Conclusion

The developed game method provides self-organization of the "predator-prey" MAS due to locally-determined coordination of agent strategies. Each agent carries out a local observation of the actions of neighboring agents. After calculating the corresponding fines for violating the terms of coordination, agents receive current losses that they use to form the dynamic vectors of mixed strategies. The method of transforming mixed strategies (6) constructed on the basis of stochastic approximation provides minimization of the functions of average losses on unit simplexes. The purposeful dynamics of mixed strategies transforms locally-coordinated actions of players into the global coordination (self-organization) of a stochastic game, which manifests itself in the form of a scheme of right-side or left-side dynamics of the MAS, when the team of victim agents behaves as a holistic organism.

Coordination of agent strategies is achieved during the process of solving a stochastic game in real time based on the collection of current information and its adaptive elaboration. The proposed method allows us to find solutions to stochastic games in pure strategies.

The effectiveness of the coordination of the system's strategies is evaluated using the characteristic functions of average losses and the coefficient of coordination. Reducing the function of average losses and increasing the coordination factor indicates the convergence of the game method.

The speed of coordinating the pure strategies of agents depends on the dimension of the stochastic game, the size of the disturbance and the parameters of the game method. With the increase in the number of players and the intensity of the disturbances, the speed and effectiveness of the game coordination of the MAS are reduced.

The reliability of the obtained results is confirmed by the repetition of the values of the calculated characteristics in the stochastic game for different sequences of random variables.

The development of the "predator-prey" game model is possible in the direction of increasing the number of predators and the use of other methods of teaching stochastic games.

References

1. Amato, C.: Decision-making under uncertainty in multi-agent and multi-robot systems: planning and learning. In: Proceedings of the Twenty-Seventh International Joint Conference on Artificial Intelligence (IJCAI 2018), pp. 5662–5666 (2018)
2. Wooldridge, M.: An Introduction to Multiagent Systems. Wiley, Hoboken (2002). 366 p.
3. Byrski, A., Kisiel-Dorohinicki, M.: Evolutionary Multi-Agent Systems: From Inspirations to Applications. Springer, New York (2017). 224 p.
4. Radley, N.: Multi-Agent Systems – Modeling, Control, Programming, Simulations and Applications. Scitus Academics LLC (2017). 284 p.
5. Ciobanu, G.: Coordination and self-organization in multi agent systems. In: Sixth International Conference on Intelligent Systems Design and Applications, pp. 16–18 (2006). https://doi.org/10.1109/ISDA.2006.127
6. Krishnamurthy, E.V.: Self-organization and autonomy in computational networks: agents-based contractual workflow paradigm. Int. J. Intell. Systx. Appl. (IJISA) 64–76 (2012). https://doi.org/10.5815/ijisa.2012.01.08
7. Kravets, P., Pasichnyk, V., Kunanets, N., Veretennikova, N.: Game method of event synchronization in multi–agent systems. Adv. Intell. Syst. Comput. **938**, 378–387 (2019)
8. Terán, J., Aguilar, J.L., Cerrada, M.: Mathematical Models of Coordination Mechanisms in Multi-Agent Systems. CLEIej Montevideo ago, vol. 16, no. 2 (2013)
9. Soria Zurita, N.F., Colby, M.K., Tumer, I.Y., Hoyle, C., Tumer, K.: Design of complex engineered systems using multi-agent coordination. J. Comput. Inf. Sci. Eng. **18**(1) (2017). https://doi.org/10.1115/1.4038158
10. Heylighen, F.: Stigmergy as a universal coordination mechanism I: definition and components. Cogn. Syst. Res. **38**, 4–13 (2016). https://doi.org/10.1016/j.cogsys.2015.12.002

11. Ye, D., Zhang, M., Vasilakos, A.V.: A survey of self-organization mechanisms in multiagent systems. IEEE Trans. Syst. Man Cybern. Syst. **47**(3), 441–461 (2017). https://doi.org/10.1109/TSMC.2015.2504350
12. Zheng, Y., Ma, J., Wang, L.: Consensus of hybrid multi-agent systems. IEEE Trans. Neural Netw. Learn. Syst. **29**(4), 1359–1365 (2018). https://doi.org/10.1109/TNNLS.2017.2651402
13. Nazin, A., Poznyak, A.: Adaptive Choice of Variants. Nauka, Moscow (1986). (in Russian). 288 p.
14. Ummels, M.: Stochastic Multiplayer Games: Theory and Algorithms. Amsterdam University Press (2014). 174 p.
15. Morice, S., Pincebourde, S., Darboux, F., Kaiser, W., Casas, J.: Predator–prey pursuit–evasion games in structurally complex environments. Integr. Comp. Biol. **53**(5), 767–779 (2013)
16. Baillard, V., Goy, A., Vasselin, N., Mani, C.S.: Potential field based optimization of a prey-predator multi-agent system. In: MATHMOD 2018 Extended Abstract Volume, 9th Vienna Conference on Mathematical Modelling, Vienna, Austria, 21–23 February, pp. 127–128 (2018)
17. Dobramysl, U., Mobilia, M., Pleimling, M., Täuber, U.C.: Stochastic population dynamics in spatially extended predator–prey systems. J. Phys. Math. Theor. **51**(6), 063001 (2018)
18. Thie, P.R., Keough, G.E.: An Introduction to Linear Programming and Game Theory. Wiley, Hoboken (2011). 480 p.
19. Neogy, S.K., Bapat, R.B., Dubey, D.: Mathematical Programming and Game Theory. Springer, Singapore (2018). 226 p.
20. Kushner, H., George Yin, G.: Stochastic Approximation and Recursive Algorithms and Applications. Springer, New York (2013). 417 p.

Modernization of the Second Normal Form and Boyce-Codd Normal Form for Relational Theory

Oleksandr Rolik, Oleksandr Amons, Kseniia Ulianytska(✉), and Valerii Kolesnik

Department of Automation and Control in Technical Systems, National Technical University of Ukraine "Igor Sikorsky Kyiv Polytechnic Institute", Kyiv 03056, Ukraine
o.rolik@kpi.ua, amons@voliacable.com, ulianitskaya.k@gmail.com, kolesnik.valerii@gmail.com

Abstract. The task of designing relational databases has always been the subject of scientific research, as it is associated with a number of interrelated steps. The result of each step is the development of models for presenting the future database with further refinement and, finally, the creation of an adequate relational database as a set of relations with the corresponding links between them. The article focuses on the normalization of databases, as one of the steps to create a datalogical model, namely, the use of the first three Normal Forms. As a result of the analysis carried out in the article, it was concluded that the definition of the Second Normal Form can be modernized and thus achieve two goals: to ensure the correct creation of potential primary keys and, thereafter, the correct external connections between relations, even before creating the data schema in a specific relational database. Moreover, to reconsider the necessity of applying the so-called "strengthened" or a higher version of the Third Normal Form, which speaks of mutual dependencies between key and non-key attributes.

Keywords: Relational database · Relational theory · Normalization · Normal forms

1 Introduction

In the modern world of information technologies, a significant place is given to database design as a foundation for developing software applications of various directions - WEB applications, application programs, analytical programs, etc.

Relational databases are built on the principles of the relational data model, which was proposed by E.F. Codd in 1970 in his article «A Relational Model of Data for Large Shared Data Banks».

Codd used the term «tuple» instead of the widely used term «record» at that time. «Tuple» is final set of interrelated admissible attribute values that collectively describe an entity (table) [1, 8]. Today, thanks to this definition, such terms as «row», «record» and «tuple» have become synonymous.

Z. Hu et al. (Eds.): ICCSEEA 2020, AISC 1247, pp. 296–305, 2021.
https://doi.org/10.1007/978-3-030-55506-1_27

The implicit component of the relational database is normalization. Normalization is a process that is explicitly or implicitly present in all components of a relational database [4].

The article provides a practical analysis of the first three Normal Forms and Boyce-Codd Normal Form (BCNF) based on the following research methods:

- building datalogical models with different options for working out normal forms;
- testing functions of Data Manipulation Language (DML) such as INSERT, UPDATE, DELETE to exclude any data anomalies in datalogical model;
- creating and working with Class Diagram based on datalogical model with different options for working out normal forms for further research.

2 Problem Statement

The issue of this article is the analysis and adjustment of one of the main components of relational databases design, namely the Normal Forms. Normal Forms are levels of normalization process.

Normalization is the process of organizing a database to reduce redundancy and improve data integrity [2].

There are three common forms of database normalization: first, second, and third normal form (abbreviated as 1NF, 2NF, and 3NF respectively). In addition, this article provides analysis of BCNF like a subtle enhancement on 3NF.

3NF level with BCNF combination is strong enough to satisfy most applications, and the higher levels are rarely used, except in certain circumstances where the data (and its usage) requires it.

In this paper, we will reconsider and analyze a few normal forms and will find that some requirements can be excluded or upgrade and therefore reduce number of normal forms at least by one. It is about second normal form and BCNF.

To study this issue within the practical course of studying relational databases, a number of relational DBs were designed and students conducted laboratory studies. All tables are normalized to 3NF (tables are in BCNF by default like it will be shown). In most tables, a composite primary key is missing. Where there was a need for unique combinations of attributes, the integrity constraint of the unique key was applied.

3 Normalization in the Design Process of Relational Databases

In the process of design of relational databases, it is essential to perform normalization of relations or normalization of tables. As stated earlier, this process more frequently includes first three normal forms.

Correct analysis of subject field and design of logically correct models on the seconds and the third stages give an opportunity to normalize a database by the use of only three normal forms. However, for all other normal forms it will true that relations will align to them only when subject field model will be fully complied with physical model of the database obtained from a certain RDBMS.

This paper considers in more details second normal form because developers consider this form as a default one. In such way, BCNF becomes also simple form, which will be considered by default.

2NF definition – A relation is in 2NF if it is in 1NF and every non-prime attribute of the relation is dependent on the whole of every candidate key. In other words, there are no partial functional dependencies between non-prime attribute and parts of composite primary key [1];

Thus, in order to be in 2NF it is required for table to comply with 1NF and all non-prime attributes must functionally completely depend on all parts of compound primary key.

A certain attribute or set of attributes of a given set A is functionally dependent on X only when each combination of values from X corresponds to a single A value. It is described as X → A [5]. Cleary 2NF can be represented in the Fig. 1:

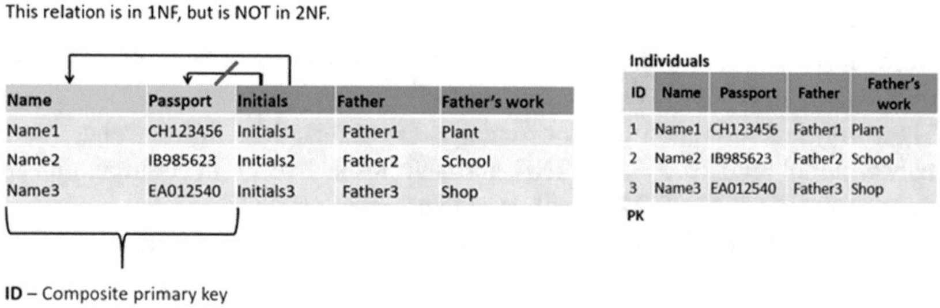

This relation is in 1NF, but is NOT in 2NF.

Name	Passport	Initials	Father	Father's work
Name1	CH123456	Initials1	Father1	Plant
Name2	IB985623	Initials2	Father2	School
Name3	EA012540	Initials3	Father3	Shop

ID – Composite primary key

a) Table without 2NF normalization

Individuals

ID	Name	Passport	Father	Father's work
1	Name1	CH123456	Father1	Plant
2	Name2	IB985623	Father2	School
3	Name3	EA012540	Father3	Shop

PK

b) Table in 2NF normalization

Fig. 1. Normalization process to 2NF

Now we adjust the understanding of 2NF at the level of initial datalogical design.

If we initially accept the fact that primary keys are indivisible values with an abstract value, then 2NF will be respected automatically.

3NF definition – A relation is in 3NF if it is in 2NF and every non-prime attribute of the relation is non-transitively dependent on every key of the relation [1].

Therefore, the relation or table is in 3NF if it is already in 2NF and each non-prime attribute depends on a primary key in non-transitional connection (indirect).

BCNF (Boyce-Codd Normal Form) is a subtle enhancement on 3NF. A relation is in BCNF if and only if for each of its dependencies X → Y, at least one of the following conditions hold:

- X → Y is a trivial functional dependency (Y⊆X)
- X is a superkey for schema R.

BCNF is based on functional dependencies that take into account all candidate keys in a relation.

To show what BCNF is, we took another table. Now Fig. 2 demonstrates the table which is not in BCNF:

ID2 – Composite PK

Parking reservation

Parking number	Start time	End time	Rate
56	12:30	12:55	Economy
56	9:45	10:30	Economy
56	18:00	19:00	Standard
12	10:00	12:00	PremiumB
12	15:15	17:15	PremiumB
12	20:35	23:55	PremiumA

ID1 – Composite PK **ID3** – Composite PK

Fig. 2. Table without BCNF normalization

A relation is in 2NF since all attributes are included in one of the potential keys, and there are no non-key attributes in relation. There are also no transitive dependencies, which corresponds to the requirements of the third normal form. Nevertheless, there is a functional dependence "Rate" → "Parking number" in which the left part (determinant) is not a potential key of the relation, that is, the relation is not in BCNF.

If we take into account the fact that 2NF is respected in the presence of only one integral primary key, and then the situation in which BCNF is not respected will not arise in the future.

As a conclusion, we can take the fact that if the definition of 2NF is changed, then BCNF can be simplified.

To prove the foregoing, let us conduct experiments with relational databases, and analyze the literature on this issue.

4 Literature Review in the Research Field

There are many articles, books and other materials devoted to Normal Forms and normalization as itself. First, there are materials made by E.F. Codd himself [1, 5, 8–10]. These materials represent normalization process, integrity constraints of relational database, database sublanguages and other – basics of relational theory.

In addition, there are many modern books devoted to relational theory and normalization process [11–13] and others.

There are materials related to the analysis of specific normal forms [14, 17], functional dependencies [15] and keys (complex primary key, foreign key) in particular [16]. In this material described terms, methods, and design questions of relational theory. All of them complement and do not contradict each other and all basics terms are unshakable.

However, there are some articles, where we can find hints of future changing of relational theory. In [16] among other things, this paper includes a discussion of the pros and cons of surrogate keys, which leads to revision of, for example, 2NF.

An extensive and detailed treatment of the problems caused by duplicates. The discussion of duplicates directly connected to integrity constraints such as primary, foreign, unique keys [14].

Eventually, internet article was found by Konrad Zdanowski which called "A Unified View on Database Normal Forms: From the Boyce-Codd Normal Form to the Second Normal Form (2NF, 3NF, BCNF)" [18]. This article refers to [12] and author of this article does not hesitate about NF definition, but only one paragraph made me think about revision some normal forms:

"In SQL one distinguishes one key in a table by labeling it as PRIMARY KEY. You can also point other keys in a table by labeling some sets of attributes as UNIQUE NOT NULL. Moreover, this really helps a database manager to process with data. In relation theory there is no need to treat primary key in a special way so let me just stress that we may have many keys in one relation. If one can extend a set of attributes X to a key in R then X is a sub key in R. It is not always possible since X may be, e.g., too big to be contained in a key which is not a superkey" [18].

Recent articles on normal forms report that appears new normal forms MRDANF [19], using basics normal forms without their revise [20, 21].

As will be shown below, working with complex primary key is very difficult. Complex primary keys are used in a very specific way, they are not being used as a foreign key in another tables.

Modern relational DBMS (such as Oracle and MS SQL Server) do not allow several primary keys in a table, so functional dependencies between primary keys are impossible by default. Even for this reason, not mention the complexity of datalogical modeling, it is worth revising the definition of 2NF and BCNF.

5 Experiment and Results

As part of the experiment, a number of RDBs were designed. Figure 3 depicts datalogical model of a sample data.

Datalogical model (Fig. 3) consists of five tables, all related, each table has an integral primary key.

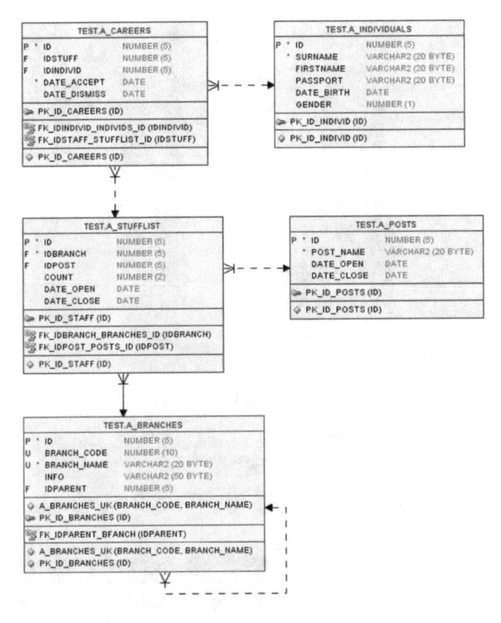

Fig. 3. Datalogical model of HR department

Three tables contain so-called candidate keys or pointer attributes of infological model (Fig. 4).

A_INDIVIDUALS	PASSPORT	Varchar2(20)	Denotative	Uniquely describes individual, candidate key
A_POSTS	POST_NAME	Varchar2(20)	Denotative	Uniquely describes post, candidate key
A_BRANCHES	BRANCH_CODE	Number (10)	Denotative	Uniquely describes branch, candidate key

Fig. 4. Candidate keys for database HR department

In the table A_BRANCHES there is one more condition which should be met and that is combination of fields BRANCH_CODE and BRANCH_NAME does not have to repeat. That means if the table has two branches names like "Tool design", their BRANCH_CODE would be different like it shown on Fig. 5.

A UNIQUE KEY has been created for this table with the next query:

ALTER TABLE "A_BRANCHES" ADD CONSTRAINT "A_BRANCHES_UK" UNIQUE ("BRANCH_CODE", "BRANCH_NAME");

The A_BRANCHES_UK is one of the integrity constraints of RDB, which actually is possible primary key of the table. The unique key is not a reference but it is not part of connections.

ID	BRANCH_CODE	BRANCH_NAME	INFO	IDPARENT
14	491	Tool Design	(null)	10
1	23	Administration	Main branch	(null)
2	14	Human Resources	(null)	(null)
3	76	Engineering	(null)	1
4	76	Production	(null)	1
5	332	Sales	(null)	4
6	332	Shipping	(null)	5
11	17	Finance	(null)	4
9	777	Purchasing	(null)	11
10	832	Research	(null)	11
7	49	Tool Design	(null)	(null)
12	17	QA	(null)	7

```
SELECT * FROM A_BRANCHES
    WHERE BRANCH_NAME = 'Tool Design';
```

ID	BRANCH_CODE	BRANCH_NAME	INFO	IDPARENT
7	49	Tool Design	(null)	(null)
14	491	Tool Design	(null)	10

Fig. 5. Table "A_BRANCHES" with query data.

We give two definitions and explanations for them:

Candidate key is not a key attribute – it is denotative attribute, which also like PRIMARY KEY can uniquely determine entity instance. Candidate key – is a logically connected attribute with some instance, like in our case PASSPORT, BRANCH_-CODE, POST_NAME, etc. In all definition of normal forms, this statement is implicitly confirmed [3, 6, 12].

In practical use, in mostly modern RDBMS a surrogate key is taken as the primary key. Primary key is generated uniquely each time a new record is inserted [22].

All tables have 2NF since DB does not have compound primary keys according to the next command:

```
ALTER TABLE "A_BRANCHES"
ADD CONSTRAINT "PK_ID_BRANCHES" PRIMARY KEY ("ID");
ALTER TABLE "A_POSTS" ADD CONSTRAINT "PK_ID_POSTS"
PRIMARY KEY ("ID");
```

If it is required to emphasize a uniqueness of one or several attributes then modern RDBMS makes those as UNIQUE KEY as shown for table A_BRANCHES.

The possible reason of application of compound key is unbundling of connection many to many (Fig. 6). Compound key in this case is applied for two fields of foreign keys in the unbundling table.

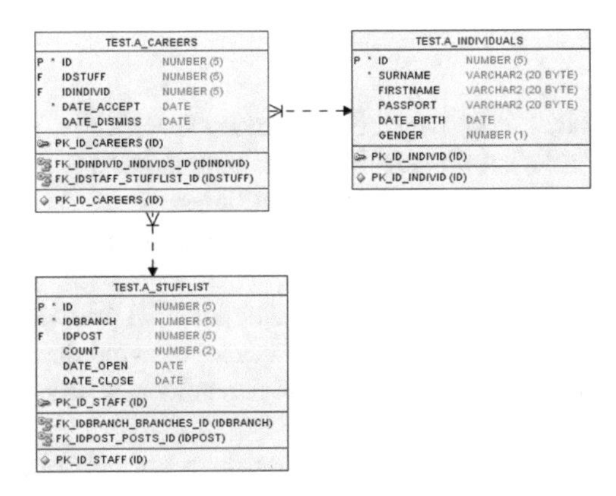

Fig. 6. Tables with many to many dependency and complex primary key

Based on mentioned information the statement of 2NF can be reconsidered what will be done in the result.

Tables of given DB are in BCNF since the possibility to create several primary keys in RDBMS does not exist from one side.

Each row in the table represents a parking reservation unit in parking. That parking has some parking place with types only Economy or Standard, another parking place with types PremiumA or PremiumB.

According to the rules of RDB development, the given model was required to be presented in the next way (Fig. 7):

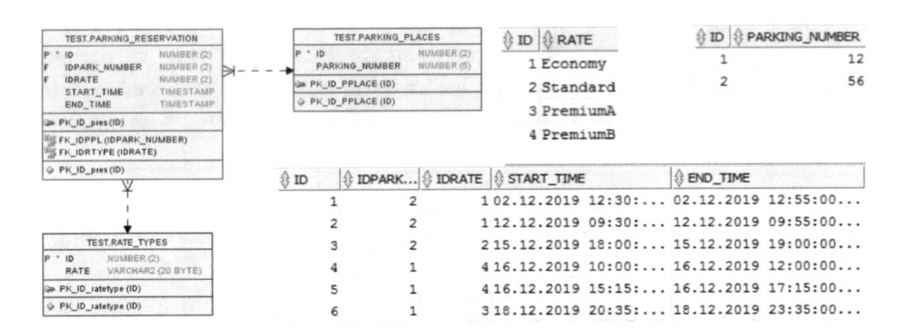

ID	RATE
1	Economy
2	Standard
3	PremiumA
4	PremiumB

ID	PARKING_NUMBER
1	12
2	56

ID	IDPARK...	IDRATE	START_TIME	END_TIME
1	2	1	02.12.2019 12:30:...	02.12.2019 12:55:00...
2	2	1	12.12.2019 09:30:...	12.12.2019 09:55:00...
3	2	2	15.12.2019 18:00:...	15.12.2019 19:00:00...
4	1	4	16.12.2019 10:00:...	16.12.2019 12:00:00...
5	1	4	16.12.2019 15:15:...	16.12.2019 17:15:00...
6	1	3	18.12.2019 20:35:...	18.12.2019 23:35:00...

Fig. 7. Normalized tables, which meet BCNF

Model with data represented on Fig. 7 correctly designed and normalized to BCNF.

With consideration of such manipulations and possible simplification of database design without compound keys or with minimal or rare availability of such keys it is possible to change formulation of 2NF and BCNF.

6 Conclusions

Let us look at so-called improved 3NF or more specifically BCNF. This form is possible only when relation already has 3NF and it has no dependencies of keys (attributes of compound key) on non-prime attributes.

Relation is in BCNF when determinants of all functional dependencies are potential keys.

In other words, this normal form suppose that potential primary key should not be a value type and moreover not compound one in order to avoid dependencies of primary key parts on non-prime attributes and to avoid dependencies inside compound primary keys.

Alternative approach for distinction of first three normal forms and Boyce-Codd normal form can be concluded in reducing the number of normal forms on one with taking into account following formulation:

Formulation of the first and the third normal form will remain the same and formulation of the second one will be changed:

Relation is in 2NF if it is in 1NF and as for primary key it uses integral non-prime for relation attribute with automatic generation of the following value.

According to this wording, the applications of the improved third normal form are negated. Therefore, on the stage of infological design within definition of restrictions for integrity it will be possible to introduce integral primary key and take non-prime attribute for this key.

The BCNF offsets its value because of that all nuances connected with big amount of potential keys are being eliminated through correct datalogical design and such integrity constraints as UNIQUE KEY and CHECK.

If to put this principle on the design stage of infological model then physically developed relational database will be free from significant problems of redundancy connected with duplication of important information (which is often included in primary and as a result in secondary keys) and also from problems of anomalies with insertion, modification and deletion.

References

1. Codd, E.F.: Normalized data base structure: a brief tutorial. In: Proceedings of 1971 ACM-SIGFIDET Workshop on Data Description, Access and Control, pp. 1–17. ACM, New York (1971)
2. The basics of relational databases. Transl. from eng: Publishing and trading house, Russian Edition (2001). 384 p.
3. Guschin, A.: Data bases, 2nd edn. corrected and add: teaching aid. Direct-Media, Moscow (2015), 311 p.
4. Date, C.J.: An Introduction to Database Systems, 6th edn. Addison-Wesley, Reading (1995)
5. Codd, E.F.: Further Normalization of the Data base Relational Model. In: Data Base Systems, pp. 33–64. Prentice-Hall, New Jersey (1972)
6. Fagin, R.A.: Normal form for relational databases that is based on domains and keys. ACM Trans. Database Syst. **6**(3), 387–415 (1981)

7. Calenko, M.: Modeling Semantics in Databases. Science, Moscow (1988)
8. Codd, E.F.: Relational completeness of data base sublanguages. In: Rustin, R.J. (ed.) Data Base Systems, Courant Computer Science Symposia Series 6. Prentice Hall, Englewood Cliffs (1972)
9. Codd, E.F.: Derivability, redundancy, and consistency of relations stored in large data banks. IBM Research Report RJ599, 19 August 1969
10. Codd, E.F., Date, C.J.: Much ado about nothing. In: Date, C.J. (ed.) Relational Database Writings 1991–1994. Addison-Wesley, Reading (1995)
11. Date, C.J.: SQL and Relational Theory: How to Write Accurate SQL Code, 2nd edn., p. 446. O'Reilly Media, Inc., Sebastopol (2012)
12. Date, C.J.: Database Design and Relational Theory, p. 449. O'Reilly Media, Inc., Sebastapol (2012)
13. Harrington, J.L.: Relational Database Design and Implementation: Clearly Explained, 3rd edn., p. 418. Morgan Kaufmann, San Francisco (2009)
14. Date, C.J.: Date on Database: Writings 2000–2006. Apress, Berkeley (2006)
15. Darwen, H.: The role of functional dependence in query decomposition. In: Date, C.J., Darwen, H. (eds.) Relational Database Writings 1989–1991. Addison-Wesley, Reading (1992)
16. Date, C.J.: Composite keys. In: Date, C.J., Darwen, H. (eds.) Relational Database Writings 1989–1991. Addison-Wesley, Reading (1992)
17. Date, C.J.: Database Design and Relational Theory: Normal Forms and All That Jazz, p. 450. APRESS, Healsburg (2019)
18. Zdanowski, K.: A Unified View on Database Normal Forms: From the Boyce-Codd Normal Form to the Second Normal Form (2NF, 3NF, BCNF), Vertabelo (2015)
19. Alotaibia, Y., Ramadan, B.: A novel normalization forms for relational database design throughout matching related data attribute. Int. J. Eng. Manuf. (IJEM), 65–72. Published Online September 2017 in MECS
20. Zhuchenkoa, A.I., Osipab, L.V., Cheropkin, E.S.: Design database for an automated control system of typical wastewater treatment processes. Int. J. Eng. Manuf. (IJEM), pp. 36–50. Published Online July 2017 in MECS
21. Rizwan Jameel Qureshi, M., Shaik, M.S., Iqbal, N.: Using fuzzy logic to evaluate normalization completeness for an improved database design. Int. J. Inf. Technol. Comput. Sci. (IJITCS), pp. 48–55. Published Online March 2012 in MECS
22. Dewson, R.: SQL Server 2008 Express for Developers: From Novice to Professional, p. 508. APRESS (2009)

Abnormal Interference Recognition Based on Rolling Prediction Average Algorithm

Ya-Hu Yang[1], Jia-Shu Xu[2], Yuri Gordienko[2], and Sergii Stirenko[2](✉)

[1] Beijing Technology and Business University, Beijing 100048, China
yyh_19940317@163.com

[2] National Technical University of Ukraine "Igor Sikorsky Kyiv Polytechnic Institute", Kyiv, Ukraine
jiashu.xu.1@gmail.com, {gord,stirenko}@comsys.kpi.ua

Abstract. Aiming at the problems that the traditional camera abnormal interference recognition methods have a single identification type, the low recognition accuracy and reliability, the generalization ability is not strong due to the prediction of flicker. A method for camera abnormal interference recognition based on the rolling prediction average algorithm is proposed. The core ideas of this method include the following steps. Firstly, the ImageNet pre-trained ResNet50 network on the self-built abnormal interference image training set is fine-tuned. Then a model for camera abnormal interference image classification and recognition is trained. Finally, the rolling prediction average algorithm is applied to the model to realize the classification and recognition of the camera abnormal interference video online or offline. The experimental results on the self-built abnormal interference image verification set show that the proposed method can correctly identify the normal images, occlusion images, blurred images and the camera rotated images. The recognition accuracy reaches 97%. Test accuracy reaches 95% when using test set videos test, which fully verifies the feasibility and effectiveness of the proposed method. Compared with other abnormal interference recognition methods, not only the accuracy rate is high, but also the false recognition rate and missed detection rate are relatively low. The proposed method can be applied to actual monitoring scenarios to assist the work of monitoring center staff.

Keywords: Intelligent video surveillance · Convolutional neural network · Deep learning · Smart city · Public safety

1 Introduction

As the front-end surveillance cameras of the intelligent video surveillance system, it is very important to effectively identify the abnormal interference such as occlusion, blur and rotation of the surveillance camera lens. The development of image processing technology provides a new method for the abnormal interference recognition of the surveillance cameras [1]. The traditional surveillance camera abnormal interference recognition methods mostly identify only one kind of abnormal interference [2–5]. However, as the number of surveillance camera has increased dramatically, the image data generated by intelligent video surveillance systems have also increased

Z. Hu et al. (Eds.): ICCSEEA 2020, AISC 1247, pp. 306–316, 2021.
https://doi.org/10.1007/978-3-030-55506-1_28

exponentially. The traditional methods that manually extract features and then use SVM (Support Vector Machine, SVM) and other classifiers to classify and identify camera interference not only consumes a lot of resources but also is difficult to meet the accuracy and reliability requirements of video surveillance.

In recent years, the deep learning technology based on massive data has carried out higher-level and more abstract expressions of raw data through nonlinear models, and has performed well in many tasks in computer vision such as image classification, target detection and recognition [6]. At present, among the image classification methods based on the Convolutional Neural Network (CNN), there are representative end-to-end image classification networks such as VGG16, VGG19, SqueezeNet, InceptionV4, DenseNet121, ResNet18 and ResNet50.

Therefore, this paper draws on the excellent image classification performance of the ResNet50 network. Firstly, we use the self-built surveillance camera image dataset to fine-tune the ImageNet pre-trained ResNet50 network to train the image classification and recognition model for surveillance camera abnormal interference. Finally, based on the trained model, the rolling prediction average algorithm is used to realize the classification and recognition of the surveillance camera abnormal interference video online and offline.

2 Background

The deeper the network, more information you will get and the features will be richer. However, according to experiments [7], as the network deepening, the optimization effect worsens, and the accuracy of the test set and training set is reduced. This is due to the deepening of the network will cause gradient explosions and gradient to disappear, resulting in deep networks failure to train. To this end, ResNet protects the integrity of the information by directly bypassing the input information to the output. The entire network only needs to learn the part of the input and output differences, which simplifies the learning objectives and difficulty.

Figure 1 shows the residual learning module of the ResNet network.

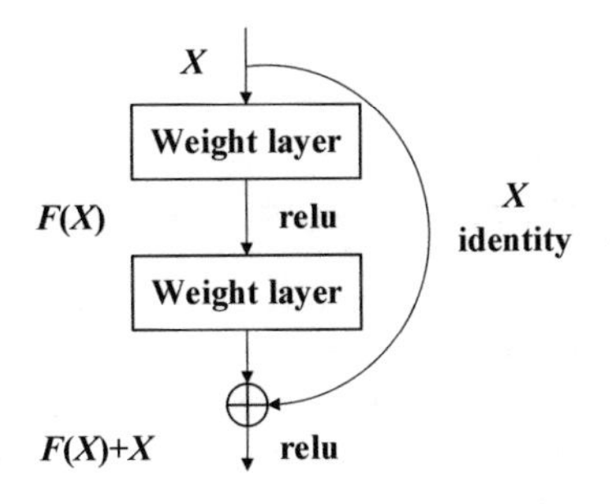

Fig. 1. ResNet residual learning model.

There are two kinds of mappings in ResNet, one is identity mapping, which is *X* shown in Fig. 1. The other is residual mapping, which is *F*(*X*) in shown Fig. 1. The target mapping to solve is *F*(*X*). According to Fig. 1

$$H(X) = F(X) + X \tag{1}$$

$$F(X) = H(X) - X \tag{2}$$

Therefore, the problem is transformed into solving the residual mapping function *F*(*X*) of the network, which simplifies the goal and difficulty of learning. Since the channels of *F*(*X*) and *X* are not all the same, the Eq. (1) has another expression, as shown in Eq. (3):

$$H(X) = F(X) + WX \tag{3}$$

Among them, *W* is the convolution operation, which is mainly used to adjust the channel dimensions of the *X*. Two residual modules are used in the ResNet network structure, as shown in Fig. 2.

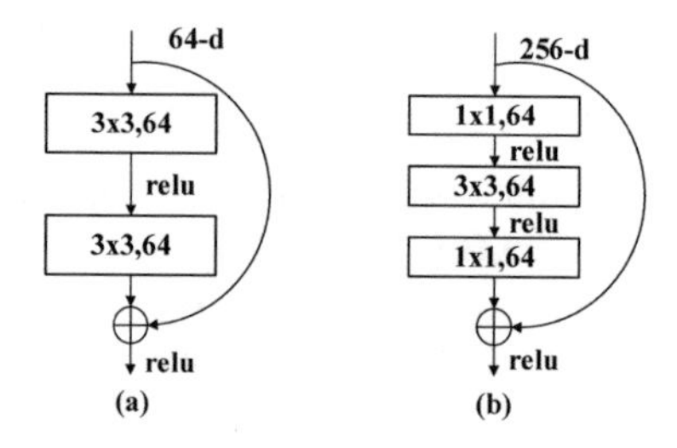

Fig. 2. (a) Two-layer and (b) Three-layer residual learning module.

The structure of Fig. 2(a) is mainly for the ResNet34 network, and the structure of Fig. 2(b) is mainly for the ResNet50, ResNet 101, and ResNet152 networks.

3 Method

3.1 Fine-Tuning ImageNet Pre-trained ResNet50 Network

It turns out that another type of migration learning, fine-tuning, can lead to a higher accuracy than migration learning by feature extraction. This paper mainly uses the self-built dataset to fine-tune the ImageNet [8] pre-trained ResNet50 network to train the image classification and recognition model of camera interference. The visualization process is shown in Fig. 3. The specific steps are as follows:

1. Delete the FC (Fully Connected, FC) layer and Softmax layer at the end of the pre-trained ResNet50 network, as shown in Fig. 3(a);

2. Replace the original FC layer and the Softmax layer with a new set of randomly initialized FC layer and Softmax layer, as shown in Fig. 3(b);
3. Freeze all CONV (convolutional) layers under the FC layer and the Softmax layer of the pre-trained ResNet50 network to ensure that any previous powerful features mastered by the pre-trained ResNet50 are not destroyed, as shown in Fig. 3(c);
4. Train ResNet50 with a small learning rate (0.0001), but only train the FC layer and the Softmax layer, as shown in Fig. 3(c);
5. Unfreeze all CONV layers in the pre-trained ResNet50 network and perform a second training, as shown in Fig. 3(d).

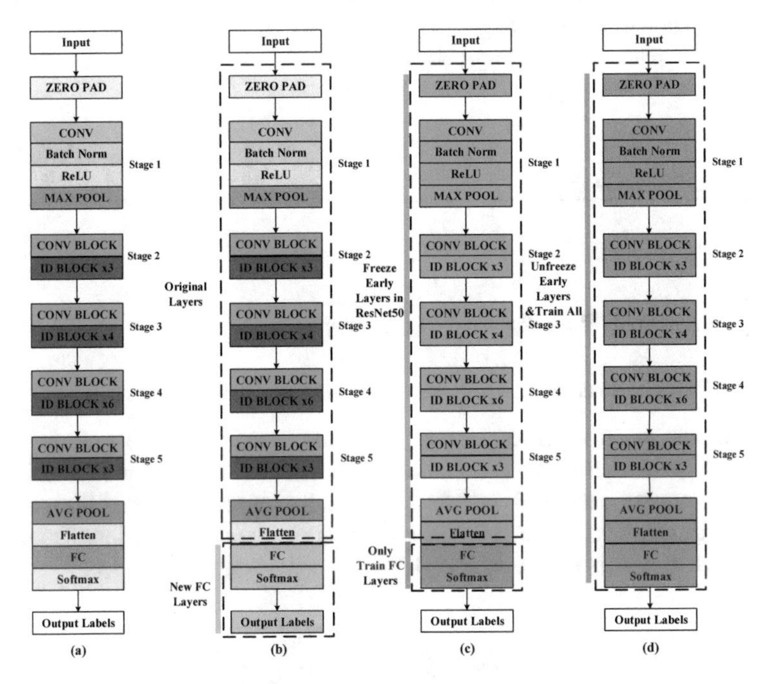

Fig. 3. Processes of fine-tuning the pre-trained ResNet50. (a) Original ResNet50 network structure, (b) Replace the original FC layer and Softmax layer, (c) Freeze the CONV layers below the FC layer and train the FC layer and the Softmax layer, (d) Unfreeze the CONV layers and train again.

3.2 Rolling Prediction Averaging Algorithm

The traditional methods encounter "predictive flicker" and may ignore context information of videos. This paper proposes a method that uses the rolling prediction average algorithm to realize the video classification and recognition of abnormal interference of surveillance cameras. It is based on the model trained for surveillance camera interference image classification and recognition with the following algorithm.

Algorithm: Realization of surveillance camera abnormal interference video classification and recognition by rolling prediction average.

input: Surveillance camera interference video file
output: Surveillance camera interference video and its corresponding label

```
1  for i in range(n):
2      For each frame, pass the frame through ResNet50 fine-tuned on our self built dataset;
3      Obtain the predictions from ResNet50 fine-tuned on our self built dataset;
4      Maintain a list of the last K predictions;
5      Compute the average of the last K predictions and choose the label with the largest
       corresponding probability;
6      Label the frame and write the output frame to disk;
7      Show the output image;
8      i += 1;
9  end
10 Release the file pointers.
```

4 Experimental Results and Analysis

a) Establishing Surveillance Camera Interference Image Dataset

In the research of surveillance camera abnormal interference recognition, there is currently no public data set for classification and recognition model training of surveillance camera abnormal interference. In this paper, the surveillance camera interference image data set was constructed. The surveillance camera interference video acquisition workflow is shown in Fig. 4.

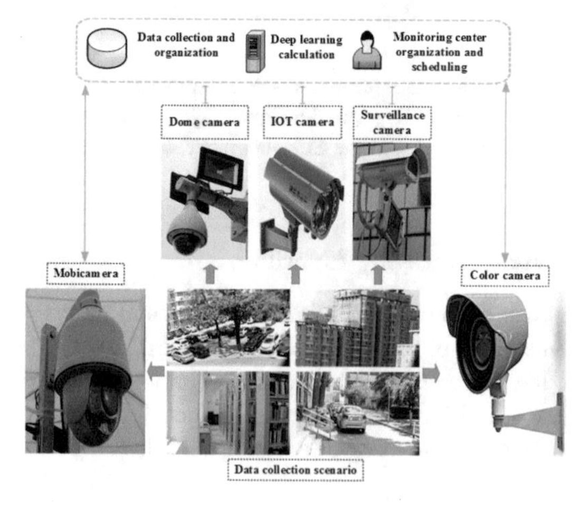

Fig. 4. Data acquisition process under video surveillance platforms.

After interference video collection, slicing and image enhancement, there are 5,400 images of 500 pixel × 281 pixel in the data set and the percentage of each category is 25%. Which illustrates that this dataset has a good balance in learning different categories. In the end, 75% was used as the training set and 25% was used as the verification set, that is, 4,050 images of the training set, and 1,350 images of the verification set. Training and verification were performed on the pre-trained ResNet50 network, and finally the trained model was combined with the rolling prediction average algorithm to perform test on 80 unknown surveillance camera interference videos.

b) **Evaluation of Surveillance Camera Interference Recognition Model**

During training, the training batch size is 64, and the ImageNet pre-trained ResNet50 network is fine-tuned, using an SGD (Stochastic Gradient Descent, SGD) optimizer with an initial learning rate of 0.0001, a momentum of 0.9, and a decline of 0.0001 in learning rate per 50 iterations. The image size of the training set and the validation set are both 500 × 281 pixels. The model is based on the deep learning framework Keras 2.2.4 and training, validation and test are performed on an Intel(R) Xeon(R) E5-2603v3@1.60 GHz processor with 256G RAM and a NVIDIA Titan XP GPU of 12G memory.

The loss function is used to estimate the degree of inconsistency between predicted value and actual value of the model. It is a non-negative real-valued function. The loss function used in this paper is the cross-entropy loss function.

The smaller the loss function, the closer the distribution of the two probabilities, the better the robustness of the model. Figure 5 shows the relationship between loss function and learning rate on ResNet34 and ResNet50 networks.

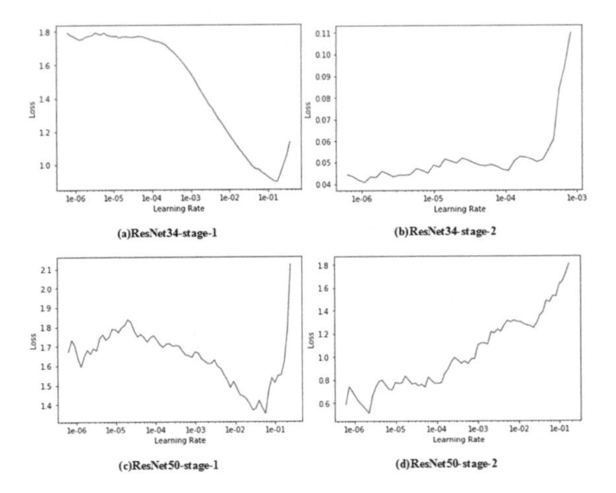

Fig. 5. The relationship between loss function and learning rate on ResNet34 and ResNet50 networks: (a) ResNet34, the initial learning rate, (b) ResNet34, the selected optimal learning rate, (c) ResNet50, the initial learning rate, (d) ResNet50, the selected optimal learning rate.

The learning rate of the ResNet50 network has a greater impact on the loss function than ResNet34 (Fig. 5). This also fully demonstrates that we can improve the fitting of the ResNet50 network by adjusting the learning rate. Evolution of accuracy and loss values for training and validation is shown in Fig. 6.

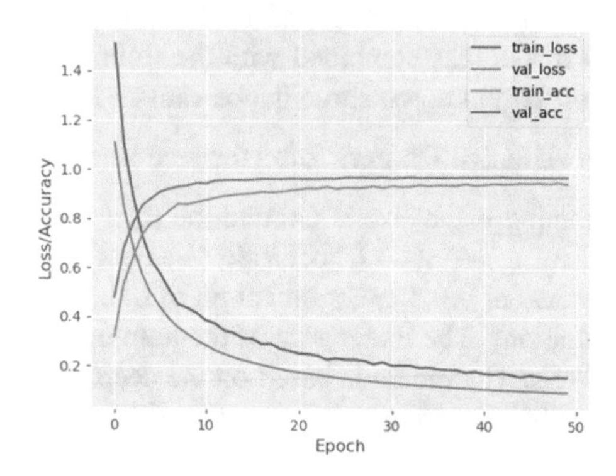

Fig. 6. Accuracy and loss function graph for training and validation processes.

From the accuracy and loss function graph of the training and validation process, the ResNet50 network after being fine-tuned by the self-built data set can accurately classify the self-built dataset, and finally achieves a classification accuracy of 97% on the validation set. And the change of the loss function is stable, and eventually reaches 0.1612 at the 50th epoch, which indicates that the predicted value has been very close to the true value.

The metrics of the model trained on the self-built dataset are shown in Table 1.

Table 1. Verification set evaluation result.

	Precision	Recall	F1-score	Support
Blurry	1.00	1.00	1.00	337
Normal	0.91	0.97	0.94	338
Occlusion	0.99	0.92	0.95	337
Scene_switching	0.97	0.98	0.98	338
Micro avg	0.97	0.97	0.97	1350
Macro avg	0.97	0.97	0.97	1350
Weighted avg	0.97	0.97	0.97	1350

The trained model has a good classification performance for the self-built data set (Table 1). The 1350 verification set images were verified, including 338 normal images, 337 blurry images, 337 occlusion images, and 338 camera rotation images. Each type of *precision*, *recall*, and *f1-score* are very close to 1, indicating that the classification performance of the model is good. The *micro avg*, *macro avg*, and *weighted avg* were both 0.97, achieving 97% verification set classification accuracy.

c) **Surveillance Camera Interference Video Recognition Results and Analysis**

The performance of the trained classification model was evaluated in detail on the validation set. Then 80 test set videos were tested in combination with the rolling prediction average algorithm. Finally, the common evaluation indexes of the video surveillance such as Accuracy (Accuracy, Acc), True Alarm Rate (True Alarm Rate, TAR) and False Alarm Rate (False Alarm Rate, FAR) were used to check the recognition performance of the proposed algorithm and the trained model on the surveillance camera interference videos. Some parts of the test set video recognition results are shown in Fig. 7. The total test set video recognition results are shown in Table 2.

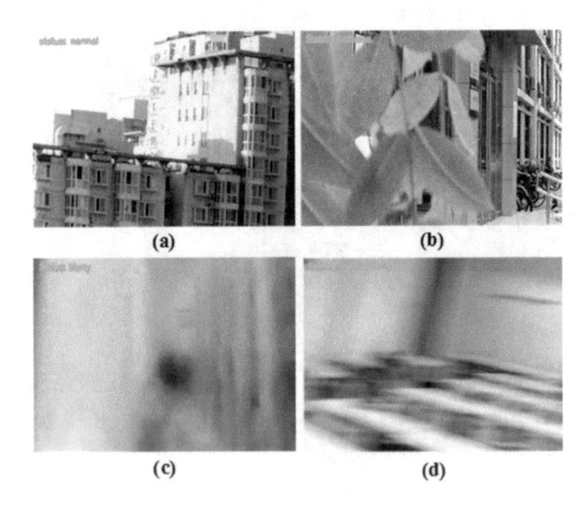

Fig. 7. Parts of the test set video recognition results. (a) Normal video recognition results, (b) Occlusion video recognition results, (c) Blur video recognition results, (d) Rotated video recognition results.

As shown in Fig. 7, including normal surveillance videos, not only the various surveillance camera interference videos are labeled with their corresponding labels, but also different interference categories in the same interference video can also be quickly classified and recognized.

Table 2. Test set videos recognition result.

Video type	Total number of tests	Total number of recognition	Unidentified number
Normal video	20	20	0
Occlusion video	20	20	0
Blurry video	20	19	1
Rotated video	20	17	3
Total	**80**	**76**	**4**

According to Table 1 and Table 2, the classification effect of normal images and occlusion images is less better than that of blurred images and scene switching images, but in video recognition, the recognition effect of normal videos and occlusion videos is better than that of blurred videos and rotated videos. It also proves that video recognition takes into account not only the apparent features (i.e., static features) but also the timing features (i.e., dynamic features), while the image classification recognition only needs to take into account the apparent features.

In order to objectively evaluate the classification and recognition results of the surveillance camera abnormal interference videos, objective evaluation indexes in [2] such as Acc, TAR and FAR are used in this paper. The calculated results of Acc, TAR and FAR of various types of videos and the overall videos are shown in Table 3.

Table 3. Objective evaluation from Acc, TAR and FAR.

Video type	Acc (%)	TAR (%)	FAR (%)
Normal video	100	0	0
Occlusion video	100	0	0
Blurred video	95	5	15
Rotated video	85	15	5
Total	**95**	**5**	**6.7**

It can be seen from Table 3 that the method achieves 95% of the surveillance camera abnormal interference video recognition accuracy, and the TAR and FAR are relatively low, respectively 5% and 6.7%.

In order to further verify the effectiveness of the proposed method, this paper performed comparative experiments on dataset with the traditional and step-by-step camera interference recognition methods. The comparison results are shown in Table 4.

Table 4. Comparison of experimental results of different methods.

Method	Acc (%)	TAR (%)	FAR (%)	Targeted video category
SVM [2]	84.20	11.78	16.03	Normal, occlusion
SAE+DBN [3]	88.46	10.65	11.63	Normal, occlusion
Improved LSD [4]	89.12	12.43	9.92	Normal, occlusion
ORB [9]	83.75	16.25	21.70	Normal, occlusion, blurry, rotated
The proposed method (Fine-tuning+RPA)	**95**	**5**	**6.70**	Normal, occlusion, blurry, rotated

According to the results of the comparative experiments, it can be seen that the surveillance camera abnormal interference categories are recognized using the proposed method are richer than that recognized using the traditional and step-by-step recognition methods of the surveillance camera abnormal interference, but the Acc is higher, and the TAR and FAR are lower respectively. It follows from: 1. By fine-tuning the ImageNet pre-trained ResNet50 network, the powerful previous knowledge of the ImageNet is used reasonably. 2. Inspired by action recognition in videos [10], this paper applies the rolling prediction average algorithm to the abnormal interference recognition of the surveillance camera, achieving a new breakthrough in interference recognition of the surveillance cameras. In general, it allows us to use this approach in the wider range of applications, like in our previous works for preliminary digital diagnostics of various diseases [11], on-demand digitalization of arts and heritage [12], optimized estimation of road traffic conditions [13], etc.

5 Conclusions

This paper proposes using the rolling prediction average algorithm, combined with the image classification and recognition model of the trained camera interference, to realize the video classification and recognition of camera interference in intelligent video surveillance. In the future, we plan to further improve recognition accuracy by combining action recognition in videos and develop an excellent abnormal interference detection system [13–18]. From the practical point of view this approach can be very promicing for on-device inference implementation of optimized networks [19] for various applications and on various platforms including standard graphic processing units (GPU) and the newly available specialized tensor processing (TP) architectures [20, 21].

References

1. Huang, K., Chen, X., Kang, Y., et al.: Intelligent visual surveillance: a review. Chin. J. Comput. **38**(6), 1093–1118 (2015)
2. Yuan, Y., Sheng, D., Xin, X., Li, C.: Support vector machine based approach for leaf occlusion detection in security surveillance video. J. Comput. Appl. **34**(7), 2023–2027 (2014). (in Chinese)
3. Wu, M., Chen, L.: Deep learning based approach for detecting leaf occlusion in surveillance videos. J. Wuhan Univ. Sci. Technol. **39**(1), 69–74 (2016)
4. Tao, L., Li, C., Hui, N.: Utility pole occlusion algorithm based on improved line segmentation detection. Comput. Eng. **43**(9), 250–255 (2017)
5. Huang, S.: Neural network based blur measurement and segmentation for partially blurred image, pp. 23–29. Anhui University, Hefei (2016)
6. Wang, L.M., Xiong, Y.J., Wang, Z., et al.: Temporal segment networks: towards good practices for deep action recognition. In: 14th European Conference on Computer Vision (ECCV 2016), pp. 20–36. Springer (2016)

7. He, K.M., Zhang, X.Y., Ren, S.Q., et al.: Deep residual learning for image recognition. In: Proceedings of the 29th IEEE Conference on Computer Vision and Pattern Recognition (CVPR 2016), pp. 770–778. IEEE (2016)
8. Deng, J., Dong, W., Socher, R., et al.: ImageNet: a Large-Scale Hierarchical Image Database. In: 2009 IEEE Conference on Computer Vision and Pattern Recognition (CVPR 2009), pp. 248–255. IEEE (2009)
9. Xiang, L.: Design and implementation of video anomaly diagnostic system, pp. 24–44. South China University of Technology, Guangzhou (2017)
10. Wang, L., Xiong, Y., Wang, Z., et al.: Temporal segment networks for action recognition in videos. IEEE Trans. Pattern Anal. Mach. Intell. **41**(11), 2740–2755 (2018)
11. Stirenko, S., Kochura, Y., Alienin, O., Rokovyi, O., Gordienko, Y., Gang, P., Zeng, W.: Chest X-ray analysis of tuberculosis by deep learning with segmentation and augmentation. In: 2018 IEEE 38th International Conference on Electronics and Nanotechnology (ELNANO), pp. 422–428. IEEE (2018)
12. Gordienko, N., Gang, P., Gordienko, Y., Zeng, W., Alienin, O., Rokovyi, O., Stirenko, S.: Open source dataset and machine learning techniques for automatic recognition of historical graffiti. In: International Conference on Neural Information Processing (ICONIP), pp. 414–424. Springer, Cham (2018)
13. Taran, V., Gordienko, Y., Rokovyi, A., Alienin, O., Stirenko, S.: Impact of ground truth annotation quality on performance of semantic image segmentation of traffic conditions. In: International Conference on Computer Science, Engineering and Education Applications (ICCSEEA), pp. 183–193. Springer, Cham (2019)
14. Mishra, P.K., Saroha, G.P.: A study on classification for static and moving object in video surveillance system. Int. J. Image Graph. Signal Process. (IJIGSP) **5**, 76–82 (2016)
15. Chandrajit, M., Girisha, R., Vasudev, T.: Motion segmentation from surveillance video using modified hotelling's T-Square statistics. Int. J. Image Graph. Signal Process. (IJIGSP) **7**, 41–48 (2016)
16. Singh, S., Saurav, S., Shekhar, C., et al.: Moving object detection scheme for automated video surveillance systems. Int. J. Image Graph. Signal Process. (IJIGSP) **7**, 49–58 (2016)
17. Singh, G., et al.: Crowd escape event detection via pooling features of optical flow for intelligent video surveillance systems. Int. J. Image Graph. Signal Process. (IJIGSP) **10**, 40–49 (2019)
18. Natalia Chaudhry, Kh., Umar Suleman, M.: IP camera based video surveillance using object's boundary specification. Int. J. Image Graph. Signal Process. (IJIGSP), **8**, 13–22 (2016)
19. Gordienko, Y., Kochura, Y., Taran, V., Gordienko, N., Bugaiov, A., Stirenko, S.: Adaptive iterative pruning for accelerating deep neural networks. In: 2019 XIth International Scientific and Practical Conference on Electronics and Information Technologies (ELIT), pp. 173–178. IEEE, September 2019
20. Kochura, Y., Gordienko, Y., Taran, V., Gordienko, N., Rokovyi, A., Alienin, O., Stirenko, S.: Batch size influence on performance of graphic and tensor processing units during training and inference phases. In: International Conference on Computer Science, Engineering and Education Applications, pp. 658–668. Springer, Cham (2019). https://doi.org/10.1007/978-3-030-16621-2_61
21. Gordienko, Y., Kochura, Y., Taran, V., Gordienko, N., Rokovyi, A., Alienin, O., Stirenko, S.: Scaling analysis of specialized tensor processing architectures for deep learning models. In: Deep Learning: Concepts and Architectures, pp. 65–99. Springer, Cham (2020). https://doi.org/10.1007/978-3-030-31756-0_3

Modified Generators of Poisson Pulse Sequences Based on Linear Feedback Shift Registers

V. Maksymovych[1], O. Harasymchuk[1], and M. Shabatura[1,2(✉)]

[1] Lviv Polytechnic National University, St. Bandery 12, Lviv, Ukraine
volodymyr.maksymovych@gmail.com,
oleh.harasymchuk@gmail.com, mandrona27@gmail.com
[2] Lviv State University of Life Safety, Lviv, Ukraine

Abstract. In this article are suggested the structures of generators of Poisson pulse sequences based on Linear Feedback Shift Registers. The results of their simulation modeling show, that statistical characteristics of output pulse sequence correspond to Poisson law of distribution. In addition, in this article have been defined the limits of control code and a repetition period of pseudorandom sequence for concrete parameters of structural scheme. The distinctive feature of the proposed generators is that they allow to expand the range of medium values of the output pulse sequence frequency. This makes it possible to extend the range of radiation intensity that is simulated to ensure compliance of the impulse flow to the Poisson distribution law.

Keywords: Imitation of radioactivity · Pseudorandom sequence · Pseudorandom number generator · Generators of Poisson pulse sequences · Linear Feedback Shift Registers · Pearson's chi-squared test

1 Introduction

The Poisson flux has got wide usage among different branches of science and technology. In particular, it is widely used in: economical, military, chemical processes modeling, imitation of radioactivity, traffic modeling of informational systems, also in a theory of massive service etc. [1–4].

The subject of generators of Poisson pulse sequences (GPPS) research and construction is engaged by scientists from different countries. There are numerous articles in literature that belong to analysis of the problem, and also spheres of GPPS usage and GPPS quality research.

It is known, that the main feature of Poisson flux, which promotes its wide spread in modeling is additively, so that the resultant flux of the sum of the Poisson fluxes is also Poisson flux, which intensity is the sum of all intensities of fluxes incoming in it.

The main issue, which is raised for specialists in modeling sphere or other researches with using of Poisson flux, is the usage of effective software and hardware generators of Poisson pulse sequence, which would ensure flux generation by law, maximally approximate to theoretical Poisson law of distribution.

Z. Hu et al. (Eds.): ICCSEEA 2020, AISC 1247, pp. 317–326, 2021.
https://doi.org/10.1007/978-3-030-55506-1_29

The main advantages of GPPS structures based on Linear Feedback Shift Registers (LFSR) are simplicity of its hardware realization and higher performance [4, 19]. However, today are known the structures that don't allow ensuring required features of output signal with minimally required hardware costs.

Concerning this, the actual task is upgrading the GPPS structures based on LFSR.

2 Related Works

For instance, the usage of GPPS for events simulation of radioactive decay, are described in scientific researches such scientists as Alejandro Veiga and Enrique Spinelli from National University of La Plata, Argentina [5], A. Linares-Barranco, D. Cascadoa, G. Jime´neza, A. Civita, B. Linares-Barrancoc (Sevilla, Spain), M. Osterb (Zurich, Switzerland). Also has been researched LFSR, as a method of pseudorandom hardware for generation the Address Event Representation Protocol (AER) [6] in real-time mode from sequence of images, that are stored in computer memory [7]. In works of Milovanovic Emina and Igor, Stojcev Mile, Nikolic Tatjana (Serbia) are described the usage of LFSR for building block in built-in-self-test (BIST) design within SoC [8]. Mohammed Abdul Samad AL-khatib and Auqib Hamid Lone in [9] propose a secure, lightweight acoustic pseudo-random number generator (SLA-LFSRPRNG). The generator is based on cryptographically secure Linear Feedback Shift Register (LFSR) and extracts the entropy from sound sources. The major attraction of proposed Pseudo Random Number Generator (PRNG) is its immunity to major attacks on pseudo-random number generators. Malik UsmanDilawar and FaizaAyub Syed in [10] analysed the network failure by assuming there occurrence as a Poisson Process was made. The accurate prediction of fault rate helps in managing the up gradation, replacement and other administrative issues of data centre components. And many other works are dedicated to consideration issues related to construction Pseudo-Random Number Generator and generators of Poisson pulse sequences constructed in different ways [11–16].

3 The Structure of GPPS Based on LFSR, the Methodology of Output Pulse Sequence Quality Evaluating

The GPPS can be realized based on pseudorandom number generator (PRNG). Currently, is known a lot of as software and hardware methods of pseudorandom numbers generation. The generators with the highest performance are hardware PRNGs based on LFSR, the designing principles, and also GPPS on their base, are examined in works [17–19].

The average frequency of output GPPS pulses, realized according to this scheme (Fig. 1), is defined by equation

$$f_{out} = \frac{\mathrm{G}}{2^{\mathrm{m}}} f_F, \quad (1)$$

where f_F is a frequency of input clock pulses, m is the quantity of LFSR binary bits, CC is comparison circuit, G is the value of control code.

The main disadvantage of suggested GPPS based on LFSR is relatively narrow range of control code values, when is ensured correspondence between output signal and Poisson law of distribution. The purpose of the research outlined in this paper is to find ways for its spreading in order ensure GPPS usage for solving wider range of tasks.

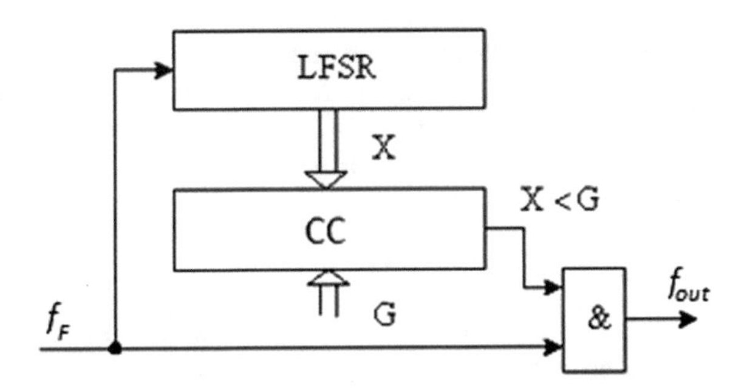

Fig. 1. The structural scheme of GPPS based on LFSR.

For evaluating the quality of LFSR and PRNG, realized on their base, is possible to use both graphical and evaluation tests [20, 21]. A methodology that has recommended itself well is the methodology of evaluating the quality of LFSR, which is based on classical methodology of verifying a hypothesis about distribution of general aggregate according to the Poisson law with using Pearson's chi-squared test (χ^2). This methodology was proposed by us and discussed in detail in [23]. A significant advantage of this method is that it takes into account the features of the hardware implementation of the generators.

For evaluation according to this methodology the flux of input pulses is divided into n similar groups, each of them consists of $i_{\max}$ pulses. Maximum quantity of groups – $n_{\max}$. The groups of input pulses correspond to the groups of output pulses with number of pulses – $k_1, k_2, \ldots, k_{n_{\max}}$.

At the same time, taking into consideration the specificity of GPPS designing, in addition the following suggestions have been proposed:

- The nominal (theoretical) average value of numbers $k_1, k_2 \ldots k_{n_{\max}}$ – k_c is fixed, independently from the value of control code G.
- The value $i_{\max}$ is changeable, it depends on G value and is defined by equation

$$i_{\max} = \frac{2^m}{G} k_c, \tag{2}$$

By using this methodology for each G value we receive the value χ_c^2. Then, according to the tables of critical dots of χ^2 distribution [24] and chosen levels of significance α (for α is chosen one of three values as usual: 0,1; 0,05; 0,01) and a number of freedom rate *st*, as a result is found the critical value $\chi^2_{\kappa p}$. If $\chi^2_c < \chi^2_{\kappa p}$ then is considered, that the pulse flux corresponds to Poisson law of distribution.

It substantially to notice, that one of the important requirements, that are necessary to follow when this technique is used – is ensuring the following inequality

$$i_{\max} \cdot n_{\max} < T_n, \tag{3}$$

where T_n is a repetition period of pseudorandom pulse sequence, that is equal to the repetition period of pseudorandom numbers at PRNG output.

If LFSR is used as PRNG, then a repetition period of pseudorandom sequence at its output, if using correct set of LFSR parameters, is equal to $2^m - 1$. That's why, taking into consideration the expressions (2) and (3), is possible to consider, that forming at the GPPS output the Poisson pulse flux will take place under the condition of compliance the inequality:

$$G > k_c \cdot n_{\max}. \tag{4}$$

Therefore, if the control code G has low values, the parameters of generator output signal will not correspond to the Poisson laws of distribution.

4 The GPPS Structure Based on LFSR with Improved Features

Structural scheme of modified GPPS is shown in Fig. 2. It differs from the previous variant (in Fig. 1) that the quantity of LFSR binary bits substantially exceeds the number of binary bits of comparison circuit CC, at input of which come only low order LFSR bits. Due to this, the repetition period of pseudorandom numbers at output of LFSR substantially exceeds the value $2^m - 1$ (here m – is quantity of CC bits), thus is possible to form the Poisson pulse flux at the output of generator with low values of control code G.

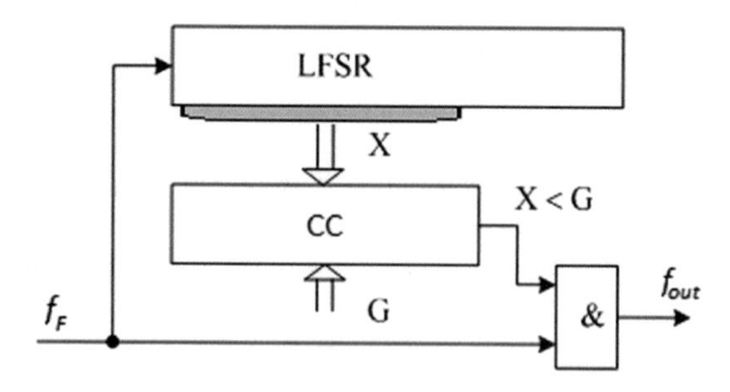

Fig. 2. The structural scheme of GPPS based on LFSR with improved features.

Further are represented the results of research of statistical characteristics of modified GPPS output signal with different parameters of LFSR structure. For comparison are shown the results of GPPS research for a case of software realization of PRNG based on standard function random for any programming language.

And it is clear, that under other equal conditions, the generators based on LFSR, as any hardware tools, substantially exceed software tools by performance and can be realized as an integrated circuit.

When GPPS is realized based on LFSR, as it is known from works [19–21], on quality (randomness) of pseudorandom evenly distributed number sequence, formed at LFSR output, they are also influence of parameters of Eq. (5), that determines the algorithm of its work:

$$Q(t+1) = T^{r}Q(t), \tag{5}$$

where $Q(t)$ and $Q(t+1)$ are the states of register of generator at the moments of time t and $t+1$ accordingly (before and after clock pulse arrival), T is a quadratic matrix of N degree

$$T_1 = \begin{vmatrix} a_1 & a_2 & \dots & a_{N-1} & a_N \\ 1 & 0 & \dots & 0 & 0 \\ 0 & 1 & \dots & 0 & 0 \\ & & \dots & & \\ 0 & 0 & \dots & 1 & 0 \end{vmatrix} \quad or \quad T_2 = \begin{vmatrix} 0 & \dots & 0 & 0 & a_N \\ 1 & \dots & 0 & 0 & a_{N-1} \\ & \dots & & & \\ 0 & \dots & 1 & 0 & a_2 \\ 0 & \dots & 0 & 1 & a_1 \end{vmatrix} \tag{6}$$

where N is a degree of irreducible polynomial

$$\mathrm{F}(x) = \sum_{i=0}^{N} a_i x^i, a_N = a_0 = 1, a_j \in \{0,1\}, j = \overline{1,(N-1)} \tag{7}$$

and r is natural number.

Thus, the parameters of such GPPS is defined not only by choosing of forming polynomial (its degree), but also the type of matrix (T_1 or T_2) and its degree r.

In Fig. 3 and Fig. 4 are shown some of the results of GPPS imitation modeling based on standard random function and GPPS on LFSR based on its various parameters.

Here is shown the dependences of value *level* from control code G, which signifies the number of values χ^2_{ccep} (the averaged value of five last current values χ^2_c) exceeding $\chi^2_{\kappa p}$.

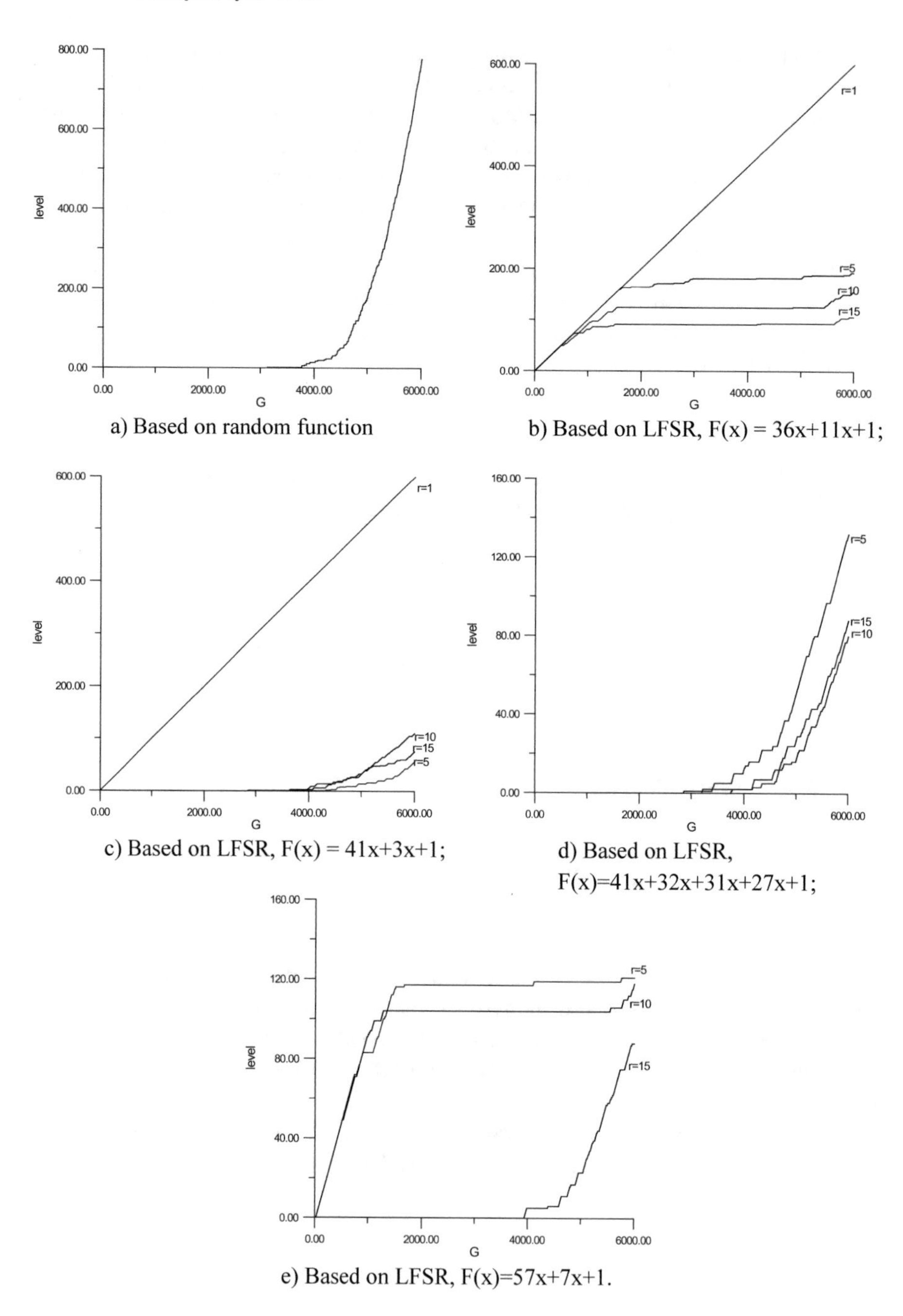

a) Based on random function

b) Based on LFSR, F(x) = 36x+11x+1;

c) Based on LFSR, F(x) = 41x+3x+1;

d) Based on LFSR, F(x)=41x+32x+31x+27x+1;

e) Based on LFSR, F(x)=57x+7x+1.

Fig. 3. The statistical characteristics of GPPS output signal (the quantity of CC binary bits – 15).

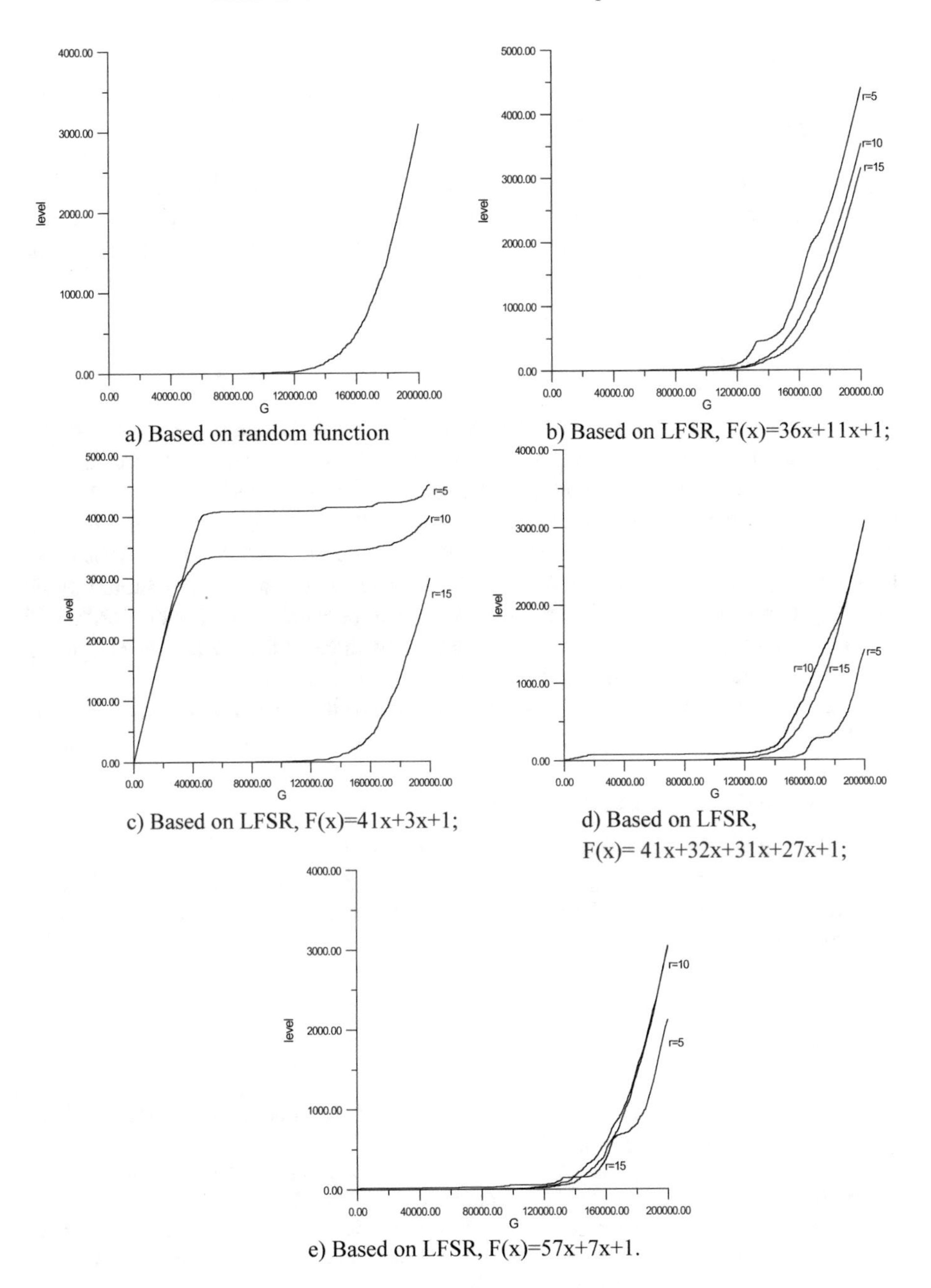

a) Based on random function

b) Based on LFSR, F(x)=36x+11x+1;

c) Based on LFSR, F(x)=41x+3x+1;

d) Based on LFSR, F(x)= 41x+32x+31x+27x+1;

e) Based on LFSR, F(x)=57x+7x+1.

Fig. 4. The statistical characteristics of GPPS output signal (the quantity of CC binary bits – 20).

The LFSR generators, that were researched, had been realized based on T_1 matrix.

The polynomial type is indicated in the signature to the corresponding figure, and a value of matrix degree r – on every figure.

The results of research have allowed to make a conclusion, that exceeding the capacity of comparison scheme CC, the capacity of polynomials LFSR and degrees of its forming matrixes, happen an extension of value range of control code G, when the output pulse sequence of generator correspond to Poisson law of distribution. The specific choice of these parameters allows ensuring required parameters of GPPS.

5 Conclusions

The suggested structure of GPPS based on LFSR, where for comparison with control code G is used only a part of low order PRNG bits, allows to ensure correspondence of output pulse sequence to Poisson law of distribution in extended range of values G, starting from the value of control code $G = 1$. At the same time, the upper limit of effective usage of value G range substantially depends on not only from choosing of forming polynomial, but also from a degree r of forming matrix. One should avoid choosing exponents $r > 15$, since the hardware implementation of such GPPS will complicate and the performance will decrease, and there will be virtually no gain in increasing the range of control code values.

The obtained results are of considerable practical value and can be used by developers of dosimetric devices in their development and testing by simulating radioactive radiation.

Further studies in this area may relate to:

1. Implementation of the considered PLD-based generators;
2. The use of the proposed generators as a basis for the construction of dosimetric detectors, taking into account their features, in particular, the dead time of such detectors.

References

1. Nikitenok, V.I.: The fast nonparametric algorithms for signal detection. In: The Technical Writer's Handbook, 131 p. BSU, Minsk (2010)
2. Krylov, V.V., Samokhvalova, S.S.: Teletraffic theory and its applications. In: The Technical Writer's Handbook, 188 p. BHV-Petersburg, St. Petersburg (2005)
3. Garasymchuk, O.I., Dudykevych, V.B., Maksymovych, V.M., Smuk, R.T.: Generators of test pulse sequences for dosimetric devices. J. Nat. Univ. "Lviv Polytechnic" Heat Power Environ. Eng. Autom. **506**, 186–192 (2004)
4. Zaharija, G., Mladenović, S., Dunić, S.: Cognitive agents and learning problems. Int. J. Intell. Syst. Appl. (IJISA) **9**(3), 1–7 (2017). https://doi.org/10.5815/ijisa.2017.03.01
5. Veiga, A., Spinelli, E.: A pulse generator with poisson-exponential distribution for emulation of radioactive decay events. https://www.researchgate.net/publication/301321983_A_pulse_generator_with_poisson-exponential_distribution_for_emulation_of_radioactive_decay_events

6. Address-Event Representation (AER) protocol. https://inilabs.com/support/hardware/aer/
7. Linares-Barranco, A., Oster, M., Cascado, D., Jiménez, G., Civit, A., Linares-Barranco, B.: Inter-spike-intervals analysis of AER Poisson-like generator hardware. Neurocomputing **70**, 2692–2700 (2007). http://www.sciencedirect.com/science/article/pii/S0925231207000951
8. Milovanovic, E.I., Stojcev, M.K., Milovanovic, I.Z., Nikolic, T.R., Stamenkovic, Z.: Concurrent generation of pseudo random numbers with LFSR of fibonacci and galois type. Comput. Inform. **34**, 941–958 (2015). https://cai.type.sk/content/2015/4/concurrent-generation-of-pseudo-random-numbers-with-lfsr-of-fibonacci-and-galois-type
9. AL-khatib, M.A.S., Lone, A.H.: Acoustic lightweight pseudo random number generator based on cryptographically secure LFSR. Int. J. Comput. Netw. Inf. Secur. (IJCNIS) **2**, 38–45 (2018). Published Online February 2018 in MECS (http://www.mecs-press.org/). https://doi.org/10.5815/ijcnis.2018.02.05
10. UsmanDilawar, M., Syed, F.A.: Mathematical modeling and analysis of network service failure in DataCentre. Int. J. Mod. Educ. Comput. Sci. (IJMECS) **6**, 30–36 (2014). Published Online June 2014 in MECS (http://www.mecs-press.org/). https://doi.org/10.5815/ijmecs.2014.06.04
11. Lowy, M.: Parallel implementation of linear feedback shift registers for low power applications. IEEE Trans. Circuits Syst.-II: Analog Digit. Signal Process. **43**(6), 458–466 (1996)
12. Wijesinghe, W., Jayananda, M.K., Sonnadara, U.: Hardware implementation of random number generators. In: Proceedings of the Technical Sessions, Institute of Physics-Sri Lanka, vol. 22, pp. 25–36 (2006). http://www.ip-sl.org/procs/ipsl063.pdf
13. Dabrowska-Boruch, A., Gancarczyk, G., Wiatr, K.: Implementation of a ranlux based pseudo-random number generator in FPGA using VHDL and impulse C. Comput. Inform. **32**, 1272–1292 (2013)
14. Maksymovych, V., Harasymchuk, O., Opirskyy, I.: The designing and research of generators of poisson pulse sequences on base of fibonacci modified additive generator. In: Advances in Intelligent Systems and Computing, AISC, vol. 754, pp. 43–53 (2019)
15. Maksymovych, V.N., Harasymchuk, O.I., Mandrona, M.N.: Designing generators of poisson pulse sequences based on the additive fibonacci generators. J. Autom. Inf. Sci. **49**(2), 1–13 (2017)
16. Maksymovych, V.N., Mandrona, M.N., Kostiv, Yu.M., Harasymchuk, O.I.: Investigating the statistical characteristics of poisson pulse sequences generators constructed in different ways. J. Autom. Inf. Sci. **49**(10), 11–19 (2017)
17. Mandrona, M.M., Maksymovych, V.M., Harasymchuk, O.I., Kostiv, Y.M.: Generator of pseudorandom bit sequence with increased cryptographic immunity. Metall. Min. Ind. **5**, 25–29 (2014)
18. Hu, Z., Legeza, V.P., Dychka, I.A., Legeza, D.V.: Mathematical modeling of the process of vibration protection in a system with two-mass damper pendulum. Int. J. Intell. Syst. Appl. (IJISA) **9**(3), 18–25 (2017). https://doi.org/10.5815/ijisa.2017.03.03
19. Garasymchuk, O.I., Maksymovych, V.M., Striletsky, Z.M.: Analysis of the technical characteristics of the pulse sequence generator with the Poisson law of distribution, constructed on the basis of the M-sequence generator. Collection of Scientific Works “Computer Technologies of Printing”, Lviv, vol. 21, pp. 110–117 (2009)
20. Garasymchuk, O.I., Maksymovych, V.M.: Algorithm for the formation of a Poisson pulse stream. J. Nat. Univ. “Lviv Polytechnic” Autom. Meas. Control **475**, 21–25 (2003)
21. Ivanov, M.A., Chugunkov, I.V.: Theory, application and quality assessment of pseudo-random sequence generators. In: The Technical Writer’s Handbook, 240 p. KUDITS - IMAGE (2003)

22. Gaeini, A., Mirghadri, A., Jandaghi, G., Keshavarzi, B.: Comparing some pseudo-random number generators and cryptography algorithms using a general evaluation pattern. Int. J. Inf. Technol. Comput. Sci. (IJITCS) **8**(9), 25–31 (2016). https://doi.org/10.5815/ijitcs.2016.09.04
23. Harasymchuk, O.I., Kostiv, Yu.M., Maksymovych, V.M., Mandrona, M.M.: Methodology for research of Poisson pulse sequence generators using Pearson's chi-squared test. Sustain. Dev. Int. J. **9**, 67–72 (2013)
24. Hurman, V.E.: Probability theory and mathematical statistics. Study allowance for high schools. Pub. No. 5, reworked and added by M., "High School", 479 p. (1977)

The Improvement of Digital Signature Algorithm Based on Elliptic Curve Cryptography

Svitlana Kazmirchuk, Anna Ilyenko(✉), Sergii Ilyenko, Olena Prokopenko, and Yana Mazur

National Aviation University, Kyiv 03058, Ukraine
{sv.kazmirchuk, ilyenko.a.v, ilyenko.s.s, bortnik.olena.v}@nau.edu.ua,
http://www.nau.edu.ua

Abstract. Electronic digital signature guarantee message integrity. In the present paper we describe existing algorithms for the formation and verification of the electronic digital signatures. Also in the article on the basis of the analysis of modern methods of formation and verification of electronic digital signatures are considered directions of improvement of the electronic digital signatures procedure using a group of points of the elliptic curves with providing the possibility of ensuring the integrity and confidentiality of information. The proposed new method of electronic digital signatures generation and verification is implemented on the Shnorr signature algorithm, that allows to recover data directly from the signature similarly to RSA-like signature systems and the amount of the recovered information will be variable. The main advantages of improvement procedure is the shorter key length with equivalent cryptographic strength, shorter length of the signature itself and reduced the total length of the transmitted data. Thus new secure digital signature scheme minimize time for formation and verification of confidential information witch depending on the used method based on the elliptic curves and adds a privacy service.

Keywords: Electronic digital signature · Message recovery · Confidentiality · Symmetric encryption · Signature · Hash

1 Introduction

In our time, the benefits of cryptographic information security are continuing to gain momentum. However, this use could be much wider. At the moment, a cryptographic tool such as electronic digital signature is most commonly used in banking records, but even this makes it very popular. Thus, to be able to provide the service of concealment, and given the fact that usually important electronic documents are signed, it can be said that it is necessary to take care of preserving the secrecy of confidential data, which is an invariable attribute of each document. Therefore, an important problem with all digital signing of information is the improvement of algorithms and methods for signing confidential data recovery from the signature itself. Improvements should be

Z. Hu et al. (Eds.): ICCSEEA 2020, AISC 1247, pp. 327–337, 2021.
https://doi.org/10.1007/978-3-030-55506-1_30

made taking into account the arithmetic of elliptic curves to construct more efficient asymmetric crypto algorithms [1, 18, 19].

While providing the integrity and authenticity properties of the message, the electronic digital signature do not regulate the confidentiality of the transmitted data. Considering the fact that usually important electronic documents are signed, it can be assumed that a separate document will contain confidential data. Therefore, the important task for digital signing of information is to improve the algorithms and methods of signature in a way that the confidentiality service of the classified information of the document is embedded [16, 17].

The purpose of this article is to conduct an analysis of practical approaches to electronic digital signature formation and verification. Based on the analysis, determine further ways to improve the method of digital signing of the electronic documents based on the complex problem of finding a discrete logarithm in a group of points of the elliptic curve, with the built-in ability to recover the confidential part of the message of any length. The areas of develop new method are to combine the performance of cryptographic circuits built on the arithmetic of elliptic curves with the functionality of RSA-like message recovery schemes, whereby the length of the recovered message is chosen based on the signer's own needs and is arbitrary, named as ECDSMR – electronic digital signature scheme with ability to recovery based on the use of elliptic curves. This research proposed an improvement of the electronic digital signatures procedure where the length of the signature is equal to the sum of the length of the end field over which the elliptic curve is determined, and the artificial excess redundancy provided to the hidden message was achieved. The major tasks include the ability to customize the method of digital signing, in particular the variable length of the recoverable part of the message, which should be determined by the needs of the message signer; selection of the required key length in view of the required level of security; providing redundant data, as appropriate.

2 Related Work

There are different classifications of modern electronic digital signature schemes. They can be classified according to the mechanism of construction (symmetric and asymmetric), with the recovery of a message or without, one-time and multiple, deterministic and probabilistic, on the problem that underlies them [12–15].

Various electronic-digital signature schemes have been proposed to address specific issues in the electronic document management field. The most successful of the cryptologists of electronic digital signature schemes were the RSA and the ElGamal signature scheme. But the first of these has been patented in the USA and a number of other countries. In the second scheme, there are many possible modifications, and all of them are quite difficult to patent. It is for this reason that ElGamal signature scheme has remained largely patent-free. Besides, this scheme has some practical advantages: the size of the blocks that are operated by the algorithms and accordingly the size of the electronic-digital signature in it were significantly lower than in RSA, with the same sustainability [18, 19].

Recent advances in the theory of computational complexity have shown that the general problem in discrete fields, which is the basis of this specified scheme of electronic-digital signature, cannot be considered to be a sufficiently strong foundation [16]. There are currently digital signature algorithms based on the problem of discrete logarithm over a finite field DSA, ElGamal, and ECDSA elliptic curves, DSTU 4145-2002, GOST R 34.10-2018 [1, 5, 16, 18, 19]. This article is focused on cryptographic algorithms based on elliptic curves [1–3, 7].

Consider an ECDSA algorithm that is similar to DSA, but is defined not over the finite field, but over the groups of points of elliptic curves [6, 8]. Define an elliptic curve E above the field F_p and the point P, lying on this curve $E(F_p)$. The point has a simple order q (this is a very important condition). The curve E and the point P, are system parameters. User generates the secret and public keys as follows: selects a random or pseudorandom integer x from the interval $[1,\ q-1]$, which will be the secret key. Then the product (multiple) $Q = xP$ is calculated, resulting in one more point lying on this curve $E(F_p)$ and forming a cyclic subgroup formed by the generating point P. Instead of using E and P as global system parameters, only one field F_p can be fixed for all users and allow each user to choose their own elliptic curve E and point P at $E(F_p)$. In this case, a certain curve equation E, the coordinates of the point P, and the order q of that point P must be included in the user's public key. If the field F_p is fixed, then the hardware and software components can be constructed in the way to optimize the calculations in that field. At the same time, there are a huge number of options for choosing an elliptical curve over the field F_p. The signature is carried out in the following sequence of steps: 1. Generate a (pseudo-) random integer $k \in [1, q-1]$; 2. Calculate $kP = (x_1, y_1)$ and put $r = x_1 \mathrm{mod} q$, where r is obtained from an integer $x_1 \in [1, p-1]$ modulo casting q.

If $r = 0$, then it is necessary to regenerate a random number and calculate r. Calculate $k^{-1} \mathrm{mod} q$ and put $s = k^{-1}(h + xr) \mathrm{mod} q$, where h is the value of the hash function of the signed message. s should not be equal to 0, otherwise the verification equation does not exist. The signature for the message is a pair of integers (r, s).

When verifying, the first step is to check whether $r \in [1, q-1]$ or $s \in [1, q-1]$, otherwise the signature is not valid. Then the following variables are calculated: $u_1 = s^{-1} h \, mod \, q$ and $u_2 = s^{-1} r \, mod \, q$. The main validation equation has the following form (similar for DSA): $u_1 P + u_2 Q = (x_0, y_0)$. The correctness of the scheme is shown by the following relation:

$$u_1 P + u_2 Q = (u_1 + xu_2)P = (s^{-1} h \, mod \, q + xs^{-1} r \, mod \, q)P = s^{-1}(h + xr) \, mod \, qP = kP.$$

Almost similar is also the algorithm GOST R 34.10-2001, described in, which has a difference only in the values for which the module-inverted elements are taken. In addition, there are schemes equivalent to the Nyberg–Rueppel (p-NEW) Signature Scheme. These schemes provide recovery of the message. One of these schemes is described in [8]. In this scheme, key generation follows the same sequence of steps as in ECDSA scheme. But the signature calculation formulas have been modified: $R = kP$, $r = m^{-1} R_x \, mod \, p$, $r' = r \, mod \, q$ and $s = k^{-1}(1 + r'x) \, mod \, q$. The signature is a pair of numbers (r, s). The message is restored from the validation equation: $m = (s^{-1}P + r's^{-1}Q_x)r^{-1} \, mod p$. Therefore, it can be seen that all circuits defined over a finite field based on the problem of discrete logarithm can be easily converted to similar ones defined over a group of points of the elliptic curve, having all the advantages described earlier [19]. In addition to digital signature schemes that perform full message recovery, there are also partial recovery algorithms. Areas of application of the algorithms are also quite wide: these schemes can be applied in applications and relationships where the complete hiding of the message in the signature is unnecessary. However, certain parts do have confidential information that should not be disclosed. Similar algorithms are used in particular in mail systems, where the confidential part includes such information as the address of the recipient, the disclosure of which is not appropriate, while the overall content of the message is not confidential. As an example, consider the security-certified scheme included in the P1363 standard - the Pintsov-Vanstone digital signature scheme [9].

First, the criteria by which the signature algorithm was chosen should be noted: crypto-stability, provided at the standard level (minimum – 80 bits of security); existential counterfeit resistance: the algorithm must maintain the likelihood of counterfeiting at a certain level (e.g. 240 operations); the minimum amount of time it takes to forge a signature may be enough to deter potential offenders; minimum size cost: the mail system can have severe restrictions on the size of the signature. In particular, barcode optical readers work more efficiently with small size data; maximum computation efficiency should be maintained at a level that allows multiple signatures/signature checks per second on modern equipment; the signature must include all the information required for verification without connecting external sources; privacy: certain parts of the message must be cryptographically protected from unauthorized disclosure. Therefore, at the stage of preparation for signature, the signing message is split into two parts $C||V$, where C reflects the data to be cryptographically protected, V is the open data. Open data integrity is also ensured as they are also signed. Generation takes the following steps: $R = kP$, where R is the point on the selected curve, that must be interpreted into the bit stream for the next transformation $c = Tr_R(C)$, where Tr_R – transformation that is parameterized by a previously calculated value R and designed to destroy any algebraic structure that C may have. As a transformation, a symmetric encryption algorithm such as AES or a simple XOR key and secret data operation can be used; $h = H(e||I||V)$, where H – cryptographic hash function, I – sender ID; $d = ah + k(mod n)$, where a – sender's private key; a pair of numbers (c, d) is a digital signature [16, 18]. The signature generation algorithm becomes much more efficient if the bit length of the data in need of confidential protection is less than or equal to the bit length R when XOR operation can be applied to that data. However, if the secret data exceeds the key length, then the use of XOR with a repetition of gamma is not acceptable for the proper level of cryptographic strength. To verify the signature, the recipient receives the following data packet: sender ID I, signature (c, d) and open data V. The following calculations are performed for verification: $U = sP - dQ$, where Q – sender's public key, and $d = H(e||I||V)$; the inversed transformation $X = Tr_{U^{-1}}(c)$ is performed; check the redundancy X and if X has the required redundancy (40 bits) accept the signature. In this algorithm, the redundancy parameter is important. The choice of its size should follow from the level of security in relation to the attack of credible forgery. Advantages include the variable of the recoverable part of the message.

Considering the principles of building a scheme, cryptographic stability is caused by a number of factors: the security of the group of points of the elliptic curve, in particular the ability to calculate a discrete logarithm in this group; resistance to hash function collision; the stability of the cipher used as a transformation; the amount of redundant data that is attached to the hidden data. A similar scheme is considered in [10], but its description is given in terms of paired cryptography, that is, cryptography based on the pairing of different groups of numbers in one algorithm. An example of such pairing is the group of points of the elliptic curve. This scheme gives the secret key s and the public key $Q = sP$, where P – generating element of the group. Suppose the following hash functions $H_1 : \{0, 1\} \rightarrow G_1$, that is, a function that displays binary data in G_1 element; $F_1 : \{0, 1\}^{k_2} \rightarrow \{0, 1\}^{k_1}$, that is, a function that displays sequence from k_2 elements to sequence from k_1 elements; $F2 : \{0, 1\}^{k_1} \rightarrow \{0, 1\}^{k_2}$ - inverse to the previous one. Let the message be $m \in \{0, 1\}^{k_2}$. Then the signature is constructed in the same way as in [6], but instead of symmetrical transformation, hashing operations are performed: $f = F_1(m)||(F_2(F_1(m))\oplus m)$, the length f is $k_1 + k_2 = q$ bits. In this case, the message is completely hidden in the signature, and it should be no longer than k_2. However, this scheme can be modified in such a way that it is possible to hide part of the message as well. As in article [11]: $f = F_1(m_2)||(F_2(F_1(m_2))\oplus m_2)$, where m_2 – part of the message that needs to be hidden and the hash functions are defined as in the previous example. The open part m_1 is taken into account further in the formation of the signature, i.e. the integrity of that part is also confirmed. That is, in this scheme it is possible to hide data in length of k_2, although the message being signed can be larger. Thus the analysis considered algorithms that, similar to the RSA algorithm, provide the ability to recover a message directly when verifying a digital signature. However, the recovery capabilities of all of these algorithms were limited: they were either not suitable for recovering long messages when multiple signatures or one having a large size (p-NEW schemes; Mohanty-Mahji scheme) or simply were not capable of recovering messages of any length (or part of them of any length): schemes offered by Zhang and Raylin Tso. The Pintsov-Vanstone scheme required the inclusion of redundancy in the hidden data, making it difficult to verify the signature using this redundancy. However, these algorithms have significant limitations: the length of the retrieved message cannot be greater than the size of the final field (or key size), and hash functions of a special appearance are required. That is way it is necessary to propose an improvement of the electronic digital signatures procedure with providing the possibility of ensuring the integrity and confidentiality of information.

3 Proposed Electronic Digital Signature Scheme with Message Recovery

This section describes the mathematical model of a digital signature with the proof of its correctness. Elliptical curve can be generated in three ways: using any curve described in the SEC standard [4] (it is only needed to agree on the size of the order of the subset of the curve points); generation of own curve by the method of complex multiplication, specifying the required parameters: field size, curve type, cofactor; or load the desired curve parameters from an existing file. This curve is denoted as $E(F_p)$, indicating that it is defined over a finite field F_p of prime characteristic $p > 3$ [2, 3, 5, 6]. For the purpose of the new ECDSMR method, before using it, it is necessary to determine the hashing and symmetric encryption algorithms that will be used. This particular implementation proposes the use of the SHA family of functions as hashing algorithms and AES (Advanced Encryption Standard) that are the applicable world-wide standards. The next step is to generate the required public and private keys. The secret key is initially generated by a cryptographically strong algorithm for pseudo-random sequencing according to standard X9.17.

It should also be noted that the key used in the developed method of digital signature should not be simple (although it may be), it is enough that it is sufficiently random. This is proven by the fact that in the group of points of the elliptic curve there are no analogues of prime numbers as they are defined in finite fields. The public key is also calculated from the secret according to the rules of scalar multiplication of the number (which is the secret key) to the point which is the base point of the curve, which is the system parameter of the method. Then the public key is calculated with the formula: $Q = x \cdot P$, where x – user's private key, $P \in E(F_p)$ – base point of the elliptic curve. After generating the key pair, the user saves his private key x in a secure location and publishes his public keys Q and the system curve parameters $E(F_p)$ that are required for further signing and message validation.

3.1 Formation of the Signature

The message M, that must be signed, is broken down into separate parts: c (confidential) - the confidentiality of which must be maintained and p (plain) - which can be transmitted over the communication channel in an open form, the length of both parts is not limited and may be arbitrary. It is up to the user to decide the question of splitting according to his own needs. If necessary c is provided with redundancy (appropriate if there is no natural redundancy in this part, for example, if only certain numeric codes are hidden, without data that could be immediately identified as fake). Redundancy is variable, the amount of which is set by the number of bytes: $b \in \left[0..\frac{bytes(p)}{2}\right]$, where $bytes(p)$ – the size of the field in bytes. The probability of an existential forgery attack depends on this size. This probability equals $p_{ef} = 2^{-b}$, where the term *ef* means existential forgery -an attack of existential forgery, as it is indicated in the English-language literature. After that a pseudorandom number $k \in [1..q-1]$ is selected. Cryptographically resistant pseudorandom number generator should also be used.

According to the rules of scalar multiplication, a point $R = k \cdot P$, $R \in E(F_p)$ is calculated in the group of points of the elliptic curve. Key data for AES encryption is generated from coordinates describing point R: symmetric encryption key

$$K = R_x mod 2^l, \tag{1}$$

where R_x is the coordinate of the x point obtained in the previous step, l means key length in bits (for AES is 128, 192, 256), that is selected depending on the required total cryptographic strength of the algorithm (for example, for $E(F_p) : p = 256$ AES key with $l = 128$ must be used); initializing vector

$$IV = R_y mod 2^{128}, \tag{2}$$

where R_y is the coordinate y of the point R. The size of the initialization vector in AES is fixed and is 16 bytes (128 bits). Than we propose to choice the last parameter – encryption mode. OFB (Output Feedback Mode) was chosen – a block clutch mode in which each subsequent encrypted block is obtained after performing the XOR operation on the initialization vector data, after executing all encryption cycles over it, and the corresponding block of open data and depends on all previous blocks That is why the first signature parameter is obtained by AES encryption:

$$r = AES\_encrypt_{K,IV}(c) \tag{3}$$

The hash value is calculated using the SHA function:

$$h = SHA(r||p), \tag{4}$$

where $||$ denotes the operation of concatenation of two binary vectors. It should be noted here that the size of the hash is selected based on the overall cryptographic strength of the ECDSMR and numerically equal to the bit length of the field characteristic p (implementations of SHA-160, SHA-256, SHA-384 with the corresponding bit lengths of the hashes were used here). The second signature parameter is calculated: $s = x \cdot h + k(mod\, q)$, where q – the order of the subset of the points of the curve $E(F_p)$. The signature is a pair of numbers (r, s). The signed message is sent via the communication channel in the form $M_{signed} = p||(r, s)$. Method of message attachment user defines himself. In addition, after creating the signature, the proposed method adds and ensures the service of data privacy, as well as increases cryptographic stability.

3.2 Verification of the Signature

On the receiving side, the recipient performs the following steps to verify the signature of the received message and to recover the information hidden in the signature. First checks whether $s \in [1..q - 1]$, otherwise the signature is recognized as invalid; calculates the hash function h for (4); by the rules of scalar multiplication in the group of points of the elliptic curve calculates

$$SPHQ = s \cdot P - h \cdot Q. \quad (5)$$

Than by rules (1) and (2), with the difference that the coordinates x and y are taken from a *SPHQ* point, calculates the key parameters for the AES cipher: initialization vector *IV* and key *K*. Next restores the encrypted part with inversed (3) function, it should be noted that the functions AES_encrypt and AES_decrypt are symmetric: $c' = AES\_decrypt_{K,IV}(r)$ Farther checks the redundancy of the binary vector provided in paragraph 3.1 of the signature generation step and concludes that the signature is valid. The proof of the correctness of the method ECDSMR. To do this, it is shown that the point on an elliptical curve.

$$SPHQ = s \cdot P - h \cdot Q = ((x \cdot h + k) mod\, q)P - h \cdot (x \cdot P) = ((x \cdot h) mod\, q)P \\ + (k mod\, q)P - ((x \cdot h) mod\, q)P = kP.$$

Therefore, all subsequent actions with the coordinates of the point *SPHQ* lead to the generation of the same key parameters for AES as in encryption, so the scheme is correct. Several nuances of the ECDSMR method should be noted. First, you need to decide on the stage of providing redundancy to the hidden data. In this implementation, it was provided by adding redundancy to a part *c* of the ANSI symbol numbered 01 as the dividing character between the information and redundant part and the required number of bytes from the hash function calculated from *c*.

Another nuance is that, given the fact that the length of the hidden part of the message can be arbitrarily selected, it is possible to consider the case where the user does not need ECDSMR signature security services. In this case, the steps of generation and verification are modified so that the algorithm is transformed into the Shnorr signature algorithm. Specifically, when generating a signature, the hash function is calculated taking into account the $r = R_x mod\, q$ modification. The signature is also a pair of numbers *(r, s)*. When verifying a signature, the hash function undergoes similar changes. In this case, only one condition needs to be checked: $SPHQ_x mod\, q = r$.

4 Results

As a result of the proposed ECDSMR method, a software implementation of the procedure for the formation and verification of a digital signature in C# was obtained using the CryptoLib library. This library fully supports cryptographic algorithms and is certified for use in Ukraine at the state level. The algorithms were tested in the Crypto++ 5.6.0 software environment on a dual-core Intel Core 1.83 GHz processor running Windows 8 32 bit x86 (Table 1).

The main aim of performing the experiment is to compare algorithms on the same grounds and draw a conclusion about improved procedure of formation and verification of electronic digital signature. During the experiment the algorithms where constructed in such a way that the same parameters are almost everywhere equivalent in value, which provides clarity of comparison. All algorithms where compared by using some experiment specification such as: *K* - length of the key; *q* – the order of the cyclic subset of the base point for algorithms based on discrete logarithm; has a length equal to the size of the finite field F_P; *H (M)* – hash function from the incoming message *M*, if there are several hash functions in the algorithm, then the sequence number is added to the mark ($H_1(M)$); *LH* (*M*) – the length of the output value of the hash function $H(M)$; *G* – the point on the elliptic curve (the length equals $2q$); L_{redund} – the length of the redundancy in bits ($0 \leq L_{redund} \leq q/2$); *S*(*M*) – symmetric cipher.

After experiment we received results which show important difference between classical and proposed method of the electronic digital signatures procedure using a group of points of the elliptic curves with providing the possibility of ensuring the integrity and confidentiality of information. Comparing the created algorithm with the classical algorithms we can draw the main differences: there are no restrictions on the length of the recoverable portion of the message. Signing is possible even at the zero length of the retrieved message: the first parameter of the signature is a cipher text calculated from the redundancy with the length of $q/2$ bits; all the advantages of using a group of points of the elliptic curve are present: shorter key length with equivalent crypto security, shorter length of the signature itself; reduced the total length of the transmitted data, that in sum is equal to $L(M) + q + L_{redund}$, where $L(M)$ – the length of the signed message (all lengths are in bits). It can be seen that even with the maximum redundancy value, the length of the transmitted data is smaller than in other algorithms by the value of $q/2$ bits, that is, half the size of the finite field over which the elliptic curve is defined.

Table 1. Comparison of modern algorithms for the formation and verification of digital signature

Algorithm	Open key, bit	Length of signature, bit	Additional algorithm parameters	Time of formation the signature, ms	Time of verification the signature, ms	Message restore/length limit
RSA	2048	$2K$	–	1,48	0,07	+/$L(M)$
DSA	2048	$2q$	$H(M)$	0,42	0,52	–
ElGamal	2048	$2K$	$H(M)$	0,83	3,84	–
Shnorr	2048	$LH(M) + q$	$H(M)$	0,63	2,35	–
NR	2048	$K + q$	–	3,78	7,78	+/$\leq$ q
ECDSA	224	$2q$	$H(M)$	2,88	8,53	–
Zhang	224	$q + G/3q$	$H_3(M) = H_1(M) + H_2(M)$	3,65	9,37	+/$\leq (q - LH_1(M))$
ECDSMR	224	$q + L_{redund}$	$H(M)$, $S(M)$	3,50	5,22	+/–

5 Conclusion

Thus, in this article we give a full description of the proposed ECDSMR method for the creation and verification of electronic-digital signature using elliptic curves. Proposed electronic digital signature scheme with message recovery is implemented on the Shnorr signature algorithm, which allows to recover data directly from the signature similarly to RSA-like signature systems and the amount of the recovered information will be variable, i.e. it will depend on the information message.

We can say that this proposed method is more simple compared to others and much more economical than computing resources. As a result of the modification, the size of the public key has decreased by 10%, that is, the minimum size of the public key is 224 bits, and the time for formation and verification decreased in the range from 1,04 times to 1,79 times depending on the method used based on the elliptic curves. Thus, in this way, the proposed algorithm for the formation and verification of an electronic-digital signature ECDSMR with the ability to provide the function of message recovery and privacy service.

Our future research will focus on building more other encryption algorithms based on elliptic curves with the ability to recover message using strong elliptic curves over finite fields of simple order.

References

1. ISO/IEC. ISO/IEC 9796-3:2006, Information technology – Security techniques – Digital signature schemes giving message recovery – Part 3: Discrete logarithm based mechanisms (2006)
2. NIST, Recommended elliptic curves for federal government use (1999)
3. NIST SP 800-131A, Transitions: Recommendation for Transitioning the Use of Cryptographic Algorithms and Key Lengths (2011)
4. SEC 2: Recommended Elliptic Curve Domain Parameters, Standards for Efficient Cryptography Group (2000)
5. FIPS PUB 186-4 – Digital Signature Standard (2013)
6. RFC 6979 – Deterministic Usage of the Digital Signature Algorithm (DSA) and Elliptic Curve Digital Signature Algorithm (ECDSA) (2015)
7. The state standard of Ukraine 4145-2002, Information Technology – Cryptographic protection of information – Digital signature based on elliptical curves. Formation and Verification (2002). (in Ukrainian)
8. Miyaji, A.: A message recovery signature scheme equivalent to DSA over elliptic curves. In: Kim, K., Matsumoto, T. (eds.) Advances in Cryptology—ASIACRYPT 1996. Lecture Notes in Computer Science, vol. 1163, pp. 1–14. Springer, Heidelberg (1996)
9. Pintsov, L.A., Vanstone, S.A.: Postal revenue collection in the digital age. In: Frankel, Y. (eds.) Financial Cryptography, FC 2000. Lecture Notes in Computer Science, vol. 1962, pp. 105–120. Springer, Heidelberg (2001)
10. Zhang, F., Susilo, W., Mu, Y.: Identity-based partial message recovery signatures (or how to shorten ID-based signatures). In: Patrick, A.S., Yung, M. (eds.) Financial Cryptography and Data Security, FC 2005. Lecture Notes in Computer Science, vol. 3570, pp. 45–56. Springer, Heidelberg (2005)
11. Tso, R., Gu, C., Okamoto, T., Okamoto, E.: Efficient ID-based digital signatures with message recovery. In: Bao, F., Ling, S., Okamoto, T., Wang, H., Xing, C. (eds.) Cryptology and Network Security, CANS 2007. Lecture Notes in Computer Science, vol. 4856, pp. 47–59. Springer, Heidelberg (2007)
12. Hu, Z., Dychka, I., Onai, M., Zhykin, Y.: Blind payment protocol for payment channel networks. Int. J. Comput. Netw. Inf. Secur. (IJCNIS) **6**(11), 22–28 (2019)
13. István, V.: Construction for searchable encryption with strong security guarantees. Int. J. Comput. Netw. Inf. Secur. (IJCNIS) **5**(11), 1–10 (2019)

14. Goyal, R., Khurana, M.: Cryptographic security using various encryption and decryption method. Int. J. Math. Sci. Comput. (IJMSC) **3**(3), 1–11 (2018)
15. Shukla, R., Bhandari, R.: A novel minimized computational time based encryption and authentication using ECDSA. Int. J. Mod. Educ. Comput. Sci. (IJMECS) **9**(5), 19–25 (2013)
16. Kazmirchuk, S., Anna, I., Sergii, I.: Digital signature authentication scheme with message recovery based on the use of elliptic curves. In: Hu, Z., Petoukhov, S., Dychka, I., He, M. (eds.) ICCSEEA 2019. AISC, vol. 938, pp. 279–288. Springer, Cham (2020). https://doi.org/10.1007/978-3-030-16621-2_26
17. Vysotska, O., Davydenko, A.: Keystroke pattern authentication of computer systems users as one of the steps of multifactor authentication. In: Hu, Z., Petoukhov, S., Dychka, I., He, M. (eds.) ICCSEEA 2019. AISC, vol. 938, pp. 356–368. Springer, Cham (2020). https://doi.org/10.1007/978-3-030-16621-2_33
18. Menezes, A.J., Vanstone, S.A., Van Oorschot, P.C.: Handbook of Applied Cryptography. CRC Press, London (1996)
19. Schneier, B.: Applied Cryptography, 2nd edn. Wiley, Hoboken (2015)

Method Simultaneous Using GAN and RNN for Generating Web Page Program Code from Input Image

Ivan Dychka, Viktor Legeza, Liubov Oleshchenko(✉), and Dmytro Bohutskyi

National Technical University of Ukraine "Igor Sikorsky Kyiv Polytechnic Institute", Kyiv, Ukraine
dychka@pzks.fpm.kpi.ua, viktor.legeza@gmail.com, oleshchenkoliubov@gmail.com, bogutskii.d@gmail.com

Abstract. The article presents an analysis of neural networks that can be used to generating web page program code from input image. As part of the research of the possibilities of modification of algorithms for generating code, the method of generating code simultaneous using the generative adversarial and recurrent neural networks based on the input image is proposed. Methods for creating source code using generative adversarial network (GAN) and recurrent neural networks (RNN) are analyzed. Using domain-specific language (DSL) for tree generation code method, it was possible to use GAN and RNN together, achieving on the average 10% productivity increase. Using GRID's mesh, the generated code is adaptable. To implement the proposed method, software with a graphical user interface is created. Classes of GAN and RNN were considered, among which the feasibility of using a neural network with long-term short-term memory (LSTM) for analysis and forecasting of tags, blocks and HTML elements of the language in the input image was selected and substantiated. The stages of the research are presented. For the experiment datasets made on the basis of existing sites from open sources for 2018–2019 are used. The estimation of the effectiveness of the proposed method is described.

Keywords: Program code generating · Web-pages · Input image · Neural network · Generative adversarial neural network · Long short-term memory · DSL

1 Introduction

The number of new pages and sites on the Internet is growing rapidly every year. As of January 2019, the number of all active websites in the world exceeded 1.9 billion, which is half a billion more than in the last 2018. Just before the start of the new year, the number of domain names in all top-level domains exceeded 342 million. The most popular top-level domains were.com, immediately followed by the national Chinese.cn domain. With the increase in the number of sites on the Internet, the threshold for entering this area also increases. Thus, for a person far from programming, creating a site comes down to an impossible task, and for a web programmer, creating sites comes

Z. Hu et al. (Eds.): ICCSEEA 2020, AISC 1247, pp. 338–349, 2021.
https://doi.org/10.1007/978-3-030-55506-1_31

down to everyday routine. In times of fast Internet, a tool is needed that will allow to quickly create ready-made web pages. After all, in order to start creating web pages, people learn for months, or even years. Therefore, there is a need to create software that will generate program code for a web page using the method of interaction of two neural networks based on the input image that the user is able to draw. Thus, this will solve the problem of quickly creating a website for both a beginner and an experienced programmer.

There are technologies for generating program code from an image, but all of them are only prototypes and have a number of drawbacks that negate the possibility of their use for both beginners and experienced developers.

1.1 Neural Networks for Encode an Input Image

In order to analyze the input image, we can use the methods of computer vision. A lot of researches has proven that deep neural networks are able to analyze and learn from hidden variables that describe objects of images with respect to the corresponding lengths of variable text values. To create own method of generating program code based on the input image, we decided to use own dataset, a generative adversarial neural network (GAN) to convert the input image to DSL tree and a recurrent neural network (RNN) to convert the DSL language to program code future web page [1].

Consider two neural networks: the first is the generator G: $Z \rightarrow Rn$ with parameters θ, the purpose of which is to generate a similar sample from data, and the second is the discriminator D: $Rn \rightarrow [0, 1]$ with parameters γ. The purpose is to give the maximum estimate on the samples from X and the minimum on the generated samples from G. The distribution generated by the generator will be denoted by p_{gen}. We also note that in the current presentation the architecture of neural networks is not fundamental, therefore, we can assume that the parameters θ and γ are simply the parameters of multilayer perceptrons [2].

As an example, we can consider the generation of realistic photos: in this case, the input for the generator can be random multidimensional noise and the output of the generator (and the input for the discriminator) is an RGB image. The output for the discriminator is the probability that the photo is real, that is, a number from 0 to 1. Our task is to learn the distribution of p_{gen} so that it describes p_{data} as best as possible. We set the error function for the resulting model. From the discriminator side, we want to recognize samples from X as correct, i.e. towards unity, and samples from G as incorrect, i.e. towards zero, so we need to maximize the following value:

$$E_X \sim p_{data}[\log D(x)] + E_X \sim p_{gen}[\log(1 - D(x))], \quad (1)$$

where

$$E_X \sim p_{gen}[\log(1 - D(x))] = E_Z \sim p_Z[\log(1 - G(z))]. \quad (2)$$

On the generator side, it is required to learn to "deceive" the discriminator, that is, to minimize by p_{gen} the second term of the previous expression. In other words, G and D play the so-called minimax game, solving the following optimization problem:

$$minGmaxDE_X \sim p_{data}[\log D(x)] + E_Z \sim p_Z[\log(1 - D(G(z))]. \quad (3)$$

The second neural network is a recurrent neural network, which is based on the LTSM method. Recurrent neural networks were based on David Rumelhart's work in 1986. Hopfield networks is a special kind of RNN were discovered by John Hopfield in 1982. In 1993, a neural history compressor system solved a "Very Deep Learning" task that required more than 1000 subsequent layers in an RNN unfolded in time [3].

Elman network is known as "simple recurrent networks" (SRN):

$$h_t = \sigma_h(W_h x_t + U_h h_{t-1} + b_h), \quad (4)$$

$$y_t = \sigma_y(W_y h_t + b_y),$$

where x_t is input layer; h_t is hidden layer vector; y_t is output vector; W, U та b are matrix and vector parameters; σ_h та σ_y are activators functions.

Long short-term memory (LSTM) is a deep learning system that avoids the vanishing gradient problem. LSTM is normally augmented by recurrent gates called "forget" gates. LSTM prevents back propagated errors from vanishing or exploding. Errors can flow backwards through unlimited numbers of virtual layers unfolded in space. LSTM can learn tasks that require memories of events that happened thousands or even millions of discrete time steps earlier. LSTM works even given long delays between significant events and can handle signals that mix low and high frequency components [4]. Many applications use stacks of LSTM RNNs and train them by Connectionist Temporal Classification (CTC) to find an RNN weight matrix that maximizes the probability of the label sequences in a training set, given the corresponding input sequences. CTC achieves both alignment and recognition. LSTM can learn to recognize context-sensitive languages unlike previous models based on hidden Markov models (HMM) and similar concepts.

Neural network training is carried out with the help of reverse distribution in time, for which an iterative gradient descent is used that changes each weight coefficient in proportion to its derivative in relation to the error.

1.2 Research Data

For training neural networks more than existing 10,000 open access web pages created for 2018–2019 were used. Using rules for recognizing HTML blocks, GAN construct the own dataset which is implemented by creating a tree, which is an algorithm written using domain-specific language.

1.3 Objective

The object of the study in this paper is the process of automation of web page creation using machine learning algorithms to generate web page code from an input image.

The purpose of the work is to develop a software method of reproducing a web page using a combination of GAN and RNN neural networks for generation program code from an input image template that will eliminate the shortcomings of existing software solutions.

2 Related Works

A neural network model developed to generate program code based on a source image includes artificial neural networks based on deep machine learning methods.

Of the existing methods for generating program code, it is worth mentioning the DeepCoder method – a system that can generate computer programs using statistical forecasts to increase traditional search methods [5].

Program code generation is possible in a way of understanding the interactions between an example of input-output through differentiable translators.

DSLs are computer languages, such as markup, modeling, or programming languages. They are designed for a special domain, but they are more syntactically strict than full-featured programming languages.

When we use the DSL language, we have the ability to limit the complexity of the programming language that we need to simulate.

All these advantages allow DeepCoder to write programs much faster than its predecessors. DeepCoder writes a program of three lines in a split second, while previous systems needed several times or tens of times more time to try all possible options [6].

Table 1 shows the speed of generating programs from three lines of code with various tasks.

Since the creation of applications is an actively researched area, the generation of something from the input image is still virtually unexplored area of research. Most existing methods take two components as a basis. The first is convolutional neurons (CNN), which perform the learning function without control, which transforms an unanalyzed input image.

Table 1. Dependence of speed on program length [6]

Timeout needed	DFS			Enumeration			λ^2
to solve	20%	40%	60%	20%	40%	60%	20%
Baseline	$163s$	$2887s$	$6832s$	$8181s$	$>10^4s$	$>10^4s$	$463s$
DeepCoder	$24s$	$514s$	$2654s$	$9s$	$264s$	$4640s$	$48s$
Speedup	**6.8**×	**5.6**×	**2.6**×	**907**×	**>37**×	**>2**×	**9.6**×

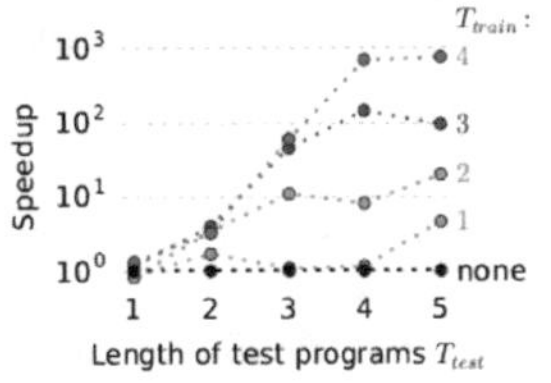

RNN makes language modeling on text from an input image [6]. But each of the existing methods has a number of common drawbacks, for example, a long working time, the lack of a user interface and the need for a powerful remote server.

3 Method of Generating Web Page Program Code from Input Image Using GAN, RNN and DSL

At the time of training, the input image is encoded by a GAN based computer vision model [7]. Proposed method allows to create DSL code consisting of GAN based LSTM layers, which turns into a dataset, which RNN will then use to recreate the program code of the web page (Fig. 1).

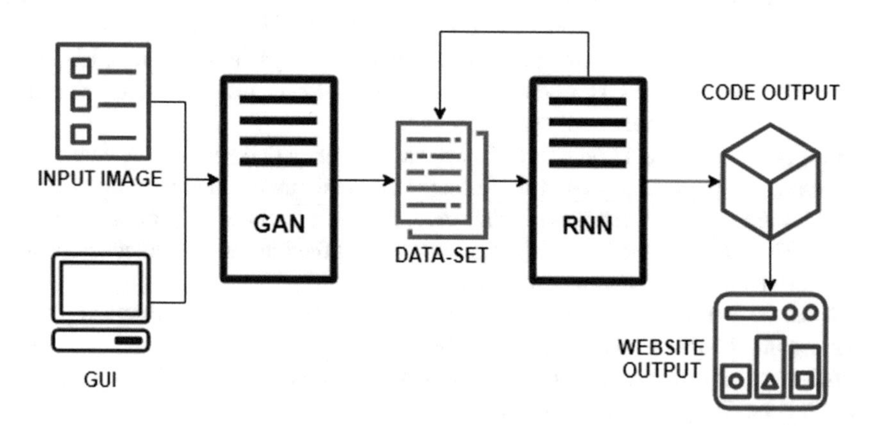

Fig. 1. Proposed method's architecture.

The final DSL token sequence is compiled into final HTML & CSS languages using traditional compiler design techniques.

The main method of operation of the native GAN & RNN method based on LSTM is shown in the following steps.

Step 1. Creating an input image (using PNG or JPEG format).
Step 2. Sending and choosing method's settings.
Step 3. GAN analyzes an input image.
Step 4. GAN generates special DSL tokens which reproduced different blocks in the input image.
Step 5. RNN takes the dataset from the GAN.
Step 6. Created DSL tokens which are in dataset decode into programming code with RNN.
Step 7. RNN makes programming code adaptive and responsive with help of own's dataset.
Step 8. Final tests are made on last code changes.
Step 9. Generated code is ready to deploy.

To implement method based on the neural network LSTM high-level programming language Python and open Keras library are used. Because the software method consists of a user interface and a server part that creates web pages using HTML and CSS languages, it was decided to implement the software as a web application connected to the server part, neural networks running on remote server. For the generative adversarial network the input image is used that the user uploaded to the GUI software. The input image can be of format PNG or JPEG. Next, the GAN analyzes the input image and forms a tree that is written in DSL. This is the first to distinguish own code generation method from existing counterparts and allow own method to work faster than competitors in the field of code generation. After the input image has been analyzed by GAN and the first dataset is formed, it is as input for the RNN. The RNN then generates the code of the future web page based on the input date. In the final stage, RNN repeatedly passes through the generated code to create an adaptive and cross-browser layout.

Unlike existing solutions, the proposed method for creating a dataset is implemented by creating a tree, which is written in DSL. It is worth considering the stages of creating DSL code using the example of creating trees or graphs. First of all, user must create an input image, which depicts the main elements that are present on the site [8]. Consider the process of creating a DSL tree in more detail. For example, a user uploaded an input image. Figure 2 illustrates scheme of user's input image and created DSL tree based on input image.

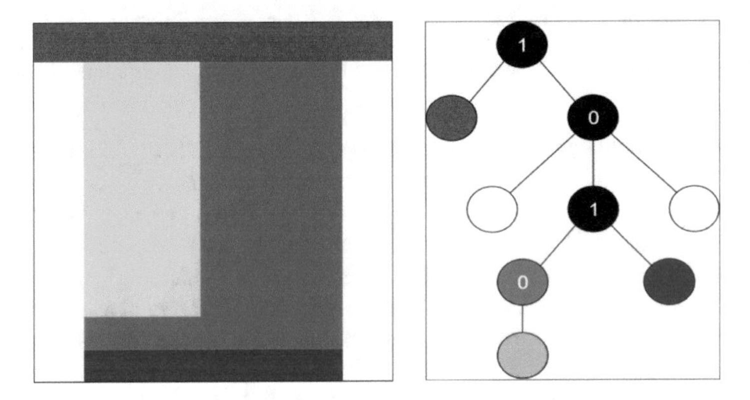

Fig. 2. Scheme of user's input image and created DSL tree based on this input image.

After uploading one image to the user interface, it enters GAN, which should form a dataset using the DSL language.

Thanks to creation own DSL based dataset with the help of GAN, the first acceleration of the algorithm was achieved. Consider the process of creating own dataset. After recognizing an input image, GAN generates a dataset as a DSL tree. It was decided to create its own parsing grid to create own dataset. Thus, there are several rules for recognizing the logical HTML elements depicted in the input image. Black circles indicate logical nodes. The value of the nodes can be 1 and 0. 1 means that the elements are arranged vertically, 0 - horizontally. These rules parse the image.

Figure 2 shows a red header (top of the web page), a white body (a body of the site), a gray container (the area that restricts the body of the site), a blue footer (the bottom of the web page) and a yellow sidebar (the side menu of the web page).

Figure 3 shows more complex scheme of user's input image in terms of image recognition by the neural network. The neural network has been trained on more than 10,000 open access web pages. Using the above rules for recognizing HTML blocks, GAN constructed DSL dataset. On this already complex tree, many different nodes are visible (Logo, Navbar, Side menu, Items, Buttons, Footer, Info etc.).

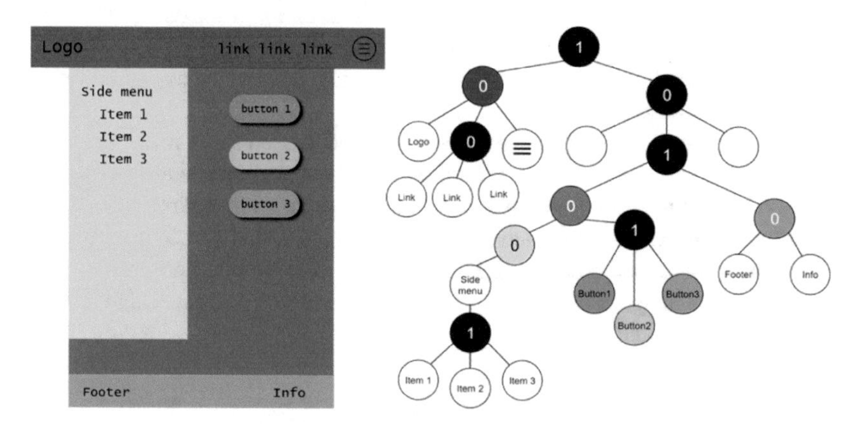

Fig. 3. Input image schema and DSL tree that is based on a user's input image.

GAN starts analyzing the image from top to bottom, left to right. It breaks each element into blocks that are interconnected. A special grid has been created on which the work takes place. For example, in the created tree there are nodes, blocks, pictures and elements. Thus, a DSL code is created, which will later be used as a dataset for RNN in which the page structure based on the image is written.

The next step of proposed method is to transfer such DSL tree to a second recurrent neural network for further code generation. At this stage of operation of the proposed method is the generation of program code using a recurrent network based on the input dateset from GAN. RNN generates HTML & CSS code based on the created GAN tree. Immediately afterwards, a new project structure is created and all dependencies in package.json are specified. At this stage we can start a local server and view the result of the method in different browsers and devices.

4 Results

4.1 Graphical User Interface and Features of Developed Software

The user of developed software creates a site image in a graphical editor, then uploads the image to the GUI and performs settings. The project structure is created and dependencies are specified. The user can download the finished result or view the code generated on the local server (Fig. 4).

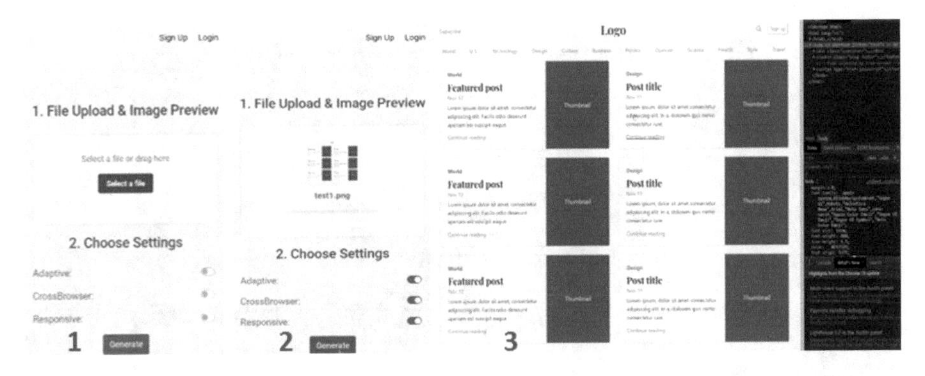

Fig. 4. Using graphical user interface of the developed software (1 - application start window, 2 - uploading the input image to the application, 3 - code generation in the right part of the window for the image in the left part of the window).

4.2 Quality Estimation of Generated Code

To train RNN, we had to create own dataset based on the available code for web pages created for the period 2018–2019. A code generator that creates a user interface based on the DSL language, is compiling it into a software HTML & CSS languages.

To evaluate the quality of the code generation method, it is necessary to investigate and determine the key parameters that are important for the algorithm itself to operate in the minimum and maximum mode [9–12]. Indicators are the total running time of the algorithm; running time based on a text-only input image algorithm; the running time of an input based algorithm consisting of text and images; code generation accuracy. To measure the accuracy of code generation, a test run of the software was performed (Fig. 5).

```
1  ...
2  768/768 [==============================] - 0s 63us/step - loss: 0.4817 - acc: 0.77
3  Epoch 147/150
4  768/768 [==============================] - 0s 63us/step - loss: 0.4764 - acc: 0.77
5  Epoch 148/150
6  768/768 [==============================] - 0s 63us/step - loss: 0.4737 - acc: 0.76
7  Epoch 149/150
8  768/768 [==============================] - 0s 64us/step - loss: 0.4730 - acc: 0.77
9  Epoch 150/150
10 768/768 [==============================] - 0s 63us/step - loss: 0.4754 - acc: 0.77
11 768/768 [==============================] - 0s 38us/step
12 Accuracy: 76.56
```

Fig. 5. Results of accuracy of code generation at 150 epochs

Figure 5 shows the results of testing the accuracy of code generation at the passage of 150 epochs. Figure 5 shows that the code generation accuracy is 77% on average. For existing Pix2Code and DeepCoder analogs, 75% and 70% accuracy values were obtained. It's also worth noting that when the GAN analyzes the input image, it can make errors and packet loss can occur [13]. Figure 6 shows the percentage of packet loss.

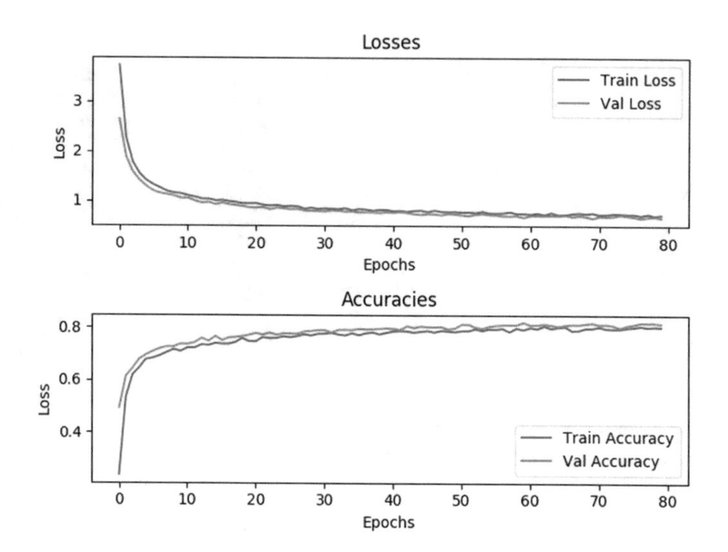

Fig. 6. Loss and accuracies while analyzing an image.

The generated code average accuracy of proposed method is greater by a few percent, which allows more accurate generation of program code, as well as to create its adaptability and cross-browser.

Next is to test the main indicator of the algorithm – the total speed of operation. In order to test the speed of the program method, tests in two loads: input image with text and input image with text and pictures are performed. Testing of the overall performance of the program method was performed using unit and manual tests. Due to the simultaneous operation of GAN and RNN on the basis of our own DSL tree creation method, we managed to increase the speed of the main method on average by 10% (Table 2).

Table 2. The speed generation of web pages using different methods

Program method of code generating	The speed programing code generation of web pages with text	The speed programing code generation of web pages with text and image
Pixcode	2.454 s	5.754 s
Deepcode	2.980 s	6.142 s
Proposed method	2.208 s	5.178 s

We managed to achieve a minimum percentage of inaccuracies in the process of generating program code. Using its own dataset, it was possible to generate the most semantic code that successfully passes ESLint and GooglePageSpeed tests.

Using flex-box layout and abandoning Bootstrap, we managed to achieve code adaptability that existing solutions do not have. Using ReactJS, a convenient GUI was created that allows to customize the method's execution (Table 3).

Table 3. Error results of proposed method

Code generating indicators	Errors (%)
Web UI (HTML/CSS)	10.21
Semantic	6.43
Adaptive	4.89
Responsive	2.41

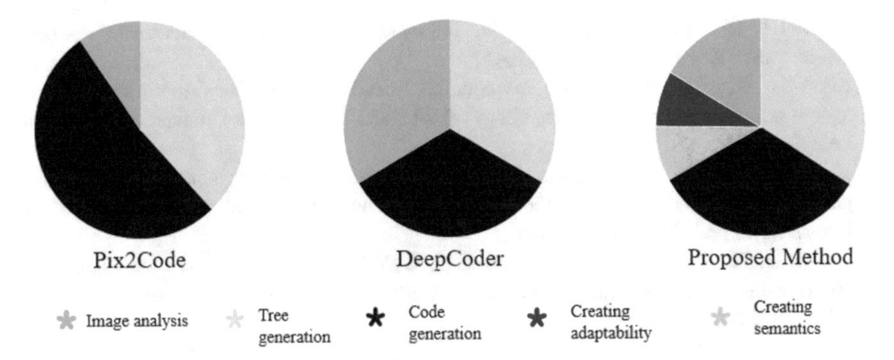

Fig. 7. Comparison of execution time to perform code generation from an input image of existing solutions and proposed method.

Maximum semantics is achieved using a dataset created using 2019 web page creation standards. Cross-browser access is achieved by generating rendering mechanisms vendor-prefixed properties. A further use of the proposed method is creating applications on iOS and Android platforms in Dart with the help of Flutter framework (Fig. 7).

5 Conclusions

The methods of creating program code based on the source image using a neural networks GAN and RNN are analyzed. Ways to improve existing methods and created improved method are analyzed. Using own DSL tree generation method, we managed to get GAN and RNN working together, thus achieving 10% performance increase. Using own GRID, it was possible to achieve adaptability of the generated code. A user interface and settings have been created, thanks to which the user can flexibly customize the operation of the algorithm.

The proposed method not only has higher performance, but also generates code for different browsers and is adaptive, this advantages were not presented by any of the analogues. Cross-browser access is achieved by generating vendor-prefixed properties that are the rendering mechanisms of modern browsers. This allows customizing the properties for each individual rendering engine to allow for inconsistencies between implementations.

The main disadvantage of the proposed method is the inability to recognize the image from the input file, which is located in the project directory.

References

1. Nguyen, T., Csallner, C.: Reverse engineering mobile application user interfaces with REMAUI (T). In: 30th IEEE/ACM International Conference Automated Software Engineering (ASE), pp. 248–259 (2015)
2. Reed, S., Akata, Z., Yan, X., Logeswaran, L., Schiele, B., Lee, H.: Generative adversarial text to image synthesis. In: Proceedings of the 33rd International Conference on Machine Learning, vol. 3 (2016)
3. Vinyals, O., Toshev, A., Bengio, S., Erhan, D.: Show and tell: a neural image caption generator. In: Proceedings of the IEEE Conference on Computer Vision and Pattern Recognition, pp. 3156–3164 (2015)
4. Zhang, H., Xu, T., Li, H., Zhang, S., Huang, X., Wang, X., Metaxas, D.: Stackgan: text to photo-realistic image synthesis with stacked generative adversarial networks. arXiv preprint arXiv:1612.03242 (2016)
5. Krizhevsky, A., Sutskever, I., Hinton, G.E.: Imagenet classification with deep convolutional neural networks. In: Advances in Neural Information Processing Systems, pp. 1097–1105 (2012)
6. Balog, M., Gaunt, A.L., Brockschmidt, M., Nowozin, S., Tarlow, D.: Deepcoder: learning to write programs. arXiv preprint arXiv:1611.01989 (2016)
7. Yu, L., Zhang, W., Wang, J., Yu, Y.: Seqgan: sequence generative adversarial nets with policy gradient. arXiv preprint arXiv:1609.05473 (2016)
8. Shetty, R., Rohrbach, M., Hendricks, L.A., Fritz, M., Schiele, B.: Speaking the same language: matching machine to human captions by adversarial training. arXiv preprint arXiv: 1703.10476 (2017)
9. Hu, Z., Dychka, I., Oleshchenko, L., Kukharyev, S.: Applying recurrent neural network for passenger traffic forecasting. In: Proceedings of International Conference on Computer Science, Engineering and Education Applications ICCSEEA 2019. Advances in Computer Science for Engineering and Education II, pp. 68–77 (2019)
10. Sharma, M., Tomer, M.S.: Predictive analysis of RFID supply chain path using long short term memory (LSTM): recurrent neural networks. Int. J. Wirel. Microw. Technol. (IJWMT) **8**(4), 66–77 (2018). https://doi.org/10.5815/ijwmt.2018.04.05
11. Pawade, D., Sakhapara, A., Jain, M., Jain, N., Gada, K.: Story scrambler - automatic text generation using word level RNN-LSTM. Int. J. Inf. Technol. Comput. Sci. (IJITCS) **10**(6), 44–53 (2018)

12. Mishra, N., Soni, H.K., Sharma, S., Upadhyay, A.K.: Development and analysis of artificial neural network models for rainfall prediction by using time-series data. Int. J. Intell. Syst. Appl. (IJISA) **10**(1), 16–23 (2018). https://doi.org/10.5815/ijisa.2018.01.03
13. Srivastava, N., Hinton, G.E., Krizhevsky, A., Sutskever, I., Salakhutdinov, R.: Dropout: a simple way to prevent neural networks from overfitting. J. Mach. Learn. Res. **15**(1), 1929–1958 (2014)

Method for Synthesis Scalable Fault-Tolerant Multi-level Topological Organizations Based on Excess Code

Heorhii Loutskii[1], Artem Volokyta[1(✉)], Pavlo Rehida[1], Oleksandr Honcharenko[1], and Vu Duc Thinh[2]

[1] National Technical University of Ukraine "Igor Sikorsky Kyiv Polytechnic Institute", Kyiv, Ukraine
georgijluckij80@gmail.com, artem.volokita@kpi.ua, pavel.regida@gmail.com, alexandr.ik97@ukr.net
[2] Ho Chi Minh City University of Food Industry, Ho Chi Minh, Vietnam
thinhvd@cntp.edu.vn

Abstract. Scaling distributed computing systems means increasing the failure probability of individual elements in the system. Fault tolerance issues are of particular importance in distributed systems design. These issues should be considered at different stages of design. In this paper, the main focus is on improving fault tolerance at the design stage of a topology. Based on the following parameters: degree, diameter, topological traffic, and ease of message routing. At our time, high performance of computing systems is achieved through the large number of computing elements. This leads to an increase of the number of failure situations. As a consequence, one of the key requirements for such systems is fault tolerance. There are several approaches to ensure fault tolerance. The simplest and effective approach is to ensure fault tolerance at the topology level. This article discusses methods for synthesizing fault-tolerant topologies. These topologies are hierarchical, scalable with a simple routing system.

Keywords: Topology synthesis · De Bruijn topology · Excess code · Fault tolerance

1 Introduction and Related Work

Modern distributed systems consist of individual elements. Clusters and distributed systems include millions of elements. This allows you to scale the system and use parallel computing. As the number of items increases, the number of potential failures increases too [1–3]. Failure can be easily localized in a small system. After that, you can stop the system and fix the failure. But, this method is not acceptable for a distributed system with many computing nodes. As a consequence, a high level of fault tolerance is one of the key requirements for distributed systems. Also, it is necessary to pay attention to routing methods while designing distributed systems. Routing should be simple and fault tolerant: provide alternative routes and use minimum of resources.

Z. Hu et al. (Eds.): ICCSEEA 2020, AISC 1247, pp. 350–362, 2021.
https://doi.org/10.1007/978-3-030-55506-1_32

There are several approaches that provides reliability and fault tolerance [4–6]. First. Fault tolerance of distributed system provided only by software that uses complex routing and virtualization algorithms. But such approach requires a lot of computing resources and makes the distributed system slow [7–10].

Second. This approach is aimed to provide hardware fault tolerance at the topology level. It improves fault tolerance with minimal usage of resources. But the fault-tolerant topology should be consistent with the next requirements: optimal topological parameters and simple routing. Synthesis of fault-tolerant topologies is an important task while designing distributed systems [11–15].

The main methodology, used on this research, is a graphs theory. The main objective of this research is to propose method of synthesis fault tolerant topologies with next properties:

- High possibility to failures' bypassing
- Simple and fault tolerant routing
- Standard logical structure, shown to user

The structure of main part of article is next. In this section, the prerequisites and the urgency of this research are highlighted, the objective and methodology are shown. In 2nd section the main parameters of directly connected networks are described. In 3rd section considered the methods of topologies synthesis, that used as basis for proposed method. The main features of these are highlighted. In 4th section the essence of proposed method is shown. The 5th section is devoted to issues of routing. The method of failures' bypassing on routing "one-to-one" are described. The main parts of possible routing algorithm are shown. The 6th section highlights the results. In this section the main parameters of topologies, created by this method, are shown. The 7th section it is conclusion.

2 Parameters of Directly Connected Networks

Directly connected networks (DCN) or point-to-point networks are popular network architectures. These networks are easy to scale. DCN consists of a set of nodes. These nodes are directly connected to other network nodes. The main component of these nodes is the router, that controls the transfer of messages between nodes. Therefore, DCN is often called routing based networks.

Directly connected topological networks are a good approach for designing scalable systems [1]. But the effectiveness of such systems depends on the parameters of the topology. These topology parameters are: number of nodes (N), degree (S), diameter (D), average diameter $(\overline{D})$, topological traffic (Q), cost (C).

The topology will be represented as a next graph $G(X, Y)$, where $X = \{0, 1, \ldots, (N-1)\}$ – set of nodes, $Y = \{0, 1, \ldots, (E-1)\}$ – set of edges.

Diameter (D) - is the minimum distance between the outermost vertices of the graph. For example, in the three-dimensional hypercube (Fig. 1), the outermost pairs of vertices are 0–7, 1–6, 2–5, 3–4. It takes 3 steps to send data from sender to recipient. Therefore, the diameter D = 3.

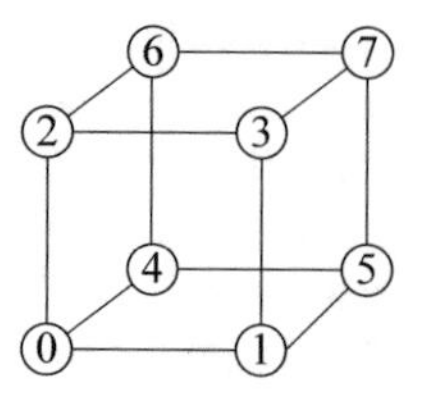

Fig. 1. Hypercube topology.

Degree (S) – is the maximum number of edges, that are incident to one vertex. For example, vertices 1 and 3 (Fig. 2), that's why degree S = 3.

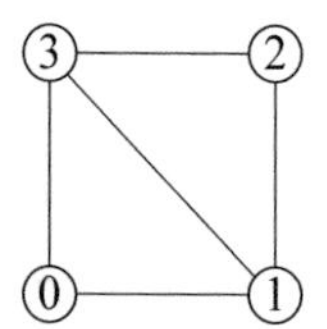

Fig. 2. An example graph with degree S = 3

The average diameter $\left(\overline{D}\right)$ – is the average distance between the vertices of the graph. The average diameter is determined on the basis of the following formula:

$$\overline{D} = \frac{\sum_{i=0}^{N} \sum_{j=i+1}^{(N+i-1) modN} d_{ij}}{N(N-1)}$$

For a regular graph, this formula is simplified. For any i the formula is determined by:

$$\overline{D} = \frac{\sum_{j=i+1}^{(N+j-1) modN} d_{ij}}{N-1}$$

For example, let's consider the calculation of the average diameter for a three-dimensional hypercube (Fig. 1). This graph is homogeneous. The average diameter is the average distance between any vertex i and all other $(N-1)$ vertices. For a node with number $i = 0$, the average distance is determined on the basis of the spanning tree (Fig. 3).

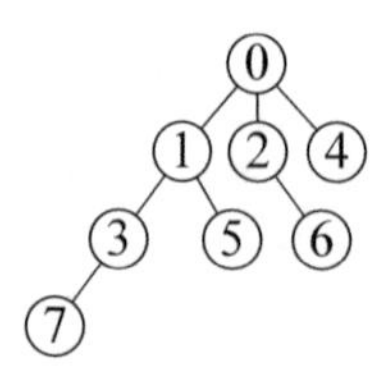

Fig. 3. Hypercube topology. Spanning tree.

Then in accordance with (2):

$$\overline{D} = \frac{1*3+2*3+3*1}{7} = \frac{12}{7}$$

Topological traffic determines the potential average load of edges by messages (packets). If each node can transfer only one packet at a time. It means next: there may be no more than N packets in the network at a time.

Each packet covers an average distance of $(\overline{D})$. Therefore, to transmit all packets without delay, you need $\overline{D}N$ edges. Graph has $R = \frac{N*S}{2}$ edges, than traffic:

$$Q = \frac{\overline{D}N}{\frac{N*S}{2}} = \frac{2\overline{D}}{S}$$

Traffic for three-dimensional hypercube (Fig. 1)

$$Q = \frac{2*12}{7*3} = \frac{24}{21} = \frac{8}{7}$$

As the number of nodes increases, the topological traffic will approach 1. Thus, the hypercube is close to the ideal value of Q.

The cost (C) of the topological organization depends on the degree and equals $C = N*S$. Therefore, degree value always imposed by significant restrictions. In practice, degree value should be less than 7, and are in the range $S = 4-6$.

The diameter D significantly affects the performance of the system and has to be optimized. Therefore, while searching for optimal solutions of topological organizations, it is necessary to search for solutions with minimum values of the multiplicative parameter $S*D$. With the same $S*D$ values, we should select the topologies with topological traffic that is closer to 1.

Currently, there are a large number of topologies. There are many distributed systems that have been successfully constructed, based on those topologies. However, with the increasing number of elements, these topologies are unacceptable because of their large degree or large diameter. Therefore, the paper proposes a new method for the synthesis of topologies with minimum values of $S*D$.

3 Topology Synthesis Methods for Fault-Tolerance Routing

There are many methods for synthesizing topologies. In the context of this article, the most interesting is synthesis, based on code transformations and Cartesian product.

Synthesis Based on Code Transformations. This approach based on the fact that neighbors of each node can be obtained by transformations of the code of a given node.

De Bruijne topology and hypercube are good examples of this synthesis method. The de Bruijne topology is based on left and right offsets with inserts 0 and 1 in the vertex code [7]. The hypercube is built on the basis of an Exchange transform:

neighbors of any vertex can be obtained by inverting the i bit of the given vertex code (where - $i \in [0, n-1], n - codelength$). The main advantage of this method is the ease of routing. Since the neighbors of the vertex are obtained through transformations, the routing is performed similarly. Also, this method makes allows to control the degree and diameter. It is possible because the topology parameters are determined by the corresponding transformations.

Synthesis Based on Cartesian Product. Two graphs are required for this method [1]. The first graph describes the links inside of the cluster, and the second graph describes links between clusters. For example, Fig. 4 shows a Cartesian product of two topologies - a square and a square. Each vertex has a dual code. The vertex code consists of two parts. The first part is the cluster number (number from the outer graph), the second part is the number inside the cluster (from the inner graph).

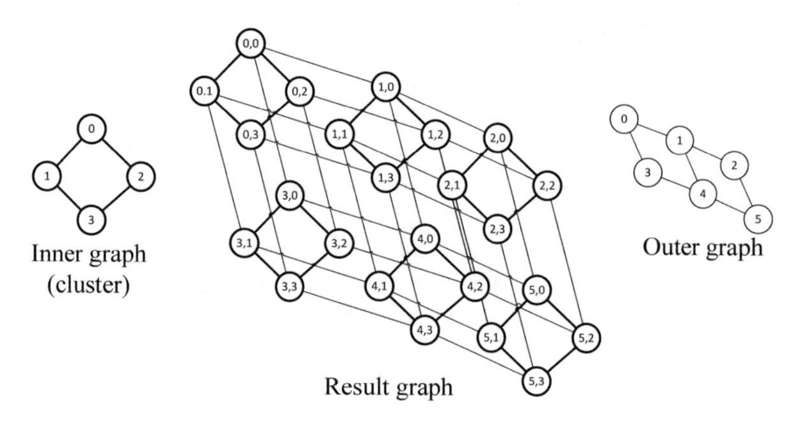

Fig. 4. Example of Cartesian product

The Cartesian product is interesting for optimizing *SD* parameters. Also, we can use features from each topology that was used in multiplication. For example, while using a hypercube as an external topology, allows to use hypercube routing for transferring data between clusters. Thus, topological traffic of the resulting topology improves. Also, features of second topology can be used inside of the cluster. The first level of the cluster - is the level of the outer graph. The second level is the inner graph level. Navigating between levels allows you to bypass faults.

Faults in Outer Connections. In this case, we take advantage of the fact that all the vertices of the sender cluster have links to the vertices of the receiver cluster. Therefore it is possible to make a transfer inside of the cluster. After that, message could be send to the destination cluster. At last, message will be transfered to the node, using the inner links inside cluster.

Faults in Inner Connections. In this case, it is necessary to use advantages of outer links. At first, send the message to the neighboring cluster. After that, deliver it to the vertex that has link with the vertex of the receiver. At last, deliver the message to the node using inner links.

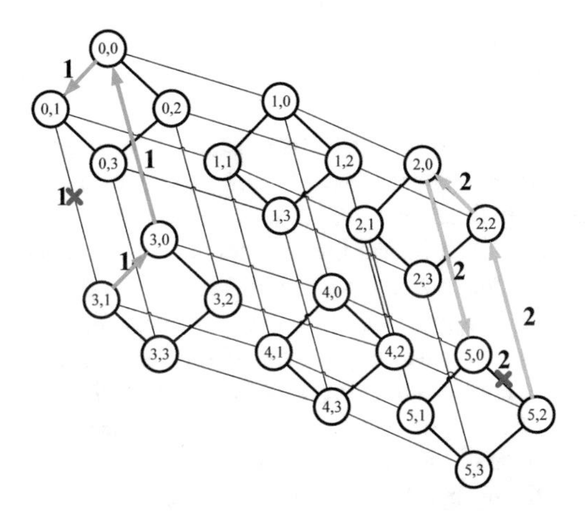

Fig. 5. Bypassing faults for inner and outer links failure

Figure 5 shows an example of this cases. Number 1 (blue line) shows an example of bypassing failure in case of outer failure. (between nodes 3.1 and 0.1). Number 2 (green line) shows the path to bypass failure in case of inner failure. (between nodes 5.2 and 5.0).

4 Proposed Method for Synthesizing Topologies Based on Excess Code

4.1 Excess Code 01T and Its Use

Excess code 01T is an extension of the binary system, that supplemented by a "minus 1" digit (−1 is represented by the letter T) [13–15]. The peculiarity of this numeral system is that some numbers in this code can be represented in different ways. This can be used to build a fault-tolerant topology. This allows the topology nodes to be clustered with the same number. As a consequence, in the event of a failure, we can replace one node with another with that number. The most effective is to use excess code in code-based synthesis.

This allows you to keep the ease of routing and increase fault tolerance. For example, we replace the node encoding system in de Bruijne's topology. Therefore, we are able to obtain a redundant de Bruijne topology. It is highly fault tolerant and supports tree-based and code-based routing. The authors discussed the usage of excess coding in synthesis of topology based on Latin square [14]. This paper examines the approach to synthesis of topologies based on Cartesian product with excess code.

4.2 The Essence and Advantages of the Proposed Method

As described previously, two topologies are required for synthesis with Cartesian Product. The internal topology is a cluster (or bottom-level topology). The external topology are connections between clusters (or top-level topology). But while designing distributed system, these features allow you to clearly separate the levels of the hierarchy.

Our method proposes the following. The cluster is a local part of the system and its internal structure is hidden from the user. Therefore, it is better to place topology with the excess code at the bottom-level. Also, it is better to use code transformations to synthesis bottom-level (which means faster execution of the task).

The first step is to select a graph for the cluster. For example, a good variant of a cluster is an excess de Bruijn graph or an excess hypercube. This allows you to get benefits of redundancy: the rate of increase of the number of nodes (3n), flexible adjustment of the topology parameters, a convenient routing, bypassing the failure connections, and the presence of nodes with the same numbers. The disadvantages are the next: are the high rate of growth of the node degree and the number of edges. But this will not be significant, because the cluster is not scalable. All its parameters are known and constant.

In the second step, we have to select the top-level topology. We can chose any standard topology (such as a ring, hypercube or mesh). Further nodes of the outer topology are replaced by clusters. The inner links are unchanged, inside the cluster. Outer links are formed by the Cartesian product. If there is a connection between clusters 1 and 2, then for each node $(1, i)$ there is an outer link with the node $(2, i)$. An example of synthesized topology is shown in Fig. 4.

All structures in our method are double-leveled. This means that top-level routing is independent of bottom-level routing. Due to that cluster cant be changed, it gives us the stability of scaling. This allows the use of redundancy possibilities (nodes with the same number are connected, we call it “quasi-quantum connection”). For the user the cluster is transparent. Clusters have the same number of nodes. This fact makes possible to design symmetrical system. This is another benefit.

5 Multilevel Routing with Bypassing Failures

Cluster routing is independent of the top-level topology. This allows you to use ready-made solutions. But it is worth considering the possibility of increasing through the interaction of levels.

The following algorithm is proposed. At first, routing on the bottom-level is performed. The message transfers to the node of the cluster, which is equivalent to the destination node in another cluster. At the next step, we use top-level routing. This option works only with outer connections between clusters.

Let’s consider options for bypassing the failure situations in the example. This is a top-level scalable ring. A redundant de Bruijn topology used as bottom-level topology inside the cluster. The redundant de Bruijn topology has three routing trees. It is known that this topology is still working even with two failures [15].

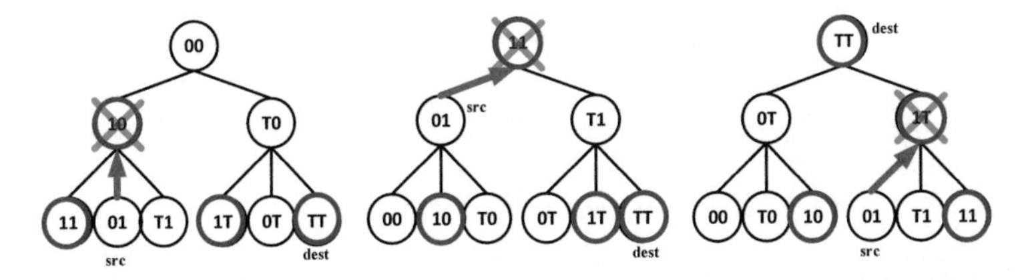

Fig. 6. Situation in excess de Bruijn, when bypassing of malfunctions are impossible

But with three failures, it is likely that bypassing the failure will be impossible. This situation is shown in Fig. 6 (node 01 is the sender, node TT is the receiver, red nodes 10, 11, 1T with failures).

This problem can be solved using outer links. Let the ring contain N clusters and cluster 1 has failures. Then we do next steps.

1. The node transfers the message to the neighboring cluster. The number of the new cluster larger by one than the sender's number. This is cluster 2 (Fig. 7).
2. A bottom-level routing algorithm is performing in cluster 2.
3. After that routing top-level algorithm is performing and message is sending to the receiver's node.

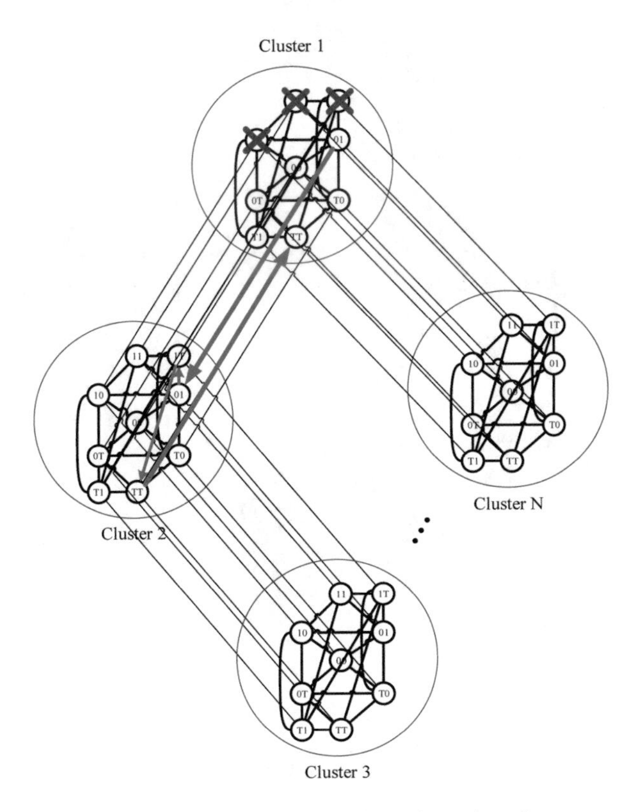

Fig. 7. Bypassing the failure through another cluster

Figure 7 presents this example for transferring a message between nodes 01 and TT in Cluster 1.

Another case. Failure in outer link. It is not possible to transfer control to the top-level algorithm control.

In this case, our algorithm includes the following steps:

1. Searches of the node with needed outer link. If such node was found, the message transfers to this node. This node starts top-level routing algorithm and transfers message to the receiver's cluster.
2. In that case, if any of nodes don't have link with the receiver cluster, that means that all nodes in this cluster are not working. Such a failure is serious and needs detailed analysis.
3. In the receiver's cluster, a bottom-level algorithm is executed. The message is transferred to the needed node.

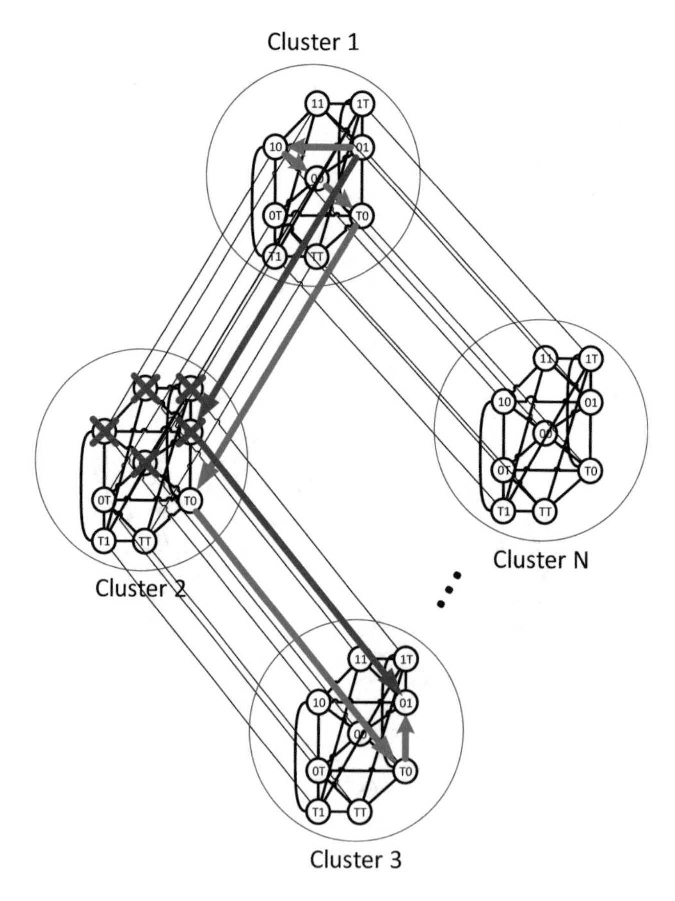

Fig. 8. Bypassing the failure in outer link.

Figure 8 presents an example of such bypassing. Node 1.01 sends a message to node 3.01. The red arrow shows the normal transmission path, but it cannot be used due to a failure. The blue arrow shows the alternative path in bypass. Searching for the node with the outer link (vertex 1, T0) in Cluster 1 is performing. After that, the message is transferred to Cluster 2. Then the message is transferred to Cluster 3. The routing method is performing inside Cluster 3, and the message transferred to node 3,01.

6 Results and Discussions

Let's consider the following topologies for comparison: hypercube, de Bruijne topology, excess hypercube, excess de Bruijne, circle, mesh, and binary tree. Table 1 presents the key parameters for topologies. Those parameters for classic topologies are well-known. For excess topologies those parameters are taken for experiments, presented in the past researches.

Table 1. Parameters for basic topologies

Parameter	Topologies						
	Hyper cube	De Bruijn	Excess hypercube	Excess De Bruijn	Mesh	Binary tree	Circle
Nodes (N)	2^n	2^n	3^n	3^n	n^2	$2^n - 1$	n
Degree (S)	n	4	2n	6	4	3	2
Diameter (D)	n	n	n	n	n/2	2n − 1	n/2
Local connectivity (L)	n	2	2n	4	4	2	2
SD	n^2	4n	$2n^2$	6n	2n	3 * (2n − 1)	n
Routing trees	2	2	3	3	–	1	–
Codes transformations	Yes	Yes	Yes	Yes	No	No	No

Table 2 presents the prognosed topological parameters for proposed multilevel topologies. Uses the Cartesian product properties, we can theoretically predict main parameters of multilevel topologies.

Hypercube, de Bruijne topology and their modifications based analogues excess code were selected as clusters. Hypercube, ring, mesh and binary tree were used as top-level topologies.

As it shown in Table 2, realizations, includes from excess versions has higher local connectivity. It allows to increase fault-tolerance by using those additional connections. But the degrees of those topologies are higher too. This is a main limitation of proposed method.

Table 2. Parameters for multi-level topologies

External topologies (rank m)	Parameters	Clusters (rank n)			
		Hypercube	De Bruijn	Excess hypercube	Excess De Bruijn
Hypercube	Nodes (N)	2^{n+m}	2^{n+m}	$3^n * 2^m$	$3^n * 2^m$
	Degree (S)	n + m	4 + m	2n + m	6 + m
	Diameter (D)	n + m			
	Local Connectivity (L)	n + m	2 + m	2n + m	4 + m
Circle	Nodes (N)	$m * 2^n$	$m * 2^n$	$m * 3^n$	$m * 3^n$
	Degree (S)	n + 2	6	2(n + 1)	8
	Diameter (D)	$n+\frac{m}{2}$			
	Local Connectivity (L)	n + 2	4	2(n + 1)	6
Mesh	Nodes (N)	$2^n * m^2$	$2^n * m^2$	$3^n * m^2$	$3^n * m^2$
	Degree (S)	n + 4	8	2(n + 2)	10
	Diameter (D)	$n+\frac{m}{2}$			
	Local Connectivity (L)	n + 4	6	2(n + 2)	8
Binary tree	Nodes (N)	$2^n(2^m - 1) \approx 2^{n+m}$	2^{n+m}	$3^n * 2^m$	$3^n * 2^m$
	Degree (S)	n + 3	7	2n + 3	9
	Diameter (D)	n + 2 m − 1			
	Local Connectivity (L)	n + 2	4	2(n + 1)	6

The combinations with excess hypercube have a very big degree. It limits a rank of clusters in range of 1–3. Unlike this, excess de Brujine's clusters haven't this limitation and, as result, are more useful. On the top level, the circle has a very bad ratio between count of nodes and diameter. It produces to very big diameters of result topologies.

As result, the most perspective solutions are next. These are: 1. Excess de Brujine with tree; 2. Excess de Brujine with hypercube; 3. Excess de Brujine with mesh.

Solutions 1 and 2 propose speed of nodes' count growth $3^n * 2^m$. We said above, that cluster it is unscalable part of system. As result, this formula may be represented as $N_1 * 2^m$, where N_1 – fixed size of cluster. It is good. The 3rd solution may propose nodes' count growth as $3^n * m^2$. It may be represented as $N_1 * m^2$. The 1st and 3rd solutions have constant degree and it is a big advantage. But their local connectivity is constant too. The degree of 2nd solution changes as 6 + m. But the diameter of 2nd solution is least relative to growth's speed: only n + m against n + 2 m − 1 on 1st solution. The 2nd solution has less coefficient on formula, but speed of growth n^2 removes all advantages of this.

7 Conclusions

A new method of synthesis of fault-tolerant topologies based on Cartesian product and usage excess code was presented in this paper. All steps of this method was illustrated. Also the multi-level routing method and method for bypassing failures were presented. Compatible usage of routing trees and multi-level topologies is shown too. Formulas for the basic topological parameters for infinite scaling are presented.

The main advantages of using redundancy for the synthesis of multi-level topologies are given. There is the ability to degree and diameter adjustment based on the choice of cluster of different rank. But there is a disadvantage: redundant topologies have a higher degree. This is a fee for increased fault tolerance.

The proposed method may be used for constructing multilevel distributed or cluster systems. It highlights a way, that allows to use main advantages of excessity without changes on logical structure of system. It allows to use existing software solution, simplify routing and increase a fault-tolerance.

Future researches may address bypassing failures in complex cases, when is not possible to use routing trees and outer links. An additional task is to specify the maximum number of failures that can potentially be bypassed. And, as alternative way, it may be considered a using of virtual excessity for the routing in standard topologies.

References

1. Ganesan, E., Pradhan, D.K.: The hyper-deBruijn networks: scalable versatile architecture. IEEE Trans. Parallel Distrib. Syst. **4**(9), 962–978 (1993)
2. Mukhin, V., Volokyta, A., Heriatovych, Y., Rehida, P.: Method for efficiency increasing of distributed classification of the images based on the proactive parallel computing approach. Adv. Electr. Comput. Eng. **18**(2), 117–122 (2018)
3. Hu, Z., Mukhin, V., Kornaga, Y., Volokyta, A., Herasymenko, O.: The scheduler for distributed computer systems based on the network centric approach to resources control. In: Proceedings of the 2017 IEEE 9th International Conference on Intelligent Data Acquisition and Advanced Computing Systems: Technology and Applications, IDAACS 2017, pp. 518–523(2017)
4. Korniyenko, B., Galata, L.: Implementation of the information resources protection based on the CentOS operating system. In: 2019 IEEE 2nd Ukraine Conference on Electrical and Computer Engineering (UKRCON), pp. 1007–1011. IEEE (2019)
5. Shmalko, O., Rehida, P., Volokyta, A., Loutskii, H.: The programming language for embedded real-time devices with reducing errors and without reducing the performance of programs. Inf. Telecommun. Sci. **9**(2), 44–53 (2018)

6. Kravets, P.I., Shymkovych, V.M., Samotyy, V.: Method and technology of synthesis of neural network models of object control with their hardware implementation on FPGA. Paper presented at the Proceedings of the 2017 IEEE 9th International Conference on Intelligent Data Acquisition and Advanced Computing Systems: Technology and Applications, IDAACS 2017, vol. 2, pp. 947–951 (2017). https://doi.org/10.1109/idaacs.2017.8095226
7. Dürr, F.: A flat and scalable data center network topology based on De Bruijn graphs. arXiv preprint arXiv:1610.03245 (2016)
8. Faizian, P., Mollah, M.A., Yuan, X., Alzaid, Z., Pakin, S., Lang, M.: Random regular graph and generalized De Bruijn graph with k-shortest path routing. IEEE Trans. Parallel Distrib. Syst. **29**(1), 144–155 (2017)
9. Aggarwal, A., Verma, R., Singh, A.: An efficient approach for resource allocations using hybrid scheduling and optimization in distributed system. Int. J. Educ. Manag. Eng. (IJEME) **8**(3), 33 (2018)
10. Kulakov, Y., Kohan, A., Kopychko, S.: Traffic orchestration in data center network based on software-defined networking technology. In: International Conference on Computer Science, Engineering and Education Applications, pp. 228–237. Springer, Cham, January 2019
11. Lu, Y., Sun, F., Zhao, Y., Li, H., Liu, H.: Distributed traffic balancing routing for LEO satellite networks. Int. J. Comput. Netw. Inf. Secur. (IJCNIS) **6**(1), 19–25 (2014). https://doi.org/10.5815/ijcnis.2014.01.03
12. Kamal, Md.S., et al.: De-Bruijn graph with MapReduce framework towards metagenomic data classification. Int. J. Inf. Technol. **9**(1), 59–75 (2017)
13. Olexandr, G., Rehida, P., Volokyta, A., Loutskii, H., Thinh, V.D.: Routing method based on the excess code for fault tolerant clusters with InfiniBand. In: Advances in Computer Science for Engineering and Education II, ICCSEEA 2019. Advances in Intelligent Systems and Computing, vol. 938. Springer, Cham (2020)
14. Honcharenko, O., Volokyta, A., Loutskii, H.: Fault-tolerant topologies synthesis based on excess code using the Latin square. In: The International Conference on Security, Fault Tolerance, Intelligence ICSFTI 2019, Ukraine, Kyiv, 14–15 May 2019, pp. 72–81 (2019)
15. Loutskii, H., Volokyta, A., Rehida, P., Goncharenko, O.: Using excess code to design fault-tolerant topologies. Techn. Sci. Technol. **1**(15), 134–144 (2019). https://doi.org/10.25140/2411-5363-2019-1(15)-134-144

Spatial Transformer Steerable Nets for Rotation and Flip Invariant Classification

Viacheslav Dudar(✉) and Vladimir Semenov

Taras Shevchenko National University of Kyiv,
64/13, Volodymyrska Street, Kyiv 01601, Ukraine
slavko123@ukr.net, semenov.volodya@gmail.com

Abstract. The paper focuses on problem of achieving approximate rotation and flip invariance of CNNs for classification. Group equivariant and steerable convolutional nets are models that could generate rotation and flip equivariant representations of input images. However, training and evaluation of such networks requires much memory and computations. We propose to use small steerable network to obtain orientation vector that is equivariant to rotations and use it to compute rotation invariant tensor, and feed it to usual CNN. We show that in case there is only one object on the scene our model shows approximate rotational and flip invariance and works faster than aforementioned approaches.

Keywords: Convolutional neural nets · Steerable nets · Rotational invariant neural net · Horizontal flip invariance of CNN · Dense steerable net · Interpretation of steerable filters · Spatial transformer network

1 Introduction

Convolutional neural networks are state of art methods in image classification, image segmentation, object detection, and many other computer vision fields [1]. From our point of view, the success of CNNs is determined by such factors:

- Built in invariance to image shifts
- CNNs are universal approximators in the class of translational invariant functions
- Weight sharing: features learned at one position are used in all other positions.

However, CNNs have no guarantees about invariance of classification to other types of geometric transformations: horizontal image flips, rotation, scaling, shear, and general local and global affine transforms [1, 12, 15].

This paper focuses on the ways to build in invariance to some geometric transformations of the input image to the model. Such models will be able to give accurate predictions on rotated/flipped images it hasn't seen at the training stage.

There are several ways to enforce invariance of the model to affine transforms. Most popular one is data augmentation [1]. At the training stage random affine transform is applied to each image in the minibatch. In this way model is trained to make correct predictions for affine transformed inputs. The problem of this approach is that large number of epochs is required to train the model to have some degree of invariance. Since general affine transform has 6 degrees of freedom, it is very unlikely

Z. Hu et al. (Eds.): ICCSEEA 2020, AISC 1247, pp. 363–372, 2021.
https://doi.org/10.1007/978-3-030-55506-1_33

that some small number of randomly transformed samples will be enough for the model to become truly invariant. That's why ways for building in invariance to affine transforms into the inner structure of convolutional neural network get a lot of attention.

Structure of the paper is as follows. In Sect. 2 we discuss existing approaches for improving generalization of neural networks to rotation/flip of inputs. Section 3 describes steerable convolutional kernels w.r.t. rotation/flip, since architecture we propose to use is based on specific steerable kernels. In Sect. 4 we describe proposed convolutional network that has desired rotational/flip invariance. Results of experiments are in Sect. 5. Finally, we draw conclusions and describe future work in Sect. 6.

2 Related Work

Here we list some of approaches that are used to build in equivariance/invariance to affine input image transforms into neural networks.

The first one is tangent propagation [2]. The idea of this method is to minimize norm of the model output gradient w.r.t. parameters of the input transformation. At the first stage gradient of input w.r.t. parameters of input transformation needs to be found. This could be done numerically, by deforming the image, and then subtracting initial image from it. At the second stage derivative of the norm of gradient w.r.t. transformation parameters is found. This stage involves differentiating Jacobian of the model, and could be done with technique similar to backpropagation, called tangent propagation. After that usual cross-entropy loss with regularization term described above is minimized. Multiplier λ at regularization term is responsible for trade-off between model fitting capability and generalization.

The next approach relies on usage of Gabor filters for CNN [3]. Suppose the variable u controls the frequency and v controls orientation. Then Gabor filter is defined as:

$$G_{u,v}(z) = \frac{\left\|k_{u,v}\right\|^2}{\sigma^2} \exp\left(-\left(\frac{\left\|k_{u,v}\right\|^2 \|z\|^2}{2\sigma^2}\right)\right)\left(e^{ik_{u,v}z} - e^{-\frac{\sigma^2}{2}}\right) \tag{1}$$

Here $k_{u,v} = k_v e^{ik_u}$, $k_v = (\pi/2)/\sqrt{2}^{v-1}$, $k_u = \pi u/U$, $\sigma = 2\pi$, $u = 0\ldots U, v = 0\ldots V$. Experiments show that behavior of some cells in visual cortex of animals can be modelled as convolution of input image with Gabor filter.

Authors propose to multiply trainable convolutional kernels by Gabor filters. So, kernels for v-th scale are such:

$$C_{i,u}^{v} = C_{i,o} \circ G_{u,v} \tag{2}$$

Authors use fixed number of orientations u for each convolution, and change the value of scale v depending on the depth of the current layer (deeper layers correspond to larger values of v).

One of advantages of such approach is that reduced number of filters is needed to be stored, and their augmentation for accounting for different orientations is done at inference time.

Usage of kernels multiplied with Gabor filters improves the model response to different orientations of features, but does not guarantee that output of the network will be invariant to rotations.

One more approach is Spatial Transformer networks [4]. Authors propose to use additional CNN that takes image as input, and outputs parameters of the affine image transform. After that estimated affine transform is applied to the input image, which is then fed to the second CNN that performs classification. Bilinear sampling is used to make the affine image transformation differentiable. Authors also experimented with use of spatial networks at several layers, and with parallel implementations, when several spatial networks find several regions on the input image which are then processed with the second network. This approach showed better results than usual CNNs on several datasets: Distorted MNIST, SVHN and CIFAR-10.

However, Spatial Transform Networks do not guarantee that if the image is rotated, then change of the estimated parameters will compensate that rotation. Thus rotational invariance could be learned by the model, but it is not build in initially.

In the paper Learning Rotation Invariant Convolutional Filters for Texture Classification [5] authors propose to rotate convolutional kernels with some angular step in order to detect features with different orientation. Filtering with rotated kernels implies that if the image is rotated then corresponding output will be cyclically shifted. The next layer of the network is orientation pooling – operation that finds maximal activation among orientation channels. Spatial pooling and fully connected layer are applied next. Output of the network is invariant to rotations of the image because of orientation pooling (exact invariance is achieved when the image is rotated by the angle that is a multiple of angular step). Experiments showed that such CNN generalizes better, especially in case of small training set.

The drawback of this approach is that the way to perform multilayer equivariant computations was not proposed.

The paper Group Equivariant Convolutional Networks [6] proposed the way to build in invariances to some geometric transformations into the structure of neural network. It was shown that usual CNN architecture can be modified to be equivariant to sets of transformations that form the discrete groups: translations, 90° rotations, symmetric reflections, and their combinations. 2 discrete groups are considered: p4 – group of translations and rotations by 90°, and p4m – p4 augmented with reflections. In the usual CNN we replace convolutional kernels with their 4 rotated copies (in case of group p4), or with 8 rotated and reflected copies (for the group p4m), and perform convolution with transformed kernels, to get 4(8) separate tensors, that detect features at different orientations [13]. If we apply transformation from the group to the input image, these tensors are modified in accordance with induced group. To make convolution from one group of tensors to another equivariant to transformations, we need to modify kernels in accordance with the group table. G-convolution is introduced – it is an extension of usual convolution that is equivariant to group operations also. For all other CNN operations, such as pooling and batch normalization, their group versions are also developed [14]. After these operations are implemented, we can take any plain

CNN architecture, replace all layers into group layers, and obtain network invariant to group operations (it has to be trained from scratch).

To make G-convolution efficient on GPUs, in practice all kernels are concatenated along output dimension and then convolution is performed. After that resulting tensor is split to components back, if needed. This speeds up G-convolution, since single convolution with many outputs is much better parallelized than several smaller convolutions.

One more approach is Steerable CNNs [7]. Here authors proposed to use steerable kernels to improve efficiency of group convolutions at the same time preserving invariance of output to 90° rotations and reflections. The network is called steerable under some group of transformations, if representations it generates change in predictable way under group transformations of input. Authors show that each steerable representation is a composition of low-dimensional independent feature types.

Similarly to G-convolutions, Steerable Convolutions could be used like usual convolutions. Just, instead of number of output features one needs to specify multiplicities of each feature type. We will use steerable convolutions throughout the paper, and will describe types of features and their interpretation in the next sections.

One more approach to mention is capsule networks [8] which combine several ideas. First of all, each feature is equipped with orientation vector, which length represents presence of the feature at position. This vector could represent direction, relative shift or other geometrical properties. Simple features are combined into more abstract features with the procedure of dynamic routing: simple features vote for presence and configuration of higher level features, and if their vote agrees with the total configuration, their contribution into total sum is enhanced. Approximate equivariance of representations is reached by the fact that everything depends on relative configurations of features: if, for example, two lower level features are rotated, then orientation vector of higher level feature is also expected to be rotated in the same way. However, there are no inner model mechanisms that could guarantee that with respect to affine transforms. That's why training set has to be augmented with affine transformed images in order to generalize well to different affine transformations.

Group and steerable networks also were generalized to 3D inputs with respect to more complicated 3D spherical rotation groups [15].

Finally, in the paper A General Theory of Equivariant CNNs on Homogeneous Spaces [9] general theory of equivariant CNNs with respect to group transforms was developed, that covered several previously developed methods.

3 Properties of Steerable Feature Types

Here we list types of steerable kernels for different transformations. Consider firstly a simple 2-element group of horizontal image flips. There are 2 types of convolutional kernels that have steerable behavior under group transformations: symmetric and asymmetric:

$$k_s = \begin{pmatrix} a & d & a \\ b & e & b \\ c & f & c \end{pmatrix}, k_a = \begin{pmatrix} -g & 0 & g \\ -h & 0 & h \\ -i & 0 & i \end{pmatrix} \tag{3}$$

We denote input image by I, convolution by $*$, horizontal image flip by f, $\pi/2$ clockwise rotation by r, and identity transformation by e. Then behavior of these 2 kernels can be summarized with the following Table 1:

Table 1. Steerable kernels for horizontal flips group

Kernel\transformation	e	f
k_s	A	fA
k_a	B	$-fB$

The table above gives a way to predict result of convolution for 2 kernel types for horizontal flips of the image. It shows, that if $I * k_s = A$ then $fI * k_s = fA$, and so on. So for symmetric kernel we just need to flip the result, and for asymmetric kernel we also need to multiply result by -1. Also note that arbitrary kernel can be represented as linear combination of symmetric and asymmetric.

For the rotation group steerable kernel types are such (k_x and k_y form a pair):

$$k_s = \begin{pmatrix} a & b & a \\ b & c & b \\ a & b & a \end{pmatrix}, k_x = \begin{pmatrix} d & e & h \\ -i & 0 & i \\ -h & -e & -d \end{pmatrix}, k_y = \begin{pmatrix} h & i & -d \\ e & 0 & -e \\ d & -i & -h \end{pmatrix},$$

$$k_a = \begin{pmatrix} g & f & -g \\ -f & 0 & -f \\ -g & f & g \end{pmatrix} \tag{4}$$

Table 2. Steerable kernels for rotation group

	e	r	r^2	r^3
k_s	A	rA	r^2A	r^3A
k_x	B	rC	$-r^2B$	$-r^3C$
k_y	C	$-rB$	$-r^2C$	r^3B
k_a	D	$-rD$	r^2D	$-r^3D$

Transformation table for these kernels (Table 2):

For the roto-reflection group D4 (rotations by $\pi/2$ and horizontal flips) there are such types of steerable kernels, (k_x, k_y) and (k_{d1}, k_{d2}) kernels form pairs:

$$k_s = \begin{pmatrix} a & b & a \\ b & c & b \\ a & b & a \end{pmatrix}, k_x = \begin{pmatrix} -d & 0 & d \\ -e & 0 & e \\ -d & 0 & d \end{pmatrix}, k_y = \begin{pmatrix} d & e & d \\ 0 & 0 & 0 \\ -d & -e & -d \end{pmatrix},$$
$$k_{xy} = \begin{pmatrix} 0 & f & 0 \\ -f & 0 & -f \\ 0 & f & 0 \end{pmatrix}, k_d = \begin{pmatrix} g & 0 & -g \\ 0 & 0 & 0 \\ -g & 0 & g \end{pmatrix}, k_{d1} = \begin{pmatrix} 0 & -h & 0 \\ -i & 0 & i \\ 0 & h & 0 \end{pmatrix}, k_{d2} = \begin{pmatrix} 0 & i & 0 \\ -h & 0 & h \\ 0 & -i & 0 \end{pmatrix} \tag{5}$$

Table 3 describes the way convolutional output changes under group transformation of input.

Table 3. Steerable kernels of the roto-reflection group D4.

	e	r	r^2	r^3	f	fr	fr^2	fr^3
k_s	A	rA	r^2A	r^3A	fA	frA	fr^2A	fr^3A
k_x	B	rC	$-r^2B$	$-r^3C$	$-fB$	$-frC$	fr^2B	fr^3C
k_y	C	$-rB$	$-r^2C$	r^3B	fC	$-frB$	$-fr^2C$	fr^3B
k_{xy}	D	$-rD$	r^2D	$-r^3D$	fD	$-frD$	fr^2D	$-fr^3D$
k_d	E	$-rE$	r^2E	$-r^3E$	$-fE$	frE	$-fr^2E$	fr^3E
k_{d1}	F	rG	$-r^2F$	$-r^3G$	$-fF$	$-frG$	fr^2F	fr^3G
k_{d2}	G	$-rF$	$-r^2G$	r^3F	$-fG$	frF	fr^2G	$-fr^3F$

Again, arbitrary convolutional kernel can be represented as linear combination of these types. The kernel types for the roto-reflection group have geometric interpretation. The first kernel k_s that detects some symmetric patterns. Convolution with such kernel is invariant under rotations and reflections. Thus it cannot detect feature orientation.

The pair k_x, k_y is generalization of Sobel filter [10]. Really, if $d = 1, e = 2$ then it becomes exactly Sobel operator that is widely used in image processing. From the Table 3 we see that pair k_x, k_y behaves like equivariant vector: if the image is rotated, then this vector is rotated by the same angle, if the image is flipped, then vector is flipped also. This is the same property as Sobel gradient operator has. Inverse statement also can be proved: if the pair of kernels behaves as equivariant vector, then these kernels have the same parameterization as k_x, k_y. Thus this pair of kernels is able to detect orientation of features.

Features k_{xy} and k_d determine relation between vertical and horizontal, first and second diagonal directions. These features are useful in case we need to determine orientation of the feature that has some axis of symmetry (for example, edge).

Given the table above, and the group table we can construct steerable convolution operation. At the first stage we take an input image and convolve it with 5 types of kernels described above to get the first column of the Table 3. After that we can obtain remaining 7 columns by changing signs and permuting tensors in accordance to the table. In this way we get 8 tensors that change each other following group table in case image is rotated, flipped, or both. To compute the next layer we apply convolution with steerable kernels to the concatenation of all 8 tensors to obtain the first tensor. To compute next tensors, we need to rearrange kernels following the group table and changing sign of some of them. In this way we get equivariant set of tensors. At each stage we can apply any nonlinearity, thus following architecture of plain CNN.

4 Proposed Architecture

Steerable nets described above provide the way to get rotational equivariant feature map with reduced number of parameters. But the computational cost of getting such representations is high, since we need to handle 8 tensors and convolve all them with permuting kernels. This leads also to increased memory consumption on GPUs. We propose to unite framework of steerable nets with the approach of spatial transformer networks to retain roto-reflection invariance but to reduce computational amount.

In case rotational invariance is needed, we propose to use small steerable net (with respect to rotational group) to get rotationally equivariant tensor. After that we convolve it with kernels k_x, k_y, and average result over spatial dimension, to get single rotational equivariant vector. After that we can rotate the last tensor to orient that vector in direction of $(0, 1)$. The tensor we obtained will be invariant to rotations of the input image. After that we can apply plain convolutional network to the tensor we got to get rotational invariant classification.

In case rotational and flip invariance is needed, we can perform procedure from the previous paragraph, but use steerable net with respect to roto-reflection group. After rotation we will obtain tensor that is invariant to rotations and equivariant to horizontal image flips (since vector $(0, 1)$ is invariant under horizontal image flips). After that we apply steerable net (flip group) to that tensor to obtain classification result that is invariant both to rotation and horizontal image flips.

All parameters of this network could be trained with backpropagation. Tensor rotation operation is differentiable in case we use bilinear interpolation.

5 Experiments

To test if prosed approach works in practice, we conducted two experiments. We trained and tested proposed network on CIFAR-10 to see if induced rotation/flip invariance can boost performance on this standard classification task. For the second experiment we trained proposed network on MNIST dataset and tested it on MNIST-rot dataset to see if the network will be able to generalize on rotated images, and compared it with standard way to achieve rotation invariance: g-convolution network.

For CIFAR-10 we used roto-reflection group (network equivariant to rotations and reflections at the first stage, and equivariant to reflections at the second stage). We used DenseNet [12] architecture for the tests to improve gradient backpropagation by connecting all layers to each other. In all cases we used 3 average poolings: from spatial size 32×32 to 16×16, then to 8×8 and to 1×1. The first network outputs tensor with spatial size 8 × 8 (the last layer of the net is average orientation pooling). This tensor is padded with zeros to increase spatial size to 12 × 12, rotated in accordance with found vector, and fed into the second flip invariant network. Second network also follows Dense Net architecture: it computes convolutions, applies ReLU nonlinearity to outputs, and concatenates them back to tensor. After that it applies orientation average pooling (for 2 group elements) and full average pooling, and feeds resulting vector to fully connected layer with softmax nonlinearity.

The Table 4 shows net configurations and results: number of convolutional blocks between poolings, structure of steerable convolution (number of k_s filters, number of k_x and k_y filters (there has to be equal number of filters), number of k_{xy} filters and number of k_d filters). For the second net configuration of convolutional block consists of number of symmetric k_s and asymmetric k_a kernels.

Table 4. CIFAR-10 classification results

Net configuration	Train error	Train accuracy	Test error	Test accuracy
1 block: [2, 2, 2, 2] 2 blocks: [8, 8] 31,168 params	0.75	73.82%	0.88	69.99%
2 blocks: [2, 2, 2, 2] 4 blocks: [8, 8] 97,608 params	0.61	78.90%	0.82	72.48%
1 block: [4,4,4,4] 4 blocks: [8, 8] 107,610 params	0.40	86.60%	0.84	73.88%
2 blocks: [4,4,4,4] 4 blocks: [8, 8] 216,426 params	0.02	99.86%	1.42	79.32%

For the second experiment we compared performance of our network with g-convolution network trained on MNIST and tested on MNIST-rot. In this way we test if network that trains to classify straight digits is able to generalize to classify rotated digits without seeing them at training time. We use DenseNet architecture (of similar structure to our proposed network) as baseline solution that is not designed to deal with rotations. We train all the networks using SGD with momentum for 100 epochs (dense connectivity makes these networks to converge very fast). Results are listed in the Table 5.

Table 5. MNIST-rot results of networks trained on MNIST

Network	Train error (MNIST)	Train accuracy (MNIST)	Test error (MNIST-rot)	Test accuracy (MNIST-rot)
DenseNet (baseline) 4 blocks: [8,8,8,8] 39956 params	0.01	99.67%	6.39	41.86%
G-conv Dense Net 4 blocks: [8,8,8,8] 49132 params	0.02	99.23%	0.84	83.42%
Proposed Net 2 blocks: [8,8] 2 blocks: [8,8] 29118 params	0.04	98.64%	0.46	87.58%

6 Conclusion

In this paper we proposed a way to unite spatial transformer networks and steerable networks to achieve rotation and flip invariance and reduce amount of computations.

Proposed approach needs small steerable network to determine orientation vector after that plain CNN or horizontal flips steerable CNN is applied which runs much faster. In this way faster training and evaluation of the network are achieved.

Experiments showed that this approach is not working well for scenes with many objects, like images from CIFAR-10. In that case single vector is not enough to describe orientation of the scene and objects there, and to achieve rotation invariance use of group convolutional net is required.

Experiment with training on MNIST and testing on MNIST-rot showed that our model generalizes well to images of rotated objects it hasn't seen at the training stage and shows better performance than g-convolutional net.

In general proposed architecture makes sense in cases there is one object of interest on input images and testing images could have much more rotational distortions that training ones.

In future we plan to try to generalize this approach to images with several objects each having its own orientation.

Acknowledgements. We gratefully acknowledge the support of NVIDIA with the donation of the Titan X Pascal GPU used for this research.

References

1. Goodfellow, I., Bengio, Y., Courville, A.: Deep Learning. MIT Press, Cambridge (2016). http://www.deeplearningbook.org
2. Simard, P.Y., Lecun, Y.A., Denker, J.S., Victorri, B.: Transformation invariance in pattern recognition—tangent distance and tangent propagation. In: Neural Networks: Tricks of the Trade. Lecture Notes in Computer Science, pp. 239–274 (1998)
3. Luan, S., Zhang, B., Chen, C., Cao, X., Han, J.: Gabor convolutional networks. arXiv:1705.01450 (2017)
4. Jaderberg, M., Simonyan, K., Zisserman, A., Kavukcuoglu, K.: Spatial transformer networks. arXiv:1506.02025 (2015)
5. Marcos, D., Volpi, M., Tuia, D.: Learning rotation invariant convolutional filters for texture classification. arXiv:1604.06720 (2016)
6. Cohen, T.S., Welling, M.: Group equivariant convolutional networks. arXiv:1602.07576 (2016)
7. Cohen, T.S., Welling, M.: Steerable CNNs. arXiv:1612.08498 (2016)
8. Sabour, S., Frosst, N., Hinton, G.E.: Dynamic routing between capsules. arXiv:1710.09829 (2017)
9. Kanopoulos, N., Vasanthavada, N., Baker, R.L.: Design of an image edge detection filter using the Sobel operator. IEEE J. Solid-State Circu. **23**(2), 358–367 (1988)
10. Krizhevsky, A., Sutskever, I., Hinton, G.E.: ImageNet classification with deep convolutional neural networks. In: Proceedings of the Advance Conference in Neural Information Processing Systems (NIPS 2012), Lake Tahoe, NE (2012)
11. Huang, G., Liu, Z., van der Maaten, L., Weinberger, K.Q.: Densely connected convolutional networks. In: Proceedings of the IEEE Conference on Computer Vision and Pattern Recognition (2017)
12. Azulay, A., Weiss, Y.: Why do deep convolutional networks generalize so poorly to small image transformations? arXiv:1805.12177 (2018)
13. Dieleman, S., De Fauw, J., Kavukcuoglu, K.: Exploiting cyclic symmetry in convolutional neural networks. In: International Conference on Machine Learning (ICML) (2016)
14. Marcos, D., Volpi, M., Komodakis, N., Tuia, D.: Rotation equivariant vector field networks. In: International Conference on Computer Vision (ICCV) (2017)
15. Weiler, M., Geiger, M., Welling, M., Boomsma, W., Cohen, T.: 3D steerable CNNs: learning rotationally equivariant features in volumetric data. arXiv:1807.02547 (2018)
16. Oyedotun, O.K., Dimililer, K.: Pattern recognition: invariance learning in convolutional auto encoder network. Int. J. Image Graph. Signal Process. (IJIGSP) **8**(3), 19–27 (2016). https://doi.org/10.5815/ijigsp.2016.03.03

Hardware/Software Co-design for XML-Document Processing

Anatoliy Sergiyenko, Maria Orlova, and Oleksii Molchanov(✉)

National Technical University of Ukraine "Igor Sikorsky Kyiv Polytechnic Institute", Kyiv 03056, Ukraine
aser@comsys.kpi.ua, oleksii.molchanov@gmail.com

Abstract. Algorithms and tools for XML-documents processing are reviewed. The document analyzer algorithms, which are implemented in the stack state machine, are considered. A need for high-speed reconfigurable hardwired XML-request analyzers is determined. The SM16 processor architecture core is developed which effectively evaluates the stack-based parsing algorithms and is implemented in the field-programmable gate array (FPGA). The processor has the stack architecture with three additional stack blocks, hash-table, and instructions that accelerate the execution of parsing operations. The hardware/software FPGA-based system, which has the main processor and tens to hundreds of SM16 executive processor elements, is proposed. This system is scalable for different amounts of packets streams. It efficiently processes XML-documents and can be easily reconfigured to the given document grammars, allowing to adapt to fast-changing modern-world communications.

Keywords: XML · Parser · FPGA · Stack processor · Grammar · Stack automaton

1 Introduction

The XML-language gained widespread adoption in the Web services due to its standardization and inherent scalability. The size of XML-documents ranges from hundreds of bytes to tens of gigabytes [1]. Most XML documents are processed in the form of requests to the Web servers. Recently, XML is used to share data along the Internet of Things, for which a special profile: Devices Profile for Web Services (DPWS) has been developed [2]. A new XML format, Efficient XML Interchange (EXI), has been introduced to increase the capacity of the transmission channels [3].

Although XML has many advantages, parsing of XML queries results in a significant slowdown in the Web services performance [1]. The XML parsing takes up a lot of memory and computing resources, consuming about 30% of processing time in the Web service applications [4]. The experience of using XML in the databases shows that XML parsing is a major bottleneck in the productivity gains and can increase the transaction costs up to 10 times or more [5].

The article overviews the XML-document parsing algorithms and discusses the ways of accelerating them. The hardware/software co-design approach is proposed, which is implemented in the field-programmable gate array (FPGA) and provides high-speed reconfiguration to the new grammar rules.

Z. Hu et al. (Eds.): ICCSEEA 2020, AISC 1247, pp. 373–383, 2021.
https://doi.org/10.1007/978-3-030-55506-1_34

The structure of this article is as follows. Section 2 reviews existing parsing algorithms. Section 3 overviews existing parsing acceleration approaches. Section 4 introduces the SM16 processor designed for XML-document processing. Section 5 proposes a hardware/software co-design system for XML-document processing. Section 6 presents the results of the SM16 configuration on an FPGA. Section 7 concludes the results of this work.

2 XML-Documents Parsing Algorithms

The common XML-document parsing algorithms fall into two categories of parsers. The event-driven parser analyses the next XML phrase and then feeds back the message to the client application that the next tag is found. Consequently, such a parser does not load the entire XML-document at a time and therefore requires a small amount of memory. An example of such a parser is the SAX parser, which is considered as the industry standard [6]. In most cases, the parser task is to confirm the document's correctness and to read its fields. Such a parser makes the document filtering.

The tree-based parser writes the entire XML-document to the server and creates a document object model (DOM) in the form of a tree [7]. DOM provides maximum flexibility, but it consumes a lot of memory and strict computational requirements.

To define the structure of an XML document, some XML-schema languages are used. The XML-schemas are usually formalized using regular tree grammar [8]. The grammar of a particular query type is expressed using XML-query languages such as XPath [9]. In general, such a grammar is presented as a tuple G = (N, T, S, P). Here, N is a finite set of the non-terminal symbols, T is a finite set of the terminal symbols, $S \subseteq N$ is a set of initial symbols, P is a set of rules in the form $X \rightarrow ar$, where $X \in N$, $a \in T$, and r is a regular expression over N. The rule says that X originates the sub-trees with the root and children that satisfy the expression r [8].

Consider an example of such a grammar presented below:

```
N = { Data, Header, Body, Content }
T = { d, h, b, c } # tag names
S = { Data }
P = # Rules written below:
Data → '<d>' Header Body* '</d>' ; # Rule1
Header → '<h>' '</h>' ;               # Rule2
Body → '<b>' '</b>' ;             # Rule3
Body → '<b>' Content '</b>' ;     # Rule4
Content → '<c>' '</c>' ;           # Rule5
```

The next example of XML-document matches the described grammar:

$$<d> \; <h> \; </h> \; <b> \; <c> \; </c> \; </b> \; <b> \; </b> \; </d> \quad (1)$$

Parsing of this document should result in a tree presented in Fig. 1. Each vertex of a graph corresponds to a particular document node, the adjacent vertices above are the parent nodes, and the lower vertices represent the child nodes.

Consider a simple XML-parsing algorithm that validates the document in terms of a given regular tree grammar, as described in [8]. The algorithm is implemented in a stack finite state machine (FSM), which has three stacks: P, Y, and S. It traverses a tree (Fig. 1) in-depth and triggers the event phrases of a document. These include the opening e (<…) and closing /e (</…) tags. The algorithm uses three stacks as follows:

Stack P stores the symbol sets from N. The top specifies the symbols from N that can be used to match the next node in the document. Stack P is initialized with set S.

Stack Y stores the sets of rules from P. Each set specifies the rules that can match the current node in the document.

Stack S stores the lists of symbol sets from N. The first symbol set denotes the symbols corresponding to the first child of the current node; the second symbol set denotes the symbols corresponding to the second child of the current node, etc.

The algorithm implemented in stack FSM described as follows.

```
Initial state: XML-document D is passed as input; stacks P, Y and S are
empty.
Add a set of initial non-terminal symbols to P;
Traverse D in depth {
  when opening tag e is matched {
     choose rules of form Xi → a(ri), where a is the name of the tag e,
     // Xi ∈ N' — non-terminal symbols from top set of symbols in stack P;
     if (N' = ∅) {show a message that D is incorrect; End;}
     push set of chosen rules to stack Y;
     push empty list to stack S;
     push set of non-terminal symbols from ri to stack P;
  }
  when closing tag /e is matched {
     pop set of rules { Xi → a(ri) } from stack Y;
     pop list of sets of non-terminal symbols {X1, X2, …, Xj } from stack S,
     //where Xj is a set of non-terminal symbols;
     collect set of non-terminal symbols X' = { Xi } such that ri accepts
     //{X1, X2, …, Xj };
     if (X'= ∅) { show a message that D is incorrect; End;}
     append X' to the end of list in to of stack S;
     pop set of non-terminal symbols from stack P;
  }
}
show message that D is correct;
End.
```

The execution of this algorithm when parsing an XML document (1) is shown in Table 1. In [8], other parsing algorithms are described. The algorithm described above and the parsing example are simplified. However, they do show the basic properties inherent to the XML-parsing algorithms, including the fact that it is advisable to use a three-stack FSM to implement an event-driven parser. The parsing effectiveness depends on the implementation of this FSM, which is discussed in the next section.

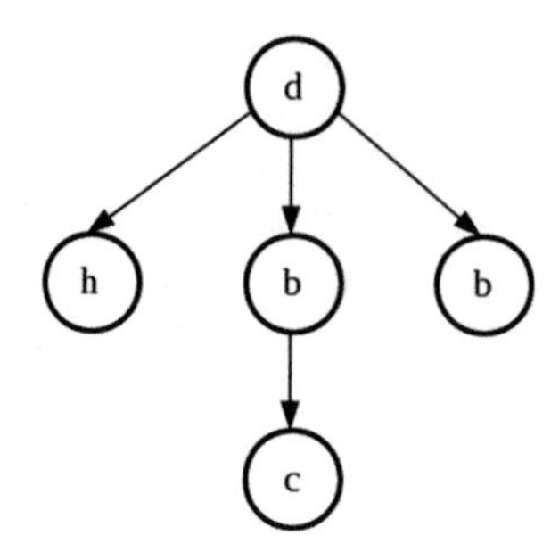

Fig. 1. Example of XML-document parse-tree

3 Acceleration of XML-Document Filtering

The XML text filtering is a difficult problem because it has to support the real-time processing of wide streams of various XML-requests. Different methods and accelerators have been proposed to improve these tasks, which are discussed below.

Table 1. Parsing steps of XML-document (1)

Step	Input tag	Stack P	Stack Y	Stack S
1		(Data)		
2	<d>	(Header, Body) (Data)	(Rule1)	()
3	<h>	(Header, Body) (Data)	(Rule2) (Rule1)	() ()
4	</h>	(Header, Body) (Data)	(Rule1)	((Header))
5	<b>	(Content) (Header, Body) (Data)	(Rule3, Rule4) (Rule1)	() ((Header))
6	<c>	() (Content) (Header, Body) (Data)	(Rule5) (Rule3, Rule4) (Rule1)	() () ((Header))
7	</c>	(Content) (Header, Body) (Data)	(Rule3, Rule4) (Rule1)	((Content)) ((Header))
8	</b>	(Header, Body) (Data)	(Rule1)	((Header), (Body))
9	<b>	(Content) (Header, Body) (Data)	(Rule3, Rule4) (Rule1)	() ((Header), (Body))
10	</b>	(Header, Body) (Data)	(Rule1)	((Header), (Body), (Body))
11	</d>	(Data)		

3.1 Software Accelerators

In [10] an approach is proposed to translate the terminal and nonterminal symbols into binary codes to reduce memory consumption. The pre-scanner in [11] builds an XML-document tree for later document splitting and processing it in parallel. In [12], the researchers split the parsing process into phases, which are pipelined. In [13, 14], an XML-filter (XFilter) is built as FSM implemented in software. A combination of these FSMs for several different XML-queries is also used. According to the LazyDFA approach, weakly deterministic FSM is dynamically constructed for the filtration process [15]. In [16], a dynamic FSM is used to find the matchings by the Aho-Korasik searching algorithm.

3.2 Hardware Filters

FPGA is an efficient solution for the hardware filtering of XML-queries. FSM that is constructed for a specific set of grammars is implemented in FPGA in [17]. In [18], the acceleration is obtained by using the associative memory to store the document tree. The systems based on stack FSM, which are compiled from the given grammars, are shown in [19–21].

An FSM skeleton is proposed in [22], which is capable of being reconfigured quickly without the FPGA project redesigning. This FSM skeleton becomes a working FSM after loading the transition conditions corresponding to a specific set of XML-requests. In [23] a new data structure is investigated, which is processed in the systolic processor configured in FPGA. The disadvantages of these approaches include high hardware redundancy and limitations of the processed document class.

3.3 Multi-processor Filtering

Several approaches use parallel architectures. In this case, the acceleration up to n times is achieved because n documents are processed simultaneously [24]. To speed up the processing, the processing units (PUs) have the hardware support [17]. The GPU accelerator is used to filter XML requests in [25], but its efficiency drops significantly when processing the short queries.

3.4 Hardware/Software Co-design Filtering

It makes sense to use FPGAs to speed up the XML-parsing in the software systems. Such FPGA processes the incoming documents and issues a set of tags for each document, which is further processed in software. Thus, up to 97% of all incoming requests can be filtered from the stream [26]. FPGAs effectively implement a tag decoder that speeds up the SAX analyzer [21].

Comparing the mentioned approaches, the following conclusions are set.

- The hardware filtering systems have the highest performance, but they are designed for a limited number of grammars, and their reconfiguration is long-lasting.
- Reconfigurable FSM-based hardware filtering systems have excessive hardware costs and focus on a particular class of grammars. An n-processor system is capable

of speeding up the processing of multiple documents when the coarse-grained parallelism with *n* threads is organized.

- The hardware/software systems distinguish the software and hardware subsystems. The software subsystem performs the preliminary and final processing of a document. The hardware subsystem recognizes tags and branches of the document tree.
- More flexible architectures that can provide both – the high throughput and the ability to quickly be reconfigured to the arbitrary XML-grammar – are required.

4 Processor for XML-Document Processing

An application-specific programmable processor can be designed considering the XML-parsing features described above. A processor for the XML-document processing, which has the stack architecture described in [27], has been developed.

Various authors have developed several stack processors, which are available for replication in FPGA [28, 29]. In [29] it is shown that a 16-bit stack processor has approximately 2.5 times shorter program length and much less hardware volume than the program for the Xilinx Microblaze processor [30] has when implementing the packet parsing (see Table 2). Besides, a stack processor program has the same syntax as the Forth language [27]. Therefore, it is advisable to develop a stack processor architecture that not only minimizes the hardware costs but also ensures the efficient execution of the XML-document processing algorithms.

The structure of the developed 16-bit SM16 processor is shown in Fig. 2. This processor has a common dual-stack architecture [27]. It is a successor of an eight-bit SM8 processor, which has been developed to implement the data communication protocols [31]. The processor includes a program counter (PC), Data RAM block, Program ROM block, an instruction register (IR), return address stack (Rstack), data stack (Dstack), ALU. The T, N registers are the top registers of the DStack. Register R is the top of the RStack, which also plays the role of a loop counter.

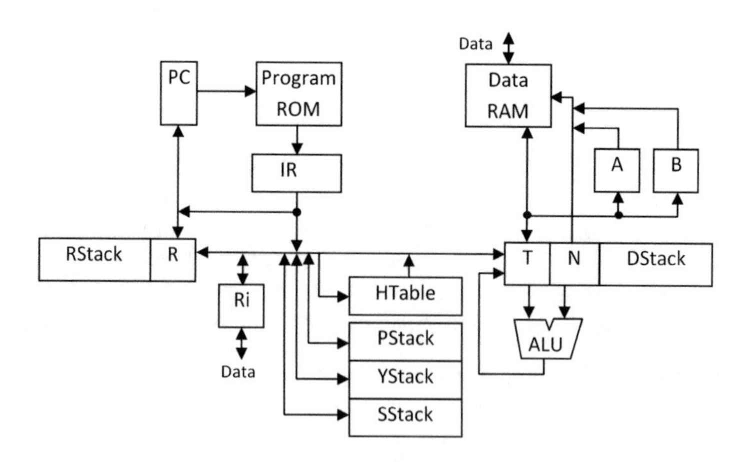

Fig. 2. SM16 processor structure

The program is loaded into the Program ROM during the FPGA configuration. The dual-port Data RAM downloads the XML-document from the external devices in the DMA mode. A and B are the index registers, and the peripheral register Ri serves for the inter-processor communications. The HTable ROM stores the hash table for the transcoding the long tags found in XML-documents into the numbers. The PStack, Ystack, and SStack stacks perform the same function as the P, Y and S stacks of a stack FSM described in Sect. 2.

The processor instructions are executed in a single cycle except for the branch instructions that are executed in two cycles. This feature makes the architecture friendly to the parsing algorithms that are branch intensive. All instructions have 16-bit width (Fig. 3). The instruction has one to three opcode fields F1, F2, F3, and a variable-length D field that stores a constant, or a jump address or shift. The processor can perform up to three operations F1, F2, F3. For example, the instruction: L1 HASH @ab+ DJNZ L1 calculates the hash code with a byte in the T register and code in the N register. If $[R] \neq 0$, then the control flow is passed to the L1 label and the register R is decremented. Otherwise, the loop is exited. At the same time, next byte at the address [A] is read into the register T, then, this address is incremented. The calculated code is used as the HTable memory access address.

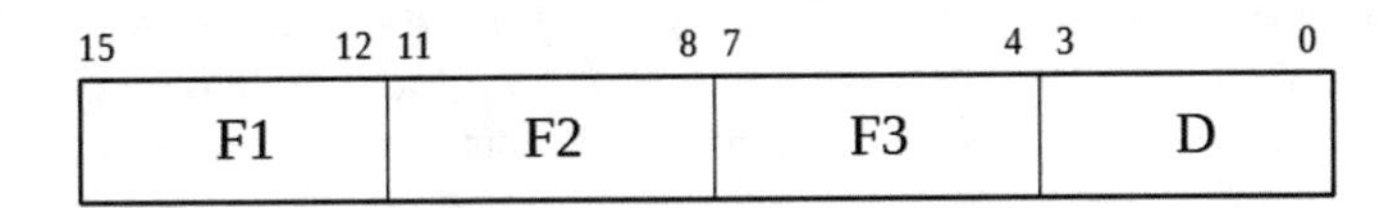

Fig. 3. SM16 instruction format

The stack processor programming is usually done in a threaded code style when the program looks like a sequence of subroutine calls. This makes it possible to obtain the programs of minimal length, which is important for the FPGA implementation. The ability to insert a return operation in most instructions and combine it with a conditional branch reduces both the subroutine length and their duration. The processor has an interrupt system as well. Because the stack processor context has minimum volume, the interrupt overhead is also negligible. Due to a large number of subroutines, conditional, and memory read instructions in the XML-parsing applications, the average run time of a single instruction is one and a half clock cycles.

5 System for Hardware/Software Processing of XML-Documents

The SM16 processor is capable of parsing XML-documents according to the algorithms described in Sect. 2. It is reprogrammed to another grammar by recompiling its description into a program that is embedded in the respective segment of the FPGA configuration file. This process takes about a minute and is much faster than the process of translating the grammar descriptions in FSM.

The insufficient performance of a single processor is compensated by the use of a parallel many-core system. Such a system consists of dozens of SM16 processing units (PUs) and a master processor that distributes the XML-documents among these PUs. For example, the Xilinx Zynq-7 FPGA, which integrates the master Cortex-A9 core, may contain from 20 to 300 PUs SM16, depending on its logical capacity.

To evaluate the performance of the system, it should be compared with the XML-query processing system proposed in [32]. This system consists of a master processor Cortex-A9 and a hardware accelerator configured in FPGA. As a result of its testing, it is established that such a hardware/software parser when parsing the XML-requests in the EXI format reduces the duration of parsing by 23% and a pure hardware parser does it by 95%. Moreover, the configured parser equipment operates at a frequency of 50 MHz and has the hardware costs of 12.8 thousand LUTs. With the same hardware costs, the proposed system includes approximately 20 PUs and provides no worse or even better throughput, but has the advantages of the fast reconfiguration and the ability to handle the large-scale tasks.

6 Results and Discussions

Table 2 presents the results of the SM16 processor configured in Xilinx Spartan-6 FPGA. The SM16 hardware costs are specified for a minimum core configuration and the core with three stacks and a hash table. The table analysis shows that the SM16 processor has the same performance as the J1 processor at a third of the higher hardware costs, it has a much higher speed than the well-known MSP430 processor and somewhat loses to the Microblaze processor. Nevertheless, it should be noted that the SM16 processor has a larger instruction set which is adapted to handle the XML-documents.

The developed assembler compiles the program text into machine codes, which are stored in the Program ROM. First, it parses a sequence of different simple operations.

Then, it tries to combine subsequences of simple operations creating composite operations. One composite operation may consist of two or three simple operations. Operations DJNZ and RET can be combined with some data transfer or ALU operations and with memory access operations.

To accelerate the simulation and debug of the developed programs, a software processor simulator is developed for SM16, as it helps to reduce the number of errors before usage of FPGA simulation tools (as it is done in [33, 34]).

Table 2. Parameters of the processors configured in FPGA

Microcontroller	Bit-length	Hardware costs (LUTs)	Max. clock frequency, MHz	Speed, MIPS
b16-small [28]	16	280	100	50 MIPS
J1 [29]	16	342	106	70 MIPS
MSP430 [35]	16	1240	65	25 MIPS
Microblaze [30]	32	2046	130	174 DMIPS
SM8 [36]	8	181	140	94 MIPS
SM16	16	477–580	105	70 MIPS

7 Conclusion

An overview of existing algorithms and XML-document processing tools has been done. Its analysis shows that the event-driven parsers implemented in FSM have the highest performance. Typically, such devices are implemented in FPGAs. However, they require a long adjustment process when the grammar is changed. There are the adjustable FSMs that are universal to a given set of grammars, but they have excessive hardware costs. The implementation of the XML-document analysis in a multiprocessor system has limited efficiency. The three-stack FSM provides the effective implementation of the most tree grammar processing algorithms.

The SM16 processor core is proposed, in which architecture meets the XML-document processing algorithms. It has the stack architecture and provides both low hardware costs, minimized program code length, and efficient stack FSM implementation in FPGAs. Three stack blocks, hash tables, and instructions, which speed up certain operations, have been added to the core to improve its performance. The instruction format is selected, which allows us both to perform several operations in parallel and to minimize the program length.

An FPGA-based hardware/software system is proposed. It has a master Cortex-A9 processor and from tens to hundreds of SM16 slave PUs. Such a system provides a high throughput of the XML parsing. The needed throughput is derived by the selection of the processing unit number because the parsing problem solving is scalable for the large packet streams. This system is not only capable of processing the XML-documents efficiently but also can be quickly reconfigured to process the documents of other grammars.

Further work includes implementation of the stack-based parsing algorithm described in Sect. 2 for SM16, implementation, and testing of hardware/software system for parsing of XML-documents based on SM16 PUs.

References

1. Head, M.R., Govindaraju, M., van Engelen, R., Zhang, W.: Benchmarking XML processors for applications in grid web services. In: Proceedings of the 2006 ACM/IEEE Conference on Supercomputing (2006)
2. Driscoll, D., Mensch, A.: Devices profile for web services ver. 1.1 (2009). http://docs.oasis-open.org/ws-dd/dpws/wsdd-dpws-1.1-spec.html
3. Schneider, J., Kamiya, T.: Efficient XML Interchange (EXI) Format 1.0 (2011). http://www.w3.org/TR/exi/. Accessed 9 Nov 2019
4. Apparao, P., Bhat, M.: A detailed look at the characteristics of XML parsing. In: BEACON-1: 1st Workshop on Building Block Engine Architectures for Computers and Networks (2004)
5. Mattias, N., Jasmi, J.: XML parsing: a threat to database performance. In: Proceedings of the 20th International Conference on Information and Knowledge Management CIKM 2003, pp. 175–178 (2003)
6. SAX Parsing Model. http://sax.sourceforge.net. Accessed 11 Nov 2019
7. Document object model (DOM) level 2 core specification. http://www.w3.org/TR/DOM-Level-2-Core. Accessed 11 Nov 2019
8. Murata, M., Lee, D., Mani, M., Kawaguchi, K.: Taxonomy of XML schema languages using formal language theory. ACM Trans. Internet Technol. **5**(4), 660–704 (2005)
9. XML Path Language Version 1.0. http://www.w3.org/TR/xpath. Accessed 11 Nov 2019
10. Chiu, K., Devadithya, T., Lu, W., Slominski, A.: A binary XML for scientific applications. In: Proceedings of the IEEE 1st International Conference on e-Science and Grid Computing (e-Science 2005), pp. 336–343. IEEE (2005)
11. Lu, W., Chiu, K., Pan, Y.: A parallel approach to XML parsing. In: Proceedings of the 7th IEEE/ACM International Conference on Grid Computing, pp. 223–230. IEEE/ACM (2006)
12. Head, M.R., Govindaraju, M.: Approaching a parallelized XML parser optimized for multi-core processor. In: Proceedings of the 2007 Workshop on Service-Oriented Computing Performance: Aspects Issues and Approaches (SOCP 2007), pp. 17–22 (2007)
13. Altinel, M., Franklin, M.J.: Efficient filtering of XML documents for selective dissemination of information. In: Proceedings of the 26th International Conference on Very Large Data Bases (VLDB 2000), pp. 53–64 (2000)
14. Diao, Y., Altinel, M., Franklin, M.J., Zhang, H., Fischer, P.: Path sharing and predicate evaluation for high-performance XML filtering. ACM Trans. Database Syst. (TODS) **28**, 467–516 (2003)
15. Green, T.J., Gupta, A., Miklau, G., Onizuka, M., Suciu, D.: Processing XML streams with deterministic automata and stream indexes. ACM Trans. Database Syst. (TODS) **29**, 752–788 (2004)
16. Silvasti, P.: XML-document-filtering automaton. In: Proceedings of the Very Large Data Base Endowment (VLDB Endowment 2008), vol. 1, no 2, pp. 1666–1671 (2008)
17. Lunteren, J.V., Engbersen, T., Bostian, J., Carey, B., Larsson, C.: XML accelerator engine. In: 1st International Workshop on High Performance XML Processing (2004)
18. El-Hassan, F., Ionescu, D.: SCBXP: an efficient hardware-based XML parsing technique. In: 5th Southern Conference on Programmable Logic (SPL), pp. 45–50. IEEE (2009)
19. Mueller, R., Teubner, J., Alonso, G.: Streams on wires—a query compiler for FPGAs. In: Proceedings of the Very Large Data Base Endowment (VLDB Endowment 2009), vol. 1, no. 2, pp. 229–240 (2009)

20. Moussalli, R., Salloum, M., Najjar, W., Tsotras, V.: Massively parallel XML twig filtering using dynamic programming on FPGAs. In: Proceedings of the IEEE 27th International Conference on Data Engineering (ICDE 2011), pp. 948–959. IEEE (2011)
21. Mitra, A., Vieira, M., Bakalov, P., Najjar, W., Tsotras, V.: Boosting XML filtering with a scalable FPGA-based architecture. In: Proceedings of the CIDR 2009 - 4th Biennal Conference on Innovative Data Systems Research (2009)
22. Teubner, J., Woods, L., Nie, C.: XLynx—an FPGA-based XML filter for hybrid XQuery processing. ACM Trans. Database Syst. (TODS) **38**(4), 1–39 (2013)
23. Woods, L., Alonso, G., Teubner, J.: Parallelizing data processing on FPGAs with shifter lists. ACM Trans. Reconfigurable Technol. Syst. (TRETS) **8**(2), 1–22 (2015). Special Section on FPL 2013
24. Letz, S., Zedler, M., Thierer, T., Schutz, M., Roth, J., Seiffert, R.: XML offload and acceleration with cell broadband engine. In: XTech: Building Web 2.0. XTech Conference (2006)
25. Moussalli, R., Halstead, R., Solloum, M., Najjar, W., Tsotras, V.: Efficient XML path filtering using GPUs. In: Proceedings of the 2nd International Workshop on Accelerating Data Management Systems (ADMS 2011) (2011)
26. Fischer, P., Teubner, J.: MXQuery with hardware acceleration. In: Proceedings of the IEEE 28th International Conference on Data Engineering (ICDE), pp. 1293–1296. IEEE (2012)
27. Koopman, P.: Stack Computers: The New Wave, 234 p. Ellis Horwood, Mountain View Press, Mountain View (1989)
28. Paysan, B.: b16-small—less is more. In: Proceedings of the EuroForth 2004, 9 July 2006
29. Bowman, J., Garage, W.: J1: a small Forth CPU core for FPGAs. In: Proceedings of the EuroForth 2010, pp. 1–4, January 2010
30. Kale, V.: Using the MicroBlaze processor to accelerate cost-sensitive embedded system development. Xilinx, WP469 (v1.0.1) (2016). https://www.xilinx.com/products/design-tools/microblaze.html#documentation. Accessed 11 Nov 2019
31. Sergiyenko, A., Molchanov, O., Orlova, M.: Nano-processor for the small tasks. In: 2019 IEEE 39th International Conference on Electronics and Nanotechnology (ELNANO), pp. 674–677. IEEE (2019)
32. Altmann, V., Skodzik, J., Danielis, P., Van N.P., Golatowski, F., Timmermann, D.: Real-time capable hardware-based parser for efficient XML interchange. In: Proceedings of the 9th International Symposium on Communication Systems, Networks & Digital Sign (CSNDSP), pp. 415–420 (2014)
33. Rani, A., Grover, N.: Novel design of 32-bit asynchronous (RISC) microprocessor & its implementation on FPGA. Int. J. Inf. Eng. Electron. Bus. (IJIEEB) **10**(1), 39–47 (2018). https://doi.org/10.5815/ijieeb.2018.01.06
34. Daghooghi, T.: Design and development MIPS processor based on a high performance and low power architecture on FPGA. Int. J. Mod. Educ. Comput. Sci. (IJMECS) **5**(5), 49–59 (2013). https://doi.org/10.5815/ijmecs.2013.05.06
35. Girard, O.: OpenMSP430. OpenCores, Rev. 1.13 (2013). http://opencores.org. Accessed 11 Nov 2019
36. Molchanov, O., Orlova, M., Sergiyenko, A.: Software/hardware co-design of the microprocessor for the serial port communications. In: Hu, Z., Petoukhov, S., Dychka, I., He, M. (eds.) Advances in Computer Science for Engineering and Education II, pp. 238–246. Springer, Cham (2020)

Load Balancing in Software Defined Networks Using Multipath Routing

Yurii Kulakov(✉), Alla Kohan, Sergii Kopychko, and Roman Cherevatenko

National Technical University of Ukraine "Igor Sikorsky Kyiv Polytechnic Institute", 37 Peremohy Avenue, Kyiv 03056, Ukraine
ya.kulakov@gmail.com, a.v.kohan433@gmail.com, kopychko.sn@gmail.com, r.cherevatenkol@gmail.com

Abstract. Currently, in connection with the advent of new network technologies, such as Software-Defined Networking (SDN), there is a need to develop new and improve well-known methods of managing network resources and traffic construction. With an increase in network dimension and significant energy consumption, the task of balancing traffic becomes relevant. In this regard, the main purpose of this work is to research and develop a method for balancing traffic in SDN networks, which allows for the most uniform loading of information transmission channels. The analysis of the features of the organization and functioning of SDN to improve the efficiency of load balancing.

A method of balancing traffic is proposed, which, due to the centralized method of generating routing information in the SDN controller and using the multi-path routing method, allows to increase the efficiency of the dynamic traffic reconfiguration procedure.

Routing information is generated using a modified wave routing algorithm. This algorithm simultaneously with the formation of a given route forms the paths from all intermediate nodes to the final node. This reduces the formation time of the formation of new tracks and eliminates the re-formation of routing information for previously formed sections of the track.

The results of modeling the process of balancing traffic when changing the load on the communication channels are presented. From the simulation results, it follows that the proposed method of balancing traffic by selecting a path with a minimum value of the path selection criterion ensures uniform network load. Choosing a path with a minimum number of channels causes a recalculation of the metric of the minimum number of paths.

Keywords: Software-defined networking · Multipath routing · Streaming algorithm · Traffic balancing

1 Introduction

Currently, in connection with the growing needs for computing power and information volumes, the urgent task is to increase the efficiency of computer networks. Modern computer networks are characterized by a large dimension and a diverse composition of equipment, including mobile communications. In this regard, the process of managing

Z. Hu et al. (Eds.): ICCSEEA 2020, AISC 1247, pp. 384–395, 2021.
https://doi.org/10.1007/978-3-030-55506-1_35

modern computer networks is becoming more complicated, in particular, the procedure for balancing traffic. To solve these problems, the software-configured networks (SDN) technology is currently used [1, 2]. In the SDN, the control plane is deleted to a separate device called the controller. The controller is the most important and fundamental part of the SDN architecture [3–6]. Currently, with the advent of new SDN technologies, there is a need to develop new methods for balancing traffic. This, due to the features of the SDN organization, will improve the efficiency of the traffic balancing procedure. Compared to a traditional network, the main advantage of load balancing in SDN is that balancing is done centrally in the SDN controller. This allows a more efficient load balancing strategy. The SDN controller updates the routing information for the SDN switches by updating their routing tables in order to select the best route in terms of minimizing power consumption and channel congestion. Compared to the distributed methods of traffic construction and its balancing, the centralized method eliminates the need to exchange service information between network switches.

Unlike conventional routing methods, in multi-path routing, several paths between network nodes are formed simultaneously [7]. This allows the dynamic balancing of channels due to rerouting.

In this regard, the main goal of this work is to research and develop a method of balancing traffic in SDN networks based on multi-path routing in order to increase the efficiency of the functioning of computer networks.

To achieve this goal, this paper provides a review of the literature on known traffic balancing methods. An algorithm for the formation of a set of minimally loaded disjoint paths is given. A criterion is proposed and justified for choosing a path that ensures the maximum possible uniform network load. The traffic balancing algorithm is presented.

2 Overview of Ways to Balance Traffic in Computer Networks

In [8], a study is made of various methods and approaches to solving problems, load balancing, and overload control. The focus is on issues such as optimizing data traffic on the network using SDN, QoS, load balancing and congestion control.

In work [9], a method for balancing the load of the controller and balancing the load of the channel in the SDN is proposed. It is shown that the load distribution problem for lines and controllers using well-known methods of path formation is an NP-complete problem.

The work [10] addressed load balancing issues in the Fat-Tree topology. In the Mininet emulator. The simulation results in the Mininet emulator showed an average load improvement of 50%, an average latency improvement of 41%, and significant improvements in terms of ping response, bandwidth usage and system bandwidth.

In work [10] a method for balancing traffic is proposed, which, due to the centralized method of generating routing information in the SDN controller and using multi-path routing, simplifies the reconfiguration of traffic and ensures the most uniform network load. The formation of routing information using the wave routing algorithm [11] allows you to simultaneously generate paths from all intermediate nodes to the final node. This substantially reduces the time complexity of forming multiple paths.

In [12], a multi-path routing algorithm was proposed, which allows increasing network performance by 10–15% by reducing the volume of service packets. This allows you to reduce energy consumption by 41% and increase the maximum use of communication channels by 60% compared to distributed methods balancing [13]. In addition, ping delay in the network decreases by 5–10%, and the number of control packets in the network decreases by 60–70% [14].

In turn, the use of multi-path routing allows quick re-routing after failures of channels and network switches [15].

In [16], a method for centralized formation of routing information in distributed data centers based on SDN technology was proposed, which eliminates the re-formation of routes between intermediate sections of the route.

The main disadvantage of the known traffic control methods is that many large flows, called elephant flows, are redirected to the same path, which leads to load imbalance and bandwidth loss [17]. In this regard, when designing traffic and balancing it is necessary to take into account the size and nature of a load of channels.

3 A Method of Balancing Traffic in the SDN Based on Multi-path Routing, Taking into Account the Nature of the Load Channels

Based on the analysis of known methods of traffic balancing and taking into account the peculiarities of the organization and functioning of software-configured networks, the following sequence of traffic balancing is proposed in this work:

1. The formation of many disjoint paths between the final vertices of the path.
2. Among the generated paths, the path selection with the minimum criterion for loading the path channels:
3. Adjustment of the path loading table.
4. Dynamic reconfiguration of paths adjacent to this path.

3.1 Formation of Many Disjoint Paths Between the End Nodes of the Path

The network load depends on the length of the paths along which information is transmitted. The presence of common channels between different paths leads to channel overloads and a decrease in the efficiency of computer networks. In this regard, when balancing traffic, it is necessary to form a lot of minimally loaded disjoint paths. For SDN networks, the following modified wave algorithm satisfies these requirements.

Algorithm for the formation of many minimally loaded disjoint paths.

1. $D_h = 0, D_p = 0, d_p = 0, d_h = 0,$
2. if $L_{h,h} < L_{p,p}$ and …< $L_{n,n.}$ then go to W_h; //Search for the transition to the
3. else go to W_p; // second peak with less workload.
4. for i=1 step1 W_{n+1} Calculate d_n, D_n;
5. if $d_h < d_p \dots d_n$ and $< d_n$ then go to d_h; // Choosing the least loaded
6. else go to d_p; // channel to go to the next vertex.
7. if $d_h = d_p$;
8. if $D_h < D_p$ then go to d_h; // Calculation of the minimum number of
9. else go to d_p; // transitions to the final vertex at $d_h = d_p$.
10. if $W_p + 1 = \varnothing$ then go to 4; // Completion of the program in the absence
11. end; // of the need for transition, due to reaching the final peak.

where:

d_p, d_h - Channel load.
D_p, D_h - Number of transitions.
W - Vertices of the graph.

As a result of the work of this algorithm, a set of minimally loaded disjoint paths is formed in the SDN controller for each pair of nodes. The smallest path is entered into a common path table between all the vertices.

For example, for the network shown in Fig. 1 minimum paths are presented in the Table 1, based on the values of which load balancing is carried out.

In Fig. 1. A graph of data transmission between vertices is presented.

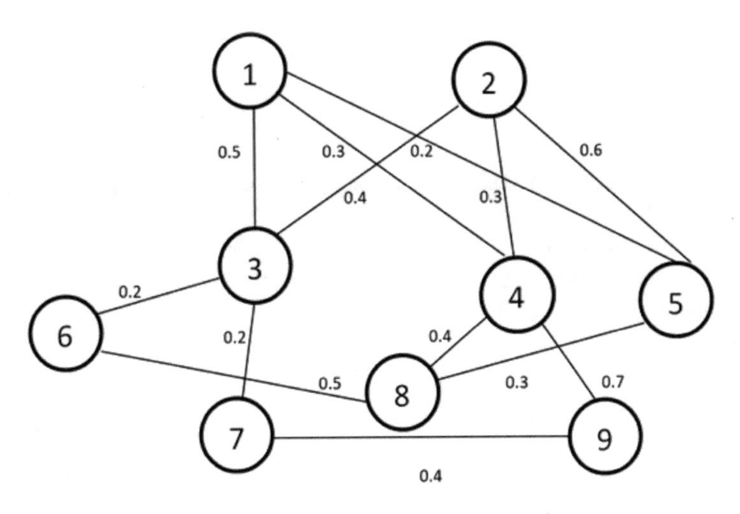

Fig. 1. Data transfer graph between vertices.

Table 1. Table of minimal paths between vertices of a graph

	R_1	R_2	R_3	R_4	R_5	R_6	R_7	R_8	R_9
R_1	0	L1.4.2	L1.3	L1.4	L1.5	L1.3.6	L1.3.7	L1.5.8	L1.4.9
R_2	L1.4.2	0	L2.3	L2.4	L2.5	L2.3.6	L2.3.7	L2.4.8	L2.3.7.9
R_3	L1.3	L2.3	0	L3.2.4	L3.1.5	L3.6	L3.7	L3.6.8	L3.7.9
R_4	L1.4	L2.4	L3.2.4.	0	L4.1.5	L4.8.6	L4.2.3.7	L4.8	L4.9
R_5	L1.5	L2.5	L3.1.5	L4.1.5	0	L5.8.6	L5.1.3.7	L5.8	L5.1.4.9
R_6	L1.3.6	L2.3.6	L3.6	L4.8.6	L5.8.6	0	L6.3.7	L6.8	L6.3.7.9
R_7	L1.3.7	L2.3.7	L3.7	L4.2.3.7	L5.1.3.7	L6.3.7.	0	L7.3.6.8	L7.9
R_8	L1.5.8	L2.4.8	L3.6.8	L4.8	L5.8	L6.8	L7.3.6.8	0	L8.4.9
R_9	L1.4.9	L2.3.7.9	L3.7.9	L4.9	L5.1.4.9	L6.3.7.9	L7.9.	L8.4.9.	0

Minimum paths in (Table 1) were obtained by analyzing all possible paths from Fig. 1.

4 Dynamic Load Balancing Based on the Nature of the Load Channels

4.1 Load Balancing Procedure

In this paper, the balancing procedure consists in choosing among a set of available paths a minimally loaded path with relatively uniform loading of channels of a given path. In order to ensure uniform loading, it is necessary to choose tracks with a minimum average load and a minimum mean square deviation of the track load from the average value. To this end, in this paper, it is proposed to use the criterion for loading path channels:

$$\mathrm{D_i} = \frac{n}{N}\left(d_i^0 + \frac{1}{n}\sum\nolimits_{j=1}^{n}\left(d_j - d_i^0\right)^2\right), \tag{1}$$

Where:

n - is the number of links (channels) of the path P_i between the initial vertex and the final vertex V_e;
N - is the number of links of the maximum path P_m between the vertices V_s and V_s;
d_i^0 - is the average load value of the channels of the path P_i;
$\mathrm{d_j}$ - channel loading $\mathrm{L_j} \in \mathrm{P_i}$.

$$d_i^0 = \frac{1}{n}\sum\nolimits_{j=1}^{n} d_j. \tag{2}$$

The ratio n/N allows one to take into account the relative path length P_i. Value d_i^0 – defines a less loaded path.

The root-mean-square deviation of the load of the path channels $\left(\left(d_j - d_i^0\right)^2\right)$ characterizes the degree of uniformity of the load of the path channels.

Thus, the coefficient Di allows you to choose the most optimal way in terms of load balancing in the network.

4.2 Load Balancing Procedure

Balancing the load of channels consists of choosing a data transmission path that optimizes the loading of network channels. In the process of balancing, service information is exchanged between the SDN controller and network switches (Fig. 2).

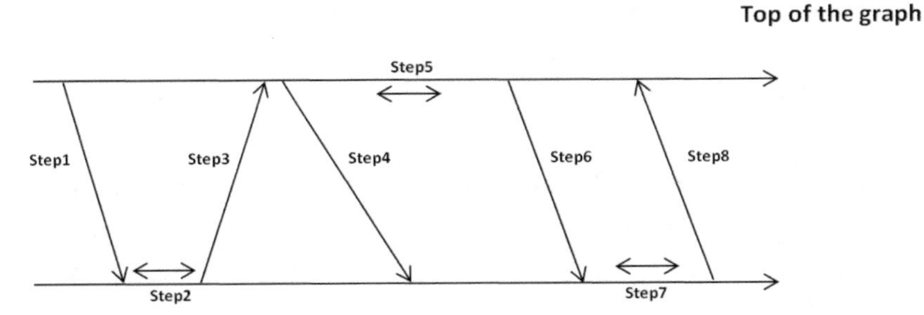

Step1 - Request for a route, **Step2**– Formation of a route, **Step3** - Transfer of a route, **Step4** - Number of a route, **Step5** - Transfer of information, **Step6** - End of transfer, **Step7** - Making corrections to a route, **Step8** - Update of a route.

Fig. 2. The process of exchanging service information between the controller and the switch.

Load Balancing Algorithm

1. At times Step1, the switch sends a routing request with its address and the recipient address.
2. In Step2, the controller, according to the table of minimum paths (Table 1), determines the existence of valid paths between the sender and receiver of information. In the absence of valid paths, the SDN creates many valid paths using a modified routing protocol.
3. At Step3, the route is transferred by the SDN controller to the top-sender of the information sender. Among the available paths, the path with the minimum load of channels is selected and its number is reported to the SDN controller.
4. Based on the number of the selected route, the SDN controller adjusts the routes adjacent to the selected route.
5. Then the information is transferred to the selected vertex.
6. After the transfer of information is completed, the used routes are updated.

4.3 Example Channel Load Balancing

As an example, we consider the modeling of the load balancing process during the transfer of information between the vertices W7 and W5 of the graph shown in Fig. 3. Using the above formula for the average value of the load of the path channels, we calculate the congestion of the channels for each of the possible paths, namely:

1) W_7, W_3, W_1, W_5.
2) W_7, W_3, W_2, W_5.
3) W_7, W_9, W_4, W_8, W_5.
4) W_7, W_3, W_6, W_8, W_5.
5) W_7, W_9, W_4, W_2, W_5.

Finding the average congestion of the path channel.

1) W_7, W_3, W_1, W_5.
1) 0.2 + 0.5 + 0.2 = 0.9.
2) 0.9/3 = 0.3.

Next, using the average load (0.3) we find the deviation of the channel load value L from the average channel load.

$P\Delta = P_i - P_{cp}$
1) 0.3 − 0.2 = 0.1.
2) 0.3 − 0.5 = 0.2.
3) 0.3 − 0.2 = 0.1.

At this stage, the mean square deviation of the channel load is calculated.

2) $0.3 + ((0.1)^2 + (0.2)^2 + (0.1)^2) = 0.320$.

Then, multiplying the above found mean square deviation is performed by dividing the number of links by the number of links of the maximum path (n/N), between the vertices Vs and Vs;

3) (0.32 * 3)/6 = 0.160.

We calculate the load factor for the remaining possible paths:

2) W_7, W_3, W_2, W_5.
1) 0.2 + 0.4 + 0.6 = 1.2/3 = 0.4.
2) $0.4 + ((0.2)^2 + (0.2)^2)/3 = 0.426$.
3) (0.426 * 3)/6 = 0.213.

3) W_7, W_9, W_4, W_8, W_5.
1) 0.4 + 0.7 + 0.4 + 0.3 = 1.8/4 = 0.45.
2) $0.45 + ((0.05)^2 + (0.25)^2 + (0.05)^2 + (0.15)2)/4 = 0.472$.
3) (0.4725 * 4)/6 = 0.315.

4) W_7, W_3, W_6, W_8, W_5.
1) 0.2 + 0.2 + 0.5 + 0.3 = 1.2/4 = 0.3.
2) $0.3 + ((0.1)^2 + (0.1)^2 + (0.2)^2)/4 = 0.315$.
3) (0.315 * 4)/6 = 0.210.

5) W_7, W_9, W_4, W_2, W_5.
1) 0.4 + 0.7 + 0.3 + 0.5 = 1.9/4 = 0.47.
2) $0.47 + ((0.07)^2 + (0.23)^2 + (0.17)^2 + (0.03)2)/4 = 0.419$.
3) (0.4919 * 4)/6 = 0.327.

1) W_7, W_3, W_1, $W_5 = 0.160$.
2) W_7, W_3, W_2, $W_5 = 0.213$.
3) W_7, W_9, W_4, W_8, $W_5 = 0.315$.
4) W_7, W_3, W_6, W_8, $W_5 = 0.210$.
5) W_7, W_9, W_4, W_2, $W_5 = 0.327$.

Figure 3 - is a diagram of the results of the effectiveness of the possible paths W7–W5.

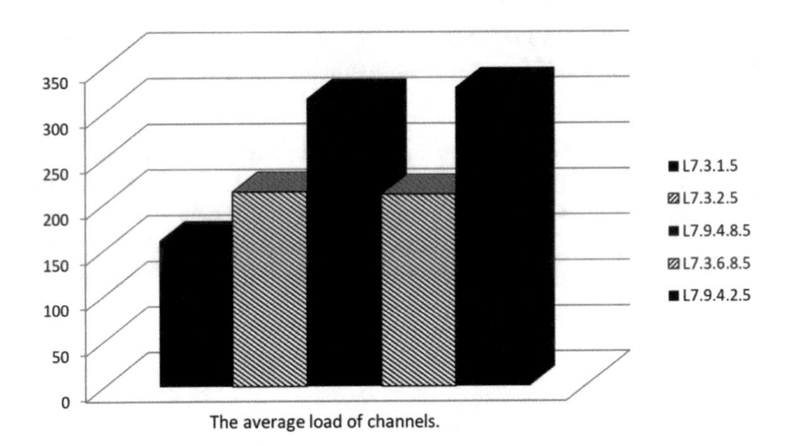

Fig. 3. The average load of channels

Thus, for the paths from vertex 7 to vertex 5, with the lowest load factor, path 1. with D1 = 0.160.

For each path presented in (Table 1), the coefficients of channel loading are determined, the value of which is presented in (Table 2).

The load factors in (Table 2) were obtained by calculating the shortest path, using the formula (1).

Table 2. Path downloads

	R_1	R_2	R_3	R_4	R_5	R_6	R_7	R_8	R_9
R_1	0	0.141	0.083	0.050	0.033	0.149	0.149	0.101	0.214
R_2	0.141	0	0.050	0.050	0.075	0.124	0.124	0.141	0.179
R_3	0.083	0.050	0	0.141	0.149	0.033	0.033	0.149	0.124
R_4	0.050	0.050	0.141	0	0.101	0.181	0.137	0.066	0.077
R_5	0.033	0.075	0.149	0.101	0	0.117	0.137	0.050	0.223
R_6	0.149	0.124	0.033	0.181	0.117	0	0.044	0.071	0.129
R_7	0.149	0.124	0.033	0.137	0.137	0.044	0	0.240	0.057
R_8	0.101	0.141	0.149	0.066	0.050	0.071	0.240	0	0.229
R_9	0.179	0.215	0.124	0.077	0.223	0.129	0.057	0.229	0

4.4 Adjustment of the Path Loading Table

We simulate the situation that the data flow between nodes loads the data channel by 0.1, and will go along the path W7, W3, W2, W5. Consequently, the load on each transition will increase by 0.1. We calculate the coefficient of channel loading efficiency taking into account the change in weight of each transition.

1) 0.3 + 0.5 + 0.7 = 1.5/3 = 0.5.
2) $0.5 + ((0.2)^2 + (0.2)^2)/3 = 0.5266$.
3) (0.5266 * 3)/7 = 0.225. – The final coefficient of download efficiency with a load of streaming information (0.1).

In this case, the load of channels having adjacent paths with data will also increase. For example, when forming a path (L7.3.2.5). The path L1.3 is not changed and its weight is 0.5. But the L3.7 path has already been used for data transfer. Therefore, it weighs 0.1 more, namely L3.7 = 0.3. Given these changes, the weight of the path is L1.3.7 will change. We calculate the channel loading efficiency coefficient taking into account the change in weight of the transition (s).

L1.3.7

1) 0.5 + 0.3 = 0.8/2 = 0.4.
2) $0.4 + ((0.1)^2 + (0.1)^2)/2 = 0.41$.
3) (0.41 * 2)/5 = 0.164.

As a result, the new load factor of our channel is D1 = 0.164. While the past was D1 = 0.149.

In this example, we calculate the new congestion of the channel for each path that has common ground with the path along which the information is being transmitted.

In this case, it is necessary to adjust 14 routes. As a result, the load of individual channels will increase.

Table 3 shows the load values of the network channels when transmitting information between the vertices W7 and W5. The paths are bold. The Di value of which has changed. The loading of channels in (Table 3) was obtained by calculating the values according to formula (1), taking into account the additional loading of channels between the vertices.

Table 3. Channel downloads for information transfer

	R_1	R_2	R_3	R_4	R_5	R_6	R_7	R_8	R_9
R_1	0	0.141	0.083	0.050	0.033	0.149	**0.164**	0.101	0.214
R_2	0.141	0	**0.062**	0.050	**0.087**	**0.145**	**0.162**	0.141	**0.202**
R_3	0.083	**0.062**	0	**0.205**	0.149	0.033	**0.042**	0.149	**0.141**
R_4	0.050	0.050	**0.205**	0	0.101	0.181	**0.179**	0.066	0.077
R_5	0.033	**0.087**	0.149	0.101	0	0.117	**0.156**	0.050	0.223
R_6	0.149	**0.145**	0.033	0.181	0.117	0	**0.072**	0.071	**0.153**
R_7	**0.164**	**0.162**	**0.042**	**0.179**	**0.156**	**0.072**	0	**0.274**	0.057
R_8	0.101	0.141	0.149	0.066	0.050	0.071	**0.274**	0	0.229
R_9	0.214	**0.202**	**0.141**	0.077	0.223	**0.153**	0.057	0.229	0

5 Results and Discussion

The proposed method of balancing traffic due to the simultaneous formation of many shortest routes between all the peaks allows you to carry out the routing procedure in a dynamic mode without recalculating the entire path. For example, you still need to go from vertex 7 to vertex 5, but the difference is that with each transition, the controller transfers information about any changes to each vertex of the path, and dynamically changes the route if necessary. This allows us to exclude the procedure of returning to the point of the beginning of movement if some channel turned out to be overloaded during the transmission of information. Consider the process of transmitting information along the path L7.3; L3.2; Assume that during transmission, the L2.5 channel was overloaded. For conducting experiments. In the absence of an alternative path, a return is made to the nearest last peak (2). In real-time, the controller changes the path to the peak we need to get to (W2, W4, W8, W5) Thus, there is no need to return to the top of the movement (7) to plan a new route. Consequently, there are savings on transitions, and hence on congestion of the channels of the entire network, which is more effective in our case to compared to known methods of balancing traffic.

If there are two types of traffic in the network: elastic and non-elastic, first of all, it is necessary to form the paths for inelastic traffic, since it places higher demands on the quality of service (QoS). In this case, load balancing is mainly carried out due to elastic traffic.

6 Conclusions

The work proposes a method of balancing traffic, which through accounting Features of the organization of SDN, in particular, due to the presence of a central controller in the network, reduces the time it takes to form multiple routes and simplify the process of balancing traffic.

A criterion is proposed and justified for choosing a path from the set of available paths, which allows for more uniform loading of information transmission channels with given QoS parameters.

From a practical point of view, the use of the method proposed in the work reduces the time of balancing traffic due to the exception procedures for recalculating routes when changing the load of individual paths.

During operation, networks can dynamically change the load of channels. This may affect the consistency of simulation results.

This method of balancing traffic allows you to provide more uniform network load, which reduces network consumption

Further improvement of traffic balancing methods is associated with the application of decision theory methods when choosing a path with the aim of predicting the nature of load changes in communication channels during data transmission. This will reduce the frequency of path reconfiguration and ensure a more uniform network load.

References

1. Isong, B., Kgogo, T., Lugayizi, F.: Trust establishment in SDN: controller and applications. Int. J. Comput. Netw. Inf. Secur. (IJCNIS) **9**(7), 20–28 (2017). https://doi.org/10.5815/ijcnis.2017.07.03. Published Online July 2017 in MECS (http://www.mecs-press.org/)
2. Sahoo, K.S., Mishra, S.K., Sahoo, S., Sahoo, B.: Software defined network: the next generation internet technology. Int. J. Wirel. Microw. Technol. (IJWMT) **7**(2), 13–24 (2017). https://doi.org/10.5815/ijwmt.2017.02.02. Published Online March 2017 in MECS (http://www.mecs-press.net)
3. Mahmood, Z.A., Nasir, A.A., Fatima, W.H.: Evaluating and comparing the performance of using multiple controllers in software defined networks. Int. J. Mod. Educ. Comput. Sci. (IJMECS) **8**, 27–34 (2019). https://doi.org/10.5815/ijmecs.2019.08.03. Published Online August 2019 in MECS (http://www.mecs-press.org/)
4. Abdolhossein, F., Keihaneh, K.: A centralized controller as an approach in designing NoC. Int. J. Mod. Educ. Comput. Sci. (IJMECS) **9**(1), 60–67 (2017). https://doi.org/10.5815/ijmecs.2017.01.07. Published Online January 2017 in MECS (http://www.mecs-press.org/)
5. Ashutosh, K.S., Naveen, K., Shashank, S.: PSO and TLBO based reliable placement of controllers in SDN. Int. J. Comput. Netw. Inf. Secur. (IJCNIS) **2**, 36–42 (2019). https://doi.org/10.5815/ijcnis.2019.02.05. Published Online January 2017 in MECS http://www.mecs-press.org
6. Gamess, E., Tovar, D., Cavadia, A.: Design and implementation of a benchmarking tool for OpenFlow controllers. Int. J. Inf. Technol. Comput. Sci. (IJITCS) **10**(11), 1–13 (2018). https://doi.org/10.5815/ijitcs.2018.11.01. Published Online November 2018 in MECS (http://www.mecs-press.org/)

7. Sharma, G., Singh, M., Sharma, P.: Modifying AODV to reduce load in MANETs. Int. J. Mod. Educ. Comput. Sci. (IJMECS) **8**(10), 25–32 (2016).https://doi.org/10.5815/ijmecs.2016.10.04. Published Online October 2016 in MECS (http://www.mecs-press.org/)
8. Matnee, Y.A., Abooddy, C.H., Mohammed, Z.Q.: Analyzing methods and opportunities in software-defined (SDN) networks for data traffic optimizations. Int. J. Recent Innov. Trends Comput. Commun. (IJRITCC) **6**(1), 75–82 (2018). http://www.ijritcc.org
9. Wang, H., Xu, H., Huang, L., Wang, J., Yang, X.: Load-balancing routing in software defined networks with multiple controllers. Comput. Netw. **141**, 82–91 (2018). https://doi.org/10.1016/j.comnet.2018.05.012
10. Ejaz, S., Iqbal, Z., Shah, P.A., Bukhari, B.H., Ali, A., Aadil, F.: Traffic load balancing using software defined networking (SDN) controller as virtualized network function (2019). https://doi.org/10.1109/ACCESS.2019.2909356. Received February 8, 2019, accepted March 21, 2019, date of publication April 4, 2019, date of current version April 18, 2019
11. Kulakov, Y., Kogan, A.: The method of plurality generation of disjoint paths using horizontal exclusive scheduling. Adv. Sci. J. **10**, 16–18 (2014). https://doi.org/10.15550/ASJ.2014.10. ISSN 2219-746X
12. Rajasekaran, K., Balasubramanian, K.: Energy conscious based multipath routing algorithm in WSN. Int. J. Comput. Netw. Inf. Secur. (IJCNIS) **1**, 27–34 (2016). https://doi.org/10.5815/ijcnis.2016.01.04 in MECS (http://www.mecs-press.org/)
13. Han, Y., Seo, S.S., Li, J., Hyun, J., Yoo, J.H. and Hong, J.W.K.: Software defined networking-based traffic engineering for data center networks. In: Asia-Pacific Network Operations and Management Symposium (2014). https://ieeexplore.ieee.org/document/6996601
14. Kumar, P., Dutta, R., Dagdi, R., Sooda, K., Naik, A.: A programmable and managed software defined network. Int. J. Comput. Netw. Inf. Secur. (IJCNIS) **12**, 11–17 (2017). https://doi.org/10.5815/ijcnis.2017.12.02. In MECS http://www.mecs-press.org/. Accessed Dec 2017
15. Moza, M., Kumar, S.: Analyzing multiple routing configuration. Int. J. Comput. Netw. Inf. Secur. (IJCNIS) **5**, 48–54 (2016). https://doi.org/10.5815/ijcnis.2016.05.07. In MECS http://www.mecs-press.org/. Accessed May 2016
16. Kulakov Y., Kohan A., Kopychko S.: Traffic orchestration in data center network based on software-defined networking technology. In: International Conference on Computer Science, Engineering and Education Applications ICCSEEA 2019: Advances in Computer Science for Engineering and Education II, pp. 228–237 (2019)
17. Shu, Z., Wan, J., Lin, J., Wang, S., Li, D., Rho, S., Yang, C.: Traffic engineering in software defined networking: measurement and management. IEEE Access **4**, 3246–3256 (2016). http://www.ieee.org/publications_standards/publications/rights/index.html
18. Abbasi, M.R., Guleria, A., Devi, M.S.: Traffic engineering in software defined networks: a survey. J. Telecommun. Inf. Technol. **4**, 3–13 (2016)

Information System Penetration Testing Using Web Attack Automated Simulation

Inna V. Stetsenko(✉) and Viktoriia Savchuk

Igor Sikorsky Kyiv Polytechnic Institute,
37 Prospect Peremogy, Kiev 03056, Ukraine
stiv.inna@gmail.com, viktoria859@gmail.com

Abstract. The usage of information systems in the day-to-day operations of entrepreneurial companies opens enlarged opportunities for communication services provision and processing of information resources between relevant departments and organizations. At the same time, it opens the way for attackers to gain access to information that poses a real threat to information security and system safety. In addition, the possibility of relevant threat implementation increases through the extended availability of computer technology. As a consequence, the complexity of information system protection against related threats and their implementation ways (attacks) is significant. Every day new vulnerabilities of software could be identified and therefore new ways for attackers appear. Thereby, the new method of penetration time estimation for the information system is proposed in this research. It is grounded on using information about vulnerabilities and presents a preliminary and detailed study of possible options for the information attack implementation using a simulation model. Stochastic Petri net with informational arcs is used for model creation. The results which are obtained for different sets of vulnerabilities show a strong dependence of penetration time on system vulnerabilities.

Keywords: Cyber attack simulation · Vulnerabilities · Stochastic petri net

1 Introduction

Security problems can be successfully solved in small computer networks using a variety of tools, including support for modelling attacks, but cannot be effectively solved for large systems. The complexity of modeling attacks in large computer networks is associated with the incompleteness and uncertainty of information which has been used by modeling tools, as well as the significant computational complexity of algorithms of constructing and analyzing attack models.

In the real world, penetration tests are conducted to secure a network. A penetration test, colloquially known as an ethical hacking, is an authorized simulated cyberattack on a computer system, performed to evaluate the security of the system [1]. The test is performed to identify vulnerabilities, including the potential for unauthorized parties to gain access to the system's features and data, and to assess the level of risk of those vulnerabilities [2].

Z. Hu et al. (Eds.): ICCSEEA 2020, AISC 1247, pp. 396–406, 2021.
https://doi.org/10.1007/978-3-030-55506-1_36

Practically, penetrating testing is a challenging task which relies on human experience and expertise in various techniques operated and an extensive number of tools used in the process. To avoid the above issues the method of penetration time estimation based on stochastic Petri net is proposed in this article and the correspondent application is developed. The main goal of the research is to get a fast way to assess the protection of the investigated information system and to identify the weaknesses of the system that decrease its penetration time.

The considering of delayed transition expands the scope of the Petri nets implementation and makes it possible not only to trace the order of events occurring in the system but to take into account the interaction of events in time. The state of the system is presented by the state of places of Petri net described by the number of tokens in every place and the state of transitions pf Petri net described by the moments of token outputs in every transition. The current state of the model corresponds to the attack phase. Then, the change of the state and the next attack generation are presented by firing Petri net transitions.

The use of Petri net is described and analyzed through case studies elucidating several properties of Petri net variants and their suitability to modeling attacks in penetration testing. The advantages of simulating attacks by Petri nets have been explored and described. The model contains the mechanisms needed to describe the algorithm of the attacker's actions, including the arbitrary selection of one of the alternatives. It allows investigating the impact of the dynamic characteristics of network traffic attacking in the form of deterministic and random time delays. The drawbacks of such models for attack research stem from the shortcomings of Petri nets. A large number of elements which is needed to build a complicated model, the absence of possibility to use information about the condition without changing it when the event occurs and the inability to create values for statistics. Petri-object simulation technology, which is used for model development in this research, due to the combination of object-oriented technology and stochastic Petri net with multichannel transitions and information arcs provides advanced possibilities.

Section 1 introduce to the formalized models of information attacks research. The second section describes related works. The next section represents the proposed method for simulation. Section 4 describes application architecture. Experimental results of simulation are given in the next section. The last section summaries the article and gives the perspective of future research.

2 Related Works

Up-to-date malware analysis and detection technics were reviewed in work [3]. Recent research has focused on the impact of network topologies on the spread of malware. In work [4] the authors obtained results for dimensionless network topology using a mathematical model. The topology was characterized by the number of nodes and power exponent parameters. In research [5] the structure of social and technology networks was investigated when they were attacked by a computer virus or worm. The authors used a simulation of the spread of infection based on a susceptible-infected-recovered epidemic model. They proposed a structural risk model. In work [6] computer virus propagation via Internet resources depending on users' connections topology is investigated using Petri-object simulation.

Among the variants of formalized models of information attacks, it is necessary to distinguish a model based on the mathematical apparatus. In work [7] it was proposed to use the stochastic Petri nets with delays of a special kind, restraining arcs and weighted transitions. The colored Petri net was used by the authors of work [8] for the development of the model of web content integrity verification. The attack model based on calculating the conditional probabilities of attack paths is proposed in work [9].

An attack on an information system is a sequence of actions of an intruder that lead to the implementation of a threat by exploiting system vulnerabilities. A threat is a potentially possible event, action, phenomenon, or process, that can produce damage to a system resource [5]. Thus, the attacker initiates some action that leads to the expected result using malware that exploits the system vulnerabilities. Logical web vulnerabilities using for remote access were considered in work [10]. To score the vulnerabilities the authors of the work used the Common Vulnerability Scoring System (CVSS).

An information attack commonly consists of three stages [11]:

1. Gathering information is the main stage. At this stage, the target of the attack is selected, information about it (OS, configuration, services) is collected, the most vulnerable places of the attacked system are identified, the impact on which leads to the desired result, and the type of attack is selected. The usage of Open Source Intelligence tools at the gathering information stage has been considered in work [12].
2. The stage of implementation of the attack. At this stage, the intruder gets unauthorized access to resources of those nodes of information system to which attack is carried out. If the nature of the impact of the attack is active [13], then this stage is completed by the implementation of the goals for which the attack was undertaken. The result of such actions may be a violation of the confidentiality, integrity, and availability of information. Besides, there may be hiding of the source and the fact of the attack, the so-called 'covering of traces'.
3. The stage of further spread of the attack - the actions that are directed on the continuation of the attack on resources of other nodes of IS. In the case of passive attacks [14], this stage is the stage of completion of the attack.

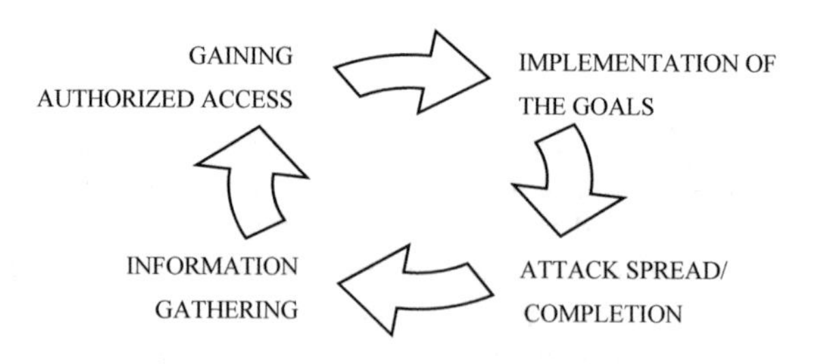

Fig. 1. The life cycle of a typical information attack

Figure 1 schematically presents the stages of the life cycle of typical information attack.

In this research, full hacker way is simulated and all possible way of system hacking is investigated. The model of information system penetration, which is proposed, is taking into account such parameters and characteristics of attack as the time of action, multiplicity, access method, list of vulnerabilities used by the attack, type of attack, and impact on the target of the attack.

3 The Proposed Method of Attack Simulation

Modeling of attacks was carried out in a specially designed environment by object-oriented programming language Java. Petri net's components (places, transitions, arcs, tokens) are represented as objects and all processes are occurring by OOP methods. The application uses a graphical shell to simplify model creation. The feature is the application convert Petri net to a Java method after the model drawing, which provides the opportunity to reuse small model parts to create a big model [15].

The screenshot of Petri net snippet for modeling hacker penetration is depicted in Fig. 2. A successful attack occurs only when the host service has an available vulnerability, which can be exploited by a hacker. The nodes (CVE-2019-10097, CVE-2019-10092, CVE-2019-10082, and CVE-2019-10098) present vulnerabilities.

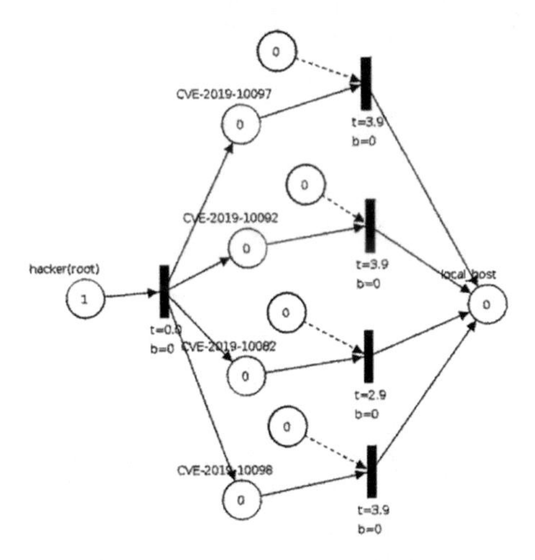

Fig. 2. Petri net snippet for modeling hacker penetration

Common Vulnerabilities and Exposures (CVE) is a list of information-security vulnerabilities and exposures—each containing an identification number, a description, and at least one public reference—for publicly known cybersecurity vulnerabilities, its format: CVE-0000-0000 [16, 17]. All of the presented CVEs are vulnerabilities of service Apache HTTP Server version 2.4.39. Apache web server has been chosen because of its popularity: it is the first in the top 5 web servers [18].

4 Application Architecture

The application, as it was intended in the beginning, is an automated analogy of the penetration test. For this reason, the first goal is to identify vulnerabilities that can be exploited to penetrate tested network. For this reason, it is necessary to know which services and networks belonging to a particular organization. To collect the parameters of an organization network use:

1. Search engines (Google).
2. Whois-services.
3. Search engines on databases of Internet registrars – Hurricane Electric BGP Toolkit, RIPE.
4. Data visualization services of the domain name of the site – Robtex (Fig. 3) [19].
5. DNSdumpster is a domain zone analysis service that contains historical data on domain zones, which greatly helps to collect data. Domaintools - similar services.

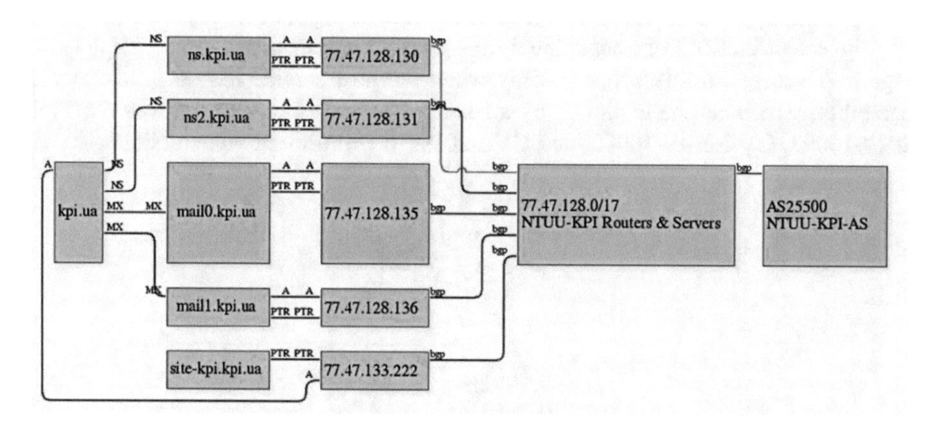

Fig. 3. The static graph over the entity and related entities that belong to the National Technical University of Ukraine "Igor Sikorsky Kyiv Polytechnic Institute" using Robtex service [19].

Web service 'whois' is a resource that allows you to find the site IP address by the domain name, as well as other important data to search for networks. Important data which can be found after 'whois' search: net name, net description, physical address, contacts and other [20]. To identify the available services, it is possible to use one of the two most well-known tools designed to make computer networks safer: Shodan or Censys [21, 22]. They have similar capabilities, support work with API. For a full search, both services require registration to get the API key. The application architecture uses Shodan API. The service has REST API, implementations of which in different programming languages presented in Github, including Java [23, 24].

The following information could get after Shodan scanning:

- IP – a unique network address of a node in a computer network built using the IP Protocol;
- Port – a numeric number that is a parameter of transport protocols (such as TCP and UDP);
- Protocol – a set of logical-level interface conventions that define communication between different programs;
- Hostname – is a symbolic name assigned to a network device that can be used to organize access to that device in various ways;
- Service – the name of a particular service;
- Product – the name of the software with which the service is implemented;
- Product version – specific software version;
- Banner – welcome information which is given by the service when a connection attempt has been done;
- CPE-Common Platform Enumeration, a standardized way of naming software applications, operating systems, and hardware platforms;
- OS – version of the operating system.

An IP address, port, and CPE are data that the applications use. IP and port are very important to know about potential vulnerability location in target information system. And CPE is the best way to determine the presence of any vulnerability in the National Vulnerability Database and in a positive case to find the identification number of this vulnerability [25]. CVE Details is a resource that helps to identify the main characteristics of the vulnerability, which are very important for the model [26]:

- vulnerability type – the type defined by CWE (Common Weakness Enumeration) – community developed list of software weakness type [27];
- authentication – the type of rules which hacker should have to exploit a vulnerability (root or user);
- gained access level – the rule which hacker can gain after vulnerability exploiting (root or user);
- access – the type of access (network or local);
- complexity – the metric describing how difficult to exploit a vulnerability.

All of these characteristics are determined by The Common Vulnerability Scoring System, which is open standard for assessing the severity of actual vulnerabilities [28]. In the model, there is a direct relationship between complexity of the vulnerability and the time delay: the lower the complexity score, the longer the hacker will exploit it. The complexity is assessed relative to its danger, that is, the more difficult it is to exploit, the less dangerous it is, because not every hacker can have enough skills. Table 1 shows full description about complexity metric with values and scores [29].

Table 1. Common Vulnerability Scoring System complexity metric description.

Value	Description	Score
High	Specialized access conditions exist. For example, the attacking party must already have elevated privileges; the attack depends on social engineering methods; the victim must perform several suspicious or atypical actions	0.35
Medium	The access conditions are somewhat specialized. For example, the attacking party is limited to a group of systems at some level of authorization, possibly untrusted; some information must be gathered before an attack can be launched	0.61
Low	Specialized access conditions or extenuating circumstances do not exist. For example, the affected product typically requires access to a wide range of users, possibly anonymous and untrusted; the attack can be performed manually and requires little skill or additional information gathering	0.71

The Selenium framework for Java has been used to automate the process of collecting information [29]. It is a framework for managing browsers using driver WebDriver for different browsers. The application uses WebDriver for scraping and parsing information on the sites, as mentioned above, the site of the National Vulnerability Database and the site of CVE data source. The information resulted from WebDriver's work is used by the Petri net model.

5 Experimental Results

The simulation was carried out on a typical architecture that consists of a web server, a file server, and a database. From the Internet, only the network, in which the web server and file server are located, is available. The architecture of the network in the experiment is depicted in Fig. 4. The network in which the webserver and the file are located has direct access to the Internet. Table 2 shows the data about hosts vulnerabilities collected in the previous stages of the application. Figure 5 shows the Petri net model of the spread of an attack on IP vulnerability data.

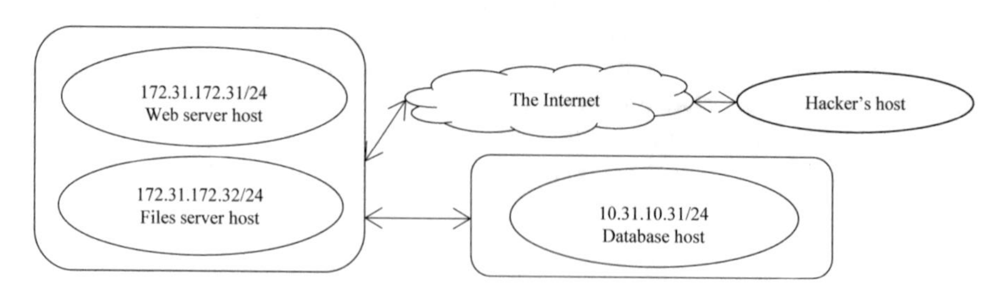

Fig. 4. The architecture of the network.

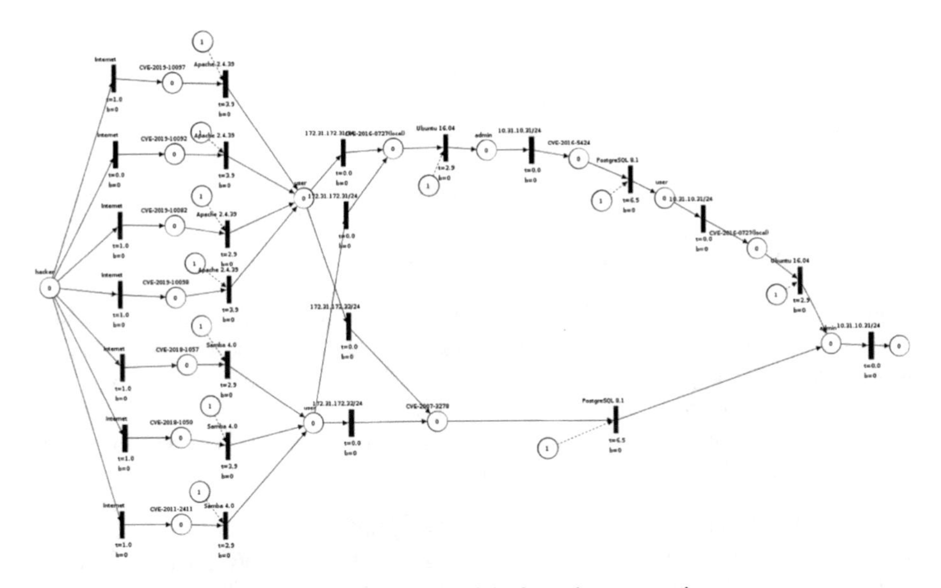

Fig. 5. The Petri net model of attacks propagation

Table 2. The data about hosts vulnerabilities obtained automatically by the application.

Host ID	CVE ID	Access type	Complexity	Authentication	Gained access
172.31.172.31/24	CVE-2019-10097	Remote	Medium	Single system	None
	CVE-2019-10092	Remote	Medium	Not required	None
	CVE-2019-10082	Remote	Low	Not required	None
	CVE-2019-10098	Remote	Medium	Not required	None
	CVE-2016-0727	Local	Low	Not required	Admin
172.31.172.32/24	CVE-2018-1057	Remote	Low	Single system	None
	CVE-2018-1050	Remote	Medium	Not required	None
	CVE-2011-2411	Remote	Low	Single system	None
10.31.10.31/24	CVE-2007-3278	Remote	Medium	Not required	Admin
	CVE-2016-5424	Remote	High	Single system	None
	CVE-2016-0727	Local	Low	Not required	Admin

Timeouts for exploiting a vulnerability has been calculated as $\left(1 - score_{complexity}\right) \cdot 10$ because 10 s is the most common order of time delays. Simulating attack many times all possible ways of successful penetration could be found and the time needed for their implementation could be obtained:

Path 1 (CVE-2019-10097, CVE-2016-0727, CVE-2016-5424, CVE-2016-0727) 17.2 s
Path 2 (CVE-2019-10092, CVE-2016-0727, CVE-2016-5424, CVE-2016-0727) 17.2 s
Path 3 (CVE-2019-10082, CVE-2016-0727, CVE-2016-5424, CVE-2016-0727) 16.2 s
Path 4 (CVE-2019-10098, CVE-2016-0727, CVE-2016-5424, CVE-2016-0727) 17.2 s
Path 5 (CVE-2018-1057, CVE-2016-0727, CVE-2016-5424, CVE-2016-0727) 16.2 s
Path 6 (CVE-2018-1050, CVE-2016-0727, CVE-2016-5424, CVE-2016-0727) 17.2 s

Path 7 (CVE-2011-2411, CVE-2016-0727, CVE-2016-5424, CVE-2016-0727) 16.2 s
Path 8 (CVE-2019-10097, CVE-2007-3278) 11.4 s
Path 9 (CVE-2019-10092, CVE-2007-3278) 11.4 s
Path 10 (CVE-2019-10082, CVE-2007-3278) 10.4 s
Path 11 (CVE-2019-10098, CVE-2007-3278) 11.4 s
Path 12 (CVE-2018-1057, CVE-2007-3278) 10.4 s
Path 13 (CVE-2018-1050, CVE-2007-3278) 11.4 s
Path 14 (CVE-2011-2411, CVE-2007-3278) 10.4 s

From the result of simulation follows that the vulnerability CVE-2007-3278 is the most dangerous; it reduces the attack time by a third compared to other attack paths. The best way, in this case, to get rid of it that means to change the database version to a newer one. The latest version of PostgreSQL is 12.1 released on 2019-11-14 [30]. Figure 6 shows the modification of the spread of attacks in the case of this version of the database service. The result of simulation:

Path 1 (CVE-2019-10097, CVE-2016-0727, CVE-2009-2943, CVE-2016-0727) 14.6 s
Path 2 (CVE-2019-10092, CVE-2016-0727, CVE-2009-2943, CVE-2016-0727) 14.6 s
Path 3 (CVE-2019-10082, CVE-2016-0727, CVE-2016-5424, CVE-2016-0727) 13.6 s
Path 4 (CVE-2019-10098, CVE-2016-0727, CVE-2009-2943, CVE-2016-0727) 14.6 s
Path 5 (CVE-2018-1057, CVE-2016-0727, CVE-2016-5424, CVE-2016-0727) 13.6 s
Path 6 (CVE-2018-1050, CVE-2016-0727, CVE-2016-5424, CVE-2016-0727) 14.6 s
Path 7 (CVE-2011-2411, CVE-2016-0727, CVE-2016-5424, CVE-2016-0727) 13.6 s

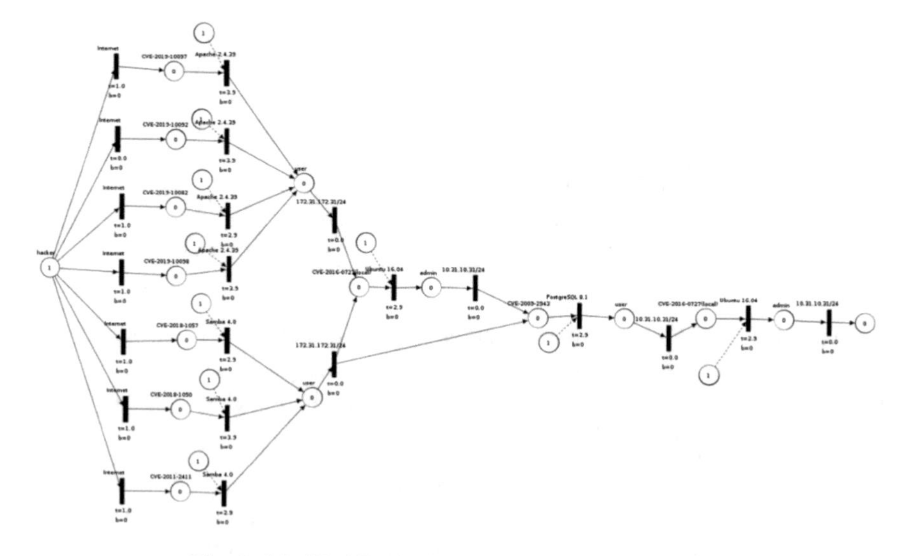

Fig. 6. Modified Petri net model of attacks propagation

Thus, minimum attack time increased by 30% in comparison with the previous version. According to the results, it can be concluded that the vulnerabilities CVE-2019-10082, CVE-2018-1057, and CVE-2011-2411 should be rejected in the first place because a hacker in the fastest way penetrate the system and reaches full control over the database. Moreover, the most dangerous vulnerability in the system is CVE-2007-3278 in the database which cuts the hacker's path to the target in half. The main parameters that affected the results are the access rights that are needed for the vulnerability and what can be obtained with its help.

6 Conclusions

A method for penetration tests without operating the information system is proposed. The system reliability is estimated using the model which is described by stochastic Petri net. Software application architecture is presented that identifies services and their vulnerabilities by using publicly available methods of passively scanning the hosts. Based on the collected and provided information about the system infra-structure, the system model is automatically generated, the simulation result of which is the time of the attack propagation. In this way, it is possible to assess the criticality of a particular vulnerability in the system and to draw conclusions about its elimination or acceptance of the risk that exists because of its presence in the system.

The application can be improved by adding other vulnerability databases. Furthermore, when developing, it could be focused not only on the above characteristics of the vulnerability but applied all the well-known Common Vulnerability Scoring System metrics.

References

1. Pandey, K.: A bug tracking tool for efficient penetration testing. Int. J. Educ. Manage. Eng. **8** (3), 14–20 (2018)
2. Henry, K.M.: Penetration testing: Protecting Networks and Systems. IT Governance LTD, Ely (2012)
3. Tahir, R.: A study on malware and malware detection techniques. Int. J. Educ. Manage. Eng. **8**(2), 20–30 (2018)
4. Yang, L., Yang, X.: The effect of network topology on the spread of computer viruses: a modeling study. Int. J. Comput. Math. **94**(8), 1–19 (2017)
5. Guo, H., Cheng, H.K., Kelley, K.: Impact of network structure on malware propagation: a growth curve perspective. J. Manage. Inf. Syst. **33**(1), 296–325 (2016)
6. Stetsenko, I.V., Lytvynov, V.V.: Computer virus propagation petri-object simulation. Adv. Intell. Syst. Comput. **1019**, 103–112 (2020)
7. Khorkov, D.A.: On the using of Petri nets for computer attack. Reports of Tomsk State University of Control Systems and Radio Electronics, vol. 1–2, no. 19, pp. 49–50 (2009). (in Russian)
8. Hijazi, S., Hudaib, A.: Verification of web content integrity: detection and recovery security approach using colored petri nets. Int. J. Comput. Netw. Inf. Secur. (IJCNIS) **10**(10), 1–10 (2018)

9. Zimba, A., Chama, V.: Cyber attacks in cloud computing: modelling multi-stage attacks using probability density curves. Int. J. Comput. Netw. Inf. Secur. (IJCNIS) **10**(3), 25–36 (2018)
10. Ndichu, S., McOyowo, S., Okoyo, H., Wekesa, C.: A domains approach to remote access logical vulnerabilities classification. Int. J. Comput. Netw. Inf. Secur. (IJCNIS) **11**(11), 36–45 (2019)
11. Lukatsky, A.: Protect Your Information: With Intrusion Detection. BPB Publications, New Delhi (2004)
12. Wiradarma, A.A., Susmita, G.M.: IT risk management based on ISO 31000 and OWASP framework using OSINT at the information gathering stage. Int. J. Comput. Netw. Inf. Secur. (IJCNIS) **11**(12), 17–29 (2019)
13. Serdyuk, V.: Information security of automated enterprise systems. Account. Comput. **1**, 104–107 (2007). (in Russian)
14. Natrov, V.V.: Network attack classification. information technology in management and modeling: a collection of reports. – Belgorod, pp. 128–132 (2005)
15. Stetsenko, I.V., Dyfuchyn, A., Leshchenko, K., Davies, J.: Web application for visual modeling of discrete event systems. In: Picking, R., Cunningham, S., Houlden, N., Oram, D., Grout, V., Mayers, J., (eds.) Proceedings of the Seventh International Conference on Internet Technologies and Applications, ITA2017, Wrexham, UK, pp. 86–91, (2017)
16. CVE official. https://cve.mitre.org. Accessed 05 Oct 2019
17. CVE format. https://nvd.nist.gov/vuln/detail/CVE. Accessed 05 Oct 2019
18. Top 5 web servers. https://binge.co/what-are-the-best-web-servers. Accessed 06 Oct 2019
19. Kpi.ua visualization of the domain name of the site. https://www.robtex.com/dns-lookup/kpi.ua. Accessed 06 Oct 2019
20. Whois online service (2019). https://whois.net. Accessed 06 Oct 2019
21. Shodandan.io. https://www.sho. Accessed 10 Oct 2019
22. Censys. https://censys.io/. Accessed 16 Oct 2019
23. Shodan REST API Documentation. https://developer.shodan.io/api/. Accessed 11 Oct 2019
24. Shodan client for Java. https://github.com/fooock/jshodan/. Accessed 16 Oct 2019
25. National Vulnerability Database. https://nvd.nist.gov/. Accessed 16 Oct 2019
26. CVE Details – the ultimate security datasource. https://www.cvedetails.com/. Accessed 16 Oct 2019
27. Common Vulnerability Scoring System. https://www.first.org/cvss/. Accessed 16 Oct 2019
28. Common Weakness Enumeration. https://cwe.mitre.org/index.html/. Accessed 10 Oct 2019
29. Selenium - Web Browser Automation. https://www.seleniumhq.org//. Accessed 10 Oct 2019
30. PostgreSQL. https://www.postgresql.org/. Accessed 16 Oct 2019

Optimal Decision of Supply Chain Based on Nested Loop Algorithm from the Perspective of System Analysis

Yao Zhang, Jinshan Dai(✉), Mengya Zhang, and Qingying Zhang

School of Logistics Engineering, Wuhan University of Technology, Wuhan 430063, China
785370647@qq.com, jinshan.dai@whut.edu.cn

Abstract. In the process of supply chain optimization, there is a phenomenon that the optimization results are always be of an unsatisfactory, because of the single selection of optimization objectives. In the process of multi-objective supply chain optimization, this paper applies the analytic hierarchy process of system analysis to decompose the system at multiple levels, establishes the supply chain optimization decision model based on nesting cycle algorithm, uses the supervised fuzzy cycle algorithm to sort the importance of each index of the supply chain, and introduces the subjective supervision operator and the self-feedback system to ensure the rationality and accuracy of the sorting results. Finally, through the appropriate index selection principle, select the appropriate index, carry on the supply chain optimization.

Keywords: System analysis · Fuzzy mathematics · Supply chain optimization · Subjective supervision operator

1 Instruction

With the development of society, competition between enterprises is intensifying. Under the current situation, it is difficult for a single enterprise rely on its own strength to adapt to the rapid changes, resulting in a gradual strengthening of cooperation between a single enterprise and upstream and downstream enterprises, and a gradually forming of supply chain management. Supply chain management optimizes the resource allocation of enterprises by optimizing the cooperation between upstream and downstream enterprises, which promotes the improvement of corporate performance [1].

As a complex, multilevel open system, the optimization of the supply chain is often a multi-objective decision making process [2, 3]. Therefore, in the general process of supply chain optimization, methods such as the analytic hierarchy process of system analysis will be implemented to disassemble the supply chain optimization goals [4]. After the analytic hierarchy process, a large number of supply chain indicators are obtained, and then key performance indicators will be selected for supply chain evaluation and optimization. In many cases, this type of supply chain evaluation and optimization method will be limited by the number of indicators, which may cause the supply chain optimization to have the benefit reversal situation, thus the optimization

Z. Hu et al. (Eds.): ICCSEEA 2020, AISC 1247, pp. 407–418, 2021.
https://doi.org/10.1007/978-3-030-55506-1_37

to be an unsatisfactory [5–7]. Therefore, this paper uses a systematic analysis method based on the indicators decomposed by the supply chain to design a nested loop algorithm, by using supervised self-feedback fuzzy decision model to analyze supply chain indicators and optimize the overall supply chain.

2 Theoretical Basis of Nested Loop Supply Chain Optimization Method

2.1 Fuzzy Mathematics and Decision Model

In all fields of science, the quantities involved can always be divided into two categories: certainty and uncertainty. For the uncertainty problem, it can also be divided into two categories: random uncertainty and fuzzy uncertainty. Fuzzy mathematics is a kind of mathematical method to study the change of the quantity which belongs to uncertainty and has fuzziness [8].

Decision making is a common activity in people's life and work. It is the process of selecting the best solution to solve the possible problems. Decisions made by fuzzy mathematical methods are called fuzzy decisions, and the purpose is to sort the solution set according to the "good to bad order", or select a "satisfactory" solution, or judge the category of "maximum possible" of each case in the solution set [9]. When making decisions, if there is a hierarchical decision Standard used as the basis for decision-making, then this process can be called as supervised decisions [10].

2.2 Self-feedback System and "Supervision"

In the case of unknown target weights and classification criteria, how to obtain the optimal fuzzy decision standard matrix, target weight, and fuzzy decision recognition matrix is the key issue. At the same time, there are often some hidden connections between the various index values within the decision consideration range. When data amount is limited, to make accurate correlation analysis is difficult. Therefore, this article introduces a self-feedback system [11].

This system is based on an algorithm that solves the above problems and uses fuzzy crossover and iterative algorithms to make full use of the implicit relationship between each element and the functional relationship between the optimal membership degree, element weights and fuzzy recognition matrix [12–14]. The establishment of fuzzy decision standard matrix and membership often contains subjectivity, and the iterative results may not be consistent with objective reality, so this article introduces a supervision factor to improve the accuracy and credibility of the model [15–17].

3 The Main Process of Nested Loop Optimization

3.1 The Main Steps of Nested Loop Optimization

The main steps for nested loop optimization of supply chain are:

Step 1. Target the object, goals and index of optimization
Step 2. Evaluate the quality and potency ratio of each index, rank them in various dimensions according to the supervised self - feedback model and adjust with subjective judgement.
Step 3. Optimizes each indicator in turn according to the path selected by the given model method.
Step 4. Check the overall index of supply chain. If it meets the demands, then stop the process and output the result. If not, then repeat **Step 2** and **Step 3** until supply chain achieves the expected optimization goals.

The main flowchart is shown in Fig. 1.

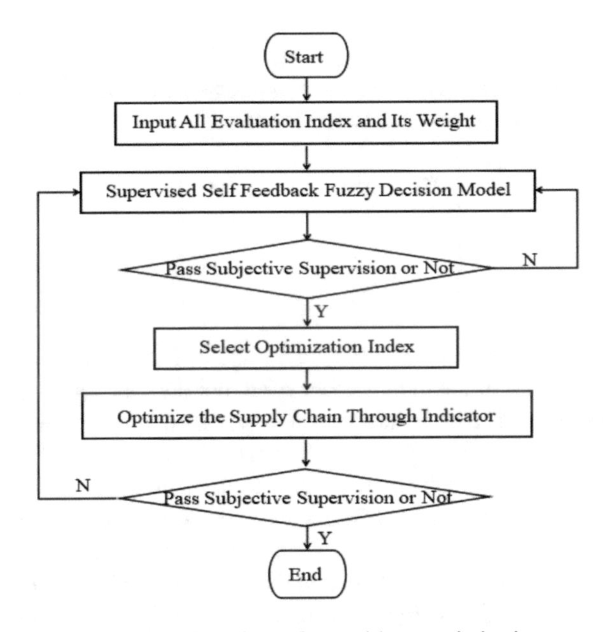

Fig. 1. The flowchart of nested loop optimization.

3.2 Establishment of Decision-Making Model

Let n schemes form a decision schemes set $D = \{d_1, d_2, \ldots, d_n\}$.

The fuzzy concept β categorized as c level, and the n samples to be identified constitute a set of samples.

$$X = \{x_1, x_2, \ldots, x_n\} \tag{1}$$

The fuzzy decision recognition matrix for n schemes is $U = (u_{nj})_{c \times n}$.
Each sample is represented by m index eigenvalues.

$$x_j = \left(x_{1j}, x_{2j}, \ldots, x_{mj}\right)^T \tag{2}$$

And the sample set follows $m \times n$ index characteristic matrix table

$$X = \begin{bmatrix} \mathrm{x}_{11} & \mathrm{x}_{12} & \cdots & \mathrm{x}_{1n} \\ \mathrm{x}_{21} & \mathrm{x}_{22} & \cdots & \mathrm{x}_{2n} \\ \vdots & \vdots & \vdots & \vdots \\ \mathrm{x}_{m1} & \mathrm{x}_{m2} & \cdots & \mathrm{x}_{mn} \end{bmatrix} = \left(\mathrm{x}_{ij}\right) \tag{3}$$

In formula (2), x_{ij} represents the eigenvalue of i in sample j.

Sample set will recognize eigenvalues of m indicators based on c level between two extremes and then $m \times c$ index characteristic matrix

$$Y = \begin{bmatrix} y_{11} & y_{12} & \cdots & y_{1n} \\ y_{21} & y_{22} & \cdots & y_{2n} \\ \vdots & \vdots & \vdots & \vdots \\ y_{m1} & y_{m2} & \cdots & y_{mn} \end{bmatrix} = (y_{ih}) \tag{4}$$

In formula (4), x_{ij} represents the standard value of i at j level.

Since the magnitude of the m indicators x_{ij} are not the same, it is necessary to wipe the effects of the index. The basic idea is as follows: for the fuzzy concept β, the higher the level is, the worse it is [18]. Due to it, the larger the eigenvalue of the indicator, the greater the level of the x_{ij}, greater than or equal to the standard value of level c and has a relative 0 of membership degree; with larger eigenvalues and lower level indicators, x_{ij} is smaller than or equal to y_{i1} at first level and has a relative 1 of membership of degree; when x_{ij} lies between y_{ic} and y_{i1}, and the relative membership of β is between 0 and 1, which is determined by linear interpolation. The opposite is true.

According to the above regulations, the formula of the relative membership of the index to the fuzzy concept β can be obtained.

$$r_{ij} = \begin{cases} 0 & , \quad x_{ij} \geq y_{ic} \ OR \ x_{ij} \leq y_{ic} \\ \frac{y_{ic} - x_{ij}}{y_{ic} - y_{i1}} & , \quad y_{ic} > x_{ij} > y_{i1} \ OR \ y_{ic} < x_{ij} < y_{i1} \\ 1 & , \quad x_{ij} \leq y_{i1} \ OR \ x_{ij} \geq y_{i1} \end{cases} \tag{5}$$

Similarly, the relative membership formula between the standard value y_{ih} of indicator i at h level and fuzzy concept β can be obtained.

$$r_{ij} = \begin{cases} 0 & , \quad y_{ic} = y_{ih} \\ \frac{y_{ic} - y_{ih}}{y_{ic} - y_{i1}} & , \quad y_{ic} > x_{ij} > y_{i1} \ OR \ y_{ic} < x_{ij} < y_{i1} \\ 1 & , \quad y_{i1} = y_{ih} \end{cases} \tag{6}$$

In formulas (5) and (6), r_{ij} is the valley floor degree of membership of sample j with i eigenvalue to β; S_{ih} is the relative degree of membership of sample h with i eigenvalue y_{ih} to β; y_{i1} and y_{ic} are standard value of indicator i at level 1 and c.

To find the best membership of degree of u_{hj} at h level in plan j and the optimal weight of object i, it needs to build a object function with subjective supervision factor of target weight in fuzzy environment:

$$\min\left\{F(u_{hj}, w_i) = (1-\alpha)\sum_{j=1}^{n}\sum_{h=1}^{c}[u_{hj}(\sum_{i=1}^{m} w_i(r_{ij} - s_{ih}))^p)^{\frac{1}{p}}]^2 + \alpha\sum_{j=1}^{n}\sum_{h=1}^{c}[(\sum_{i=1}^{m} w_i(r_{ij} - s_{ih}))^p)^{\frac{1}{p}}]^2\right\}$$

Its mathematical physics meaning is: the average value of the weighting of the sum of weighted generalized weight distance high squares and the sum of generalized weighted (α is weight) distance squares of all n decision schemes to c standard levels is the smallest. α is called subjective supervision factor,$\alpha \in [0, 1]$. The larger the α, the greater the subjective influence. If $\alpha = 0$, the formula (6) becomes the minimum sum of weighted generalized distance squares from all n decision schemes to c standard levels. In such circumstance, the objective function is not affected by subjective factors. Introduce Lagrangian constants λ_1 and λ_2 to build a Lagrangian formula and turn the problem of finding extreme with equality constraints into a problem of finding extreme by beam without equality constraints. Let $p = 2$, then the object function is

$$\min\left\{L(u_{hj}, w_i, \lambda_1, \lambda_2) = (1-\alpha)\sum_{j=1}^{n}\sum_{h=1}^{c}[u_{hj}^2\sum_{i=1}^{m}[w_i(r_{ij} - s_{ih})]^2\right.$$

$$\left. + \alpha\sum_{j=1}^{n}\sum_{h=1}^{c}[(\sum_{i=1}^{m}[w_i(r_{ij} - s_{ih})]^2 - \lambda_1(\sum_{h=1}^{c} u_{hj} - 1) - \lambda_2(\sum_{i=1}^{m} w_i - 1)\right\}$$

Find the partial derivative and make the derivative zero

$$\frac{\partial L}{\partial u_{hj}} = 2(1-\alpha)u_{hj}\sum_{i=1}^{m}[w_i(r_{ij} - s_{ih})]^2 - \lambda_1 = 0 \tag{7}$$

$$\frac{\partial L}{\partial w_i} = (1-\alpha)\sum_{j=1}^{n}\sum_{h=1}^{c} 2w_i[u_{hj}^2 w_i(r_{ij} - s_{ih})]^2 + \alpha\sum_{j=1}^{n}\sum_{h=1}^{c} 2w_i[u_{hj}^2 w_i(r_{ij} - s_{ih})]^2 - \lambda_2 = 0 \tag{8}$$

To simplify (8)

$$u_{hj} = \left(\sum\nolimits_{k=1}^{c} \frac{\sum\nolimits_{i=1}^{m} \left[w_i \left(r_{ij} - s_{ih} \right) \right]^2}{\sum\nolimits_{i=1}^{m} \left[w_i \left(r_{ij} - s_{ik} \right) \right]^2} \right)^{-1} \tag{9}$$

$$w_i = \left(\sum\nolimits_{k=1}^{c} \frac{\sum\nolimits_{j=1}^{n} \sum\nolimits_{h=1}^{c} \left[(1-a)u_{hj}^2 + a \right] \left(r_{ij} - s_{ih} \right)^2}{\sum\nolimits_{j=1}^{n} \sum\nolimits_{h=1}^{c} \left[(1-a)u_{hj}^2 + a \right] \left(r_{kj} - s_{kh} \right)^2} \right)^{-1} \tag{10}$$

According to (9) and (10), α is the determined factor of subjective weight supervision. When $\alpha = 0$, it reflects the law of the internal mechanism of the objectives in the decision-making system. It is a objective weight method. If $\alpha \neq 0$, it is affected by subjective factors. Then it is a combination of subjective and objective factors. Apparently, different form of α in line with separate weight α like a controller which targets the objects and balance the weight. α can be determined by the knowledge, experience or wishes of decision-making expert and the two numerical values or clustering validity indicators.

3.3 Addition of Subjective Supervision Factor

Because there are some hidden connections in these decision objectives, and it is difficult to make accurate correlation analysis under the condition of small samples, this paper adds a self-feedback system to the fuzzy decision model by cross-iterative algorithm in order to fully identify the internal connections of these targets [19]. On the basis of making the ranking fully carried out, the subjective supervision factor is added to make the calculation accord with the statistical law after adjusting the weight [20]. By constantly adjusting the subjective supervision operator, we can find out the target weight which can not only reasonably show the improved difference, but also make the recalculated results conform to the statistical law. The iterative process is as follow:

(1) Establishment of Factor Set
The factor set u is a set of factors that affect the evaluation object, represents as $U = \{u_1, u_2, \ldots, u_n\}$.
(2) Establishment of Evaluation Set
Evaluation set V is a collection of judgement on possibility of objects, represents as $V = \{v_1, v_2, \ldots, v_m\}$. Take this paper as a object, then the evaluation set V represents as $V = \{$Very important, important, average, unimportant, basically no impact$\}$.
(3) Establishing of Weight Set
Among the evaluation factors, the importance of each evaluation factor is different, but for this paper, in the absence of data, it is impossible to operate in practice if the influencing factors are used as the evaluation goal of the factor set in a general way.

Therefore, this paper will draw up the initial weight, in the continuous iteration to obtain the weight set of the goal.

For each factor u_i the corresponding weight given is w_i^t. The weight set will be W^t , $W^t = \{w_1^t, w_2^t, \ldots, w_n^t\}$.

To bring the above data into the self-feedback system, the algorithm steps are as follows:

Step 1. Given the target initial weight as $W^0 = (w_i^0)$, subjective supervision factor as α and iterative calculation accuracy as ε.

Step 2. Substituting (w_i^0) into Eq. (7) to calculate the fuzzy decision recognition matrix $U^0 = (u_{hj}^0)$.

Step 3. Substitute u_{hj}^0 into (8) to calculate the target weight $W^1 = (w_i^1)$.

Step 4. Substitute (w_i^1) into (7) to calculate the fuzzy decision recognition matrix $U^1 = (u_{hj}^1)$.

Step 5. Compares and judges. If $\left|u_{hj}^1 - u_{hj}^0\right| \leq \varepsilon$ and $\left|w_i^1 - w_i^0\right| \leq \varepsilon$, iteration ends, otherwise repeat **step 2** and **step 3**. If $\left|u_{hj}^l - u_{hj}^{l-1}\right| \leq \varepsilon$ and $\left|w_i^l - w_i^{l-1}\right| \leq \varepsilon$, the loop iteration ends. The approximate optimized result are $U^1 = (u_{hj}^1)$ and $W^1 = (w_i^1)$. l means the times of cyclic iterations. If not, then the iterative computation continues.

3.4 Optimization of Index Selection

Three types of indicator selection theories are as follow:

(1) The barrel effect. That is, in the process of optimization, first consider the "short board" of all indicators to ensure that the worst indicator can be optimized at first, so that the supply chain can rapidly increasing supply chain Elasticity, and improving the overall level of the supply chain.

(2) The value theory. That is, in the process of optimization, consideration is given to the cost and the degree of optimization or effectiveness of the indicators that can be brought by the cost paid. A higher value ratio indicates that it is more reasonable and economical to optimize the indicator.

(3) Comprehensive theory. That is, in the process of optimization, it is necessary to ensure the improvement of the "short plate", so as to quickly improve the overall level of the supply chain; on the other hand, the cost problem to be generated in the process of optimizing the index should be fully considered, so as to select a more economical and reasonable index. The main optimization goal is selected by a balance of two aspects.

4 Calculation Example

Based on the current situation of company D, applying the analytic hierarchy process to analyze the status of its supply chain, the indicators we find out are in Table 1.

Table 1. Analysis of the indicators obtained by company D

Indicators to be optimized
1. Transportation cost
2. Inventory cost
3. NDC delivery flexibility
4. RDC distribution flexibility
5. NDC inventory capacity
6. RDC inventory capacity
7. Demand satisfaction rate
8. Customer coverage

The eight goals are summarized and the data is standardized to obtain the standard index eigenvalue matrix as:

$$R^T = \begin{bmatrix} 0.500 & 0.500 & 0.721 & 0.272 & 1.000 & 1.000 & 0.274 & 0.372 \\ 0.872 & 0.912 & 0.257 & 0.481 & 0.170 & 0.781 & 1.000 & 0.547 \\ 0.733 & 0.781 & 0.314 & 0.640 & 0.217 & 0.312 & 0.657 & 0.230 \\ 0.367 & 0.694 & 1.000 & 0 & 0.324 & 0.297 & 0.874 & 1.000 \\ 0.362 & 1.000 & 0.982 & 0.500 & 0 & 0.853 & 0.254 & 0.146 \end{bmatrix}$$

The larger and better the indicators take the standardized function as follows:

$$r_{ij} = \frac{x_{ij} - \bigwedge_j x_{ij}}{\bigvee_j x_{ij} - \bigwedge_j x_{ij}}$$

The larger and better the indicator, the standardized function is as follows:

$$r_{ij} = \frac{\bigvee_j x_{ij} - x_{ij}}{\bigvee_j x_{ij} - \bigwedge_j x_{ij}}$$

Due to the inaccuracy of the evaluation criteria for individual elements, and to simplify the calculation, the standard decision mode matrix is determined by linear interpolation at equal intervals. The standard fuzzy index matrix is:

$$S = \begin{bmatrix} 1 & 0.75 & 0.50 & 0.25 & 0 \\ 1 & 0.75 & 0.50 & 0.25 & 0 \\ 1 & 0.75 & 0.50 & 0.25 & 0 \\ 1 & 0.75 & 0.50 & 0.25 & 0 \\ 1 & 0.75 & 0.50 & 0.25 & 0 \end{bmatrix}$$

Bring the above data into a self-feedback system using a matrix $[5, 4, 3, 2, 1]$ multiply the final fuzzy recognition matrix of each target each by each, and get the weights of the modified targets after normalization. Get the weights of the modified targets after normalization and use them to modify the elasticity index. The algorithm steps are as follows:

Step 1. Given the target initial weight as $W^0 = (w_i^0)$, subjective supervision factor as α and iterative calculation accuracy as ε.
Step 2. Substituting (w_i^0) into Eq. (7) to calculate the fuzzy decision recognition matrix $U^0 = (u_{hj}^0)$.
Step 3. Substitute u_{hj}^0 into (8) to calculate the target weight $W^1 = (w_i^1)$.
Step 4. Substitute (w_i^1) into (7) to calculate the fuzzy decision recognition matrix $U^1 = (u_{hj}^1)$.
Step 5. Compares and judges. If $\left|u_{hj}^1 - u_{hj}^0\right| \leq \varepsilon$ and $\left|w_i^1 - w_i^0\right| \leq \varepsilon$, iteration ends, otherwise repeat **step 2** and **step 3**. If $\left|u_{hj}^l - u_{hj}^{l-1}\right| \leq \varepsilon$ and $\left|w_i^l - w_i^{l-1}\right| \leq \varepsilon$, the loop iteration ends. The approximate optimized result are $U^1 = (u_{hj}^1)$ and $W^1 = (w_i^1)$. l means the times of cyclic iterations. If not, then the iterative computation continues.
Step 6 Using a matrix $[5, 4, 3, 2, 1]$, multiply the final fuzzy recognition matrix of each target item by item and normalize to get the weight of each bomb target.

After three iterations, the sorting results are as follow (Table 2):

Table 2. Sorting order of index importance after iteration

First iteration	Second iteration	Third iteration
1. Transportation cost	1. Demand satisfaction rate	1. RDC distribution flexibility
2. Inventory cost	2. Transportation cost	2. NDC delivery flexibility
3. NDC delivery flexibility	3. Customer coverage	3. Demand satisfaction rate
4. RDC distribution flexibility	4. NDC capacity	4. Transportation cost
5. RDC capacity	5. RDC distribution flexibility	5. Customer coverage
6. NDC capacity	6. NDC delivery flexibility	6. RDC capacity
7. Demand satisfaction rate	7. RDC capacity	7. NDC capacity
8. Customer coverage	8. Inventory cost	8. Inventory cost

D Company uses the comprehensive theory as the supply chain indicator selection theory. It only selected the most significant indicator in each iteration of the three iterations, and optimized them in turn, and the result is shown in Fig. 2.

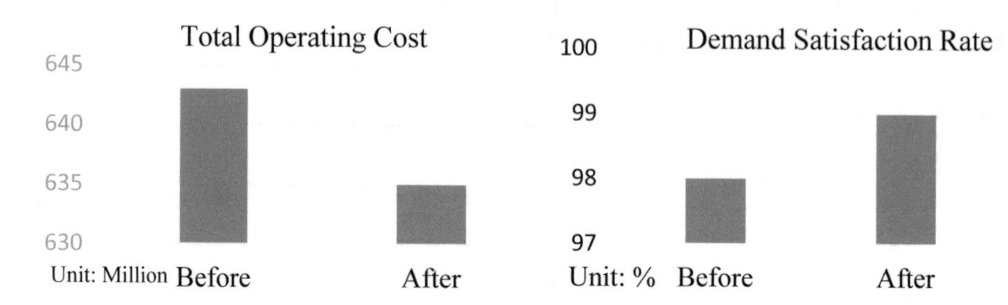

Fig. 2. Comparison of Total operating costs and demand satisfaction rate.

From Fig. 2, it can be seen that the cost of the company after optimization has decreased by 9 million, and the service satisfaction rate has increased from 98% to 99%. It can be seen that the optimization effect of this model is good.

5 Conclusion

The Supply chain optimization model based on nested loop algorithm in this paper is start from the system analysis idea, using the analytic hierarchy process to split the indicators of the multi-objective decision-making supply chain optimization process, and then perform the nested loop optimization process on it. In nested loop optimization, inner loop-iterative calculation in the supervised self-feedback fuzzy decision model, ranking the importance of each index and supervising judgment of the rationality of the ranking; outer loop integration-the degree of optimization of the body plan Supervision mainly controls the end of the nested loop by judging whether the supply chain optimization as a whole has reached the target at this time.

In the inner loop of importance ranking, this article introduces subjective supervision operators and self-feedback systems. The purpose is to improve the accuracy and credibility of the model, so that it can accurately optimize the supply chain and improve the core competitiveness of the supply chain.

To a certain extent, this model solves the problem that the measurement angle is fixed, the improvement effect and the corresponding influence are not clear in the general process of supply chain optimization. And apply the subjective supervision operators and self-feedback system can, in certain level, increase the rationality and helping optimize the supply chain from multiple angle which can lower the cost of supply chain optimization. In future development, the change of self-feedback system and supervision operator can increase the accuracy and making the whole model specific for every different scene.

Acknowledgment. This paper is supported by (1) Ministry of Communications Science and Technology Project “Research on Multimodal Transport of Wuhan Shipping Center”; (2) Wuhan University of Technology Educational Reform project “The Exploration and Practice of the Course Reform of Port Enterprise Management Facing the Needs of Compound Talents in the New Era (2018418110)”.

References

1. Battini, D., Peretti, U., Persona, A., et al.: Sustainable humanitarian operations: closed-loop supply chain. Int. J. Serv. Oper. Manage. **25**(1), 65–79 (2016)
2. Rohmer, S.U.K., Gerdessen, J.C., Claassen, G.D.H.: Sustainable supply chain design in the food system with dietary considerations: a multi-objective analysis. Eur. J. Oper. Res. **1**, 273–275 (2019)
3. Anitha, P., Patil, M.M.: A review on data analytics for supply chain management: a case study. Int. J. Inf. Eng. Electron. Bus. (IJIEEB) **10**(5), 30 (2018)
4. Jiaguo, L., Yuexiang, Z., Jun, L., et al.: Analysis on the selection model of supply chain elastic comprehensive optimization path. J. Harbin Univ. Technol. **5**, 101–106 (2014)
5. Bhinge, R., Moser, R., Moser, E., et al.: Sustainability optimization for global supply chain decision-making. Procedia CIRP **26**, 323–328 (2015)
6. Tang, O., Musa, S.N.: Identifying risk issues and research advancements in supply chain risk management. Int. J. Prod. Econ. **133**(1), 25–34 (2011)
7. Haghighi, P.S., Moradi, M., Salahi, M.: Supplier segmentation using fuzzy linguistic preference relations and fuzzy clustering. Int. J. Intell. Syst. Appl. (IJISA) **6**(5), 76 (2014)
8. Mendel, J.M.: Uncertain Rule-based Fuzzy Systems: Introduction and New Directions, p. 684. Springer, Cham (2017)
9. Fahmi, A., Abdullah, S., Amin, F., et al.: Weighted average rating method for solving group decision making problem using triangular cubic fuzzy hybrid aggregation. Punjab Univ. J. Math. **50**(1), 23–34 (2018)
10. Karatzinis, G., Boutalis, Y.S., Kottas, T.L.: System identification and indirect inverse control using fuzzy cognitive networks with functional weights. In: 2018 European Control Conference (ECC), pp. 2069–2074. IEEE (2018)
11. Zhike, X., Liuxin, G.: Construction of multi object, multi index and multi-level fuzzy pattern recognition model. Pract. Understanding Math. **43**(04), 143–147 (2013)
12. Christyawan, T.Y., Haris, M.S., Rody, R., et al.: Optimization of fuzzy time series interval length using modified genetic algorithm for forecasting. In: 2018 International Conference on Sustainable Information Engineering and Technology (SIET), pp. 60–65. IEEE (2018)
13. Kherad, M., Vahdat-Nejad, H., Araghi, M.: Trasfugen: traffic assignment of urban network by an approximation fuzzy genetic algorithm. Int. J. Model. Simul. Sci. Comput. **9**(04), 1850034 (2018)
14. Cheowsuwan, T., Arthan, S., Tongphet, S.: System design of supply chain management and thai food export to global market via electronic marketing. Int. J. Mod. Educ. Comput. Sci. **9** (8), 1 (2017)
15. Fan, G., Zhong, D., Yan, F., et al.: A hybrid fuzzy evaluation method for curtain grouting efficiency assessment based on an AHP method extended by D numbers. Expert Syst. Appl. **44**, 289–303 (2016)
16. Lakshmi, T.M., Martin, A., Begum, R.M., et al.: An analysis on performance of decision tree algorithms using student's qualitative data. Int. J. Mod. Educ. Comput. Sci. (IJMECS) **5**(5), 18 (2013)

17. Mirshekaran, M., Piltan, F., Esmaeili, Z., et al.: Design sliding mode modified fuzzy linear controller with application to flexible robot manipulator. Int. J. Mod. Educ. Comput. Sci. (IJMECS) **5**(10), 53 (2013)
18. Feng, F., Jun, Y.B., Liu, X., et al.: An adjustable approach to fuzzy soft set based decision making. J. Comput. Appl. Math. **234**(1), 10–20 (2010)
19. Hou, Z., Wang, J.: A new cluster validity criterion for the Cross Iterative Fuzzy Clustering Algorithm. In: Fifth World Congress on Intelligent Control and Automation (IEEE Cat. No. 04EX788), vol. 3, pp. 2322–2326. IEEE (2004)
20. Li, X., Zhang, G., Qi, F., et al.: Grey cluster estimating model of soil organic matter content based on hyper-spectral data. J. Grey Syst. **26**(2), 28 (2014)

The Possibilistic Gustafson-Kessel Fuzzy Clustering Procedure for Online Data Processing

Zhengbing Hu[1] and Oleksii K. Tyshchenko[2](✉)

[1] School of Educational Information Technology, Central China Normal University, 152 Louyu Road, Wuhan 430079, China
hzb@mail.ccnu.edu.cn

[2] Institute for Research and Applications of Fuzzy Modeling, CE IT4Innovations, University of Ostrava, 30. dubna 22, 701 03 Ostrava, Czech Republic
lehatish@gmail.com

Abstract. The script is engaged in the intricacy of clustering data streams. The advised plan considers the possibilistic fuzzy clustering on the grounds of the Gustafson and Kessel techniques for the case when data should be grouped in the following manner. Another aspect should also be considered, which has to do with the fact that classes possess a random configuration (which cannot typically be classified linearly) and are mutually intersecting.

Keywords: Fuzzy clustering · Hypersphere · Membership function · Ellipsoidal cluster · Distance measure · Fuzzifier · Goal function

1 Introduction

Clustering is practiced regularly as a mechanism for the exploratory data examination, which suggests that an academic specialist constantly challenges some requisite complications correlated with the fact that some data attributes are beforehand undiscovered [1, 2]. Certain features may routinely hold a relationship and a range of extreme values, some clumps and density of classes, a measure of groups' intersection, and nesting. Complications to be coupled together with this sort of ambiguity are exceptionally critical when it proceeds to implement a data analytical approach for the case of online sequential data processing [3]. Aside from methodological challenges related to the incapability of "retrieving" a long representation for a revolved analytical approach with other pattern's elements; there exist some hassles of a primitive nature due to properties' shifts in time of a researchable object. This manuscript generally reports several techniques endeavored at supervising the principal kinds of indetermination in initial data. Argument' extension applied for making these schemes build prototypes with the smallest quantity of manually tuned parameters that can softly accommodate to properties' altering of the data under examination.

The majority of online clustering procedures originate from offline versions of the same clustering approaches [4]. As a matter of discussion, most of the online

Z. Hu et al. (Eds.): ICCSEEA 2020, AISC 1247, pp. 419–428, 2021.
https://doi.org/10.1007/978-3-030-55506-1_38

modifications of the clustering methods are grounded on the Euclidian distance measure that finally leads to clusters of a spherical shape. At the same time, after the recurrent computation for a covariance matrix was offered, it has become possible to handle the Mahanalobis distance in the recurrent modifications of the clustering techniques.

In practical applications, data frequently provide some overlying clumps of data points, given that a particular data object in equal measure appertains to two or more groups. Admittedly, that being the case, conventional Kohonen neural networks confirm the fact that they are hollow [5–7]. In this synopsis, the clarification may be gained through soft computing techniques. To start with, the Fuzzy C-Means (FCM) approach [8, 9] is associated as a compelling instrument for this kind of assignment. The FCM procedure collapses in handling with massively noisy information. To clarify the point, membership grades are treated as normalized intervals between the current units and the prototypes/centroids. The representations that are located distant from all of the prototypes and may, consequently, be outliers are proportionately weighted. It implies that the membership grades for the before-mentioned data examples are low and do not offer as much as the normal data to the centroids' means and covariance matrices. To defeat this FCM's disadvantage, the primary intention is to soften the stipulation given by the relative typicality (membership degrees). As things now stand, the central attention should be placed on adjusted recurrent tweaks of FCM operating in the online scheme [10].

One of the primary flaws of the method is its sensitivity to the initial conditions (values) of its arguments and the potential match of several centroids in the same position. These procedures mostly rely on the Euclidian distance, and accordingly, the distinguished groups are characterized by (hyper)spherical configurations. By entering the Mahalanobis distance into the procedure, it enables the discovery of the (hyper) ellipsoidal structure of clusters [11]. It is comparatively evident that the presence of complicated nonconvex shapes in a group's architecture may affect the excess of spherical clumps and inevitably to the processing decrease.

For the foregoing reasons, the paper's discovery provides the possibilistic fuzzy clustering approach on the grounds of the Gustafson and Kessel techniques for online data processing of clusters characterized by hyprellipsoidal forms and various volumes.

The document contains seven parts where Sect. 2 draws a brief review of a typical fuzzy clustering goal function and the methods based on it. Sect. 3 clarifies the primary scheme by Gustafson-Kessel. Section 4 concentrates on a possibilistic case of the procedure by Gustafson&Kessel. Section 5 introduces adaptive variants of the procedure by Gustafson-Kessel. Section 6 presents the simulation outcome. Section 7 arrives at a conclusion for the paper and resumes the examination.

2 The Primary Fuzzy Clustering Objective Function

Let us examine the fundamental clustering objective function [12]

$$J(q_j, c_j) = \sum_{k=1}^{N} \sum_{j=1}^{m} q_j^{\beta}(k) d^2\left(x(k), c_j\right) \tag{1}$$

following the stipulation

$$\sum_{j=1}^{m} q_j(k) = 1.$$

Since the expression (1) holds the distance metric $d\left(x(k), c_j\right)$, at this point, it becomes compelling to highlight the fact that the distance function is subject to optimization. Consequently, its tweaking specifically inside the procedure is unworkable. The metric's adjustment produces a novel algorithm.

The most widespread clustering approaches (FCM, K-means) are achieved through the instrumentality of the Euclidean metric

$$d^2\left(x(k), c_j\right) = \left(x(k) - c_j\right)^T \left(x(k) - c_j\right) = \left\|x(k) - c_j\right\|^2.$$

It is worthwhile noting that the degree of membership q_j is identical for all objects which are situated at regular intervals from a centroid. In most scenarios, clumps' edges discovered by specific techniques invariably possess a hypersphere form. This experience consistently limits their usability to applicative purposes when a contour of groups is dramatically distinct from the hypersphere; or when a value scales up some particular elements in characteristic vectors that have a distinct measuring scale. An earlier hypothesis about a spherical configuration of the groups makes a global minimum of the goal function inapproachable.

3 Methods Stemming from the Gustafson-Kessel Procedure

The topical Gustafson-Kessel scheme [13, 14] implies a trade-off that enables processing ellipsoidal bunches of varying elongation without delivering any valuable corrections to the procedure. With this background, a scaling matrix is injected in the equation for estimating a distance between feature vectors. The matrix adjusts the scale of the space in every particular position (coordinate point)

$$d_G^2\left(x(k), c_j\right) = \left(x(k) - c_j\right)^T G\left(x(k) - c_j\right) = \left\|x(k) - c_j\right\|_G^2. \tag{2}$$

By optimization (minimization) of the cost function (1) with allowances made for the metrics (2), the topical minimization techniques which bear in mind not only the estimation of clumps' centroids c_j, $1 \leq j \leq m$ and degrees of membership q_j, but additionally the computation of a covariance matrix R_j for every individual clump and a scaling matrix G_j of the clump

$$\begin{cases} q_j(k) = \dfrac{\left(\left(x(k) - c_j(k)\right)^T G_j\left(x(k) - c_j(k)\right)\right)^{1/(1-\beta)}}{\sum\limits_{l=1}^{m}\left((x(k) - c_l(k))^T G_l(x(k) - c_l(k))\right)^{1/(1-\beta)}}, \\ c_j = \dfrac{\sum\limits_{k=1}^{N} q_j^{\beta}(k)x(k)}{\sum\limits_{k=1}^{N} q_j^{\beta}(k)}, \\ R_j = \sum\limits_{k=1}^{N} q_j^{\beta}(k)\left(x(k) - c_j\right)\left(x(k) - c_j\right)^T, \\ G_j = \left|R_j\right|^{1/n} R_j^{-1}. \end{cases} \tag{3}$$

Now it can be seen that the achieved matrix G_j by this means is positive definite and symmetric. Besides, the measuring scale it initiates to be established for every coordinate by a quadratic root of the conformable characteristic vector is favorable [15] with respect to the figure of merit (1). The implications of the eigenmodes in the pattern G_j can modify the appearance of the analogous class that raises the possibility of utilizing this method for a broad spectrum of applicative purposes.

4 A Possibilistic Version of the Gustafson-Kessel Method

As it has been earlier stated in Sect. 3, procedures that hold the restrictions (2) are termed probabilistic and own some meaningful impediments like weak robustness to information outliers interchangeably with adequately overlying batches. The stipulation (2) is excessively stringent for numerous experiential assignments: extreme data values may possess a grade of membership (usually moderately close to zero for all groups), and points at the heart of overlaying clumps should be specified as having grades of membership approaching one for each overlapping bunch. Methods that do not employ this limitation are also known as possibilistic ones.

To come up to a possibilistic variant of the algorithm by Gustafson and Kessel, the goal functional relation is utilized

$$J(q_j, c_j) = \sum_{k=1}^{N}\sum_{j=1}^{m} q_j^{\beta}(k)\left\|x(k) - c_j\right\|_G^2 + \sum_{j=1}^{m} s_j \sum_{k=1}^{N}\left(1 - q_j(k)\right)^{\beta}. \tag{4}$$

Following up the minimization of the previous expression, an algorithm is put down in a slightly different way [16]

$$\begin{cases} q_j(k) = \left(1 + \left(\frac{(x(k) - c_j)^T G_j (x(k) - c_j)}{s_j}\right)^{1/(1-\beta)}\right)^{-1}; \\ c_j = \frac{\sum_{k=1}^{N} q_j^\beta(k) x(k)}{\sum_{k=1}^{N} q_j^\beta(k)}; \\ R_j = \sum_{k=1}^{N} q_j^\beta(k) (x(k) - c_j)(x(k) - c_j)^T; \\ G_j = |R_j|^{1/n} R_j^{-1}; \\ s_j(k) = \frac{\sum_{k=1}^{N} q_j^\beta(k) (x(k) - c_j)^T G_j (x(k) - c_j)}{\sum_{k=1}^{N} q_j^\beta(k)}. \end{cases} \quad (5)$$

5 Adaptive Variants of the Procedure by Gustafson-Kessel

Considering computational flows handling with counting the covariance matrix and its (pseudo) transposition, it drives to a high degree of computational complexity factor of the method (that further grows swiftly following a unit's capacity); to underlying volatility of the procedure to outliers (when this is the case, the matrix turns out to be ill-conditioned); and to the group data processing.

Through the use of the recipe by Sherman and Morrison for accumulative transposition of the initial matrix and the nonlinear programming method by Uzawa, Arrow, and Hurwicz for achieving the recursive minimization scheme of the expression (4) provides procuring an adjusted model of the plan by Gustafson and Kessel that is fitted for running totally in a recurrent style.

The most specific issue regarding this flow consists in a recursive transposition of the covariance matrix $R_j^{-1}(k+1)$ with reference to the scheme by Morrison and Sherman and recursive counting of its determinant $|R_j(k+1)|$.

An adjusted variant of the possibilistic method undergoes the identical weaknesses (as opposed to their batch cases) like perishability to data outliers and struggles with the apportionment of profoundly protruding classes. Furthermore, the before-mentioned flows cannot be wholly named unsupervised procedures because a scholar should beforehand apprehend some clumps handling the data contraction. To neutralize these blemishes, it appears rational to pull the possibilistic method by Gustafson and Kessel (5) to an adaptive variant. An extraction scheme for this procedure is related to a derivation idea for the flow (6) besides the fact that the statement (4) is managed as the goal function. Conclusively, an adjusted possibilistic method by Gustafson-Kessel can be achieved in the frame of

$$\begin{cases} q_j(k+1) = \dfrac{\left\|x(k+1) - c_j(k)\right\|_{G_j(k)}^{2/(1-\beta)}}{\sum_{l=1}^{m} \left\|x(k+1) - c_l(k)\right\|_{G_l(k)}^{2/(1-\beta)}}; \\ R_j(k+1) = R_j(k) + q_j^{\beta}(k+1)\left(x(k+1) - c_j(k)\right)\left(x(k+1) - c_j(k)\right)^T; \\ R_j^{-1}(k+1) = R_j^{-1}(k) \\ \qquad - \dfrac{q_j^{\beta}(k+1)R_j^{-1}(k)\left(x(k+1) - c_j(k)\right)\left(x(k+1) - c_j(k)\right)^T R_j^{-1}(k)}{1 + q_j^{\beta}(k+1)\left(x(k+1) - c_j(k)\right)^T R_j^{-1}(k)\left(x(k+1) - c_j(k)\right)}; \\ \left|R_j(k+1)\right| = \left|R_j(k)\right|\left(1 + q_j^{\beta}(k+1)\left\|x(k+1) - c_j(k)\right\|^2\right); \\ G_j(k+1) = \left|R_j(k+1)\right|^{1/n} R_j^{-1}(k+1); \\ c_j(k+1) = c_j(k) + \eta(k) q_j^{\beta}(k+1) G_j(k+1)\left(x(k+1) - c_j(k)\right). \end{cases} \tag{6}$$

By going over the relationship (that may look slightly unwieldy), its composition is similar to the statement (6).

$$\begin{cases} q_j(k+1) = \left(1 + \left(\dfrac{\left(x(k+1) - c_j(k)\right)^T G_j(k)\left(x(k+1) - c_j(k)\right)}{s_j(k)}\right)^{1/(1-\beta)}\right)^{-1}; \\ R_j(k+1) = R_j(k) + q_j^{\beta}(k+1)\left(x(k+1) - c_j(k)\right)\left(x(k+1) - c_j(k)\right)^T; \\ R_j^{-1}(k+1) = R_j^{-1}(k) - \dfrac{q_j^{\beta}(k+1)R_j^{-1}(k)\left(x(k+1) - c_j(k)\right)\left(x(k+1) - c_j(k)\right)^T R_j^{-1}(k)}{1 + q_j^{\beta}(k+1)\left(x(k+1) - c_j(k)\right)^T R_j^{-1}(k)\left(x(k+1) - c_j(k)\right)}; \\ \left|R_j(k+1)\right| = \left|R_j(k)\right|\left(1 + q_j^{\beta}(k+1)\left\|x(k+1) - c_j(k)\right\|^2\right); \\ G_j(k+1) = \left|R_j(k+1)\right|^{1/n} R_j^{-1}(k+1); \\ c_j(k+1) = c_j(k) + \eta(k) q_j^{\beta}(k+1) G_j(k+1)\left(x(k+1) - c_j(k)\right); \\ s_j(k+1) = \dfrac{\sum_{p=1}^{k+1} q_j^{\beta}(p)\left(x(p) - c_j(k+1)\right)^T G_j(k+1)\left(x(p) - c_j(k+1)\right)}{\sum_{p=1}^{k+1} q_j^{\beta}(p)}. \end{cases} \tag{7}$$

It is crucial to state that the possibilistic methods hold some defects. To start with, there is no barrier for cluster centers to collapse closer (since the restraint is absent) or to spread out unhampered by restriction. As a consequence, a decision of these methods may be portrayed by arbitrary uncertainty, i.e., tiny disturbances in the original sample (and through the actualization of centroids equally well) can bring about a substantial contrast in the target separation. The chief advice for supervising this state is to employ several repetitions of the probabilistic procedure [12, 16, 17] for initialization of the possibilistic flow.

It should be noted from the outset that the adjusted possibilistic procedure (7) matches the adjusted probabilistic method by Gustafson and Kessel (6) except for a representation of the membership function $q_j(k+1)$ and an argument value of the fuzzy boundary $s_j(k+1)$. This experience makes it suitable for employing these methods collectively to enhance stability.

6 Simulation

To verify the workability of the produced idea, one should apply the clustering validity measures to assess outcomes accurately. With that in mind, an artificially developed data set was put into use. The revealed set of samples was actuated several times, according to the processing plan, to aggregate the output. The current section presents only a brief simplistic illustration. The produced data set comprised of 650 points. The outcomes following on from the validity techniques are aggregated in Table 1.

To begin with, all these schemes are enabled at random fashion, so there may be numerous operating concerns. By the same token, the gained outcomes are inconsiderably stipulated by the data structure, and there is no universal efficacy index that can suit flawlessly well to the clustering complexity.

When it comes to examining the case, there is a fragment of the whole processing flow in Fig. 1 manipulating distinct initializations. What can be indeed observed is that there exists a changing size of groups to be distinguished. In this experimental research, upcoming points (objects) are put on the 2D plane. Within some time, the clusters are being formed. As one can see, the shapes of the formed clusters are quite tricky. The significant barrier of this class of methods is that the prototypes' actuation is enabled in a random manner. It is profoundly guided to run every method numerous times to achieve reliable outcomes. To eliminate the difficulty explained above, the cluster centroids are initialized by irregularly chosen data objects.

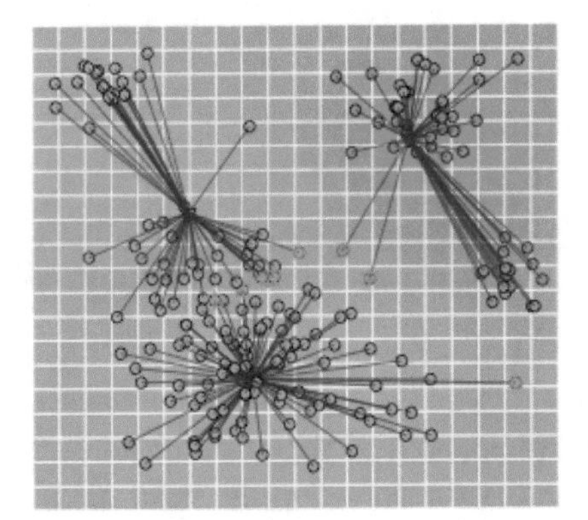
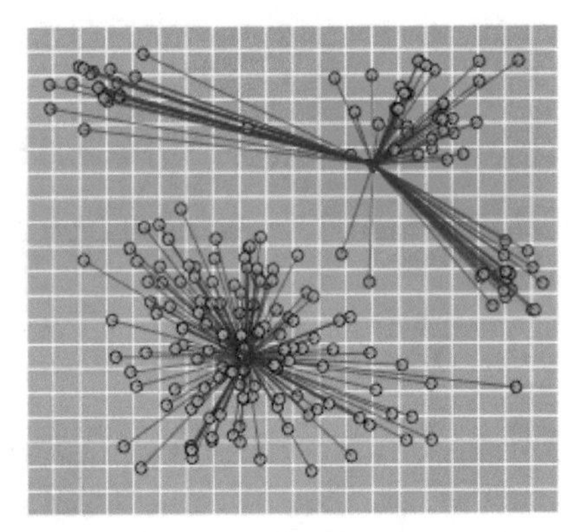

Fig. 1. The outcomes acquired by multiple initializations based on the synthetic data

In regards to the achieved issues, the hard clustering method may be quite the appropriate solution to a case, although it is highly likely that this method meets the initialization problem. When it comes to the fuzzy procedures, the only distinction is the shape of the clusters obtained; at the same time, the GK procedure can detect elongated clusters much better. For this reason, GK confirms insignificantly better decisions grounded on the validity measures. For illustration, it was established in [18, 19] that the real data are portrayed by the Mahalanobis norm rather than the Euclidean norm.

The issues of implementing distinct techniques are exhibited in Fig. 1. Affectedly formed objects are spread around two and three centers, respectively, unpredictably.

The most compelling point about this data set is that edge points of the ellipsoidal groups are positioned closer to a core of a bordering cluster (than to the native one). This observation clarifies the issue of why the FCM method does not supply the solicited separation. Meanwhile, pivots and distribution elements (the covariance matrixes) caused by the scheme by Gustafson and Kessel are remarkably close to the originally provided items.

XB strives to count a degree of the total diversity inside clumps and the division of clumps. DI is expected formerly to recognize compressed and well-divided clusters. PC considers a degree to which the groups intersect. Considering the estimates achieved through the two known clustering indexes (PC and XB indexes), the improved revised GK flow produces the best outcomes for this data set.

That is a well-known fact that clustering procedures are usually responsible for either some data preprocessing or being a part of some more advanced decision-making systems. Based on the presented results, the clustering quality goes high by approximately 10% compared to the most popular FCM algorithm. At the same time, the more entangled and tricky data a user may encounter, the more significant this difference may become. Although the developed approach is more complicated and sophisticated from the computational viewpoint, the provided clustering results will payback to a great extent.

Table 1. The clustering results

	PC	XB	DI
K-Means	0.8421	0.3164	0.4643
FCM	0.8873	0.2852	0.5317
Modified GK	0.9347	0.1538	0.6976

7 Conclusion

This script chiefly covered the concept of the possibilistic fuzzy clustering on the grounds of the Gustafson and Kessel techniques for managing the main kinds of ambiguities in data. Parameters' generalization employed for getting these systems to make it admissible to build prototypes with a minimum quantity of by-hand tuned arguments that can react to the properties' altering of the data understudy in a flexible manner. Viewed in this way, as the plurality of the practical assignments is identified by ellipsoidal configurations of clusters, the stated scheme is a more beneficial tool. This proficient way of treating the objects may finally improve the quality of decision making since a recommender system based on this clustering preprocessing technique provides a higher and more sophisticated quality of making decisions.

Acknowledgment. RAMECS and self-determined research funds of CCNU from the colleges' primary research and operation of MOE (CCNU19TS022) preserved fractionally this experimental effort.

Oleksii K. Tyshchenko was additionally maintained by the National Science Agency of the Czech Republic within the project TACR TL01000351.

References

1. Bouveyron, C., Celeux, G., Murphy, T.B., Raftery, A.E.: Model-Based Clustering and Classification for Data Science: With Applications in R. Cambridge Series in Statistical and Probabilistic Mathematics. Cambridge University Press, Cambridge (2019)
2. Palumbo, F., Montanari, A., Vichi, M.: Data Science: Innovative Developments in Data Analysis and Clustering. Studies in Classification, Data Analysis, and Knowledge Organization. Springer, Cham (2017)
3. Nasraoui, O., Ben N'Cir, C.-E.: Clustering Methods for Big Data Analytics: Techniques, Toolboxes and Applications. Unsupervised and Semi-Supervised Learning. Springer, Cham (2019)
4. Höppner, F., Klawonn, F., Kruse, R., Runkler, T.: Fuzzy Cluster Analysis: Methods for Classification, Data Analysis and Image Recognition. Wiley, Chichester (1999)
5. Du, K.-L., Swamy, M.N.S.: Neural Networks and Statistical Learning. Springer, London (2014)
6. Kohonen, T.: Self-Organizing Maps. Springer, Berlin (1995)
7. Hu, Z., Bodyanskiy, Ye.V., Tyshchenko, O.K., Boiko, O.O.: A neuro-fuzzy Kohonen network for data stream possibilistic clustering and its online self-learning procedure. Appl. Soft Comput. J. 68, 710–718 (2018)
8. Bezdek, J.C., Keller, J., Krishnapuram, R., Pal, N.: Fuzzy Models and Algorithms for Pattern Recognition and Image Processing. The Handbook of Fuzzy Sets. Springer, Kluwer (1999)
9. Bezdek, J.-C.: Pattern Recognition with Fuzzy Objective Function Algorithms. Plenum Press, New York (1981)
10. Bodyanskiy, Ye.V., Tyshchenko, O.K., Mashtalir, S.V.: Fuzzy clustering high-dimensional data using information weighting. In: Rutkowski, L., et al. (ed.) Lecture Notes in Artificial Intelligence, Proceedings of the 18th International Conference on Artificial Intelligence and Soft Computing (ICAISC 2019), Zakopane, Poland, 16–20 June 2019, Part I, vol. 11508, pp. 385–395 (2019)
11. Mao, J., Jain, A.K.: A self-organizing network for hyperellipsoidal clustering. IEEE Trans. Neural Netw. **7**, 16–29 (1996)
12. Gan, G., Ma, Ch., Wu, J.: Data Clustering: Theory, Algorithms and Applications. SIAM, Philadelphia (2007)
13. Gustafson, D.E., Kessel, W.C.: Fuzzy clustering with fuzzy covariance matrix. In: Proceedings of the IEEE CDC, San Diego, pp. 761–766 (1979)
14. Krishnapuram, R., Jongwoo, K.: A note on the Gustafson-Kessel and adaptive fuzzy clustering algorithms. IEEE Trans. Fuzzy Syst. **7**(4), 453–461 (1999)
15. Aggarwal, C.C., Reddy, C.K.: Data Clustering: Algorithms and Applications. CRC Press, Boca Raton (2014)
16. Xu, R., Wunsch, D.C.: Clustering. IEEE Press Series on Computational Intelligence. Wiley, Hoboken (2009)

17. Hu, Z., Bodyanskiy, Ye.V., Tyshchenko, O.K.: Self-learning and adaptive algorithms for business applications. In: A Guide to Adaptive Neuro-fuzzy Systems for Fuzzy Clustering under Uncertainty Conditions. Emerald Publishing Limited, UK (2019)
18. Babichev, S., Škvor, J., Fišer, J., Lytvynenko, V.: Technology of gene expression profiles filtering based on wavelet analysis. Int. J. Intell. Syst. Appl. (IJISA) **10**(4), 1–7 (2018)
19. Izonin, I., Trostianchyn, A., Duriagina, Z., Tkachenko, R., Tepla, T., Lotoshynska, N.: The combined use of the wiener polynomial and SVM for material classification task in medical implants production. Int. J. Intell. Syst. Appl. (IJISA) **10**(9), 40–47 (2018)

Distribution of Resources Between Composite Applications in a Hyperconverged System

Serhii Bulba, Nina Kuchuk(✉), and Anna Semenova

National Technical University “KhPI”, Kyrpychova Str., 2, Kharkiv, Ukraine
nina_kuchuk@ukr.net

Abstract. The analysis of the features of the functioning of systems with hyperconverged architecture is carried out. The necessity of solving the problem of efficient distribution of computing resources is shown. It is considered in the centralized management of composite applications. The analysis of the features of the functioning of composite applications in hyperconverged systems is carried out. A formal description of the resource allocation process of a hyperconverged system is proposed. On its basis, a mathematical model for the distribution of computing resources between composite applications has been developed. The approach to entering virtual time when processing a package of composite applications is described. Using this approach allowed to modify the developed mathematical model. A computational experiment was conducted. The results of the experiment showed that with increasing application complexity, the efficiency of computing resource allocation increases. This occurs using the proposed approach compared to the standard ones.

Keywords: Computing resource · Composite application · Hyperconverged system · Time windows

1 Introduction

1.1 Motivation

Informational systems and infrastructures are changing to more efficient and safer ones. So converged systems replaced cloud technologies. And hyperconverged systems (HCS) replaced convergent ones. A distinctive feature of HCS is the hypervisor. HCS hypervisor for access to applications in most cases uses composite applications. This improves transaction efficiency. But this raises the problem of the centralized distribution of resources between composite applications.

1.2 Analysis of Related Works

Features of the functioning of HCS are considered in [1]. Various methods for distributing a computing resource are widely represented in the scientific literature, for example, in [2–8]. In the article [2], the authors consider the distribution of resources with a large number of parallel tasks in a multiprocessor system. In [3], a special classifier is introduced for resource allocation. The methods proposed in [4] are focused on cloud technologies. In [5], resource allocation is focused on IoT Networks.

Z. Hu et al. (Eds.): ICCSEEA 2020, AISC 1247, pp. 429–439, 2021.
https://doi.org/10.1007/978-3-030-55506-1_39

In articles [6–8], special quality criteria are introduced for distribution. For example, in [8] - this is energy efficiency.

The areas of use the composite applications are described in detail in [9, 10]. A number of works present the specifics of the functioning of composite applications. So, access to data is described in [11]. Transaction security issues are described in [12]. The possibilities of clustering are disclosed in [13–15]. However, in these works, the problem of efficient distribution of computing resources in the centralized management of composite applications is not considered. To solve this problem, many different methods are also proposed. For example, multiagent systems [16], evolutionary search methods [17], and intelligent systems based on ant algorithms [18, 19] are considered. Existing models do not consider the possibility of managing time windows. However, for composite applications this factor is significant. Time window accounting is especially important when running multiple composite applications at the same time. Therefore, the development of a method for allocating resources between composite applications in a hyperconverged system that takes into account time windows is relevant.

1.3 Goals and Structure

An important task is the efficient allocation of resources between HCS composite applications. However, existing methods do not take into account time windows and HCS features. As a result, when several composite applications are executed in parallel, the efficiency of resource allocation is significantly reduced. The purpose of the report is to develop a functioning model of HCS composite applications. This model should take into account the time windows allocated to applications when accessing a computing resource. The use of this approach will increase the efficiency of using HCS computing resources.

The material is structured as follows. Section 2 describe mathematical model of resource allocation process between composite applications. Section 3 describe an approach that allows to introduce virtual time into the model. Section 4 is devoted to the results of a study of the effectiveness of the proposed method.

2 Mathematical Model of Resource Allocation Between Composite Applications

2.1 Features of Composite Applications Functioning in HCS

The composite application integrates several existing features into a new program. Composite applications can be built on any technology or architecture. The composite application consists of functionality. They were integrated from different sources. In Fig. 1 is a schematic representation of a part of a composite relationship. It is available to the operator. Different functional modules are presented under the various modules of the application. They may be located on remote distributed services. They provide the work of the operators.

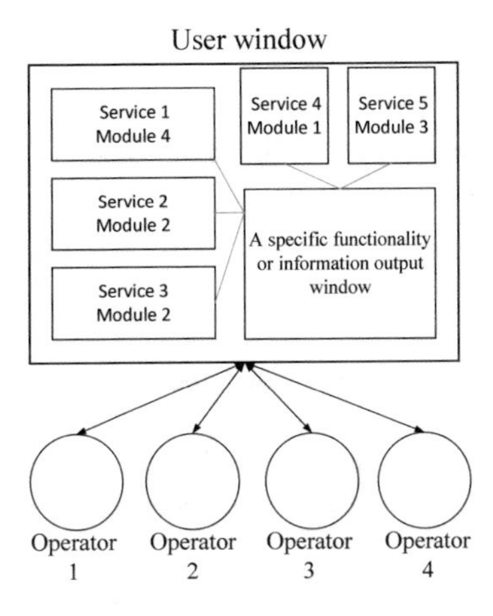

Fig. 1. Presentation of temporary windows

In Fig. 2 the structure of the composite application is presented. It operates on the basis of territorially distributed services, databases and computing resources.

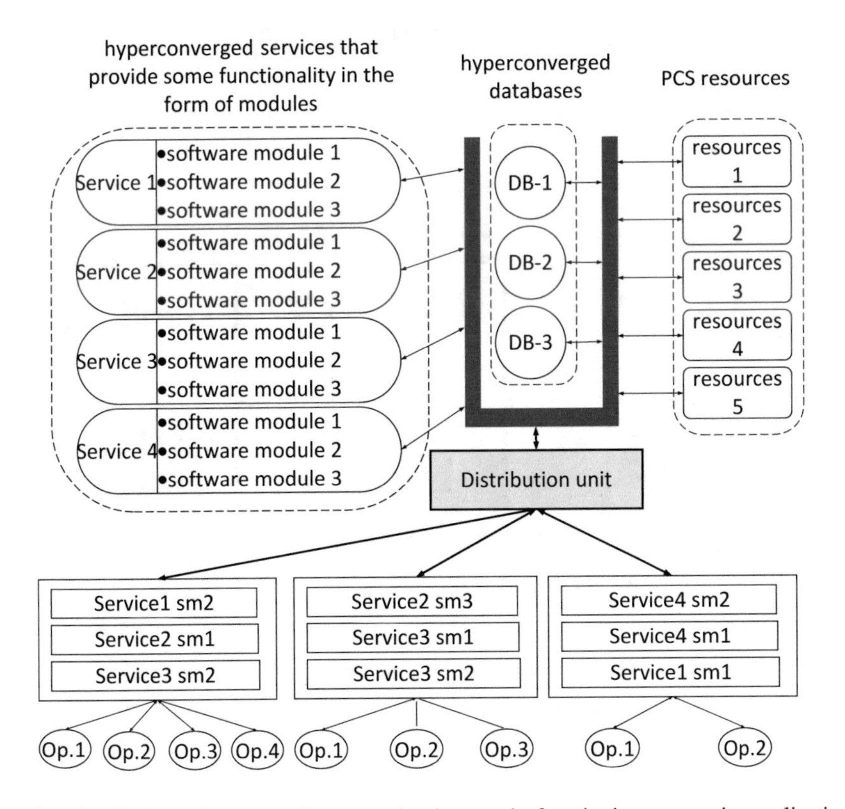

Fig. 2. Distribution of resources between simultaneously functioning composite applications

2.2 Formalization of the Resource Allocation Process in HCS

Consider the set available for CP resources P. It consists of S subsets different types of resources. We decompose the set of resources as follows:

$$P = \bigcup_{s=1}^{S} P_s; \quad \forall s_i \neq s_j, s_i, s_j \in \overline{1,\ S} \quad \Rightarrow \quad P_{s_i} \bigcap P_{s_j} = \emptyset, \tag{1}$$

where s – conditional number of a specific type of resource, $s \in \overline{1,\ S}$; P_s – set computing units (CU) type s.

In turn, each resource type s has a certain number of computing units N_s. If a separate block with a conditional number i denote as p_{si}, then

$$P_s = \bigcup_{i=1}^{N_s} p_{si}, \quad s \in \overline{1, S}, \tag{2}$$

where i – block number in set P_s.

Based on expressions (1) and (2), we carry out the decomposition of the set P:

$$P = \bigcup_{s=1}^{S} \bigcup_{i=1}^{N_s} p_{si}, \quad card\, P = \sum_{s=1}^{S} card\,(p_s). \tag{3}$$

An example the interaction of computing units is given in Fig. 3. The presented computing units are of different types.

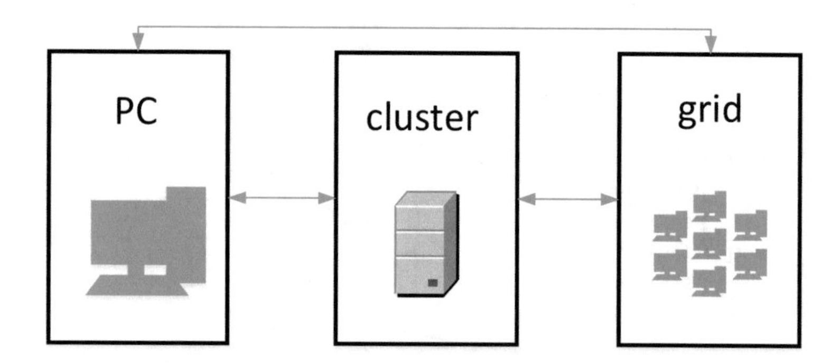

Fig. 3. An example the interaction of computing units

Consider the distribution resources during one cycle of information processing lasting T_{comm}, $\tau_{\mathrm{comm}} = [0,\ T_{\mathrm{comm}}]$. To plan the execution of composite applications, time intervals must be considered. During which a computing unit is needed p_{si} not available to package CP. Set available time intervals on CU with number i type s define as the union of time intervals during the considered time period T_{comm}:

All tasks of the CP package must begin no later than T_{comm}, that is

$$t_{mn}^{(0)} \in \tau_{\text{comm}},\ \forall n \in \overline{1, n_m};\ \forall m \in \overline{1, N_{DS}}, \tag{4}$$

where $t_{mn}^{(0)}$ – relative start time n-th task of m-th CP,

N_{DS} – total amount CP, requiring resource allocation in the time interval is considered, n_m – number of independent task of m-th CP.

Structure m-th CP defined by an oriented acyclic graph

$$AG_m = \langle DS_m, E_m \rangle, \tag{5}$$

where DS_m – set of the tasks m-th CP, $DS_m = \bigcup_{n=1}^{n_m} DS_{mn};\ card\ DS_m = n_m$, E_m – set of directed edges of the graph, $E_m = \left\{ e_{ij}^{(m)} | i,j \in DS_m \right\}$.

An example is shown at Fig. 4.

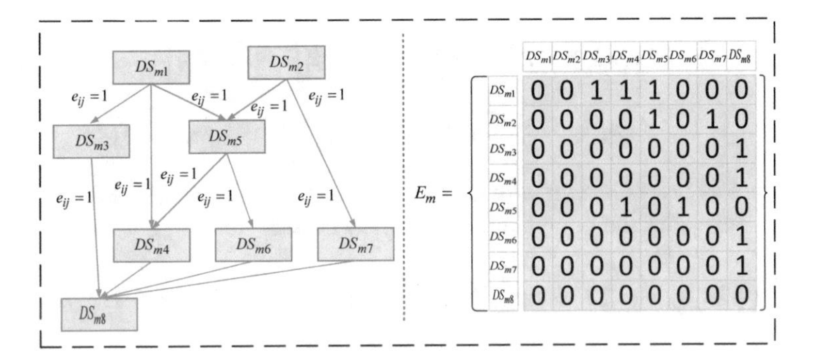

Fig. 4. Task graph CP

Let the set R specifies the resource allocation option $\Re$ between CP, then

$$R = \{\Re\} = \left\{ \bigcup_{m=1}^{N_{DS}} \bigcup_{n=1}^{n_m} R_{mn} \right\}, \tag{6}$$

where R_{mn} – tuple, specify distribution $\Re$ for n-th task of m-th CP thus:

$$R_{mn} = \eta(DS_{mn}, CP_{mn}) = \left\langle P_{si}(DS_{mn}),\ t_{mn}^{(0)} \right\rangle, \tag{7}$$

where CP_{mn} – a set of computing blocks, able to complete tasks DS_{mn},

η – displaying on tuple sets such that

$$\eta : (DS, P) \to (P, T); \tag{8}$$

$DS = \bigcup_{m=1}^{N_{DS}} DS_m$, P – set resources available to the package CP,

$P_{si}(DS_{mn})$ – i - th CU type s, what will carry out n-th task of m-th CP, starting time $t_{mn}^{(0)} \in \tau_{comm}$.

In addition to the start time for each task, we will consider the relative time of its end. For most tasks, the following condition must be met: $t_{mn}^{(1)} \in \tau_{comm}$.

We introduce additional conditions for the implementation of the resource allocation process.

An elementary resource (CU) cannot be used simultaneously to calculate multiple tasks. The fulfillment of this condition is possible due to further decomposition of the sets DS and P. Formally, this requirement is as follows:

$$P_{si}(DS_{m_1n_1}) = P_{si}(DS_{m_2n_2}) \Rightarrow \left[t_{mn}^{(0)}, t_{mn}^{(1)}\right] \cap \left[t_{mn}^{(0)}, t_{mn}^{(1)}\right] = \varnothing; \tag{9}$$

$$\forall m_1 m_2 \in \overline{1, N_{PS}}; \forall n_1 \in \overline{1, n_{m_1}}; \forall n_2 \in \overline{1, n_{m_2}}. \tag{10}$$

Sequence of tasks for each m-th CP corresponds to the graph AG_m. Each task can be completed only after the completion of the predecessors, thus

$$\left(\forall i | e_{mn_i, mn_f} = 1\right) \Rightarrow \; t_{mn_f}^{(0)} \geq \max_i \left(t_{mn_i}^{(1)}\right), \; \forall n_f \in DS_m, \; \forall m \in \overline{1, N_{DS}}. \tag{11}$$

It is necessary to take into account the communication costs between the tasks of one CP. To do this, a matrix of objects that transmit data is introduced. All within one CP:

$$D_m = \left\{d_{nn_f}^{(m)} | n, n_f \in \overline{1, n_m}\right\}, \tag{12}$$

where d_{nn_f} – amount of data that is sent from job number n to task with number n_f.

Accordingly, we define the matrices. They will form the time required for the transfer of data during distribution $\Re$:

$$\Theta_m(\Re) = \left\{\theta_{nn_f} | n, n_f \in \overline{1, n_m}\right\}, \tag{13}$$

where $\Theta_{nn_f}^{(m)} = \Theta_{nn_f}^{(mn)}(\Re, d_{nn_f}^{(m)})$ – transmission time of generated data from n to n_f.

Then condition (11) can be modernized as follows:

$$\left(\forall i | e_{mn_i, mn_f} = 1\right) \Rightarrow t_{mn_f}^{(0)} \geq \max_i \left(t_{mn_i}^{(1)} + \Theta_{n_i n_f}^{(m)}\right), \; .\forall n_f \in DS_m, \forall m \in \overline{1, N_{DS}}. \tag{14}$$

The full execution time of a composite application package is defined as

$$T_n = \max_{m \in \overline{1,N_{DS}}} \left(\max_{n \in \overline{1,n_m}} t_{mn}^{(1)} \right). \quad (15)$$

Then, if the following condition is satisfied:

$$T_n \le T_{\text{comm}}, \quad (16)$$

then the CP package is fully implemented with the selected distribution $\Re$ during one cycle of information processing. Otherwise, the selected time interval is not enough.

Let $N_-(\Re)$ – subset of CP, which when allocating resources $\Re$ do not have time to complete all the tasks. Then

$$N_-(\Re) = \left\{ m | \max_{n \in \overline{1,n_m}} t_{mn}^{(1)} > T_{comm} \right\}. \quad (17)$$

So, the above expressions (1)–(17) constitute a mathematical model of the distribution of resources between composite applications. This happens during one cycle of information processing. But the developed model does not take into account time windows. That is, the resource is allocated to the CP until the corresponding transaction is completed.

3 Introduction of Virtual Time

When allocating resources for composite applications, time windows must be considered. The main problem is the lack of a continuous time variable. Therefore, it is proposed to introduce virtual time. Such a variable will combine all the application activity intervals into one inextricable section. This will allow a continuous countdown.

For distribution $\Re$ introduce $\gamma_{si}(\Re)$ – time interval display, considered, τ_{comm}, into the union of sets intervals $\tau_{si}(\Re)$:

$$\gamma_{si}(\Re) : \tau_{comm} \to \tau_{si}(\Re), \ s = \overline{1,S}, \ i = \overline{1,N_S}. \quad (18)$$

The set $\tau_{si}(\Re)$ formed as follows: each of its elements is a virtual continuous time interval:

$$\tau_{si}(\Re) = \bigcup_{P_{si}(DS_{mn}) \neq \varnothing} \tau_{si}(DS_{mn}), \quad (19)$$

$$\sum_{m=1}^{N_{DS}} \sum_{n=1}^{n_m} |\tau_{si}(DS_{mn})| \le T_{comm}, \quad (20)$$

Where

$$s = \overline{1,S},\ i = \overline{1,N_S},\ \forall n \in \overline{1,n_m};\ \forall m \in \overline{1,N_{DS}}.$$

Total time allocated to complete the task n CP m calculated as

$$\tau_{mn} = \sum_{s=1}^{S} \sum_{i=1}^{N_S} |\tau_{si}(DS_{mn})|. \tag{21}$$

Continuous counting of interval combining time $\tau_{si}(DS_{mn})$ we will carry out using additional variables $t_{si,mn}$.

A general view of the process allocating time windows to a specific task of a fixed composite application is presented schematically and in detail in Fig. 5.

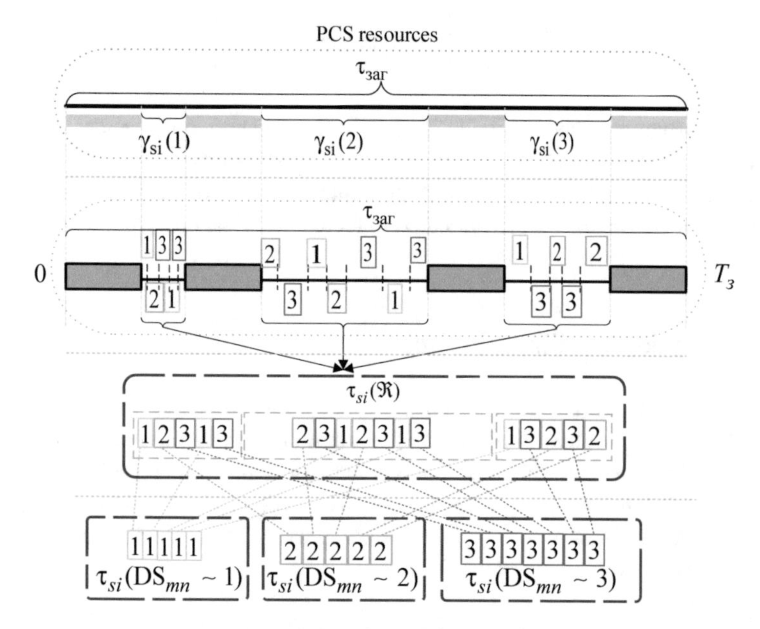

Fig. 5. Presentation of time windows

Therefore, the introduction of virtual time using expression (18) allows us to take into account time windows. This significantly reduces resource costs when running composite applications.

4 Discussion and Conclusion

To analyze the possibilities of the proposed method, a computational experiment was conducted. The calculations were performed on a personal computer with a processor *Intel(R) Core(TM) i7*-4500 with clock rate 1.80 GHz 2.40 GHz, and random access memory 6 Gb, in to the 64 bit operating system Windows 8. The package of composite applications was formed of 6 applications with the variable structure. The experimental results are shown in Fig. 6.

With increasing application complexity, the efficiency of computing resource allocation increases. This occurs using the proposed approach compared to the standard ones.

The necessity of solving the problem of efficient distribution of computing resources is shown. It is considered in the centralized management of composite applications. The analysis of the features of the functioning of composite applications in hyperconverged systems is carried out. A formal description of the resource allocation process of a hyperconverged system is proposed. On its basis, a mathematical model for the distribution of computing resources between composite applications has been developed. The approach to entering virtual time when processing a package of composite applications is described. Using this approach allowed to modify the developed mathematical model.

A feature of the proposed method is the ability to take into account time windows when allocating resources. The scientific novelty of the method is the possibility of continuous counting of virtual time for each composite application.

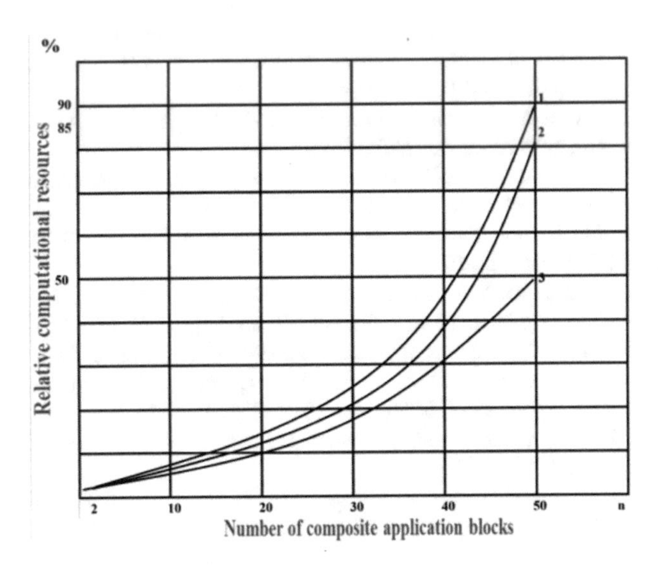

1 – standard algorithm, 2 – application of a mathematical model
3 - application of modified mathematical model

Fig. 6. Dependency computational resources

References

1. Merlac, V., Merlac, V., Smatkov, S., Kuchuk, N., Nechausov, A.: Resources distribution method of university e-learning on the hypercovergent platform. In: Conference Proceedings of 2018 IEEE 9th International Conference on Dependable Systems, Service and Technologies, DESSERT 2018, Kyiv, pp. 136–140 (2018). https://doi.org/10.1109/dessert.2018.8409114
2. Mohseni, Z., Kiani, V., Rahmani, A.M.: A task scheduling model for MultiCPU and multi-hard disk drive in soft real-time systems. Int. J. Inf. Technol. Comput. Sci. (IJITCS) **11**(1), 1–13 (2019). https://doi.org/10.5815/ijitcs.2019.01.01
3. Zhou, L.-Y., Boateng, L.K., Amoh, D.M., Okine, A.A.: Improving the reliability of churn predictions in telecommunication sector by considering customer region. Int. J. Inf. Technol. Comput. Sci. (IJITCS) **11**(6), 1–8 (2019). https://doi.org/10.5815/ijitcs.2019.06.01
4. Kuchuk, G., Nechausov, S., Kharchenko, V.: Two-stage optimization of resource allocation for hybrid cloud data store. In: International Conference on Information and Digital Technologies, Zilina, pp. 266–271 (2015). https://doi.org/10.1109/DT.2015.7222982
5. Mohammed, A.S., Saravana, B.B., Saleem Basha, M.S.: Fuzzy applied energy aware clustering based routing for IoT networks. Adv. Inf. Syst. **3**(4), 140–145 (2019). https://doi.org/10.20998/2522-9052.2019.4.22
6. Kuchuk, N., Mozhaiev, O., Mozhaiev, M., Kuchuk, H.: Method for calculating of R-learning traffic peakedness. In: 2017 4th International Scientific-Practical Conference Problems of Info communications Science and Technology, PIC S and T 2017, Proceedings, pp. 359–362 (2017). https://doi.org/10.1109/INFOCOMMST.2017.8246416
7. François-Xavier, D., Sandrine, A.: Generalized greedy alternatives. In: Applied and Comp. Harmonic Analysis, Elsevier (2018). https://doi.org/10.1016/j.acha.2018.10.005
8. Kuchuk, G., Kovalenko, A., Komari, I.E., Svyrydov, A., Kharchenko, V.: Improving big data centers energy efficiency. traffic based model and method. In: Kharchenko, V., Kondratenko Y., Kacprzyk J. (eds.) Green IT Engineering: Social, Business and Industrial Applications, Studies in Systems, Decision and Control, vol. 171. Springer, Cham (2019). https://doi.org/10.1007/978-3-030-00253-4_8
9. Bulba, S.: Composite application distribution methods. Adv. Inf. Syst. **2**(3), 128–131 (2018). https://doi.org/10.20998/2522-9052.2018.3.22
10. Kuchuk, N., Mohammed, A.S., Shyshatskyi, A., Nalapko, O.: The method of improving the efficiency of routes selection in networks of connection with the possibility of self-organization. Int. J. Adv. Trends Comput. Sci. Eng. **8**(1), 1–6 (2019). https://doi.org/10.30534/ijatcse/2019/0181.22019
11. Kuchuk, G.A., Akimova, Y.A., Klimenko, L.A.: Method of optimal allocation of relational tables. In: Engineering Simulation, 2000, vol. 17, no. 5, pp. 681–689 (2010)
12. Semenov, S., Sira, O., Kuchuk, N.: Development of graphicanalytical models for the software security testing algorithm. Eastern-Eur. J. Enterpr. Technol. **2**, (4(92)), 39–46 (2018). https://doi.org/10.15587/1729-4061.2018.127210
13. Fränti, P.: Efficiency of random swap clustering. J. Big Data **5**(1), 1–29 (2018). https://doi.org/10.1186/s40537-018-0122-y
14. Fahim, A.: Homogeneous densities clustering algorithm. Int. J. Inf. Technol. Comput. Sci. (IJITCS) **10**(10), 1–10 (2018). https://doi.org/10.5815/ijitcs.2018.10.01
15. Tripathy, R.: Fuzzy clustering of sequential data. Int. J. Intell. Syst. Appl. (IJISA) **11**(1), 43–54 (2019). https://doi.org/10.5815/ijisa.2019.01.05

16. Javed, R., Anwar, S., Bibi, K., Usman Ashraf, M., Siddique, S.: Prediction and monitoring agents using weblogs for improved disaster recovery in cloud. Int. J. Inf. Technol. Comput. Sci. (IJITCS) **11**(4), 9–17 (2019). https://doi.org/10.5815/ijitcs.2019.04.02
17. Moghissi, G.R., Payandeh, A.: A parallel evolutionary search for shortest vector problem. Int. J. Inf. Technol. Comput. Sci. (IJITCS) **11**(8), 9–19 (2019). https://doi.org/10.5815/ijitcs.2019.08.02
18. Ankita, S.K.S.: An automated parameter tuning method for ant colony optimization for scheduling jobs in grid environment. Int. J. Intell. Syst. Appl. (IJISA) **11**(3), 11–21 (2019). https://doi.org/10.5815/ijisa.2019.03.02
19. Narendrababu Reddy, G., Phani Kumar, S.: MACO-MOTS: modified ant colony optimization for multi objective task scheduling in cloud environment. Int. J. Intell. Syst. Appl. (IJISA) **11**(1), 73–79 (2019). https://doi.org/10.5815/ijisa.2019.01.08

Improvement of Merkle Signature Scheme by Means of Optical Quantum Random Number Generators

Maksim Iavich[1(✉)], Avtandil Gagnidze[2], Giorgi Iashvili[1], Tetyana Okhrimenko[3], Arturo Arakelian[4], and Andriy Fesenko[5]

[1] Scientific Cyber Security Association, Caucasus University, Tbilisi, Georgia
m.iavich@scsa.ge
[2] East European University, Tbilisi, Georgia
taniazhm@gmail.com
[3] National Aviation University, Kyiv, Ukraine
[4] University of Georgia, Tbilisi, Georgia
[5] Taras Shevchenko National University of Kyiv, Kyiv, Ukraine
aafesenko88@gmail.com

Abstract. The corporation Google, NASA and have teamed up with D-Wave, the global manufacturer of quantum CPUs. Quantum computers will have the opportunity to hack almost every cryptosystem, which is used in the majority of products, for example, they will be able to break RSA. RSA crypto scheme is the part of many products, which are used in different areas on the various platforms. Nowadays, this crypto scheme is the part of the basic commercial systems; the number of these systems is actively increasing. Hash-based digital signature schemes offer an alternative, which can be considered safe against the attacks of quantum computers. These digital signature schemes use the cryptographic hash function. The security of the hash based digital signature schemes is based on the collision resistance of the hash function, which they use. In 1979, Ralph Merkle proposed Merkle signature scheme. Merkle signature scheme has an efficiency problems, so it cannot be used in practice. World scientists are working on improving the scheme. One of the improvements is integrating PRNG (pseudo random number generator) not to calculate and store the large amount of one-time keys pairs. This approach cannot be considered secure, because according to our research quantum computers are able to crack PRNG, which were considered safe against attacks of classical computers. In the article it is offered to use the concrete pseudo random number generator and the true random number generator for generating the seed. As a pseudo random number generator it is offered to use an algorithm based on a hash function, as the whole the algorithm is based on it. NIST has recommended two continuous hash based PRNGs: HASH_DBRG and HMAC_DBRG. In the article it is offered to use HASH_DBRG as it is more efficient. As the true random number generator it is offered to use physical quantum random number generator (QRNG) for generating the seed for HASH_DBRG. The implementation algorithm of the scheme is offered and the security of this scheme is analyzed. The offered scheme is rather efficient, because it does not store all the signature keys. The practical experiment shows that the algorithm is also rather fast. The offered scheme is secure and can be used as post-quantum security tool.

Z. Hu et al. (Eds.): ICCSEEA 2020, AISC 1247, pp. 440–453, 2021.
https://doi.org/10.1007/978-3-030-55506-1_40

Keywords: RSA · Hash function · Pseudo random number generator · Digital signatures

1 Introduction

The corporation Google, NASA and USRA (the Universities Space Research Association) have teamed up with D-Wave, the global manufacturer of quantum CPUs. Quantum computers will have the opportunity to hack almost every cryptosystem, which is used in the majority of products, for example, they will be able to break RSA. RSA crypto scheme is the part of many products, which are used in different areas on the various platforms. Nowadays, this crypto scheme is the part of the basic commercial systems, the number of these systems is actively increasing. Hash-based digital signature schemes offer an alternative, which can be considered safe against the attacks of quantum computers. RSA crypto system is actively used in different operating systems, such as Microsoft, Apple, Novell and Sun. RSA crypto scheme is used in hardware too. It is used in secure telecommunication, Ethernet, network and smart cards. This system is often used in the cryptographic hardware. It must be emphasized that RSA algorithm is used in the secure internet communications, such as SSL, S/WAN and S/MIME. This scheme is used in almost every organization; RSA is used in banks, government structures, financial organizations, banks, libraries and in educational institutions.

As we can see, RSA is the mostly often used encryption technology, so we can consider it as the most commonly used asymmetric crypto scheme. This scheme is developing together with the development of World Wide Web. So the hacking of RSA will cause chaos, it will lead to the breaking of almost all the products around the world.

1.1 Literature Review

In [1–3] one time signature schemes are offered. In this signature schemes each pair of the keys can be used only for signing one document, so they are not efficient in practice. In [4–6] Merkle signature scheme is offered as a solution to this problem and the efficiency of the scheme is discussed. It is shown, that this crypto system has efficiency problems, and cannot be used in practice. In [7–9] different improvements of the efficiency of Merkle are offered. The authors of [9] offer to integrate pseudo random number generator into the scheme not to calculate and store large amount of one-time keys pairs. In [10] pseudorandom number generators that are widely used in Cryptography are described and analyzed pseudorandom number generators that are widely used in Cryptography. The authors of [12] show a polynomial quantum time attack on PRNG Blum-Micali pseudorandom number generator, which is considered safe from threats of standard computers. In [13–17] true random number generators are described and different was of their practical use are offered.

1.2 Digital Signatures

Digital signatures are a key technology, which make various IT systems and Internet secure. They provide integrity, non-repudiation of information and authenticity. Digital signatures are very actively used in the authentication and identification protocols. So having secure digital signatures is obligatory for computer science and cyber security. RSA, ECDSA and DSA are digital signature algorithms, which are actively used today in practice. These algorithms are not quantum secure because their security is based on the complexity of large number factorization and discrete logarithms computation. Hash-based digital signature schemes offer an alternative, which can be considered safe against the attacks of quantum computers. These digital signature schemes use the cryptographic hash function. The security of hash based digital signature schemes is based on the collision resistance of the hash function, which they use.

1.2.1 Hash Based Digital Signatures

One-time signature scheme – "Lamport One Time Signature Scheme" was offered in [1, 2].

The signature key X of this system consists of 2n lines of the length n, and the lines are selected randomly.

$$X = (x_{n-1}[0],\ x_{n-1}[1], \ldots,\ x_1[0],\ x_1[1],\ x_0[0],\ x_0[1]) \in \{0,1\}^{n,2n}$$

Verification key Y of this scheme consists of 2n lines of the length n, and the lines are selected randomly.

$$Y = (y_{n-1}[0],\ y_{n-1}[1], \ldots,\ y_1[0],\ y_1[1],\ y_0[0],\ y_0[1]) \in \{0,1\}^{n,2n}$$

To calculate the key we use one way function f:

$$f : \{0,1\}^n \to \{0,1\}^n;$$

$$y_i[j] = f(x_i[j]),\ \ 0 <= i <= n-1,\ j = 0,1$$

1.2.2 Document Signature

Message m of arbitrary size, is transformed into size n by means of the hash function:

$h(m) = hash = (hash_{n-1}, \ldots, hash_0)$

Function h is a cryptographic hash function:

$$H : \{0,1\}^* \to \{0,1\}^n$$

Signature occurs as follows:

$$sig = (x_{n-1}[hash_{n-1}], \ldots, x_1[hash_1],\ x_0[hash_0]) \in \{0,1\}^{n,n}$$

If the i-th bit in the message is equal to 0, i-th string in this signature is assigned to $x_i[0]$. If the i-th bit in the message is equal to 1, i-th string in this signature is assigned to $x_i[1]$.

The length of this signature is n^2.

1.2.3 Signature Verification

For signature verification sig = (sig_{n-1}, …, sig_1, sig_0), the hash of the message is calculated.

hash = ($hash_{n-1}$, …, $hash_1$, $hash_0$) and the following equality must be verified:

$$(f(sig_{n-1}),\ldots,\ f(sig_1),\ f(sig_0)) = (y_{n-1}[hash_{n-1}],\ldots,y_1[hash_1],\ y_0[hash_0])$$

If it is true, then the signature is correct.

This system has a very big disadvantage; it has a huge size of the keys.

To get the security O (2^{80}), the whole size of the private and public keys has to be 51200 (160 * 2 * 160) bits, it is 50 (51200/1024) times greater than in the RSA scheme.

Is must be also mentioned, that the signature size in “Lamport One Time Signature Scheme” is much greater than in RSA scheme.

Winternitz One time Signature Scheme was offered to decrease the size of the signature [3, 18].

One time signature schemes are inefficient in practice, because each pair of the keys can be used only for signing one document. Ralph Merkle offered a solution to this problem in 1979. He offered to use the binary hash tree to decrease the arbitrary, concrete number of one-time verification keys to the single public key. This key is the root of the binary hash tree.

2 Merkle Signature Scheme

2.1 Key Generation

The size of the tree must be H >= 2 and using one public key 2^H documents can be signed. 2^H signature and verification key pairs are generated; X_i, Y_i, 0 <= I <= 2H. X_i is the signature key and Y_i is the verification key.

To get the leaves of the tree, signature keys are hashed using the hash function:

$$h : \{0,1\}^* \rightarrow \{0,1\}^n \tag{4}$$

To get the parent node, the concatenation of two previous nodes is hashed. The root of the hash tree is the public key of the signature - public.

Figure 1. illustrates the tree with H = 3; a[i,j] are the nodes of the tree;

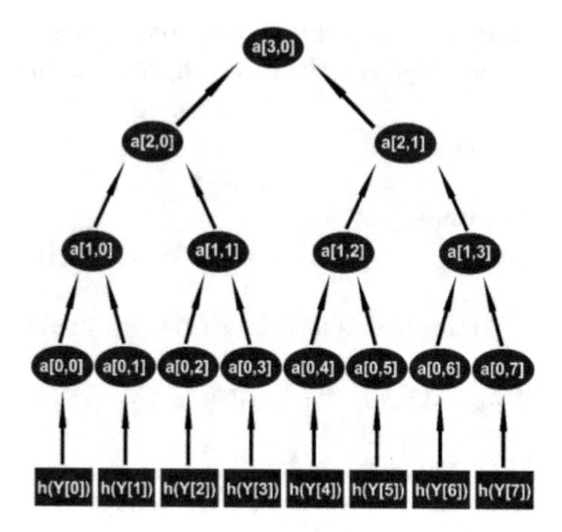

Fig. 1. Merkle tree with H = 3

2.2 Message Signature

To sign a message of any size, it is transformed to size of n by means of hashing.

h(m) = hash, to sign the message, is used an arbitrary one-time key X_{arb}. The signature is a concertation of one-time signature, one-time verification key, index of a key and all brother nodes according to the selected arbitrary key with the index "*arb*".

$$Signature = (sig||arb||Y_{arb}||auth_0, \ldots, auth_{H-1}) \tag{5}$$

2.3 Signature Verification

To verify the signature, the one time signature is checked using the selected verification key, if the verification has passed successfully, all the needed nodes are calculated using "auth", index "arb" and Y_{arb}. If the root of the tree matches the public key, then the signature is correct.

This crypto system has efficiency problems, so it cannot be used in practice [4–6, 19].

$\mathbf{2^H}$ pairs of one-time keys must be calculated in order to generate the public key. Storing such a big number of key is problematic in practice.

3 Merkle with PRNG

World scientists are working on improving the scheme [7–9]. One of the improvements is integrating PRNG (pseudo random number generator) into the scheme. It is performed not to calculate and store large amount of one-time keys pairs.

3.1 PRNGs in Cryptography

Pseudorandom number generators are widely used in Cryptography [10, 20]. This type of PRNGs are called cryptographically secure pseudorandom number generators CSPRNGs. Blum Blum Shub and the Blum and Micali generators are often used in the various cryptography applications. These CSPRNGs are based on number theory. Blum Blum Shub works as follows: the output bits are received from the following recursive formula:

$$X_{i+1} = X_i^2 \ mod\ N \tag{1}$$

for $N = pq$. N is the product of two primes numbers p and q, which are congruent to *3 mod 4*.

As the internal state is used X_i - the i-th number. This algorithm gets X_0 and N as inputs and it outputs X_i in the i-th output bit. The initial state X_0 should be received from the true random number generator. This PRNG has various properties since concrete common computational complexity assumptions hold. In this PRNG, even if the hacker got the internal state X_i at the stage I, the next bits of the message are not leaked.

To guess X_{i-1} from X_i is the computationally hard problem, because the quadratic residuosity problem cannot be solved in the polynomial time. Afterwards, the works showed that hacking Blum Blum Shub is the same as factoring [11]. So, the hacker that knows N will be able to use Shor's algorithm to factor the integers and to break the security of Blum Blum Shub generator by means of quantum computer.

A polynomial quantum time attack on PRNG Blum-Micali was also shown [12]. This PRNG is considered safe from threats of standard computers. This attack uses Grover algorithm along with the quantum discrete logarithm, and is able to restore the values at the generator output for this attack.

Integrating all CSPRNGs cannot be considered secure, because according to our research quantum computers are able to crack PRNG, which were considered safe against attacks of classical computers.

3.2 Secure CSPRNGs for Merkle

It is not enough when the random numbers are only uniform. The numbers must be unpredictable and the generator must limit the damage, if the key is compromised. CSPRNGs must be unpredictable. Also, if the hacker intercepts a part of the sequence, he must not be able to predict the previous values of the generator with a probability more than 0.5.

In practice, the hacker must not be able to intercept a subsequence from the generator and receive the previous and the next bits of the output better than by random guessing in polynomial time.

This concept follows from Yao's definition of the indistinguishable sources [13]. The big part of PRNGs cannot be considered as CSPRNGs. For example, Mersenne Twister, Matsumoto and Nishimura showed, that the internal state in this PRNG can be recovered from the output [14].

As a CSPRNG in Merkle we offer an algorithm based on a hash function, as the whole the algorithm is based on it. NIST has recommended two continuous hash based PRNGs: HASH_DBRG and HMAC_DBRG. We offer to use HASH_DBRG as it is more efficient.

We offer to use physical quantum random number generator (QRNG) for generating the seed for HASH_DBRG.

4 Radioactivity-Based Quantum Random Number Generators

In 1961, the researchers offered to use quantum phenomena as a source of randomness [15]. Afterwards the researchers began to work actively on it. Radioactive decay was the rather accessible source of the true randomness. Geiger-Muller tubes were already enough sensitive to grab and enhance α, β and γ radiation, rather qualitative radioactive samples were already available. Almost all QRNGs based on radioactivity were using the detection of β radiation.

In a Geiger-Muller detector, each particle makes an ionization event that is enhanced in the Townsend avalanche. We get a device that creates a pulse for every detected particle. Any concrete atom's probability to decay in the time interval *(t, t + dt)* can be presented as exponential random variable:

$$Pr(t)dt = \lambda_n e^\wedge(-\lambda_n t)dt \tag{2}$$

where λ_n is a decay constant.

The time period between two detected pulses is the random exponential variable, if the sample saves a lot of original atoms and the detector system is not changed in this time interval.

The times values are independent from the previous ones. The number of the pulses, which occur in a concrete time, follow a Poisson distribution. It can be shown experimentally, that these pulses occur at the independent times. The probability of finding n pulses in some period is:

$$P_n(T) = (\lambda T)^n/n! * e - \lambda T \tag{3}$$

where T is the number of seconds and λ is a mathematical mean of pulses, which were detected in one second. This number corresponds to the exponential distribution parameter.

QRNGs, which are based on radioactive decay, have a lot of general features. Most of them use digital counters to convert the pulses, received from the detector into the random values. The counter increments the values of the output by 1 as soon as it receives a pulse as the input and the value of the counter can be reset to zero. Timing with a digital clock is also very important element.

If f is a frequency of the clock, a fast clock with $f > \lambda$ generates a big amount of pulses and a slow clock with $f < \lambda$ outputs a pulse rarely, so many counts can be registered in detector.

In order to get the random number, the randomness in the time of arrival generator must be converted into random values. It can be done in the various ways.

For instance, the generators of Vincent and Isida and Ikeda use the fast clock counter. The counter is read and then it is reset to zero, when we receive a count on the detector. The counter value at the detection time is used to get the random number. The distribution of the values must be corrected, because it is not uniform.

A slow clock can also be used to understand when to read the value of the counter. For instance in the Schmidt generator, the detector pulses enhance the value of the counter. When the slow clock makes a new pulse, we get the random value by reading the counter value and after that, the count resets to zero once more.

The output of the generator corresponds to the number of particles received in the every clock period. In order to generate values from 0 to $V - 1$, we use modulo V operation. If $V = 2$, we receive the binary RNG (random number generator). The modulo M addition of multiple outputs must be taken, as the distribution of the values is not uniform, so we will receive a distribution with a very little bias. This procedure is named contraction.

4.1 Time of Arrival Generators

Today, almost all the existing QRNGs are based on the quantum optics. Most parameters of the quantum states of the light have rather good randomness property. It gives us the opportunity of good choice of the implementations. Light emitting diodes, light from lasers or any single photon sources is a very convenient and rather affordable alternative of the radioactive material to get a source of quantum randomness and there exist a lot of different available detectors.

Time of arrival generators, are representatives of optical quantum random number generators (OQRNGs). These OQRNGs are based on the same basic principles as the QRNGs that get the radioactive decay [16].

QRNG that uses time has a rather bad source of photons, it has timing circuitry and detector, which trace the time of every detection or the number of clicks in a concrete period of the time. The detector receives photons from the LED incoherent light and from the coherent states of the laser in time, that is distributed exponentially $\lambda e^{-\lambda t}$, where λ is the average photons number in a second.

The time between two detections is exponential as it is the difference of two random exponential variables. We can compare the differences of the time between the arrivals of consecutive pulses; we will get two time differences t_1 and t_2, so we can compare them also. In order to get the random bit, if $t_1 > t_0$ we assign 1 and if $t_0 > t_1$ we can assign 0. We take the integers with the number of the counted clock periods c_0 and c_1 instead of the real times t_0 and t_1.

We can get the case, where $t_0 = t_1$, with some negligible probability for an ideal constant time measurement and of course this case must be taken into the consideration. We must find two coherent measures $c_0 = c_1$ for which we will get the same time. So, in the basic scheme we generate 0 or 1 after comparing two time intervals, and if the output is not defined we have to discard the results. If we are considering this equality as the valid result, we must have a special analysis of the probabilities of every outcome and we need the approach of their assignment to the random binary bit.

In 2007 the first optical QRNG was offered, which uses time detection. It gets the photons from the LED arriving at a PMT after it compares the arrivals times.

We offer to use this time arrival generator as a seed of HASH_DBRG.

5 The Proposed Scheme

5.1 Key Generation

The size of the tree must be H >= 2 and using one public key 2^H document can be signed. We use time arrival generator to generate a seed. The PRNG HASH_DBRG takes the seed as the input and are generated signature and verification keys; X_i, Y_i, 0 <= i <= 2H.

X_i - is the signature key, Y_i - is the verification key. To get the leaves of the tree, signature keys are hashed using the hash function:

$$h : \{0,1\}^* \rightarrow \{0,1\}^n \tag{4}$$

To get the parent node, the concatenation of two previous nodes is hashed. The root of the tree is the public key of the signature – public.

5.2 Message Signature

To sign a message of any size, it is transformed to size of n by means of hashing.

h (m) = *hash*, to sign the message, is used an arbitrary one-time key X_{arb}. This key is calculated by means of PRNG HASH_DBRG using the same seed got from time arrival generator.

The signature is a concertation of one-time signature, one-time verification key, index of a key and all brother nodes according to the selected arbitrary key with the index "*arb*".

$$Signature = (sig||arb||Y_{arb}||auth_0, \ldots, auth_{H-1}) \tag{5}$$

5.3 Signature Verification

To verify the signature, the one-time signature is checked using the selected verification key, if the verification has passed, all the needed nodes are calculated using "auth", index "arb" and Y_{arb}. If the root of the tree matches the public key, then the signature is correct.

5.4 Security

Nothing is changed in the classical version of Merkle. In order to improve the efficiency the hash based PRNG is integrated, the seed of PRNG is integrated by means of Quantum Random Number Generator. The offered PRNG is not vulnerable to attacks of quantum computers.

As we can see, the proposed implementation of improved Merkle is secure.

5.5 Pseudo Code of OQRNG Simulation

1. Importing necessary labs
2. Defining the function
3. Defining the string
4. For each loop
   4.1. generations detection time for the first photon
   4.2. generating detection time for the second photon
   4.3. generation two pseudo random time periods
   4.4. calculate the difference between them
5 if the first period is more adding the bit 1 to the string
6 if the second period is more adding the bit 0 to the string
7 checking if time intervals are equal
   6.1 if they are equal, calling the function again
8 Returning the string value

5.6 Pseudo Code of the Scheme

1. Importing necessary libs
2. Importing HASH_DRBG class
3. Importing the class of OQRNG
4. Define class
5. Defining "alt_hashes(hashes)" method
6. Set list "arr "
7. If hashes == "", raise Exception
8. Foreach loop
   8.1. Sorting hashes and appending into arr
9. Length_of_block == length of arr
10. While loop, if length is odd, copy last element in list
    10.1. Append it into arr list
11. Set list "another_arr "
12. Foreach loop
    12.1. For loop with range from 0 to length of "arr" and iteration by 2
        12.1.1. Define variable with "sha512()" value
        12.1.2. Hash elements that are in "arr" list
        12.1.3. Append them into new "another_arr" list
        12.1.4. Return this list in hex
13. Set list "hash_arr"
14. Generating the seed by OQRNG
15. Foreach loop
    15.1. Passing OQRNG to HASH_DRBG
    15.2. Generating the signature keys

16. Create message put it in “st” variable
17. Convert “st” value in binary
18. Generating the signature keys once more with HASH_DRBG with the same seed
19. Adding the keys to temporal tuple
20. Generate “one-time signature”
    20.1. If st == 0
        20.1.1. Choose “First_signature_key” bit
    20.2. Else
        20.2.1. Choose “Second _signature_key” bit
21. First_pub_key = hash(hash_arr[0])
22. Second_pub_key = hash (hash_arr[1])
23. Clearing the temporal tuple
24. Encryption
    24.1. Concatenate “one-time signature” with message’s hash
25. Verification of “one-time signature”
    25.1. If bit of “one-time signature” == 0
        25.1.1. Compare with “First_secret_key” bit
    25.2 Else
        25.2.1. Compare with “Second _secret_key” bit
26. Verification of “signature”
    26.1. Concatenate siblings with each other
    26.2. If this equals to public key
        26.2.1. Sign is correct
    26.3. Else
        26.3.1. Sign is not correct

6 Results and Discussion

As the result, we receive the secure improvement of Merkle digital signature scheme, which is secure against the attacks of quantum computers. This scheme uses pseudo random number generator not to calculate and store large amount of one-time keys pairs. As a pseudo random number generator an algorithm based on a quantum resistant hash function is used. This approach is rather interesting, because the whole Merkle algorithm used the quantum resistant hash function. As the pseudo random number generator HASH_DBRG is used. It must be emphasized that it is the NIST standard. This PRNG is considered rather secure and efficient. Physical quantum random number generator (QRNG) for generating the seed for HASH_DBRG is used in the offered algorithm. As the QRNG we use time of arrival generator, which is the representative of optical quantum random number generators (OQRNGs). It is a very interesting approach, because most parameters of the quantum states of the light have rather good randomness property. So we have a very big good choice of the implementations. It must be emphasized, that light emitting diodes, light from lasers or any single photon sources is a very convenient and very affordable alternative of the radioactive material to get a source of quantum randomness. Also a lot of detectors like these already exist.

The integration of such OQRNG is rather efficient, because it is used only for generating the seed for HASH_DBRG. We also offer the simulation of OQRNG. It must be emphasized that this approach is rather unique, because there are no other implementation with OQRNG.

We implemented the our algorithm with Python 3 and performed the tests the laptop with I5 CPU and 8 Gb of RAM. As the message with took the sequence of bits of the 128 bits length.

Public key generation time 0.019363815214316526

The signature time 0.0007876256585058229

Verification time 0.011725638324516278

We have got very good efficiency results. It must be emphasized, that they are almost the same, if we compare them with the classical Merkle scheme implementation.

7 Conclusions

We have received efficient and secure digital signature scheme, which is secure against quantum computers attacks. As it is not efficient to store all the signature keys, our scheme does not store all of them. It only stores the short seed of the time of arrival generator, which is a secure physical quantum random number generator and is the representative of optical quantum random number generators.

This seed is used for HASH_DBRG pseudo random number generator, which uses quantum secure hash function. This pseudo random number generator is efficient and secure. This PRNG is the NIST standard.

Our practical results show that the scheme is rather efficient. The results also show that speed of the algorithm implementation of the scheme is rather good and it is almost the same as in the case of the classical algorithm representation.

The integration of OQRNG does not affect the efficiency, because it is used only for generating the seed for HASH_DBRG.

As we can see, our approach saves the capacity of the scheme and does not affect the speed. It must be also mentioned that our approach increases the security of the scheme and the scheme can be used as a post-quantum security tool.

It would be very interesting to work on increasing the efficiency of quantum random number generator in the future works.

Acknowledgement. This work was carried out as a part of PHDF-19-519 and the grant financed by Caucasus University as well as part of Young Scientist Project № 0117U006770 of the Ministry of Education and Science of Ukraine.

References

1. Ajtai M.: Generating hard instances of lattice problems. In: Complexity of Computations and Proofs, Quad. Mat., vol. 13, pp. 1–32. Dept. Math. Seconda Univ. Napoli, Caserta (2004). Preliminary version in STOC 1996

2. Babai, L.: On Lovász lattice reduction and the nearest lattice point problem. Combinatorica **6**, 1–13 (1986)
3. Goldreich, O., Goldwasser, S., Halevi, S.: Collision-free hashing from lattice problems. Technical report TR96-056, Electronic Colloquium on Computational Complexity (ECCC) (1996)
4. Buchmann, J., García, L.C.C., Dahmen, E., Döring, M., Klintsevich, E.: CMSS – an improved merkle signature scheme. In: Barua, R., Lange, T. (eds.) Progress in Cryptology – INDOCRYPT 2006. Lecture Notes in Computer Science, vol. 4329. Springer, Heidelberg (2006)
5. Gagnidze, A., Iavich, M., Iashvili, G.: Improvement of hash based digital signature. In: CEUR Workshop Proceedings (2018)
6. Buchmann, J., Dahme, E., Schneider, M.: Merkle tree traversal revisited. In: 2nd International Workshop on Post-Quantum Cryptography, LNCS, vol. 5299, pp. 63–77. Springer, Heidelberg (2008)
7. Iavich, M., Gagnidze, A., Iashvili, G., Gnatyuk, S., Vialkova, V.: Lattice based Merkle. In: CEUR Workshop Proceedings, vol. 2470, pp. 13–16 (2019)
8. Gagnidze, A., Iavich, M., Iashvili, G.: Analysis of post quantum cryptography use in practice. Bull. Georgian Natl. Acad. Sci. **11**(2), 29–36 (2017)
9. Buchmann, J., Coronado, C., Dahmen, E., Döring, M., Klintsevich, E.: CMSS – an improved Merkle signature scheme. In: Progress in Cryptology – INDOCRYPT 2006, LNCS, vol. 4329, pp. 349–363. Springer, Heidelberg (2006)
10. Gagnidze, A., Iavich, M., Iashvili, G.: Novel version of merkle cryptosystem. Bull. Georgian Natl. Acad. Sci. **11**(4), 28–33 (2017)
11. Tynymbayev, S., Gnatyuk, S.A., Aitkhozhayeva, Y.Z., Berdibayev, R.S., Namazbayev, T. A.: Modular reduction based on the divider by blocking negative remainders. In: News of the National Academy of Sciences of the Republic of Kazakhstan, Series of Geology and Technical Sciences, vol. 2, no. 434, pp. 238–248 (2019)
12. Elloá, B.G., Francisco, M.A., Bernardo Jr, L.: Quantum permanent compromise attack to blum-micali pseudorandom generator. In: The 7th International Telecommunications Symposium (ITS 2010) (2010)
13. Yao, A.C.: Protocols for secure computations. In: Proceedings of SFCS 1982 Proceedings of the 23rd Annual Symposium on Foundations of Computer Science, pp. 160–164, 03–05 November 1982
14. Matsumoto, M., Nishimura, T.: Mersenne twister: a 623-dimensionally equidistributed uniform pseudo-random number generator. ACM Trans. Model. Comput. Simul. (TOMACS) **8**(1), 3–30 (1998). Special issue on uniform random number generation TOMACS Homepage archive
15. Isida, M., Ikeda, H.: Random number generator. Ann. Inst. Statist. Math. (Tokyo) **8**, 119–126 (1956)
16. Gnatyuk, S., Kovtun, M., Kovtun, V., Okhrimenko, A.: Development of a search method of birationally equivalent binary Edwards curves for binary Weierstrass curves from DSTU 4145-2002. In: Proceedings of 2nd International Scientific-Practical Conference on the Problems of Infocommunications. Science and Technology (PIC S&T 2015), Kharkiv, Ukraine, 13–15 October 2015, pp. 5–8 (2015)
17. Gaeini, A., Mirghadri, A., Jandaghi, G., Keshavarzi, B.: Comparing some pseudo-random number generators and cryptography algorithms using a general evaluation pattern. Int. J. Inf. Technol. Comput. Sci. (IJITCS) **8**(9), 25–31 (2016)

18. Dychka, I., Tereikovskyi, I., Tereikovska, L., Pogorelov, V., Mussiraliyeva, S.: Deobfuscation of computer virus malware code with value state dependence graph. Adv. Intell. Syst. Comput. **754**, 370–379 (2018)
19. Gupta, L., Garg, H., Samad, A.: An improved DNA based security model using reduced cipher text technique. Int. J. Comput. Netw. Inf. Secur. (IJCNIS) **11**(7), 13–20 (2019)
20. Dawood, O.A., Rahma, A.M., Hossen, A.M.: The new block cipher design (Tigris Cipher). Int. J. Comput. Netw. Inf. Secur. (IJCNIS) **7**(12), 10–18 (2015)

Improved Post-quantum Merkle Algorithm Based on Threads

Maksim Iavich[1], Sergiy Gnatyuk[3](✉), Arturo Arakelian[2], Giorgi Iashvili[1], Yuliia Polishchuk[3], and Dmytro Prysiazhnyy[3]

[1] Scientific Cyber Security Association, Caucasus University, Tbilisi, Georgia
m.iavich@scsa.ge
[2] University of Georgia, Tbilisi, Georgia
[3] National Aviation University, Kyiv, Ukraine
s.gnatyuk@nau.edu.ua, d.prysiazhnyy@gmail.com

Abstract. Today scientists are actively working on the creation of quantum computers. Quantum computers will be able to solve the problem of factoring the large numbers. So, quantum computers can break the crypto system RSA, which is used in many products. Hash based digital signatures are the alternative to RSA. These systems use cryptographic hash function. The security of these systems depends on the resistance to collisions of the hash functions that they use. The paper analyzes hash based digital signature schemes. It analyzes the improvements of the scheme. Fractal Merkle algorithm is also analyzed. This algorithm can be considered as the static one, because it does not depend on the number of the threads in CPU. Authors have offered the post-quantum algorithm, which uses the threads of CPU in the parallel mode. The mathematical model of this algorithm and the pseudo code of its implementation are offered in the paper. This algorithm was analyzed and is shown that this algorithm provides rather good speed up, and can be implement to provide post-quantum security in modern information and communication systems.

Keywords: Information security · Cryptography · Post-quantum era · Hash function · Collision · Merkle algorithm · CPU

1 Introduction

Nowadays there is active work on the development of quantum computers. These computers, by means of Shor's algorithm (for example) will be able to crack easily systems that are based on factorization of integers. Accordingly, the most common cryptosystem RSA, that uses public key, will be vulnerable [1].

1.1 Related Literature Review

From viewpoint of quantum computing RSA and other public cryptography algorithms are vulnerable [1–3] and they cannot ensure high security level in modern and future information-communication systems [4]. Alternatives of RSA for post-quantum era are hash-based cryptosystems [4]. Their safety is based on safety of hash functions and it is

Z. Hu et al. (Eds.): ICCSEEA 2020, AISC 1247, pp. 454–464, 2021.
https://doi.org/10.1007/978-3-030-55506-1_41

not dependable on computing ability of intruder. Hash-based one-time signatures were suggested in [5, 6].

1.2 Lamport-Diffie One-Time Signature Scheme

Lamport-Diffie one-time signature scheme was offered [7]. For the signature key *X* in this system, 2n random lines of size n are generated.

$$X = (x_{n-1}[0],\ x_{n-1}[1], \ldots, x_0[0],\ x_0[1]) \in \{0,1\}^{n,2n} \tag{1}$$

Verification key:

$$Y = (y_{n-1}[0], y_{n-1}[1], \ldots, y_0[0], y_0[1]) \in \{0,1\}^{n,2n}. \tag{2}$$

It is calculated as following:

$$Y_i[j] = f(x_i[j]),\ 0 <= i <= n-1,\ j = 0, 1, \tag{3}$$

where *f* is one way function:

$$f : \{0,1\}^n \rightarrow \{0,1\}^n. \tag{4}$$

Signature of the Message
To sign the message m, we hash:

$$h(m) = hash = (hash_{n-1}, \ldots, hash_0), \tag{5}$$

where h is a cryptographic hash function:

$$h : \{0,1\}^* \rightarrow \{0,1\}^n. \tag{6}$$

The signature is calculated as following:

$$Sig = (x_{n-1}[hash_{n-1}], \ldots, x_0[hash_0]) \in \{0,1\}^{n,n}. \tag{7}$$

The size of the signature is n^2.

Message Verification
To verify the signature sig the message is hashed:

$$hash = (hash_{n-1}, \ldots, hash_0). \tag{8}$$

After that the following equality is verified:

$$(f(sig_{n-1}), \ldots,\ f(sig_0)) = (y_{n-1}[hash_{n-1}], \ldots, y_0[hash_0]). \tag{9}$$

If the equation is true, then the signature is correct.

1.3 Winternitz One-Time Signature Scheme

In the Lamport scheme key generation and signature generation are efficient, but the signature size is equal to n2.

To reduce it we use Winternitz one-time signature scheme [7]. In this scheme several bits of the hashed message are simultaneously signed by one line of the key.

The Winternitz parameter is the number of bits of the hashed message that will be signed simultaneously. It is chosen as w >= 2.

After that we calculate:

$$p_1 = n/w$$

$$p_2 = (\log_2 p_1 + 1 + w)/w,\ p = p_1 + p_2. \tag{10}$$

We generate randomly the signature keys:

$$X = (x_{p-1}[0], \ldots, x_0) \in \{0,1\}^{n,p}. \tag{11}$$

We compute the verification key:

$$Y = \left(y_p - 1[0], \ldots, y_0\right) \in \{0,1\}^{n,p}, \tag{12}$$

where

$$y_i = f^{2^\wedge w-1}(x_i), 0 <= i <= p-1. \tag{13}$$

Signature of the Message

The length of the signature and the verification key is equal to **np** bits, one-way function *f* is used $p(2^w - 1)$ times.

To sign the message it is hashed: hash = h(m). The minimum number of zeros is added to the hash, so that the hash would be a multiple of w. Afterwards it is divided into p_1 parts of size w.

$$\text{hash} = k_{p-1}, \ldots, k_{p-p1}. \tag{14}$$

The checksum is following:

$$c = \sum\nolimits_{i=p} -p1^{p-1}(2^w - ki). \tag{15}$$

As c <= $p12^w$, the length of its binary representation is less than

$$\log 2\ p_1 2^w + 1 \tag{16}$$

We add to this binary representation the minimum number of zeros, so that it would be a multiple of w, and divide it into p2 parts of the length w.

$$c = k_{p2-1}, \ldots, k_0, \tag{17}$$

And the message signature is calculated as follows:

$$sig = (f^\wedge k_p - 1(x_{p-1}), \ldots, f^\wedge k_0(x_0)). \tag{18}$$

The size of the signature is equal to p_n.

Signature Verification
To verify the signature sig = (sig_{n-1}, …, sig_0) bit string k_{p-1},…, k_0 are calculated. Then the following equality is verified:

$$(f^\wedge(2^w - 1 - k_{p-1}))(sig_{n-1}), \ldots, (f^\wedge(2^w - 1 - k_0))(sig_0) = y_{n-1}, \ldots y_0. \tag{19}$$

Such systems are ineffective, because of transferring public key many times. Merkle cryptosystem eliminates this problem. In this system one public key can be used many times to encrypt many different messages.

2 Merkle Crypto-System

The Merkle signature scheme allows you to sign multiple messages with the same key [8, 9]. This system uses one-time signatures and a binary tree whose root is a public key.

Key Generation
The size of the tree must be H >= 2 and using one public key 2H document can be signed. Are generated signature and verification keys; Xi, Yi, 0 <= i <= 2H. Xi- is the signature key, Yi- is the verification key. To get the leaves of the tree, signature keys are hashed using the following hash function:

$$h : \{0, 1\}^* \rightarrow \{0, 1\}n \tag{20}$$

To get the parent node, the concatenation of two previous nodes is hashed a[i, j] are the nodes of the tree:

$$a[1, 0] = h(|a[0, 0] \parallel a[0, 1]). \tag{21}$$

The root of the tree is the public key of the signature - pub.

Message Signature

To sign a message of any size, it is transformed to size of n by means of hashing, h (m) = hash, to sign the message, is used an arbitrary one-time key X_{any}, and the signature is a concertation of: one-time signature, one-time verification key, index of a key and all brother nodes according to the selected arbitrary key with the index "any".

$$\text{Signature} = \left(\text{sig}||\text{any}\left|\right|Y_{\text{any}}||\text{auth}_0, \ldots, \text{auth}_{H-1}\right). \quad (22)$$

Signature Verification

To verify the signature, an one-time signature is checked using the selected verification key, if the verification has passed, all the a[i, j] are calculated using "auth", index "any" and Y_{any}. If the root of the tree matches the public key, then the signature is correct.

Here we offer the following pseudo code of its initial implementation:

2.1 Pseudo Code

1. Importing necessary libs
2. Define class
3. Defining "alt_hashes(hashes)" method
4. Set list "arr"
5. If hashes == " ", raise Exception
6. Foreach loop
   6.1. Sorting hashes and appending into arr
7. Length_of_block == length of arr
8. While loop, if length is odd, copy last element in list
   8.1. A = Append it into arr list
9. Set list "another_arr"
10. Foreach loop
    10.1. For loop with range from 0 to length of "arr" and iteration by 2
        10.1.1. Define variable with "sha512()" value
        10.1.2. Hash elements that are in "arr" list
        10.1.3. Append them into new "another_arr" list
        10.1.4. Return this list in hex
11. Set list "hash_arr"
12. Foreach loop
    12.1. Generate Hex and put it into "hash_arr" list
13. Create message put it in "st" variable
14. Convert "st" value in binary
15. First_secret_key = hash_arr[0]
16. Second_secret_key = hash_arr[1]
17. Generate "one-time signature"
    17.1. If st == 0
        17.1.1. Choose "First_secret_key" bit
    17.2. Else
        17.2.1. Choose "Second _secret_key" bit

18. First_pub_key = hash(hash_arr[0])
19. Second_pub_key = hash (hash_arr[1])
20. Encryption
    20.1. Concatenate "one-time signature" with message's hash
21. Verification of "one-time signature"
    21.1. If bit of "one-time signature" == 0
        21.1.1. Compare with "First_secret_key" bit
    21.2. Else
        21.2.1. Compare with "Second _secret_key" bit
22. Verification of "signature"
    22.1. Concatenate siblings with each other
    22.2. If this equals to public key
        22.2.1. Sign is correct
    22.3. Else
        22.3.1. Sign is not correct

In order to improve the efficiency of this cryptosystem Fractal Merkle algorithm was offered [10–12].

2.2 Fractal Merkle Algorithm

Starting with a Merkle tree of height S, further notation to deal with subtrees was introduced. First a subtree with height s < S must be chosen. We let the altitude of a node n in Tree be the length of the path from n to a leaf of Tree (therefore, the altitude of a leaf of Tree is zero). Consider a node n with altitude at least s. We define the h-subtree at n to be the unique subtree in Tree which has n as its root and which has height s. For simplicity in the suite, we assume s is a divisor of S, and let the ratio, R = S/s, be the number of levels of subtrees. We say that an s-subtree at n is "at level j" when it has altitude js for some j ∈ {1, 2, … S}. For each j, there are 2S−ih such s-subtrees at level j. We say that a series of s-subtrees Tree j (j = 1 … R) is a stacked series of s-subtrees, if for all j < R the root of Tree j is a leaf of Tree j + 1 [13–15]. We illustrate the subtree notation and provide a visualization of a stacked series of s-subtrees in Fig. 1.

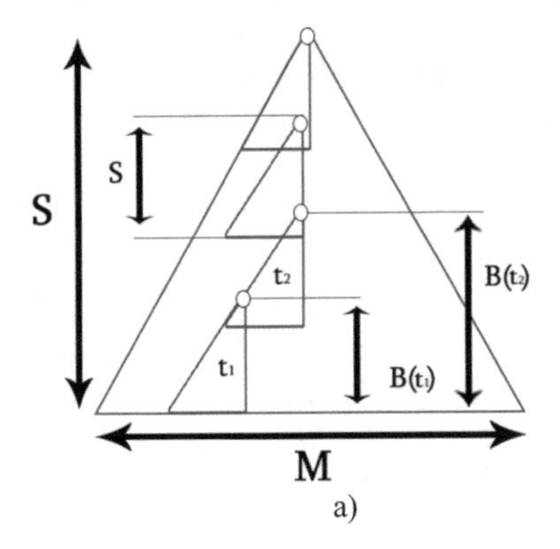

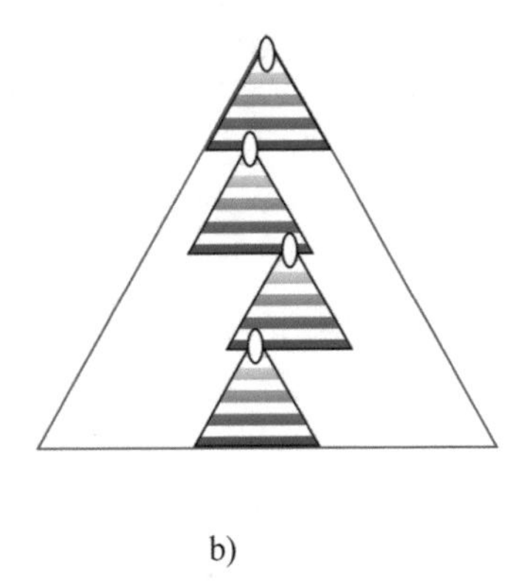

Fig. 1. The subtree notation (a) and a visualization of a stacked series of s-subtrees (b)

Figure 1 a) The height of the Merkle tree is S, and thus, the number of leaves is M = 2S. The height of each subtree is s. The altitude B(t1) and B(t2) of the subtrees t1 and t2 is marked.

Figure 1 b) Instead of storing all tree nodes, we store a smaller set - those within the stacked subtrees. The leaf whose pre-image will be output next is contained in the lowest-most subtree; the entire authentication path is contained in the stacked set of subtrees [16].

3 Algorithm Based on Threads

The offered algorithm can be considered as the static one, because it does not depend on the number of the threads in CPU. We offer the algorithm, which uses the threads of CPU in the parallel mode.

Key Generation

The size of the tree must be H >= 2 and using one public key 2H document can be signed. Are generated signature and verification keys; Xi, Yi, 0 <= i <= 2H. Xi- is the signature key, Yi- is the verification key. To get the leaves of the tree, signature keys are hashed using the hash function h:{0,1}*→{0,1}n.

To get the parent node, the concatenation of two previous nodes is hashed; a[i, j] are the nodes of the tree;

$$a[1,0] = h(a[0,0] \| a[0,1]). \tag{23}$$

The tree is divided into quantity of threads in CPU. In loop, which length equals to threads' quantity, we are calculating parent nodes. Let's say we have "t" threads and "d" nodes. The quantity of nodes "d" is divided on quantity of threads "t", "d/t". Evert d/t nodes are passed to the separate thread. The parent nodes are received by means of (23), so we receive t sets of parent nodes. These sets are concatenated and afterwards divided into t more sets. We continue this process until we get to the root of the tree.

The root of the tree is the public key of the signature - pub.

Message Signature

To sign a message of any size, it is transformed to size of n by means of hashing: h (m) = hash, to sign the message, is used an arbitrary one-time key Xany, and the signature is a concertation of: one-time signature, one-time verification key, index of a key and all brother nodes according to the selected arbitrary key with the index "any":

$$\text{Signature} = (\text{sig}\|\text{any}\|\text{Y any}\|\text{auth0}, \ldots, \text{authH} - 1). \tag{24}$$

Signature Verification

To verify the signature, an one-time signature is checked using the selected verification key, if the verification has passed, all the a[i, j] are calculated using "auth", index "any" and Yany. If the root of the tree matches the public key, then the signature is correct.

Here we offer the pseudo code of its initial implementation.

3.1 Thread Pseudo Code

1. Importing necessary libs
2. Set queue "q"
3. Define class
4. Defining "alt_hashes(hashes)" method
5. Set list "arr"
6. If hashes == " ", raise Exception
7. Foreach loop
    7.1. sorting hashes and appending into arr
8. Length_of_block == length of arr
9. While loop, if length is odd, copy last element in list
    9.1. append it into arr list
10. Set list "another_arr"
11. Foreach loop
    11.1. For loop with range from 0 to length of "arr" and iteration by 2
        11.1.1. Define variable with "sha512()" value
        11.1.2. Hash elements that are in "arr" list
        11.1.3. Apennd them into new "another_arr" list
        11.1.4. put hex into q
12. Set list "hash_arr"
13. Foreach loop
    13.1. Generate Hex and put it into "hash_arr" list
14. Create message put it in "st" variable
15. Convert " st" value in binary
16. hash_arr3 equals to hash_arr
17. length_of_arr equals to the length of hash_arr list
18. add first element of hash_arr to auth_list
19. While loop, Length_of_block > 1
    19.1. If length_of_arr! = 2
        19.1.1. halfLength equals to length_of_arr/2
        19.1.2. Define j equals to 0
        19.1.3. Define new list hash_arr3
        19.1.4. For loop which range is 2
            19.1.4.1. Define new list new_hash_arr
            19.1.4.2. Call method in thread
            19.1.4.3. j equals to halfLength
            19.1.4.4. halfLength equals to length_of_arr
            19.1.4.5. join all threads
            19.1.4.6. append result from queue to hash_arr3
        19.1.5. reset thread_hash_arr list
        19.1.6. For loop which range is length of hash_arr3
            19.1.6.1. concatenate hash_arr3's elements
        19.1.7. Append them to auth_list
        19.1.8. Reset hash_arr3
        19.1.9. Reset queue

19.1.10. length_of_arr equals to length_of_arr/ 2
19.2. else
19.2.1. call thread_method
19.2.2. reset thread_hash_arr
19.2.3. reset queue
19.2.4. set value of thread_hash_arr[0] to thread_pub variable
20. First_secret_key = hash_arr[0]
21. Second_secret_key = hash_arr[1]
22. Generate "one-time signature"
22.1. If st == 0
22.1.1. Choose "First_secret_key" bit
22.2. Else
22.2.1. Choose "Second _secret_key" bit
23. First_pub_key = hash(hash_arr[0])
24. Second_pub_key = hash (hash_arr[1])
25. Encryption
25.1. Concatenate "one-time signature" with message's hash
26. Verification of "one-time signature"
26.1. If bit of "one-time signature" == 0
26.1.1. Compare with "First_secret_key" bit
26.2. Else
26.2.1. Compare with "Second _secret_key" bit
27. Verification of "signature"
27.1. Concatenate siblings with each other
27.2. If this equals to public key
27.2.1. Sign is correct
27.3. Else
27.3.1. Sign is not correct

4 Results Analysis and Discussion

During experimental study we have compared the implementation of threads algorithm with a classic one. The test was performed on PC with two thread CPU. The length of the message was 128 bits.

The classical algorithm results:
Key generation time – 0.049351,
Signature time – 0.0002425,
Verification time – 0.0038651.

Proposed algorithm with threads:
Key generation time – 0.013841,
Signature time – 0.0002425,
Verification time – 0.0038651.

In these experimental conditions, we can see that proposed algorithm is faster than classical one in 3.57 times from viewpoint of key generation.

5 Conclusions

After results analyzing we can declare that our proposed algorithm with threads gives a good speed up in the key generation stage (in comparison with existed algorithm).

Proposed algorithm is very relevant, because nowadays CPU's are very fast and dynamic. Their development is going very fast. This algorithm is dynamic; its speed depends on the CPU and thread quantity.

Future research work can be related to improving proposed algorithm from view-point of security and constructing post-quantum encryption system.

Acknowledgement. This work was carried out as a part of PHDF-19-519 and the grant financed by Caucasus University as well as part of Young Scientist Project № 0117U006770 of the Ministry of Education and Science of Ukraine.

References

1. Iavich, M., Gagnidze, A., Iashvili, G., Gnatyuk, S., Vialkova, V.: Lattice based Merkle. In: CEUR Workshop Proceedings, vol. 2470, pp. 13–16 (2019)
2. Gagnidze, A., Iavich, M., Iashvili, G.: Improvement of hash based digital signature. In: CEUR Workshop Proceedings (CEUR-WS.org) (2018)
3. Buchmann, J., Dahmen, E., Schneider, M.: Merkle tree traversal revisited. In: 2nd International Workshop on Post-Quantum Cryptography - PQCrypto 2008, LNCS, vol. 5299, pp. 63–77. Springer, Heidelberg (2008)
4. Gagnidze, A.G., Iavich, M.P., Iashvili, G.U.: Analysis of post quantum cryptography use in practice. Bull. Georgian Natl. Acad. Sci. **11**(2), 29–36 (2017)
5. Ajtai, M.: Generating hard instances of lattice problems. In: Complexity of computations and proofs, Quad. Mat., vol. 13, pp. 1–32. Dept. Math., Seconda Univ. Napoli, Caserta (2004). Preliminary version in STOC 1996
6. Babai, L.: On Lovász lattice reduction and the nearest lattice point problem. Combinatorica **6**, 1–13 (1986)
7. Goldreich, O., Goldwasser, S., Halevi, S.: Collision-free hashing from lattice problems. Technical report TR96-056, Electronic Colloquium on Computational Complexity (ECCC) (1996)
8. Coron, J.S., Dodis, Y., Malinaud, C., Puniya, P.: Merkle-Damgård revisited: how to construct a hash function. In: Shoup, V. (eds.) Advances in Cryptology – CRYPTO 2005. Lecture Notes in Computer Science, vol. 3621. Springer, Heidelberg (2005)
9. Li, H., Lu, R., Zhou, L., Yang, B., Shen, X.: An Efficient Merkle-Tree-Based Authentication Scheme for Smart Grid. IEEE (2013)
10. Jakobsson, M., Leighton, T., Micali, S., Szydlo, M.: Fractal Merkle tree representation and traversal. In: Joye, M. (ed.) Topics in Cryptology – CT-RSA 2003. CT-RSA 2003. Lecture Notes in Computer Science, vol. 2612. Springer, Heidelberg (2003)
11. Buchmann, J., Dahmen, E., Schneider, M.: Merkle tree traversal revisited. In: Buchmann, J., Ding, J. (eds.) Post-Quantum Cryptography. PQCrypto 2008. Lecture Notes in Computer Science, vol. 5299. Springer, Heidelberg (2008)
12. Hu, Z., Gnatyuk, S., Okhrimenko, T., Kinzeryavyy, V., Iavich, M., Yubuzova, Kh.: High-speed privacy amplification method for deterministic quantum cryptography protocols using pairs of entangled qutrits. In: CEUR Workshop Proceedings, vol. 2393, pp. 810–821 (2019)

13. Hu, Z., Gnatyuk, S., Kovtun, M., Seilova, N.: Method of searching birationally equivalent Edwards curves over binary fields. In: Advances in Intelligent Systems and Computing, vol. 754, pp. 309–319 (2019)
14. Gaeini, A., Mirghadri, A., Jandaghi, G., Keshavarzi, B.: Comparing some pseudo-random number generators and cryptography algorithms using a general evaluation pattern. Int. J. Inf. Technol. Comput. Sci. (IJITCS) **8**(9), 25–31 (2016)
15. Dawood, O.A., Rahma, A.M., Hossen, A.M.: The new block cipher design (Tigris Cipher). Int. J. Comput. Netw. Inf. Secur. (IJCNIS) **7**(12), 10–18 (2015)
16. Gupta, L., Garg, H., Samad, A.: An improved DNA based security model using reduced cipher text technique. Int. J. Comput. Netw. Inf. Secur. (IJCNIS) **11**(7), 13–20 (2019)

Optimization of Stock Basing on Improved Grey Prediction Model: A Case Study on Garment Supply Chain

Jinshan Dai, Yao Zhang, Mengya Zhang(✉), and Qingying Zhang

School of Logistics Engineering,
Wuhan University of Technology, Wuhan 430063, China
Jinshan.dai@whut.edu.cn, kmno4031l@163.com

Abstract. Grey prediction model is the core of grey system theory. It has been widely used because of the lack of strict requirements and restrictions on data. However, when the general grey prediction model is be applied to predict the random oscillation sequence, the accuracy is not ideal. For this reason, this paper selects a small sample with a large amplitude of oscillation characteristics of the initial product demand on the market for research. The grey prediction method is be improved innovatively, and the envelope curve of grey oscillation interval can be used to predict the value interval of demand, then to determine the partition function of demand. Basing on the actual operating data from garments enterprise, the author applied the improved grey prediction method and used the Monte Carlo method to simulate the demanding amount for next three cycles. It makes the setting and optimization of safety stock more accurate and effective data support, reduces the cost redundancy caused by information uncertainty, and improves the service level.

Keywords: Grey prediction model · Optimization of safety stock · Random oscillation sequence · Inventory strategy

1 Introduction

Inventory strategy refers to the decision-making behavior that enterprises hold a certain amount of inventory in order to make the supply chain system run stably. Although ensuring the implementation of inventory strategy will lead to the rise of inventory cost, but from the perspective of the operation of the whole supply chain is economic and effective. The main content of inventory strategy is the setting of safety stock, which depends on supply-and-demand, production-and-transportation lead-time, product value, and service-level [1]. The determination of the inventory cache value should been placed on the historical sales data to calculate the theoretical safety stock, and then optimize the adjustment according to the actual situation of the company [2, 3].

The setting and optimization of safety stock need to go through the following steps: Goal Planning - Data Collection - benchmark model construction - demand analysis and prediction - safety stock setting - simulation and comparison.

Z. Hu et al. (Eds.): ICCSEEA 2020, AISC 1247, pp. 465–474, 2021.
https://doi.org/10.1007/978-3-030-55506-1_42

In the sale process of supply chain for finished goods, the setting of safety stock is been generated for many reasons such as the scale effect of batch purchase, lead-time of goods preparation, uncertainty of sales volume, imbalance of supply and demand, which make it difficult to reach the balance of supply and demand [4]. At the same time, for the initial products put on the market, the demand data is a small sample time series, and the data has a large range of shocks due to the uncertainty of information. General data analysis methods are often insufficient to support the analysis and prediction of demand data in the early stage of safety stock setting [5], which leads to the error of safety stock setting, the redundancy of cost and the reduction of service level.

2 Methodology for Determination of the Demand Distribution Function Basing on Improved Grey Prediction Model

2.1 Methodology for Determination of the Demand Distribution Function

Grey prediction model is the core of grey system theory [6, 7]. It mainly aims at the grey uncertainty prediction problems existing in the real world. It uses a small amount of effective data and grey uncertainty data to reveal the future development trend of related systems through the accumulation of sequences [8]. Because the grey prediction model has no strict requirements and restrictions on data, it has been widely used. For the products put on the market at the initial stage, the demand data is a small sample time series, and the data has a large range of shocks due to the uncertainty of information. For small sample data, gray prediction method is often been used to improve the accuracy of prediction. However, when the conventional grey prediction model is been used to predict the random oscillation sequence, the accuracy is not ideal [9, 10].

The reason is that the demand of products put on the market at the initial stage is random oscillation, while the equivalent grey prediction model is usually in the form of index, which is monotonous. The simulation data calculated by the index expression does not conform to the characteristics of random oscillation of the demand of products put on the market at the initial stage. Therefore, this paper innovatively models and forecasts from the perspective of "scope". Based on the data processing of the model, it uses the shock sequence interval grey prediction modeling method of envelope curve to determine the demand distribution function, and then determines the safety inventory strategy based on the demand distribution function and other parameters [11]. Through the improved grey prediction model, the value range of demand is provided [12]. It can be considered that the demand follows the uniform distribution within the value range (the demand takes an integer).

2.2 Data Pre-processing Basing GM (1, 1)

Step 1: Pre-Processing of Data
To weaken the stochasticity of the original time series, the original time series needs to be data processed before building the grey model, and the data processed time series is generation sequence.

① Suppose that $X^{(0)} = \{X^{(0)}(1), X^{(0)}(2), X^{(0)}(3), \ldots X^{(0)}(n)\}$ is the original data of an index which needs to be forecast,and the stepwise ratio of the calculating sequence $\lambda(t) = \frac{X^{(0)}(t-1)}{X^{(0)}(t)}, t = 2, 3, \cdots, n$. If the majority of the stepwise ratios fall in the coverable area, then we can build the *GM* (1, 1) model use it to grey forecast [9].

Otherwise, we need to pre-process the data appropriately. The processing methods could be root extraction, taking logarithm, and smoothing. The smoothing design of the pre-processed date is three-point smoothing, and the data could been processed as follows:

$$X^{(0)}(t) = \left[X^{(0)}(t-1) + 2X^{(0)}(t) + X^{(0)}(t+1)\right]/4 \tag{1}$$

$$X^{(0)}(1) = \left[3X^{(0)}(1) + X^{(0)}(2)\right]/4 \tag{2}$$

$$X^{(0)}(n) = \left[X^{(0)}(n-1) + 3X^{(0)}(n)\right]/4 \tag{3}$$

② After pre-processing, we operated an accumulative generating processing to the data, i.e. Taking the first data of the original sequence as the first date of the generation sequence, and adding the second data of the original sequence to its first data, as the second data of the generation sequence. Following this rule, we could obtain the generation sequence:

According to $X^{(1)}(k) = \sum_{n=1}^{k} X^{(0)}(n)$, we obtained a new sequence as follows:

$$X^{(1)} = \left\{X^{(1)}(1), X^{(1)}(2), X^{(1)}(3), \ldots X^{(1)}(n)\right\} \tag{4}$$

Comparing with the original sequence, the stochasticity of the new sequence is been largely decreased, and its smoothness increased largely.

Step 2: Differential Equation for Trends
The trends of new sequence could been approximately expressed by differential equation as [11]:

$$\frac{dX^{(1)}}{dt} + aX^{(1)} = u \tag{5}$$

In the equation: a is developing grey number; u is the endogenous controlling grey number.

Step 3: Solving of the Differential Equation
Let $Y_n = [X^{(0)}(2), X^{(0)}(3), \ldots, X^{(0)}(n)]^T$, $\hat{\alpha}$ the vector of the parameter left for evaluating.

$$B = \begin{bmatrix} -\frac{1}{2}(X^{(1)}(2) + X^{(1)}(2)) & 1 \\ -\frac{1}{2}(X^{(1)}(2) + X^{(1)}(3)) & 1 \\ \cdots & \cdots \\ -\frac{1}{2}(X^{(1)}(n-1) + X^{(1)}(n)) & 1 \end{bmatrix}, \tag{6}$$

Therefore, the model could be expressed as $Y_n = B\hat{\alpha}$ and by least square method. It can been obtained that $\hat{\alpha} = (B^T B)^{-1} B^T Y_n$. By solving the differential equation, the discrete time response functions of the grey forecast could obtain.

$$\hat{X}^{(1)}(t+1) = \left[X^{(0)}(1) - \frac{u}{a}\right] e^{-at} + \frac{u}{a},\ t = 0, 1, 2\ldots, n-1 \tag{7}$$

$\hat{X}^{(1)}(t+1)$ is the accumulated forecast value obtained, and the following equation can be obtained by reducing the forecast value:

$$\hat{X}^{(0)}(t+1) = \hat{X}^{(1)}(t+1) - \hat{X}^{(1)}(t) \tag{8}$$

Note: If the data has been pre-processed, it need to been converted with the correspondent methods to obtain the actual forecast value.

2.3 Improving Method to Grey Forecast of the Envelope Curve Based on the Oscillatory Area

To the data of time series with a few samples, the uncertainty of the information would cause a large amplitude of oscillation [13]. Therefore, in order to increase the validity of the forecasting, the forecasting model takes the fixed value from grey system. The traditional target should been applied to build model and forecast from the perspective of "area", and the procedures are as follows:

Step1: The Determination of the Oscillatory Sequence Envelope Curve
Calculating and obtaining the envelope curve functions from both upper bound and lower bound according to the designing principle of the oscillatory sequence envelope curve.

Step2: Expanding the Area of Oscillatory Sequence
Obtaining the upper bound sequence *S* and lower bound *U* by expanding the area of the oscillatory sequence *X*.

Step3: Grey Forecast Model of the Area of the Oscillatory Sequence
Building the grey forecast model of the area of sequenced *X(⊗)*, and then calculating the area with *S* as the adjacent grey layer. *W* is the ordinate of the neutrality line pivot point of the adjacent grey layer, as shown in the Fig. 1.

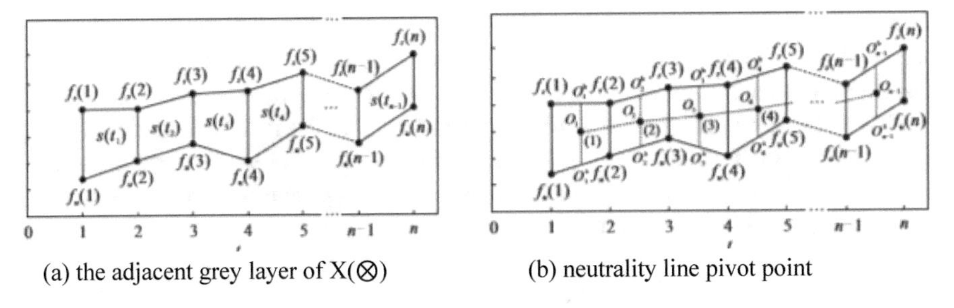

(a) the adjacent grey layer of X(⊗) (b) neutrality line pivot point

Fig. 1. Diagram of area grey number forecast model

The area sequence A and coordinates sequence W of sequence of area grey number (⊗) respectively are

$$A = (s(t_1), s(t_2), \ldots, s(t_{n-1})) \tag{9}$$

$$W = (w(t_1), w(t_2), \ldots, w(t_{n-1})), \text{ and,} \tag{10}$$

$$s(t_p) = \frac{[f_s(p) - f_u(p)] + [f_s(p+1) - f_u(p+1)]}{2}, p = 1, 2, \ldots, n-1$$

$$w(t_p) = \frac{[f_s(p) + f_u(p)] + [f_s(p+1) + f_u(p+1)]}{4}, p = 1, 2, \ldots, n-1$$

Building the GM(1,1) models of the area sequence **A** and coordinates sequence **W** separately, it can been obtained that

$$\hat{f}_u(k) = P_1 e^{-a_w(k-2)}(1 - P_2^{k-2}) - P_3 e^{-a_w(k-2)}(1 - P_4^{k-2}) + (-1)^k . \hat{f}_u(2) \tag{11}$$

$$P_1 = \frac{2(1 - e^{a_w})[w(t_1) - \frac{b_w}{a_w}]}{1 + e^{a_w}}, P_2 = -e^{a_w} \; P_3 = \frac{(1 - e^{a_s})[s(t_1) - \frac{b_s}{a_s}]}{1 + e^{a_s}}, P_2 = -e^{a_s}$$

The parameters 'a' and 'b' are the model parameters of the area sequence A and coordinates sequence of W respectively. $\hat{f}_u(k)$ and $\hat{f}_s(k)$ defined the areas of the value.

3 Inventory Strategy Setting Based on the Improving of Grey Forecast Model

After obtaining the probability distribution function of the demanding amount by using the improved grey forecast method, the safe inventory setting could be taken next [5, 14, 15].

According to the large scale of fluctuation of the demand to products put in the market in the initial period, the steady increasing trend of its selling process, and the relevant documents of inventory strategy setting, the (s, S) inventory model could been taken. For example, setting the weekly inventory limit as S, and if the inventory is less than s during the inventory checking, the relevant agent shall apply for products allocating, and adding products to the amount of S before the next cycle starts [13].

Due to the demanding amount to the products put in the market in initial period is integer with a small scale, the discrete (s, S) inventory model should been taken.

In the beginning of each cycle, if the inventory is Q, then the profit would be C_1 within the cycle. k is ordering cost.

$$C_1 = \sum_{0}^{Q} kr \cdot p(r) + \sum_{Q+1}^{\infty} kQ \cdot p(r) \tag{12}$$

Due to the fee c, charged for storing each unsold product within each cycle, the inventory cost within each cycle would be C_2.

$$C_2 = \sum_{0}^{Q} c \cdot (Q - r) \cdot p(r) \tag{13}$$

Meanwhile the cost for lacking product supplying is C_3.

Considering all the fore mentioned factors, the net interest would be C.

$$C = C_1 - C_2 - C_3 = \sum_{0}^{Q} [(k + c)r - cQ] \cdot p(r) + \sum_{Q+1}^{\infty} kQ \cdot p(r) - C_3. \tag{14}$$

When inventory amount is lower than s and reallocating to S, the net interest C(s) of next cycle would be:

$$\mathrm{C(S)} = \sum_{0}^{S} [(k + c)r - cS] \cdot p(r) + \sum_{S+1}^{\infty} kS \cdot p(r) - C_3 \tag{15}$$

On the contrary, keeping inventory amount as S, without reallocating products, and the net interest C(S) of next cycle would be:

$$\mathrm{C(s)} = \sum_{0}^{s} [(k + c)r - cs] \cdot p(r) + \sum_{s+1}^{\infty} ks \cdot p(r) \tag{16}$$

If the largest interest made from its terminal warehouses and retailing stores whose products are been delivered by the terminal warehouses, it shows that S and s need to meet the inequation:

$$\mathrm{C(s)} \geq \mathrm{C(S)}, \tag{17}$$

$$\text{i.e.} \sum_{0}^{s} [(k+c)r - cs] \cdot p(r) + \sum_{s+1}^{\infty} ks \cdot p(r) \geq \sum_{0}^{S} [(k+c)r - cS] \cdot p(r) + \sum_{S+1}^{\infty} kS \cdot p(r) - C_3 \tag{18}$$

$$\sum_{0}^{s} c(S-s) \cdot p(r) - \sum_{s+1}^{S} [k(r+s) + c(r-S)] \cdot p(r) + \sum_{S+1}^{\infty} k(s-S) \cdot p(r) + C_3 \geq 0 \tag{19}$$

According to these two inequations, the values of s and S of the inventory model (s, S) could be obtained.

4 Simulating Contrast of the Inventory Setting of Clothing Company Based on the Monte Carlo Method

An example of products put in the market in initial period of D clothing company is been taken at here, and the rationale of the fore mentioned safe inventory setting method will be verified by Monte Carlo method [16].

Given that a single cost of the clothing is 300 RMB, cost for storing each clothing is 36 RMB, loss for lacking supplying of each single clothing is 180RMB, and the ordering fee is not considered. Suppose that in one of those warehouses, its products delivering amount for the previous 16 cycles respectively are in Table 1. The small scale of samples and large scale of amplitude of this sequence is applicable to the mentioned safe inventory setting method.

Table 1. The Sale Information of Company X

Month	7	8	9	10	11	12	1	2
2017	244	205	269	248	212	166	243	244
2018	132	216	175	203	220	174	159	131

At first, according to the improved grey forecasting model, we obtain the upper bound function S, and lower bound function U as follows:

$$S = (187, 183, 179, 175, 170, 166, 162, 158, 153, 149, 145, 141, 136, 132, 128, 124)$$

$$U = (280, 275, 269, 263, 258, 253, 247, 242, 236, 231, 225, 220, 214, 209, 203, 198)$$

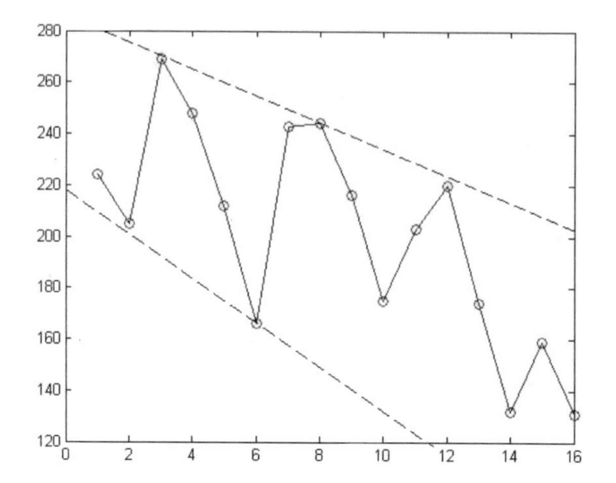

Fig. 2. Schematic diagram of grey oscillation envelope curve

Relevant parameters are

$$P_1 = 89.73,\ P_2 = -0.02,\ P_3 = 234.13,\ P_4 = -1.03$$

$$a_w = 0.15,\ b_w = 93.220,\ a_s = 0.025,\ b_s = 236.440$$

According to the Fig. 2. The demand value areas for next three cycles are obtained like $[24, 109]$, $[29, 102]$, $[41, 113]$.

It is regarded as the demanding amount for each cycle is corresponding to the evenly distributed area, then the (s, S) inventory strategy is taken to get $s = 62, S = 95$.

By using the Monte Carlo method to simulate the demanding amount for next three cycles, and calculating the total profit respectively in accuracy to the hundreds. Finally, the total profit of a normal inventory strategy obtained is 48,300 RMB, and that of the improved inventory strategy is 62,700 RMB, with an increasing rate of 29.8%.

5 Conclusions

This paper selects a small sample with a large amplitude of oscillation characteristics of the initial product demand on the market for research. The grey prediction method is been improved innovatively, and the envelope curve of grey oscillation interval is used to predict the value interval of demand, then to determine the partition function of demand. It makes the setting and optimization of safety stock more accurate and effective data support, reduces the cost redundancy caused by information uncertainty, and improves the service level.

Applying the improved grey prediction method basing on the actual operating data from garments enterprise, and using the Monte Carlo method to simulate the demanding amount for next three cycles, considerable economic benefits can be achieved which has showed aforementioned. Moreover, it will make sense especially for the initial products at the context of demand data at the form of small sample. The findings of this research can lead to a better solution for setting and optimization of safety stock than traditional ways.

Acknowledgment. This paper is supported by (1) Ministry of Communications Science and Technology Project "Research on Multimodal Transport of Wuhan Shipping Center" (2018-MS3-093); (2) Wuhan University of Technology Educational Reform project "The Exploration and Practice of the Course Reform of Port Enterprise Management Facing the Needs of Compound Talents in the New Era (2018418110)".

References

1. Anitha, P., Patil, M.M.: A review on data analytics for supply chain management: a case study. Int. J. Inf. Eng. Electron. Bus. (IJIEEB) **10**(5), 30–39 (2018)
2. Haghighi, P.S., Moradi, M., Salahi, M.: Supplier segmentation using fuzzy linguistic preference relations and fuzzy clustering. Int. J. Intell. Syst. Appl. (IJISA) **6**(5), 76–82 (2014)
3. Li, Q.: A VMI model in supplier-driven supply chain and its performance simulation. Int. J. Inf. Eng. Electron. Bus. (IJIEEB) **2**(2), 17–23 (2010)
4. Oliva, R., Watson, N.: Cross-functional alignment in supply chain planning: a case study of sales and operations planning. J. Oper. Manag. **29**(5), 434–448 (2011)
5. Manzini, R., Gebennini, E.: Optimization models for the dynamic facility location and allocation problem. Int. J. Prod. Res. **46**(8), 2061–2086 (2008)
6. Kayacan, E., Ulutas, B., Kaynak, O., et al.: Grey system theory-based models in time series prediction. Expert Syst. Appl. **37**(2), 1784–1789 (2010)
7. Chen, C., Huang, S.: The necessary and sufficient condition for GM(1, 1) grey prediction model. Appl. Math. Comput. **219**(11), 6152–6162 (2013)
8. Cui, J., Liu, S., Zeng, B., et al.: A novel grey forecasting model and its optimization. Appl. Math. Model. **37**(6), 4399–4406 (2013)
9. Tien, T.: A research on the grey prediction model GM(1, n). Appl. Math. Comput. **218**(9), 4903–4916 (2012)
10. Hsu, L.: Applying the Grey prediction model to the global integrated circuit industry. Technol. Forecast. Soc. Chang. **70**(6), 563–574 (2003)
11. Adhikari, R., Agrawal, R.K.: An introductory study on time series modeling and forecasting (2009). http://arxiv.org/abs/1302.6613
12. Zhang, L., Xie, L., Li, W., Wang, Z.: Security solutions for networked control systems based on des algorithm and improved grey prediction model. Int. J. Comput. Netw. Inf. Secur. (IJCNIS) **6**(1), 78–85 (2014)
13. Morita, H., Kase, T., Tamura, Y., et al.: Interval prediction of annual maximum demand using grey dynamic model. Int. J. Electr. Power Energy Syst. **18**(7), 409–413 (1996)

14. Brunaud, B., Lainezaguirre, J.M., Pinto, J.M., et al.: Inventory policies and safety stock optimization for supply chain planning. AIChE J. **65**(1), 99–112 (2019)
15. Gueant, O., Lehalle, C., Tapia, J.F., et al.: Dealing with the inventory risk: a solution to the market making problem. Math. Financ. Econ. **7**(4), 477–507 (2013)
16. Engle, R.F., Granger, C.W.: Co-integration and error correction: representation, estimation, and testing. Econometrica **55**(2), 251–276 (1987)

Computer Science for Medicine and Biology

Smart Prothesis: Sensorization, Humane Vibration and Processor Control

Alexander M. Sergeev[1(✉)], Anatoly A. Solovyev[1,2], and Nikita S. Kovalev[1]

[1] Mechanical Engineering Research Institute of the Russian Academy of Sciences, 4, Malyi Kharitonyevsky Pereulok, Moscow 101990, Russian Federation
amserg.imash@yandex.ru
[2] Moscow Institute of Physics and Technology (State University), 9, Institutskiy per., Dolgoprudny, Moscow Region 141700, Russian Federation

Abstract. This article describes the research phase aimed at the creation the sensorized human upper limb smart prosthesis system. The main focus of the article is on obtaining programming tactile sensation that are organic to humans and which arise with the application of Force Sensing Resistors (FSRs). These sensors act as the artificial tactile mechanoreceptors. The micro vibrating motors incorporated in vibro-bracelet placed on the upper arm is used to transmit the sensation of touch to an amputee. The signal transmission is provided by Bluetooth technology. Arduino was used as the tool for research work.

Keywords: Smart prosthesis · Impulse · Feedback · Amputee · Receptor · Touch · Sensorization · Force Sensing Resistors (FSRs) · Vibrating motor · Vibro-bracelet · Microcontrollers · PWM signals

1 Introduction

The idea of the need to integrate sensing elements into human hand prosthesis is not new. One of the first scientist who expressed it was "the father of cybernetics" Norbert Wiener.

Answering his own question whether "an artificial hand" can touch objects like "a natural one", Wiener responded affirmatively. "It is easy" - he wrote - "to put pressure gauges into the artificial fingers, and these can communicate electric impulses to a suitable circuit. This can in its turn activate devices acting on the living skin, say, the skin of the stump. For example, these devices may be vibrators. Thereby we can produce a vicarious sensation of touch, and we may learn to use this to replace the missing natural tactile sensation" [1].

The bioelectric (myoelectric) prosthesis was embedded with the sensitive elements, but much later the start of its production. Wiener recognizing the achievements of Soviet specialists in this field stressed that "Such artificial hands have already been made in Russia, and they have even permitted some hand amputees to go back to effective work" [ibid].

Z. Hu et al. (Eds.): ICCSEEA 2020, AISC 1247, pp. 477–486, 2021.
https://doi.org/10.1007/978-3-030-55506-1_43

Indeed, in the USSR attempts to create the upper limb prosthesis controlled by biopotential signal the remaining muscles of the anatomical intact part of an arm, were undertaken extremely actively in time for Wiener wrote his book. In 1956, the Central Research Institute of Prosthetics and Prosthetic Design of the Ministry of Social Security of the Russian Soviet Federative Socialist Republic (CRIPPD MSS RSFSR), together with the Mechanical Engineering Research Institute of the USSR Academy of Sciences with the participation of the Research Institute of Applied Mathematics, became the first mastered this area. In 1958, the CRIPPD team including Ya. S. Jacobson, A.E. Kobrinsky, B.P. Popov, E.P. Polyan, Ya. L. Slavutsky and A. Ya. Sysin developed the muscle biocurrents controlled forearm prosthesis with an electric servo and device for sensing grip force.

It should be noted that 1970 was a milestone for the prosthetic industry. This year in the USSR the First International Symposium on Prosthetics and Prosthetic Engineering was held. By the beginning of the Symposium, the research and production team had created a forearm prosthetic design "for the amputee's grip force sense", equipped with pressure sensor as the initial link and a vibrator as the final link. In this case the vibration frequency determined the grip force. The device made it possible to sense the object grip force "with the precision inherent in a healthy person (ranging from 75 to 1500 g)" [2].

In this context should be stressed the contribution of the specialists of Kharkov Research Institute of Prosthetics and Traumatology named after prof. M.I. Sitenko the development of feedback prostheses. In the late 60 s as part of a study *"sensing"* prostheses they used electric pulse frequency feedback [4].

Based on the various research groups data, the practice of orthopedic centers in Russia and abroad information, and disability statistics, we concluded that the continuation R&D of forearms and hands "sensing" prostheses is necessary. The term *smart prosthesis* was chosen to designate mechatronic non-invasive prostheses created using haptic feedback via state of art sensor, vibration and wireless technologies.

2 The Smart Prosthesis System as an Agent in the Process of the Amputee Tactile Perception

At some point in the study of sensors with the capabilities of high-quality information transfer to the amputee about the nature of his via prosthetic contact with the environment, the term "sensing" began to be used in the scientific papers. It is clear that the for object of research does not always find an adequate characteristic concerning its properties. In addition, a man-made device that imitates a product of living nature, when it reaches similarity, is often described either in terms with the addition of "artificial" or with the use of those verbal characteristics that previously belonged to live matter. The same happened with the term and ultimately with the concept of sensing.

No matter how perfect the prosthesis of the upper limb or its part is, it's still a rehabilitation tool – kind *an agent* - in the process of human perception. We have a deal with the problem connecting live with not live, organic with not organic in prosthetics in our research. Something the same is in robotics, with the difference that there is the *contact* the artificial antropomorphic object with man, *not inextricable link* with him like in the area of concern to us [5, 6].

Our research is aimed at the rehabilitation of people who have lost one of their hands. Their peculiarity is that they can return to work after successful prosthesis fitting. This distinguishes mentioned category from those who suffered more serious injuries and lost, for example, both upper limbs. In this case, other agents (rehabilitation systems) are applicable, in particular the use of the Inductive Tongue Control System [7].

Therefore, it is very hard to endow with the *sense* the technical rehabilitation aids (TRAs). It is a conventional term for artificial organs and processes connected with the physical, mental and emotional human spheres. So, it would be better to use the term "sensing" in the *system 'prosthesis – person'*.

The term "sense organs" is not an exception meaning - a specialized peripheral anatomical and physiological system, which, thanks to its receptors, ensures the receipt and primary analysis of information about the condition of the body's external and internal environment. Receptors are the primary link here. Earlier we have already dwelt on the role of various skin receptors kinds responsible for touch [8]. In turn, touch is habitually regarded as a set of sensations born in the cerebral cortex, which are a subjective reflection of objective reality. In the context of the present research we are interested in the sensations of vibration and pressure, for which the receptors named the Vater-Pacinian corpuscles are responsible.

Today in prosthetics the compensatory principle is implanting high-precision sensors into an artificial hand and efforts to replace the skin receptors with these sensors. Thus, sensors that can create a sense of pressure and/or vibration in an amputee are *quasi receptors*. Therefore, we can name the integration prosthesis with sensors that provide an amputee the sensations as *sensorization*.

3 The Main Trends in Sensorization

Solving the problems related to the hand prosthesis sensorization, two interrelated directions in the world practice should be taken into account. The first trend concerns the reception and transmission of a signal of *grip force* (compression), as well as its differentiation. The force sensors are responsible for this process. In a number of sources these sensors are referred to as press sensors. The second one is related to *touch*, in transmission characteristics of which the *touch sensors* are involved. Some sources name it *contact sensors*.

According to the latest publications, for a certain number of sensorized prostheses developers, the topic of *touching* within the concept of an amputee *quality of life* has high priority. The psycho-emotional state of a person who has lost a hand is considered by them as the dominant factor of project activity [9].

In order to increase the "sensitivity" of sensors, on the one hand, and to combine touch and pressure devices, on the other, developers are searching for structures and materials that can respond to stimuli with an effect as close to the skin receptors as possible. Particularly intensive work is underway to create artificial skin and electronic skin (e-skin) [10–13].

Despite the researches intensification it should be noted that most of them are far from practical application type. Meanwhile the urgent need for high-quality prosthesis presupposes the search for structures and technologies, the innovative nature of which does not prevent their production. In this regard the smart prosthesis developed by MERI of RAS team can be described as novel, simple, cost-effective and user-friendly due to an integrated approach to the experimental part of the study.

4 Sensorization of the Smart Prosthesis: Experimental Study

It is known that the generator potential (depolarization) can be detected in the unmyelinated ending of the Vater-Pacinian corpuscle when it is compressed. This results in short burst of impulses in the sensory fibre, which adapts in 1–2 s to zero or very low frequency. According to modern concepts, the main function of Pacinian corpuscles in the skin is vibration detection. The frequency of vibrations to which they respond are in the range of 70–1000 Hz. They are most sensitive in the middle of the range of 200–400 Hz, where skin deformation of only 1 μm is a sufficient stimulus.

Our skin contains two types of receptors that allow to feel the vibration. These are the tactile bodies of Meisner, "specializing" in slow vibrations, and the already mentioned Vater-Pacinian corpuscles which are responsible for determining high-frequency vibrations. Most smartphones vibrate at a frequency of 130–180 Hz, which falls on the range between the sensitivity peaks of these two types of receptors. This is probably why smartphones vibrate calls to use both types of receptors. But some think that the Pacinian corpuscles take a greater part in the formation of the sensation of vibration.

For our experiment to give a sensation of touch in the "hand prosthesis - amputee" system and to replace Pacinian corpuscles that are found in the glabrous skin of the fingertips a Force Sensing Resistors (FSRs) were chosen. The principle of the FSR is to change the resistance value depending on how much pressure is being applied to the active area. The more force applied, the less resistance from the FSR. Figure 1 shows the construction of the FSR.

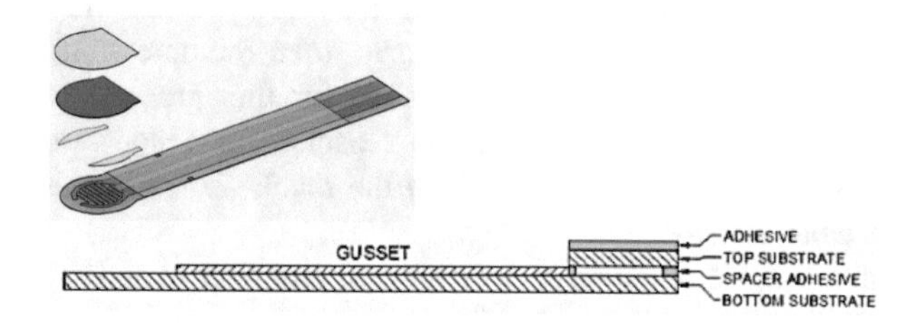

Fig. 1. FSR construction

For a simple force-to-voltage conversion, the FSR device is tied to a measuring resistor in a voltage divider (Fig. 4, on the left). The output to this setup can be described with the following equation:

$$V_{out} = \frac{R_m V_+}{(R_m + R_{fsr})}$$

In the equation R_m, R_{fsr} values represent the resistances of the measuring resistor RM and FSR, respectively. V_{out} is the voltage outputted from the sensor, V_+ is the voltage inputted to the FSR.

The measuring resistor, RM, is chosen to obtain the desired sensitivity range for the force under current limiting. Depending on the impedance requirements of the measuring circuit, the voltage divider may be supplemented with an operational amplifier. The output FSRs voltage changes proportional to the applied force magnitude: it increases with increasing force and decreases with decreasing force (Fig. 2, on the right).

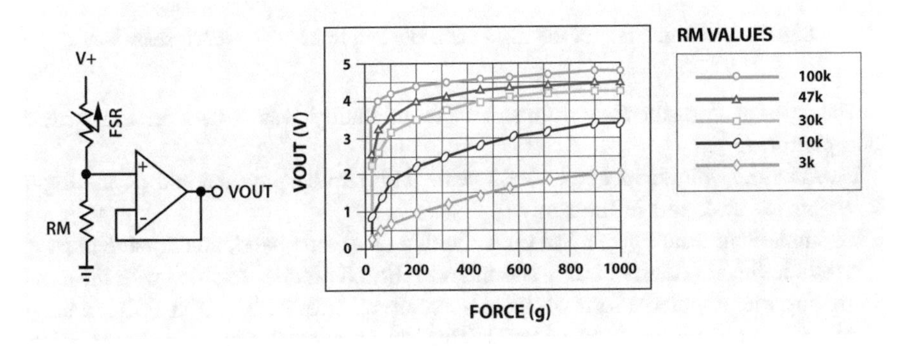

Fig. 2. Voltage divider circuit (on the left) and the graph of dependence the FSR output voltage V_{out} on the applied force magnitude (on the right)

In our experiment a 10 kΩ measuring resistor was selected for the voltage divider. At the same time FSRs reliably work in the range from 20 g to 800 g.

For study, a myoelectric prosthesis developed by the Mechanical Engineering Research Institute of the Russian Academy of Sciences together with Federal State Budgetary Institution "Federal Bureau of Medical and Social Expertise" was used.

The aim was to confirm the hypothesis about the possibility of restoring trans-radial amputee tactile sensations using the vibrotactile method. The theoretical basis of the work was the concept of "double replacement". The core of the concept is the replacement skin mechanoreceptors with pressure sensors and the replacement of natural touch stimuli (various solids) with micro-vibration motors (Fig. 3).

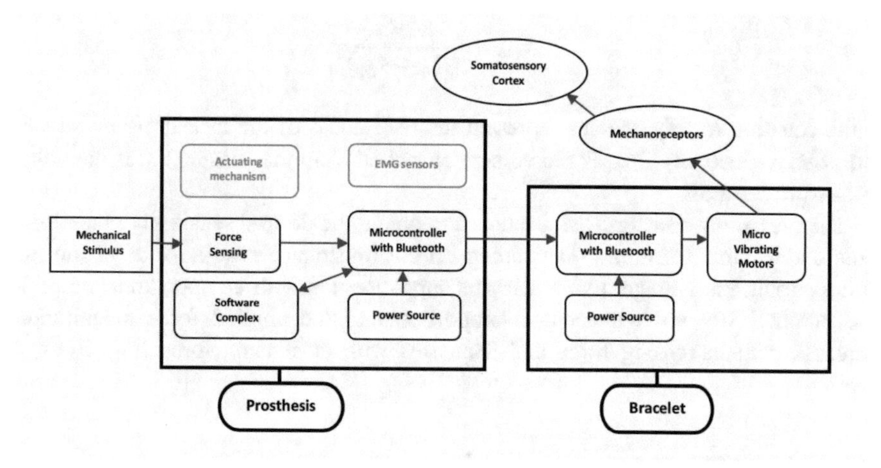

Fig. 3. Block diagram of the main elements and links of the touch sensation

The system "prosthesis-vibrobracelet" functionality was tested on the Arduino Nano platform.

Five healthy volunteers aged 34–42 years with similar physical and psychological characteristics took part in the study.

To implement a sure grip of an object with a prosthetic hand, an amputee must use at least three fingers (thumb, index and middle). Based on this, and also with the aim of minimizing the amputee efforts on the perception of information from FSRs, a sensor system of three FSRs has been adopted. Thus, to improve the functionality of a hand prosthesis in three artificial fingers we incorporated FSRs to sense the touch of the object and detect the grip force the amputee hand.

The FSRs placement on the myoelectric prosthetic hand is presented in Fig. 4.

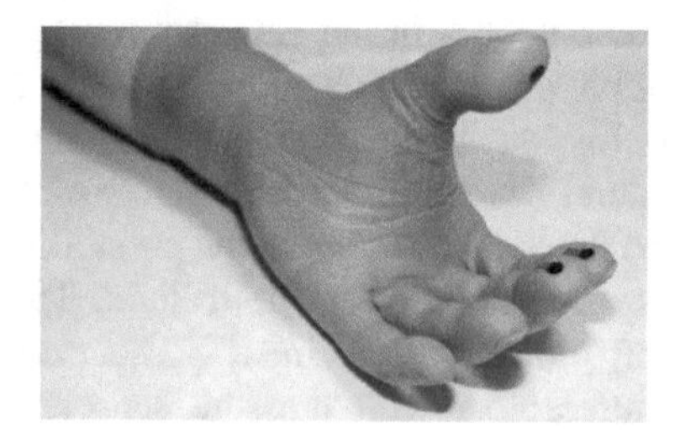

Fig. 4. FSRs location

The information from each FSR is transmitted via Bluetooth to the microprocessor installed in the prosthetic socket, encoded and then transmitted via the Bluetooth microcontroller to the vibrating devices placed into the vibro-bracelet on the upper arm (Fig. 5).

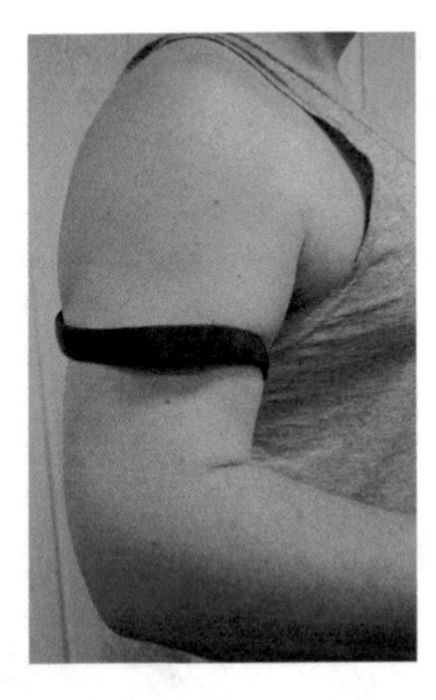

Fig. 5. The vibration bracelet

Vibro-bracelet includes three vibrating motors, a microprocessor, a microcontroller with Bluetooth and a power source. The coded signals from the sensor system through vibration motors are transmitted to the mechanoreceptors of the patient's upper arm skin to create amputee sense of contact with the object. Each finger corresponds to a separate vibration motor.

The used vibration motors operate at a nominal voltage of 5 V and a current of 40 mA in accordance with the Arduino Nano microcontroller PWM ports specification. The PWM signal is calibrated programmatically so that when a maximum force is applied to the sensor, the increasing duty cycle does not exceed the 100% threshold, where the pulse width is 2 ms.

The results of the studies are shown in Figs. 6 and 7. Figure 6 shows the minimum average values of the V_{out} voltage and the nominal current I_{vibro} of the vibrating motor that the amputee confidently distinguish as the different in value vibration signals.

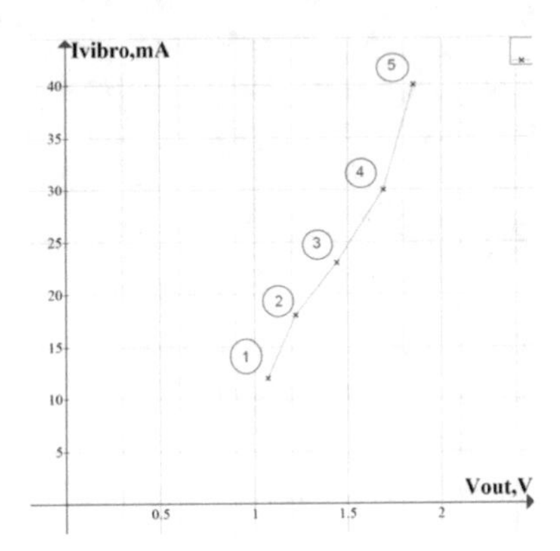

Fig. 6. Dependence of average values FSR V_{out} voltage and nominal current of the I_{vibro} vibrating motor

Figure 7 illustrates the dependence of the voltage V_{vibro} on the vibration motor in the forces range from 20 to 300 g.

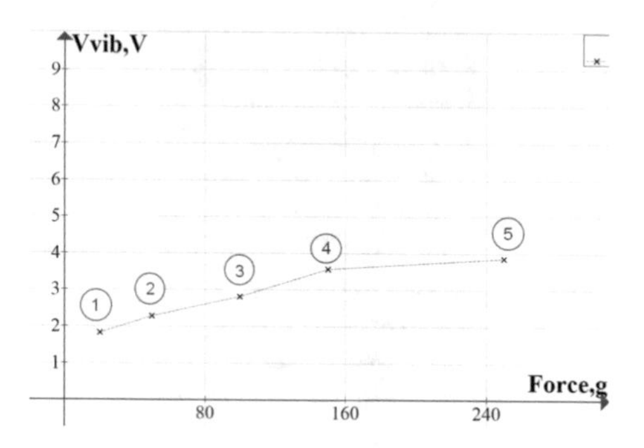

Fig. 7. The vibromotor voltage in force range from 20 to 300 g.

Our experiment performed on the created test complex the elements of which (Fig. 8) allows to identify all the features of the signal received from the prosthesis interaction with a stimulus. Depend on the applied force to the resistive sensors located on the finger tips, the corresponding PWM signals transmit to the individual vibration motor. Thus, every time when a force is applied the vibration motor acts on the upper arm area receptors and this signal will decoded then by amputee. The aim is to adapt optimally the signal for the amputee, who would realize the grip force degree without being overloaded physically and psychologically.

Fig. 8. 1, 2 - Arduino microcontroller: 3 - Bluetooth (master); 4 - Bluetooth (slave); 5,6 - *Force Sensing Resistors* (type 1, type 2); 7 - vibrating motors; 8 - power supply

Note that microcontrollers Arduino receive analogue signals from external sources and converted it digital ones through the general-purpose input/output (GPIO) ports. The microcontroller ports are designed to work with logical zero and logical unit levels and allow digitize analog signals with level limitation and generate various shapes signals using PWM and a low-pass filter. For digitizing the vibration signal is using the ADC.

In future, it is assumed that microelectromechanical systems (MEMS), including microcontrollers and sensors will be used for the sample being introduced into production for the microminiaturization of the component base.

5 Conclusions

We have tested our technology, but our research in sensorization will be ongoing. The noninvasive prosthetic hand development is our ultimate goal. We believe that amputee comfort is the main criterion for the prosthesis. Therefore, we used the Bluetooth technology in a way that had not previously been used in feedback experiments by other teams. We also designed the spiral bracelet instead of the usual cuff which allows to embed the vibration system into it more efficiently.

The next stage of our work is aimed at individualizing the vibrotactile effect via programming method.

Not only hardware and software but the advances in higher amount of user training allow bringing this safe and economic devices to amputee for fulfilling communications with real world. We aware that the hand smart prosthesis creation necessitates interdisciplinary cooperation and close interaction with potential users of these devices [14].

References

1. Wiener, N.: God and Golem Inc. A Comment on Certain Points where Cybernetics Impinges on Religion, pp. 74–75. The M.I.T Press, Cambridge (1964)
2. Popov, B.P.: Modern directions in upper limb prosthetics. In: Sat: Materials Symposium on Prosthetics and Prosthetic Design, p. 23. CRIPPD (TsNIIP), Moscow (1970). (In Russian)
3. Schneider, A.Yu.: Feedback devices in bioelectrically controlled prostheses. In: Sat: Materials Symposium on Prosthetics and Prosthetic Design, pp. 64–70. CRIPPD (TsNIIP), Moscow (1970). (In Russian)
4. Linetsky, M.L., Gibner, V.M., Zakarlyuka, Z.D.: The tasks of “sensing” bioelectric prostheses. In: Sat: Materials Symposium on Prosthetics and Prosthetic Design, pp. 70–74. CRIPPD (TsNIIP), Moscow (1970). (In Russian)
5. Alvarez-Dionisi, L.E., Mittra, M., Balza, R.: Teaching artificial intelligence and robotics to undergraduate systems engineering students. Int. J. Mod. Educ. Comput. Sci. **7**, 54–63 (2019)
6. Olaronke, I., Oluwaseun, O., Rhoda, I.: State of the art: a study of human-robot interaction in healthcare. Int. J. Inf. Eng. Electron. Bus. (IJIEEB) **3**, 43–55 (2017)
7. Johansen, D., Popović, D.B., Struijk, L.N.S.A., Sebelius, F., Jensen, S.: A novel hand prosthesis control scheme implementing a tongue control system. Int. J. Eng. Manuf. (IJEM) **5**, 14–21 (2012)
8. Skvorchevsky, A.K., Solovyev, A.A., Sergeev, A.M., Kovalev, N.S., Korovkin, Yu.V.: Development of a system for a human wrist smart prosthetic based on the concept of double substitution. Med. High Technol. **3**, 14–26 (2018).p. 17. (In Russian)
9. Heeti Al, A.: Prosthetic hands get smart - and a sense of touch. (Prostheses are morphing into mind-controlled extensions of the human body that let their wearers feel what they’re touching). C/Net, 3 April 2018. https://www.cnet.com/news/prosthetic-hands-get-a-sense-of-touch/

10. Hansen, A.: Artificial skin technology mimics touch sensations and reflexes. Scope. Stanford Medicine, 31 May 2018. https://scopeblog.stanford.edu/2018/05/31/artificial-skin-technology-mimics-touch-sensations-and-reflexes/
11. Quick, D.: Skin-mounted electronics that can be applied and worn like a temporary tattoo, 15 August 2011. https://newatlas.com/skin-mounted-electronics/19517/
12. Saraf, R., Maheshwary, V.: High-resolution thin-film device to sense texture by touch. Science **312**(5779), 1501–1504 (2006). https://digitalcommons.unl.edu/cgi/viewcontent.cgi?referer=https://www.google.ru/&httpsredir=1&article=1009&context=chemeng_nanotechnology
13. Wu,Y., Liu, Y., Zhou, Y., et al.: A skin-inspired tactile sensor for smart prosthetics. Sci. Robot. **3**(22), eaat0429 (2018). https://robotics.sciencemag.org/content/3/22/eaat0429.full
14. Biddiss, E., Chau, T.: Upper limb prosthesis use and abandonment: a survey of the last 25 years. Prosthetics Orthot. Int. **31**(3), 236–257 (2007). p. 254. https://journals.sagepub.com/doi/pdf/10.1080/03093640600994581

Detection of Ventricular Late Potentials in Electrocardiograms Using Machine Learning

Xavier Fagan[1], Kateryna Ivanko[2(✉)], and Nataliia Ivanushkina[2]

[1] Ecole Centrale de Lyon, Lyon, France
xavier.fagan@ecl16.ec-lyon.fr
[2] Igor Sikorsky Kyiv Polytechnic Institute, Kyiv, Ukraine
koondoo@gmail.com, n.ivanushkina@gmail.com

Abstract. The first signs of myocardial electrical instability reflect the depletion of regulatory systems at the cellular level. These changes at the first stage may not manifest themselves clinically as functional and anatomical changes. The development of methods for detecting early signs of cardiac abnormalities makes it possible to prevent life threatening pathological processes. Such a task includes eliciting a violation of the electrical homogeneity of the myocardium based on the registration of ventricular late potentials (VLP) by high resolution electrocardiography. The goal of this paper is to evaluate the performance of the Simson method, widely accepted as the standard method for detecting VLP and to compare it to other methods based on time-frequency analysis. The simulation of VLP with different signal-to-noise ratio conducted in this study allows us to generate a variety of VLP of different shapes, which correspond to the states of norm and pathology. Comparison of Simson method, acknowledged as the standard method for VLP detection, time-frequency and wavelet analysis as well as combinations of their features is performed in order to determine whether VLP presence in ECG. Machine learning approach is used to find the efficiency of each set of features. A new method for VLP detection based on wavelet analysis is proposed as a suitable replacement for the Simson method. Using the proposed features made it possible to separate healthy and sick patients with an overall accuracy of 99%, wherein 98% of the cases with VLP were correctly identified outperforming the Simson method.

Keywords: ECG · Ventricular late potential · Time-frequency analysis · Machine learning

1 Introduction

One of the main challenges in ECG analysis is to identify small patterns associated with cardiac electrical instabilities. Prediction of the development of potentially life-threatening arrhythmias becomes possible with an early diagnostic, but this requires the revealing of small and specifically localized changes in ECG. In this context, late potentials are of a great interest: they consist of delayed action potentials that appear at the end of the P-wave (for Atrial Late Potentials, ALP) and at the end of the QRS

Z. Hu et al. (Eds.): ICCSEEA 2020, AISC 1247, pp. 487–497, 2021.
https://doi.org/10.1007/978-3-030-55506-1_44

complex (for Ventricular Late Potentials, VLP (Fig. 1)). It is a true challenge to identify VLP for many reasons: first of all, because of their low amplitude equivalent to the noise amplitude in the ECG, and because of the limited time in which they can be identified (less than 50 ms in general). Also, VLP shape differs from one person to another, which makes it harder to detect these small patterns. Different ALP and VLP detection methods are reported previously [1–5]. The most common one is the Simson method [1]. It requires pre-processing of the three Frank orthogonal leads and calculation of the vector magnitude. Then three parameters are calculated and their value should allow distinguishing a healthy patient from a sick one. But this method is not satisfactory for some reasons: first of all, it cannot be used from the conventional leads, so other papers focused on how to extract the information contained in Frank leads only from conventional leads [2]. Then, this method is not efficient enough if VLP amplitude is low. For this reason other methods have been suggested by the researchers, such as: short-time Fourier transform, Wigner-Vile distribution [3]; bispectral estimation [4]; ant colony optimization algorithm [5].

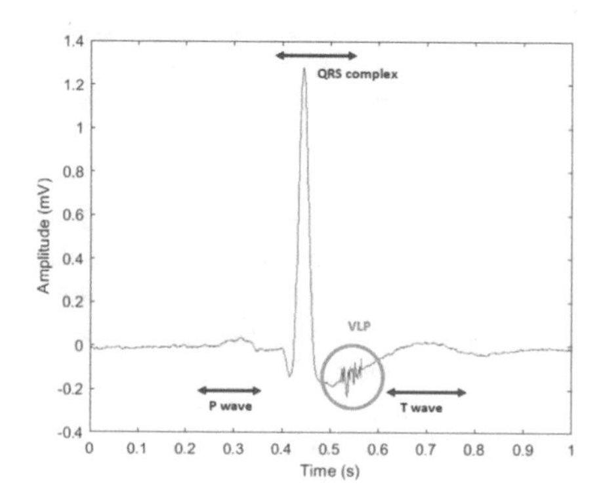

Fig. 1. Example of VLP in a ECG. The VLP were amplified in order to be seen here

In order to make it possible detection of anomalies such as late potentials, it is necessary to perform a high-resolution ECG (HR ECG). This term groups a set of acquisition and computer techniques used to obtain a detailed ECG. It requires sophisticated equipment, since the degree of precision achieved is very important. The environment of acquisition must be controlled, because small disturbances such as electric noise, breathing or muscle movement can cover VLP and make them undetectable [6].

2 Materials and Methods

The data used in this study were provided by the PhysioNet resource [7]. 100 signals were downloaded from the PTB (Physikalisch-Technische Bundesanstalt) Diagnostic ECG Database [8], which includes recordings from patients with a variety of risk factors for sudden cardiac death, as well as healthy controls. This particular database was chosen because it is one of the rare ones containing the Frank leads: the Simson method is defined for those leads, and using other leads would make the Simson method implementation unusable. Also, the high sampling frequency (Fs = 1 kHz) plays a significant role: because of the low amplitude of VLP and short time window of their localization, it is impossible to detect them in a low-resolution ECG (for instance if Fs = 250 Hz). Also, the noise in the signals from the PTB database was not pre-filtered, which makes it possible to compare signals with "real" noise instead of a simulated one and makes the model more "robust" for application in real-life situations. There is not much information about how to characterize VLP. It is known that VLP consist of a series of action potentials appearing in the end of the QRS complex, due to cardiac irregularities. It is also acknowledged that VLP have the frequency components in the band between 40 and 250 Hz and low amplitude (a few dozens of μV). Testing of digital signal processing techniques for late potentials extraction requires signals with various VLP morphologies. Generation of artificial VLP based on their real properties and modification of the model parameters allowed us to create the database, which reflects the states of normal and pathologic conditions. Since VLP can have a wide range of properties, the work in this paper is based on the idea of using the widest description of the VLP as possible, in order to have robust results. This is why VLP considerably varying in their amplitude, frequency range and duration. Signal-to-noise ratio also very affects the results. To simulate VLP in the part of the signals, late potentials were obtained as the sum of the harmonic components with different amplitudes, frequencies and phases. In other cases, VLP activity was simulated by solving the Hodgkin-Huxley equations and the modulated sequences of action potentials were added to the terminal part of QRS-complex [9, 10].

An important factor in detecting VLP is the ratio VLP-signal/noise. In order to compare VLP and noise amplitude, the standard deviation is calculated for both of them. Then, for each Frank lead the two values of standard deviation are divided and the three ratios obtained are averaged in order to obtain a global ratio. The main issue is that VLP are close in amplitude to the noise. In this work the ratio VLP-signal/noise is computed by calculating the standard deviation of the noise after using the Butterworth filter, which allows comparing the VLP only to the noise in the same frequency range as VLP. For the VLP, the standard deviation is calculated before adding them to the signals, in order to obtain data for the VLP alone. In the database the global ratio varies from 10 to 0.5 and the ratio between the VLP amplitude and ECG amplitude (computed by dividing the amplitude of the VLP by the amplitude of the R-peak) is mostly around 1%, but in some cases it goes up to 5%.

Signal averaging is used in order to reduce the noise present in the signal. This approach is based on the fact that noise is considered as a random component while VLP are deterministic. By summing a great number of cardiac cycles (as a general rule, 150–300), the random components will be decreased and the deterministic ones will be conserved, hence the ratio VLP/noise is enhanced and VLP are easily detected [6]. In reality, when using the Simson method the minimum number of beats required in order to perform a signal averaging is 300 (which approximates to a 5-min ECG signal recording). Once the pre-processing stage is performed [11, 12], the noise level is low enough to make the VLP detectable. In this phase it is important to choose the best tool in order to reveal the presence of VLP or their absence. The idea is to acquire parameters of the ECG, which value differs if the VLP are present. Since a well-known description on VLP is in time and frequency domains, in most cases a time-frequency analysis is performed, and the tool considered as a standard for detection of VLP is the Simson method. In this paper other methods issued from time-frequency analysis are studied in order to be compared to Simson method. The methods analysed in this paper are classic time-frequency tools, such as spectrograms, scalograms and wavelet decomposition [13, 14].

3 Identification of VLP Features

To compare the performance of the Simson method, time-frequency and wavelet analysis as well as combinations of their features for VLP detection in ECG, machine learning approach was used to find the efficiency of each set of features. We expected to propose new features for VLP detection as a suitable replacement for the Simson method. The methods implemented and presented in this section, are grouped by the processing techniques required for ECG.

A. Methods Issued from the Vector Magnitude

As discussed previously, the vector magnitude (V_M) is required in order to perform the Simson method. To extract the vector magnitude from the ECG, for each orthogonal lead (X, Y and Z) the beats having the higher correlation index (at least equal to 0.98) are summed together and averaged, as shown in Fig. 2a. Then the three average beats are filtered through a Butterworth fourth-order band-filter, with cut frequencies of 40 and 250 Hz in order to keep only the frequencies to which the VLP belong. Finally the Vector Magnitude is obtained as a sum of the three filtered beats as follows: $V_{\mathrm{m}} = \sqrt{X_{\mathrm{f}}^2 + Y_{\mathrm{f}}^2 + Z_{\mathrm{f}}^2}$.

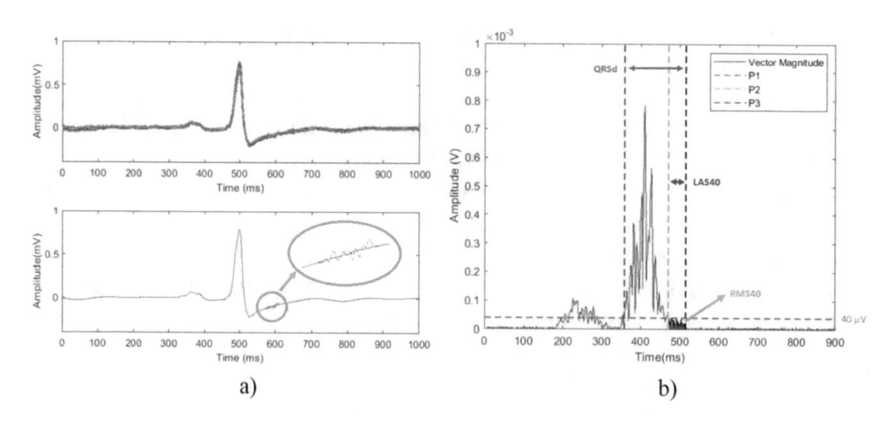

Fig. 2. a) Accumulation of beats and average beat with a zoom on the VLP region. b) Vector Magnitude and how to extract the three features used in the Simson method.

For the *Simson method*, it is necessary to identify when the QRS complex begins (the point will be called P1 here) and when it ends (P3). Then an intermediate point called P2 corresponds to the moment when the QRS complex goes below 40 μV. Then three features are extracted, as shown in Fig. 2b: the duration of the QRS complex (QRSd, associated to the segment P1-P3), the RMS value of the last 40 ms of the QRS complex RMS40 and the time while the end of the QRS complex stays below 40 μV, called LAS40 and equal to the segment P2-P3 [1].

To extract the *spectrogram features*, the spectrogram of the vector magnitude was calculated. The parameters for the spectrogram used in this study were: hamming windows of duration 5 ms each, no windows overlap and 50 frequency divisions. Then a submatrix was extracted in order to keep only the times and frequencies for VLP detection, i.e. between 40 Hz and 250 Hz and for time, starting from P2 and lasting 50 ms. Still, this method gives 242 features (11 times and 22 frequencies). Since the frequency distribution of the VLP is unknown (other than F_{min} and F_{max}), it is possible to reduce the number of features by averaging all the values corresponding to an instant time moment t_i. This makes it possible to reduce the dimension of feature vector to 11 features (Fig. 3).

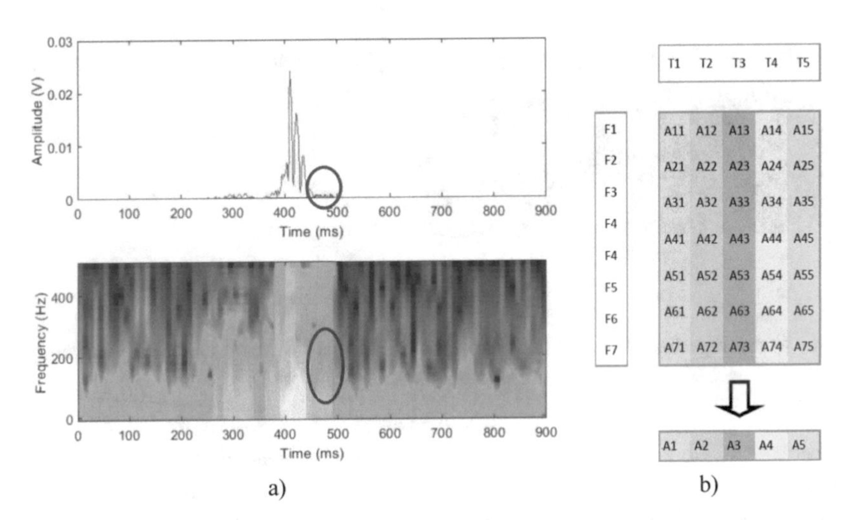

Fig. 3. a) Vector Magnitude and its spectrogram. The VLP zone is circled in red. b) Process of averaging in the spectrogram matrix.

B. Methods issued from the average beat

The main flaw of the using vector magnitude is that it requires combining of 3 Frank leads. For the database studied, the ratio noise/signal in the Y lead was sensibly higher than in the two others, so often the VLP in the Y lead would not appear in the Vector Magnitude and the noise would be amplified. In order to try different solutions, it is possible to take only one lead and calculate the average beat. The idea of this approach is to make a pre-processing similar to what was done previously for the vector magnitude without mixing the three leads: the average beat is obtained by superimposing all the beats in a signal (synchronized on the R-peaks, the most easily identifiable part of the beat), and then averaging those that have a high correlation index (R = 0.98). It was revealed in this study, that by its shape, the obtained signal is more suitable to a wavelet analysis than the Vector Magnitude.

By using continuous wavelet decomposition, it is possible to obtain *scalogram features*. The concept is the same as for spectrogram, but scales take the place of frequencies. In this case, the symmetrical wavelet of the 4th order was used as it is generally applied for ECG processing and it enhances most the differences between a signal with VLP and without. As for the spectrogram, the matrix in the time-scales domain of the VLP is extracted. The time segment chosen is the same as for spectrogram, from P2 to 50 ms after P2. The scales domain is the one that correspond to the frequency domain 40–250 Hz, in this case it was between scales 3 and 18. Figure 4 shows how the features were extracted. Implementation of this method gives 800 features (since there are 16 scales and 50 time positions): as for the spectrogram, by averaging scale-wise, this number is reduced to 50.

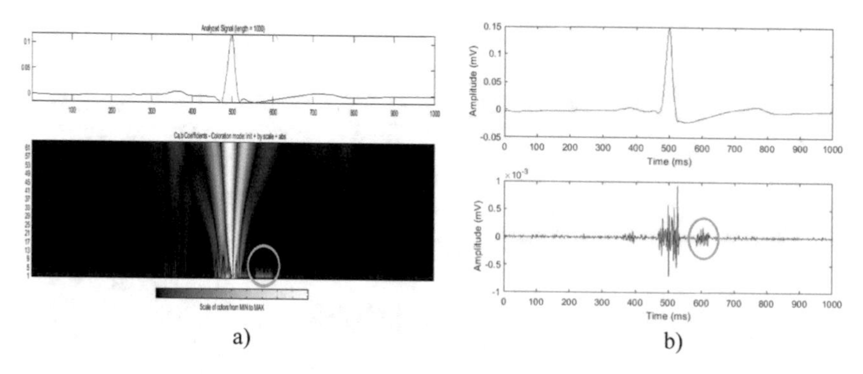

Fig. 4. a) Average beat and its scalogram. The VLP zone is circled in green. b) Average beat and its first level of detail D1. The VLP zone is circled in green.

In order to obtain features from the *multi-scale wavelet decomposition*, the ECG signals were also decomposed using symmetrical wavelet function of the 4th order. By comparing the ECG signals in norm with those, which contain VLP, it is observed that by going till the 5th level of decomposition, the VLP signals can be identified on the D1 level of detail. The results are shown in Fig. 4b.

The most immediate element to notice is that VLP cannot be seen on the average beat, but by plotting the first level of detail D1 they become easily noticeable for the same signal. The shape of absolute value of D1 is similar to a Vector Magnitude, as Fig. 5a shows, but the ratio VLP/noise is sensibly higher here, and the last part of the QRS complex (before the VLP) is cut using the wavelets: hence, it is easier to isolate VLP from the QRS complex and better results can be expected with this plot than with the Vector Magnitude. Two points will be defined following Fig. 5b: P2 as the end of the QRS peak and P3 as the end of the VLP peak. For P2, we start from the maximum and P2 is the first point below a threshold value (linked to the mean value and the standard deviation of the noise) and the same for P3, but starting at the end of the signal and finding the first point above the threshold. The proposed in this study method allows the differentiation to be made between a healthy or a sick patient: if the VLP is absent, P2 and P3 are very close, so the length of the segment P2-P3 will be used as a feature.

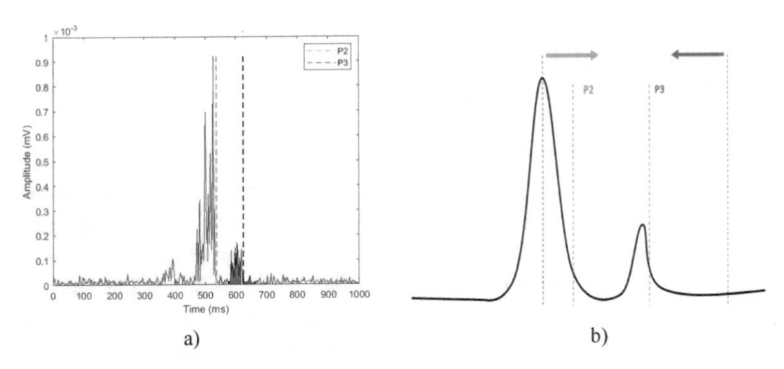

Fig. 5. a) Absolute value of D1 (red) and the two segments used for the RMS ratio (blue and orange). b) Plot of how P2 and P3 are obtained from the plot of the first level of detail D1.

The second proposed in this study feature is the ratio between the RMS values of two segments: one from 40 ms after P2 to 80 ms after it and one from P3 to 40 ms after it:

$$r = \frac{RMS_{VLP}}{RMS_{NOISE}} = \sqrt{\frac{\sum_{i=P2+40}^{P2+80} (D1_i)^2}{\sum_{i=P3}^{P3+40} (D1_i)^2}} \tag{1}$$

The idea behind this choice is that for a sick patient, it will be the ratio VLP/noise and for a healthy one it will be the ratio between two segments of noise because P2 and P3 coincide and P3 is, by definition, the last point before the noise zone. Hence, for a sick patient this ratio should be higher than for a healthy one.

4 Results and Discussion

In this part the Simson method is applied to the PTB database, then the same is done with all the other methods in order to compare their performance with Simson in the same context. In this paper, the purpose of Machine Learning is to analyse the efficiency of all the methods described, and it is of great help especially for the methods with a great number of features. For instance, for the 242 features of the spectrogram method, it is not possible to determine its efficiency by plotting two or three features.

Hence, the Machine Learning helps comparing all the methods, and a more in-depth study will be done for the most interesting ones. Another point is about how to evaluate the accuracy of the methods: in this context the overall accuracy is not enough. It is important to have a low rate of "false negative", i.e. sick patients who are tested and found healthy. This issue is more important than the opposite, because we can imagine that for some sick patients more thorough exams can be done, and this will discard the "false positive" results. This is why the results will not only be based on the overall accuracy, but also on the rate of sick patients correctly identified.

First of all, the results by using the standard Simson method were analysed. For this method, VLP are identified if at least two of these criteria are verified: QRSd is above 114 ms; RMS40 is below 40 μV; LAS40 is above 38 ms.

Applying this method, the results are plotted with three features for Simson as shown in Fig. 6, a. The plot shows that 80% of the patients are correctly identified, but only 66% of the patients with VLP are identified. Also, success rate depends on the amplitude of the simulated VLP: Fig. 6, b is plotted using the ratio VLP/noise. This figure shows for instance that if the ratio VLP/noise is under 2, the success rate is around 50% and if the ratio goes below 1, the success rate is 0%. It is also possible to use Machine Learning with the Simson features in order to have a reference for

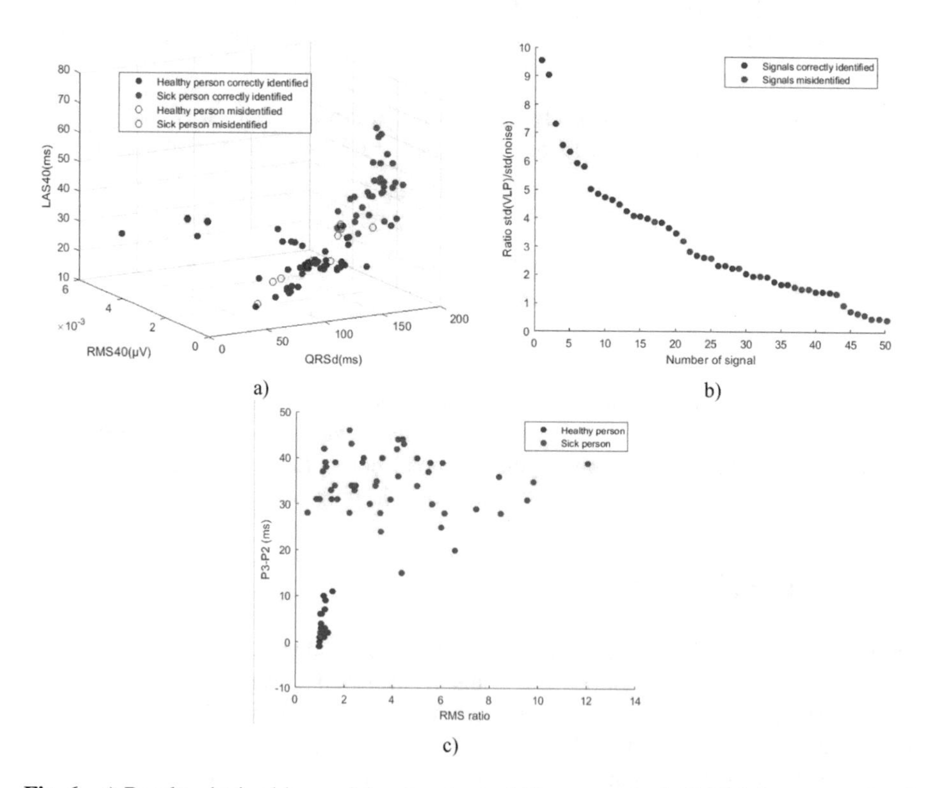

Fig. 6. a) Results obtained by applying the standard Simson method. b) Link between ratio of VLP/noise and success rate of the standard Simson method. c) Results obtained by applying the wavelet method

comparing the results given by Machine Learning on the other methods. For Simson, we obtained a maximum accuracy of 84%. As far as the spectrogram and the scalogram are concerned, there have too many features, so Machine Learning is the only way to analyse the performance of these methods.

In the study, 22 classifiers of Matlab system were used, among which are classifiers implementing linear and quadratic discriminant analysis, different variations of k-nearest neighbors method, k-means method, different variations of support vector machines, decision trees, etc. Table 1 shows the results with the best accuracy percentage along with the number of features. To assess the accuracy of the classifiers the cross-checking was used. Cross-validation was carried out with a breakdown of the data into 5 parts, 4 of which were used for training, and the 5th was used for testing. The procedure repeated five times and each of the five parts was used for testing. We can see that every method gives a higher success rate than Simson, with the best result for the scalogram (without averaging in the frequency domain). But considering the very high number of features, finally the best compromise is the spectrogram (average). Moreover, the averaging has a positive effect for the spectrogram, but it gives worse results for the scalogram. On the other hand, the number of features is reduced, which is an improvement. For spectrogram and scalogram features, the results are probably not as satisfactory as expected because the number of signals in the database is too low with respect to the number of features.

Table 1. Comparison of machine learning results for different methods of VLP detection

Method	Number of features	Best accuracy
Simson method	3	84%
Spectrogram	242	87%
Spectrogram (average)	11	89%
Scalogram	800	93%
Scalogram (average)	50	85%
Wavelet	2	99%
Simson + spectrogram (average)	14	93%

Table 2. Comparison of machine learning results for Simson method and wavelet method

Classification method	Simson method	Simson + spectrogram	Wavelet
Fine tree	76% (76%)	85% (84%)	98% (96%)
Linear discriminant	84% (74%)	77% (68%)	98% (96%)
Quadratic discriminant	80% (76%)	84% (96%)	99% (98%)
Linear SVM	84% (74%)	85% (76%)	98% (96%)
Cubic SVM	77% (74%)	93% (90%)	98% (96%)
Cosine KNN	84% (74%)	86% (74%)	98% (96%)

The method giving the best results is the method based on wavelet decomposition, using the first level of detail D1. Comparing it to the other methods in Table 1 we can see, that it gives the best results and has a very little number of features, so this method will be analysed more in detail to see if it can replace Simson method. Figure 6c shows the results of classification for wavelet method: as described previously, only two features are used, the length of the segment P2-P3 and the ratio VLP/noise (which becomes noise/noise for a healthy patient). Two features are enough to separate healthy and sick patients with an efficiency of 99% and 98% of the VLP are correctly identified. The values correspond to the prevision: VLP make the segment P2-P3 longer, and VLP contains more energy than noise, so the RMS ratio is higher if VLP are present. But the RMS ratio is not enough to detect VLP: for some signals, the amplitude of the VLP is so low that the ratio VLP/noise is similar to the results for healthy people. But by using the length of the P2-P3 segment it becomes possible to separate the two groups of signals. Moreover, the features of healthy patients are localised, whereas the sick ones occupy a much larger zone, because a "wide" model of VLP was taken, with different amplitudes and frequencies. It is possible to identify a "healthy area" and consider all the results outside it as sick. To conclude the comparison with Simson, Table 2 shows the accuracy rate of some Machine Learning features: the methods compared are Simson, Simson combined with the spectrogram (average) and the wavelet method. Between parenthesis is indicated the rate of sick patients correctly identified, since the importance of this result was discussed previously. Simson method can be improved quite significantly by associating it to another one: the results are better than for each method taken separately. But the proposed in this study method, based on wavelet decomposition features, still outperform the Simson one by using a limited number of features.

5 Conclusions

The novel approach to VLP detection, based on wavelet decomposition features, was proposed in the study. Comparing to other considered methods, it is the best performing one, simple in implementation and still has room for improvement, which makes it an ideal replacement for the Simson method in detection of ventricular late potentials. Using the proposed 2 features made it possible to separate signals in norm and pathology with an overall accuracy of 99%, wherein 98% of the cases with VLP were correctly identified. Creating a database of real HR ECG signals with synthetic VLP that includes pathological alterations in morphology of the terminal part of QRS complex is a promising direction for further investigation and improvement of methods and algorithms for diagnosis of early symptoms of cardiac electrical instability.

References

1. Breithardt, G., et al.: Standards for analysis of ventricular late potentials using high-resolution or signal-averaged electrocardiography. Circulation **83**(4), 1481–1488 (1991)
2. Maheshwari, S., Acharyya, A., Schiariti, M., Puddu, P.E.: Frank vectorcardiographic system from standard 12 lead ECG: an effort to enhance cardiovascular diagnosis. J. Electrocardiol. **49**, 231–242 (2016)
3. Laciar, E., Orosco, L.: Analysis of ventricular late potentials in high resolution ecg records by time-frequency representations. LAAR **39**, 255–260 (2009)
4. Sharmila, K., et al.: Use of higher order spectral analysis for the identification of sudden cardiac death. In: 2012 IEEE International Symposium on Medical Measurements and Applications Proceedings (2012)
5. Subramanian, S., Gurusamy, G., Selvakumar, G.: A new ventricular late potential Classification system using ant colony optimization. J. Comput. Sci. **8**(2), 259–264 (2012)
6. Majeed, I., et al.: High Resolution Electrocardiography. Pak. J. Physiol. **1**(1-2) (2005)
7. Goldberger, A.L., et al.: PhysioBank, PhysioToolkit, and PhysioNet: components of a new research resource for complex physiologic signals. Circulation **101**(23), e215–e220 (2000)
8. Oeff, M., Koch, H., Bousseljot, R., Kreiseler, D.: The PTB diagnostic ECG database. National Metrology Institute of Germany (2012)
9. Ivanushkina, N.G., Ivanko, K.O., Timofeyev, V.I.: Analysis of low-amplitude signals of cardiac electrical activity. Radioelectron. Commun. Syst. **57**(10), 465–473 (2014)
10. Matveyeva, N.A., Ivanushkina, N.G, Ivanko, K.O.: Combined method for detection of atrial late potentials. In: Proceedings of the IEEE International Conference "Electronics and Nanotechnology", pp. 285–289 (2013)
11. Ivanushkina, N.G., Ivanko, K.O., et al.: Analysis of electrocardiosignals for formation of the diagnostic features of post-traumatic myocardial dystrophy. Radioelectron. Commun. Syst. **60**(9), 405–412 (2017)
12. Yadav, A., Grover, N.: A robust approach for r-peak detection. Int. J. Inf. Eng. Electron. Bus. (IJIEEB) **9**(6), 43–50 (2017)
13. Ardhapurkar, S.S., Manthalkar, R.R., Gajre, S.S.: Interpretation of normal and pathological ECG beats using multiresolution wavelet analysis. Int. J. Inf. Technol. Comput. Sci. (IJITCS) **5**(1), 1–14 (2012)
14. Khazaee, A.: Automated cardiac beat classification using RBF neural networks. Int. J. Mod. Educ. Comput. Sci. (IJMECS) **5**(3), 42–48 (2013)

Significant Parameters of the Keystroke for the Formation of the Input Field of a Convolutional Neural Network

Ivan Dychka[1], Ihor Tereikovskyi[1(✉)], Liudmyla Tereikovska[2], Anna Korchenko[3], and Volodymyr Pogorelov[3]

[1] National Technical University of Ukraine "Igor Sikorsky Kyiv Polytechnic Institute", Kiev, Ukraine
dychka@scs.ntu-kpi.kiev.ua, terejkowski@ukr.net
[2] Kyiv National University of Construction and Architecture, Kiev, Ukraine
tereikovskal@ukr.net
[3] National Aviation University, Kiev, Ukraine
annakor@ukr.net, volodymyr.pogorelov@gmail.com

Abstract. The article deals with improvement of personality and emotion recognizers based on the keystroke pattern. The article shows that improvement of those recognizers can be implemented using neural network solutions. Based on the analysis of literature sources, we have determined the potential of using convolutional neural networks. The difficulties of using such networks are largely related to justification of a keystroke parameter list, whereby the input field of the convolutional neural network is determined. We suggest determining the said list with respect to feasibility of using the keystroke parameters that have a positive impact on efficiency of the neural network model. In addition, we assume that for admissible resource intensity the efficiency can be assessed experimentally using indices such as recognition accuracy or training and validation sample losses. The experiments allowed plotting the values of efficiency indices versus the number of training iterations for different options of keystroke parameters. The experimental results have shown that, for an input field formed using one keystroke parameter, the highest recognition accuracy is demonstrated by a convolutional neural network with the input field formed using the key hold time. At the same time, the combination of this parameter with other parameters does not result in any significant positive changes. We have justified the need for further research on improvement of keystroke parameters preprocessing procedures in order to increase their informational value. We have also determined that it is necessary to develop a method for determining the architectural parameters of a convolutional neural network designed to recognize the user personality and emotions based on their keystroke.

Keywords: Keystroke recognition · Keystroke parameters · Recognition accuracy · Convolutional neural network · Input field

Z. Hu et al. (Eds.): ICCSEEA 2020, AISC 1247, pp. 498–507, 2021.
https://doi.org/10.1007/978-3-030-55506-1_45

1 Introduction

In the modern context, ever-growing use of biometrics analysis tools is a significant trend in developing the technologies of covert user monitoring in different-purpose information systems [10, 19, 20]. These technologies are successfully used in on-line monitoring of emotional state of the staff operating critical infrastructure facilities, in automatic check of learning content perception by students of distance learning systems, as well as in bank information system to prevent fraud in borrowing loans [2, 5–7]. Most of the mentioned tools are based on analysis of behavioural biometrics. With regard to the difficulty of forgery, inalienability from the user personality, possibility to register parameters using only standard peripheral equipment, as well as extensive use of password and process data in the form of a character set in information systems, the keystroke-based covert monitoring tools are the most promising ones. It should be noted that the term keystroke (KS) is used to refer to a biometric behavioural personality characteristic that determines the peculiarities of keyboard typing [3, 10, 12]. The specific feature of KS analysis is the need to analyse big multidimensional data, which considerably impairs the efficiency of statistic models traditionally used in monitoring tools. Therefore, improvement of the said tools over the last few years was attributed to the use of neural network models due to the fact that they are well-proven in solving similar tasks [8, 9]. Although a considerable number of researches deals with development of neural network tools for KS analysis, the practical experience and the results of works [11, 13, 19] indicate that it is necessary to further improve them, which determines the relevance of research in this field.

2 Analysis of Literature Sources in the Field of Research

The analysis of research and practice works [14–17] reveals that main effort in the field of developing the neural network tools for KS analysis are aimed at improving the efficiency of the said tools due to adaptation of the view and the design parameters of neural network models to the set problem statement. For example, neural network tools for KS recognition based on classical neural networks in the form of a two-layer perceptron, PNN, Hopfield network, Kohonen topographic map, and a radial basis function network have been developed in the relatively early research works [4, 10, 11]. It should be noted that classical neural networks have not been widely spread, because limited capabilities of their architecture prevent reaching the acceptable recognition accuracy. More advanced works [21] are influenced by new theoretical advances in the field of neural networks. For example, works [12, 22] suggest performing the KS analysis using a modern recurrent neural network of LSTM type. The suggestion is based on the proved LSTM ability to efficiently analyse textual information with an unlimited number of characters. In the meantime, it is indicated that the suggested solution is limited due to the fact that the LSTM network design is associated with complexity of training sample formation. More promising suggestions are proposed in works [18, 22]. In these works, KS analysis is implemented using a convolutional neural network (CNN) designed in a similar manner to the model of visual image recognition by animals. Even use of early CNN modifications allowed achieving the KS-based user recognition accuracy at a level of 93–94%. Similar results have been obtained in work [8]. Here, KS analysis was performed using a LeNet CNN in order to improve stability of user password protection. It should be noted that the CNN input was KS time parameters. Article [14] also deals with solving the task of improving neural network systems for KS analysis by employing modern CNN types. It has been established that the difficulties in solving this task are related to the fact that KS parameters must be transformed into a form suitable to be input into the CNN. A transformation procedure has been proposed based on presentation of KS parameters of fixed-size input text as a square multichannel image. Each encrypted character in the text is attributed to an individual image point and described by a corresponding ASCII code and KS parameters such as key hold time (KHT) and time between key presses (TBKP). In addition, the abscissa of an encrypted character corresponds to the position of this character in the text. The ordinate corresponds to the position of the previous text character on the keyboard. If the number of text characters exceeds the number of characters on the keyboard, the image is complemented with rows corresponding to the space character from the top along the axis of ordinates in order to preserve its square shape. If, in contrast, the number of text characters is less than the number of characters on the keyboard, the image is complemented with columns of space characters from the right in order to preserve the shape. The suggested transformation procedure is illustrated in Fig. 1, which presents the fragmentary mapping of text “HEY THERE” encrypted using KHT. Each image point corresponding to the encrypted value of a text character in Fig. 1 is described by two digits recorded in the corresponding cell. The first digit is the ASCII code of an input character, while the second digit is the KHT. For example, the “Y” character corresponds to the image point at the intersection of the third column and the sixth row. The corresponding cell contains values 121 (ASCII code) and 51 (KHT). It has been shown by experiments that use of the suggested procedure provided the error of user emotion and personality recognition at the level of the best modern KS analysis systems. Thus, the analysis of literature sources demonstrates that CNN are currently the most advanced type of neural network models in terms of KS recognition. Notably, in most of the sources analysed, the input field of a CNN was formed using processed time parameters of KS that include: KHT, TBKP, and the ratio of these parameters (PR). In addition, the values of key hold time dynamics (KHTD) and dynamics of time between key presses (TBKPD) can be used. These parameters are calculated using the following equations:

$$\tau_r(i) = t_u(i) - t_d(i), \tag{1}$$

$$\tau_b(i, i-1) = t_u(i) - t_d(i-1), \tag{2}$$

$$q_{br}(i, i-1) = \frac{\tau_b(i, i-1)}{\tau_r(i)}, \tag{3}$$

$$v_r(i, i-1) = \frac{\tau_r(i) - \tau_r(i-1)}{\tau_r(i)}, \tag{4}$$

$$v_b(i, i-1) = \frac{\tau_b(i) - \tau_b(i-1)}{\tau_b(i)}, \tag{5}$$

where τ_r is KHT, t_d, t_u is time of key press and release, τ_b is TBKP, i is the number of key use when typing the text, q_{br} is PR, v_r, v_b are KHTD and TBKPD.

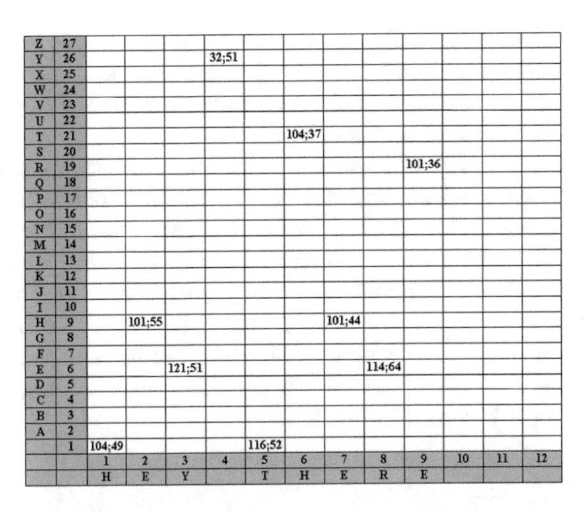

Fig. 1. Mapping of encrypted text "HEY THERE"

It is also arguable that feasibility of using certain KS parameters in CNN-based KS analysis tools is not sufficiently justified. At the same time, theoretical studies in the field of neural networks show that accuracy and resource intensity of a network largely depend on the list of parameters analysed. For example, use of minor parameters can considerably reduce the recognition accuracy due to the noise component introduced into the input data stream. On the other hand, failure to use significant parameters leads to the similar negative result. Therefore, the objective of this work is to determine the keystroke parameters used in formation of the input field of a convolutional neural network designed to recognize the user personality.

3 Research Environment

Similarly to the data in [1, 2, 5, 9], determination of the CNN input parameters was based on the hypothesis on feasibility of using the KS parameters that have positive impact on efficiency of the neural network model. *It has also been assumed that for admissible recourse intensity the efficiency can be assessed using indices such as recognition accuracy (*Accuracy*), as well as training and validation sample losses (*Loss). The values of these indices have been calculated as follows:

$$Accuracy = \frac{N_{right}}{N} \times 100\%, \tag{6}$$

$$Loss = \frac{1}{N} \sum_{t=1}^{N} e^{T}(t, Q)\ W(\theta)\ e(t, Q), \tag{7}$$

where N_{right} is the number of correctly recognized examples, N is the total number of examples, $e(t, Q)$ is n_y-by-1 error vector at a given time t, parametrized by the parameter vector Q, n_y is the number of neural network outputs, $W(Q)$ is the weighting matrix, specified as a positive semidefinite matrix.

Impact of the used KS parameters on efficiency was assessed experimentally, because neural network theory does not provide a corresponding analytic solution. Based on the results of [21], a CNN of SqueezeNet type was used in the experiments. Its main advantages include small memory footprint, high recognition speed, availability of a pretrained model, possibility of implementation using proved tools, and sufficient recognition accuracy. The SqueezeNet model adapted to the task of recognizing personalities of 10 users based on their KS has been implemented using the application package MATLAB R2018b. The structural chart of SqueezeNet visualized using MATLAB tools is shown in Fig. 2.

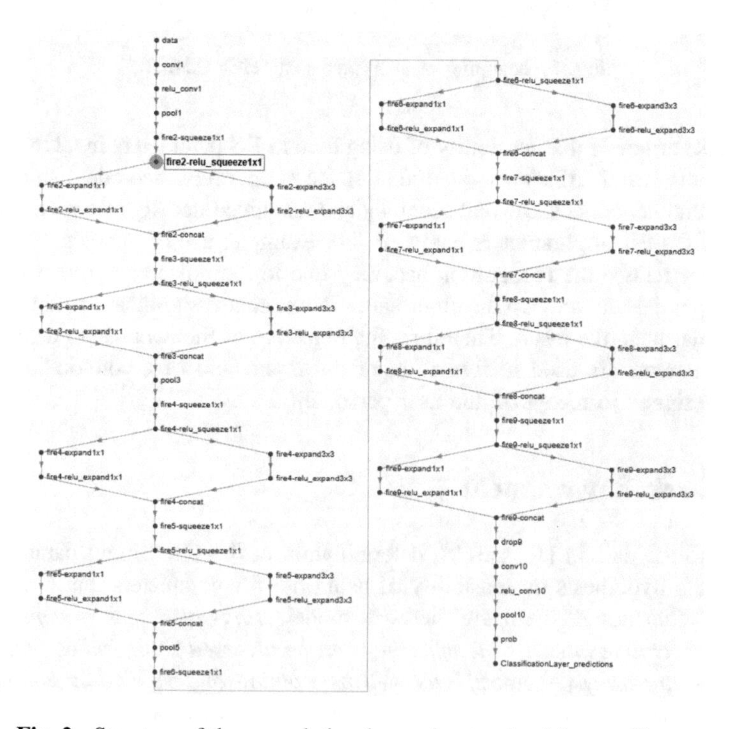

Fig. 2. Structure of the convolutional neural network of SqueezeNet type

The experiments were conducted on a personal computer with CPU Intel Core i7-8700 (3.2–4.6 GHz), 16 GB RAM, graphics card nVidia GeForce GTX 1660Ti, under operating system Microsoft Windows 10.

4 Experiments and Discussion of the Results

Following the first set of experiments, values of the efficiency indices have been plotted against the number of training stages for each individual KS parameter (KHT, TBKP, KHTD, TBKPD). For example, Figs. 3 and 4 show plots of these dependencies based on training and validation data for KHT and TBKP. In order to improve clearness of the results obtained for the training sample, smoothed values were calculated for the accuracy index (*Accuracy_smoothed*) and the loss index (*Loss_smoothed*), with their respective plots also shown in Fig. 3 and Fig. 4.

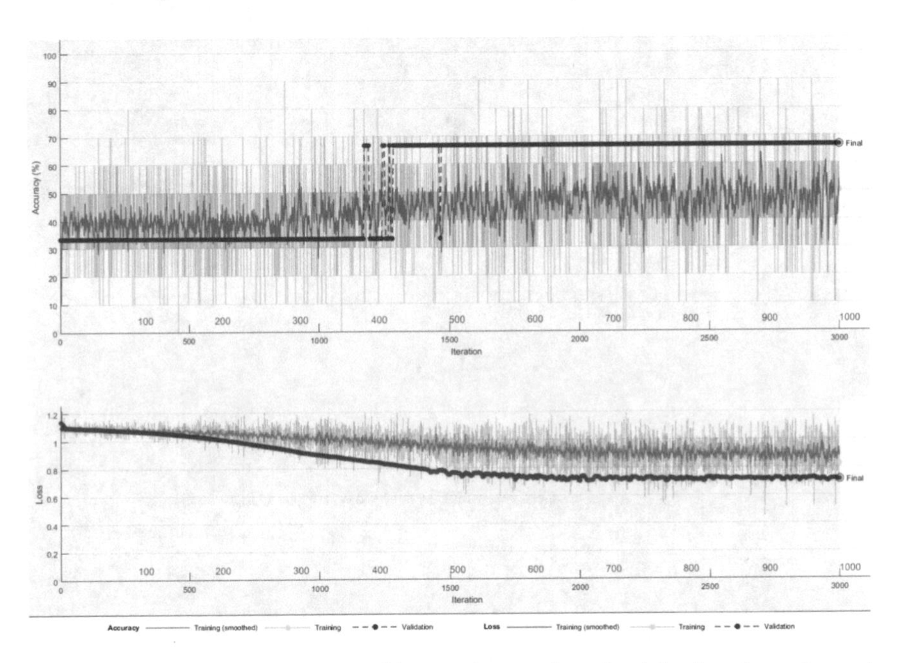

Fig. 3. Plot of recognition accuracy and loss against number of training iterations when using KHT

According to the recommendations of [20–22], the training process lasted 1,000 stages. However, the experiments have shown that the efficiency indices stabilize after 400-500 learning stages, which is demonstrated by plots in Fig. 3 and Fig. 4. Therefore, part of the experiments was limited to 500 stages in order to speed up the research. Since the CNN input field can be presented as a multichannel raster image, the second part of the experiments dealt with different options of simultaneous use of several KS parameters combined. For example, Fig. 5 shows plots of efficiency indices for the combination of two parameters (KHT and KHTD) at 500 training stages.

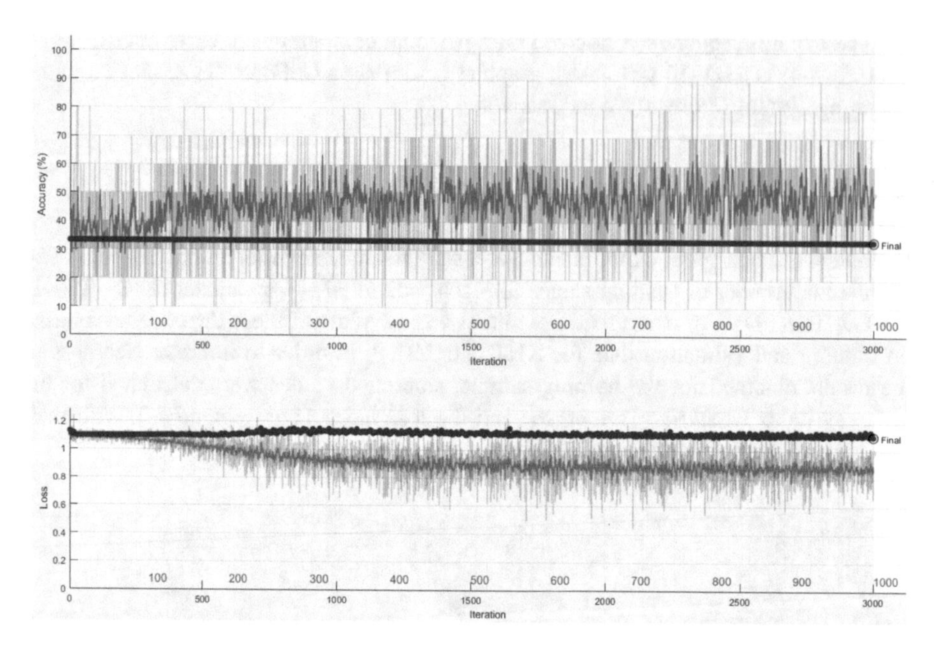

Fig. 4. Plot of recognition accuracy and loss against number of training stages when using TBKP

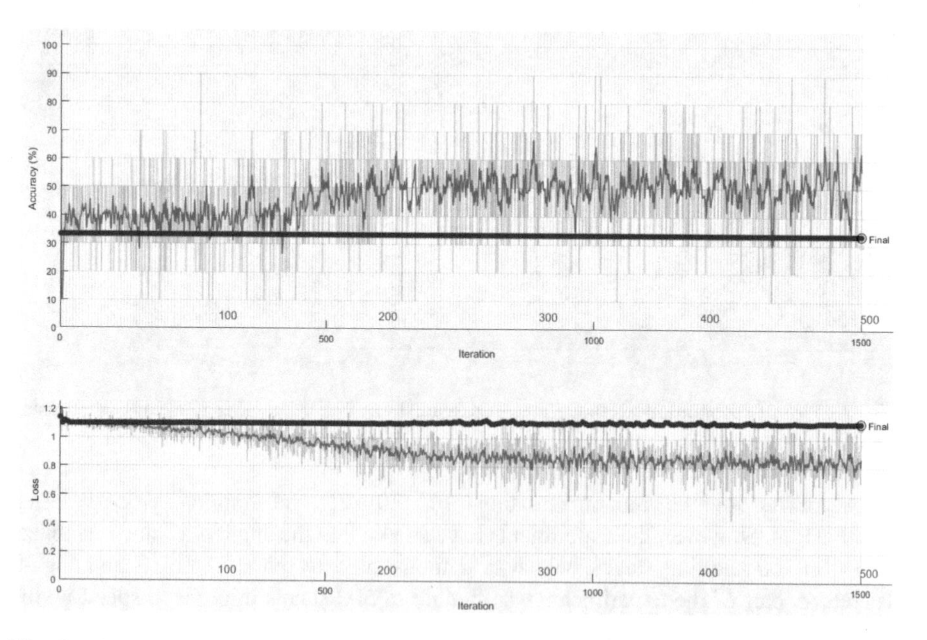

Fig. 5. Plot of recognition accuracy and loss against number of training stages when using KHT and KHTD

Key results of the experiments are summarized in Table 1, which shows the efficiency indices of a neural network model when using different KS parameters.

Table 1. Values of efficiency indices for different KS parameters

Parameters used	Training sample accuracy (%)	Validation sample accuracy (%)	Training sample loss (%)	Validation sample loss (%)
KHT	71	67	0.76	0.76
TBKP	70	33	1.01	1.17
TBKP, KHTD	40	33	0.78	1.18
TBKPD	50	32	0.83	0.84
PR	67	33	0.95	4.5
KHT, KHTD	68	53	0.67	1.16
KHT, TBKPD	71	66	0.8	0.8
TBKP, KHTD	70	34	1	1.17
KHT, TBKP, KHTD	70	35	0.68	1.17

According to the analysis of data presented in Table 1, if the input field is formed using one KS parameter, the highest recognition accuracy is demonstrated by the CNN with the input field formed based on KHT. At the same time, the combination of this parameter with other parameters does not result in any significant positive changes. It should be noted that the result obtained is consistent with the findings of work [10], in which KS was analysed using a fairly simple probabilistic neural network, and is somewhat in contrast to the findings of work [19], in which KS was analysed using a bimodal statistical model after complex preprocessing. This demonstrates the need for further research on improvement of KS parameters preprocessing procedures in order to increase their informational value. In addition, developing a method for determination of architectural parameters of a CNN designed for KS analysis is of interest.

5 Conclusion

It has been demonstrated that improvement of mathematical support of tools for user personality and emotions recognition based on keystroke through the use of neural network solutions is an important direction for the development of such tools. Based on the analysis of literature sources, we have determined the potential of using convolutional neural networks. The difficulties of using such networks are largely related to justification of a keystroke parameter list, whereby the input field of the convolutional neural network is determined. We suggest determining the said list with respect to feasibility of using the keystroke parameters that have a positive impact on efficiency of the neural network model. In addition, we assume that for admissible resource intensity the efficiency can be assessed experimentally using indices such as recognition accuracy or training and validation sample losses. The experiments allowed plotting the values of efficiency indices versus the number of training iterations for different options of keystroke parameters. The experimental results have shown that, for an input field formed using one keystroke parameter, the highest recognition accuracy is demonstrated by a convolutional neural network with the input field formed using the key hold time. At the same time, the combination of this parameter with other parameters does not result in any significant positive changes. We have justified the need for further research on improvement of keystroke parameters preprocessing procedures in order to increase their informational value. It is also necessary to develop a method for determining the architectural parameters of a convolutional neural network designed to recognize the user personality and emotions based on their keystroke.

References

1. Aitchanov, B., Korchenko, A., Tereykovskiy, I., Bapiyev, I.: Perspectives for using classical neural network models and methods of counteracting attacks on network resources of information systems. News Natl. Acad. Sci. Republic Kazakhstan Ser. Geol. Tech. Sci. **5** (425), 202–212 (2017)
2. Akhmetov, B., Tereykovsky, I., Doszhanova, A., Tereykovskaya, L.: Determination of input parameters of the neural network model, intended for phoneme recognition of a voice signal in the systems of distance learning. Int. J. Electron. Telecommun. **64**(4), 425–432 (2018)
3. Akhmetov, B., Lakhno, V., Malyukov, V., Sarsimbayeva, S., Zhumadilova, M., Kartbayev, T.: Decision support system about investments in smart city in conditions of incomplete information. Int. J. Civ. Eng. Technol. **10**(2), 661–670 (2019)
4. Akhmetov, B., Lakhno, V., Akhmetov, B., Alimseitova, Z.: Development of sectoral intellectualized expert systems and decision making support systems in cybersecurity. In: Advances in Intelligent Systems and Computing, vol. 860, pp. 162–171 (2019)
5. Alghamdi, S.J., Elrefaei, L.A.: Dynamic user verification using touch keystroke based on medians vector proximity. In: 2015 7th International Conference on Computational Intelligence, Communication Systems and Networks (CICSyN), pp. 121–126. IEEE (2015)
6. Dychka, I., Tereikovskyi, I., Tereikovska, L., Pogorelov, V., Mussiraliyeva, S.: Deobfuscation of computer virus malware code with value state dependence graph. In: Advances in Intelligent Systems and Computing, vol. 754, pp. 370–379 (2018)
7. Gnatyuk, S.: critical aviation information systems cybersecurity. In: Meeting Security Challenges Through Data Analytics and Decision Support. NATO Science for Peace and Security Series, D: Information and Communication Security, vol. 47, no. 3, pp. 308–316. OS Press Ebooks (2016)
8. Gnatyuk, S., Sydorenko, V., Aleksander, M.: Unified data model for defining state critical information infrastructure in civil aviation. In: Proceedings of the 2018 IEEE 9th International Conference on Dependable Systems, Services and Technologies (DESSERT), Kyiv, Ukraine, 24–27 May 2018, pp. 37–42 (2018)

9. Hayreddin, Ç., Shambhu, U.: Sensitivity analysis in keystroke dynamics using convolutional neural networks. In: 2017 IEEE Workshop on Information Forensics and Security (WIFS) 4–7 December 2017, pp. 1–6 (2017)
10. Hu, Z., Tereykovskiy, I., Zorin, Y., Tereykovska, L., Zhibek, A.: Optimization of convolutional neural network structure for biometric authentication by face geometry. In: Advances in Intelligent Systems and Computing, vol. 754, pp 567–577 (2018)
11. Malik, J., Girdhar, D., Dahiya, R., Sainarayanan, G.: Reference threshold calculation for biometric authentication. IJIGSP **6**(2), 46–53 (2014)
12. Kobojek, P., Saeed, K.: Application of recurrent neural networks for user verification based on keystroke dynamics. J. Telecommun. Inf. Technol. **N3**, 80–90 (2016)
13. Lakhno, V.A., Tretynyk, V.V.: Information technologies for maintaining of management activity of universities. In: Advances in Intelligent Systems and Computing, vol. 754, pp. 663–672 (2018)
14. Liu, M., Guan, J.: User keystroke authentication based on convolutional neural network. In: Communications in Computer and Information Science 2019, vol. 971, pp. 157–168 (2019)
15. Lin, C.-H., Liu, J.-C., Lee, K.-Y.: On neural networks for biometric authentication based on keystroke dynamics. Sens. Mater. **30**(3), 385–396 (2018)
16. Oyedotun, O.K., Dimililer, K.: Pattern recognition: invariance learning in convolutional auto encoder network. Int. J. Image Graph. Signal Process. (IJIGSP) **8**(3), 19–27 (2016)
17. Saket, M., Soumyajit, G., Vikram, P.: Deep secure: a fast and simple neural network based approach for user authentication and identification via keystroke dynamics. In: Conference: IWAISe, International Joint Conference on Artificial Intelligence (IJCAI), Melbourne, Australia, pp. 34–40 (2017)
18. Sassi, A., Ouarda, W., Amar, C., Miguet, S.: Sky-CNN: a CNN-based learning approach for skyline scene understanding. Int. J. Intell. Syst. Appl. (IJISA) **11**(4), 14–25 (2019)
19. Savinov, A.N.: Matematicheskaya model mehanizma raspoznavaniya klaviaturnogo pocherka na osnove Gaussovskogo raspredeleniya. In: Savinov, A.N., Sidorkina, I.G. (eds.) Izvestiya Kabardino-Balkarskogo nauchnogo centra RAN, Vyp. I, pp. 26–32. Kabardino-Balkarskij nauchnyj centr RAN, Nalchik (2013)
20. Tereikovskyi, I., Chernyshev, D., Tereikovska, L.A., Mussiraliyeva, S., Akhmed, G.: The procedure for the determination of structural parameters of a convolutional neural network to fingerprint recognition. J. Theor. Appl. Inf. Technol **97**(8), 2381–2392 (2019)
21. Deng, Y., Zhong, Y.: Keystroke dynamics advances for mobile devices using deep neural network. GCSR **2**, 59–70 (2015)
22. Xiaofeng, L., Shengfei, Z., Shengwei, Y.: Continuous authentication by free-text keystroke based on CNN plus RNN. Proc. Comput. Sci. **147**, 314–318 (2019)

Two-Layer Perceptron for Voice Recognition of Speaker's Identity

Zhengbing Hu[1], Ihor Tereikovskyi[2(✉)], Oleksandr Korystin[3], Victor Mihaylenko[4], and Liudmyla Tereikovska[4]

[1] School of Educational Information Technology, Central China Normal University, Wuhan, China
hzb@mail.ccnu.edu.cn

[2] National Technical University of Ukraine "Igor Sikorsky Kyiv Polytechnic Institute", Kiev, Ukraine
terejkowski@ukr.net

[3] Scientifically Research Institute of the Ministry of Internal Affairs, Kiev, Ukraine
alex@korystin.pro

[4] Kyiv National University of Construction and Architecture, Kiev, Ukraine
{kpm_knuba, tereikovskal}@ukr.net

Abstract. The article is devoted to the problem of ensuring reliable authentication of users of information systems for various purposes. The prospects of solving this problem through the use of voice signal analysis tools to recognize the speaker's personality are shown. The main advantages of such tools include the increased durability of the biometric access code, the use of common registration tools, as well as the possibility of implementation of hidden monitoring of the user's identity. The relevance of research in the direction of developing low-resource means of recognizing the speaker's personality by voice fragments of a fixed duration, using only available computing power on the spot, is substantiated. Based on the analysis of literary works, the prospects of using neural network solutions are shown, the creation of which is complicated by the existing uncertainty in choosing the type of neural network model, as well as in determining the set of input parameters. As a result of the studies, it was determined that in the task of recognizing the speaker's identity by voice fragments of a fixed duration, it is advisable to use a type of neural network model such as a two-layer perceptron, the input parameters of which are associated with small-cepstral coefficients characterizing each of the quasi-stationary fragments of the analyzed voice signal, and the output parameters match of recognizable speakers. By computing experiments, it is proved that each of the quasistationary fragments should be described using 20 chalk-cepstral coefficients. At the same time, the recognition accuracy of the speaker using a two-layer perceptron is at the level of the best modern means of this purpose and is 8% higher than the recognition accuracy using a convolutional neural network such as LeNet. The need for further research in the direction of adapting the parameters of the two-layer perceptron to the recognition conditions under the influence of various kinds of interference was also established.

Z. Hu et al. (Eds.): ICCSEEA 2020, AISC 1247, pp. 508–517, 2021.
https://doi.org/10.1007/978-3-030-55506-1_46

Keywords: Speaker recognition · Voice signal · Two-layer perceptron · Cepstral coefficients · Neural network model

1 Introduction

At the moment, one of the most important tasks in the development of information systems for various purposes is to ensure reliable authentication of users both at the entrance to information system and during the interaction of the user and the system [3, 5, 8]. Since classical authentication systems based on passwords' using and/or hardware keys have a number of drawbacks, biometric authentication methods are increasingly used. However, most of them have a significant drawback associated with the constancy of the user's biometric code (face image, fingerprints). Such a code can be read by an attacker, duplicated, and then used for unauthorized entry into an information system. Unlike similar systems, voice authentication is more reliable due to the ability to use a password-based method for verifying friendly handshakes, which allows you to vary the duration and content of the voice signal. In addition, voice authentication is implemented using commonly used means of recording sound and can be used in the presence of interference that impedes the registration of visual information (poor lighting, glasses, or headdress) [16, 17]. Well-known examples of voice authentication using are monitoring the presence of critical infrastructure objects at the workplace, means of controlling military vehicles that "understand" only the voice of their commander, and monitoring the passage of test tasks by students of the distance education system [11, 12]. Note that the development of voice authentication tools is largely reduced to solve the speaker's identity recognition problem, solved within the framework of the theory of automatic recognition of voice signals. One of the characteristic features of voice recognition tools is their high resource consumption, which is explained by the high computational complexity of the algorithms used. This imposes significant restrictions on the scope of automatic speech recognition, where the described problems are especially acute due to the strictly limited computing resources. A well-known way to overcome this limitation is to transfer resource-intensive computing from user devices to servers in the cloud. The user application only sends voice signals there and receives responses using an Internet connection. The Siri systems from Apple and Google Voice Search from Google [4] work according to this scheme. However, for such an implementation certain conditions are necessary, which in some cases is quite problematic to provide for technical, regulatory or financial reasons. This determines the relevance of research in the direction of developing low-resource means of recognizing the speaker's identity, exploiting only available computing power on the spot.

2 Analysis of Literature Sources in the Field of Research

The results of the analysis of scientific and practical works indicate that at present the widespread identity recognition systems of a speaker are based on mathematical models based on the Bayesian approach [2], hidden Markov processes [7], Support

Vector Machines [6, 15], and also theories of neural networks [9, 14, 18]. At the same time, parameters characterizing a certain portion of the voice signal recorded with a sampling frequency of 8 to 96 kHz are supplied to the input of the recognition system.

In researches [9, 19], the possibility of effective recognition of the speaker's identity on fixed sections of the voice signal lasting 5–15 s was shown. The prevailing approach to the procedure for processing a voice signal is to use short-term analysis. That is, the voice signal is divided into time windows of a fixed size (quasi-stationary fragments), on which the signal parameters do not change. Typically, the size of a quasi-stationary fragment is selected within 10–30 ms [22]. For a more accurate representation of a signal between fragments, an overlap is made equal to half the length of the window. Then, feature extraction algorithms are applied to each fragment, as a rule, the so-called Mel-frequency cepstral coefficients (MFCC) are used as a rule [10, 20]. The calculation of the MFCC consists in the sequential implementation of the following steps:

- The input discrete voice signal is filtered using the expression:

$$x(n) = (x(n) - 0,9x(n-1))\left[0,54 - 0,46\cos\left((i-6)\frac{2\pi}{180}\right)\right], \tag{1}$$

  where x (n) is the discrete value of the amplitude of the voice signal at the n registration.
- The discrete Fourier transform calculates the spectrum of the voice signal:

$$X(k) = \sum_{n=0}^{N-1} x(n)e^{-i2\pi kn/N}, \tag{2}$$

  where N is the number of registrations.
- To smooth the spectrogram at the boundaries of quasi-stationary fragments to the values obtained by expression (3), the Heming window function is used:

$$X(k) = X(k)H(k), \tag{3}$$

  where $H(k) = [0,54 - 0,46\cos(2\pi k/N - 1)]$ is the Heming function.
- Using the expressions (8–11), the filter parameters are calculated:

$$B(b) = 1125\ln(1 + b/700), \tag{4}$$

$$B^{-1}(b) = 700\left(e^{b/1125} - 1\right), \tag{5}$$

$$f(m) = \left(\frac{N}{F}\right)B^{-1}\left(B(f_1) + m\frac{B(f_h) - B(f_1)}{M+1}\right), \tag{6}$$

$$H_m(k) = \begin{cases} 0, \text{ if } k < f(m-1) \\ \frac{k-f(m-1)}{f(m)-f(m-1)}, \text{ if } f(m-1) \le k < f(m) \\ \frac{f(m+1)-k}{f(m+1)-f(m)}, \text{ if } f(m) \le k \le f(m+1) \\ 0, \text{ if } k > f(m+1) \end{cases}, \ 1 \le m \le M, \tag{7}$$

– where m is the filter number, M is the quantity of filters (MFCC), F is the sampling rate.

Using the expression (12, 13), the MFCC values are calculated:

$$S(m) = \ln\left(\sum_{k=0}^{N-1} X(k)^2 H_m(k)\right), \ 0 \le m \le M, \tag{8}$$

$$C(n) = \sum_{n=0}^{M-1} S(m)\cos(\pi n(m+0,5)), \ 0 \le n \le M. \tag{9}$$

Also, an analysis of literary sources suggests that the main efforts in the field of consciousness of low-resource client means of recognizing the speaker's identity are associated with the development of neural network models. The input of such a model is associated with the MFCC, and the output of the model is associated with the identity of the announcer. Moreover, in the analyzed literature there is no consensus on what kind of neural network model is appropriate to use for recognition. The question of how much MFCC is sufficient to describe a single quasi-stationary fragment of a voice signal is also not adequately covered. Moreover, it is indicated in theoretical works that the correct choice of type and parameters of the neural network model is the most important factor determining the effectiveness of its application. Therefore, the aim of this paper is to determine the type and parameters of a neural network model designed to recognize the speaker's identity in fixed areas of the voice signal.

3 Development of a Neural Network Model

In accordance with the widespread methodology for constructing neural network information protection tools [1, 13, 21], the first stage of research is associated with the choice of the most effective type of neural network model (NNM), which is determined by expressions of the form:

$$\max_{V_i} = \{ V_1, V_2, .. V_I \}, \tag{10}$$

$$V_i = \sum_{k=1}^{K} z_k R_k(n_i), \quad n_i \in N_d, \ i = 1, ..I, \ \alpha_k \in Z. \tag{11}$$

where I is the number of acceptable types of NNM; V_i is the efficiency function of the i type of NSM; z_k = [0..1] is the weight coefficient of the k performance criterion; n_i is the i type of the NNM; Z is the set of weighting factors, N_d is the set of permissible types of NNM; K is the number of performance criteria; $R_k(n_i)$ is the value of the k criterion for the i type of NNM.

The choice of acceptable types of NNMs is implemented based on the results of [1, 8] and modern achievements in the use of neural networks for voice recognition [9, 14, 18]. The formation of a list of acceptable types of NNM allowed us to write the set of N_d in the form:

$$N_d = \{n_{BLP}, n_{DNN}, n_{CNN}, n_{LSTM}, n_{MK}\}, \tag{12}$$

where n_{BLP} is a two-layer perceptron, n_{DNN} are deep neural networks with direct signal propagation, n_{CNN} are convolutional neural networks, n_{LSTM} are LSTM networks, n_{MK} is a Kohonen map.

It is also determined on the basis of the formed requirements for the NNM that with neural network recognition of the speaker's identity on fixed sections of the voice signal, significant criteria for the effectiveness of the NNM type are determined by: R_1 - the ability to learn from noisy data, R_2 - the amount of memory required to train, R_3 - the learning time, R_4 - recognition accuracy, R_5 - decision time, R_6 - the ability to extrapolate learning outcomes, R_7 - the ability to interpret the output in the form of probability, R_8 - recognition of fixed fragments of the voice signal. The equivalence of these criteria is accepted $z_k = 0{,}125$. The values of performance criteria for each type of neural network (presented in Table 1) are determined by experts using data [1, 8]. The determination of the values of the sets N_d, R_k and Z made it possible to calculate the efficiency function for each of the admissible types of NNM. For calculations, expressions (10–12) were used. The results are shown in the Table 1.

Table 1. Values of performance indicators for different types of NNMs

NNMType	Effectiveness criteria								Value of the efficiency function
	R_1	R_2	R_3	R_4	R_5	R_6	R_7	R_8	
n_{BLP}	1	0,7	1	1	1	1	1	1	0,9625
n_{DNN}	1	0,8	0,8	1	0,9	1	1	1	0,9375
n_{CNN}	1	1	0,7	1	0,9	1	1	0,5	0,8875
n_{LSTM}	0,8	0,5	0,8	0,9	0,8	1	0,8	1	0,825
n_{MK}	0,5	0,4	1	0,5	1	0.9	0,8	0,8	0,7375

Based on the data of the Table 1 it is determined that the two-layer perceptron, for which the efficiency function reaches a maximum value of 0.9625, is the most effective type. The authors note that the definition of a two-layer perceptron as the most effective NNM type in the task of recognizing of fixed fragments of a voice signal with limited computing resources is also confirmed by the data [14].

According to the results of analysis [1], the structural parameters of a two-layer perceptron are determined by the number of input (N_{in}), hidden (N_h) and output neurons (N_{out}). For the task of recognition, the quantity of output neurons is equal to the number of recognizable speakers, and the number of hidden neurons can be calculated using the expression:

$$N_h = \eta\sqrt{P/N_{out}}, \tag{13}$$

where P is the number of training examples, $\eta = [1..10]$ is the proportionality coefficient.

By analogy with [9], it is planned to use the cepstral coefficients (MFCC) calculated by expressions (1–9) for each of the quasi-stationary fragments of the voice signal as input parameters of the two-layer perceptron. According to the recommendations of [4, 14], the duration of each of the quasi-stationary fragments is taken to be t_{st} = 16 ms. This is explained by the possibility of efficiently applying the fast Fourier transform to calculate the spectrum of a quasi-stationary fragment at a sampling frequency of 16000 Hz, which is typical for recording voice signals in available databases. The authors note that there is no consensus on how much MFCC should be used to determine the identity of the speaker in the analyzed literature sources.

4 Experiments and Discussion of the Results

The purpose of computer experiments was to determine the effectiveness of speaker personality recognition using a two-layer perceptron using different amount of MFCCs characterizing each of the quasistationary fragments of the voice signal.

The experiments were performed for $MFCC = \{1, 2, 3, 5, 10, 15, 20, 26\}$, which corresponds to $N_{in} = \{875, 1750, 2625, 4375, 8750, 13125, 17500, 22750\}$. The

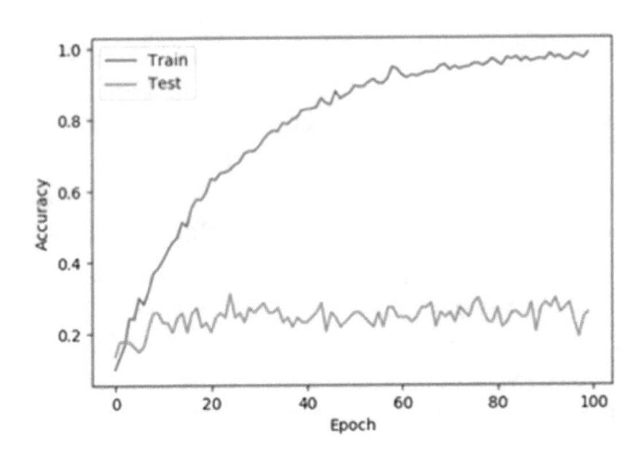

Fig. 1. Recognition graph of accuracy versus of training eras' quantity for training and educational data with MFCC = 1

calculation of input parameters' quantity was implemented taking into account the half overlap of quasi-stationary fragments with a voice signal duration of 7.008 s. To create a training and test sample, the Speaker_recognition database was used, available at www.kaggle.com. From the presented database, voice recordings of 10 announcers (5 men and 5 women) were used. Each speaker in English in the studio voiced 100 texts

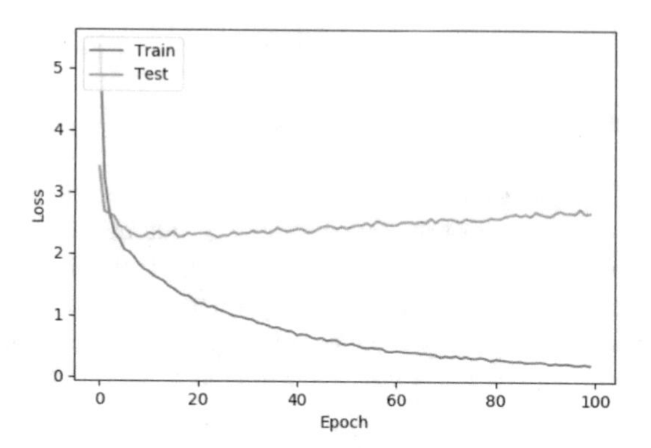

Fig. 2. Dependence of the Loss indicator on training epochs' quantity for training and educational data with MFCC = 1

of different contents. For the formation of both the training and test samples, only the first 7.008 from each of the records was used. The volume of the generated training sample was 900 examples, and the volume of the training sample was 100 examples. Thus, in this series of experiments, the number of output neurons is N_{out} = 10, and the number of hidden neurons calculated using expression (4) is N_h = 256. The model of a two-layer perceptron with a sigmoidal function of activation of hidden and output neurons is implemented in software using the Python programming language and the Keras library. Also, creating the program, the following libraries were used: Numpy - for processing arrays, Soundfile - for processing audio files and Matplotlib - for visualizing the results. The experiments were carried out on a personal computer with an Intel Core i7-8700 processor (3.2–4.6 GHz), 16 GB RAM, and an nVidia GeForce GTX 1660Ti graphics card that was running Microsoft Windows 10. The effectiveness of the two-layer perceptron was evaluated using the indicators of recognition Accuracy and Loss in the training and test samples at 100 epochs of training. For an example Fig. 1 and Fig. 2 show the graphs of the dependence of recognition accuracy and losses on the training and test samples on the number of training eras describing one quasi-stationary fragment by one MFCC.

As a result of the experiments, the ones shown in Fig. 3 and Fig. 4 plots of dependencies of recognition accuracy and losses on a test sample on the number of MFCCs used to describe a single quasi-stationary fragment of a voice signal. As follows from the experimental results, the maximum recognition accuracy and minimum losses are achieved in the case of the description of one quasi-stationary fragment of 20 MFCC. The number of input parameters is 17500.

It should be noted that the achieved recognition accuracy of the speaker's identity using a two-layer perceptron 0.94 corresponds to the best modern solutions that were used in the same conditions [4, 9, 14, 18]. Also, to verify the correct choice of the type

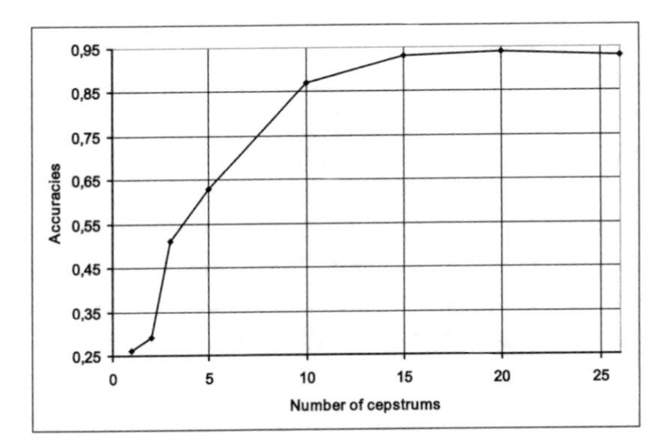

Fig. 3. Recognition graph of accuracy versus of the MFCC quantity

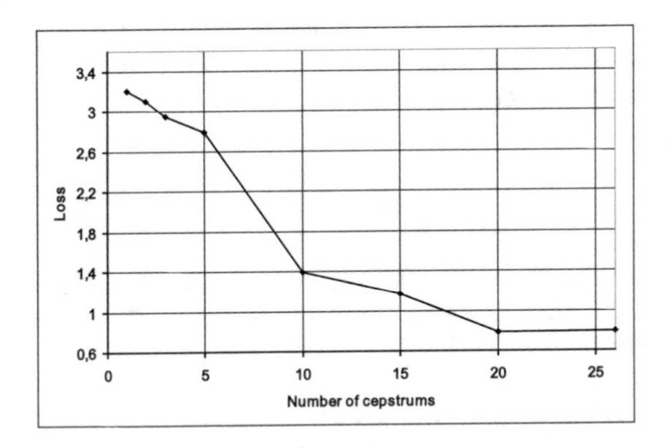

Fig. 4. Dependence graph of loss through MFCC quantity

of neural network model, computer experiments were conducted to recognize the speaker's identity using a LeNet. The experiments showed that LeNet is approximately 10% inferior to the two-layer perceptron in recognition accuracy on test data. At the same time, the LeNet training duration was approximately 1.5 times greater than the training duration of a two-layer perceptron. Thus, the experimental results confirm the feasibility of recognizing the speaker's identity by voice using the developed two-layer perceptron. Since during the experiments the influence on the recognition efficiency of the number of hidden neurons, as well as the type of activation function, was not considered, it is the solution of this issue that determines the paths for further research.

5 Conclusion

The relevance of research in developing low-resource means of recognizing the speaker's identity by voice fragments of a fixed duration, exploiting only available computing power on the spot, is substantiated. Based on the analysis of literary works, the prospects of using neural network solutions are shown, the creation of which is complicated by the existing uncertainty in choosing the type of neural network model, as well as in determining the set of input parameters. As a result of the studies, it was determined that in the task of recognizing the speaker's identity by voice fragments of a fixed duration, it is advisable to use a type of neural network model such as a two-layer perceptron, the input parameters of which are associated with Mel-frequency cepstral coefficients characterizing each of the quasi-stationary fragments of the analyzed voice signal. It was also proved using computer experiments that, unlike the generally accepted methodology for processing voice information, it is advisable to describe each of the quasistationary fragments using 20 Mel-frequency cepstral coefficients. At the same time, the recognition accuracy of the speaker using a two-layer perceptron is at the level of the best modern means of this purpose and is 8% higher than the recognition accuracy using a convolutional neural network such as LeNet. However, for the effective use of the neural network for recognition of the speaker's personality, it is necessary to level out the restrictions primarily associated with filtering the noise component. Therefore, the direction of further research may be the adaptation of the parameters of a two-layer perceptron to recognition of the speaker's identity under the influence of stationary and non-stationary noise.

References

1. Aitchanov, B., Korchenko, A., Tereykovskiy, I., Bapiyev, I.: Perspectives for using classical neural network models and methods of counteracting attacks on network resources of information systems. News Natl. Acad. Sci. Republic Kazakhstan ser. Geol. Tech. Sci. **5** (425), 202–212 (2017)
2. Jadhav, A.N., Dharwadkar, N.V.: A Speaker recognition system using Gaussian mixture model, EM algorithm and K-means clustering. Int. J. Mod. Educ. Comput. Sci. (IJMECS) **10** (11), 19–28 (2018)
3. Akhmetov, B., Lakhno, V., Malyukov, V., Omarov, A., Abuova, K., Issaikin, D., Lakhno, M.: Developing a mathematical model and intellectual decision support system for the distribution of financial resources allocated for the elimination of emergency situations and technogenic accidents on railway transport. J. Theor. Appl. Inf. Technol. **97**(16), 4401–4411 (2019)
4. Akhmetov, B., Tereykovsky, I., Doszhanova, A., Tereykovskaya, L.: Determination of input parameters of the neural network model, intended for phoneme recognition of a voice signal in the systems of distance learning. Int. J. Electron. Telecommun. **64**(4), 425–432 (2018)
5. Altincay, H.: Speaker identification by combining multiple classifiers using Dempster-Shafer theory of evidence. Speech Commun. **41**(4), 531–547 (2003)
6. Campbell W., Sturim D., Reynolds D.: Support vector machines using GMM supervectors for speaker verification. IEEE Signal Process. Lett. **13**(5), 308–311 (2006b)
7. Drugman, T., Dutoit, T.: On the potential of glottal signatures for speaker recognition. In: Interspeech, pp. 2106–2109 (2010)
8. Dychka, I., Tereikovskyi, I., Tereikovska, L., Pogorelov, V., Mussiraliyeva, S.: Deobfuscation of computer virus malware code with value state dependence graph. In: Advances in Intelligent Systems and Computing, vol. 754, pp. 370–379 (2018)
9. Variani, E., Lei, X., McDermott, E., Moreno, I.L., Gonzalez-Dominguez, J.: Deep neural networks for small footprint text-dependent speaker verification. In: 2014 IEEE International

Conference on Acoustics, Speech and Signal Processing (ICASSP), pp. 4052–4056. IEEE (2014)
10. Nijhawan, G., Soni, M.K.: A new design approach for speaker recognition using MFCC and VAD. IJIGSP **5**(9), 43–49 (2013)
11. Gnatyuk, S.: Critical aviation information systems cybersecurity. In: Meeting Security Challenges Through Data Analytics and Decision Support. NATO Science for Peace and Security Series, D: Information and Communication Security, vol. 47, no. 3, pp. 308–316. IOS Press Ebooks (2016)
12. Gnatyuk, S., Sydorenko, V., Aleksander, M.: Unified data model for defining state critical information infrastructure in civil aviation. In: Proceedings of the 2018 IEEE 9th International Conference on Dependable Systems, Services and Technologies (DESSERT), Kyiv, Ukraine, 24–27 May 2018, pp. 37–42 (2018)
13. Hu, Z., Tereykovskiy, I., Zorin, Y., Tereykovska, L., Zhibek, A.: Optimization of convolutional neural network structure for biometric authentication by face geometry. In: Advances in Intelligent Systems and Computing, vol. 754, pp 567–577 (2018)
14. Ding Jr., I., Yen, C.-T., Hsu, Y.-M.: Developments of machine learning schemes for dynamic time-wrapping-based speech recognition. Math. Probl. Eng. 56–68 (2013)
15. Karam, Z., Campbell, W.: A new kernel for SVM MLLR based speaker recognition. In: Proceedings of Interspeech 2007, Antwerp, Belgium, August 2007, pp. 290–293 (2007)
16. Lakhno, V.A.: Algorithms for forming a knowledge base for decision support systems in cybersecurity tasks. In: Advances in Intelligent Systems and Computing, vol. 938, pp. 268–278 (2020)
17. Lakhno, V.A., Kasatkin, D.Y., Blozva, A.I., Gusev, B.S.: Method and model of analysis of possible threats in user authentication in electronic information educational environment of the university. In: Advances in Intelligent Systems and Computing, vol. 938, pp. 600–609 (2020)
18. McLaren, M., Lei, Y., Scheffer, N., Ferrer, L.: Application of convolutional neural networks to speaker recognition in noisy conditions. In: 15th Annual Conference of the International Speech Communication Association, Singapore, 14–18 September 2014, pp. 686–690. ISCA (2014)
19. Singh, S., Kumar, A., Kolluri, D.R.: Efficient modelling technique based speaker recognition under limited speech data. Int. J. Image Graph. Signal Process. (IJIGSP) **8**(11), 41–48 (2016)
20. Sorokin, V.N.: Speaker verification using the spectral parameters of voice signal. J. Commun. Technol. Electron. **55**(12), 156–157 (2010)
21. Tereikovskyi, I., Chernyshev, D., Tereikovska, L.A., Mussiraliyeva, S., Akhmed, G.: The procedure for the determination of structural parameters of a convolutional neural network to fingerprint recognition. J. Theor. Appl. Inf. Technol. **97**(8), 2381–2392 (2019)
22. Zhang, W.-Q., Deng, Y., He, L., Liu, J.: Variant time-frequency cepstral features for speaker recognition. In: Interspeech, pp. 2122–2125 (2010)

Deduplication Method for Ukrainian Last Names, Medicinal Names, and Toponyms Based on Metaphone Phonetic Algorithm

Zhengbing Hu[1], V. Buriachok[2], and V. Sokolov[2(✉)]

[1] Central China Normal University, Wuhan, China
hzb@mail.ccnu.edu.cn
[2] Borys Grinchenko Kyiv University, Kyiv, Ukraine
{v.buriachok,v.sokolov}@kubg.edu.ua

Abstract. This paper attempts to optimize the phonetic search processes for fuzzy matching tasks, such as deduplication of data in various databases and registers to reduce the number of errors in personal data entry (for instance, last names). The analysis of the most common last names in the territory of Ukraine shows that the majority of these last names are of Ukrainian and Russian origin (which are also reduced to phonetic rules of the Ukrainian language). The rules for pronouncing and writing last names in Ukrainian are fundamentally different from the basic algorithms for English and quite different for the Russian language, so the phonetic algorithm should take into account the peculiarities of the formation of Ukrainian last names. The use of the phonetic algorithm gives significant advantages in search and deduplication in comparison with already known algorithms: calculation of Levenshtein, Damerau-Levenshtein, Hamming, Jaro or Jaro-Winkler distance, Q-gram index, etc. [1]. The task of searching by last name was previously formalized in English [2, 3], Russian [4, 5] and some other languages, but for the Ukrainian language such an attempt was made for the first time. The paper presents the results of the experiment on the formation of phonetic indices, as well as the results of increasing productivity when using the generated indices. A method of tailoring the search to other domains and several related languages is presented separately, for example, the search for medicines. Also, search optimization by place names in Ukrainian and Russian was separately worked out. Since in Ukraine there is an abrupt change in the names of cities and streets, the latest relevant data was collected to obtain an up-to-date list of names. Among the existing phonetic search algorithms for the Cyrillic language group, the Metaphone has proven itself in the best way.

Keywords: Deduplication · Fuzzy coincidence · Phonetic rule · Phonetic algorithm · Ukrainian last name · Ukrainian surname · International nonproprietary name · Medicine · Medication · Drug · Toponym · Metaphone

1 Introduction

Different mechanisms and approaches can be used to search for fuzzy matching between words and phrases: calculation of Levenshtein, Damerau-Levenshtein, Hamming, Jaro or Jaro-Winkler distance, Q-gram index, etc. [1], but none of them are taken

Z. Hu et al. (Eds.): ICCSEEA 2020, AISC 1247, pp. 518–533, 2021.
https://doi.org/10.1007/978-3-030-55506-1_47

into account peculiarities of sound perception of unfamiliar words (in this case last names) by a person. Although these algorithms are universal and justified in the analysis of long letters with finite alphabets (including for the analysis of similarity and search for mutations in DNA and RNA) and allow to determine n-fold errors in permutations, typing and omission of letters, but can not fully determine phonetic errors that occur in languages with old grammars due to the cancellation or variability of pronunciation and written reproduction. Therefore, the task of simplifying and unifying sound perception. The task of searching by last name and deduplication are reduced to the formation of search indexes by phonetic key and has been formalized earlier in English [2, 3], Russian [4, 5] and some other languages.

The academic value of this method is formation of quick suggests in advising information systems based on the phonetic algorithm. Such search and deduplication require minimal hardware resources due to pre-formed indexes. In addition, the adaptation of the method for a specific natural language allows you to minimize the number of indexes and their length, as well as expand the variety of suggests. This set of benefits reduces the number of human input errors.

The content of the paper is organized as follows. Section 2 "Research Methodology" contains the research objectives, its aim, and the reasoning for the error correction. In Sect. 3 "Related Works" provides an analysis of the sources for phonetic, semantic, full-text search, and other sorting algorithms. Section 4 "Analysis of the Sample of Last Names" shows the breadth of the distribution of Ukrainian last names in Ukraine and the appropriateness of applying special phonetic rules when working with Ukrainian last names. Preliminary processing of experimental data is given in Sect. 5 "Experiment Conditions". Section 6 "Ukrainian Last Names Indexing," Sect. 7 "Multilanguage Pharmacy Names Indexing," and Sect. 8 "Multilanguage Toponym Indexing" provide modifications of the phonetic algorithm for different datasets. The gain from applying the phonetic algorithm is given in Sect. 9 "Performance Gains." The paper end with Sect. 10 "Conclusions and Future Work" and Acknowledgments.

2 Research Methodology

The research objectives is to develop and implement a phonetic search optimization technology to detect fuzzy coincidence when deduplicating data to databases.

The aim of this research is to develop and implement an optimization technology using phonetic search to detect a fuzzy match at data deduplication in databases. The optimization should be determined based on a decrease in the volume of search indexes compared with the full sample to the use of a phonetic algorithm. A search based on the Metaphone phonetic algorithm should be applied to reduce the number of errors when indexing Ukrainian last names, names of medication, and toponyms.

To accomplish the aim, the following tasks have been set:

- To investigate the frequency of using of Ukrainian last names at Ukraine.
- To construct a phonetic algorithm using a sample of Ukrainian last names.
- To conduct an experimental research and implement an optimization technology for the phonetic algorithm for indexes using a sample of Ukrainian last names.

- To conduct an experimental research and to implement an optimization technology for search queries related to medicinal products and the names of cities and streets of Ukraine when two related languages are mixed.

In addition, phonetic algorithms can be used to define *n*-multiple errors in typos and misspellings, but none of them:

- Firstly, does not take into consideration the peculiarities of sound perception of unfamiliar words (in this case, last names) by human.
- Secondly, cannot fully define determining phonetic errors (such errors occur in languages with old grammar and caused by cancellation of or variability in pronunciation and written reproduction).

3 Review of the Literature

Several phonetic algorithms and their modifications are already known, for example, Soundex [6], NYSIIS [7], Daitch-Mokotoff Soundex [8], Metaphone [9], Double Metaphone [10], Russian Metaphone [4] and Polyphon [5], Caverphone [11]. However, for the Ukrainian language the author is not aware of an attempt to construct algorithms for phonetic search. The analysis of healthcare data takes a significant part of the work of medical information systems; various algorithms are used for processing, for example, the wrapper method (WFS) using particle swarm optimization (PSO) [12]. These methods do not take into account data preprocessing taking into account the phonetic features of natural speech. Other types of search, for example, semantic search [13], full-text search [14], or any other sorting algorithm [15], do not fully solve the problem of the difference between written and oral speech, as well as the variability in the writing of oral speech in, especially for languages with old grammar.

Researchers in [16] take an overview of search engine evolution from primitive to the present, but such a universal approach does not give significant gains when searching through partially structured datasets. Paper [17] presents a string searching algorithm that incorporates a certain degree of intelligence to search for a string in a text. In the search of a string, the algorithm relies on a chance process and a certain probability at each step. But such an algorithm requires significant computing resources and does not allow shifting some of the calculations to client workstations. Therefore, the use of the phonetic algorithm is optimal for generating prompts in search boxes, since it requires minimal computational costs and small data packets that are transmitted over the network. To construct groups of deduplicated data, one can use the methods proposed in [18]. A parallelization of the search process allows you to speed up the primary process of indexing datasets and their deduplication [15].

4 Analysis of the Sample of Last Names

After normalization process of the last names, statistics were collected by the most common last names. Based on these statistics, a list of last names with a frequency of use of more than 0.3% of the total number of last names (from 6,100 times) was formed. Table 1 provides examples of the most commonly used last names with a frequency of more than 0.8% (for transliteration into the Latin script was used BGN/PCGN Romanization in the 1965 edition). The female version of the last name is given through a slash [19].

Table 1. The most common Ukrainian last names.

Frequency of application, ‰	Last name
3.3	Mel'nyk
3.0	Shevchenko
2.6	Boyko
2.5	Kovalenko, Bondarenko
2.3	Tkachenko
2.2	Koval'chuk, Kravchenko
2.0	Ivanov/Ivanova
1.9	Oliynyk
1.8	Koval', Shevchuk
1.7	Polishchuk
1.4	Tkachuk, Bondar, Marchenko
1.3	Lysenko, Moroz, Savchenko, Rudenko, Petrenko
1.2	Kravchuk, Klymenko, Popov/Popova
1.1	Pavlenko, Savchuk, Kuz'menko, Levchenko
1.0	Ponomarenko, Vasylenko, Voloshyn/Voloshyna, Kharchenko, Koval'ov/Koval'ova, Karpenko, Sydorenko, Havrylyuk, Mel'nychuk, Khomenko, Pavlyuk, Shvets', Popovych
0.9	Romanyuk, Chornyy/Chorna, Panchenko, Lytvynenko, Mazur, Kushnir, Yurchenko
0.8	Dyachenko, Martynyuk, Kostyuk, Tkach, Petrov/Petrova, Semenyuk, Prykhod'ko, Kostenko, Honcharenko, Kulyk, Kolomiyets', Bilous, Nazarenko, Volkov/Volkova, Kravets', Kozak, Kovtun

The dependence of the frequency F (measured in moles) of the distribution of the last names on the number of last names can be described by a step function obtained from the approximation of the graph (in Fig. 1 the theoretical distribution is shown by the dashed line and the real is solid):

$$F = 2\pi n^{-e}, \quad (1)$$

where n is the number of last names with the same prevalence.

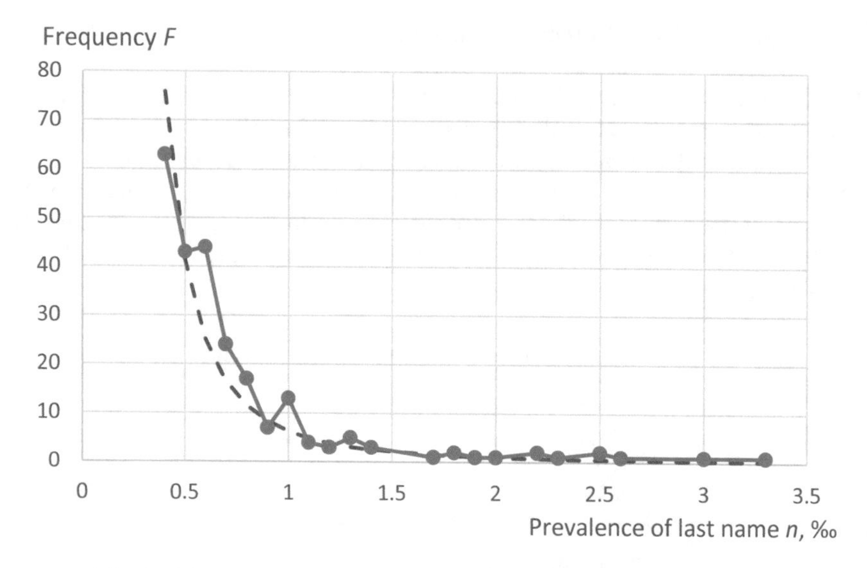

Fig. 1. The dependence of the number of last names on its distribution.

The percentage of Ukrainian last names among the most commonly used (with a frequency of use more than 0.3‰) is 88%, and all others are Russian, Belarussian and related (see Fig. 2).

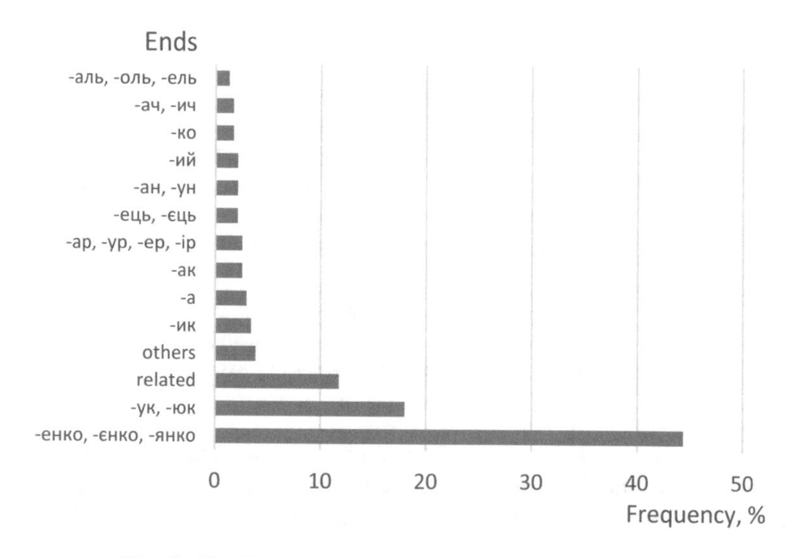

Fig. 2. The frequency with which different endings occur.

Therefore, the use of phonetic rules of the Ukrainian language is justified.

5 Experiment Conditions

A database of 17,631,472 records, representing approximately 41% of the population of Ukraine according to the State Statistics Service of Ukraine for 2018 (excluding the temporarily occupied territory of the Autonomous Republic of Crimea and Sevastopol) was used for data analysis [20].

All calculations and simulations were performed in a PostgreSQL 10.5 database environment using JetBrains DataGrip 2019.1.4 software on Amazon Web Services virtual hardware with 4 × 2.3 GHz processor and 16 GB memory size. All test tables were in the first normal form (1NF).

The data is entered in the database according to simplified validation rules (for example, a ban on entering digits); therefore mistakes in last names are possible. Additional data retrieval helps solve problems with 8,685 records (0.04%). After removal of all unnecessary elements (dashes, spaces, other service symbols) and replacement of misspelled ones (especially often at the beginning of the last name), it is necessary to bring the Latin characters in Cyrillic (the rules of casting are shown in Fig. 3). Also in the first stage the apostrophe is removed (see Listing 1), since it carries only a partial phonetic load.

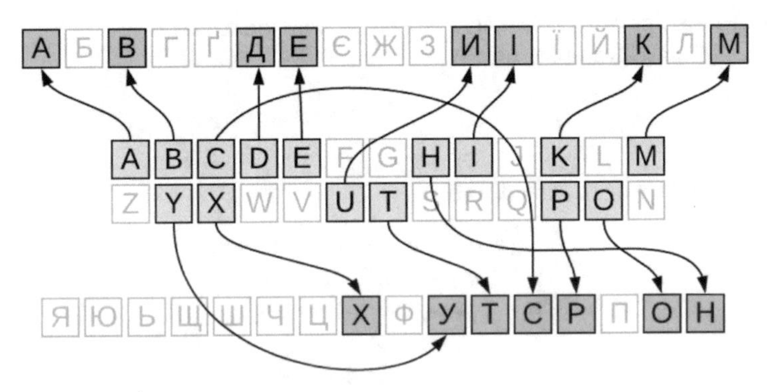

Fig. 3. Similarity in the form of Cyrillic and Latin ligatures.

Listing 1. Formation of "clean" data.

```
-- Remove all special symbols
UPDATE db SET
nm=regexp_replace(regexp_replace(regexp_replace(
replace(lower(nm), ' ', ''), '(-{2,})+', '-' , 'g'), '^-
|-$|''$', ''), '[^абвгґдеєжзиіїйклмнопрстуфхцчшщьюяa-z-
]+', '','g');

-- Remove Latin symbols, replace letters with similar
spelling (including capitalization)
UPDATE db SET nm = translate(nm, 'abcdehikmoptuxy',
'авсдеnікмортиху') WHERE nm IN (SELECT nm FROM lastnames
WHERE nm ~ '[a-z]');
```

Double and triple last names do not need to be cast because of their uniqueness. The number of such last names is 19,290 (3.6%). For all other last names, you can apply phonetic rules of regulation. It should be noted that the number of letters in the last names could be two or more.

6 Ukrainian Last Names Indexing

There are already known realizations of the Metaphone implementation algorithm for English, Russian [4] and other languages, but for the Ukrainian language, no implementation is known and the effect of phonetic search optimization to speed up the search and deduplication processes are unknown. Semantic search algorithms for last names cannot give a significant improvement in performance, since last names act as identifiers and do not carry a separate meaningful load.

For the implementation of the Metaphone phonetic algorithm, a set of phonetic rules of name normalization was developed based on the existing rules of grammar [21] and reference data on Ukrainian names [22]. This algorithm includes the following sequence of actions:

1. Remove the apostrophe (implemented in the previous step).
2. Exclude Ghe with upturn *-ґ-*:

-ґ- → *-г-*

3. Bring vowels to their sound forms:

{*-е-*, *-є-*, *-йе-*, *-іе-*, *-іо-*} → *-е-*
{*-а-*, *-я-*, *-іа-*, *-ія-*} → *-а-*
{*-і-*, *-ї-*, *-и-*} → *-и-*
{*-у-*, *-ю-*} → *-у-*
{*-о-*, *-йо-*} → *-о-*

4. Bring the short U *-у* ("ў") at the end of the word:

-у → *-в* (at the end of the word)

5. Replace letter combinations formed by *-ськ-*:

{*-ськ-*, *-сськ-*} → -1-
{*-зьк-*, *-гськ-*, *-жськ-*, *-зськ-*} → -2-
{*-цьк-*, *-дськ-*, *-тськ-*, *-кськ-*, *-чськ-*, *-цськ-*} → -3-

6. Delete the soft sign -ь-.
7. Replace consonants that are assimilated by sonant (affricate consonant and Ghe with upturn are not accounted for because of extraordinary rarity):

   {*п-* → *б-*, *х-* → *г-*, *т-* → *д-*, *ш-* → *ж-*, *с-* → *з-*} before {*-б*, *-г*, *-д*, *-ж*, *-з*}

8. Combine surd slit consonants:

   -хв- → *-ф-*

9. Replace consonants assimilated into consonant groups:

   {*-сч-*, *-жч-*, *-шч-*, *-щч-*} → *-щ-*

10. Simplify consonant groups:

    -стн- → *-сн-*
    -здн- → *-зн-*
    -слн- → *-сн-*
    -стл- → *-сл-*
    -шчн- → *-шн-*

11. Simplify another groups:

    -цв- → *-ц-*

12. Replace double letters with one (both consonants and vowels).
13. Compress endings longer than three characters:

-авко → *-A*	{*-ейко*, *-ейка*} → *-H*	*-ишко* → *-O*
{*-айко*, *-айка*} → *-B*	{*-енка*, *-енко*} → *-I*	*-ович* → *-P*
-айло → *-C*	*-ечко* → *-J*	*-онко* → *-Q*
-анко → *-D*	*-ешко* → *-K*	*-очко* → *-R*
-ашко → *-E*	*-ийло* → *-L*	*-уник* → *-S*
-евич → *-F*	*-иско* → *-M*	{*-унко*, *-унка*} → *-T*
-евка → *-G*	*-ишин* → *-N*	{*-ушко*, *-ушка*} → *-U*

This rule of thumb is important, for example, a soft sign -ь- is only removed after replacing the letters received from *-ськ-*. When implementing the above rules, we get the algorithm shown in Fig. 4. Others absorb some rules, so query optimization is applied to the database. An example implementation of a phonetic algorithm using regular expressions is shown in Listing 2.

Listing 2. Formation of indexes by phonetic rules (with comments).

```
-- 2. Exclude Ghe with upturn -ґ-
UPDATE db SET nm=replace(nm,'ґ','г');

-- 3. Bring vowels to their sound forms
UPDATE db SET nm=translate(nm,'іїєяю','иieay');
UPDATE db SET nm=replace(nm,'йе','e');
UPDATE db SET nm=replace(nm,'ie','e');
UPDATE db SET nm=replace(nm,'io','e');
UPDATE db SET nm=replace(nm,'ia','a');
UPDATE db SET nm=replace(nm,'йо','o');

-- 4. Bring the short U -y ("ў") at the end of the word
UPDATE db SET nm=regexp_replace(nm,'y$','в','g');

-- 5. Replace letter combinations formed by -ськ-
UPDATE db SET
nm=regexp_replace(nm,'(д|т|к|ч|ц)ськ','3','g');
UPDATE db SET nm=replace(nm,'цьк','3');
UPDATE db SET nm=regexp_replace(nm,'(г|ж|з)ськ','2','g');
UPDATE db SET nm=replace(nm,'зьк','2');
UPDATE db SET nm=regexp_replace(nm,'с+ьк','1','g');

-- 6. Delete the soft sign -ь-
UPDATE db SET nm=replace(nm,'ь','');

-- 7. Replace consonants that are assimilated by bell
UPDATE db SET
nm=regexp_replace(nm,'п(б|г|д|ж|з)','б\1','g');
...

-- 8. Combine deaf slit consonants
UPDATE db SET nm=replace(nm,'хв','ф');

-- 9. Replace consonants assimilated
UPDATE db SET nm=regexp_replace(nm,'(с|ж|ш|щ)ч','щ','g');

-- 10. Simplify consonant groups
UPDATE db SET nm=regexp_replace(nm,'с(т|л)н','сн','g');
UPDATE db SET nm=replace(nm,'здн', 'зн');
...

-- 11. Simplify another groups
UPDATE db SET nm=replace(nm,'цв','ц');
-- 12. Replace double letters with one
UPDATE db SET nm=regexp_replace(nm,'(\w)\1+','\1','g');

-- 13. Compress endings longer than three characters
UPDATE db SET nm=regexp_replace(nm,'авко$','A','g');
...
```

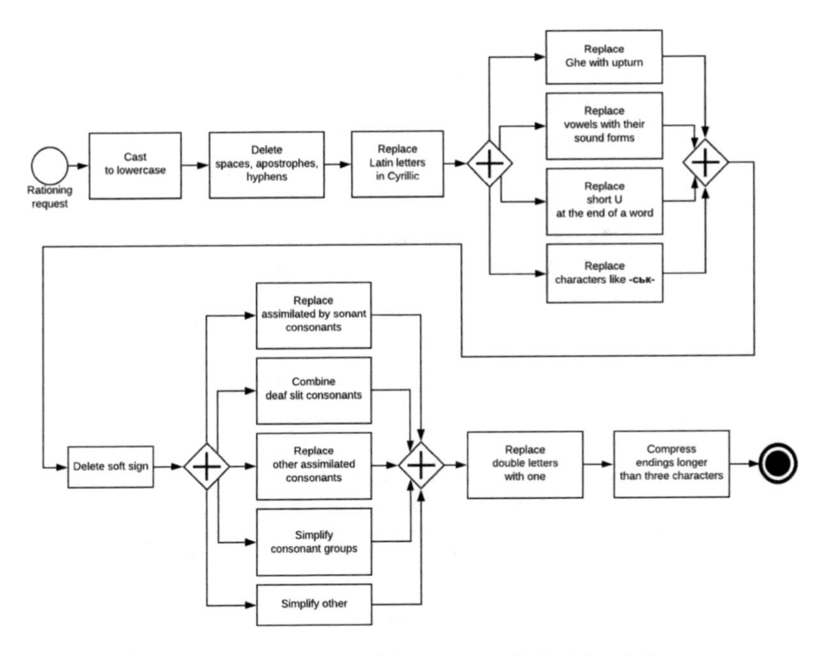

Fig. 4. BPMN diagram of the process of obtaining indexes.

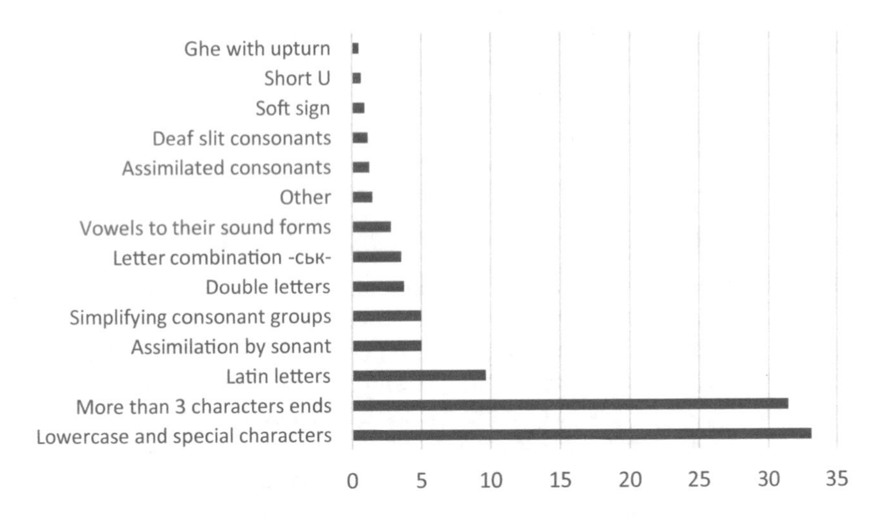

Fig. 5. Diagram of time usage for different types of last name modifications.

Due to the use of regular expressions (POSIX) and the large data, primary indexing can take considerable time, so optimizing the indexing process can save 15% of CPU time. Figure 5 shows the time costs for each operation, ranked by the execution time.

7 Multilanguage Pharmacy Names Indexing

Industries, additional conditions may arise, for example, when searching for medicines under an international nonproprietary name (INN) and a trademark, there is a need to search simultaneously in two languages: Ukrainian and Russian. However, in these related languages, there are different rules of assimilation by sonant and surd consonant, so the rules of index construction are enlarged and at the same time include only common phonetic features. Therefore, the sequence of actions for medicines will look this way:

1. Bring to the lowercase.
2. Replace dots, commas, dashes, showers, other special characters (®, ™, &, *, etc.) and double spaces with spaces.
3. Replace Latin characters with Cyrillic characters (Fig. 1).
4. Divide the names of several words into separate cells.
5. Delete all names with a length of less than four characters.
6. Remove the apostrophe, the soft sign -ь- and the hard sign -ъ-.
7. Exclude Ghe with upturn *-ґ-*:

$$\text{-ґ-} \rightarrow \text{-г-}$$

8. Bring vowels to their sound forms:

$$\{\text{-е-, -є-, -э-, -йе-, -ие-, -ie-, -io-, -ио-}\} \rightarrow \text{-е-}$$
$$\{\text{-а-, -я-, -ia-, -iя-, -иа-, -ия-}\} \rightarrow \text{-а-}$$
$$\{\text{-і-, -ї-, -и-, -ы-}\} \rightarrow \text{-и-}$$
$$\{\text{-у-, -ю-}\} \rightarrow \text{-у-}$$
$$\{\text{-о-, -йо-, -ё-}\} \rightarrow \text{-о-}$$

9. Replace double letters with one (both consonants and vowels).

According to the verification of the Morion Handbook [23], 83,637 medicines were obtained. From these data, trade and non-proprietary active substance names in the Ukrainian and Russian languages were separated. The result is 133,598 words, which transformed to 23,198 unique words. After creating the phonetic indexes according to the algorithm, 16,049 keys were obtained and the coefficient of optimization according to the Eq. (2) by the number of rows was 30.8%.

The application of this algorithm allows to automate the process of finding medicines in medical information systems for both pharmaceutical workers and users in situations where the drug is recorded on hearing or illegible recorded in a prescription.

8 Multilanguage Toponym Indexing

When forming indexes for toponyms, a thorough cleaning of the database with settlements and streets is required. Both registries must be separated. The cleaning sequence consists of the following steps:

1. Replace Latin characters with Cyrillic.
2. Delete all numerals that are indicated in Arabic and Roman numerals.
3. Delete anniversary indication.
4. Delete all abbreviations.
5. Remove the names of street types from the names: highway, path, road, highway, descent, street, promenade, house, square, etc.
6. Remove the designation of generally accepted infrastructure objects and relief elements from the names: market, station, village, farm, driveway, mountain, rampart, ravine, park, forest, grove, stop, booth, corner, bridge, river, island, bay, coast, beach, ford, etc., but retain rare names: roadside, forestry, station, meadow, natural boundary, beam, club, bathhouse, canal, cape, pond, rowing, pier, dock, etc.
7. Remove mention of family relationships, such as "family" or "brothers."
8. Delete all conjunctions and prepositions shorter than four letters.
9. Delete all names less than four letters long.
10. Replace hyphens with spaces if the parts of the word are homogeneous, and delete if they are the same name.

If the name of the settlement appears in the street name, then the previously obtained index is used, for example, "Red Kyiv." For phonetic search, the variability of forms in the same case or in different cases matters will have the same phonetic index. Comparison at the intersection of sets—directories of names with last names and many streets—does not give an adequate result, since among the streets names and second names can be in the plural or not in the nominative case.

34,962 names were collected in Ukrainian and Russian from 17,481 settlements. After removing hyphens and other data cleaning actions, 28,534 unique names were obtained. The use of the phonetic algorithm reduced the number of indices to 23,364. Similar actions for the formation of indices were carried out for street names. Table 2 shows the performance indicators of the use of phonetic indices. The data on the streets and cities of Ukraine are obtained from an open source—the State Register of Voters [24].

Table 2. Efficiency of phonetic indices for toponyms.

	Qty	Total toponyms in Ukrainian and Russian	Unique names	Phonetic indices	Reduction amount index, %
Settlements	17,481	40,210	28,534	23,364	41.9
Streets	32,804	161,934	44,906	32,955	79.6

The effectiveness of using phonetic indices for toponyms is much lower than for last names or medications, but gives a tangible result in increasing the speed of search and deduplication. The decrease in the number of indexes for streets is twice as much as for settlements, because the names of settlements are less varied.

Since the search and filtering operations take up considerable processor time in information systems, the optimization of these processes significantly saves computing power and production resources.

9 Performance Gains

The performance of indexes can be tested in two ways:

- The ratio of the number of indexes to the complete sequence or the average value of the number of elements per index.
- Decrease in index volume (number of characters).

You can use optimization coefficients by number of rows and by volume of characters to calculate the payoff:

$$\begin{cases} K_{\text{opt}}^{\text{qty}} = \left(1 - \frac{I^{\text{qty}}}{N^{\text{qty}}}\right) \cdot 100\%, \\ K_{\text{opt}}^{\text{vol}} = \left(1 - \frac{I^{\text{vol}}}{N^{\text{vol}}}\right) \cdot 100\%, \end{cases} \tag{2}$$

where I^{qty} and I^{vol} are indices by number and volume; N^{qty} and N^{vol} complete in order of number and volume.

The results of the payoff calculation are shown in Table 3. Depersonalized data obtained from the Helsi database [25].

Table 3. Calculation of winnings.

	Full size	Index size	Optimization factor, %
By number, row			
Structured sampling	547,825	434,495	20.7
Complete consistency	17,631,472		97.5
By size, symbol			
Structured sampling	9,213,759	8,358,969	9.3
Complete consistency	262,767,707		96.8

When implementing the algorithm for constructing phonetic indexes, there is a problem of integration with existing search engine optimization systems: semantic or other universal algorithms based on finding differences in sequences. In the case of last names, the use of semantic search does not make much sense, so universal algorithms for finding differences remain. Multiple algorithms do not significantly increase search quality, but require a multiple increase in hardware resources (and, consequently, can increase delays).

The order of application of phonetic algorithms is constant: first clearing of data from superfluous characters and other "garbage," and only after that carrying out of all transformations.

10 Conclusions and Future Work

Metaphone phonetic algorithm for Ukrainian can be applied to the tasks of search engine optimization by last name, data deduplication processes (combined multiple accounts), form autofill, fill-in tips. In addition, data can be widely used to reduce the number of errors when filling out electronic forms (questionnaires) in government and commercial information systems. The maximum gain in search speed when using phonetic indexes is $(97.5 \pm 0.1)\%$. The error is calculated by the number of foreign and double last names. This algorithm, along with Levenshtein distance calculation algorithms and other typing methods, can greatly improve the quality of search results:

- The sample of last names of Ukrainian origin amounted to approximately 41% of the population. Due to the high percentage of Ukrainian last names among the most frequently occurring ones (88%), we can argue that the use of phonetic search makes it possible to cover most search queries.
- Based on the statistical data and rules of formation of Ukrainian last names, we have constructed an algorithm of phonetic conversion for indexes for last names of Ukrainian origin. A given deduplication algorithm has made it possible to reduce the cyclical sorting procedure by three times.
- Theoretical results have been confirmed in the course of a field experiment using the samples of Ukrainian last names, which has allowed us to boast about saving server resources that host databases and optimizing search processes with an optimization coefficient of 30.8%.
- We have also tested algorithm performance for the mixed search for medicines and toponyms in the Ukrainian and Russian languages simultaneously.

In the study, the phonetic algorithm was applied to real datasets. Since words of foreign origin come across among these data, the phonetic rules are only partially valid. In addition, the limitation for using this algorithm may be unprepared datasets, for example, containing the Latin alphabet, special characters or numbers.

In the future, the authors plan to develop similar rules for Ukrainian names and patronymics, as well as to improve the identification of last names that may be in different genera (male and female). And to conduct a detailed study on the possibilities of simultaneous use of universal algorithms, based on the search for differences in sequences, and phonetic rules of index construction.

Acknowledgments. This scientific work was partially supported by RAMECS and self-determined research funds of CCNU from the colleges' primary research and operation of MOE (CCNU19TS022). In addition, the authors of the paper thank the management of the medical information system of Helsi LLC [20] for access to depersonalized medical data, a database of medical preparations and information resources for analysis and creation of a phonetic algorithm.

References

1. Branting, L.K.: A comparative evaluation of name-matching algorithms. In: 9th International Conference on Artificial Intelligence and Law (ICAIL 2003), pp. 224–232 (2003). https://doi.org/10.1145/1047788.1047837
2. Snae, C.: A comparison and analysis of name matching algorithms. Int. Sch. Sci. Res. Innov. **1**(1), 107–112 (2007)
3. Peng, T., Li, L., Kennedy, J.: A comparison of techniques for name matching. GSTF J. Comput. **2** (2012). https://doi.org/10.1037/e527372013-010
4. Karakhtanov, D.S.: Implementation of the Metaphone algorithm for Cyrillic surnames using the PL/SQL language. Young Sci. **8**(19), 162–168 (2010). [Publication in Russian]
5. Paramonov, V.V., Shigarov, A.O., Ruzhnikov, G.M., Belykh, P.V.: Polyphon: an algorithm for phonetic string matching in Russian language. Inf. Softw. Technol. 568–579 (2016). https://doi.org/10.1007/978-3-319-46254-7_46
6. Baruah, D., Kakoti Mahanta, A.: Design and development of Soundex for Assamese language. Int. J. Comput. Appl. **117**(9), 9–12 (2015). https://doi.org/10.5120/20581-3000
7. Silbert, J.M.: The world's first computerized criminal justice informationsharing system the New York State identification and intelligence system (NYSIIS). Criminology **8**(2), 107–128 (1970). https://doi.org/10.1111/j.1745-9125.1970.tb00734.x
8. Zahoransky, D., Polasek, I.: Text search of surnames in some Slavic and other morphologically rich languages using rule based phonetic algorithms. IEEE/ACM Trans. Audio Speech Lang. Process. **23**(3), 553–563 (2015). https://doi.org/10.1109/taslp.2015.2393393
9. Philips, L.: Hanging on the metaphone. Comput. Lang. **7**(12), 39–43 (1990)
10. Parmar, V.P., Kumbharana, C.K.: Study existing various phonetic algorithms and designing and development of a working model for the new developed algorithm and comparison by implementing it with existing algorithm(s). Int. J. Comput. Appl. **98**(19), 45–49 (2014). https://doi.org/10.5120/17295-7795
11. Koneru, K., Pulla, V.S.V., Varol, C.: Performance evaluation of phonetic matching algorithms on English words and street names—comparison and correlation. In: 5th International Conference on Data Management Technologies and Applications (2016). https://doi.org/10.5220/0005926300570064
12. Saw, T., Myint, P.H.: Feature selection to classify healthcare data using wrapper method with PSO search. Int. J. Inf. Technol. Comput. Sci. (IJITCS) **11**(9), 31–37 (2019). https://doi.org/10.5815/ijitcs.2019.09.04
13. Kumar, V., Tripathi, A.K., Chandra, N.: An efficient and optimized sematic web enabled framework (EOSWEF) for Google search engine using ontology. Int. J. Inf. Eng. Electron. Bus. (IJIEEB) **11**(5), 40–45 (2019). https://doi.org/10.5815/ijieeb.2019.05.06
14. Wirawan, K.T., Sukarsa, I.M., Bayupati, I.P.A.: Balinese historian chatbot using full-text search and artificial intelligence markup language method. Int. J. Intell. Syst. Appl. (IJISA) **11**(8), 21–34 (2019). https://doi.org/10.5815/ijisa.2019.08.03
15. Maurya, M., Singh, A.: An approach to parallel sorting using ternary search. Int. J. Mod. Educ. Comput. Sci. (IJMECS) **10**(4), 35–42 (2018). https://doi.org/10.5815/ijmecs.2018.04.05
16. Kathuria, M., Nagpal, C.K., Duhan, N.: Journey of web search engines: milestones, challenges & innovations. Int. J. Inf. Technol. Comput. Sci. (IJITCS) **8**(12), 47–58 (2016). https://doi.org/10.5815/ijitcs.2016.12.06

17. Gurung, D., Chakraborty, U.K., Sharma, P.: An analysis of the intelligent predictive string search algorithm: a probabilistic approach. Int. J. Inf. Technol. Comput. Sci. (IJITCS) **9**(2), 66–75 (2017). https://doi.org/10.5815/ijitcs.2017.02.08
18. Vijayarani, S., Muthulakshmi, M.: An efficient string matching technique for desktop search to detect duplicate files. Int. J. Inf. Technol. Comput. Sci. (IJITCS) **9**(7), 69–76 (2017). https://doi.org/10.5815/ijitcs.2017.07.08
19. Buriachok, V., et al.: Implantation of indexing optimization technology for highly specialized terms based on Metaphone phonetical algorithm. East. Eur. J. Enterp. Technol. **5**(2), 43–50 (2019). https://doi.org/10.15587/1729-4061.2019.181943
20. State Statistics Service of Ukraine: Population (estimated) as of January 1, 2018 and average in 2017 (2018). http://www.ukrstat.gov.ua/operativ/operativ2017/ds/kn/kn_u/kn1217_u.html. Accessed 24 Jan 2019. [Publication in Ukrainian]
21. Cabinet of Ministers of Ukraine: Ukrainian grammar (2019). https://mon.gov.ua/storage/app/media/zagalna%20serednya/05062019-onovl-pravo.pdf. Accessed 24 Jan 2019. [Publication in Ukrainian]
22. Red'ko, Y.K., Varchenko, E. (eds.): Directory of Ukrainian last names, p. 256. Soviet school, Kyiv (1968). [Publication in Ukrainian]
23. Program complex "Pharmacy:" Information web service (2019). https://pharmbase.com.ua/ru/project/web-content/. Accessed 24 Jan 2019. [Publication in Russian]
24. State Register of Voters: Approved datasets (2019). https://www.drv.gov.ua/ords/svc/f?p=111:4:12186060626032::NO. Accessed 24 Jan 2019. [Publication in Ukrainian]
25. Helsi: Electronic medical system for patients and doctors (2019). https://helsi.me. Accessed 24 Jan 2019. [Publication in Ukrainian]

Method of Recognition and Indexing of People's Faces in Videos Using Model of Machine Learning

Nataliya Kobets[1,2] and Tetiana Kovaliuk[1,2(✉)]

[1] Borys Grinchenko Kyiv University, 18/2 Bulvarno-Kudriavska Str., Kiev 04053, Ukraine
nmkobets@gmail.com, tetyana.kovalyuk@gmail.com
[2] Taras Shevchenko National University of Kyiv, 64/13, Volodymyrska Str., Kiev 01601, Ukraine

Abstract. This paper describes a machine learning approach for visual object detection in particular people's faces. Developed an algorithm based on the Viola-Jones algorithm. Images are used in the integral representation, which allows you to quickly calculate the necessary objects; signs of Haar are used, with the help of which the search for the needed faces and their features; boosting is used to select the most suitable features for the desired object in a given part of the image. Features of implementation of AdaBoost algorithm for solution a problem of construction of the classifiers cascade for effective detection of objects on images are considered in this paper. The algorithm is tested on an image containing several persons. Graphs of dependence of quality of detection on quantity of iterations are received.

Keywords: Face recognition · Clustering · Viola-Jones algorithm · Adaboost

1 Introduction

Terabytes of digital information, including photos and video content, appear in the world at any moment. Today, YouTube reports that more than 100 h of video are uploaded to its servers every minute [1]. Video is becoming more prevalent and video collection of public services and personal archives are constantly growing. Cisco predicts that by 2020, 82% of all consumer web traffic will be video. Video is the most popular trend in the presentation of information, as evidenced by such statistics [2]:

- almost 82% of Twitter users watch video content on this platform;
- people spend over 500 million hours watching videos on YouTube around the world;
- users watch about 10 billion videos daily on Snap chat;
- 92% of mobile users are actively involved in sharing videos with others;
- 87% of online marketers use video content;
- social videos generate 1200% more promotions than images and text content combined;
- video content leads to a 157% increase in organic traffic from search results;

Z. Hu et al. (Eds.): ICCSEEA 2020, AISC 1247, pp. 534–544, 2021.
https://doi.org/10.1007/978-3-030-55506-1_48

- placing videos on landing pages increases conversion rates by 80%.

The volume of video content proves that video search technology will become the core technology of the new digital world. Video data must be processed and their content must be indexed for information retrieval. For processing video data, computer vision methods are used.

The classic task in computer vision and image processing is a recognition task that determines whether video data contains a specific object, feature, or activity. Recognition problems can be divided into several processes:

- recognition: one or more predefined or studied objects or classes of objects can be recognized, usually together with their two-dimensional position in the image or three-dimensional position in the scene;
- identification: an individual instance of the object, such as a human face or fingerprints, or the vehicle number is recognized;
- detection: video data is checked for a certain condition. For example, detecting possible incorrect cells or tissues in medical images.

The scene recovery task is to reproduce a three-dimensional scene model from a picture of the scene or video data. In the simplest case, the model can be a set of points of three-dimensional space. The task of image recovery is to remove noise. The simplest approach to solving this problem is a variety of types of filters, such as low or medium frequency filters. One of the main tasks of computer vision, namely the recognition of people on video, as well as the indexing of video, that is, processing video to provide it with certain descriptors that indicate the presence of a particular person at a particular time, considered in this work.

The purpose of this work is to improve the quality and minimize errors in the recognition and indexing of persons present in the video. To achieve this purpose it is necessary to solve the following tasks:

- analyze known results of solving the object recognition problem on video;
- develop a solution method that combines different methods and algorithms;
- develop software that implements the above algorithms;
- to conduct research of efficiency of the developed algorithm.

2 Related Work

Many scientific papers of domestic and foreign researchers are devoted to the problem of detection of face and facial features [3–7]. A practical facial feature extraction system is described in [3]. The proposed system combines facial feature extraction and gender classification with robustness to images of different quality. Principal Component Analysis (PCA) technique has been used in the system of face recognition [4]. The system has implemented a face recognition system by using PCA, which is eigenvector based multivariate analyses. By implementing PCA, the face recognition system supplies the user with a lower-dimensional picture. A face detection with auto positioning for robotic vision has implemented in real-time system on robots.

The vision system supplies the visual communication between robot and human [5]. A real-time group face-detection system is presented in [6]. Algorithm based on analyzing facial properties and features in order to perform face detection for checking students' attendance in real time. The discrete complex fuzzy transform (DCFT) and the DCFT based facial image recognition method is presented in [7]. Authors study the DCFT to search different features of the images and increasing recognition ability. The presented DCFT consists of histogram extraction, peak points of histogram calculation and images construction. The face images data sets and local binary pattern are used to create facial image recognition method.

Viola-Jones [8, 9] algorithm is used to look for fast face detection. Four major ideas that works in real time are Haar features, integral image, AdaBoost and cascade structure. The Stanford/Technicolor/Fraunhofer HHI Video Semantic Indexing System is based on state-of-the-art image processing and machine learning techniques [10]. The system automatically annotates video shots with semantic concepts. The system processes a large amount of video data, representing them as separate fragments and recognizing various objects in them.

There is already a lot of research and practical development in the field of computer vision. But there are not many real systems that can effectively solve the problem. Finding ways to solve this problem is a very urgent problem.

3 Problem Statement

Let there be some video material with the image of people's faces appear at certain times. Some people may appear on the video several times. The background elements (the scene) can change and does not affect the quality of recognition the persons.

It is necessary to develop a software system that is able to process this video to extract information about people and the timing of their appearance on this video. The system does not have information about how many different people will be shown on the video and the moments of their appearance.

4 Description of the Proposed System

The developed system consists of four main subsystems:

- the subsystem of initial video processing and face recognition, which is engaged in video processing to extract from it all possible facial images that are found in it;
- the data clustering subsystem that selects the best facial images from the set of all examples of persons obtained in the previous step (which contain a large number of versions of the same person);
- the video processing subsystem for indexing people present on it. This subsystem use the best examples of facial images from the previous step, processes the video to determines the moments of appearance of each face;

- the subsystem for generating reports and output data based on the results of the indexing subsystem, enters new facial images and information about the indexed video into the database.

To solve the problem of identifying and indexing the presence of facial images on video, you need to have examples of the faces of these people. These examples can be obtained from the video itself. The clustering algorithm is used to select facial images. Each of the facial images is described by a numerical characteristic. This characteristic is used as a sign of clustering.

Clustering is the task of splitting multiple objects into groups called clusters. Similar objects should appear inside each group, and objects of different groups should be as different as possible. The main difference between clustering and classification is that the list of groups is not clearly defined and is determined during the operation of the algorithm.

The application of cluster analysis is reduced to the following steps in general [16]:

- defining a sample of objects for clustering;
- determining the number of variables by which the objects in the sample will be evaluated. If necessary, the values of the variables are normalized;
- calculating the values of similarity of objects;
- use of cluster analysis method to create groups of similar objects (clusters);
- presenting the results of the analysis.

After receiving and analyzing the results, it is possible to adjust the selected metric and clustering method to obtain the optimal result. The k-means algorithm, or fast cluster analysis, is the most common non-hierarchical method. It is necessary to have a hypothesis about the most likely number of clusters to use this method, in contrast to hierarchical methods that do not require previous assumptions about the number of clusters. The process of solving this task has three phases, each of which uses a specific algorithm for processing video.

The first stage of the system is to split the input video into separate frames and process each frame using cascading recognition algorithms (Viola-Jones method) to search all possible variants of faces. The result of the first stage is a database containing all the variants of facial images that have been identified on the video.

The next stage of the system is the processing of facial image examples obtained in the previous step. The clustering algorithm is applied to solve this problem. The numerical characteristic of each facial image is calculated. All images of the same face have approximately the same meaning of this characteristic and are grouped around the average value for each example. A database containing one of the most representative facial images of each person present on video is the result of the second stage.

The SURF algorithm is used in the last phase of the system [12]. This algorithm processes the video, looking for the appearance of specific images and remembering the moment of their appearance. The result of the program is a report containing a list of times when each of the face present on the video. The processed video contains selected and signed facial images.

5 Cascade Recognition Algorithm

Viola-Jones algorithm allows you to detect objects on real time image. Consider the following principles of the algorithm on which the method is based: [13, 14]

- images are used in the integral representation, which allows you to quickly calculate the necessary objects;
- signs of Haar are used, with the help of which the search for the needed object (in this context, the face and its features) occurs;
- adaptive boosting is used to select the most suitable features for the desired object in a given part of the image;
- all signs go to the input of the classifier, which gives the result "true" or "false";
- cascades of features are used to quickly drop windows where a face is not found.

5.1 Integrated Image Representation

The weight of each pixel represents its brightness. The integral value for each pixel is the sum of all values above and to the left of it, plus its own weight. The image can be integrated with several integer operations during the traversal process from the upper left corner to the right and down. The integral representation of an image is a matrix that is the same size as the original image. Each element of it stores the sum of the intensities of all pixels to the left and above this element [15]. Matrix elements are calculated according to (1):

$$L(x,y) = \sum_{i,j=0}^{i \le x, j \le y} I(i,j) \tag{1}$$

where $I(i,j)$ is the brightness of the pixel of the original image.

Each element of the matrix $L(x,y)$ represents the sum of the pixels in the rectangle from (0, 0) to (x,y) and the value of each pixel (x,y) is equal to the sum of the values of all pixels to the left and above the given pixel (x,y). The calculation of the matrix takes linear time, proportional to the number of pixels in the image, so the integrated image is calculated in one pass. The calculation of the matrix is possible according to (2):

$$L(x,y) = I(x,y) - L(x-1,y-1) + L(x,y-1) + L(x-1,y) \tag{2}$$

5.2 Signs of Class

As signs for the recognition algorithm, the authors proposed Haar signs based on Haar wavelets. Haar signs are rectangular areas that are composed of several adjacent rectangular areas marked as light or dark. The presence of the Haar signs is determined by subtracting the average value of the dark pixel region from the average values of the area of bright pixels. If the difference exceeds the threshold (determined in the learning process), then they say that the sign is existing.

Formally, a sign is a mapping $f : X \Rightarrow D_f$, where D_f is the set of admissible values of the sign. If the signs $f_1, \ldots, f_n$ are given, then the vector of signs $x = \{f_1(x), \ldots, f_n(x)\}$ is called the characteristic description of the object $x \in X$. Descriptions of signs can be identified with the objects themselves. Moreover, the set $X = D_{f_1} \ldots D_{f_n}$ is called a feature space.

The attributes are divided into the following types depending on the set D_f:

- binary sign, $D_f = \{0, 1\}$;
- nominal attribute: D_f is a finite set;
- ordinal attribute: D_f is a finite ordered set;
- quantitative attribute: D_f is a set of real numbers.

The signs of Haar give a point value of the difference in brightness along the X and Y axis, respectively. Therefore, a common Haar sign for face recognition is a set of two adjacent rectangles that lie above the eyes and cheeks. The value of the characteristic is calculated according to $F = X - Y$, where X is the sum of the brightness values of the points closed by the light part of the sign, and Y is the sum of the brightness values of the points closed by the dark part of the sign.

5.3 The Principle of the Scanning Window

Let's present an image with a two-dimensional matrix of pixels $w \times h$, in which each pixel has a certain value. The image is scanned by a search box represented by a rectangle $rect = \{x, y, w, h, \alpha\}$, where x, y are the coordinates of the center of the i-th rectangle; w is the width of the rectangle; h is the height of the rectangle; α is the angle of the rectangle to the vertical axis of the image.

The algorithm for scanning a window with signs looks like this:

- there is an investigated image, a scan window is selected, the used features are selected;
- the scanning window begins to move sequentially in the image in increments of 1 window cell (for example, the size of the window itself is 24 × 24 cells);
- in each window, approximately 200,000 options for the location of signs are calculated by changing the scale of the signs and their position in the scanning window;
- scanning is performed sequentially for various scales;
- the scanning window is scaled, i.e. the cell size is changed;
- all found signs fall into the classifier

5.4 Classifier Machine Learning Method

A finite set of images for which classes are known is specified (for example, it may be the class "nose position"). This set is called a training sample. The class affiliation of other objects is unknown. It is necessary to build an algorithm that can classify any object from the original set. To classify an object is to indicate the number (or name) of the class to which the object belongs. The classifier in classification problems is an approximating function that makes a decision which particular object belongs to the class.

Consider the classification problem. Let there be a plurality of object descriptions X and a set Y of class numbers. The mapping $Y^* : X \Rightarrow Y$ determines the relationship between the sets X, Y. The training sample is presented $X_m = \{(x_1, y_1), \ldots, (x_m, y_m)\}$. Let the function $f(X)$ of the sign vector X give the answer for any possible observation X and be able to classify the object $x \in X$.

The classification algorithms can be represented in the form $a(x) = C(b(x))$, where the function $b : X \Rightarrow R$ is called the algorithmic operator, $C : R \Rightarrow Y$ is the decision rule. The composition T of the algorithms $a_t(x) = C(b_t(x)), t = \overline{1 \ldots T}$ is the superposition of the algorithmic operators $b_t : X \Rightarrow R$, the correcting operation $F : R^T \Rightarrow R$, and the decision rule $C : R \Rightarrow Y$ C: R → Y:

$$a(x) = C(F(b_1(x), \ldots, b_r(x))) \tag{3}$$

The result is determined by a simple vote, by which the example belongs to the class that was issued by most models of the cascade.

Simple voting (4):

$$F(b1(x), \ldots, b_T(x)) = \frac{1}{T} \sum_{t=1}^{T} b_t(x) \tag{4}$$

Weighted voting (5):

$$F(b1(x), \ldots, b_T(x)) = \sum_{t=1}^{T} a_t b_t(x) \tag{5}$$

General Composition Building Algorithm

Let us consider the problem of classification into two classes $Y = \{-1, +1\}$. The decision rule $C(b) = sign(b)$ and basic algorithms $b_t : X \Rightarrow \{-1, 0, +1\}$ are specified. Basic algorithms $b_t(x) = 0$ means refusal. The algorithmic composition has the form (6):

$$a(x) = C(F(b_1(x), \ldots, b_T(x))) = sign\,(\sum_{t=1}^{T} a_t b_t(x)),\ x \in X \tag{6}$$

Composition quality functionality Q_t is defined as the number of errors allowed in a training sample (7):

$$Q(b, W^l) = Q_T = \sum_{t=1}^{l} y_l \sum_{t=1}^{T} a_t b_t(x_i) < 0 \tag{7}$$

where $W^l = (w_1, \ldots, w_l)$ is a vector of object weights.

When adding the term $a_t b_t(x)$ to the composition, only the basic algorithm b_t and its coefficient a_t are optimized, and all the previous terms $a_1 b_1(x), \ldots, a_{t-1} b_{t-1}(x)$ are assumed to be fixed. The threshold loss function in the Q_T functional is approximated by a continuously differentiable upper bound.

The iterative process can be represented as follows (8):

$$\begin{aligned} b_1 &= \arg\min_{b} Q(b, X^l) \\ b_2 &= \arg\min_{b} Q(F(b_1, b), X^l) \\ b_t &= \arg\min_{b,F} Q(F(b_1, \ldots, b_{t-1}), X^l) \end{aligned} \tag{8}$$

The threshold loss function can be approximated as sigmoid (9), logarithmic (10), exponential (11) and other functions:

$$S(z) = 2(1 + e^{z})^{-1} \tag{9}$$

$$L(z) = \log_2(1 + e^{-z}) \tag{10}$$

$$E(z) = e^{-z} \tag{11}$$

The Viola-Jones algorithm uses a threshold decision rule, where the operator is first built at zero and then the optimal value is selected. The process of sequential learning of basic algorithms is used to build the composition.

The algorithm stops under the following conditions:

- the set number of basic algorithms is built;
- the specified accuracy on the training set is achieved;
- the precision achieved on the control sample cannot be improved over the last few steps with a certain parameter of the algorithm.

AdaBoost algorithm is used to select the most suitable features and classifier learning [17, 18]. We describe the algorithm in steps.

The training sample X^l and the parameter T are input. We get the basic algorithms and their weights $a_t b_t, t = \overline{1 \ldots T}$ at the output. We define two quality metrics (12):

$$N(b; W^l) = \sum_{i=1}^{l} w_i [b(x_i) = -y_i], \; P(b; W^l) = \sum_{i=1}^{l} w_i [b(x_i) = y_i] \tag{12}$$

Step 1. To initiate object weights: $w_i = 1/l, i = \overline{1..l}$

Step2. For all values $t = 1, .., T$ of parameter T, train the basic algorithm (13):

$$b_t = \arg\min_{b} N(b; W^l) \;\; a_t = \frac{1}{2} \ln \frac{1 - N(b_t; W^l)}{N(b_t; W^l)} \tag{13}$$

Step 3. To update the weights of all objects: $w_i = w_i \exp(-\alpha_t y_i b_t(x_i), i = \overline{1 \ldots l}$.

Step 4. To normalize the weight of objects: $w_0 = \sum_{j=1}^{l} w_j$; $w_i = w_i / w_0$, $i = \overline{1..l}$

6 Algorithm Results

For training, 960 images were used, 320 of which were face images, 640 were face-free images. Face samples from the AT&T database, which contains images of 40 people using 10 different variations of head rotation and emotions, were used in the work. For testing, 2 images from each set were randomly selected, thus, in the test sample there were 80 images of people who are not in the training sample and 160 images of non-faces. A complete library of 400 images was used for direct detection. Samples not containing persons - 800. Since the AdaBoost learning algorithm is aimed at creating the best separation of data into two classes, an increase in the number of operations can achieve the best result. The algorithm is tested on an image containing multiple faces (Fig. 1).

The following graphs demonstrate the quality of the algorithm results. The research results correspond to the examples given in the technical literature [16].

Fig. 1. The result of the algorithm's operation on the test image

Figure 2 demonstrate the true positive rate (TPR) that is the ratio of correctly classified elements to the number of elements in this class, in this case the ratio of the number correctly detected persons to the total number of persons.

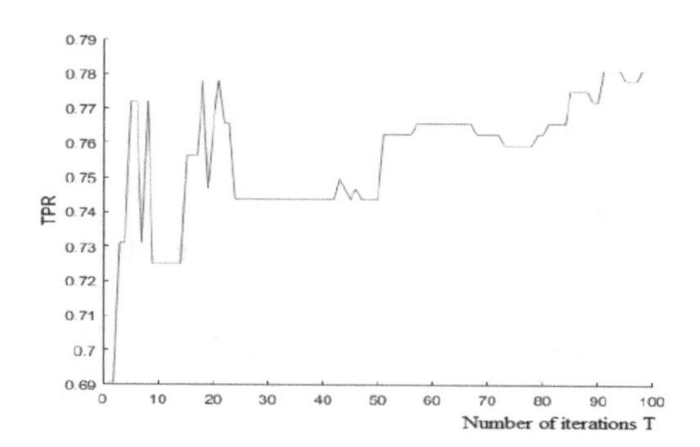

Fig. 2. TPR dependency on the number of AdaBoost iterations

Figure 3 demonstrate the false positive rate (FPR) that is the ratio of quantity is wrong of classified elements to the number of elements of this class, in the case of the ratio of the number of persons that have not been detected to the total number of persons.

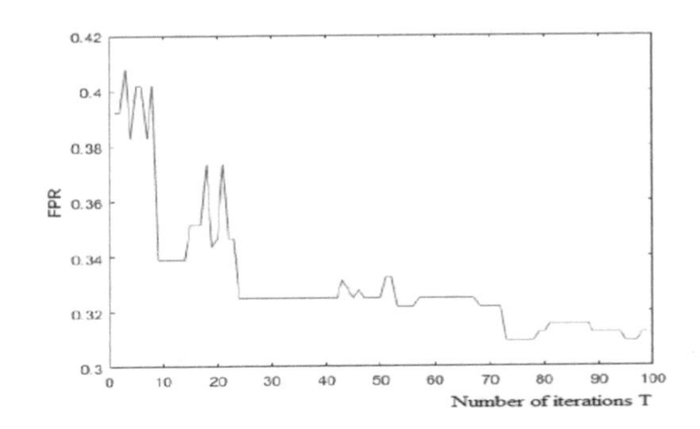

Fig. 3. FPR dependency on the number of AdaBoost iterations

7 Conclusion

The paper proposes a method for solving the problem of recognizing and indexing people present on video. Viola-Jones's image recognition method, which is based on the use of a filter cascade to detect the presence of an object in an image, has been studied in this work. Graphs of dependence of quality of detection on quantity of iterations are received. The algorithm is tested on an image containing several persons. The research results correspond to the examples given in the technical literature.

The developed software system for recognizing and indexing people on video will be used in the educational process at the university to registration of student attendance at lectures. Images of students' faces will be created using a video camera. Then images will be recognized and indexed with data entered in the attendance journal.

References

1. YouTube statistics. Access mode: http://www.youtube.com/yt/press/uk/statistics.html/. Accessed 10 Nov 2019
2. 7 tools for creating high-quality video content. https://blog.uamaster.com/7-tools-for-video-content-creation/. Accessed 05 Oct 2019
3. Tin, H.H.K.: Perceived gender classification from face images. Int. J. Mod. Educ. Comput. Sci. (IJMECS) **1**, 12–18 (2012)
4. Javed, A.: Face recognition based on principal component analysis. Int. J. Image Graph. Signal Process. (IJIGSP) **2**, 38–44 (2013)
5. Mariappan, M., Fang, T.W., Nadarajan, M., Parimon, N.: Face detection and auto positioning for robotic vision system. Int. J. Image Graph. Signal Process. (IJIGSP) **12**, 1–9 (2015)

6. Tamimi, A.A., AL-Allaf, O.N.A., Alia, M.A.: Real-time group face-detection for an intelligent class-attendance system. Int. J. Inf. Technol. Comput. Sci. (IJITCS) **06**, 66–73 (2015)
7. Tuncer, T., Dogan, S., Akbal, E.: Discrete complex fuzzy transform based face image recognition method. Int. J. Image Graph. Signal Process. (IJIGSP) **4**, 1–7 (2019)
8. Viola, P., Jones, M.J.: Rapid object detection using a boosted cascade of simple features. In: Proceedings IEEE Conference on Computer Vision and Pattern Recognition (CVPR 2001), pp. 1–9 (2001)
9. Viola, P., Jones, M.J.: Robust real-time face detection. Int. J. Comput. Vision **57**(2), 137–154 (2004)
10. de Araujo, A.F., Silveira, F., Lakshman, H., Zepeda, J., Sheth, A., Perez, P., Girod, B.: The Stanford/Technicolor/Fraunhofer HHI video semantic indexing system. https://www-nlpir.nist.gov/projects/tvpubs/tv12.papers/stanford.pdf. Accessed 10 Nov 2019
11. Cluster Analysis: basic Concepts and Algorithms. https://www-users.cs.umn.edu/~kumar001/dmbook/ch8.pdf. Accessed 05 Oct 2019
12. Oyallon, E., Rabin, J.: An analysis of the SURF method. https://www.ipol.im/pub/art/2015/69/article_lr.pdf. Accessed 15 Oct 2019
13. Wang, Y.-Q.: An analysis of the viola-jones face detection algorithm. IPOL J. Image Process. Line. **4**, 128–148 (2014)
14. Velykiy, Ya.O.: Analysis the principle of objects recognition on the basis of Viola – Jones. Open Information and Computer Integrated Technologies No. 68 (2015)
15. Galeev, S.: Modification of Viola-Jones algorithm for face recognition in real time. Modern technology and technologies, No. 5 (2017). http://technology.snauka.ru/2017/05/13529/. Accessed 12 Dec 2019
16. Naumov, N.: The Viola-Jones method as a basis for face recognition. https://habr.com/ru/post/133826/. Accessed 10 Nov 2019
17. Sherstobitov, A.I., Fedosov, V.P., Prihodchenko, V.A., Timofeev, D.V.: Face recognition on groups photos with using segmentation algorithms. Izvestiya SFedU. Engineering Sciences, pp. 66–72 (2013)
18. Šochman, J., Matas, J.: AdaBoost. center for machine perception. Czech Technical University, Prague (2010). http://cmp.felk.cvut.cz/~sochmj1/adaboost_talk.pdf/. Accessed 10 Nov 2019

Operational Intelligence Software Concepts for Continuous Healthcare Monitoring and Consolidated Data Storage Ecosystem

Solomiia Fedushko(✉) and Taras Ustyianovych

Lviv Polytechnic National University, Lviv, Ukraine
{solomiia.s.fedushko,
taras.ustyianovych.dk.2017}@lpnu.ua

Abstract. Patient health research has always remained as an important issue in medicine, because, in the absence of certain data, the lacks of digitized information researches and thorough studies have become a complex or even impossible processes. Software for real-time monitoring of a person's physical, psychological state is a necessary component that will allow him to gain additional insights about the patient's health, lifestyle and prevent certain diseases; carry out an exploratory data analysis based on the existing information. Storing data on a lot of locations, databases and services has been an issue that reduces performance and productivity, so an all-in-one software with all the necessary functions and single data warehouse is still a need for modern medical institutions. Intelligent systems and applications themselves play a major role in this process as well; contribute to detailed time-series analysis, metrics for human health assessment development, data integration, and systematization. Information gathering and data consolidation should be done using a variety of sensors, Internet of Things devices, etc. This, in turn, will greatly facilitate the detection of human health anomalies, help conduct additional researches, A/B testings. The software for operational intelligence monitoring is a need and should be an all-in-one solution that allows executing plenty of processes in real-time, contain built-in functions that may be modified and/or added for particular purposes. Developing and using of appropriate systems and frameworks saves time and cost while conducting various patient studies, and will also consolidate information and knowledge into single storage, real-time monitoring service with appropriate components.

Keywords: Health monitoring · Data consolidation · Analysis · Intelligent systems · Medical information · Big data · Data-driven medicine

1 Introduction

The integration of data analysis and various systems is an important and key component of healthcare and medicine fields of science development in the early 21st century. The usage of analytics, including statistical analysis methods, in medicine has always been important to scientists and physicians themselves [1]. Data visualization can help identify possible problems, diseases causes and defeat them. An excellent example may

Z. Hu et al. (Eds.): ICCSEEA 2020, AISC 1247, pp. 545–557, 2021.
https://doi.org/10.1007/978-3-030-55506-1_49

be the 1854 outbreak of cholera in London, when John Snow convinced authorities and critics that the disease spread from a water pump on Broad Street, leading to the now-legendary infographic map, showing the incidences of cholera clustered around the pump [2]. However, if the amount of data is much higher than the traditional size, then it should be processed in cloud environments, local or web-based applications, which will more effectively facilitate the retention and distribution of data for further analysis. The article aims to solve the problem of Big Data analytics in Healthcare for medical institutions, centers, etc., ways to develop application concept and data workflow for medical professionals in order to optimize their activities and performance. An application that should help physicians to solve and automate certain work tasks must include all or part of the functionality described in the Main part of the research and presented in the application architecture scheme. Most similar medical solutions have lacked the integrity and consolidation of all components, which is why there is a problem with the data formats diversity. Thus, Extract-Transform-Load (ETL) processes are implemented, which negatively affects performance and takes a long time to perform calculations, data transformation, etc. Existing solutions are not optimized enough, which causes data workflow, data integrity to deteriorate and potential data loss may occur, which is unacceptable in the medical and healthcare sectors.

The purpose of the research is to develop the concept and architecture of a solution for Operational Intelligence medical data monitoring on the patient's health condition, perform automated "binding" of the patient to the health care provider and doctor based on a number of factors; making personal data-driven recommendations to each user using intelligent systems; alerts on patients' health, medical records, and ways to prevent health anomalies.

The main goal is to allow low and critical level states and countries to apply information technology to medical data analysis. For this reason, the developed application should be accessible to low and high budget level healthcare institutions.

The research objectives are important due to the fact that Operational Intelligence Monitoring in healthcare and medicine will help reduce the time and cost of patients treating and physical examination and providing them with effective health recommendations. It will also allow patients to be distributed among healthcare providers based on smart data analysis. The software is designed to solve research and business problems, improve decision-making of a particular medical facility through advanced analysis and consolidated data warehouse. Simplification of various processes in a healhcare institution using information technologies for health monitoring, information analysis and visualization, data storage. The need to create a concept for the application leads to developing an appropriate software for automation healthcare facility functions, implementing complex monitoring and data collection mechanism.

The developed application will contain consolidated information and interactive dashboards to visually see all the information and make an effective decision about medical institution activity. It is a data repository that will integrate monitoring functions.

The developed application assumes API availability for quick data transfer from storages/repositories to the consolidated information source. An user-friendly interface should be taken care of. Having pre-designed (built-in) functions that calculate specific medical analysis and patient health status metrics can help optimize various types of calculations. Based on mathematical and statistical functions, different metrics will be computed. Service Reliability and Data security methods play a crucial role in the software life cycle. That is why we should take care of this and use methods for sensitive data detecting and clustering.

Data ingestion will be performed through continuous information streaming through APIs, connections between patient status data collection devices, as well as digitizing existing healthcare institution data, extracting them from medical facility databases. The data sources should unlimited and can be added by developing certain types of data inputs between the application and data source. This research can lead to a solution to real-time patient health monitoring; receive information about their condition, and alert physicians to potential health problems before their occurrence. Thus, the application will improve medical workers' performance and medical institutions will allow optimizing their activities, making it more flexible.

Complex metrics and health indicators will be calculated in a short time. In general, health care providers benefit from this, as they will be informed of their patients' status and have all information consolidated; algorithms that will objectively determine patients' distribution and particular pool for each doctor, taking into account various characteristics and ratios.

2 Analysis of the Recent Research and Publications

Researches of the various processes and routine tasks automation in the field of medicine have always remained a major factor in improving certain healthcare provider quality metrics, supporting decision-making and applying information technology in the field. The more data and information is accumulated in a particular healthcare facility, the more challenges arise with knowledge storage, processing, and analytics. Thereby, it is considered in the article to review scientific works on the latest developments implementation, production methods development and improvement, and to analyze existing solutions for monitoring in the medical field. A lot of use-cases were analyzed in the study of K. Priyanka, Kulennavar N. to provide insights of big data analytic in healthcare and medicine [4].

Monitoring application for medicine should consist of correlated components and functions in order to save time and costs for a particular healthcare provider or institution (Table 1).

Table 1. Table of analysis of the recent research and publications.

Direction of research	Scientists
Decision-making tasks in medicine	Elwyn G., Dehlendorf C., Epstein R. M. et al., Lown B. A., Clark W. D., Hanson J. L. [5], Frosch D., Thomson R. et al., Hargraves I., LeBlanc A., Shah N. D., Montori V. M. [6]
Investigation of ARAMIS (arthritis, rheumatism, and aging medical information system)	Singh G., Hunt R., B. Lazebnik L., C. Marakhouski Y., Manuc M, Gn R, S. Aye K., S. Bordin D., V. Bakulina N., S. Iskakov B, A Khamraev A, M Stepanov Y., Ally R, Garg A. [7]
Development of automated microscopy system	O. Berezsky, L. Dubchak, O. Pitsun [8], Dercksen V. [9]
Framework to evaluate the artificial intelligence (AI) readiness for the healthcare sector in developing countries	Vuong, Q.-H.; Ho, M.-T.; Vuong, T.-T.; La, V.-P.; Ho, M.-T.; Nghiem, K.-C.P.; Tran, B.X.; Giang, H.-H.; Giang, T.-V.; Latkin, C.; Nguyen, H.-K.T.; Ho, C.S.; Ho, R.C. [10], Gao, J.; Huang, X.; Zhang, L., Alberto Malva, Fabio Arpinelli, Giuseppe Recchia, Claudio Micheletto, Robert Alexander [11]
Telecare medical information systems	Qi Jiang, Zhiren Chen, Bingyan Li, Jian Shen, Li Yang, Jianfeng Ma [12], Arshad H, Nikooghadam M., Awasthi AK, Srivastava K., Das AK [13], Li, X., Wu, F., Khan, M. K., Xu, L., Shen, J., Jo, M. [14], R. Amin, S. H. Islam, P. Gope, K. R. Choo and N. Tapas [15], A. Chaturvedi, D. Mishra, S. Mukhopadhyay, Dongqing Xu, Jianhua Chen, Shu Zhang, Qin Liu [16]
Analysis of patterns between patients' disease and disease prognosis	Goldacre M., Kurina L., Yeates D., Seagroatt V., Gill L. [3]
Concept of establishing telecare medicine information systems (TMIS) for patients	Madhusudhan, R. & Nayak, C.S. [17], Chen H.M., Lo J.W., Yeh C.K., Debiao H., Jianhua C., Rui Z., Li C.T., Lee C.C., Weng C.Y., Chen S.J. [18]
Design and implementation of application system for hospital information system	Kim C.S., Kang S.S. [19], Ayatollahi H, Langarizadeh M, Chenani H. [20], Choi J., Kim J, Piao M, Jingwu W., Kwon YD, Yoon S.S., Chang H.
Tool for real-time monitoring patients	Tyagi, N., Yang, K., Gersten, D., Yan, D. [21], Sherlaw-Johnson, C., Morton, A., Robinson, M. B., Hall, A., Laffey, T. J., Cox, P. A., Schmidt, J. L., Kao, S. M., & Readk, J. Y., Fraser, H. S., Jazayeri, D., Mitnick, C. D., Mukherjee, J. S., Bayona, J. [22], Powell, W. C., Moore, T.
Big data medical system	Wu, J., Tan, Y., Chen, Z., Zhao, M. [23], Sebaa, A., Chikh, F., Nouicer, A., Tari, A., Wu, J., Tan, Y., Chen, Z., Zhao, M., Hayter, G. [24], Melnykova N., Shakhovska N., Greguš ml M., Melnykov V. [25], Melnykova, N., Melnykov, V., Vasilevskis, E. [26], Y. Bodyanskiy, I. Perova, O. Vynokurova, and I. Izonin [27], Syerov Y., Shakhovska N., Fedushko S. [28]

(*continued*)

Table 1. (*continued*)

Direction of research	Scientists
Medical diagnosis support system	Miyasa, K., Iizuka, Y., Katayama, A., & Ishikawa, R. [29], Sharma, P., Itu, L. M., Flohr, T., Comaniciu, D., Muraoka, S., Shimada, T., Sho, N. O. J. I. [30], Park, S. H., Han, K., Shortliffe, E. H., Sepúlveda, M. J.
Data privacy and security systems	Lemmey, T., Vonog, S. [31], Abouelmehdi, K., Beni-Hessane, A., Khaloufi, H., Luo, E., Bhuiyan, M. Z. A., Wang, G., Rahman, M. A., Wu, J., Atiquzzaman, M. [32], Bertino, E., Ferrari, E., Shakeel, P. M., Baskar, S., Dhulipala, V. S., Mishra, S., Jaber, M. M. [33]
Research of modern information technologies, medical and sports systems	Vorobel R. A., Gritsik V. V. [34], Zanevsky I. P. [35] Smerdov A. A., Yatsymirskyy M. M.

3 Data-Driven Solution for Operational Intelligence Medicine

3.1 Application Concepts and Architecture

The powerful computing capabilities of today's computer systems, as well as the latest advances in networking and online communications, allow performing complex data collection, integration, and consolidation into a single repository for further analysis. This is especially essential for the medical field, when the numbers of tools, sensors not only that accumulate information about the patient's condition but also help identify potential health issues and make certain predictions. The value of this data is of paramount importance to the healthcare facilities themselves for monitoring purposes.

Data migration requires a minimum amount of time, which is caused by very low response times, and costs because of open source solutions for data collection and ingestion. In this way, real-time or operational intelligence monitoring will allow the patient to be controlled, advised, set requirements for certain medical indicators, and send notifications in the case of their non-compliance or violation.

An application development that has all the needed components and functionality is important enough to be an all-in-one solution for healthcare and medicine. Also, it should be capable of establishing and maintaining connection to IoT devices. So, the software will consist of various components functions, and add-ons, which is very important to enhance the overall application usability and performance [36, 40].

For this reason, an application or online service should perform the functions described in Table 2 below:

Table 2. Health monitoring application main functions.

№	Function	Description
1	Connection with data collection devices	A kind of connector/data input to a sensor, light sensors in order to perform automation of data collecting; receiving feedback and information about the patient's well-being; analysis results; real-time data transfer
2	Data storage and compressing	A crucial role has an algorithm for data compressing to minimize its size paying attention to application usage scale and amount of data that will be indexed in the storage in order to perform further consolidation and analysis
3	Data analysis and visualization	This function allows the user to visualize information observing the metrics over a particular period, detect outliers in the data, gain knowledge and insights about samples' health status
4	Search/Query language	Data processing should be done using a query language, which is very crucial because that can help perform more of the described functions as well as data wrangling, to transform the data into a consistent format. Usage of a query language allows doing a lot of computations to gain insights, find correlations, etc.
5	Dynamic notifications to physicians/patients	This demand for a function may occur if alert conditions are violated when a patient abnormal health indicator is observed, which is critical and may affect his or her health status. This is a very important component that will help identify and prevent health problems in a timely manner. The number of conditions under which notifications can be sent is unlimited and can be supplemented by the needs of a health care provider or patient request
6	Automatic patients' partition to doctors	A built-in algorithm should determine which physician needs a particular patient and vice versa, calculate a particular doctor or medical institution staff workload, and automatically bind a patient with the physician. Then a load of each doctor can be seen and it is easy enough to split among them tasks, patients' treatment and more
7	Data privacy and security	In healthcare, it is very essential to perform sensitive data masking, because it is a need to take care of data security. The application must provide data encoding; the ability to provide high-security mode enabling for the purpose of completely or partially data hiding [37, 38]
8	Service scalability	Increase in the number of users, datasets demand for complex decisions making to optimize the system workload. So, a need for distributed clusters development that can split the load and prevent data loss and truncation occurs

Developing application architecture is one of the main research goals. Considering the functions that the solution must perform, it should include data migration, validation, compression, saving for processing stages.

Data flow might occur in real-time with a very low response time between components, as the information should arrive fairly quickly to be informed of the patient's health status, etc.

The factors that help implement information technologies and appropriate software architecture for Big Data processing into healthcare are described in the study of Yichuan Wang [39]. In Fig. 1 the software architecture and its key components are displayed.

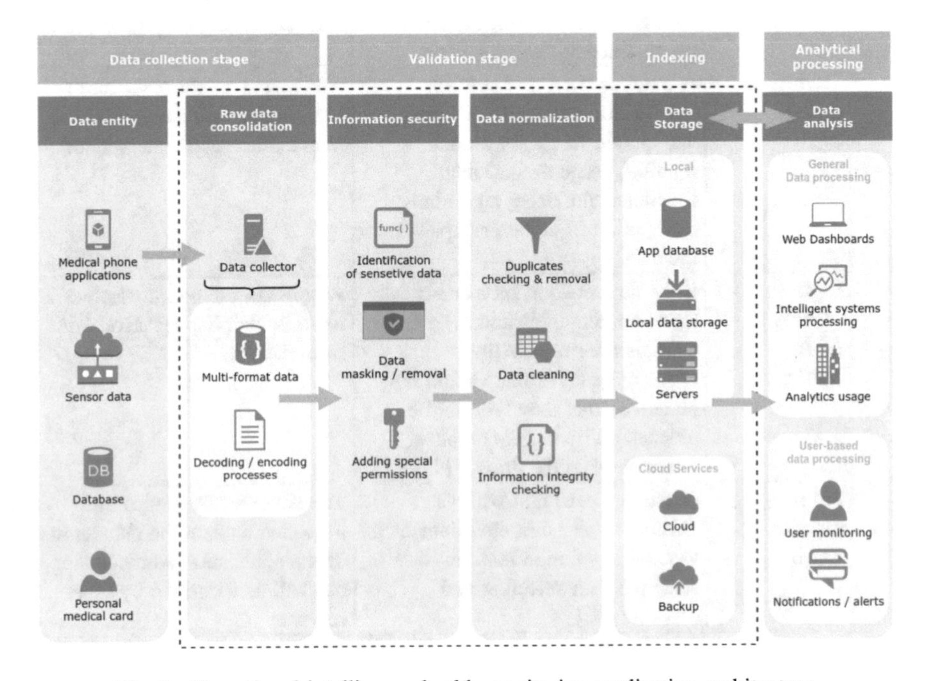

Fig. 1. Operational intelligence health monitoring application architecture.

Given that the app will contain patients, doctors, and healthcare providers' personal information, an important component is data privacy and security. This one, no less important than the other components, occurs during the data validation phase.

The main task is to detect and restrict access to sensitive data or even mask or remove it entirely from the data set. Certain methods for masking and de-identifying sensitive data are used in cloud computing to provide data privacy.

The developed protocol for identification and masking of personal data uses a kind of discovery tool algorithm with the possibility of its improvement or altering using human-computer interaction. This can be applied to the solutions' architecture described above.

Systems for dealing with sensitive and secure data in a technology environment should be implemented using particular methods [37]. Additionally, below are described some possible ways to mask personal data before its storing and indexing (Table 3):

Table 3. Methods for sensitive data processing.

№	Methods used	Usage conditions	Advantages
1	Regular expressions	Static data format; conviction that a particular factor that exists in the dataset and its format will not be modified in the near future	Lightweight solution; guaranteed masking of components that are considered necessary to be hidden
2	Trained model	A huge data set that contains information about personal data and certain labels for model training; usage of statistical functions, clustering algorithms, recognition of certain samples in the data	Automated solution; the ability to hide content without additional human interaction/intervention
3	High security mode restriction	If the dataset contains a persistent structure, the application administrator selects the appropriate fields that should be hidden so that they will not be indexed and will get completely removed from the dataset [41]	A constant method to remove fields containing sensitive data from datasets
8	3rd party services usage	Create a connection with an existing application environment to another solution that provides sensitive data detecting and masking [42]	The data will be hidden and identified without the intervention of the application administrator but will be identified by other services

3.2 Data Processing Capabilities

An application feature should be the ability to automatically “attach” a patient to a particular doctor, taking into account a variety of factors affecting both the first party (patient) and the latter (health care provider or physician). Such characteristics may be doctor’s profile compliance with the patient’s health complaints; doctor’s experience; disease severity; medical worker statistics; doctor’s workload, etc. These and other factors can be used to assist the patient in the selection of a needed specialist and will benefit those professionals who have insufficient or low numbers of patients but have the appropriate experience to perform medical treatment or consulting.

In addition, assisting in the selection of a physician for a particular patient, as well as the patients pool/partition formation for medical institution specialist can be implemented through intelligent systems usage, machine learning. It can simplify the processes described above as well but may require additional infrastructure load, tagged datasets, constant validation and algorithms accuracy checking, since the data would be ingested in real-time.

Computations of certain medical indicators, service quality in the medical facility with the help of mathematical and statistical functions will save time and money for determining the results of certain medical tests, analyzes, and conducting research in order to find variations between groups. Thus, the calculation of indicators, the service reliability level in a medical institution should be as built-in commands or functions belonging to the application. It will allow automating certain medical research processes, substituting only parameters with a data set, and to carry out complex data calculation, to carry out analytics, etc. Such indicators or metrics are as follows: Average Treatment time, Treatment Costs, Patient satisfaction (during and after the treatment), Rush hours identification, Costs per treatment (adjust costs and prices for specific services), another metrics that measure health status KPIs etc. One of the possible methods to measure patients' satisfaction level is to use an open-source Apdex Score standard commonly used in Application performance monitoring processes to determine how satisfied a user is with a particular online service and whether it works reliably. It ranges from 0 to 1, where 1 is the best score. Below is the equation for determining the Apdex Score for calculating Patient Satisfaction level for a doctor or a healthcare institution:

$$Apdexscore = \frac{\sum S + (\sum T)/2}{\sum F} \tag{1}$$

where

S is the total number of satisfied patients with the service;
T is the sum of persons who are tolerated with medical service;
F is the total number of patients, who are not satisfied with service at all and have got not their best treatment ever.

In order to determine Apdex score levels (satisfying, tolerating, frustrating), you need to have a pre-prepared data sample from patients who have either specified a certain level or calculate it independently. To do this, a threshold needs to be set and the level can be computed and identified. If a particular indicator is lower or equal than threshold, level is satisfying, if a indicator is higher than the threshold and lower the threshold multiplied by 4, then the level is toleration, in all other cases - the Apdex score level is frustrating. However, when, for example, medical treatment has been failed despite various indicators, the Apdex score level is assigned to frustrating.

An open ESRD QIP dataset - Total Performance Scores - Payment Year 2015 was used for Apdex score validation in healthcare, which contains a total performance evaluation by medical facilities across the USA, the value of which ranges from 0 to 100, where 100 is the highest [43]. The Apdex score was calculated for 10 randomly selected States to identify which health care facilities have higher Apdex scores and, accordingly, better performance. However, the rules for setting Apdex Score levels have been slightly modified for qualitative calculations. The threshold in our case is the

Total Performance Score 80th percentile. If the Performance score is higher or equal to a threshold, the Apdex Score level is set to "Satisfying"; if the performance score is less than threshold/4, then the Apdex Score is "Frustrating". In all other cases, the level is "Tolerating". As it is shown in the Figure below, the highest Apdex Score has Michigan State and it is equal to 0.60, which is a pretty good score but not the best. The lowest scores have New York and the Florida States, their Apdex is 0.534 and 0.536 accordingly.

With the help of the Apdex Score, it is possible to observe how well a healthcare facility meets patients' needs and expectations and makes it easy to identify causality between components. It can also lead to better patients understanding, drawing attention to feedback from customers and others. This is one of the key metrics that tell about performance evaluation and should be easily computed and monitored in the operational intelligence application ecosystem. So, feedback collecting should be one of the application features that leads to further medical facility improvements and activity analysis (Fig. 2).

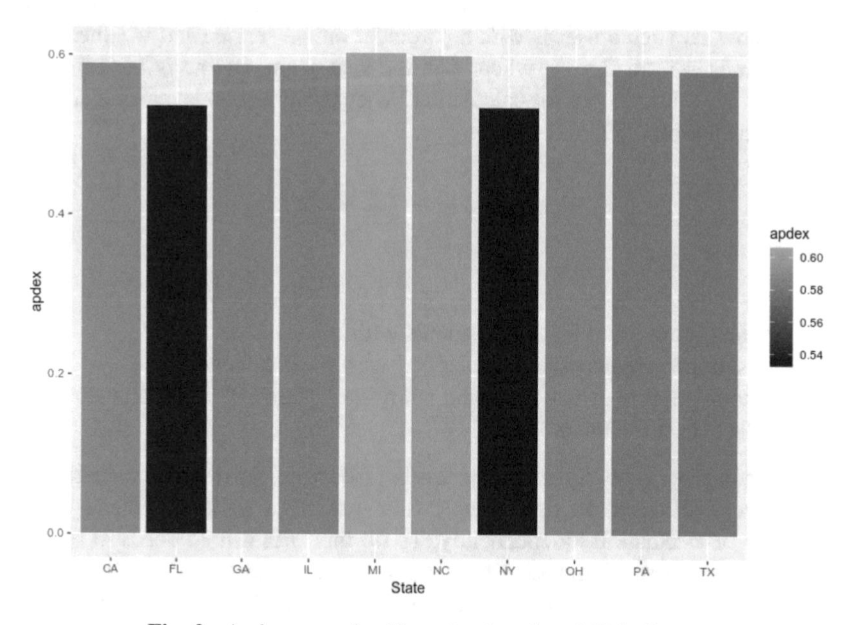

Fig. 2. Apdex score for 10 randomly selected U.S. States.

4 Conclusion

Operational intelligence in medicine is a solution for continuous data flow from certain documents, databases and information sources. Application usage for operational intelligence monitoring purposes is a very essential medical institution concept that allows increasing various KPIs, healthcare provider service level indicators, etc. Applying IoT technologies and Big Data analytics into medicine is a very crucial process, which might improve a lot of methodologies and help reduce costs and time. But it should be done carefully and special attention must be payed to a particular healthcare facility problem and issue to bring business value. Apart from that, the solution can serve for monitoring these metrics and indicators, consolidate all the data into single storage, and do specific computations. Strategy for implementing Big Data analytics and operational intelligence healthcare monitoring ought to be studied as well. So, further researches will be done to develop scalable infrastructure and software and/or framework for using and applying the application to solve real-life problems in the field of healthcare and medicine. The most essential software component should be real-time data streaming and processing functions for notifications about a particular issue etc. as well as applying intelligent algorithms and frameworks for natural language processing/understanding and using it inside a particular ecosystem, which is a very crucial and quickly-evolving trend in the field of healthcare [44, 45]. Operational intelligence healthcare will keep physicians, health institution managers or nurses with up-to-date statistics about patients' health status, medical treatment or results. The data will be available and stored on different storage services (cloud, local, hard-disks, etc.). Custom visualization and dashboard panels can be split by a few tiers of information detail (overview, intermediate or search, and detailed dashboards). The Consistent data management is a need for modern healthcare and medical service providers due to the fact that handling of increased amounts of medical data requires new approaches, usage of Big Data and Data Science technologies. So, that will lead to new researches in healthcare, well-done patient-physician interaction, data-driven decisions for medical treatment improvement, workload balance among medical institution workers, and overall management.

References

1. Kononenko, I.: Machine learning for medical diagnosis: history, state of the art and perspective. Artif. Intell. Med. **23**(1), 89–109 (2011)
2. The 1855 Map That Revolutionized Disease Prevention & Data Visualization: Discover John Snow's Broad Street Pump Map. http://www.openculture.com/2019/07/the-1855-map-that-revolutionized-disease-prevention-data-visualization.html
3. Goldacre, M., Kurina, L., Yeates, D., et al.: Use of large medical databases to study associations between diseases. QJM: J. Assoc. Phys. **93**(10), 669–675 (2000). PubMed PMID: 11029477
4. Priyanka, K., Kulennavar, N.: A survey on big data analytics in healthcare. Int. J. Comput. Sci. Inf. Technol. (MECS) **5**(4), 5865–5868 (2014)
5. Lown, B.A., Clark, W.D., Hanson, J.L.: Mutual influence in shared decision making: a collaborative study of patients and physicians. Health Expect. **12**(2), 160–174 (2009)
6. Hargraves, I., LeBlanc, A., Shah, N.D., Montori, V.M.: Shared decision making: the need for patient-clinician conversation, not just information. Health Aff. **35**(4), 627–629 (2016)
7. Hunt, R., Lazebnik, L.B., Marakhouski, Y.C., et al.: International consensus on guiding recommendations for management of patients with nonsteroidal antiinflammatory drugs induced gastropathy-ICON-G. Eur. J. Hepatogastroenterol. **8**(2), 148–160 (2018)

8. Berezsky, O., Dubchak, L., Pitsun, O.: Access distribution in automated microscopy system. In: 14th International Conference The Experience of Designing and Application of CAD Systems in Microelectronics, CADSM 2017, Polyana, Ukraine, pp. 241–243 (2017)
9. Dercksen, V.J., Hege, H.-C., Oberlaender, M.: The filament editor: an interactive software environment for visualization, proof-editing and analysis of 3D neuron morphology. Neuroinformatics **12**(2), 325–339 (2013)
10. Vuong, Q., Ho, M., et al.: Artificial intelligence vs. natural stupidity: evaluating ai readiness for the vietnamese medical information system. J. Clin. Med. **8**, 168 (2019)
11. Malva, A., Arpinelli, F., Recchia, G., et al.: Artificial intelligence applied to asthma biomedical research: a systematic review. Eur. Respiratory J. **54**, PA1482 (2019). https://doi.org/10.1183/13993003.congress-2019.pa1482
12. Jiang, Q., Chen, Z., Li, B., Shen, J., Yang, L., Ma, J.: Security analysis and improvement of bio-hashing based three-factor authentication scheme for telecare medical information systems. J. Ambient Intell. Humaniz. Comput. **9**(4), 1061–1073 (2017). https://doi.org/10.1007/s12652-017-0516-2
13. Das, A.K.: A secure user anonymity-preserving three-factor remote user authentication scheme for the telecare medicine information systems. J. Med. Syst. **39**, 30 (2015)
14. Li, X., Wu, F., et al.: A secure chaotic map-based remote authentication scheme for telecare medicine information systems. Future Gener. Comput. Syst. **84**, 149–159 (2018)
15. Amin, R., Islam, S.H., Gope, P., et al.: Anonymity preserving and lightweight multimedical server authentication protocol for Telecare medical information system. J. Biomed. Health Inf. **23**(4), 1749–1759 (2019)
16. Xu, D., Chen, J., Zhang, S., et al.: J. Med. Syst. **42**, 219 (2018)
17. Madhusudhan, R., Nayak, C.S.: A robust authentication scheme for telecare medical information systems. Multimed. Tools Appl. **78**(11), 15255–15273 (2018). https://doi.org/10.1007/s11042-018-6884-6
18. Li, C.T., Lee, C.C., Weng, C.Y., Chen, S.J.: A secure dynamic identity and chaotic maps based user authentication and key agreement scheme for e-healthcare systems. J. Med. Syst. **40**(11), 233 (2016)
19. Kim, C.S., Kang, S.S.: Design and implementation of RFID application system for hospital information system. J. Korean Soc. Med. Inform. **11**(4), 399–407 (2005)
20. Ayatollahi, H., Langarizadeh, M., Chenani, H.: Confirmation of expectations and satisfaction with hospital information systems: a nursing perspective. Healthc. Inform. Res. **22**(4), 326–332 (2016). https://doi.org/10.4258/hir.2016.22.4.326
21. Tyagi, N., Yang, K., et al.: A real time dose monitoring and dose reconstruction tool for patient specific VMAT QA and delivery. Med. Phys. **39**(12), 7194–7204 (2012)
22. Fraser, H.S., Jazayeri, D., Mitnick, C.D., et al.: Informatics tools to monitor progress and outcomes of patients with drug resistant tuberculosis in Peru. In: AMIA Symposium, American Medical Informatics Association, p. 270 (2002)
23. Wu, J., Tan, Y., et al.: Data decision and drug therapy based on non-small cell lung cancer in a big data medical system in developing countries. Symmetry **10**(5), 152 (2018)
24. Hayter, G.: U.S. Patent No. 10,111,608. Patent and Trademark Office (2018)
25. Melnykova, N., Shakhovska, N., Gregušml, M., Melnykov, V.: Using big data for formalization the patient's personalized data. Proc. Comput. Sci. **155**, 624–629 (2019). https://doi.org/10.1016/j.procs.2019.08.088
26. Melnykova, N., Melnykov, V., Vasilevskis, E.: The personalized approach to the processing and analysis of patients' medical data. CEUR Workshop. **2255**, 103–112 (2018)

27. Bodyanskiy, Y., Perova, I., Vynokurova, O., Izonin, I.: Adaptive wavelet diagnostic neuro-fuzzy network for biomedical tasks. In: 14th International. Conference on Advanced Trends in Radioelecrtronics, Telecommunications and Computer Engineering, Ukraine, pp. 711–715 (2018)
28. Syerov, Y., Shakhovska, N., Fedushko, S.: Method of the data adequacy determination of personal medical profiles. In: Hu, Z., Petoukhov, S.V., He, M. (eds.) Advances in Artificial Systems for Medicine and Education II, AIMEE2018 2018. Advances in Intelligent Systems and Computing, vol. 902, pp. 333–343. Springer, Cham (2020). https://doi.org/10.1007/978-3-030-12082-5_31
29. Miyasa, K., Iizuka, Y., et al.: U.S. Patent No. 10,068,056. Washington, DC: U.S. Patent and Trademark Office (2018)
30. Muraoka, S., Shimada, T., Sho, N.: U.S. Patent No. 10,278,661. Washington, DC: U.S. Patent and Trademark Office (2019)
31. Lemmey, T., Vonog, S.: Management of data privacy and security in a pervasive computing environment. U.S. Patent Application No. 10/121,015 (2018)
32. Luo, E., Bhuiyan, M., et al.: Privacyprotector: privacy-protected patient data collection in IoT-based healthcare systems. Commun. Mag. **56**(2), 163–168 (2018)
33. Shakeel, P.M., Baskar, S., et al.: Maintaining security and privacy in health care system using learning based deep-Q-networks. J. Med. Syst. **42**(10), 186 (2018)
34. Hrytsyk, V.V., Hrytsyk, V.V.: Information-quality assessment of data process and transfer in condition of noise and distortion of messages in computer vision tasks. Inform. Technol. Syst. **10**(1), 179–190 (2007)
35. Zanevsky, I.: Computer mathematics and modeling of complex systems in sport. Teorìâ Ta Metodika Fìzičnogo Vihovannâ (8), 17–23 (2017)
36. Shanmugasundaram, G., Sankarikaarguzhali, G.: An investigation on IoT healthcare analytics. Int. J. Inform. Eng. Electron. Bus. (MECS) **9**(2), 11–19 (2017)
37. Todeschini, E., Deloge, S.P., Anderson, D.: System and method for securing sensitive data. U.S. Patent No. 9,195,844. Washington, DC: U.S. Patent and Trademark Office (2015)
38. Manogaran, G., Varatharajan, R., Lopez, D., et al.: A new architecture of internet of things and big data ecosystem for secured smart healthcare monitoring and alerting system. Future Gener. Comput. Syst. **82**, 375–387 (2018)
39. Wang, Y., Kung, L., Byrd, T.A.: Big data analytics: understanding its capabilities and potential benefits for healthcare organizations. Technol. Forecast. Soc. Chang. **126**, 3–13 (2018)
40. Hossain, S., Muhammad, G.: Cloud-assisted industrial internet of things (IIoT) – enabled framework for health monitoring. Comput. Netw. **101**(4), 192–202 (2016). https://doi.org/10.1016/j.comnet.2016.01.009
41. Luo C., Fylakis A., Partala J., Klakegg S., Goncalves J., Liang K., Kostakos V.. A data hiding approach for sensitive smartphone data. In: Proceedings of the 2016 ACM International Joint Conference on Pervasive and Ubiquitous Computing, pp. 557–568 (2016)
42. Dynamic de-identification of data. https://patents.google.com/patent/US8881019B2/en
43. Centers for Medicare & Medicaid Services. ESRD QIP - Total Performance Scores - Payment Year 2015. (2015) https://catalog.data.gov/dataset/esrd-qip-total-performance-scores-payment-year-2015
44. Iroju, O.G., Olaleke, J.O.: A systematic review of natural language processing in healthcare. Int. J. Inform. Technol. Comput. Sci. (MECS) **8**, 44–50 (2015)
45. Olaronke, I., Oluwaseun, O.: An ontology based remote patient monitoring framework for Nigerian healthcare system. Int. J. Mod. Educ. Comput. Sci. (MECS) **8**, 17–24 (2016)

Relations of the Genetic Code with Algebraic Codes and Hypercomplex Double Numbers

Sergey V. Petoukhov(✉)

Mechanical Engineering Research Institute, Russian Academy of Sciences, M. Kharitonievsky pereulok, 4, Moscow, Russia
spetoukhov@gmail.com

Abstract. The article shows materials to the question on algebraic features of the genetic code. Presented results testify in favor that the genetic code is an algebraic code related with a wide class of algebraic codes, which are a basis of noise-immune coding of information in communication technologies. Algebraic features of the genetic code are associated with hypercomplex double (or hyperbolic) numbers and with doubly stochastic matrices. The article also presents data on structural relations of some genetically inherited macrobiological phenomena with double numbers and with their algebraic extensions. The received results confirm that multidimensional numerical systems is effective for modeling and revealing the interconnections of structures of biological bodies at various levels of their organisation. This allows one to think that living organisms are algebraically encoded entities. New classes of mathematical models and technical applications are possible on the basis of described results.

Keywords: DNA · Genetic code · Algebraic codes · Noise-immunity · Double numbers · Dyadic groups · Hamming distance

1 Introduction

Living bodies are a huge number of molecules interconnected by quantum-mechanical and stochastic relationships. These sets of molecules have an amazing ability to inherit the biological characteristics of organisms to next generations. G. Mendel, in his experiments with plant hybrids, found that the transmission of traits during the crossing of organisms occurs in accordance with certain algebraic rules, despite the colossal heterogeneity of the molecular structure of their bodies. These algebraic rules of polyhybrid crossbreeding in genetics textbooks since 1906 are presented in the form of Punnett squares resembling mathematical square matrices in their structure. Mendel also proposed a model for explanation of the observed rules, introducing the idea of binary-oppositional forms of the existence of factors of inheritance of traits: dominant and recessive forms.

This article continues the search for algebraic models of the natural features of genetic structures and of inherited macrobiological phenomena. As known, the key difference between living and inanimate objects is as follows: inanimate objects are controlled by the average random movement of millions of their particles, while in a living organism, genetic molecules have a dictatorial effect on the entire living

Z. Hu et al. (Eds.): ICCSEEA 2020, AISC 1247, pp. 558–570, 2021.
https://doi.org/10.1007/978-3-030-55506-1_50

organism. By this reason, the author focuses on the presentation and studing of the system of genetic alphabets and of the genetic code in the form of mathematical matrices constructed on binary-oppositional features of DNA alphabets.

On this way the article addresses the issues of coding information in the genetic system. In a broad sense, code is usually understood as correspondence between two sets of characters. For example, from this point of view a usual phone book can be considered as a coding system, in which its phone numbers encodes names of people. But this article considers analogies of the genetic code with more complex kinds of codes termed as algebraic codes and algebra-geometric codes, which are widely used in modern communication technologies for algorithmic providing a noise-immunity transfer of information. Obviously, the genetic coding of information has noise immunity, which allows the transfer of genetic information from ancestors to descendants along the generation chain in very difficult and different living conditions of organisms. The study of possible algorithms for noise-immunity transfer of genetic information is an important scientific task, the successful solution of which can give a lot of useful for engineering, medical, biotechnological and othersciences. It is about unraveling the bioinformational patents of living matter.

This article draws special attention to structural analogies of the molecular system of genetic coding with one of the known types of multidimensional hypercomplex numbers commonly called double numbers (although other their names are also used in the literature: hyperbolic numbers, Lorentz numbers, split-complex numbers, etc.). The set of all two-dimensional double numbers forms algebra over the field of real numbers [1, 2]. The algebra is not a division algebra or field since it contains zero divisors. Hyperbolic numbers $aj_0 + bj_1$ (where j_0 is the real unit; j_1 is the imaginary unit; a and b are real numbers) are well known in mathematics and theoretic physics and they have matrix form of their representations by the following bisymmetric matrices: $G_2 = [a, b; b, a]$, where a and b are real numbers (Fig. 1). Such matrix represetations of double numbers we will be termed simply and briefly as matrix double numbers. In such bisymmetric matrices, components a and b are located along two diagonals in the cruciform shape, which is met below in this article many times. Any bisymmetric matrix of any order with real entries a and b has an orthogonal set of eigenvectors with real eigenvalues.

$$\begin{vmatrix} a, b \\ b, a \end{vmatrix} = a* \begin{vmatrix} 1, 0 \\ 0, 1 \end{vmatrix} + b* \begin{vmatrix} 0, 1 \\ 1, 0 \end{vmatrix} = a*j_0 + b*j_1;$$

*	1	j_1
1	1	j_1
j_1	j_1	1

Fig. 1. The decomposition of the bisymmetric matrix [a, b; b, a] into two sparse matrices, where the sparse matrices j_0 and j_1 are correspondingly matrix representations of the real unit and the imaginary unit of algebra of 2-dimensional double numbers. The multiplication table of these units is shown at right.

An important special case of double numbers and their matrix representations $[a, b; b, a]$ are hyperbolic rotations, which satisfied the condition $a^2 - b^2 = 1$. Hyperbolic rotations are represented in the special theory of relativity, in Minkowski geometry and in some other physical and mathematical fields. The algebra of double numbers can be extended to algebras of 2^n-dimensional double numbers having also representations by bisymmetric ($2^n * 2^n$)-matrices. If all the components of such matrices are non-negative quantities, then these matrices are doubly stochastic matrices (under an appropriate matrix normalization), which have many applications in theory of games, optimization, etc. The author's results of analysing genetic systems from the standpoint of such doubly stochastic matrices will be published later.

The article shows connections of genetic code stuctures with such bisymmetric (2 * 2)-matrices and their unions into square matrices of higher orders. Described and some author's results testify in favor of the statement that the genetic code is an algebraic code and is connected with a wide class of algebraic and algebraic-geometric codes used in modern technologies of noise-immune communication [3, 4]. These results add previous results of matrix analysis of genetic structures [5–15].

2 The Genetic Code, DNA Alphabets and Genetic Matrices

In DNA molecules genetic information is written in sequences of 4 kinds of nucleobases: adenine A, cytosine C, guanine G and thymine T. They form a DNA alphabet of 4 monoplets. In addition, DNA alphabets of 16 doublets and 64 triplets also exist. As known, the set of these 4 nucleobases A, C, G and T is endowed with binary-oppositional indicators:

- 1) in the double helix of DNA there are two complementary pairs of letters: the letters C and G are connected by three hydrogen bonds, and the letters A and T by two hydrogen bonds. Given these oppositional indicators, one can represent C = G = 1 and A = T = 0;
- 2) the two letters are keto molecules (G and T), and the other two are amino molecules (A and C). Given these oppositional indicators, one can represent A = C = 1, G = T = 0.

Taking this into account, it is convenient to present DNA alphabets of 4 letters, 16 doublets and 64 triplets in the form of square tables, the columns of which are numbered in accordance with oppositional indicators "3 or 2 hydrogen bonds" (C = G = 1, A = T = 0), and the rows in accordance with oppositional indicators "amino or keto" (C = A = 1, G = T = 0). In such tables, all letters, doublets and triplets automatically occupy their strictly individual places (Fig. 2).

These three tables (Fig. 2) are not only simple tables but they are members of the tensor family of matrices: the second and the third tensor (Kronecker) powers of the matrix [C, A; G, T] generate similar arrangements of 16 doublets and 64 triplets in matrices $[C, A; G, T]^{(2)}$ and $[C, A; G, T]^{(3)}$ as shown in Fig. 2.

The genetic code is called a "degenerate code" because 64 triplets encode 20 amino acids and stop-codons so that several triplets can encode each amino acid at once and

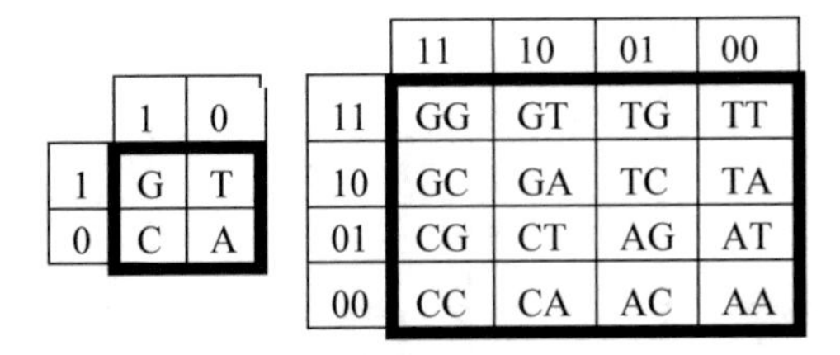

	1	0
1	G	T
0	C	A

	11	10	01	00
11	GG	GT	TG	TT
10	GC	GA	TC	TA
01	CG	CT	AG	AT
00	CC	CA	AC	AA

	111	110	101	100	011	010	001	000
111	GGG	GGT	GTG	GTT	TGG	TGT	TTG	TTT
110	GGC	GGA	GTC	GTA	TGC	TGA	TTC	TTA
101	GCG	GCT	GAG	GAT	TCG	TCT	TAG	TAT
100	GCC	GCA	GAC	GAA	TCC	TCA	TAC	TAA
011	CGG	CGT	CTG	CTT	AGG	AGT	ATG	ATT
010	CGC	CGA	CTC	CTA	AGC	AGA	ATC	ATA
001	CCG	CCT	CAG	CAT	ACG	ACT	AAG	AAT
000	CCC	CCA	CAC	CAA	ACC	ACA	AAC	AAA

Fig. 2. The square tables of DNA-alphabets of 4 nucleotides, 16 doublets and 64 triplets with a strict arrangement of all components. Each of tables is constructed in line with the principle of binary numeration of its column and rows on the basis of binary-oppositional indicators of nucleobases A, C, G and T.

each triplet necessarily encodes only a single amino acid or a stop-codon. The (8×8)-matrix of 64 triplets (Fig. 2) was built formally without any mention of amino acids and stop-codons. Nothing data preliminary exist on a possible correspondence between triplets and amino acids. How can these 20 amino acids and stop-codons be located in this matrix of 64 triplets? There are a huge number of possible options for the location and repetition of separate amino acids and stop-codons in 64 cells of this matrix. More precisely, the number of these options is much more than 10^{100} (for comparison, the entire time of the Universe existance is estimated in modern physics at 10^{17} s). But Nature uses - from this huge number of options - only a very specific repetition and arrangement of separate amino acids and stop-codons, the analysis of which is important for revealing the structural organization of the informational foundations of living matter.

Figure 3 shows the real repetition and location of amino acids and stop-codons in the Vertebrate Mitochondrial Code, which is the most symmetrical among known dialects on the genetic code. This genetic code is called the most ancient and "ideal" in genetics [16] (other dialects of the genetic code have small differences from it).

The location and repetition of all amino acids and stop-codons in the matrix of 64 triplets have the following feature (Fig. 3):

- Each of sixthteen (2 * 2)-subqudrants, forming this genetic matrix and denoted by bold frames, is bisymmerical: each of its both diagonals contains an identical kind of amino acids or stop-codon.

	111	110	101	100	011	010	001	000
111	GLY GGG	GLY GGT	VAL GTG	VAL GTT	TRP TGG	CYS TGT	LEU TTG	PHE TTT
110	GLY GGC	GLY GGA	VAL GTC	VAL GTA	CYS TGC	TRP TGA	PHE TTC	LEU TTA
101	ALA GCG	ALA GCT	GLU GAG	ASP GAT	SER TCG	SER TCT	STOP TAG	TYR TAT
100	ALA GCC	ALA GCA	ASP GAC	GLU GAA	SER TCC	SER TCA	TYR TAC	STOP TAA
011	ARG CGG	ARG CGT	LEU CTG	LEU CTT	STOP AGG	SER AGT	MET ATG	ILE ATT
010	ARG CGC	ARG CGA	LEU CTC	LEU CTA	SER AGC	STOP AGA	ILE ATC	MET ATA
001	PRO CCG	PRO CCT	GLN CAG	HIS CAT	THR ACG	THR ACT	LYS AAG	ASN AAT
000	PRO CCC	PRO CCA	HIS CAC	GLN CAA	THR ACC	THR ACA	ASN AAC	LYS AAA

Fig. 3. The location and repetition of 20 amino acids and 4 stop-codons (denoted by bold) in the matrix of 64 triplets $[C, A; G, T]^{(3)}$ (Fig. 2) for the Vertebrate Mitochondrial Code. The symbol "Stop" refers to stop-codons.

If each amino acid and stop-codon is represented by some characteristic parameter (for example, the number of carbon atoms in these organic formations or numbers of protons in its molecular structure, etc.), then a numerical (8 * 8)-matrix arises (Fig. 4) with bisymmetric (2 * 2)-subquadrants representing double numbers $aj_0 + bj_1$ described above in the Sect. 1 (Fig. 1). In other words, this phenomenologic arrangement of amino acids and stop-codons in the matrix of 64 triplets is associated to the multiblock union of matrix presentations of 16 two-dimensional double numbers. This arrangement

2	2	5	5	11	3	6	9
2	2	5	5	3	11	9	6
3	3	5	4	6	6	0	9
3	3	4	5	6	6	9	0
6	6	6	6	0	6	5	6
6	6	6	6	6	0	6	5
5	5	5	6	4	4	6	4
5	5	6	5	4	4	4	6

Fig. 4. The numeric analogue of the symbolic (8 * 8)-matrix of amino acids and stop-codons from Fig. 3 for the case of representing each of amino acids by numbers of its carbon atoms (stop-codons are conditionally represented by zero).

can be also considered as the ensemble of 16 doubly stochastic (2 * 2)-matrices with their individual number factors.

Demonstrated by the matrices in Fig. 3 and 4, the connection of the genetic code with double numbers supplements the following statement of the author, presented in a number of his publications [5–15]. The genetic code is not just a mapping of one set of elements to other sets of elements by type, for example, of a phone book in which phone numbers encode names of people. But the genetic code is inherently an algebraic code, akin to those algebraic codes that are used in communication theory for noise-immune transmission of information. Algebraic features of the genetic code participate in noise-immune properties of this code and of the whole genetic system.

One can explain the meaning and possibilities of algebraic codes by the example of transmitting a photograph of the Marsian surface from Mars to Earth using electromagnetic signals. On the way to the Earth, these signals travel millions of kilometers of interference and arrive at the Earth in a very weakened and distorted form. But, in a magical way, based on these mutilated signals on Earth, a high-quality photograph of the surface of Mars is recreated. The secret of this magic lies in the fact that from Mars not the information signals about this photo are sent, but algebraically encoded versions of these signals that is quite other signals. At receivers on Earth, these algebraically encoded signals are algebraically decoded into signals, which recreate the original photographic image of the surface of Mars. It should be emphasized that algebraic coding of information in noise-immune communication actively uses the mathematical apparatus of matrices, which is also used in quantum informatics and quantum mechanics as matrix operators.

By analogy with this example from the theory of noise-immune communications, the author's statement, that the genetic code of amino acids is algebraic one, entails the following author's hypothesis: the molecular-genetic system is a system of certain algebraic codes, which serves to provide noise-immune transmission of genetic and - in addition - some important pra-genetic information along the chain of generations. The author's works are aimed at studying algebraic properties of the genetic coding system for revealing hidden biological information algebraically encoded in the molecular genetic system [5–15].

What possible reasons of such character of the coding of amino acids and stop-codons, which is built in accordance with a matrix having 16 bisymmetric sub-quadrants? The author seeks an answer to this question analysing, firstly, relations of the genetic code with foundations of algebraic codes in the theory of noise-immune coding of information and, secondly, the role of doubly stochastic matrices in the structural organization of the genetic phenomena. The next Section describes some of the search results connected with notions of dyadic groups, Hamming distances and matrices of dyadic shifts.

3 The Genetic Code, Dyadic Groups and the Matrix of Dyadic Shifts

Many structural features of genetic coding systems indicate the important role of dyadic groups of binary numbers [5, 9, 17].

Binary numbers and their dyadic groups occupy a very important place in modern science and technology including computers, digital noise-immune communication, systems of artificial intelligence, etc. DNA alphabets, having 4 monoplets, 16 doublets and 64 triplets, are connected namely with dyadic groups of binary n-bit numbers (n = 2,3,4, …), each of which contains 2^n members [18]. For example, the dyadic group of 3-bit numbers contains 8 members:

$$000, 001, 010, 011, 100, 101, 110, 111 \quad (1)$$

The operation of the bitwise modulo-2 addition serves as the group operation in dyadic groups. It is denoted by the symbol ⊕ and has the following rules: $0 \oplus 0 = 0$; $0 \oplus 1 = 1$; $1 \oplus 0 = 1$; $1 \oplus 1 = 0$. Modulo-2 addition is utilized broadly in the theory of discrete signal processing and algebraic coding as a fundamental operation [3].

The modulo-2 addition of any two binary numbers from a dyadic group always results in a new number from the same group. For example, modulo-2 addition of two binary numbers 110 and 101 from (1), which are equal to 6 and 5 respectively in decimal notation, gives the result $110 \oplus 101 = 011$, which is equal to 3 in decimal notation. The number 000 serves as the unit element of this group: for example, $010 \oplus 000 = 010$. The reverse element for any number in this group is the number itself: for example, $010 \oplus 010 = 000$. Series (1) is transformed by modulo-2 addition of all its members with the binary number 001 into a new series of the same numbers: 001, 000, 011, 010, 101, 100, 111, 110. Such changes in the initial binary sequence, produced by modulo-2 addition of its members with any of binary numbers from (1), are termed dyadic shifts [3, 18]. Figure 5 shows an example of bisymmetric matrices of dyadic shifts where each of entries is received by modulo-2 addition of binary numerations of its column and row [3]. Such bisymmetric matrices are used in the theory of algebraic coding of noise-immune information transfering. To emphasize a close relation of the matrix of dyadic shifts with the genetic matrix of triplets in Figs. 2 and 4, appropriate triplets from Fig. 2 can be shown in Fig. 5 in all cells of the dyadic shift matrix.

	111	110	101	100	011	010	001	000
111	000 (0)	001 (1)	010 (2)	011 (3)	100 (4)	101 (5)	110 (6)	111 (7)
110	001 (1)	000 (0)	011 (3)	010 (2)	101 (5)	100 (4)	111 (7)	110 (6)
101	010 (2)	011 (3)	000 (0)	001 (1)	110 (6)	111 (7)	100 (4)	101 (5)
100	011 (3)	010 (2)	001 (1)	000 (0)	111 (7)	110 (6)	101 (5)	100 (4)
011	100 (4)	101 (5)	110 (6)	111 (7)	000 (0)	001 (1)	010 (2)	011 (3)
010	101 (5)	100 (4)	111 (7)	110 (6)	001 (1)	000 (0)	011 (3)	010 (2)
001	110 (6)	111 (7)	100 (4)	101 (5)	010 (2)	011 (3)	000 (0)	001 (1)
000	111 (7)	110 (6)	101 (5)	100 (4)	011 (3)	010 (2)	001 (1)	000 (0)

Fig. 5. The bisymmetric (8 * 8)-matrix of dyadic shifts. In each matrix cell, there is shown a binary number and its decimal value (in brackets).

One can see from Fig. 5 that this matrix of dyadic shifts consists of 16 bisymmetrical (2 * 2)-subquadrants in some analogy with the genetic matrices in Fig. 3 and 4. Each of these subquadrants is a matrix representation of 2-dimensional double number. It is obvious that this matrix has analogies with the phenomenological distribution of amino acids and stop-codons in a matrix of 64 triplets, which also consists of 16 bisymmetrical (2 * 2)-subquadrants.

If any system of elements demonstrates its connection with dyadic shifts, it indicates that the structural organization of the system is connected with the logic modulo-2 addition. The works of [5, 6, 14] show additionally that the structural organization of the molecular-genetic system is connected with dyadic shifts and correspondingly with modulo-2 addition. Dyadic shifts are also involved in the patterns of molecular-genetic alphabets. The author pays special attention to non-trivial structures of interrelated and structured alphabets of DNA and RNA since these alphabets are key elements in the noise-immune transmission of genetic information.

4 The Genetic Code and the Matrix of Hamming Distances

In the theory of algebraic coding of information in modern communication technologies, the concept of code distance plays an extremely important role, and code distance most often means the so-called Hamming distance [19]. The Hamming distance between two symbol strings of equal length is the number of positions at which the corresponding symbols are different. In other words, it measures the minimum number of substitutions required to change one string into the other, or the minimum number of errors that could have transformed one string into the other. For example, let us consider the triplet AAC, which is represented in Fig. 2 by the binary sequence 001 (in relation to the binary indicators "3 or 2 hydrogen bonds") and by the binary sequence 111 (in relation to the binary indicators "amino or keto"). It is obvious that the Hamming distance between these two representations of the triplet AAC is equal to 2. By analogy, calculations of Hamming distances between such two binary representations of each of triplets in the matrix in Fig. 2 lead to the bisymmetric matrix of Hamming distances for triplets in Fig. 6.

	111	110	101	100	011	010	001	000
111	0	1	1	2	1	2	2	3
110	1	0	2	1	2	1	3	2
101	1	2	0	1	2	3	1	2
100	2	1	1	0	3	2	2	1
011	1	2	2	3	0	1	1	2
010	2	1	3	2	1	0	2	1
001	2	3	1	2	1	2	0	1
000	3	2	2	1	2	1	1	0

Fig. 6. The bisymmetric matrix of Hamming distances for two kinds of binary representations of 64 triplets in the genetic matrix in Fig. 2.

From Fig. 6 one can show that the bisymmetric (8 * 8)-matrix of Hamming distances consists of 16 bisymmetrical (2 * 2)-subquadrants, each of which is a matrix representation of 2-dimensional double number. It is obvious that this matrix has analogies with the phenomenological distribution of amino acids and stop-codons in a matrix of 64 triplets, which also consists of 16 bisymmetrical (2 * 2)-subquadrants.

5 Hypercomplex Double Numbers and Inherited Macrobiological Phenomena

The organism is a single inherited whole. All its inherited physiological subsystems must be structurally coupled with genetic coding, otherwise they cannot be encoded and inherited (doomed to extinction). For this reason, it makes sense to study relationship of structures of inherited macrobiological phenomena with double numbers represented in the structures of the genetic coding system. This Section shows some data on such relationships.

For 150 years, in biology, inherited helical bio-lattices, associated with Fibonacci numbers, are studied under the title "phyllotaxis laws" [20]. (Fibonacci series: $F_n = F_{n-2} + F_{n-1}$: 0, 1, 1, 2, 3, 5, 8, 13, 21, …). For example, the numbers of left and right spirals in the heads of the sunflower are equal to the neighboring members of the Fibonacci series. Such phyllotaxis phenomena exist in plant and animal bodies at various levels and branches of biological evolution. The ratios F_{n+1}/F_n denote the order of symmetry of phyllotaxis lattices. In the process of growth of some organisms, their phyllotaxis lattices are transformed with the transition to other Fibonacci relations F_{k+1}/F_k (Fig. 7).

Fig. 7. Examples of phyllotaxis configurations in biological bodies.

Ukrainian Prof. O. Bodnar paid his attention that such growth transformations of phyllotaxis lattices correspond to hyperbolic rotations known in the special theory of relativity as Lorentz transformations [21, 22]. On this basis, he declared that living matter is structurally related to Minkowski geometry.

Another example of the biological realization of structures associated with double numbers is given by the heritable organization of locomotion in animal organisms. As known, living bodies have innate abilities for locomotion. For example, newborn turtles and crocodiles, when they hatched from eggs, crawl by quite coordinated movements to water; millipedes manage the coordinated movement of dozens of their legs, using inherited algorithms, etc. One of the most respectable Russian journal

"Uspekhi Fizicheskih nauk" published a large article by Prof. V. Smolyaninov "Spatio-temporal problems of locomotion control" with results of his 20 years of research on locomotion of a wide variety of animals and humans [23]. According to his results, spatio-temporal organization of locomotion control is related – by a special manner - with hyperbolic rotations and with Minkovsky geometry. On this basis, Smolyaninov put forward his "Locomotor theory of relativity" and wrote about a relativistic brain and relativistic biomechanics.

Additional structural connections of double numbers with the genetic code system and inherited macrobiological phenomena are shown in the following author's results in his preprint [24, 25]:

- Modeling the basic psychophysical law of Weber-Fechner, which has a logarithmic character, by means of hyperbolic rotations;
- Modeling of percentages (or frequencies) 2% and 3% of hydrogen bonds 2 and 3 and also their hydrogen *n*-plets in long DNA sequences on the basis of the 2^n-dimensional double numbers $[3\%, 2\%; 2\%, 3\%]^{(n)}$ where (n) refers to tensor powers;
- Using 2^n-dimensional double numbers $[\%0, \%1; \%1, \%0]^{(n)}$ to model percentages (or frequencies) %0 and %1 in a special binary representation of long Russian literary texts (by L.N. Tolstoy, F.M. Dostoevsky, A.S. Pushkin and others) on the basis of a known phonetic separation of all letters of the Russian alphabet into two classes of equivalency, denoted by 0 and 1;
- Connections of some bisymmetric genetic matrices representing 2^n-dimen- sional double numbers with matrices of the golden section f = 1,618… .
- Some similarities between a cruciform character of bisymmetric matrices of double numbers $[a, b; b, a]^{(n)}$ and a cruciform character of many inherited constructions of physiological macrosystems including sensory-motion systems (for example, in human brain, the left hemisphere serves the right half of the body and the right hemisphere serves the left half of the body).

6 Results and Discussions

Results described in this article testify in favor of the following:

- The genetic code is an algebraic code associated with the wide class of algebraic codes from theory of noise-immunity coding of information;
- The use of algebras of multidimensional numerical systems is effective both for revealing the interconnections of structures of biological bodies at various levels of their organization, and for understanding the noise-immune properties of genetic informatics. This allows one to think that living organisms are algebraically encoded entities.

One can think that the algebraic features of genetic systems are based on some quantum-mechanical, resonance and stochastic mechanisms of self-organisation of living bodies related with bisymmetric matrix operators and doubly stochastic matrices.

These mechanisms should be studied in future to understand the following key difference between living and inanimate objects, which was emphasized by creators of quantum mechanics P. Jordan and E. Schrödinger [26]: inanimate objects are controlled by the average random movement of millions of their particles; in contrast inside a living organism genetic molecules, which are only one of types of its molecules, have a dictatorial effect on the entire living organism.

The notion of "number" is the main notion of mathematical natural sciences. Pythagoras has formulated the idea: "*Numbers rule the world*" since he noted that numbers can dictate different geometric shapes. In view of this idea, natural phenomena should be explained by means of systems of numbers. As W. Heisenberg noted, modern physics is moving along the same Pythagorean path [27]. B. Russell stated that he did not know any other person who could exert such influence on the thinking of people as Pythagoras [28]; correspondingly there is no more fundamental scientific idea in the world than this idea about a basic meaning of numbers. Our proposed approach using multidimensional numeric systems can be considered as a further development of this fundamental idea of Pythagoras in connection with the genetic system and inherited biological structures.

The matrix approach to structural ensembles of the genetic system, proposed and developed by the author (under the unifying name "matrix genetics"), provides the possibility of mathematical analysis of this system for revealing hidden interrelationships in it and for deeper understanding living matter. On the basis of described results new classes of biomathematical models and technical applications, including devises and methods of medical engineering, biotechnologies, topics of works [29–34], etc. are possible.

References

1. Harkin, A.A., Harkin, J.B.: Geometry of generalized complex numbers. Math. Mag. **77**(2), 118–129 (2004)
2. Kantor, I.L., Solodovnikov, A.S.: Hypercomplex Numbers. Springer, Berlin, New York (1989). ISBN 978-0-387-96980-0
3. Ahmed, N.U., Rao, K.R.: Orthogonal Transforms for Digital Signal Processing. Springer, New York (1975)
4. Seberry, J., Wysocki, D.J., Wysocki, T.A.: Some applications of Hadamard matrices. Metrika **62**, 221–239 (2005)
5. Petoukhov, S.V.: Matrix genetics, algebrases of genetic code, noise immunity. Moscow, RCD, 316 p. (2008, in Russian)
6. Petoukhov, S.V.: Matrix genetics and algebraic properties of the multi-level system of genetic alphabets. Neuroquantology **9**(4), 60–81 (2011)
7. Petoukhov, S.V.: The system-resonance approach in modeling genetic structures. Biosystems **139**, 1–11 (2016)
8. Petoukhov, S.V.: Symmetries of the genetic code, Walsh functions and the theory of genetic logical holography. Symmetry: Cult. Sci. **27**(2), 95–98 (2016b)
9. Petoukhov, S.V.: The genetic code, 8-dimensional hypercomplex numbers and dyadic shifts (2016c). https://arxiv.org/pdf/1102.3596v11.pdf

10. Petoukhov, S.V.: Genetic coding and united-hypercomplex systems in the models of algebric biology. Biosystems **158**, 31–46 (2017)
11. Petoukhov, S.V.: The Genetic Coding System and Unitary Matrices. Preprints 2018, 2018040131 (2018a). http://www.preprints.org/manuscript/201804.0131/v2
12. Petoukhov, S.V.: Structural connections between long genetic and literary texts. Preprints 2018, 2018120142 (2018b). https://doi.org/10.20944/preprints201812.0142.v2. https://www.preprints.org/manuscript/201812.0142/v2. Accessed 15 Feb 2019
13. Petoukhov, S.V.: Connections between long genetic and literary texts. the quantum-algorithmic modelling. In: Hu, Z., Petoukhov, S., Dychka, I., He, M. (eds.) Advances in Computer Science for Engineering and Education II, ICCSEEA 2019. Advances in Intelligent Systems and Computing, vol. 938, pp. 534–543. Springer, Cham (2020). https://doi.org/10.1007/978-3-030-16621-2_50
14. Petoukhov, S.V., He, M.: Symmetrical Analysis Techniques for Genetic Systems and Bioinformatics: Advanced Patterns and Applications. IGI Global, USA (2010)
15. Petoukhov, S.V., Petukhova, E.S., Svirin, V.I.: Symmetries of DNA alphabets and quantum informational formalisms. Symmetry: Cult. Sci. **30**(2), 161–179 (2019)
16. Frank-Kamenetskii, M.D.: The most important molecule. Nauka, Moscow (1988, in Russian)
17. Hu, Z.B., Petoukhov, S.V., Petukhova, E.S.: I-Ching, dyadic groups of binary numbers and the geno-logic coding in living bodies. Progr. Biophys. Mol. Biol. **131**, 354–368 (2017)
18. Harmuth, H.F.: Information theory applied to space-time physics. The Catholic University of America, Washington, DC (1989)
19. Sagalovich, Yu.L.: Introduction to Algebraic Codes. Moscow Physical-Technical Institute, Moscow (2011, in Russian)
20. Jean, R.: Phyllotaxis: A Systemic Study in Plant Morphogenesis. Cambridge University Press, Cambridge (2006)
21. Bodnar, O.Ya.: Geometry of phyllotaxis. Reports of the Academy of Sciences of Ukraine, no. 9, pp. 9–15 (1992)
22. Bodnar, O.Ya.: Golden Ratio and Non-Euclidean Geometry in Nature and Art. Publishing House "Sweet", Lviv (1994)
23. Smolyaninov, V.V.: Spatio-temporal problems of locomotion control. Uspekhi Fizicheskikh Nauk **170**(10), 1063–1128 (2000)
24. Petoukhov, S.V.: The genetic code, algebraic codes and double numbers. Preprints 2019, 2019110301 (2019). https://www.preprints.org/manuscript/201911.0301/v1
25. Petoukhov S.V.: Hyperbolic numbers in modeling genetic phenomena. Preprints 2019, 2019080284, 36 p. (2019). https://doi.org/10.20944/preprints201908.0284.v2. https://www.preprints.org/manuscript/201908.0284/v2
26. McFadden, J., Al-Khalili, J.: The origins of quantum biology. Proc. Roy. Soc. A: Math. Phys. Eng. Sci. 1–13 (2018). https://royalsocietypublishing.org/doi/full/10.1098/rspa.2018.0674
27. Heisenberg, W.: Physics and Philosophy. Harper & Row, New York (1958)
28. Russell, B.: A History of Western Philosophy. Touchstone, New York (1945)
29. Angadi, S.A., Hatture, S.M.: Biometric person identification system: a multimodal approach employing spectral graph characteristics of hand geometry and palmprint. Int. J. Intell. Syst. Appl. (IJISA) **3**, 48–58 (2016)
30. Sahana, S.K., Al-Fayoumi, M., Mahanti, P.K.: Application of modified ant colony optimization (MACO) for multicast routing problem. Int. J. Intell. Syst. Appl. (IJISA) **4**, 43–48 (2016)
31. Algur, S.P., Bhat, P.: Web video object mining. Int. J. Intell. Syst. Appl. (IJISA) **4**, 67–75 (2016)

32. Hata, R., Akhand, M.A.H., Islam, M.M., Murase, K.: Simplified real-, complex-, and quaternion-valued neuro-fuzzy learning algorithms. Int. J. Intell. Syst. Appl. (IJISA) **10**(5), 1–13 (2018). https://doi.org/10.5815/ijisa.2018.05.01
33. Awadalla, M.H.A.: Spiking neural network and bull genetic algorithm for active vibration control. Int. J. Intell. Syst. Appl. (IJISA) **10**(2), 17–26 (2018). https://doi.org/10.5815/ijisa.2018.02.02
34. Abuljadayel, A., Wedyan, F.: An approach for the generation of higher order mutants using genetic algorithms. Int. J. Intell. Syst. Appl. (IJISA) **10**(1), 34–35 (2018)

Use of Sustainable Innovation in Information System: Integrate Individual's Spirit and Team Climate

Xiaofen Zhou[1](✉), Yi Zhang[2], and Xin Li[2]

[1] College of Logistics, Wuhan Technology and Business University, Wuhan 430065, China
103343280@qq.com

[2] School of Business Administration and Tourism Management, Yunnan University, Kunming 650504, China

Abstract. Based on the theories of employee entrepreneurship and team climate, the influencing factors of the aspiration to use the information system are studied, and a structural equation research model including two dimensions of individual and team is constructed. And 201 valid questionnaires were used to verify the model. The research found that the enterprising spirit in the entrepreneurial spirit of employees can directly affect the individual's aspiration to use it continuously, while pioneering spirit and cooperation can indirectly affect the aspiration to innovation through the mediating effect of attitude to innovations. In addition, the stronger the support for innovation and task orientation in the team, the more it can stimulate employees' innovation willingness. On this basis, the paper puts forward effective suggestions for the cultivation of employees' sustainable innovation willingness. Enterprises should create more team atmosphere, cultivate employees' innovative attitude and the characteristics of entrepreneurship, give employees autonomy in their work, and maximize the influence of employees.

Keywords: Employee entrepreneurship · Team climate · Innovation aspiration · Sustained use

1 Introduction

The research on the innovative use behavior of information system comes from the discussion of the use phase of information system by Saga et al. [1]. After the development of Jasperson [2] and Mills [3] and others, the innovative use of IS is widely regarded as a "Post-Adoptive" behavior, and its value will only occur in the Post-Adoptive stage.

Although scholars have carried out large number of researches on "Post-Adoptive behavior", there are some shortcomings to overcome. Existing literature generally studies the impact of individual traits on individual behaviors from an individual perspective [3], or studies that of organizational support structure on individual behaviors from a team perspective [4, 5]. Although this part of the research confirms that individual and team environments can affect individual behavior, few studies have

Z. Hu et al. (Eds.): ICCSEEA 2020, AISC 1247, pp. 571–584, 2021.
https://doi.org/10.1007/978-3-030-55506-1_51

combined the two to study it further. Keline (2012) believes that an organization is a system composed of multiple layers, and innovative use behavior is an emergence phenomenon, which is not only affected by individual characteristics but also by the team climate.

Therefore, this article attempts to explore the impact of the two on the innovation aspiration from both an individual and a team perspective. First, employee attributes are the number one factor in promoting the innovative use of IS. Some scholars have pointed out that the individual's pioneering spirit is one of the factors affecting individual creativity, and the stronger the individual's creativity, the easier it is to cause innovative behavior [6]. Soto also believes that entrepreneurship is a state of mind and an important aspect of individual traits [7]. Zhuhai believes that entrepreneurship is not a talent possessed by a certain group of people, but exists universally in everyone [8]. Therefore, this article attempts to consider entrepreneurship as a trait commonly possessed by employees within the team to illustrate the impact of entrepreneurship on the innovation aspiration. In addition, in the research on team climate, West believes that team climate is both subjective and objective, and divides team climate into four dimensions including task orientation, vision, support for innovation and participation security [9]. And he believes that task orientation involves team responsibility and work methods, which is a prerequisite for achieving excellent work performance; while support for innovation is an organization support method that involves the team giving more autonomy to employees [9].

Therefore, this article will focus on the two aspects of support for innovation and task orientation in the theory of team climate, and regard the entrepreneurial spirit of employees as part of the employee's traits, and discuss the role of the two on the future innovation aspiration of employees from the perspective of individuals and teams.

2 Theoretical Basis and Hypotheses

2.1 Employee Entrepreneurship and Future Innovation Aspiration

The term "Entrepreneurship" is also translated as entrepreneurial spirit, which is derived from thinking about economic growth in the field of economics. Subsequently, academia conducted in-depth research on entrepreneurship, forming two different perspectives [8, 9]. The first view is that entrepreneurship is a concept of "spiritual dimension", a combination of thinking and behavior based on innovation [9]. Birkinshaw believes that entrepreneurship is a function, i.e., entrepreneurs use creative methods to develop products and establish new methods to achieve industrial integration [9]. Another point of view is that entrepreneurship is a concept of "practical level", which is a management method that embodies the creativity of entrepreneurs [10], and an innovative strategic decision-making. It consists of three dimensions: advanced behavior, risk taking and innovation.

This shows that the academic community generally regards entrepreneurship as a characteristic possessed by managers (entrepreneurs), and believes that pioneering spirit is an important part of entrepreneurship. However, some scholars have proposed that defining entrepreneurship as a characteristic peculiar to managers is questionable, and believes that entrepreneurship is not limited to entrepreneurs themselves, but also possesses the characteristics of ordinary employees [9, 11]. It is also believed by Morris that entrepreneurial spirit exists in the employees of the enterprise, which is manifested in the process of employees using existing resources to create value [12]. Ding Hao and others also considered that the entrepreneurial spirit is not a characteristic of the founders of the enterprise when researching business model innovation. It is widespread among employees and manifests as a spirit of pursuit. And Wang Bingcheng and others also believe that the spirit of cooperation and enterprising spirit is an important manifestation of employee entrepreneurship [11]. To sum up, based on the research results of Ding Hao, Wang Bingcheng, etc., and also refers to the viewpoint of Morris et al., this paper holds that entrepreneurship is a trait commonly possessed by enterprise employees, including three dimensions of pioneering spirit, cooperation spirit, and enterprising spirit [9, 11, 13].

Firstly, pioneering spirit refers to the aspiration to adopt new methods and processes to complete work tasks in the use of information systems, and involves the expansion of usage methods and functions [2]. The innovation aspiration is expressed as the degree to which individuals desire to work in the area of innovation, and is a concrete extension of the individual's "will" in the area of innovation. Secondly, the spirit of cooperation refers to the degree to which individuals share their views and ideas with others during the use of the information system. Skyes believes that Peer Advice helps employees share new ideas and ideas with each other through informal communication channels, and it is easy to form alternative thinking within the team to solve problems encountered in work [4]. Finally, the enterprising spirit is a kind of advance action, which is manifested in the attitude of employees who are not satisfied with the status quo and actively explore new functions and methods of information systems to complete work tasks more efficiently [8]; The more enterprising individuals are able to master the more functions and methods of IS, and the more likely they are to complete tasks with new methods and workflows in future work. Therefore, this article makes the following hypotheses:

H1: Employee entrepreneurship can influence the aspiration to sustained innovation. And individuals with a stronger pioneering spirit are more likely to have an aspiration to innovate.

H2: Employee entrepreneurship can influence the aspiration to sustained innovation. And individuals with a stronger cooperation spirit are more likely to have an innovation aspiration.

H3: Employee entrepreneurship can influence the aspiration to sustained innovation. And the more enterprising individuals are more likely to have an innovation aspiration.

2.2 Team Climate Theory and Aspiration to Sustained Innovation

The study of team climate comes from the discussion of "the field" theory in the field of psychology. Levin introduced the concept of psychological climate in 1951, and was subsequently introduced into the field of behavioral science to describe the complex relationship between the environment and people. Currently, the academic community believes that team climate is both subjective and objective [14–16]. From the objective point of view, the team climate is used to distinguish the distinctive characteristics of one team from another, and it is an objective thing [15]. The subjective point of view is that team climate is an employee's perception or evaluation of specific situational factors such as team goals and team structure, and is a subjective feeling [14, 16, 17]. Litwin and others believe that team climate is a set of attributes that can be perceived by individuals. It can have a significant impact on human behavior, and divides team climate into nine dimensions including support, reward, and risk [14].

Although the academic community has different understandings of team climate, they all affirm that team climate can affect employee behavior, and most scholars have studied team climate from a subjective perspective, and believe that team climate is aggregated from different dimensions. West divides the team climate into four dimensions: Task Orientation, Vision, Support for innovation, and Participative Safety [7, 17]. West believes that task orientation refers to achieving excellent job performance. It is closely related to Vision or Outcomes. It emphasizes team or individual responsibility and achieves excellence through the rational use of existing procedures and methods. Support for innovation is expressed as the degree of support for new working methods and new perspectives [18], which is affected by both the formal rules and regulations of the organization and human factors in the team [7]. In explaining the job characteristic model, Hackman et al. proposed that job autonomy is used to describe the degree of freedom that a job gives an employee in completing a job task [19]. It is affected by human factors and organizational systems and can ultimately affect employees' working style and behavioral will. Therefore, this article will refer to the research conclusions of West et al., And take task orientation and support for innovation as team climate factors that influence employees' future innovation willingness, and put forward the following hypotheses:

H4: Team climate can affect employees' innovation aspiration. And support for innovation can have a significant positive impact on employees' aspiration to sustained innovation.

H5: Team climate can affect employees ' innovation aspiration. And task orientation can have a significant positive impact on employees' aspiration to sustained innovation.

2.3 Mediating Role of Attitude to Innovation

Rational behavior theory [20] and planned behavior theory [21] believe that the individual's perception of the outside world is an important factor leading to individual behavior, and will also stimulate innovative behavior through the mediating role of attitude. In this article, the attitude to innovation is understood as an evaluation of the innovation aspiration, and it is an individual's subjective perception of the innovation aspiration. According to Wang Jingyi's point of view, the spirit of enterprising is manifested in the attitude of employees to their responsibilities, and through continuous efforts to surpass one's own aspiration to act [8]. Therefore, the entrepreneurial spirit of employees can be regarded as a manifestation of personal attitude, which can directly affect the future innovation aspiration. Pioneering spirit is embodied in a spirit of pursuit, which is people's attitude to the problem of innovation, and can make attitudes affect individual behavior [22]. On such basis, this article believes that pioneering spirit and cooperation can not only directly affect the individual's aspiration to continue to innovate, but also indirectly affect the innovation aspiration through the mediating role of attitude to innovations. Therefore, the following hypotheses are made:

H6a: attitude to innovation has a mediating effect on the impact of pioneering spirit on the aspiration to sustained innovation.
H6b: attitude to innovation has a mediating effect on the impact of the spirit of cooperation on the aspiration to sustained innovation.

In summary, the research framework of this paper is shown in Fig. 1.

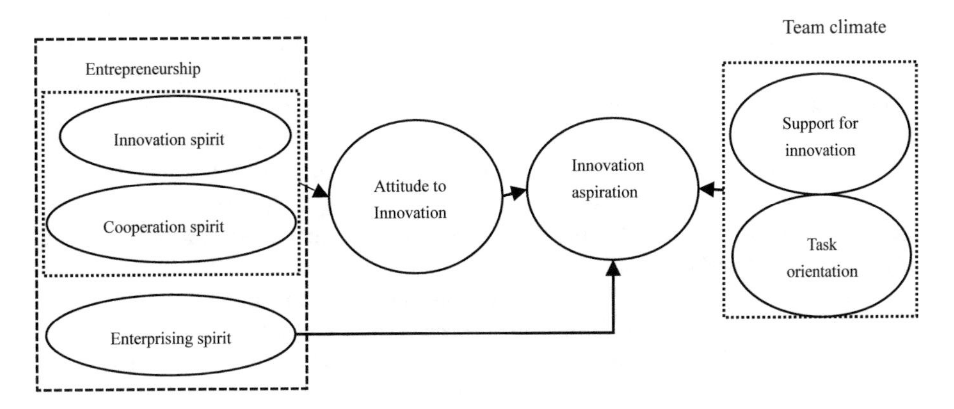

Fig. 1. Research framework

3 Questionnaire Design and Sample Features

3.1 Questionnaire Design

In this study, questionnaires were used to measure the questions to be studied. The questionnaire is mainly composed of three parts: employee entrepreneurship, team climate, future innovation aspiration and attitude to innovation. The measurement items of each part refer to the classic literature at home and abroad, and are measured using the 7-Lickert scale.

In the measurement of employee entrepreneurship. The three dimensions of pioneering spirit, cooperation spirit and enterprising spirit were measured. Because the measured samples are mainly from China, the measurement of cooperative spirit and aggressive spirit in this study mainly refers to the questionnaires of Xu Jianping [23] and Cai Hua and Maruping et al. For pioneering spirit, reference is mainly made to the research results of Ke W [24] and others. 20 indicators were used to measure employee entrepreneurship, and 8 best items were selected for data analysis.

In the measurement of team climate, the scale developed by Anderson and West [7] were mainly referred to and appropriate amendments were made in combination with the research scenario of this article. Eighteen indicators were used to measure support for innovation and task orientation, and six best items were selected to measure the two dimensions of support for innovation and task orientation.

In the measurement of future innovation aspiration and attitude to innovation, the research results of Maruping, Ajzen [25, 26] and Shen etc. were mainly referred to, 9 indicators were selected for measurement, and 7 best questions were selected for experiment.

3.2 Sample Features

In order to verify the rationality of the structural model, this experiment selected a large company in Yunnan Province who implemented the IS system to conduct a questionnaire survey. Because the innovative behavior of information systems is a "Post-Adoptive behavior" [2], the subject of this experimental investigation was employees of the company who could use IS. The selection criteria are: (1) respondents can independently use the information system to complete basic tasks; (2) their main tasks must be completed with the help of IS. With the help of the company's management, a paper questionnaire was distributed to the company's employees. Respondents were informed that the survey was anonymous and that the questionnaire should be submitted directly to the investigator to ensure that the supervisors did not affect his response to the questionnaire.

In this experiment, a total of 300 questionnaires were distributed. A total of 245 questionnaires were recovered after the recovery was completed, and the effective recovery rate was 81.67%. Among them, 32 questionnaires were not accepted due to incomplete information or data loss, so a total of 201 valid questionnaires were selected for acceptance. Of the 201 valid questionnaires, most were under 34 (86.68%), 81 (40.29%) were male, and 120 (59.7%) were female. About 38.03% of the respondents had a bachelor degree or above. The statistical results are shown in Table 1.

Table 1. Sample features of valid questionnaires

Basic information	Description	Frequency	Percentage	Basic information	Description	Frequency	Percentage
Age	≤ 24 yrs.	106	52.73%	Gender	Male	81	40.29%
	25–34 yrs.	68	33.83%		Female	120	59.7%
	35–44 yrs.	23	11.44%	Major	Economic management	23	11.44%
	≥ 45	4	1.99%		Engineering	92	45.77%
Working years	<1 yrs.	114	56.71%		other	86	42.78%
	1–3 yrs.	62	31.84%	Education background	College or below	120	59.7%
	>3 yrs.	25	12.44%		Bachelor	78	38.80%
					Post-graduate or above	3	1.49%

Note: The data comes from the collation of the questionnaire.

4 Reliability and Validity Analysis and Structural Model Testing

In this study, the reliability and validity were tested based on the structural equation model (SEM), and SMART PLS 3.0 software was used for data analysis. It is divided into two parts: the first step is mainly to test the rationality of the scale, namely: reliability and validity test; the second step involves structural model test, i.e., hypothesis test.

4.1 Scale Reliability and Validity Test

Generally speaking, the reliability test of the scale requires testing the size of Cronbach's α, but some studies have shown that the size of the α coefficient is affected by the number of measurement items. In addition, Campbell (1993) also believes that the level of α coefficient can represent the stability of the test results. Therefore, this paper uses Construct Reliability and Average Variance Extraction (AVE) to evaluate the reliability of the scale.

The values of factor load, CR and AVE are given in Table 2. As can be seen from the table, the combined reliability (CR) values are all in the range of 0.8094–0.9608, which is greater than 0.8; and the smallest average variance extraction (AVE) is greater than 0.5 ($0.5863 > 0.5$); in addition, all factors load is between 0.7256–0.9660, which is greater than 0.7. This indicates that the reliability of the scale has passed the test [27].

The validity test consists of convergence validity and discriminant validity. According to Fronell and Larcker, the test of Convergent Validity needs to meet two conditions: a. The factor load value of each measurement is >0.707; b. The average variance extraction amount is >0.5. The two conditions are true at the same time, indicating that the convergence validity has passed the test.; It can be obtained from Table 2 that the minimum factor load and AVE values are 0.7613 and 0.5863, respectively, which meets the requirements for judging the convergence validity, which indicates that the convergence validity of the scale has passed the test.

Table 2. Sample features of valid questionnaires

Factors	No.	Factor load	CR	AVE	Scale source
Pioneering spirit (PII)	PII1	0.8910	0.8822	0.7149	Ke W, Tan C, Sia C [24]
	PII2	0.7657			
	PII3	0.8744			
Cooperation spirit (COO)	COO1	0.7393	0.8094	0.5863	XU Jianping [23] CAI Hua [28] Maruping [29]
	COO2	0.7613			
	COO3	0.7954			
Enterprising spirit (ITE)	ITE1	0.9570	0.9608	0.9245	
	ITE2	0.9660			

(*continued*)

Table 2. (*continued*)

Factors	No.	Factor load	CR	AVE	Scale source
Support for innovation (SI)	SI1	0.7373			Anderson, West [7]
	SI2	0.8177			
	SI3	0.9065			
Task orientation (TO)	TO1	0.8522	0.8286	0.6181	
	TO2	0.7256			
	TO3	0.7756			
Attitude to innovation (ATTI)	ATTI1	0.9263	0.9559	0.8784	Ajzen [25, 26] Shen [30]
	ATTI2	0.9490			
	ATTI3	0.9362			
Innovation aspiration (IA)	FIA1	0.9085	0.9558	0.8440	Maruping [29]
	FIA2	0.9136			
	FIA3	0.9251			
	FIA4	0.9273			

Note: PII-Pioneering spirit; COO-Cooperation spirit; ITE-Enterprising spirit; SI-Support for innovation; TO-Task orientation; ATTI-Attitude to innovation; FIA-aspiration to sustained innovation; CR-combined reliability; AVE-average variance extraction

Table 3. Correlation coefficient and AVE square root

Indicators	1	2	3	4	5	6	7
1. Support for innovation	**0.8234**						
2. Pioneering spirit	0.2762	**0.8455**					
3. Cooperation spirit	0.0810	0.1401	**0.7657**				
4. Innovative attitude	0.0783	0.1568	0.2241	**0.9372**			
5. Innovation aspiration	0.2459	0.1319	0.2203	0.3896	**0.9187**		
6. Task orientation	0.5439	0.1160	0.0912	0.1388	0.3574	**0.7862**	
7. Enterprising spirit	0.0999	0.3196	0.2179	0.6509	0.3799	0.1584	**0.9615**

Note: The bold type on diagonal lines indicates the square root of AVE; the rest indicate the correlation coefficient

The discriminant validity test can be determined by examining the relationship between AVE and the correlation coefficient. According to the viewpoints of Fronell and Shook, if the AVE value of all indicators is greater than the square of the correlation coefficient, it indicates that the discriminant validity is passed. In Table 3, the bold font on the main diagonal indicates the square root of AVE, so only the minimum square root of AVE is greater than the maximum correlation coefficient between factors indicates that the discriminant validity has passed the test. From Table 3, the smallest square root of AVE is greater than the correlation coefficient (0.7657 > 0.5439), so the discriminant validity passes the test.

In summary, the reliability and validity of the scale have passed the test.

4.2 Structural Model Test and Results Interpretation

4.2.1 Direct Effect Test

This paper uses Smart pls 3.0 to calculate the structural model, and sets the gender, age, major, and education background of the interviewees as the control variables. The structural model results of this study are obtained. As shown in Fig. 2.

It can be seen from the figure that pioneering spirit (β = 0.022, P > 0.05) and cooperation spirit (β = 0.108, P > 0.05) cannot directly affect the innovation aspiration, so there is insufficient evidence to support H1 and H2. The enterprising spirit (β = 0.178, P < 0.05) significantly affects innovation aspiration, and therefore supports H3.

In addition, the degree of support for innovation (β = 0.215, P < 0.001) and task orientation (β = 0.254, P < 0.001) from the team can significantly affect innovation aspiration, and therefore support H4 and H5.

Finally, it can be seen from the figure that pioneering spirit and cooperation spirit cannot directly affect innovation willingness, but it seems that it can indirectly affect innovation aspiration through the mediating role of attitude to innovation (β = 0.212, P < 0.01). Therefore, it can be preliminarily judged that the attitude to innovation has a completely mediating effect in the influence of pioneering spirit and cooperation spirit on the innovation aspiration, i.e., the H6 hypothesis holds.

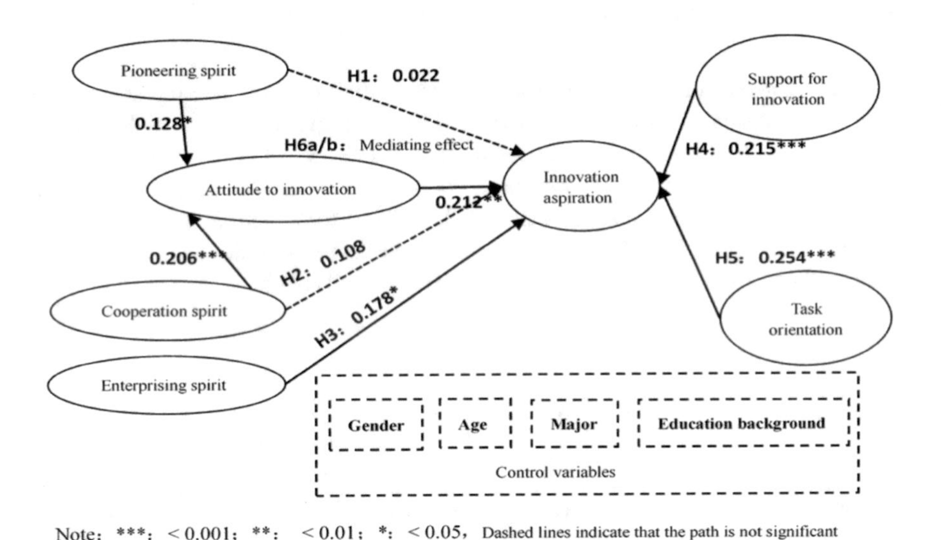

Fig. 2. Structural model test results

In order to further test the existence of mediating effect, this paper uses Bootstrapping function in SMART PLS3.0 software to further verify the mediating effect.

4.2.2 Mediating Effect Test

As can be seen from Fig. 2, in the structural model, there are two meditating paths. Namely, pioneering spirit (β = 0.128, $P < 0.05$; β = 0.212, $P < 0.01$) and cooperation spirit (β = 0.206, $P < 0.001$; β = 0.212, $P < 0.01$, respectively, indirectly affect the future innovation aspiration through attitude to innovation. In Smart pls3.0, BCA-Bootstrap was selected to test the mediating effect at a significance level of 0.05. The calculation results are shown in Table 4.

Table 4. Bootstrap analysis of the mediating effect test

Path	Indirect effect		95% confidence interval	
	Estimated value	Percentage of total effect	Lower limit	Upper limit
PII-> ATTI-> FIA	0.128 × 0.212 = 0.0271	43.64%	0.002	0.076
COO-> ATTI-> FIA	0.206 × 0.212 = 0.0436	38.03%	0.012	0.089

Note: PII-Pioneering spirit; COO-Cooperation spirit; ATTI-Attitude to innovation; FIA-aspiration to sustained innovation.

It can be seen from Table 4 that the first meditating path (pioneering spirit → attitude to innovation → aspiration to sustained innovation) accounts for 43.64% of the total effect, while the second one accounts for 38.03%. Meanwhile, at the 95% confidence level, the confidence interval does not include 0, which indicates that the two intermediate paths meet the significance of 0.05. Therefore, there is evidence that the H6 hypothesis holds, i.e., the complete mediating effect is significant.

In summary, the results of this study are shown in Table 5.

Table 5. Summary of structural model test

Model	Hypothesis	Path	Path coefficient/Confidence interval	Result
Employee entrepreneurship and aspiration to sustained innovation	H1	PII->IA	0.022	Not supportive
	H2	COO->IA	0.108	Not supportive
	H3	ITE->IA	0.178*	Supportive
Team climate for sustained innovation	H4	SI->IA	0.215**	Supportive
	H5	TO->IA	0.254*	Supportive

(*continued*)

Table 5. (*continued*)

Model	Hypothesis	Path	Path coefficient/Confidence interval	Result
Mediating role of attitude to innovation	H6a	PII->ATTI->IA	[0.002, 0.076]	Supportive
	H6b	COO->ATTI->IA	[0.012, 0.089]	Supportive

Note: ***: <0.001; **: <0.01; *: <0.05; PII-Pioneering spirit; COO-Cooperation spirit; ITE-Enterprising spirit; SI-Support for innovation; TO-Task orientation; ATTI-Attitude to innovation; IA-Innovation aspiration

5 Research Conclusions and Significance

5.1 Research Conclusions

Based on the theory of team climate and employee entrepreneurship, this article regards team climate as an organizational factor that affects employees' innovation aspiration, and uses employee entrepreneurship as an individual factor. A multi-level research model for the use of information system innovation is established, and 213 effective samples are used to verify the structure model. The result shows:

The enterprising spirit ($\beta = 0.178$, $P < 0.05$) can directly affect the innovation aspiration. This result may be due to different perceptions of employees' innovation aspiration. Generally speaking, enterprising spirit is the motivation for employees to work hard to explore new system functions and use methods [31]. Employees with high spirits are more willing to learn new knowledge in the work and complete tasks through better methods.

The pioneering spirit ($\beta = 0.022$, $P>0.05$) and cooperation spirit ($\beta = 0.108$, $P > 0.05$) cannot directly affect the future innovation aspiration, but they can influence the innovation aspiration through the mediating role of attitude to innovations (see Fig. 2 and Table 4). Both the theory of planned behavior and rational behavior show that attitude is a pre-factor affecting individual behavior, and individual behavior must indirectly affect behavior aspiration through the mediating effect of attitude [32], which is consistent with the findings of this paper. In addition, the data shows that attitude to innovation produces a completely meditating effect in the impact of the pioneering spirit and cooperation spirit on the innovation aspiration. This further shows that the attitude of employees is an important factor affecting individual behavior [33]. Managers need to pay attention to the cultivation of employee attitudes when creating a team climate that advocates innovation.

Team climate is also an important factor influencing employees' innovation aspiration. Support for innovation from the team ($\beta = 0.215$, $P < 0.001$) and task orientation ($\beta = 0.254$, $P < 0.001$) can directly affect the innovation aspiration, which shows that the stronger the employee's work autonomy within a team, the more uniquely, innovatively the tasks will be completed. Since work autonomy enables individuals to have more control over work methods and working hours, and the team's task orientation allows employees to adjust their work capabilities and working methods. It is important to make them realize that if they want to complete the task better, they must use new methods and processes to stimulate the individual's desire to use IS innovatively.

5.2 Theoretical Significance

The theoretical significance of this article lies in a more in-depth discussion of the influencing factors of innovative use behavior of IS.

Firstly, from a research perspective. The current research generally regards "entrepreneurship" as a characteristic possessed by entrepreneurs. Based on the opinions of Wang Bingcheng and Ding Hao [9, 11], this paper also draws on the research results of Morris and others, focusing on "entrepreneurship" to employees, and believes that entrepreneurship is not only a unique characteristic of managers, but also common among all employees.

Secondly, from the perspective of the research object. Most of the current researches study the influence of current factors on the innovation behavior of current information systems. However, in this article, we explore whether the current factors will affect the individual's future behavioral willingness. Therefore, there is innovation in research objects.

Lastly, it complements the existing theoretical research. Existing researches generally study the impact on individual innovation behavior from the perspective of individual traits or teams. Few literatures consider both factors. This article also considers the impact of factors at the individual and group levels on the aspiration to use IS innovatively, supplementing the shortcomings of theoretical research.

5.3 Practical Significance

By constructing a multi-layered research model, this paper finds that for ordinary employees, pioneering spirit and the ability to innovate cannot directly influence the individual's future innovation aspiration, but can influence the future innovation behavior through the mediating role of attitude to innovation. Besides, the team climate can directly affect the future innovation aspiration. This requires that managers not only pay attention to the cultivation of employee characteristics, but also create a good team climate. Secondly, the mediating role of attitude to innovations shows that in terms of the cultivation of personal traits, attitude to innovations and entrepreneurial spirit of employees are equally important. The two complement each other and develop together. In terms of building a team climate, managers should actively grant employees autonomy and create a goal-oriented work team, thus to form an individual to drive the team, and the team to drive the individual's chain of action to promote the maximum use value of the information system.

Acknowledgement. This research was financially supported by the Distinguished Young and Middle-aged Program for Scientific and Technology Innovation in Higher Education of Hubei, T201938, "Smart Logistics Park Service Innovation and Support". And supported by Hubei Education Science Planning Project (2018GB121).

References

1. Saga, V.L., Zmud, R.W.: The nature and determinants of IT acceptance, routinization, and infusion. Pittsburgh Carnegie Mellon University (1994)
2. Jasperson, J., Carter, P.E., Zmud, R.W.: A comprehensive conceptualization of post-adoptive behaviors associated with information technology enabled work systems. MIS Q. **29**(3), 525–557 (2005)
3. Mills, A., Chin, W.: Conceptualizing creative use: an examination of the construct and its determinants. In: Proceedings of the AMCIS 2007 Proceedings Paper 289 (2007). http://aisel.aisnet.org/amcis2007/289
4. Sykes, T.A.: Support structures and their impacts on employee outcomes: a longitudinal field study of an enterprise system implementation. MIS Q. **39**(2), 473–495 (2015)
5. Liang, H., Peng, Z., Xue, Y.: Employees' exploration of complex systems: an integrative view. J. Manag. Inf. Syst. **32**(1), 322–357 (2015)
6. Soto, J.H.: Socialism, economic calculation, and entrepreneurship (New thinking in political economy). Edward Elgar, UK (2010)
7. Anderson, N.R., West, M.A.: Measuring climate for work group innovation: development and validation of the team climate inventory. J. Organ. Behav. **19**(3), 235–258 (1998)
8. Wang, J.: Structural model construction and case study about entrepreneurship. Techno-Econ. Manag. Res. **264**(7), 45–51 (2018). (in Chinese)
9. Ding, H., Wang, B., Duan, H.: Research on business model innovation, legality of innovation and employee entrepreneurship of small and micro-tech enterprises. Sci. Technol. Progress Policy **30**(21), 80–85 (2013). (in Chinese)
10. Kodithuwakku, S.S., Rosa, P.: The entrepreneurial process and economic success in a constrained environment. J. Bus. Ventur. **17**(5), 431–465 (2002)
11. Wang, B., Ding, H., Duan, H.: An empirical study on the relationship among business model innovation, employee entrepreneurship and personality traits. Ind. Technol. Econ. (6): 106–160 (2013). (in Chinese)
12. Morris, M.H.: Entrepreneurial Intensity: Sustainable Advantages for Individuals, Organizations and Societies. Quorum Books, London (1998)
13. Shi, P., Xu, L.: A study of the three levels of entrepreneurship and their enlightenment. Foreign Econ. Manag. **28**(2), 44–51 (2006)
14. Muchinsky, P.M.: An assessment of the litwin and stringer organization climate questionnaire: An empirical and theoretical extension of the sims and lafollette study. Pers. Psychol. **29**(3), 371–392 (2010)
15. Forehand, G.A., Haller, G.V.: Environmental variation in studies of organizational behavior. Psychol. Bull. **62**(6), 361 (1964)
16. Xie, H., Ma, Q.: A study on the effect of organizational climate on employee's informal knowledge sharing behavior. Stud. Sci. Sci. **25**(2), 306–311 (2007)
17. Li, J.: A study on library service innovation based on team climate. Res. Libr. Sci. (3), 22–24 (2011). (in Chinese)
18. West, M.A., Anderson, N.: Innovation, cultural values, and the management of change in British hospitals. Work Stress **6**(3), 293–310 (1992)

19. Hackman, J.R., Oldham, G.R.: Work Redesign. Addison-Park, Menlopark (1980)
20. Fishbein, M., Ajzen, I.: Belief, Attitude, Intention and Behavior: An Introduction to Theory and Research. Addison-Wesley (1975)
21. Taylor, S., Todd, P.: Decomposition and crossover effects in the theory of planned behavior: a study of consumer adoption intentions. Int. J. Res. Mark. **12**(2), 137–155 (1995)
22. Sun, J., Sun, L.: Sense of Innovation. Shanghai Science Press, Shanghai (2010). (in Chinese)
23. Xu, J., Wang, C.: Regional culture's characteristics of entrepreneurial spirits: an empirical research from Zhejiang. Sci. Sci. Manag. S.&. T. **29**(12), 141–145 (2008). (in Chinese)
24. Ke, W., Tan, C., Sia, C.: Inducing intrinsic motivation to explore the enterprise system: the supremacy of organizational levers. J. Manag. Inf. Syste. **29**(3), 257–289 (2012)
25. Ajzen, I.: Attitudes Personality and Behavior, vol. 43, no. 3, pp. 228–233. Open University Press, UK (1988)
26. Ajzen, I., Fishbein, M.: Factors influencing intentions and the intention-behavior relation. Hum. Relat. **27**(1), 1–15 (1974)
27. Fornell, C., Larcker, D.F.: Evaluating structural equation models with unobservable variables and measurement error. J. Mark. Res. **18**(1), 39–50 (1981)
28. Cai, H., Yu, Y., Jiang, T.: Measurement and analysis of private entrepreneurship. Stat. Decis. (16), 163–165 (2009). (in Chinese)
29. Maruping, L.M., Magni, M.: Motivating employees to explore collaboration technology in team contexts. MIS Q. **39**, 1–16 (2015)
30. Shen, A.X., Cheung, C.M., Lee, M.K., et al.: How social influence affects we-intention to use instant messaging: the moderating effect of usage experience. Inf. Syst. Front. **13**(2), 157–169 (2011)
31. Obeng, A.Y., Mkhize, P.L.: Impact of IS strategy and technological innovation strategic alignment on firm performance. Int. J. Inf. Technol. Comput. Sci. (IJITCS) **9**(8), 68–84 (2017)
32. Marhraoui, M.A., El Manouar, A.: IT innovation and firm's sustainable performance: the intermediary role of organizational agility – an empirical study. Int. J. Inf. Eng. Electron. Bus. (IJIEEB) **10**(3), 1–7 (2018)
33. Lavy, I., Rashkovits, R.: Motivations of information system students in final project and their implications to technology and innovation. Int. J. Mod. Educ. Comput. Sci. (IJMECS) **8**(2), 1–13 (2016)

Health Status Recognition System for Communication Equipment Based on Data Mining

Yongjun Peng[1,2], Rui Guo[2], Anping Wan[3(✉)], Ningdong You[4], and Lihua Wu[5]

[1] Department of Information Communication, National University of Defense Technology, Wuhan 430010, China
[2] The PLA Rocket Force Command College, Wuhan, China
[3] Department of Mechanical Engineering, Zhejiang University City College, Hangzhou 310015, China
anpingwan@zju.edu.cn
[4] Wuhan Natural Resource and Planning Information Center, Wuhan, China
[5] Wuhan Institute of Digital Engineering, Wuhan, China

Abstract. Communication equipment system is complicated, poor working conditions, high load for a long time running, the system will often appear all sorts of fault, the control system of chain reaction will lead to the whole production line to stop operation, this paper proposes a communication equipment system based on data mining health status identification system, using a comprehensive feature selection method was carried out on the working condition of data mining analysis, key parameters, the parameters that influence the stability of communication equipment as a communication equipment health assessment indicator; Communication equipment based on certain health state evaluation indexes, the working state of clustering analysis, the characteristics of the working condition of each cluster, obtain state distribution of historical condition, define history condition of running state categories; Then using ARIMA Communication equipment health algorithm opposition access module to determine the eigenvalues of the training model, predict the trend of change of parameters, with supplemental predicted state recognition. At last, the communication equipment health status recognition system is developed, and applied to the engineering practice, proves the validity and practicability of the system.

Keywords: Data mining · Communication equipment · Health operating conditions · Clustering analysis

1 Introduction

Communication equipment systems possess the disadvantages of complicated process, poor working environment and long-term high-load operation [1, 2]. The Communication equipment features strong coupling, nonlinearity and large hysteresis, along with physical and chemical changes [3, 4]. At the present stage, the variable setting in the practical

Z. Hu et al. (Eds.): ICCSEEA 2020, AISC 1247, pp. 585–595, 2021.
https://doi.org/10.1007/978-3-030-55506-1_52

grinding process is generally adjusted by operators based on experience, resulting in high subjectivity and arbitrariness. Therefore, it is an urgent problem to be solved how to accurately establish the model of raw material grinding process in communication equipment to optimize the control of key parameters during the process [5].

The model of raw material grinding process for communication equipment has been studied in-depth both at home and abroad. Cai X [6] established a soft-sensing model of the material thickness by using the method of least square support vector machine to achieve the indirect measurement of material thickness and the adjustment of parameters in the Communication equipment system. Lin X [7] established a communication equipment grinding model using wavelet neural network and realized the optimal parameter setting through ant colony algorithm. Umucu Y [8] built the model of cement granularity by using multilayer perceptron neural network and radial basis function neural network and obtained a higher prediction accuracy. Wang Kang [9] constructed a recursive neural network model of slag powder production process with data-driven idea, based on which, using adaptive dynamic programming, the tracking controller with control constraints was designed and applied to the production process of slag powder. Lin Xiaofeng [10] established the production target prediction model of communication equipment raw material grinding process using wavelet neural network, and then combined the case-based reasoning technology with the particle swarm optimization algorithm and the target prediction model to achieve the optimal setting of key parameters during the grinding process. Yan Wenjun [11] constructed the communication equipment control loop model by using least square method, extracted multi-loop switch control rules according to the field operation experience and the abnormal operating condition characteristics and finally achieved the optimization control for the circuits.

Summarizing various models of the communication equipment described above, we find that most researchers only explored the interrelationship between single indicators of communication equipment operation. However, communication equipment is a multivariable coupled and nonlinear system, and the variables affect each other, which makes it difficult to establish a complete mechanism model of the production process. With the development of information and automation technology, especially the wide application of sensors and data acquisition devices on complex products, life cycle data of communication equipment can be recorded in real time and feature 4 V characteristics of big data. Among them, the data during operation have the largest growth rate. These data have implied product characteristics of service performance and evolution in time and space [12]. Big data analysis focuses on improving the efficiency of processing massive data with existing data mining methods through distributed or parallel algorithms. It has already been usefully explored and initially applied in many fields [13].

In this paper, using ARIMA Communication equipment health algorithm opposition access module to determine the eigenvalues of the training model, predict the trend of change of parameters, with supplemental predicted state recognition. At last, the communication equipment health status recognition system is developed, and applied to the engineering practice, proves the validity and practicability of the system.

2 Communication Equipment Intelligent Control Model

2.1 Determination Model of Health Status Indicators

Health status indicators refer to a series of equipment operating parameters that can characterize the health status of the system. The housing vibration is an important factor to reflect the stable operation and good production condition of Communication equipment system, which is the parameter indicator under key monitoring in practical system operation. Therefore, the communication equipment system health status can be described as a continuous status in which the communication equipment housing vibration amplitude maintains at a reasonable range within a certain time period under the premise of unchanged rated output.

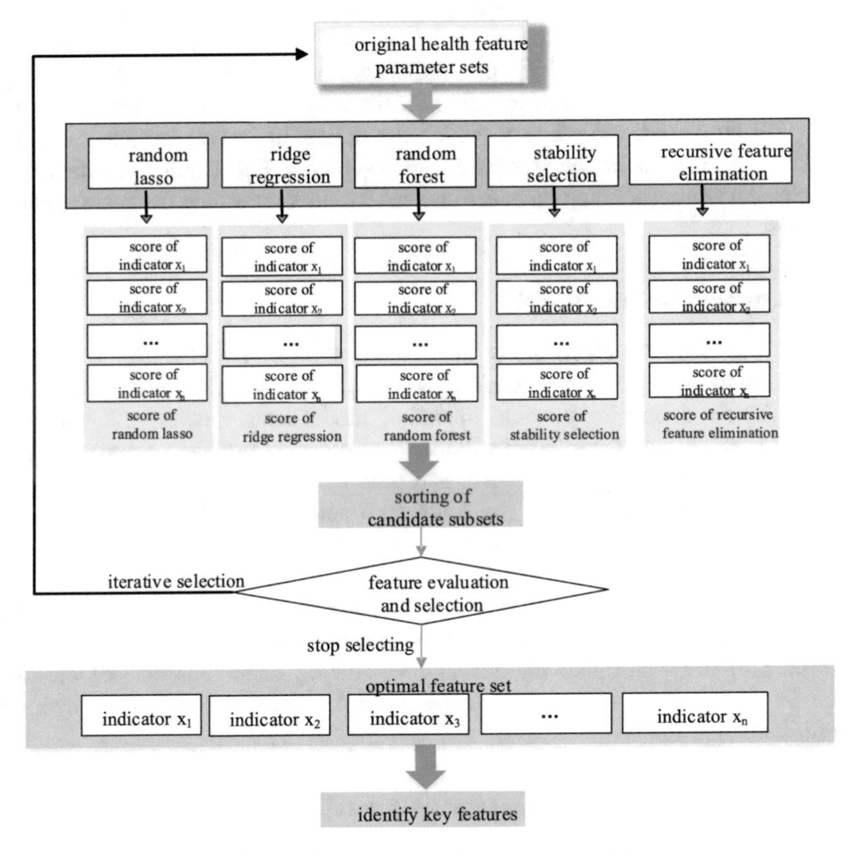

Fig. 1. Key feature mining process model

Vibration amplitude of the housing is a monitoring variable, the stability of which needs to be achieved by adjusting other adjustable parameters. Moreover, communication equipment system is a high coupling system so the vibration will change with a variety of factors. In order to find key parameters that have great influence on the vibration and determine characteristics related to the steady status, the feature selection is necessary. Common feature selection methods include random lasso, ridge regression, random forest, stability selection, recursive feature elimination, etc. These methods have their own advantages and disadvantages. Ridge regression and random lasso need to adjust the parameters to achieve the control of the sparseness of model coefficients. Random forest often emerges overfitting problem. Stability selection is based on subsampling so the results obtained through different samplings are different. The stability of the recursive feature elimination depends on the choice of the underlying model.

To optimize the accuracy of the steady status judgment results, this paper combines the above five algorithms to avoid limitations and disadvantages of certain single method. The principle is, by solving the relationship between input and output, to apply five methods respectively to obtain stable features. And then, the importance of each feature will be scored, according to which the degree of feature importance will be assessed. The feature selection process model is shown in Fig. 1. Finally, obtained key parameters and corresponding numerical ranges together constitute the health status of the system.

2.2 Clustering Analysis Model

The operating condition clustering of healthy operation status refers to the clustering analysis of different operating modes for the sample data based on determined health status indicators to obtain possible operating mode categories. At the very start, according to the parameter value distribution and practical production experience, the sample data is pretreated and screened, and then the screening result is taken as the input of the cluster calculation. The K-means method is used to obtain operating conditions cluster in the dataset. The centroids of K clusters in the dataset are found respectively, and the points in the dataset are assigned to the centroid nearest to it, which determines their categories according to corresponding centroids. The iteration method needs to be adjusted constantly according to running results to determine the value of K. After confirming the cluster of operating conditions, all the operating conditions in the sample data need to be marked to establish a healthy operating condition library.

2.3 Stable Operating Condition Library Establishment Model

According to the definition of data status in the clustering grouping, the category annotation of the existing operating condition record will be completed. The stable operating condition label is set to 0 while the unstable label is set to 1. The stable operating conditions are extracted to establish a stable operating condition library.

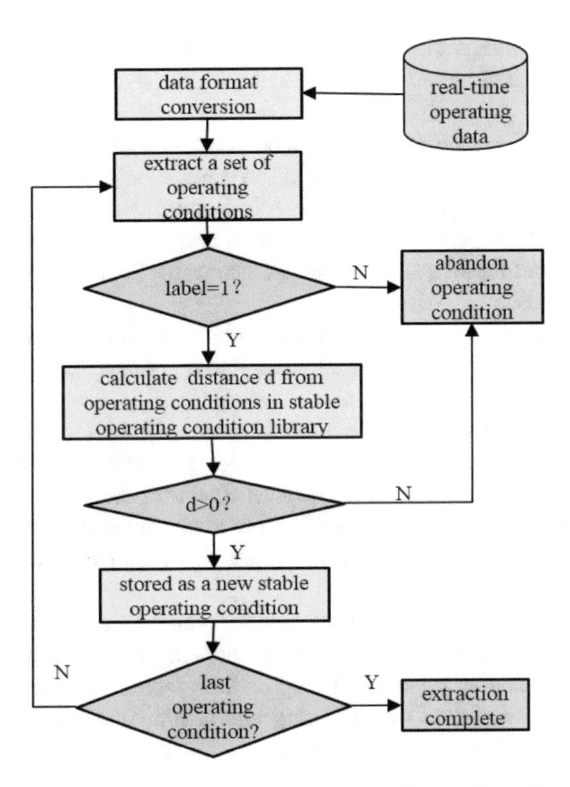

Fig. 2. Establishment process model of stable condition library

The process of establishing the stable operating condition library is shown below in Fig. 2: Every operating condition contains the controllable variables X, the stable characterization variables Y and the category labels. For each operating condition, we calculated the similarity distance between the parameters of X and existing operating conditions in the operating condition library. If the distance equals zero, the condition is seemed to have already been stored and no longer needs to be recorded repeatedly. Otherwise, the condition will be stored in the stable operating condition library in the form of vectors along with corresponding time stamp. Obviously, the stable operating condition library is not fixed and needs to be regularly trained and updated.

2.4 Real-Time Health Status Evaluation Model

Real-time health status evaluation refers to the process of obtaining health status evaluation results in real time based on the comparison between the real-time system operating data and healthy operating condition library. In the practical production process, it is reasonable not to directly use transient single-point data as a basis for judgments but to clean null values and abnormal values from the operating data, and then perform evaluation after a certain period of average calculation. According to determined indicators of health characteristic, the average value, the variance and the number of outliers of real-time running data of each parameter in the collection window are calculated respectively. The results are taken as characteristic variables of the stable operating condition judgment to be compared with the status in the healthy operating condition library.

Step 1 Take T as a starting time at some point and take Δt seconds as sampling interval (based on computer performance in the production site) to collect the real-time operating data of each parameter of the health characteristic indicators.

Step 2 Clean null values and abnormal values for the collected data of each parameter and record the frequency of occurrence respectively. Perform the successive technical accumulation.

Step 3 Calculate the mean value and the variance of the data after processing null values and abnormal values of each parameter with n * Δt seconds as a period. (Where n is the multiple of the sampling window period).

Step 4 Determine the number of occurrences of null values and abnormal values in the judgment cycle (where outliers refer to values exceeding the critical range of the health characteristic parameters). If it is greater than the preset counting threshold, the current operating condition will be considered as abnormal or not suitable for real-time evaluation of health status. Therefore, it is necessary to remind production staff on site of inspection.

Step 5 Respectively take the mean and the variance of each parameter in the sampling period as characteristics of instant operating condition and compare them with the healthy operating condition library. When there is one status similar to the abnormal operating conditions or not in line with healthy conditions, it is necessary to remind production staff on site of inspection.

2.5 Health Status Prediction Evaluation Model

Health status prediction evaluation refers to the operating condition prediction and the health status evaluation within one certain period in the future based on the real-time operation data evaluation. One concept is here to be clear firstly – stationary sequence. For a sequence {X (t)}, if its values fluctuate within a finite range, the mean and the variance are both constant while the autocovariance and the autocorrelation coefficient of sequence variables also keep unchanged after a k-period delay, it can be defined as a stationary sequence, otherwise, as a non-stationary sequence.

Due to the harsh operating conditions of the communication equipment system, the housing vibration is affected by external environmental factors and other attribute parameters. The operating condition sequence belongs to non-stationary sequence. Therefore, the Autoregressive Integrated Moving Average (abbreviated as ARIMA) Model Algorithm is used for time series modeling of communication equipment system operating condition prediction in this paper. The essence of ARIMA model is to add a difference operation before the Autoregressive Moving Average Model (ARMA) operation as shown below:

$$\begin{aligned} x_t = &\phi_0 + \phi_1 x_{t-1} + \phi_2 x_{t-2} + \ldots + \phi_p x_{t-p} + \\ &\varepsilon_t - \theta_1 \varepsilon_{t-1} - \theta_2 \varepsilon_{t-2} - \ldots - \theta_q \varepsilon_{t-q} \end{aligned} \tag{1}$$

The model indicates that the value of variable x at time t is a multivariate linear function of the x-value of previous p periods and the disturbance ε of previous q periods. The error term is the current random disturbance ε_t, which is a zero-mean white noise sequence. Therefore, based on the real-time operating data, the historical time series can be used to predict system operating conditions at a certain time t in the future and to further achieve the predictive evaluation of the operating health status.

3 Application Cases and Analysis

The system has been put into operation in a powder factory in Henan province. The field application proves that the system provides more economical and effective decision-making for the intelligent control of the communication equipment and it runs stably.

3.1 Data Acquisition and Preprocessing

The system adopts OPC protocol to realize data communication and sends the connection requests to the communication equipment central control system server for real-time data collection. The system can set the type of parameter collected and sampling interval.

There are 65 kinds of data signal parameters collected in the practical operation of the field system. After attribute screening, data outlier processing and null processing, a subset of 30 attributes of main process and performance parameters of communication equipment systems are obtained. Taking into account process conditions, manual settings of production staff and controllability of parameters in actual operation of the communication equipment system, the data are processed by further simplification and dimensionality reduction. Finally, the remaining 12 main characteristic parameters are as follows: feeding amount, powder specific surface area, material layer thickness, mill outlet temperature, mill inlet temperature, mill inlet pressure, separator revolving speed, mill pressure difference, main ventilator revolving speed, cold air valve opening, hot air valve opening, circulating air valve opening. Some features and values are shown in Table 1:

Table 1. Feature selection of partial data

Feeding amount (t/h)	Powder specific surface area	Housing vibration (mm)	Host current (a)	Material layer thickness (mm)	Mill inlet pressure (pa)	Mill inlet temperature (°C)
176	431	5.6	284.9	151.3	−510	199.9
164	425	5.7	254	153.8	−510	189.2
180	391	5.2	251.2	151.9	−484	177.7
180	404	4.8	275.2	151.1	−438	180
170	416	6.5	254.1	145.3	−410	201.7
180	411	5.8	290	149	−599	182

It can be seen from the table that dimensions of parameters are inconsistent and ranges of values vary greatly. In order to eliminate the influence of the range of values and the dimensional differences on the analysis results, the data needs to be normalized. To avoid existence of a maximum or a minimum in the data that affects subsequent analysis, zero-mean method is adopted as normalization method, which is more stable than min-max normalization. Part of data after normalization are as shown in Table 2:

Table 2. Partial data after normalization

Feeding amount	Powder specific surface area	Housing vibration	Host current	Material layer thickness	Mill inlet pressure	Mill inlet temperature
1.24	1.57	−1.55	−0.14	1.13	−0.21	0.26
−0.47	1.17	−1.46	−1.61	1.24	−0.21	−0.41
1.81	−1.07	−1.95	−1.75	1.16	0.14	−1.13
1.81	−0.21	−2.34	−0.6	1.13	0.75	−0.98
0.38	0.58	−0.67	−1.61	0.88	1.13	0.38
1.81	0.25	−1.36	0.1	1.04	−1.41	−0.86

3.2 Health Characteristics Indicator Determination

In the application filed, 12 main parameters of the communication equipment system under operation status are collected for 15 working days with 2 s as the sampling period.

The operation data of three working days were randomly selected as the sample, and the score of attribute indicator of each parameter was obtained by using the health evaluation indicator mining module. The comprehensive score record is shown in Table 3. In addition to the vibration amplitude of housing, the outlet temperature of the mill, the thickness of the material layer and the pressure difference of the mill are finally determined to be the health status characteristics of the communication equipment system.

Table 3. Score results of candidate characteristics

Feature name	Random lasso	Ridge regression	Random forest	Stability selection	Recursive feature elimination	Average
Feeding amount	0.1	0	0.03	0.08	0.18	0.08
Powder specific surface area	0.31	0.39	0.07	0.0	0.09	0.17
Material layer thickness	0.6	1.0	1.0	0.8	0.71	0.82
Mill outlet temperature	0.21	0.45	0.32	0.66	0.42	0.41

(continued)

Table 3. (*continued*)

Feature name	Random lasso	Ridge regression	Random forest	Stability selection	Recursive feature elimination	Average
Mill inlet temperature	0.0	0.0	0.23	0.0	0.14	0.07
Mill inlet pressure	0.11	0.0	0.43	0.24	0.13	0.18
Separator revolving speed	0.06	0.0	0.27	0.0	0.59	0.18
Mill pressure difference	0.5	0.79	0.67	0.95	0.95	0.77
Cold air valve opening	0.29	0.0	0.0	0.0	0.09	0.08
Hot air valve opening,	0.21	0.0	0.01	0.12	0.0	0.07
Circulating air valve opening	0.6	0.21	0.14	0.24	0.33	0.3
Main ventilator revolving speed	0.01	0.1	0.01	0.0	0.0	0.02

4 Comparative Analysis of Health Status Recognition System Operating Effects

4.1 Field Operation Data Selection

The communication equipment may occur violent vibration during the operation, which may result in abnormal situations. As an uncontrollable factor, the violent vibration will affect the evaluation of the control data. Therefore, the data of continuous operation without violent vibration are selected as the evaluation data to exclude the influence of the vibration on this evaluation.

The communication equipment Health Status Recognition System is used in a grinding production line in ZD Group. After three days of commissioning operation, working conditions data are obtained. Two sets of data in the process of intelligent control are randomly taken as the control group and another four sets of data under manual control are taken as the comparison group. The time window of the data is one hour, and the control effect of the intelligent system is analyzed.

4.2 Characteristics Evaluation of Power Consumption

According to the four health characteristic indicators such as housing vibration amplitude, mill outlet temperature, material layer thickness and mill pressure difference, the real-time health status evaluation module is used to calculate the mean value, the variance and the number of outliers for each parameter during the actual operation within the sampling window. The result obtained, as a stable operating condition judgement characteristic variable, is compared with the condition in the healthy operating condition mode library. In the specific operation, $\Delta t = 2$ s, $n = 15$. The production staff can judge whether the system is healthy according to the operating status signal provided by the system. The health status prediction and evaluation module is used to predict the health status of the system in a certain time series in the future.

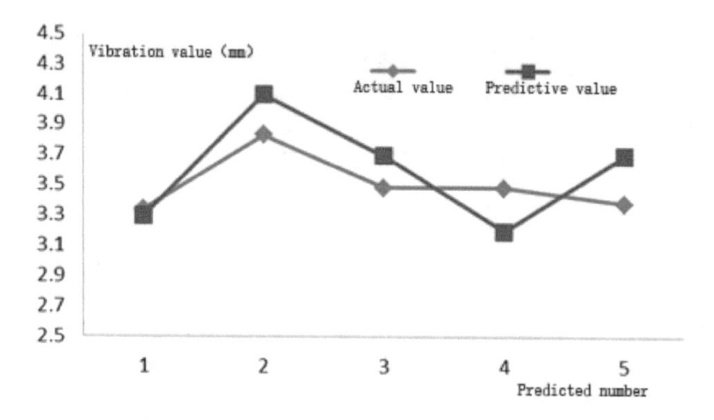

Fig. 3. Predicted values compared to actual values

Taking the housing vibration amplitude as an example, the process of judging the steady status by using time series is illustrated. The predictive model can give a continuous 5-min forecast, standard error and confidence interval. The relationship between predicted and actual values is shown in Fig. 3. It can be seen from the figure that the prediction error is low, and the predicted value basically reflects the changing trend of the value, which indicates the prediction effect of the model is good.

5 Conclusions

In this paper, firstly we present a model and a system for the health status identification and intelligent control of the communication equipment based on data mining, and use a comprehensive feature selection method to excavate and analyze the operating condition data to get key parameters that affect the stable operation of the communication equipment, which are determined as health status evaluation indicators. Next, we perform cluster analysis of operating conditions to obtain the state distribution in the historical operating conditions and define the healthy operating condition categories in the historical conditions. Finally, we make use of ARIMA algorithm to establish the system healthy status feature training model to predict the changing trend of the system operating parameters, realizing the health status identification. At last, the communication equipment health status recognition system is developed, and applied to the engineering practice, proves the validity and practicability of the system.

Acknowledgments. This research is financially supported by National Natural Science Foundation of China (51705455), and China Postdoctoral Science Foundation funded project (2017M621916, 2018T110587).

References

1. Woywadt, C.: Grinding process optimization-featuring case studies and operating results of the modular vertical roller mill. In: IEEE-IAS PCA Cement Industry Technical Conference (2017)
2. Lin, X., Zhang, M.: Modelling of the vertical raw cement mill grinding process based on the echo state network, pp. 2498–2502 (2016)
3. Lin, X., Kong, W.: Adaptive dynamic programming in raw meal fineness control of communication equipment grinding process based on extreme learning machine. J. Syst. Simul. **28**(11), 2764–2770 (2016)
4. Altun, D., Benzer, H., Aydogan, N., et al.: Operational parameters affecting the vertical roller mill performance. Miner. Eng. **103**, 67–71 (2017)
5. Lin, X., Liang, J.: Modeling based on the extreme learning machine for raw cement mill grinding process. In: Deng, Z., Li, H. (eds.) Lecture Notes in Electrical Engineering, pp. 129–138 (2015)
6. Cai, X., Meng, Q., Luan, W.: Soft sensor of communication equipment material layer based on LS-SVM. In: International Conference on Measurement Information and Control, pp. 22–25 (2013)
7. Lin, X., Qian, Z.: Modeling of communication equipment raw meal grinding process and optimal setting of operating parameters based on wavelet neural network. In: IEEE International Joint Conference on Neural Networks (IJCNN), pp. 3015–3020 (2014)
8. Umucu, Y., Deniz, V., Bozkurt, V., et al.: The evaluation of grinding process using artificial neural network. Int. J. Miner. Process. **146**, 46–53 (2016)
9. Wang, K., Li, X.-L., Jia, C., Song, G.-Z.: Optimal tracking control for communication equipment process based on adaptive dynamic programming. Acta Automatica Sinica (10), 1542–1551 (2016)
10. Lin, X., Qian, Z., Liang, J.: Modeling and optimal setting of communication equipment raw grinding process. Control Instrum. Chem. Ind. 2 (2016)
11. Yan, W., Qin, W.: Modeling and control optimization in cement vertical roller mill process. Control Eng. China (06), 929–934 (2012)
12. Su, X., Liu, T., Cao, H., et al.: A multiple distributed BP neural networks approach for short-term load forecasting based on hadoop framework. Proc. CSEE (17), 4966–4973 (2017)
13. Fusco, G., Colombaroni, C., Isaenko, N.: Short-term speed predictions exploiting big data on large urban road networks. Transp. Res. Part C Emerg. Technol. 73, 183–201 (2016)

Computer Science and Education

Technology-Enhanced Financial Education and Sustainability Goals

Margherita Mori(✉)

University of L'Aquila, L'Aquila, Italy
margherita.mori@univaq.it

Abstract. This paper aims at providing evidence on how technological progress may allow to scale up financial education, in sight of boosting financial resilience. What makes this tough task a must – rather than an option – is the vital role that financial literacy and competence can play to eradicate financial exclusion: it fuels serious worries that encourage to minimize financial vulnerability and maximize financial well-being; in turn, financial inclusion can significantly contribute to the Sustainability Goals that have been adopted by the United Nations in 2015. These thoughts sound like an invitation to analyze the state-of-the-art and sort out unexploited opportunities. To this end, a review of most recent literature paves the way to a closer look at innovative technologies to be usefully applied to the education industry, in an attempt at furthering financial capabilities, both locally and on a global scale. Interdisciplinary issues to be addressed encompass web-based learning platforms, technology-enhanced pedagogical models and educational games; focusing on the audience, it must be accounted for the recent trend towards widening the scope of financial education beyond traditionally acknowledged borders, as shown by the increasing recourse to initiatives aimed at the younger and the elderly, in line with the lifelong learning approach. The significance, advancing features and value of this research conceptual paper can be identified with its results in terms of best practices that are worth disseminating, to the benefit not only of the targeted groups – including unbanked and underbanked households – but of society as a whole.

Keywords: Financial education · Financial inclusion · Lifelong education · Sustainable development · Technological progress

1 Introduction

Financial education has gained momentum as a strategic tool to reduce – and ultimately eradicate – financial vulnerability, not only in fragile areas but also in industrialized countries: supporting arguments stem from the 2030 Agenda that was agreed upon by the United Nations (UN) in 2015 and that sheds light on the intrinsic value of a set of Ps, meaning People, Planet, Prosperity, Peace and Partnership; the 17 Sustainable Development Goals (SDGs), that make of this Agenda a high-profile to-do list, deal with economic growth, sustainable communities, reduced inequalities, innovation and infrastructure, as well as multi-stakeholder partnerships aimed at mobilizing and sharing

Z. Hu et al. (Eds.): ICCSEEA 2020, AISC 1247, pp. 599–608, 2021.
https://doi.org/10.1007/978-3-030-55506-1_53

knowledge, expertise, technology and financial resources. Yet, much room remains for furthering financial capabilities, in an attempt at fighting financial exclusion.

Based on what is still on the cards, this paper takes inspiration from the role of financial education within the framework of sustainable development and aims at providing evidence on how technological progress in both the financial and the education industry may be usefully resorted to, in order to improve financial literacy and competence: they are vital to minimize financial exclusion, which is closely tied to financial vulnerability, while financial inclusion can help to fulfill the SDGs; to provide ground for their achievement, it seems appropriate to begin by overviewing the state-of-the-art and proceed to sort out unexploited opportunities, that should be taken not to leave anything unattended, as well as challenges to be met. The resulting analysis sounds like a contribution to advancements in financial education from both a theoretical and a practical point of view, with utmost attention to be paid to academic implications, such as those pertaining to training the trainers.

From a methodological perspective, this is a research conceptual paper that promises to benefit from an interdisciplinary approach – namely: a combination of financial, technological, pedagogical and social issues – and that is structured to flow from identifying basic educational needs on financial topics to disseminating best practices; along this pathway, special emphasis is put on innovations in technology that are worth applying to the education industry, in order to hopefully boost financial capabilities, both locally and on a global scale. Focusing on the audience, the proposed investigation is set to draw upon the recent trend towards expanding the scope of financial education beyond traditionally acknowledged borders, as shown by the increasing recourse to lifelong learning initiatives aimed at the younger and the elderly.

2 Literature Review

Given the cross-cultural approach of this paper, most relevant literature can be accessed by surveying the debate on several topics, including the SDGs and their socio-economic dimensions, the role of financial inclusion and financial education, evolutionary trends in finance, developments in pedagogical methods and technological progress, to mention just a few of them: it is therefore challenging to provide a unified and unique framework for analysis, which can be considered as the starting point to contribute to advancements in the pertaining areas of research and to stimulate discussion on them; to make this challenge truly appealing, the combination of their implications has not been thoroughly discussed yet and actually needs to be included in the strategic toolkit for sustainable development. In turn, its milestone can be identified with the Resolution that was adopted by the UN General Assembly to set up the 2030 Agenda [22], and particularly with its Goal 4, designed to "ensure inclusive and equitable quality education and promote lifelong learning opportunities for all".

While "integration of the SDGs into national and local budgeting processes remains in its infancy" [19], social implications have been widely explored with regard to sustainable banking [16] and its forward steps encourage to devise ways to promote sustainability in other contexts: for instance, useful insights can be gained by considering the efforts undertaken by schools and universities in the United States to infuse sustainability topics into their curricula, as an integral part of a well-rounded education [21]; it is not less relevant to identify influencing factors for enhancing online learning usage models [1], as online education has become the core of educational settings worldwide, thanks to the advent of technology, and this trend has impacted financial education too. To give due credit to technological innovations, their applications have been closely scrutinized, to the benefit of digital firms [3] and – generally speaking – of all organizations that have implemented business intelligence techniques [20], including companies involved in Financial Technology (or Fintech).

Turning to the financial system, much attention has been devoted to the banking sector and its growing involvement in creating value in the technology-driven world: as it has been argued, "tech-savvy customers, exposed to advanced technologies in their day-to-day lives, expect banks to deliver seamless experiences" and banks have been pushed by these expectations to modernize their industry landscape, in order to activate services like mobile banking, e-banking and real-time money transfers, as well as to empower customers to avail themselves of most of these services at their fingerprints anytime anywhere [14]; by the way, a colossal increase in the processing capability of computers and the expansion of internet usage has led to huge amounts of digital information being generated and stored, thus enabling the banking workforce to move away from repetitive, process-driven tasks towards the more strategic and innovative areas of activities that can be expected to ultimately drive the industry forward [13]. Not surprisingly, artificial intelligence (AI) is reshaping the financial industry and especially the traditional approach to services provided by banks, that tend to focus on operational risk management gains as a result of fraud detection or improved know-your-customer processes, and on opportunities for cost reduction like chatbots, not to mention robo-advisors with their potential and risks [10].

3 Data Analysis

3.1 Some Relevant Figures

Evolving trends bring about challenges and opportunities, that lead to evoke the need for consumers (both households and firms, mostly micro-, small- and medium-sized enterprises) to fully understand the potential and the limitations of financial services through proper financial education, whenever a gap remains unbridged between the appropriate and actual level of financial competence and literacy. According to a realistic view, this is not a rare occurrence: a significant share of the world population can be labelled as underbanked, due to supply and demand factors that are both responsible for an inadequate recourse to formal financial services by their respective, potential target markets; to make things worse, about 1.7 billion adults – more than half of those who work in the world – have been estimated to remain unbanked in the last comprehensive survey by the World Bank [8], which means that they do not even hold an account at a financial institution or through a mobile money provider.

These undesirable behaviors are most acute among low-income people in emerging and developing economies, though surveys by the Federal Deposit Insurance Corporation show that the United States too suffers from disappointing findings as far as banking services demanded by – and supplied to – households, especially those with volatile income [11]: approximately 8.4 million – corresponding to 6.5% of – US households, composed of 14.1 million adults and 6.4 million children, were reportedly unbanked in 2017; in other words, no one in these households had a checking or savings account. At the same time, 24.2 million – corresponding to 18.7% of – US households, made up of 48.9 million adults and 15.4 million children, were categorized as underbanked, meaning that the each of them had a checking or savings account but also used products or services from an alternative financial services provider (such as money orders, check cashing, international remittances, payday loans, refund anticipation loans, rent-to-own services, pawn shop loans or auto title loans).

What sounds consoling is that visible improvements have been recently recorded by the American Bankers Association Foundation while benchmarking the relationship between older Americans and banks [4]: to mention just a few data, financial exploitation prevention training is now largely standard, being offered for frontline staff and customer service representatives at all banks surveyed in 2019 and required for customer-facing staff at 9 in 10 banks of all sizes, up from 7 in 10 in 2017; furthermore, banks are increasingly reporting elder financial exploitation to Adult Protective Services (APS), provided that 81% of banks surveyed listed reporting to APS as one of their key actions in dealing with suspected elder fraud, up from 62% in 2017. Therefore, it makes sense to pursue sustainable development by undertaking projects – such as those centered around financial education – aimed at including people who would otherwise run the risks associated with resorting to informal, less reputable financial channels.

3.2 Evidence from the Financial Industry

Apart from statistics, valuable information and data can be obtained by looking at the financial arena as a huge laboratory with both physical and virtual features, despite being experiments usually reserved to applied sciences and hence excluded from the methodological set available to finance; even surfing the web is likely to prove a rewarding exercise, as it allows to disclose initiatives that are based upon the financial footprint of several SDGs and that have been increasingly developed at both national and international level. As a quick reference, the project denominated "Banking the Unbanked" by Bank of Alexandria SAE (shortly AlexBank) consists of three integrated lines of intervention (micro-deposits, local support to micro business and a mobile wallet platform) that aim at supporting customers who would have no access to traditional banking, in order to improve their living and working conditions [2].

Furthermore, this Banks's Group – Intesa Sanpaolo – has been cooperating in the "Loan for Hope" initiative that was launched in 2010 and renewed in 2015 to assist households in temporary difficulties with "social microfinance", while "business microfinance" has been adapted to start-up businesses, young entrepreneurs and – in more general terms – small- and micro-sized enterprises [15]. Consideration has also been devoted to differently-abled people (such as wheelchair-bound and visually-impaired persons) who need customized barrier-free access to banking and still deserve to be better taken care of: it is meaningful that in 2018 Italy's Museum of Saving promoted a competition of ideas – in cooperation with Intesa Sanpaolo Innovation

Center and Fondazione La Stampa Specchio dei Tempi – to stimulate solutions aimed at fostering economic autonomy of people with cognitive disabilities and hence to improve financial inclusion of these consumers by addressing their specific needs, especially in terms of accessibility, language and communication modes [18].

As far as developments under way, technological advances can be expected to accelerate the trend towards better serving the underbanked and targeting the unbanked, as it can be argued by focusing on digital disruptions to the financial system that imply valuable opportunities, besides serious threats. For example, the World Food Programme – or WFP, the food-assistance branch of the UN and the single largest organization providing humanitarian cash transfers – makes recourse to various forms for their settlement from traditional banknotes, bank transfers or value vouchers to more innovative digital platforms such as smart cards, mobile money and blockchain technology; much depends on the local context, most likely ranging from the immediate aftermath of a natural disaster to a protracted refugee crisis; as such, the shift from the concept of food aid to that of food assistance since the late 2000s unveils the trade-off between cash transfers, that are effective where food is available in the markets but beyond people's financial reach, and physical food deliveries, that help respond better in emergency situations where basic market infrastructure is not functioning [23].

4 Discussion

4.1 Investing in Intangible Capital

Along the pathway to sustainable and inclusive growth, investment in intangible capital stands as an essential ingredient: success stories include the "Financial Education in Schools" program that has been run by Italy's Central Bank for several years, in line with international best practices [5]; in addition, this Bank has released a series of videos – entitled "Economy and finance - it's never too late" – to foster financial education among adults by covering a diversified array of topics, such as investment risk and yield, wealth and indebtedness, insurance companies and pension funds [17]. Institutional activities centered around promoting financial education have been undertaken and carried out in many other countries, as well as at a supranational level.

As a reinforcement, it is useful to refer to the initiatives promoted by the Financial Literacy and Education Commission, that was established in the United States under the Fair and Accurate Credit Transactions Act of 2003, and by the National Centre for Financial Education that is in charge of implementing India's Strategy for Financial Education. National policies aimed at improving the level of the population's financial literacy have also been decisive for China, where subjects related to financial issues and specific programs have been included in school curricula: contents are adapted by age and an easy language is used by resorting to several tools (including comics, films and videogames) that also combine technology and social media; the different campaigns in the field of financial education are designed to reach all the regions of this large country, characterized by a great diversity of population [6].

Besides public institutions, financial intermediaries have increasingly been involved in leveraging their biggest asset – their core business – to address some of the world's toughest challenges and in realigning their activities to deliver shared value. Efforts to capitalize on new opportunities encompass those related to the "BNL Edu-Care" project: it has been developed within the framework of sustainable development strategies of the BNP Paribas Group in Italy with the aim of contributing to inclusive and quality education, as well as promoting lifelong learning; this initiative includes free training courses in financial education that have been offered for years to encourage greater awareness of the variety of personal choices available [7].

4.2 Financial Literacy as a Life Skill

In line with best practices, several variables need to be kept under control, with their list being spearheaded by age-appropriate learning objectives and main categories of financial services and products to be dealt with (such as bank accounts, investment in financial assets, retirement accounts, insurance and education savings). The wide range of available options enables to design financial education projects directed to kids, based upon gamification, while distance education (particularly web-based learning) may be conveniently resorted to in order to improve financial literacy and competence among adults: the school system is likely to provide reliable support at any level, including colleges and universities, within a lifelong learning framework, which should pose unprecedented academic challenges, among others; actually, the elderly may be seen as making up a market segment that educational services can be usefully delivered to, in this case focused on financial topics and based on a practical – rather than theoretical – approach, as non-credit courses tend to allow for.

Many people feel that they are ill-equipped to make sound financial decisions and would reportedly appreciate participating in initiatives designed to improve their financial literacy, whether in person or online, especially if the content is perceived as being interesting and relevant; even more conspicuous results might be achieved by those – households and entrepreneurs alike – who are not aware of their need for financial education, particularly if they are affected by financial fragility. Focusing on employers, they are likely to take advantage of employees becoming less stressed about their financial health, which is likely to result in increased productivity, reduced absenteeism and turnover, higher engagement and job satisfaction, fewer requests for pay increases and lower healthcare costs, not to mention the advantages of setting themselves apart in the marketplace by offering holistic financial wellness benefits.

Helping young people in their transition to the job market too can fall within the scope of promoting financial education, that may be considered as the best driver to end inequality and pursue sustainable development: no surprise, provided that financial health is vital for escaping the poverty trap and can contribute to sustainable growth, while financial literacy is increasingly recognized as a life skill for the 21st century; to this end financial institutions – banks in the first place – are well positioned to play a pivotal role, as it can be argued by taking a look at the activities under way at the Center for Financial Education and Capability that has been set up by Banco Bilbao Vizcaya Argentaria S.A. (shortly BBVA), recently named as the world's best bank for financial inclusion. To better perform the underlying tasks, special attention should be paid to make financial education effective, which implies well-trained educators, tested program materials, timely instruction, relevant subject matters and evidence of impact.

5 Results

5.1 The Role of Technological Innovations

The need to invest in intangible capital – in an attempt at reaching higher and higher degrees of financial literacy and competence – should be properly combined with the need to take advantage of technological innovations, in finance as well as in the education industry. To make this scenario even more attractive, investing in education can be assumed to yield the highest return rate from a social point of view and the financial sphere of the economy makes no exception to this rule: accordingly, there is no choice but to meet the opportunities and take the challenges ahead, including those that pertain to AI in banking – and, in more general terms, in finance – and that encompass deep and machine learning; it is not a case that financial services companies are redefining the way in which they work and create innovative products, while the banking industry is set to be dominated by "intelligent" institutions.

For instance, it must be accounted for the more and more widely endorsed attitude towards going cashless, though payment behaviors significantly differ from country to country. Some characteristics of cash make it unique and still desirable: cash payments do not require any other service providers or technical infrastructure, allow for transparent expenditure overview and feature ease and speed of use; however, the advantages of cashless transactions have been increasingly acknowledged, with supporting arguments mostly tied to the positive consequences of digital disruption, that are far from being exhausted, especially in the financial system.

Further developments are supposed to provide an enhanced experience for customers and employees while delivering real business value on every dimension, since AI is underpinned by a series of interrelated technologies around machine learning and natural language. However, AI is not just better technology, which Fintech companies have been promoting heavily in the last few years, and the risk of a disruptive use of technology within the financial sector should not be underestimated, in terms of cybersecurity, data privacy, money laundering and consumer protection-related issues; furthermore, interacting with new technologies may prove hard especially for senior citizens, not only in developing but also in industrialized countries, thus leading to focus on demographic implications of financial education, as well as on social ones, such as those that call for neutrality and unbiased learning outcomes.

5.2 Technology and Financial Educators

Technology has fueled remarkable advancements not only in the global financial system but also in the education industry, as shown by the increasing recourse to web-based learning platforms, technology-enhanced pedagogical models, educational games and simulations, among others; accordingly, financial competence and literacy have been evolving thanks to the design and use of high-quality digital tools and have become more effective, in particular by facilitating positive financial behaviors and smart decisions. Meanwhile, financial education has started to be perceived as a life-long process, which has prompted to develop awareness-raising activities directed towards the younger – besides the older – generation and to involve schools in the launch of training initiatives based on interactive materials.

To this end, gamification is set to yield valuable results, since people – especially youths – are inclined to learn more about financial literacy by playing a game and it can provide the potential to cooperate or compete with friends while learning, thus making playing a popular way to add social interaction to learning. Success stories include a library of ten game-based personal finance programs, collectively known as the Northwest Youth Financial Education project [9]: students have responded exceptionally well to learning personal finance concepts through games and the feedback from instructors who have been involved in implementing these programs proves their ease of use, even without technological expertise; in more general terms, findings suggest that game-based experiential financial education has good chances to result in high levels of engagement, knowledge gain and intended behavioral changes.

All in all, digital technologies have been supporting a shift of cultural practices in teaching financial education, as well as in learning its concepts at all stages of people's life cycle: for instance, digitalization has contributed to a more active learning experience based on a participatory pedagogical approach and on a convenient mix of formal and informal learning; meanwhile, one-to-one technology-enhanced learning has been favored by being personal, portable, wirelessly-networked technologies a reality in the lives of learners in many countries, which acts as a stimulus to combine tech and personal experiences on an individual basis, and to even improve financial competence through mobile applications [12]. Based on progress under way, digital financial literacy can be assumed to become an increasingly important aspect of education for the Digital Age, thus leading to prioritize the need to develop programs for vulnerable groups, including the elderly, the less educated, owners of micro- to medium-sized enterprises, startup firms and – due to the persisting gender gap – women.

6 Conclusions

The analysis carried out so far allows to emphasize the role that financial education can play to improve financial inclusion and well-being, while reducing – and ultimately eradicating – financial exclusion and vulnerability, in sight of globally fulfilling the SDGs. Reliable estimates show that much room remains for progress, not only in fragile economies but also in industrialized countries, which sounds like an incentive to take the opportunities and meet the challenges ahead: they imply unprecedented efforts to apply technological innovations to the education industry within a lifelong learning framework that should be centered around basic finance; best practices that are worth disseminating encompass the launch of initiatives that aim at furthering financial capability and entail the recourse to technology-enhanced methods.

The wide choice of encouraging results leads to highlight the need to combine evolutionary trends in the three industries involved, that encompass technology, education and finance, and to recommend a cross-cultural approach that is intended to drive beyond the boundaries of these research areas, as the relevance of social implications suggests. For instance, based upon lessons learned, innovative pedagogical models that take advantage of increased digitalization can maximize the impact of financial education policies and related initiatives for financial inclusion, with its scope being recently expanded to cover digital financial inclusion; there it follows that academic issues should be addressed – possibly by undertaking joint efforts, as soon as possible – to advance knowledge at the crossroads of the disciplines pertaining to the above mentioned areas, both in theory and practice.

It is not only a matter of promoting and performing research activities on them and their combination: to draw conclusions that include specific guidelines, it seems not less relevant to propose courses aimed both at training the trainers in financial education and at boosting financial competence among people who would not be otherwise targeted, by universities in the first place; for example, financial education can be infused into curricula at all levels, as an integral part of a well-rounded education, provided that financial literacy is increasingly acknowledged as a life skill. Further developments can be envisaged that involve the elderly, women and minorities – to mention just a few segments of financial services consumers who tend to prevail among the unbanked and underbanked – since coping with their unsatisfied financial needs promises to turn beneficial not only to them, but to society on a global scale.

References

1. Ahuja, S., Kaur, P., Panda, S.N.: Identification of influencing factors for enhancing online learning usage model: evidence from an indian university. Int. J. Educ. Manage. Eng. (IJEME) **9**(2), 15–24 (2019)
2. AlexBank: creating shared value – driving financial inclusivity across Egypt. Bank of Alexandria S.A.E., Cairo (2017). https://www.alexbank.com/document/documents/ALEX/Retail/Sustainability/SusRep2017_En.pdf. Accessed 31 Oct 2019
3. Al-Samawi, Y.: Digital firm: requirements, recommendations, and evaluation the success in digitization. Int. J. Inf. Technol. Comput. Sci. (IJITCS) **11**(1), 39–49 (2019)
4. American Bankers Association Foundation: 2019 Older Americans Benchmarking Report: Findings from a Survey of Banks. Washington, DC (2019)
5. Bank of Italy: The Bank of Italy launches the 2015/2016 "Financial Education in Schools" Programme. Rome, 12 October 2015. https://www.bancaditalia.it/media/notizia/the-bank-of-italy-launches-the-2015-2016-financial-education-in-schools-programmme/?com.dotmarketin%E2%80%A6. Accessed 05 Nov 2019
6. BBVA: Why does China top the ranking in financial literacy? Bilbao, 12 June 2017. https://www.bbva.com/en/china-top-ranking-financial-literacy/. Accessed 31 Oct 2019
7. BNP Paribas: Financial education for an educated choice, Rome (2016). http://www.bnpparibas.it/en/bnp-paribas/corporate-social-responsibility/. Accessed 31 Oct 2019
8. Demirgüç-Kunt, A., Klapper, L., Singer, D., Ansar, S., Hess, J.R.: The Global Findex Database 2017: Measuring Financial Inclusion and the Fintech Revolution. World Bank Group, Washington, DC (2018)
9. Erickson, E., Hansen, L., Chamberlin, B.: A model for youth financial education in extension involving a game-based approach. J. Ext. **57**(4) (2019)
10. Facundo, A., Schmukler, S.L., Tessada, J.: Robo-Advisors: Investing through Machines. World Bank Group, Washington, DC (2019)
11. FDIC: National survey of unbanked and underbanked households. Washington, DC, October 2018

12. Government of Canada: Improving financial literacy through mobile technology: small change pilot program outcomes, 5 February 2018. https://www.canada.ca/en/financial-consumer-agency/programs/research/small-change-pilot.html. Accessed 07 Nov 2019
13. Jubraj, R., Graham, T., Ryan, E.: Redefine Banking with Artificial Intelligence. Accenture, Dublin (2018)
14. Kurshid, A.: Why banks need artificial intelligence, November 2019. https://www.wipro.com/business-process/why-banks-need-artificial-intelligence/. Accessed 05 Nov 2019
15. Intesa Sanpaolo Group: Financial inclusion of vulnerable people: micro-finance projects. Turin, 27 March 2019. http://www.group.intesasanpaolo.com/scriptIsir0/si09/sostenibilita/eng_inclusione_finanziaria.jsp?tabId=microcredito&tabParams=eyd0YWJJZCc6J21pY3JvY3JlZGl0byd9#/sostenibilita/eng_inclusione_finanziaria.jsp%3FtabId%3Dmicrocredito%26tabParams%3Deyd0YWJJZCc6J21pY3JvY3JlZGl0byd9. Accessed 31 Oct 2019
16. Lange, K., Schmitt, E.M.: The Social Dimension of Sustainable Banking. Institute for Social Banking, Berlin (2019)
17. Marcocci, M.: Economia e finanza. Non è mai troppo tardi. Aziendabanca, 8 May 2017. https://www.aziendabanca.it/notizie/educazione-finanziaria-video-serie-banca-d-italia-orizzonti-tv. Accessed 30 Oct 2019. (in Italian)
18. Museo del Risparmio: Al via un contest a premi per agevolare l'autonomia economica delle persone con disabilità cognitiva, press release. Turin, 18 July 2018. www.museodelrisparmio.it/wp-content/uploads/2018/07/Comunicato-Stampa-progetto-Eureka.pdf. Accessed 31 Oct 2019. (in Italian)
19. Okitasari, M., Sunam, R., Mishra, R., Masuda, H., Morita, K., Takemoto, K., Kanie, N.: Governance and National Implementation of the 2030 Agenda: Lessons from Voluntary National Reviews. United Nations University Institute for the Advanced Study of Sustainability (UNU-IAS), Tokyo (2019)
20. Oppong, S.O., Asamoah, D., Oppong, E.O., Lamptey, D.: Business decision support system based on sentiment analysis. Int. J. Inf. Eng. Electron. Bus. **11**(1), 36–49 (2019)
21. Stone, J.A.: A sustainability theme for introductory programming courses. Int. J. Modern Educ. Comput. Sci. **11**(2), 1–8 (2019)
22. UN: Transforming our world: the 2030 agenda for sustainable development. Resolution 70/1 adopted by the General Assembly, New York, NY, 25 September 2015
23. WFP: Cash transfers for fast and effective assistance, Rome, May 2018. https://docs.wfp.org/api/documents/WFP-0000070759/download/. Accessed 31 Oct 2019

Construction of Individual Learning Scenarios

Valentyn Tomashevskiy[1], Iryna Pohrebniuk[2], Nataliia Kunanets[3], Volodymyr Pasichnyk[3], and Nataliia Veretennikova[3(✉)]

[1] Informatics and Computer Science Department, National Technical University of Ukraine "Igor Sikorsky Kyiv Polytechnic Institute", Kiev, Ukraine
simtom@i.ua

[2] Economic, Management and IT Department, Berdyansk University of Management and Business, Berdyansk, Ukraine
iryna.pohrebniuk@gmail.com

[3] Information Systems and Networks Department, Lviv Polytechnic National University, Lviv, Ukraine
nek.lviv@gmail.com, vpasichnyk@gmail.com, nataver19@gmail.com

Abstract. It is proposed technologies for improving digital learning at higher educational institutions, based on the concept of learning considering students' abilities. Technologies of construction of individual learning scenarios based on a student model and gap maps of user knowledge are analyzed. The student's model presents the basic parameters of their level of preparation and individual cognitive features. It allows to realize a number of new technologies for forming individual calendar plans of theme repetition, constructing individual trajectories of processes of forgetting the acquired knowledge by a student and developing individual scenarios of adaptive testing. A key feature of the proposed approach is its practical implementation in a real functioning open adaptive learning system, which has been successfully tested and used in several Ukrainian universities.

Keywords: Student model · Gap map · Digital learning · Adaptive testing

1 Introduction

Rapid introduction of new information technologies is taking place in many spheres of society. Digitization is a global trend affecting all areas of the modern economy. Several countries, including Singapore, the United Kingdom, New Zealand, the UAE, Estonia, Japan, Israel have become leaders in the digital economy, prioritizing digital transport and education. It is identified a list of digital technologies in the Davos Economic Forum 2019 "Globalization 4.0: Shaping Global Architecture in the Age of the Fourth Industrial Revolution", including cloud and mobile technologies, blockchain, virtualization technologies, identification, artificial intelligence, biometric technologies, augmented reality technologies, additives (3D printing), etc.

In this context, there is an urgent need to develop individualized approaches to higher education institutions, creating individual learning scenarios that take into account the student's level of knowledge and provide competency-building according to the standard of specialty.

Z. Hu et al. (Eds.): ICCSEEA 2020, AISC 1247, pp. 609–620, 2021.
https://doi.org/10.1007/978-3-030-55506-1_54

There is no exception to the system of higher education, which requires new approaches and digitization. The speakers of the forum emphasized the need to improve the ways and forms of the educational process. These include mastering digital competencies with an emphasis on teamwork and creativity, learning through games that develop critical thinking, and supporting students' initiatives beyond the curriculum. Digitization of education is a major trend in the development of education systems in almost every country in the world and covers all levels from primary education to PhDs.

The purpose of the paper is to analyze technologies for constructing individual learning scenarios based on student's model and gap maps of user's knowledge.

To achieve this goal, the methods of information and mathematical modeling were used, in particular the Petri net, which allowed to simulate parallel (synchronous and asynchronous) processes in the system of adaptive learning, as well as the mindmapping method, the Rasch model, etc.

1.1 Literature Review

According to Harasim, the invention of the Web technologies made online education increasingly accessible, open, flexible; allowed new pedagogical models to emerge and reasoned the revolution in digital knowledge age that enabled greater and faster human communication and collaboration and led to fundamentally new forms of economic activity that produced the knowledge economy and required basic changes in education [1].

One of the first studies to investigate trends in distance education was carried out by Berge and Mrozowski [2], who examined research literature over a ten-year period from 1990 to 1999.

A series of studies were conducted by Zawacki-Richter to explore the distance education research domain. Zawacki-Richter developed a categorization of research areas in distance education and identified the most important and the most neglected research areas [3].

Zawacki-Richter and Anderson went one step further and provided a comprehensive survey on the state of online distance education as an independent field of inquiry, while also offering a clear orientation for future research [4].

2 Students' Models Based on the Parameters of the Education Level and Cognitive Characteristics

The prerequisites for the emergence and development of digital learning based on the concept of learning have appeared with the introduction of electronic tools and the latest technologies in higher educational institutions and it takes into account an ability of the individual, satisfies their needs to obtain an appropriate level of education. Digital learning is an innovative educational information service that enables anyone to study remotely.

This approach requires the development of information and digital technologies that become a platform for change and transformation in the education area. A radical digital transformation of learning processes needs to be done. It is expected that teachers acquire digital competencies and new skills [5].

It is believed that improving distance learning systems is based on the digitalization of learning technologies and the use of a student-centered approach model. Furthermore, such model considers the adaptation mechanism to the requirements of the digital economy in the formation of innovative learning and testing scenarios. In this regard, the development of adaptive intelligent learning systems based on artificial intelligence and digital technologies is long-term and urgent [6].

An entrant has an opportunity to study on digital form of education at any age for any period, passing the appropriate tests in each semester (amount of credits). If an exam (test) is not completed in time, the student is given another attempt. If a student does not take the tests of three subjects per semester, then they are deducted. There is a minimum required list of subjects (credits) in each semester. The number of disciplines in the curriculum is divided into two blocks:

- A block of compulsory subjects defined by higher education standards.
- A block of student-selected subjects where the student can choose such subjects that are the most important for a professional future.

Today, there are a lot of Web-based e-learning systems in the IT market, including Blackboard, WebCt, Moodle, IBM LearningSpace, but they do not use a student-centered approach, which reduces the quality of the learning process and does not allow organizing adaptive learning.

Modern platforms for providing distance education services are formed on the principles of integration of distance learning systems with models, methods, technologies of expert systems within a single architecture, which combines interacting logic and linguistic, mathematical, imitation and other types of models. It promotes the emergence of new adaptive and intelligent learning environments.

Remembering and forgetting processes play an important role in learning. Memory is one of the most important mental processes that implements the knowledge acquisition. The law of forgetting a meaningful material is represented by an approximating logarithmic function of the form $f(t) = a\ln(t) + b$, where t is time elapsed from the moment of complete mastery of material; a, b are parameters that characterize individual student memory characteristics and are determined by the method of least squares for individual statistics based on 3–4 tests over a period.

An adaptive system creates a recurrence schedule for topics, building individual forgetting curves for each student. This promotes an individual approach to sending a reminder to repeat certain material.

For mastering teamwork competencies, projects are being formed within studying a few professional disciplines, the realization of which is possible by virtual teams, formed randomly from course listeners, who will perform tasks in the remote access mode. Each team is assigned a mentor consultant who can consult using modern information technologies and video conference tools. The mentor may be a representative of a company that cooperates with the department at university [7].

3 Building Gap Maps of User's Knowledge

A new technology of mind mapping has emerged at the intersection of psychology and computer science, as a new way of analytically representing information based on the graphical display of associative or logical connections. The technique of creating mind mapping is an alternative method of visual presentation of theoretical material. Formally, mind mapping is a model of the teacher's knowledge of a topic or course in general and it is a kind of reflection of the subject area. The use of mind mapping allows you to cover the whole situation as well as to hold a large amount of information in order to find connections between individual elements, to remember information and to recall it after a long period of time.

The effect of the mind mapping method is intensified using digitization technologies that contribute to the formation of an electronic platform for educational materials, which allows the creation of a database of electronic textbooks, including interactive software products with virtual 3D representation of individual components, digital copies of training and teaching manuals. Visualization of educational material helps to create visual images, and the process of their remembering is much more efficient. This approach generates the need to improve memory models for using them in knowledge control processes.

Formalizing the mind mapping method in semantic data models helps to automate knowledge control processes by comparing students' knowledge with reference mind mapping and building knowledge gap maps. It is a gap map of knowledge (Fig. 1) that is a qualitative indicator of students' learning.

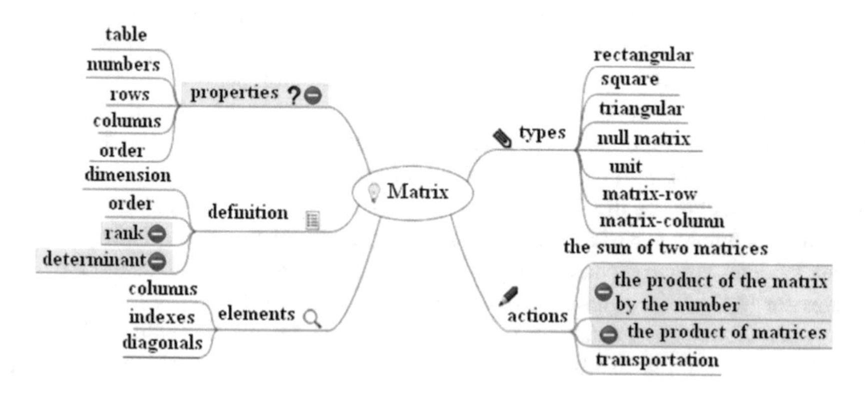

Fig. 1. Gap map of student's knowledge in the topic of the Matrix

4 Technology of Constructing an Individual Scenario of Adaptive User Testing

Adaptive testing means a wide-ranging class of testing techniques that involve changing the sequence of submitting tasks in the actual testing process using the student's answers to the provided tasks. The adaptive approach is based on the individualization of the test selection process, which ensures the generation of effective tests by optimizing the complexity of the tasks according to the level of students' readiness [8]. In other words, the basic idea of adaptive testing is that the test tasks need to be adapted to the level of student preparation, and the selection of tasks follows from the assumption that weak students should not be given difficult tasks because they are more likely to be unable to complete them correctly. Also, it is ineffective to give easy tasks strong students. The scenario of adaptive testing means an individual set of test tasks with different levels of difficulty; each test is selected for each student, depending on their answer to the previous question.

In general, the algorithm of computer adaptive testing consists of the following steps:

1. The corresponding task by parameter is selected from the task bank.
2. The selected task is given to the student who answers it correctly or incorrectly.
3. The test ability rating is updated based on answers.

The previous three steps are repeated until, according to a certain criterion, the assessment of the measured quality is considered satisfactory and the tests are finished [9].

The models based on probabilistic criteria that are relevant to modern Item Response Theory testing have been successfully used in adaptive testing. In our opinion, the one-parameter Rasch model is most suitable for parameter estimation of the knowledge level and task complexity in the adaptive remote test [10]:

$$P_j(\theta) = \frac{e^{1,7(\theta-\beta_j)}}{1+e^{1,7(\theta-\beta_j)}};$$

$$P_i(\beta) = \frac{e^{1,7(\theta_i-\beta)}}{1+e^{1,7(\theta_i-\beta)}},$$

where θ (latent variable that determines the level of student preparation) and β (latent variable that determines the test complexity) are independent variables for the first and second functions respectively.

The one-parameter Rasch model is one of a family of logistic curves. Both latent parameters θ and β are usually evaluated during the test development, and they are respectively referred to as logits of the knowledge level and logits of the task complexity level.

The logit of the knowledge level is the natural logarithm of the ratio of the student's correct answers to all the tasks of the test, to the ratio of incorrect ones.

The logit of the task complexity level is the natural logarithm of the ratio of the incorrect answers to the task to the ratio of correct answers to this task by the number of students.

The estimation of these parameters is carried out with the assumption of normality of empirical test data distribution across a set of both students and test tasks. The values of latent variables are also considered normally distributed [11].

Since there are some data processing difficulties in computer-based Rasch model testing, it is necessary to modify them.

Modification of the Rasch model is suggested in [12]. It is said that:

1. A test result matrix is used to evaluate the parameters of the task complexity of the preparation level, and it is presented with a table, whose rows are students, the columns are the task numbers. The results for each row and column are summarized below and the matrix is sorted and analyzed. There are difficulties applying adaptive tests, selected from the bank of test tasks by a certain criterion, because the matrix is not filled completely. Each student is tested by their individual scenario. The same question may have a different ordinal number, the number of questions in the test is also different for each student, so that the test tasks have a different number of answers to them. Therefore, the analysis of the preparation level and the task complexity should be done separately. The test task number is defined as the test number in the test task bank.

2. The proportion of correct and incorrect answers are calculated by the number of questions answered by the student and not by the total number of questions in the test:

$$p_i = \frac{x_i}{n}$$

Where p_i is a proportion of correct answers of the i student; x_i is a number of correct answers; n is a number of tasks that the i student answers to; i is a student number.

Similarly, it is for the task complexity regarding the number of students who responded to it, rather than the total number of students:

$$p_j = \frac{R_j}{N},$$

where p_j is the proportion of correct answers to the j task; R_j is a number of students completing j task correctly; N is the number of students who completed the j task; j is the test reference number.

It is necessary to store statistics for each task to evaluate objectively the task complexity.

3. The Rasch model for the selection of test tasks is applied from the moment of occurrence of incorrect or correct answer, and the parameter of knowledge level is calculated for this question. In cases when the student answers the test questions correctly or incorrectly, than it is occured the change of the difficulty level in one or another way:

- if the answer to the question is correct, the probability of answering the task of a higher level of difficulty is equal to 0.7 and it happens the move to a higher level of difficulty if the level of difficulty is not the highest; in the case of the highest level of difficulty, the student remains at that level;
- if the answer to the question is correct, the probability of answering at a higher level of difficulty is equal to 0 and it happens the move to a lower level of difficulty if the level of difficulty is not the lowest; in the case of the lowest level of difficulty, the student remains at that level.

4. If the student answers all the questions of the test correctly (incorrectly), the above mentioned model is applied.

It is impossible to evaluate objectively the complexity of tasks in small samples, so the algorithm of the transition from the dichotomous scale of knowledge levels to the interval scale of logits is further proposed.

Estimates of latent complexity parameters for assignments usually lie in the interval (−5; 5) and have several decimal places. They can also take negative values, which allow us to develop formulas for the transformation of logit scales. Linear transformations are more preferable because they retain the interval character of the scale. The most common is the transformation proposed by Chopin:

$$\theta_i = 50 + 4,55\theta$$

$$\beta_j = 50 + 4,55\beta$$

As a result of these transformations, we obtain positive values of the parameters θ and β, which are located in the interval (30; 70), and they are rounded to integers.

Consequently, from these formulas it is possible to deduce the inverse to move from the ordinary scale of tasks with different levels to logits.

The difference between the interval limits (30, 70) is up to 40. It is supposed that the teacher chooses 4 levels of difficulty for the test tasks ($z_j = 1,4$).

Then the easiest difficulty level z_j will correspond to the interval from (30; 40), z_j – (40; 50), z_j – (50; 60), z_j – (60,70).

It follows the next:

$$\beta = \frac{\beta_j - 50}{4,55}$$

It is suggested taking j as the left boundaries of z_j intervals to scale the usual levels of complexity in logits, i.e. the easiest task in the logit scale by the formula will have the following meaning:

$$\beta = \frac{30 - 50}{4,55} = 4,3056$$

Thus, it is necessary to determine only the values of the logits of the students' knowledge level taking into account the second point, in case of the statistics lack (small number of students, large number of test tasks) after the distribution of test tasks by the teacher by difficulty levels ($z_j = \overline{1,z}$) and turning them into logits of task complexity.

5 Implementation of a Remote Adaptive Learning System

The proposed DAOS (Distance Adaptive Open System) remote adaptive learning system is considered as a complex, multi-component system that is open to necessary changes and improvements. The effectiveness of the recommendation is based on the use of a systematic approach to the formation of an adaptive distance learning system,

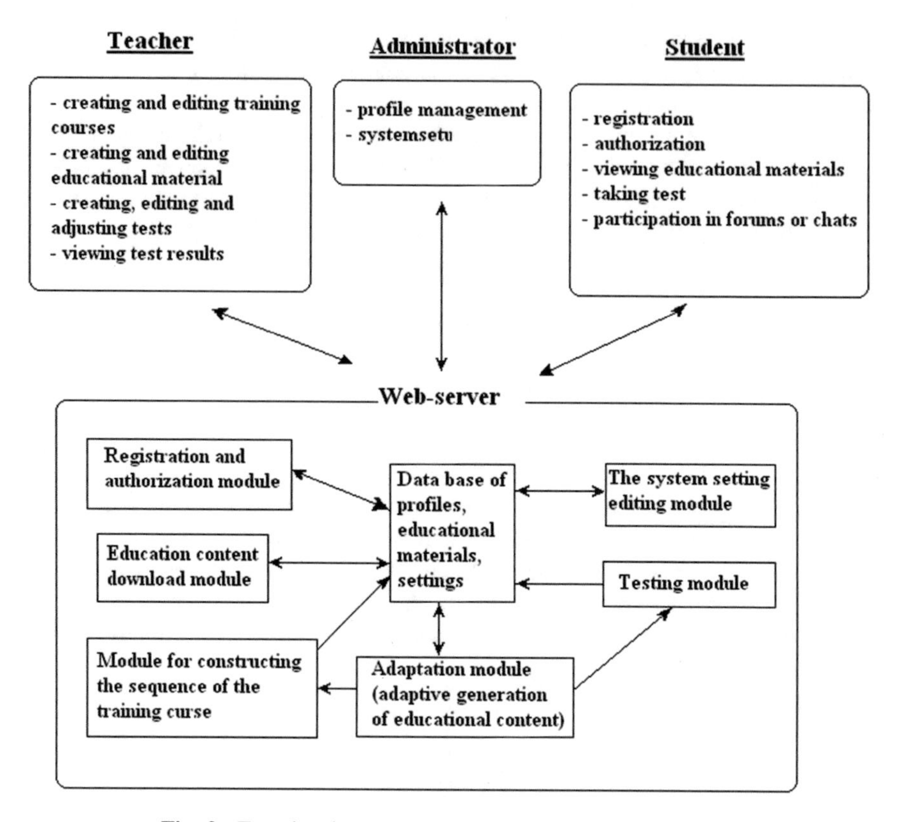

Fig. 2. Functional model of remote adaptive training system

taking into account the dependence of the system components that are subsystems and components, which will contribute to the formation of requirements for it, the development of its functional tasks, the creation of a basis for project development, as well as the choice of user interface design [13, 14] (Fig. 2).

The DAOS system is based on a three-level architecture and contains the following components:

- a client is an arbitrary browser;
- an application server is based on Java technologies to support cross-platform;
- a database with MySQL management system.

An actor, as an external entity to the system, actively interacts with the system and is logically related to multiple roles.

The design of the system considers that one person can play several roles, which are specified in their model, which contains the specification of each actor and the features of their interaction with the system.

The administrator-actor performs the following functions:

1. Administrative tasks such as management of an actor-listener, test creation, data exporting or importing, introduction of supporting theoretical material.

2. Testing support.
3. Support for system interactivity and user authentication.
4. Output of results of intermediate messages.

The application server contains the following functionality:

1. Construction of rules of test passing on the basis of the loaded models (priority, choice of test types, choice of difficulty level, transition from one level of difficulty to another one, consideration of information about the results of already passed test tasks).
2. Processing and storing the results of study and repetition of the accompanying theoretical material and test results in the database.
3. Formation of individual scenarios for studying theoretical material based on test results and forgetting curves.
4. Generation of performance reports, chart and spreadsheet plotting.

The database is intended for storage of:

- information about users;
- information about models (training scenarios);
- test results;
- system settings;
- accompanying theoretical material and tests with answer options (Fig. 3).

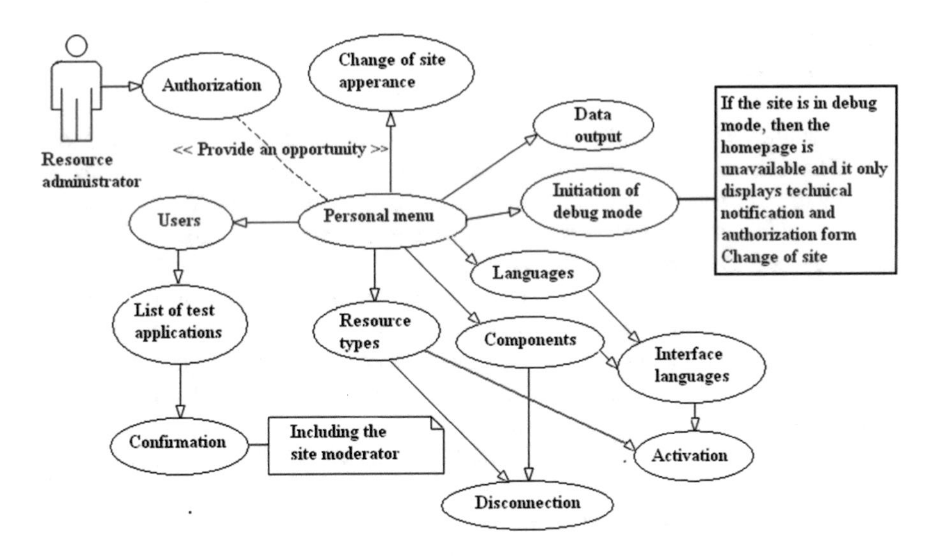

Fig. 3. Precedent chart for an actor of recommendation system administrator

Work with the system begins with registration and subsequent authorization, after that a teacher and a student can log in. Teachers' profiles are entered into the system by the administrator. After the creation of Teacher profile (an author of the learning courses), a teacher receives a personal account. A new Student user can be registered directly on the site. When access to information is granted, system users can view a list of learning courses, tests and related training material.

The teacher creates educational material and uploads it into the system through a specially designed interface. The interface is different for viewing and adding learning content for different user groups.

The student must choose the faculty, educational qualification level (bachelor, master), direction of preparation, teacher, learning course and type of educational content (lectures, laboratory classes, practical classes, tests, etc.) to study the courses or to pass the tests. The Student actor group is granted access rights to view course materials, and the Teacher actor group has an access and editing rights. An actor of Teacher group gets access to a database of learning materials and statistics after logging in. Statistical information is stored in the test system for each answer to a test question. The algorithm of student work with the system is as follows:

Step 1. Login Authorization.
Step 2. Selection of the learning course.
Step 3. Performing the testing process.
Step 4. Notification of completion of the test response process.

Closing the software or disconnecting means that you will not pass the test.

The actor of the Teacher group has an opportunity to receive test tasks from the system server, to edit them or to develop a new test and to save it in the database.

6 Testing the Results

Previous studies of the Adaptive Distance Learning model have been conducted using the STELLA simulation system and described in [15]. A group of 30 students was divided into three categories according to the average success grade. The first category includes students whose success rate is 60% and above (66% in the group), the second one has the success rate from 30% to 60% (21% in the group), and the third category has the success rate less than 30% (13% in the group). The passing score for the discipline was 60 during the three weeks of study.

The obtained results confirm that the more information a student learns, the more they forget. Studies have shown that the average achievement of all students in adaptive learning is increased by more than 60%.

In the classical scheme, students in the first category lose 9% of their knowledge, students in the second - 10%, in the third - 14%. But, during learning with the adaptive scheme, the results are completely opposite. The knowledge level of students in the first category increases by 6–20%, the second - by 95–100%, and the third in general by 300%.

Based on the Internet technologies and developed models of individual learning scenario based on knowledge gaps, a remote adaptive learning system using Google Web Toolkit framework was created, which provides fast server operation and support of an arbitrary browser introduced at the National Transport University (Kyiv).

This system of adaptive learning is implemented at the International University of Finance (Kyiv) in the distance learning package called Virtual University [16].

7 Conclusion

The proposed approach will make it possible to reconcile individual educational scenarios with the requirements of standards for students' acquisition of certain competences and skills. At the same time, the basic level of knowledge obtained by a student at the previous levels (courses) of study as a starting point for replenishing the baggage of knowledge and skills is a significant factor in the formation of such scenarios.

The research made it possible to create a system of adaptive distance learning in the form of a multilevel Petri net that takes into account cognitive features (individual forgetting time) and the preparation level of a student. According to the results of modeling of classical and adaptive distance training courses, it is determined that the model of adaptive training course, taking into account the parameter of forgetting time, allows to increase the level of preparation of a student (by 9%), while slightly increasing the time to study the course (4%) due to the repetition of educational theoretical frames. These models are implemented using hierarchical time color-coded Petri nets in CPN Tools environment and imitate the mechanisms of adaptation of learning scenarios for each student, depending on their individual learning characteristics (time of learning material forgetting and test results), forming frames for repetition of material.

References

1. Harasim, L.: Shift happens: online education as a new paradigm in learning. Internet High. Educ. **3**(1), 41–61 (2000)
2. Berge, Z.L., Mrozowski, S.: Review of research in distance education, 1990 to 1999. Am. J. Dist. Educ. **15**(3), 5–19 (2001)
3. Zawacki-Richter, O.: Research areas in distance education: a Delphi study. Int. Rev. Res. Open Dist. Learn. **10**(3), 1–17 (2009)
4. Zawacki-Richter, O., Anderson, T.: Online Distance Education: Towards a Research Agenda. AU Press, Edmonton (2014)
5. Khan, I.H.: A unified framework for systematic evaluation of ABET student outcomes and program educational objectives. Modern Educ. Comput. Sci. **11**, 1–6 (2019). https://doi.org/10.5815/ijmecs.2019.11.01
6. Matsyuk, O., Nazaruk, M., Turbal, Y., Veretennikova, N., Nebesnyi, R.: Information analysis of procedures for choosing a future specialty. In: Shakhovska, N., Medykovskyy, M.O. (eds.) CSIT 2018. AISC, vol. 871, pp. 364–375. Springer, Cham (2019)
7. Bomba, A., Kunanets, N., Nazaruk, M., Pasichnyk, V., Veretennikova, N.: Model of the data analysis process to determine the person's professional inclinations and abilities. In: Hu, Z., Petoukhov, S., Dychka, I., He, M. (eds.) ICCSEEA 2019. AISC, vol. 938, pp. 482–492. Springer, Cham (2020)
8. Assessing Student Learning Outcomes for Information Literacy Instruction in Academic Institutions Paperback by Elizabeth Fuseler Avery (ed.) (2004)

9. Computer Adaptive Testing of Student Knowledge. https://www.researchgate.net/publication/49619355_Computer_Adaptive_Testing_of_Student_Knowledge
10. Estrellado, M.E.L., Sison, A.M., Tanguilig III, B.T.: Test bank management system applying Rasch model and data encryption standard (DES) algorithm. Modern Educ. Comput. Sci. **10**, 1–8 (2016)
11. Nguyen, L., Do, P.: Learner model in adaptive learning. World Acad. Sci. Eng. Tech. **21**, 395–400 (2008)
12. Pohrebniuk I.: Model of knowledge assessment in adaptive testing [in Ukrainian]. In: Sixth Scientific and Practical Conference with International Participation "Mathematical and Simulation Modeling of Systems", Chernihiv, pp. 378–382 (2011)
13. Fedoruk, P.: The automation of adaptive processes in the system of distance education and knowledge control. Int. J. Inf. Tech. Knowl. **1**, 376–380 (2007)
14. Mestadi, W., Nafil, K., Touahni, R., Messoussi, R.: Knowledge representation by analogy for the design of learning and assessment strategies. Modern Educ. Comput. Sci. **6**, 9–16 (2017). https://doi.org/10.5815/ijmecs.2017.06.02
15. Tomashevsky, V., Novikov, Yu., Kamenskaya, P.: Models of Adaptive Learning Processes. Scientific Papers [Peter Mohyla Black Sea State University]: Computer Technology, vol. 134, no. 121, pp. 36–50 [in Ukrainian]. http://nbuv.gov.ua/UJRN/Npchduct_2010_134_121_6
16. Kut, V., Kunanets, N., Pasichnik, V., Tomashevskyi, V.: The procedures for the selection of knowledge representation methods in the "virtual university" distance learning system. In: International Conference on Computer Science, Engineering and Education Applications ICCSEEA 2018: Advances in Computer Science for Engineering and Education, pp. 713–723 (2018)

Mathematical Models of Formation and Functioning of Teams of Software Systems Developers

Mykola Pasyeka[1(✉)], Ivanna Dronyuk[2], Nadia Pasyeka[3], Vasyl Sheketa[1], Nadia Lutsan[3], and Oksana. Kondur[3]

[1] Ivano-Frankivsk National Technical University of Oil and Gas, Ivano-Frankivsk, Ukraine
pms.mykola@gmail.com, vasylsheketa@gmail.com
[2] Department of Automated Control Systems, Lviv Polytechnic National University, Lviv, Ukraine
ivanna.m.droniuk@lpnu.ua
[3] Vasyl Stefanyk Precarpathian National University, Ivano-Frankivsk, Ukraine
pasyekanm@gmail.com, lutsan.nadia@gmail.com, leuro@list.ru

Abstract. A systematic review was conducted and a number of original results were presented, which are not exhaustive in the study of mathematical models of formation and functioning of members of software development teams using cloud technology. However, they reflect the general methodology of construction and study of applied mathematical models of dynamic functioning of firm's developers of software systems, and also can be effectively used in solving problems of wide class of management of social and economic systems. From the point of view of the theory of research of mathematical models it is necessary to recognize that various results of research of commands of developers of the program systems received in psychology and sociology, today in formal models find insufficiently full reflection. For many models, there are certain difficulties in obtaining analytical solutions. For example, such a common in practice class of formation and existence of teams, taking into account the psychological of each team member in program engineering teams, is not yet a subject of deep systemic, formal, theoretical research.

Keywords: Group dynamics · Mathematical models · Teams · Software systems · Program systems

1 Introduction

The activity of the team of software systems developers is essentially regulated by the established norms. The teams, unlike organizations, have no formal hierarchy.

Whitworth and Biddle (2007) studied social interactions in teams. Conclusion that such as collective ownership, short iterations, pair programming and daily stand up meetings increased perceived social support and motivation in software development teams. Law and Ho (2004) conducted a case study on social factors in an XP environment. They observed that colocation improved team communication, and that

Z. Hu et al. (Eds.): ICCSEEA 2020, AISC 1247, pp. 621–630, 2021.
https://doi.org/10.1007/978-3-030-55506-1_55

customer onsite and an iterative planning strategy led to increased customer satisfaction at the same time Robinson and Sharp (2005) used an ethnographic approach to analyse the social implications of the practices found that four practices - namely pair programming, test first development, simple design, refactoring - all incorporated pairing. Pairing involves two and more technical developers sitting side by set task from the development of software systems. According to the study, functioning of commands can be described as an intensive but also is potentially stressful form of technical conversation. Let's consider the main characteristics of the team of software systems developers that distinguish it from a group, team or organization:

- a common goal;
- joint work on the project;
- a common understanding and perception of interests;
- independence of software development activities;
- collective and mutual responsibility for the results of joint activities;
- specialization and interchangeability of roles (as well as optimal distribution of functions and volumes and synergy in the interaction of team members);
- balance of the software development team (certainty of mutual expectations of its members).

Since the software development teams are autonomous by definition, unlike the theory of organizational systems management, the governing body - the principal - is not singled out when considering team models. However, this does not prevent us from interpreting the target function or efficiency of a team as a whole as a target function of some centres, which solves the tasks of team management of its formation and functioning [1–4, 20–22]. It is possible to consider that the set of admissible actions reflect "who can do what", target functions - "who wants what", awareness - "who knows what". By analogy with these definitions we can single out the following components of almost any model of software development teams:

1. Team composition (many members of the team). In order to describe the team, it is necessary to specify its composition at least. In most models, the composition of the software development team is considered fixed, although there are other studies of "teams" with dynamically changing composition [5].
2. The states of the team members (together with the functions they perform (roles) and the scope of tasks) have many acceptable states. Sometimes, in describing a model, formulas are formulas that reflect the relationship between the states of team members and the laws of variability of states over time [6].
3. Depending on whether the team members choose their own states independently or are considered to be externally defined (as a result of solving optimization tasks or established by a certain governing body), models that take into account the activity of team members and models with passive team members are selected accordingly. The activity of team members of software development systems is usually described within the framework of theoretical and gaming models [7].
4. The outcome of the software development team, which depends on the states of the team members (their individual actions) [8].

5. The intended functions of the members of the software development teams may depend on their individual actions (states) or the result of joint activities. And the target functions of different agents can be either the same (then there is one target function that reflects the unified effectiveness of the team) or different [9].
6. The awareness of the members of the software development teams (the information they possess under significant external and internal influences) can be either the same or different. Besides, it can be trivial (when there is a general knowledge - a fact when everybody knows that it is known to everybody or nontrivial (in this case it is necessary to take into account the effects of reflexion - the idea of team members' ideas about each other) [10, 15, 16].

Let's consider the model of carrier growth of a member of the team of software systems developers.

2 Dynamic Career Management Model for Members of Software Development Teams

Career is "the way to success, leading the society, in the official capacity". Career types are organizational and between organizational, vertical, horizontal, step-by-step, specialized, non-specialized. To increase the career growth of a member of the software development team, let's consider the task of management from the point of view of an individual (individual career), as well as from the point of view of the organization (staff promotion) [11].

For the fixed *i*-th member of the software development team we will introduce an oriented graph (*V*, *E*) where the vertices of the graph correspond to the possible positions he can occupy in the organization, and the vertices v_{ij} are ordered in the sense that the arcs go only from the vertices with a lower first index to the vertices with a higher first index. The first index $i \in I = \{1, 2, \ldots, m\}$ represents the number of the level of the hierarchy, and the other index of the hierarchy $j \in J(i)$ – where many positions at the *i*-th level of the career hierarchy [12, 13]. The arc length $t_{i,j}^{k,i} \geq 0$, $t_{i,j}^{k,i} = +\infty$ with $k < i$ from the top *i*, *j* from the top *k*, *l*, we will consider that the indicator reflects the time to be worked in the position *j* of the level *k*, in order to take up the position l at the level *k*.

For each pair of vertices *i*, *j* and *k*, *l*, $k > i$, you can find the length of $T_{i,j}^{k,i}$ the shortest path connecting these corresponding vertices:

$$\tau_{i,j}^{k} = \min_{i \in J(k)} T_{i,j}^{k,i}. \tag{1}$$

The value (1) can be interpreted as the minimum time required to reach the *k*-th level of the hierarchy starting from the *j*-th position in the *i*-th level of the hierarchy. At the same time, the value:

$$\tau_0^k = \min_{i \in J(k)} T_0^{k,i} \tag{2}$$

reflects the minimum time required to "start" from the very beginning of a professional career and reach the k-th level of the hierarchy in the company [14].

Let's consider the marker chain, the tops of which correspond to the levels of the hierarchy of positions in the considered organization, i.e. belong to the ordered set $I = \{1, 2, \ldots, m\}$.

Let's add $m + 1$ top, which reflects the release from the organization, and we will consider that we know the transition probabilities, where: p_{ij} – the probability that in the next period a member of the software development team will remain at the same (i-th) level, p_{ij} – the probability that he will move to the j-th level, $j > i$, p_{im+1} the probability that he will be released (the probability of transition p_{im+1}, $m + 1$, we will consider equal to one. Suppose that once an employee is released from this organization, he or she will not return to it again).

The probabilities p_{ij}, j < i, we will consider the values to be equal to zero because the decrease in the position is impossible. Calculating the minimum paths in the column (Fig. 1), we obtain: $\tau_0^1 = 1$, $\tau_0^2 = 2$, $\tau_0^3 = 3$, $\tau_0^4 = 4$.

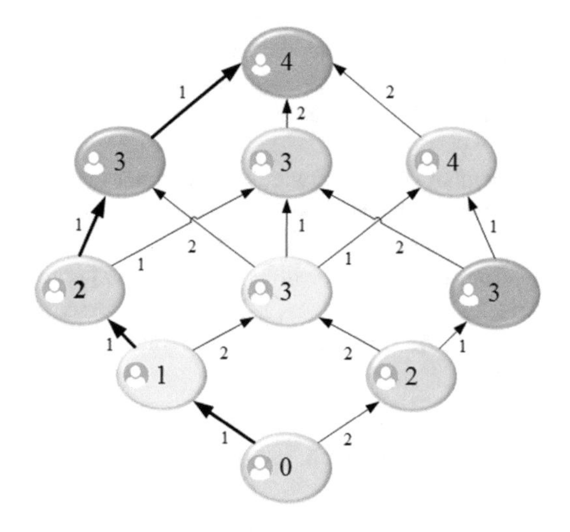

Fig. 1. Individual graphical model of career management

The graph shows four levels of the hierarchy ($m = 4$), where the numbers at the arcs represent their length, and the numbers inside the circles of the vertex represent the length of the minimum path from the entrance (zero vertex) to this vertex. The thickest lines represent the shortest path from the entrance to the highest level.

Recall that the value τ_0^i characterizes the achievement of the maximum hierarchy level, $i = \overline{1, 4}$, planned by the i-th member of the software development team [18, 19]. Thus, we can consider the Markov chain (Fig. 2).

Let the values of transition probabilities be described using the matrix in Table 1.

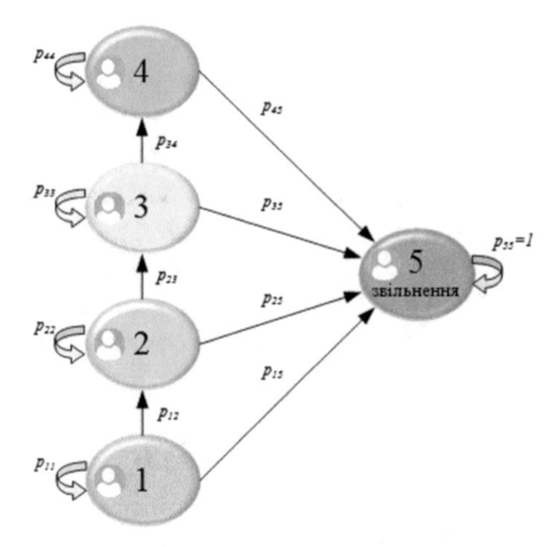

Fig. 2. Markov career development chain model

Table 1. Transition probability matrix

	1	2	3	4	5 (Liberation)
1	0,70	0,20	0,00	0,00	0,10
2	0,00	0,80	0,10	0,00	0,10
3	0,00	0,00	0,70	0,10	0,20
4	0,00	0,00	0,00	0,90	0,10
5 (Liberation)	0,00	0,00	0,00	0,00	1,0

Let's denote $p(0) = (0, 0, \ldots, 1, \ldots, 0, 0)$, where $m + 1$ is a dimensional stochastic vector, all components of which are equal to zero except one (not equal to 1). It is a component that corresponds to the level of hierarchy l, on which a member of the team of software systems developers is located or comes to work. Transient probability matrix is $P = \| p_{ij} |$.

Then the dynamics of $p(t)$ of the states of the mark chain will correspond to the condition: $p(t) = p(0)P^t, t = 1, 2, \ldots,$
where

$$\begin{aligned} p(1) &= (0.7, 0.2, 0, 0, 0.1), \\ p(2) &= (0.49, 0.3, 0.02, 0, 0.19), \\ p(3) &= (0.24, 0.339, 0.065, 0.006, 0.35), \\ p(4) &= (0.058, 0.22, 0.088, 0.032, 0.602). \end{aligned} \quad (3)$$

In fact, $p_i(t)$ is the probability that at the moment t an employee will be at the i-th level of the hierarchy, $i \in I$. Let's consider the i-th member of the software development team, who gets to the lowest level of the hierarchy in the organization. Then the solution of the problem of individual career takes the form of $(\tau_0^k)_{i \in I}$, and is a set of minimum terms of time for which a member of the team of software systems developers plans to reach the appropriate level of the hierarchy. Performance of the task on staff promotion can be represented in the form of the matrix $p_{ij} = p_j(\tau_0^k), i, j \in I$ the rows of which contain the probability that at this point τ_0^k the employee will be at the j-th level of the hierarchy. It is possible to introduce various aggregated criteria for the consistency of an individual's plans with career development proposals in the organization. For example, the probability of unsuccessful career (from the employee's point of view) as the maximum probability that the level of hierarchy at which a certain member of the software development team will be located is much lower than he or she expected:

$$Q = \max_{i=\overline{1,m}} \sum\nolimits_{j<i} p_{ij}(\tau_0^k). \tag{4}$$

With (4) we get: $Q = \max\{0, 0.49, 0.58, 0.4\} = 0.58$.

That is, the probability of unsuccessful career and a member of the software development team whose career plans are described by the graph (Fig. 1) in the organization is described by the Markov chain (Fig. 2) is 0.58. This value is rather high and it is unlikely that the employee in question will decide to apply for a job in this organization. From the perspective of career management (more precisely, career offerings) in such a situation it is necessary to reduce the likelihood of unsuccessful career development, perhaps primarily by reducing the likelihood of exemption from different levels of administrative hierarchy. Let the turnover of staff be reduced and the new transition probabilities are described in the matrix in (Table 2).

Table 2. Transition probability matrix with regard to staff turnover

	1	2	3	4	5 (Liberation)
1	0,50	0,49	0,00	0,00	0,01
2	0,00	0,50	0,49	0,00	0,01
3	0,00	0,00	0,50	0,49	0,02
4	0,00	0,00	0,00	0,95	0,05
5 (Liberation)	0,00	0,00	0,00	0,00	1,00

Then we recalculate the dynamics *p(t)* of the states of the Markov chain, which will correspond to the condition: $p(t) = p(0)P^t, t = 1, 2, \ldots,$
where

$$\begin{aligned} p(1) &= (0.5, 0.49, 0, 0, 0.01), \\ p(2) &= (0.25, 0.49, 0.24, 0, 0.02), \\ p(3) &= (0.063, 0.245, 0.36, 0.288, 0.044), \\ p(4) &= (0.004, 0.031, 0.105, 0.701, 0.16). \end{aligned} \tag{5}$$

With (5) we get: $Q = \max\{0, 0.25, 0.31, 0.14\} = 0.31$.

Taking into account the management actions, the probability of an unsuccessful career has been almost halved to 0.31. At the same time, the probability of dismissal in four periods is equal to 0.16. Probably, such conditions of career growth for many members of software development teams can look rather attractive.

So, the research of career management tasks is formulated as the task of coordinating the interests of team members in the firm. It is shown that mutually beneficial decisions can be made on the basis of comparing the results of solving the task of individual career planning (which is reduced to the task of searching for the shortest way in the network) and the task of promotion of team members in the career, i.e. the task of building and studying the properties of the brand chain. In our research we consider theoretical-play and optimization models of management of development of members of teams of software systems developers in firms, that is, the impact on the employees of the firm, which is carried out in order to increase the efficiency of their activities in terms of the interests of the firm.

The system of classifications of management tasks by members of the teams of developers of software systems is introduced, from the point of view of the company the following tasks are singled out: selection of team members, formation, distribution of functional duties and dismissals. From a personal point of view, it is proposed to consider the task of adaptation, motivation, training and promotion of team members.

3 Result

Having made average mathematical calculations behind commands of developers of software Web-systems with use of cloud technologies which partially were formed using the information on psychotypes of their members, and also partially such commands which were formed behind traditional approaches, and also carried out set tasks for check of a hypothesis by a capacitor method. Let's carry out grouping of commands behind the type of pathways in the process of formation at corresponding clusters for carrying out of the comparative cluster analysis.

On the basis of empirical data, we'll get information about the weighted average percentage of the program tasks, both experimental and reference commands. Experimental teams of software systems developers using cloud technology are such groups, which are formed taking into account the psychotypes of their members, and the reference ones are those, which are formed on the basis of traditional methods. The result is calculated by a mathematical formula (6).

$$\Delta = \left(\sum\nolimits_{i=1}^{n} EG_i - \sum\nolimits_{i=1}^{n} CG_i\right)/n = \frac{1107,06 - 964,17}{14} = 10,21\%, \tag{6}$$

where Δ is the difference in job lengths between the two clusters;

n - number of commands for commands of the first cluster group that took part in the experiment; EG_i - experimental groups of the cluster; CGi - control groups of the cluster.

Efficiency of the teams of developers of software Web-systems using charm technologies for the two clusters (taking into account the psychotypes in the formation of teams and without taking into account) (Fig. 3).

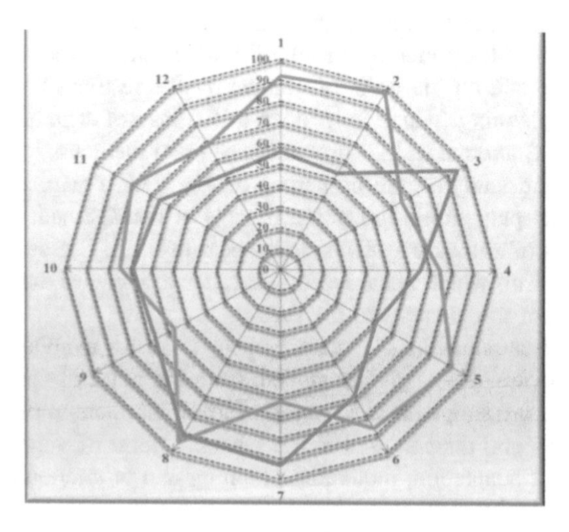

Fig. 3. Cluster analysis of the fragmentation of software systems behind the principle of team formation.

4 Discussion and Conclusion

This study considers a set of mechanisms of organizational management of the innovative development of software system developers, in particular, mechanisms of financing, management of organizational projects, institutional management, staff motivation and personnel development management. Many classes of models are considered for the first time: self-development model, games with variable composition, multi-criteria model of stimulation, models of needs hierarchy and career management. Their further development is a promising task for future research.

It should be noted that mathematical modelling was used as the main instrument of research into the development of mechanisms of management of innovative development of the firm and its personnel. On the one hand, their use of mathematical models allows us to obtain reasonable conclusions and establish a quantitative relationship between significant phenomena and processes. On the other hand, it should be remembered that, building any mathematical model, they introduce a number of assumptions and the results of the analysis of these models are fair only within their limits.

Therefore, it is probably not necessary to consider the results of mathematical modelling as the final result, and as a certain "algorithm", substituting the numerical

values corresponding to a particular real situation, you can get an exhaustive answer to the question of how to manage the innovative development of software development teams and firms in which they exist. Advantages of mathematical modelling is that it allows: to explain the observed phenomena and connections between them; to predict the future development of the processes of mutual relations between the members of the team of software systems developers, that is to choose the optimal variants of development.

References

1. Teh, A., Baniassad, E., Rooy, D.V., Boughton, C.: Social psychology and software teams: establishing task-effective group norms. IEEE Softw. **29**(4), 53–58 (2012)
2. Barsade, S.G., Knight, A.P.: Group affect. Ann. Rev. Organ. Psychol. Organ. Behav. **2**(1), 21–46 (2015). https://doi.org/10.1146/annurev-orgpsych-032414-111316
3. Barsade, S.G., Gibson, D.E.: Group affect its influence on individual and group outcomes. Curr. Dir. Psychol. Sci. **21**(2), 119–123 (2012). https://doi.org/10.1177/0963721412438352
4. Collins, A.L., Lawrence, S.A., Troth, A.C., Jordan, P.J.: Group affective tone: a review and future research directions. J. Organ. Behav. **34**(S1), 43–62 (2013). https://doi.org/10.1002/job.1887
5. Derex, M., Beugin, M.P., Godelle, B., Raymond, M.: Experimental evidence for the influence of group size on cultural complexity. Nature **503**, 389–391 (2013). https://doi.org/10.1038/nature12774
6. Ellis, A.P.J., Pearsall, M.J.: Reducing the negative effects of stress in teams through cross-training: a job demands-resources model. Group Dyn. Theory Res. Pract. **15**(1), 16–31 (2011). https://doi.org/10.1037/a0021070
7. Knight, A.P., Eisenkraft, N.: Positive is usually good, negative is not always bad: the effects of group affect on social integration and task performance. J. Appl. Psychol. **100**(4), 1214–1227 (2015). https://doi.org/10.1037/apl0000006
8. Medykovskyj, M., Pasyeka, M., Pasyeka, N., Tyrchyn, O.: Scientific research life cycle performance of information technology. In: XIIth International Scientific and Technical Conference «Computer Science & Information Technologies» (CSIT'2017), Lviv, Ukraine, 5–8 September 2017, pp. 425–428 (2017)
9. Palla, G., Barabási, A.L., Vicsek, T.: Quantifying social group evolution. Nature **446**, 664–667 (2007). https://doi.org/10.1038/nature05670
10. Pasyeka, N., Mykhailyshyn, H., Pasyeka, M.: Development algorithmic model for optimization of distributed fault-tolerant web-systems. In: IEEE International Scientific-Practical Conference «Problems of Infocommunications. Science and Technology» Kharkiv, 9–12 October, pp. 663–669 (2018)
11. Pasyeka, N.M., Pasyeka, M.S.: Construction of multidimensional data warehouse for processing students' knowledge evaluation in universities. In: 13th International Scientific and Technical Conference, 23–26 February 2016, Lviv, pp. 822–824 (2016)
12. Ram, L.V.: Guidelines on teambuilding. New Straits Times-Management Times, 10 May 2003
13. Rosen, C.C.H.: The influence of intra team relationships on the systems development process: a theoretical framework of intra-group dynamics. In: Proceedings of the 17th Workshop of the Psychology of Programming Interest Group, pp. 30–42. PPIG, Sussex, UK (2005)
14. Rashkevych, Y., Peleshko, D., Pasyeka, M.: Optimization search process in database of learning system. In: IEEE International Workshop on Intelligent Data Acquisition and Advanced Computing Systems: Technology And Application, pp. 358–361 (2003)

15. Rashkevych, Y., Peleshko, D., Pasyeka, M., Stetsyuk, A.: Design of web-oriented distributed learning systems. Upravlyayushchie Sistemy i Mashiny, no. 3–4, 72–80 (2002)
16. Romanyshyn, Y., Sheketa, V., Poteriailo, L., Pikh, V., Pasieka, N., Kalambet, Y.: Social-communication web technologies in the higher education as means of knowledge transfer. In: Proceedings of the IEEE 2019 14th International Scientific and Technical Conference on Computer Sciences and Information Technologies (CSIT), 17–20 September 2019, Lviv, Ukraine, vol. 3, pp. 35–39 (2019)
17. Rosen, C.C.H.: The influence of intra team relationships on the systems development process: a theoretical framework of intra-group dynamics. In: 17th Workshop of the Psychology off Programming Interest Group, Sussex University (2005)
18. Sy, T., Côté, S., Saavedra, R.: The contagious leader: impact of the leader's mood on the mood of group members, group affective tone, and group processes. J. Appl. Psychol. **90**(2), 295–305 (2005). https://doi.org/10.1037/0021-9010.90.2.295
19. Dingsøyr, T., Lindsjørn, Y.: Team performance in agile development teams: findings from 18 focus groups. In: International Conference on Agile Software Development. Springer, Heidelberg, pp. 46–60, June. 2013
20. Qureshi, R., Basheri, M., Alzahrani, A.A.: Novel framework to improve communication and coordination among distributed agile teams. Int. J. Inf. Eng. Electron. Bus. (IJIEEB) **10**(4), 16–24 (2018). https://doi.org/10.5815/ijieeb.2018.04.03
21. Cadavid, H.F.: Continuous delivery pipelines for teaching agile and developing software engineering skills. Int. J. Modern Educ. Comput. Sci. (IJMECS) **10**(5), 17–26 (2018). https://doi.org/10.5815/ijmecs.2018.05.03
22. Anwer, F., Aftab, S.: Latest customizations of XP: a systematic literature review. Int. J. Modern Educ. Comput. Science IJMECS **9**(12), 26–37 (2017). https://doi.org/10.5815/ijmecs.2017.12.04

Open Data Platform Architecture and Its Advantages for an Open E-Government

Mohammad Alhawawsha and Taras Panchenko(✉)

Taras Shevchenko National University of Kyiv, Kyiv, Ukraine
Taras.Panchenko@gmail.com
https://csc.knu.ua

Abstract. E-Government is a set of pervasive technologies and automated processes now. The open data plays a crucial role in the successful implementation of this concept. The Open Data Platform (ODP) architecture is described here as the framework for the open data access systems implementation, including specific requirements. The proposed architecture and its components were discussed in this paper in detail for its availability, productivity, and reliability. The open data subsystem based on the architecture presented here was developed for the Jordan Government and was successfully implemented and tested. Thus, this architecture showed its viability.

The focus of the paper is the detailed analysis of the proposed ODP architecture and its characteristics. The ODP is a significant system for the mature e-Government. We propose here the architecture for it with usage-proven characteristics. This fact adds the value to the e-Government framework stability, and significant characteristics and improves the overall quality of the system.

Keywords: E-government · Open data · Digital transformation · Digitalization · System architecture · High availability · High performance · Open Data Platform

1 Introduction. Literature Review and the Task

Digital transformation of the society, the states, the cities, and the citizens is a common process nowadays. It affects almost all areas of our life. Digitalization takes place even in e-health and e-government, data becomes open and widely accessible. All these changes require mature infrastructure which can cope with challenges like reliability, availability, performance, accessibility, and security.

The topicality, approaches, architectures, solutions, and challenges of E-Government and SMART-Government were described and discussed in [1–12]. The need for open E-Government varies from different countries, which is addressed partially in [1–5]. There are also a lot of publications, which highlight different aspects of eGovernment, including success stories and challenges during the development and implementation [6–11].

A lot of non-trivial tasks, which must be solved during the qualitative implementation of the e-Government concept, should be carefully depicted and examined in beforehand – for example, security issues when dealing with personal data, and proper

Z. Hu et al. (Eds.): ICCSEEA 2020, AISC 1247, pp. 631–639, 2021.
https://doi.org/10.1007/978-3-030-55506-1_56

availability of the data themselves. However, one of the very beginning and significant tasks is the Open Data concept [12] and its proper, good implementation. It means that Data Portal (or similar system) should be:

- reliable enough (including fault tolerance),
- available (accessible) in different conditions (high availability),
- productive, have enough Service Level Agreement (SLA) points (to comply with applicable requirements), should provide the performance required,
- usable (useful and straightforward to understand for the end-users, including applicable accessibility requirements), and
- prevent unauthorized access to the sensitive part of data (internal representations), as well as the internal part of the system at all, from non-authorized entities (including users).

We do not mention security nor confidentiality here because the topic is 'Open Data', which means the openness of the data presented to end-users. Some restrictions must take place – concerning internal parts of the whole system – however, there is no even authentication (identification) and/or authorization to access the open part of the system and the data.

Data should be presented in a clear structure. We mean the data itself, also proper formats and forms of representation, as well as catalog system – again, for the end-user to find the desired data with less efforts. Nevertheless, these questions are out of the scope of this research.

So, here we concentrate on Open Data Framework (ODF) and Open Data System (ODS) development, particularly on its architecture, the block-scheme, and its properties. We discuss expectations, requirements to such a framework, and its behavior under different conditions. We will also use the Open Data Platform (ODP) to address our development.

The research objective is to develop the architecture of the ODP, which complies with the requirements described above. The methodology is the analytical-comparative – which relies on contemporary technologies and the best practices, adapting them for the best result obtaining.

So, next, we will show the proposed architecture, clarify some aspects and connections of the sub-blocks, and then provide the analysis of the pros and cons of that architecture.

2 Open Data Platform Architecture

Considering the non-functional requirements to the system given in the introduction section, and using the ongoing trend to micro-services architecture, we present here the architecture for the Open Data Platform and then will discuss it.

Figure 1 shows the proposed structural scheme of ODP components. It includes architectural, workflow, dataflow, and connectivity aspects.

It is evident that ODP should be web-based because users will access open data via the internet. So, there should be Front-End part with data access servers and report preparation and generation services. This block includes caching service to accelerate repetitive data access requests processing. Also, cache usage can improve availability by improving the overall performance of back-end service by minimizing request count to that part, and also it provides data if the back-end is not accessible for some reason. Here back-end part is the Data Storage and data access servers and services.

We suppose that the front-end querying engine could be a combination of SQL Server and/or OLAP (On-Line Analytical Processing – in contrast to Transactional Processing, OLTP) Server for performance optimization, especially for long-processing queries with aggregations over large data arrays. This part of the whole infrastructure is the main storage of Data Marts, also known as Data Warehouse – the only source of truth for all the information circulating in the system (ODS).

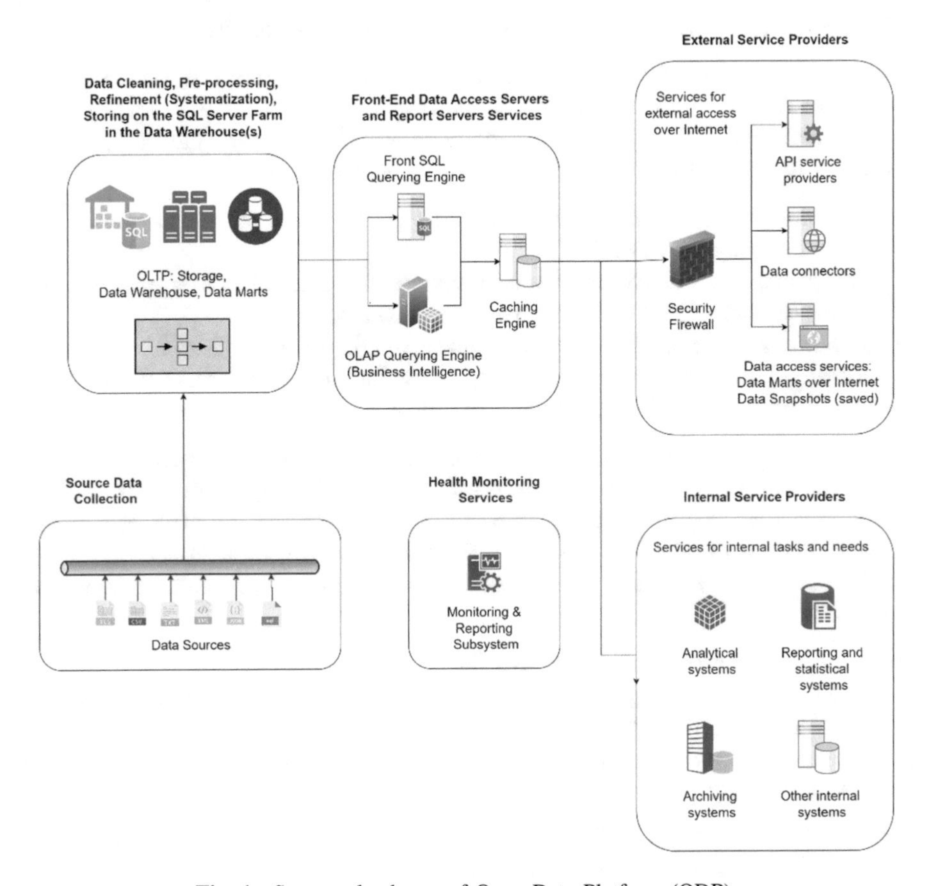

Fig. 1. Structural scheme of Open Data Platform (ODP).

On the other hand, ODP has many diverse data sources as inputs, so there will be needed the enterprise bus-based solution (micro-service) with a collection of data transformers, which convert data from input format to internal representation. We call this component as Source Data Collection because of a variety of input data formats from different sources, which should be converted to the shared data storage (data warehouse).

As in most data processing systems, we need Data Warehouse with the following components:

- data cleaning, which means seeking for outliers, corner data, inappropriate and out-of-the-range data, joining different notations of the same items, and other similar tasks,
- data pre-processing, which aggregates data, checking its consistency, accuracy, making possible cross-validations,
- refinement and systematization, putting the data into proper form and
- storing the resulting cleaned and refined data to the SQL Server Farm (Data Warehouse), where the data will be extracted (queried) from lately.

All these components (cleaning, pre-processing, refinement) are essential to prepare qualitative data for further use by the community.

External Service Providers are the front-end part of the whole Platform, including the Open Data Portal implementation. These services include:

- Application Programming Interfaces (APIs) for the users or "robots" (other programs querying the open data),
- data connectors, which include data sources for external software, connection strings (if appropriate) and other data access means,
- data access services, which consist of data marts (in the form of downloadable files primarily, but can also provide querying engine and other access means), data snapshots (primarily for versioning), and other conventional means.

APIs recommended being primarily constructed using a service-oriented architecture based on Simple Object Access Protocol (SOAP) (Fig. 2), with Web Services Description Language (WSDL) for their definition and discovery. This architecture provides high availability and scalability for this subsystem and its users.

Internal Service Providers give access to the data for other services for internal use. It contains:

- archiving subsystem and the storage (for history versioning),
- analytical and reporting subsystems – for other government institutions, agencies and "internal" data customers ("insiders"), who need the data,
- other systems, which provide access to the data needed and according to the required SLA.

Analytical and reporting subsystem aimed to accelerate access to complex data extracts, which requires aggregations of many primary data lines and multiple data formats of representation.

Furthermore, the last, but not least, component of the ODP are Health Monitoring Services, which include a row of infrastructure solutions aimed to:

- monitor the functionality of all the components,
- alert the situations when some event triggers,
- alarm some resource saturation at the early stages,

- prevent critical issues by informing administrators on critical situation changes (loading, performance, accessibility)
- visualize the "health" of the whole complex,
- predict the bottlenecks in performance and other corner situations,
- automatize maintenance tasks (archiving, adjusting hardware virtualization parameters)

It is not depicted on the scheme, but it goes without saying, that all the steps should be carefully logged to track all the changes, access, and potentially tricky situations.

As we know from big projects and companies like Google and Facebook, Grammarly, and many other web-oriented products, one of the biggest teams is the infrastructure team (development, support, and monitoring).

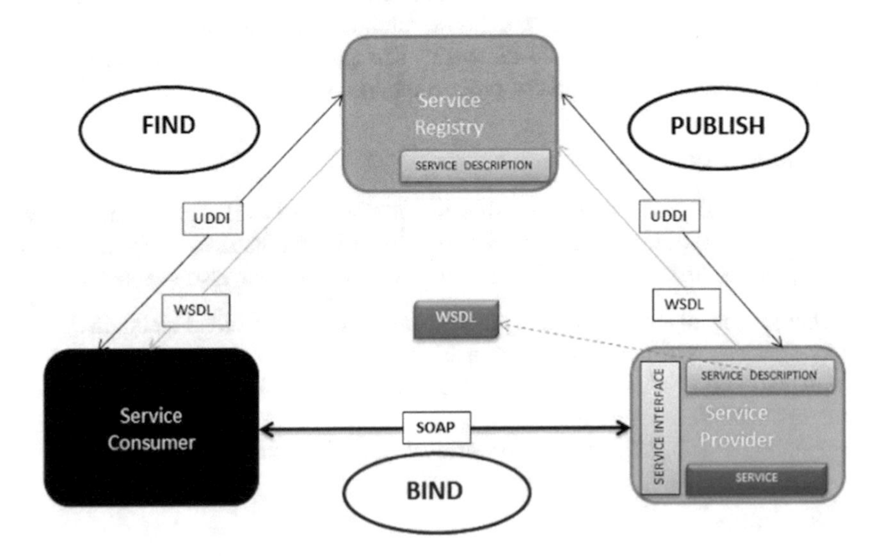

Fig. 2. The SOAP web services architecture, SOAP + WSDL + UDDI (Ref.: https://dzone.com/articles/soap-web-services-using-cxfjibx-jax-ws).

3 Open Data Platform Properties. The Result and the Discussion

The main properties are straightforward outcomes of the architecture proposed. Let us summarize them. Also, we will provide a brief analysis of the questions discussed.

3.1 Availability

The proposed architecture of ODP is fault-tolerant because of:

- separate blocks are independent and scalable enough,
- caching and querying engines help to optimize the workload of the system,
- multi-layer structure (separate blocks are stacked to implement the whole dataflow) and microservice architecture (every particular service and block construction) supports the right independency level as well as scalability capability,
- separation of inner and outer services customers,
- monitoring subsystem helps to identify errors and to solve critical situations quickly.

The data will remain accessible (for some time due to configuration) via cache even if the querying engine and/or data warehouse (storage) become unavailable. Also, the usage of OLAP querying and multi-dimension storage engine helps to improve the data availability through the better query processing speed.

3.2 Productivity

The high level of productivity (in other words, Service Level Agreement – SLA – concerning system performance) should be assured by the hardware (or virtual/cloud configuration) at first, but the following architectural solutions also support it:

- caching mechanism and multi-layer structure resolves high load situations,
- microservices provides extended scalability,
- combination of SQL and OLAP query engines provides flexibility in building more efficient data flow inside the ODP,
- monitoring, again, helps to identify critical deviations early.

3.3 Reliability

Reliability, in our case, is based on:

- microservice architecture (thus, high ODP's components independency level),
- monitoring with the reactive subsystem, which prevents system disasters, with the help of the system's health tracking dashboards, informers and closed-loop configured reactors,
- caches, which extend availability in some cases,
- multi-tier architecture, which helps to avoid complex interconnections and thus localize problems (firstly, concerning productivity) fast,
- separation of inner and outer services customers,
- firewall, which helps to filter traffic and avoid unauthorized access to the system by many parameters, for example, frequency and diversity of requests.

Nevertheless, of course, the reliability of data (the core ODP's value) as a whole depends mostly on the quality of input data.

3.4 Security

Security, as mentioned before, is not in the focus of this work, because of the open nature of the ODP and the data inside, but let us check the main criteria briefly:

- confidentiality, which should be addressed at the pre-processing stage only, because of just there any personal data should be excluded,
- integrity should be supported by additional limitations applied for the internal staff to have no access to the workflow and dataflow,
- availability was discussed above,
- authenticity is not applicable because in most cases user accessing open data will be an unauthorized entity,
- accountability is all about log and historical snapshots, described above,
- reliability was also shown before,
- non-repudiation could be guaranteed just in case of physically limited access to the ODP infrastructure.

3.5 Summary and Analysis

The proposed ODP and its architecture differ a lot from known previously (for example, form [6–11]). The described Platform follows micro-service architecture principles to enhance availability and scalability, and a row of technologies (OLAP, Reporting and Monitoring subsystems) to increase productivity and reliability. This architecture is a new one for the task of Open Data for the e-Government and is published for the first time.

Here we discussed different characteristics of the ODP of the proposed structure (architecture). This discussion is independent of the underlying implementation technologies and particular solutions – it concerns just principal questions and system properties.

We have not discussed why and how we have come to this architecture because it includes our experience in many different IT-projects being combined with the non-functional requirements, and available technologies usage with best practices following. So, it could be a separate discussion with a huge literature review section, but our positive experience of the developed ODP usage gives us the right to make a conclusion about this architecture effectiveness (which was proven by the implementation and the experience).

The disadvantage of this architecture is that it is huge (and, probably, expensive), combining many technologies and techniques (caching, reporting, OLAP, and so on) – but we try to take all these challenges under control by using the monitoring subsystem. Moreover, particular subsystems could be delivered as fault-tolerant (it should be simple to double them as microservices), but in this case, the system implementation and support (hardware or "cloud" resources primarily) at all will become even more expensive.

The experimental implementation of the Open Data System for the Jordan Government confirms the feasibility of the solution and the architecture proposed. Some similar elements of the architecture can also be seen in Facebook & Google infrastructure projects, so this supports the method proposed here and approaches the architecture presented.

4 Conclusion

In this paper, we described and discussed the proposed Open Data Platform (ODP) Architecture as the foundation and the framework for open data access for the e-Government concept implementation. We discussed its properties, advantages and disadvantages, showed how the high indicators could be reached, and how to get the most of the described infrastructure. However, we should notice that, of course, the reliability of data (the core ODP's value) as a whole depends mostly on the quality of input data.

The proposed Open Data Platform Architecture has the following characteristics: availability, productivity, relative reliability (regarding to the input data) and partial security (because this is not the goal of such a system due to the open character of the data inside the system), and comply with non-functional requirements described at the beginning of the research.

Here we have not concentrated on the specific implementations of the ODP, yet mention it, because this experience will be discussed in a separate paper. Additionally, we should mention that part of the open data system for the Jordan Government was implemented over the ODP using the proposed architecture. That specific implementation has shown that ODP is robust enough, providing high availability and performance during the load tests. Thus, we can conclude that the proposed architecture for ODP is mature enough and is highly recommended to use in other similar systems and implementations.

Foundations for this research and the theoretic roots can be found partially in [13–19].

To summarize, the main contribution is the proposed Architecture for ODP and its analysis. It has good quality indicators, which was confirmed by the following experimental implementation of the Open Data System for the Jordan Government. This case indicated the viability of the architecture, as well as the overall quality (concerning parameters discussed here) and the compliance with non-functional requirements.

Application value is straightforward – now we have the first successful architecture implementation, and we are going to reproduce it in other similar systems. The question of security and the privacy can also be addressed to this architecture in the next researches in a broader context.

References

1. Alhawawsha, M.: A comparison of the E-government system architecture in Jordan with the E-government system of the United States. Sci. J. **34**(2/2), 18–24 (2017). https://doi.org/10.15587/2312-8372.2017.100190. Technology Audit and Production Reserves
2. Alhawawsha, M.: Developing of the e-government system based on java for online voting. Sci. J. **3**(35), 9–13 (2017). https://doi.org/10.15587/2312-8372.2017.104033. Technology Audit and Production Reserves

3. Alhawawsha, M., Glybovets, A.: E-government versus SMART government: Jordan versus the United States. Eureka J. **31**(3), 3–11 (2017). https://doi.org/10.21303/2504-5571.2017.00338
4. Alhawawsha, M., Glybovets, A.: E-government developed system prototype about USA and Jordan. Geom. Model. Inf. Technol. **N1**(3), 26–42 (2017)
5. Alhawawsha, M.: Identification of the issues in the E-governance in Jordan and USA. Inf. Syst. Technol. **6**(1/1), 37–40 (2017)
6. Saleh, Z.I., Obeidat, R.A., Khamayseh, Y.: A framework for an E-government based on service-oriented architecture for Jordan. Int. J. Inf. Eng. Electron. Bus. (IJIEEB), **5**(3), 1–10 (2013). MECS Press. https://doi.org/10.5815/ijieeb.2013.03.01
7. Osho, L.O., Abdullahi, M.B., Osho, O., Alhassan, J.K.: Effective networking model for efficient implementation of e-governance: a case study of Nigeria. Int. J. Inf. Eng. Electron. Bus. (IJIEEB), **7**(1), 18–28 (2015). MECS Press. https://doi.org/10.5815/ijieeb.2015.01.03
8. Balakumaran, P.J, Ramamoorthy, H.V.: Evolving an E-governance system for local self-government institutions for transparency and accountability. Int. J. Inf. Eng. Electron. Bus. (IJIEEB), **5**(6), 40–46 (2013). MECS Press. https://doi.org/10.5815/ijieeb.2013.06.05
9. Olaniyan, J.O., Ademuyiwa, A.J.: Comparative analysis of personnel distributions in the local government service in ekiti-state, Nigeria, for service delivery. Int. J. Math. Sci. Comput. (IJMSC), **5**(2), 44–53 (2019). MECS Press. https://doi.org/10.5815/ijmsc.2019.02.04
10. Mohammad, H., Almarabeh, T., Ali, A.A.: E-government in Jordan. Eur. J. Sci. Res. **35**(2), 188–197 (2009)
11. Majdalawi, Y.K., lmarabeh, T., Mohammad, H., Quteshate, W.: E-government strategy and plans in Jordan. J. Softw. Eng. Appl. **8**, 211–223 (2015). https://doi.org/10.4236/jsea.2015.84022
12. AlSukhayri, A.M., Aslam, M.A., Arafat, S., Aljohani, N.R.: Leveraging the Saudi linked open government data: a framework and potential benefits. Int. J. Modern Educ. Comput. Sci. (IJMECS), **11**(7), 14–22, (2019). MECS Press. https://doi.org/10.5815/ijmecs.2019.07.02
13. Panchenko, T.: Compositional methods for software systems specification and verification, Ph.D. Thesis, Kyiv (2006). 177 p
14. Panchenko, T.: Application of the method for concurrent programs properties proof to real-world industrial software systems. In: Proceedings of ICTERI, pp. 119–128 (2016)
15. Ostapovska, Y.A., Panchenko, T.V., Polishchuk, N.V., Kartavov, M.O.: Correctness property proof for the banking system for money transfer payments, Probl. Program. **6**(2–3), 119–132 (2018)
16. Panchenko, T., Ivanov, I.: A formal proof of properties of a presentation system using Isabelle. In: Proceedings 2017 IEEE First Ukraine Conference on Electrical and Computer Engineering, UkrCon, pp. 1155–1160. IEEE (2017)
17. Panchenko, T., Fabunmi, S.: Quality of concurrent shared memory programs. In: Proceedings 2018 11th International Conference on the Quality of Information and Communications Technology, QUATIC, pp. 299–300. IEEE (2018)
18. Ivanov, I., Panchenko, T., Nikitchenko, M., Sunmade. F.: On formalization of semantics of real-time and cyber-physical systems. In: Proceedings International Conference on Computer Science, Engineering and Education Applications, pp. 213–223. Springer, Cham (2018)
19. Panchenko, T., Shyshatska, O., Omelchuk, L., Rusina, N., Fabunmi, S.: Compositional-nominative approach to the client-server systems properties proofs within different formal execution models. In: Proceedings of 2019 IEEE 2nd Ukraine Conference on Electrical and Computer Engineering, UKRCON, pp. 1127–1132 (2019). https://doi.org/10.1109/ukrcon.2019.8880029

The Problem of Double Costs in Blockchain Systems

Nikolay Poluyanenko[1], Alexandr Kuznetsov[1(✉)], Konstantin Lisickiy[1], Serhii Datsenko[1], Oleksandr Nakisko[2], and Serhii Rudenko[2]

[1] V. N. Karazin, Kharkiv National University, 4 Svobody Sq., Kharkiv 61022, Ukraine
nlfsr01@gmail.com, kuznetsov@karazin.ua, konstantin.lisickiy@mail.ru, sergdacenko@gmail.com

[2] Kharkiv Petro Vasylenko National Technical University of Agriculture, Alchevskikh str., 44, Kharkiv 61002, Ukraine
nakisko307@gmail.com, oblikua7@gmail.com

Abstract. The paper investigates the probability of one of the most famous attacks - double spending attacks on the blockchain system, which are based on consensus mechanisms based on the proof of the work done using hash functions. Unlike previously known works describing the analytical expression for the probability of success of a double-spend attack, this paper proposes a model of independent players that is different from existing models. In our opinion, this model more adequately describes the processes that occur during the formation of a chain of blocks, but leads to other quantitative results. An expression is given that characterizes the probability of an attacker conducting a successful double-spend attack on a blockchain system depending on the number of confirmations used, as well as the hashrate by an honest network and the attacker. Quantitative results are presented, as well as a comparison of the results obtained with the most famous works in this area. Based on the results obtained, recommendations are given for determining the "safe" number of confirmations for successfully resisting a double spending attack on the blockchain system.

Keywords: Security of blockchain systems · Double spend attack · Blockchain technology · Double expenses · Attack on distributed systems · Independent players model

1 Introduction

As a rule, all "classical" payment systems are centralized; having an administrative link that provides control over the legitimacy of any operation. At the same time, the basis for making decisions on the legitimacy of the payment is the information of the administrator himself, and not the information that is submitted by the payer. Therefore, the payer is only able to form an application for the repeated spending of the same funds,

Z. Hu et al. (Eds.): ICCSEEA 2020, AISC 1247, pp. 640–652, 2021.
https://doi.org/10.1007/978-3-030-55506-1_57

and the administrative link will confirm only the first incoming application and reject the subsequent ones, which blocks the possibility of double spending the same funds.

In blockchain systems, the absence of an administrative resource is assumed, and therefore, the possibility of double spending the same funds becomes likely [1]. To protect against a double spending attack, sellers can take various protection measures, for example watch [2], the most effective of which is to wait for a transaction with payment to be included in one of the blocks of the blockchain registry. In this case, the node forming the block will not allow inclusion in the transaction block trying to re-spend the previously spent funds. And, even if such a block is formed by an attacker's node, honest network nodes will reject it and the block will not be added to the blockchain registry of honest users.

The process of including a transaction in a new unit is called transaction confirmation. Inclusion in one block corresponds to one confirmation. The formation and addition to the register of the blockchain chain of more blocks referring to a block with a transaction corresponds to the confirmations. However, if an attacker has sufficiently large resources (possesses high-performance equipment that can provide a hash rate for the attacker, he still has the chance of successfully spending double waste by forming an alternative registry blockchain chain.

The success of a double spend attack depends on the attacker's resources (hash) and the number of confirmations. The likelihood of an alternative chain forming exponentially decreases with an increase in the number of confirmations and a decrease in the attacker's hashrate. The more confirmations a transaction has, the less likely it is that the transaction is canceled due to the replacement of the existing chain with an alternative one formed by the attacker. However, on the other hand, the more the seller waits for confirmations, the longer the time spent on the transaction itself.

Therefore, transactions with zero confirmation potentially have a high risk of becoming a victim of a double-spending attack, and transactions awaiting a large number of confirmations suffer losses due to delays in their conclusion. Therefore, the issue of finding the optimal number of confirmations at which the risk of double spending attack will be below some acceptable level and the waiting time will be the minimum necessary is an urgent task.

There is an opinion ([3–5] and others) that if a consensus mechanism is used based on a proof of work (PoW - Proof-of-work) based on a hash function and the attacker has 10% of the computing power (hashrate) from the general network, and 6 confirmations are expected - the probability of success of such an attack will be 0.1%. The given assessment is based on the "player ruin" model (see [6] which does not use independent events for an honest network and an attacker. However, when using a different model, it is possible to expect a different probability of a successful attack and therefore different from the above probability and, as a result, a different minimum required number of confirmations.

2 The Investigated Model

Player ruin model uses two elementary events:

- the elementary event "a block is formed by an honest network and the attacker did not form a block" with probability p;
- the elementary event "the block is not formed by an honest network and the attacker formed a block" with probability q;
- moreover, the probability of forming a block by an honest network (p) is related to the probability of forming a block by an attacker (q) by the ratio $p+q=1$.

In our opinion, it is more rational to use another model – the model of independent players. If we leave the introduced notation (probabilities p and q) and refuse to fulfill the condition $p=1-q$, then as a result of each attempt (or a series of attempts during a given time interval), the space of elementary outcomes contains the following events:

- the elementary event "a block is formed by an honest network and the attacker did not form a block" with probability $p(1-q)$;
- the elementary event "the block is not formed by an honest network and the attacker formed a block" with probability $(1-p)q$;
- the elementary event "the block is not formed by an honest network and the attacker did not form a block" with probability $(1-p)(1-q)$;
- the elementary event "the block is formed by an honest network and the attacker formed a block" pq.

The set of all elementary outcomes makes up a complete group of events:

$$p(1-q)+(1-p)q+(1-p)(1-q)+pq=1.$$

In order to be able to compare with the results obtained in the works of Satoshi Nakamoto and Meni Rosenfeld, we will also use some simplifications:

1. The propagation time of the block over the network is negligible, i.e. information exchange between nodes occurs almost instantly (synchronization time is zero);
2. The hash of the attacker, the hash of the honest network and the complexity of mining does not change with time throughout the race;
3. Except for the attacker, all other network users act strictly in accordance with the rules of the blockchain network protocol;
4. The victory of the attacker will be considered the formation of the required number of confirmation blocks earlier or simultaneously (it is believed that the attacker formed one block in advance) with an honest network or, otherwise, the subsequent formation by the attacker of a chain of blocks of equal length with an honest network.

An attacker can win at the start of the race. Moreover, he needs to form N or more blocks by the time when the honest network will form N no more than blocks. If he does not succeed, then he still has the opportunity to catch up with an honest network on $N+j$ a block, where j is the number of blocks formed by an honest network in addition to the necessary N blocks.

We obtain an expression characterizing the probability of the formation of blocks depending on N,j and the number of unsuccessful attempts to form a block with an honest network (k).

3 The Probability of the Attacker's Victory

Based on the above group of events, we obtain the following possible probabilities and combinations in which the attacker wins (for $N = 1, j = 0$):

1. At the first attempt (in this case, when $t = N + j + k = 1 + 0 + 0 = 1$). There can only be one combination - when both an honest network and an attacker simultaneously find blocks. In this case, the probability of the attacker's victory will be determined as: $PI_{N=1,j=0,k=0} = p \cdot q$;
2. At the second attempt ($t = N + 1 = 2$). In this case, there may be two different situations:

   – an honest network finds a block in the first test, but does not find in the second (the probability of which is equal $p \cdot (1 - p)$). In this case, if an attacker finds a block during the first test, we come to step 1, therefore, he can only find the block in the second test (probability).
   – an honest network does not find a block in the first test, but only in the second test (the probability of which is equal $(1 - p) \cdot p$). In this case, the attacker can find the block only during the first test (probability $q \cdot (1 - q)$), only during the second test (probability $(1 - q) \cdot q$), and in both tests (probability $q \cdot q$);

   The total probability of this event:

$$\begin{aligned} PI_{N=1,j=0,k=1} &= [p \cdot (1-p) \cdot (1-q) \cdot q] \\ &+ [(1-p) \cdot p \cdot q \cdot (1-q) + (1-p) \cdot p \cdot (1-q) \cdot q + (1-p) \cdot p \cdot q \cdot q] \\ &= p \cdot (1-p) \cdot [(1-q) \cdot q] + (1-p) \cdot p \cdot [q \cdot (1-q) + (1-q) \cdot q + q \cdot q]. \end{aligned}$$

Carrying out similar considerations, we further obtain the probability of an attacker winning an honest network (PI):

$$\begin{aligned} PI = p^N \cdot \sum_{k=0}^{\infty} \left((1-p)^k \cdot \binom{t-1}{N-1} \cdot \left[\binom{t-1}{N} \cdot q^N \cdot (1-q)^k + \left\{ 1 - \sum_{i=0}^{N-1} \binom{t}{i} \cdot q^i \cdot (1-q)^{t-i} \right\} \right] \right) \\ + \sum_{j=1}^{\infty} \left[p^{(N+j)} \cdot q^{(N+j)} \cdot \sum_{k=1}^{\infty} \left(sum_p \cdot (1-p)^k \cdot (1-q)^k \right) \right], \end{aligned} \tag{1}$$

where:

$$t = N + j + k,$$

$$sum_p = \sum_{ip_{(N+j)}=1}^{k} \left(\sum_{ip_{(N+j-1)}=ip_{(N+j)}+1}^{k+1} \left(\cdots \sum_{ip_2=ip_3+1}^{N+j+k-2} \left(\sum_{ip_1=ip_2+1}^{N+j+k-1} (sum_q) \right) \right) \right);$$

$$sum_q = \sum_{iq_{(N+j)}=1}^{k+1} \left(\sum_{iq_{(N+j-1)}=iq_{(N+j)}+1}^{k+2} \left(\cdots \sum_{iq_{(N+1)}=iq_{(N+2)}+1}^{k+N-1} \left(\sum_{iq_N=\max\left(iq_{(N+1)}+1, ip_{(N-1)}\right)}^{k+N} \left(\cdots \sum_{iq_3=\max(iq_4+1, ip_2)}^{N+j+k-2} \left(\sum_{iq_2=\max(iq_3+1, ip_1)}^{N+j+k-1} (1) \right) \right) \right) \right) \right).$$

Please note that if $N = 1$, then the external summation sum_q for begins with $iq_{(N+j)} = ip_{(N+j-1)}$, and if $j = 0$, then in this case $sum_q = 1$.

4 Results of Calculating the Probability of a Successful Double Spend Attack

Expression (1) gives us the opportunity to build a probability table for an attacker to successfully conduct a double spend attack on a blockchain system based on PoW consensus, depending on N, j and k for various values of p and q. Moreover, the values of p and q can be arbitrary (that is, the condition $p + q = 1$ is not imposed). But when receiving expression (1), by analogy with the works of Satoshi Nakamoto [7], Meni Rosenfeld [3] and others, we mean that the attacker has an additionally formed block due to which his chain becomes longer than the blockchain chain formed by an honest network.

Tables 1, 2, 3, 4, 5, 6, 7, 8, 9 and 10 show the results of calculating the probabilities for various j and $k(PI_{j,k})$ with $N = 1$, 3, 5 and probability values q from 0.01 to 0.3. The results are obtained by the formula (1) with calculated accuracy. The results are rounded to 6 decimal places. As you can see from the tables, the values $PI_{j,k}$ quickly decrease with increasing indices j and k, which allows for calculations to be limited to the first few values from an infinite sum of expression (1).

Expression (1) allows you to calculate the exact value of the probability. However, for values q close to 0.5 (starting from 0.45), the terms with the indices $j, k > 30$ begin to make a significant contribution, which leads to the need to take into account the corresponding terms in expression (1). Moreover, the amount grows at an exponential rate, which leads to significant time costs for its accurate determination.

However, as can be seen from expression (1), the value sum_p does not depend on p and q, this allows one-time calculation sum_p (tabulation) for specific N, j, and k, and use of the obtained values in further calculations of probabilities for any p and q.

Table 1. The probability of an attacker successfully conducting a double spend attack $N = 1,\ p = 0,99,\ q = 0,01$ for various j and k ($PI = 0,010201$).

k	j		
	0	1	2
0	0,009900	–	–
1	0,000295	0,000001	$<10^{-6}$
2	0,000005	$<10^{-6}$	$<10^{-6}$
3	$<10^{-6}$	$<10^{-6}$	$<10^{-6}$

Table 2. The probability of a successful attack by an attacker double spending $N = 1,\ p = 0,9,\ q = 0,1$ at for various j and k ($PI = 0,120879$).

k	j					
	0	1	2	3	4	5
0	0,090000	–	–	–	–	–
1	0,025200	0,000729	0,000066	0,000006	0,000001	$<10^{-6}$
2	0,003897	0,000262	0,000041	0,000006	0,000001	$<10^{-6}$
3	0,000506	0,000059	0,000014	0,000003	0,000001	$<10^{-6}$
4	0,000060	0,000011	0,000004	0,000001	$<10^{-6}$	$<10^{-6}$
5	0,000007	0,000002	0,000001	$<10^{-6}$	$<10^{-6}$	$<10^{-6}$
6	0,000001	$<10^{-6}$	$<10^{-6}$	$<10^{-6}$	$<10^{-6}$	$<10^{-6}$
7	$<10^{-6}$	$<10^{-6}$	$<10^{-6}$	$<10^{-6}$	$<10^{-6}$	.$<10^{-6}$.

Table 3. The probability of a successful attacker carrying out a double spend attack $N = 1,\ p = 0,8,\ q = 0,2$ for various a j and k ($PI = 0,285714$)

k	j						
	0	1	2	3	4	5	6
0	0,160000	–	–	–	–	–	–
1	0,083200	0,004096	0,000655	0,000105	0,000017	0,000003	$<10^{-6}$
2	0,023808	0,002621	0,000734	0,000185	0,000043	0,000009	0,000002
3	0,005745	0,001049	0,000453	0,000164	0,000052	0,000015	0,000004
4	0,001280	0,000336	0,000207	0,000101	0,000042	0,000015	0,000005
5	0,000273	0,000094	0,000078	0,000050	0,000026	0,000012	0,000005
6	0,000057	0,000024	0,000026	0,000021	0,000013	0,000007	0,000004
7	0,000012	0,000006	0,000008	0,000008	0,000006	0,000004	0,000002
8	0,000002	0,000001	0,000002	0,000003	0,000002	0,000002	0,000001
9	$<10^{-6}$	$<10^{-6}$	0,000001	0,000001	0,000001	0,000001	0,000001
10	$<10^{-6}$	$<10^{-6}$	$<10^{-6}$	$<10^{-6}$	$<10^{-6}$	$<10^{-6}$	$<10^{-6}$

Table 4. The probability of an attacker successfully conducting a double spend attack $N = 1,\ p = 0,7,\ q = 0,3$ for various j and k ($PI = 0,493530$)

k	j						
	0	1	2	3	4	5	6
0	0,210000	–	–	–	–	–	–
1	0,151200	0,009261	0,001945	0,000408	0,000086	0,000018	0,000004
2	0,059913	0,007779	0,002859	0,000943	0,000288	0,000083	0,000023
3	0,020197	0,004084	0,002316	0,001099	0,000458	0,000173	0,000061
4	0,006351	0,001715	0,001387	0,000893	0,000485	0,000233	0,000101
5	0,001930	0,000630	0,000688	0,000577	0,000396	0,000234	0,000123
6	0,000576	0,000212	0,000300	0,000317	0,000269	0,000192	0,000120
7	0,000171	0,000067	0,000119	0,000155	0,000158	0,000135	0,000099
8	0,000050	0,000020	0,000044	0,000069	0,000084	0,000083	0,000071
9	0,000015	0,000006	0,000015	0,000028	0,000041	0,000047	0,000046
10	0,000004	0,000002	0,000005	0,000011	0,000018	0,000024	0,000027
11	0,000001	$<10^{-6}$	0,000002	0,000004	0,000008	0,000012	0,000015
12	$<10^{-6}$	$<10^{-6}$	0,000001	0,000001	0,000003	0,000005	0,000007
13	$<10^{-6}$	$<10^{-6}$	$<10^{-6}$	$<10^{-6}$	0,000001	0,000002	0,000004
14	$<10^{-6}$	$<10^{-6}$	$<10^{-6}$	$<10^{-6}$	$<10^{-6}$	0,000001	0,000002
15	$<10^{-6}$	$<10^{-6}$	$<10^{-6}$	$<10^{-6}$	$<10^{-6}$	$<10^{-6}$	0,000001
16	$<10^{-6}$	$<10^{-6}$	$<10^{-6}$	$<10^{-6}$	$<10^{-6}$	$<10^{-6}$	$<10^{-6}$

Table 5. The probability of an attacker successfully conducting a double spend attack $N = 3,\ p = 0,9,\ q = 0,1$ for various j and k ($PI = 0,0025$)

k	j			
	0	1	2	3
0	0,000729	–	–	–
1	0,001006	0,000018	0,000002	$<10^{-6}$
2	0,000516	0,000018	0,000002	$<10^{-6}$
3	0,000169	0,000009	0,000001	$<10^{-6}$
4	0,000042	0,000003	0,000001	$<10^{-6}$
5	0,000009	0,000001	$<10^{-6}$	$<10^{-6}$
6	0,000002	$<10^{-6}$	$<10^{-6}$	$<10^{-6}$
7	$<10^{-6}$	$<10^{-6}$	$<10^{-6}$	$<10^{-6}$

Table 6. The probability of an attacker successfully conducting a double spend attack $N = 3,\ p = 0,8,\ q = 0,2$ for various j and k ($PI = 0,04$)

k	j						
	0	1	2	3	4	5	6
0	0,004096	–	–	–	–	–	–
1	0,010322	0,000315	0,000050	0,000008	0,000001	$<10^{-6}$	$<10^{-6}$
2	0,009634	0,000554	0,000121	0,000026	0,000005	0,000001	$<10^{-6}$
3	0,005728	0,000491	0,000142	0,000039	0,000010	0,000003	0,000001
4	0,002624	0,000304	0,000113	0,000039	0,000013	0,000004	0,000001
5	0,001014	0,000150	0,000069	0,000029	0,000011	0,000004	0,000001
6	0,000348	0,000063	0,000036	0,000018	0,000008	0,000004	0,000001
7	0,000109	0,000023	0,000016	0,000010	0,000005	0,000003	0,000001
8	0,000032	0,000008	0,000006	0,000005	0,000003	0,000002	$<10^{-6}$
9	0,000009	0,000002	0,000002	0,000002	0,000001	0,000001	$<10^{-6}$
10	0,000002	0,000001	0,000001	0,000001	0,000001	$<10^{-6}$	$<10^{-6}$
11	0,000001	$<10^{-6}$	$<10^{-6}$	$<10^{-6}$	$<10^{-6}$	$<10^{-6}$	$<10^{-6}$
12	$<10^{-6}$	$<10^{-6}$	$<10^{-6}$	$<10^{-6}$	$<10^{-6}$	$<10^{-6}$	$<10^{-6}$

Table 7. The probability of an attacker successfully conducting a double spend attack $N = 3,\ p = 0,6,\ q = 0,4$ for various j and k ($PI = 0,5$)

k	j						
	0	1	2	3	4	5	6
0	0,013824	–	–	–	–	–	–
1	0,056402	0,002389	0,000573	0,000138	0,000033	0,000008	0,000002
2	0,084935	0,006306	0,002064	0,000660	0,000206	0,000063	0,000019
3	0,082104	0,008393	0,003632	0,001501	0,000590	0,000221	0,000080
4	0,061875	0,007793	0,004327	0,002244	0,001084	0,000491	0,000211
5	0,039890	0,005754	0,003994	0,002543	0,001481	0,000797	0,000400
6	0,023173	0,003618	0,003073	0,002356	0,001628	0,001025	0,000597
7	0,012526	0,002019	0,002061	0,001873	0,001515	0,001104	0,000736
8	0,006434	0,001027	0,001241	0,001320	0,001234	0,001031	0,000781
9	0,003186	0,000485	0,000685	0,000842	0,000903	0,000857	0,000731
10	0,001536	0,000216	0,000352	0,000496	0,000603	0,000645	0,000617
11	0,000725	0,000091	0,000170	0,000272	0,000374	0,000447	0,000476
12	0,000338	0,000037	0,000078	0,000141	0,000217	0,000289	0,000340

Table 8. The probability of an attacker successfully conducting a double spend attack $N = 5,\ p = 0,9,\ q = 0,1$ for various j and k ($PI = 0,00007$)

k	j			
	0	1	2	3
0	0,000006	–	–	–
1	0,000019	$<10^{-6}$	$<10^{-6}$	$<10^{-6}$
2	0,000020	$<10^{-6}$	$<10^{-6}$	$<10^{-6}$
3	0,000012	$<10^{-6}$	$<10^{-6}$	$<10^{-6}$
4	0,000005	$<10^{-6}$	$<10^{-6}$	$<10^{-6}$
5	0,000002	$<10^{-6}$	$<10^{-6}$	$<10^{-6}$
6	0,000001	$<10^{-6}$	$<10^{-6}$	$<10^{-6}$
7	$<10^{-6}$	$<10^{-6}$	$<10^{-6}$	$<10^{-6}$

Expression (1) is well suited for values $q<0,2$, since it suffices to calculate only the first few terms in expression (1) (it is enough to calculate the first few sum_p with indices j and k). The probabilities $PI_{N,j,k}$ with increasing indices quickly tend to zero and already at when $j, k > 10$ they become negligible.

Table 9. The probability of an attacker success $<10^{-6}$ fully conducting a double spend attack $N = 5,\ p = 0,8,\ q = 0,2$ for various j and k ($PI = 0,006$)

k	j						
	0	1	2	3	4	5	6
0	0,000105	–	–	–	–	–	–
1	0,000608	0,000013	0,000002	$<10^{-6}$	$<10^{-6}$	$<10^{-6}$	$<10^{-6}$
2	0,001160	0,000047	0,000009	0,000002	$<10^{-6}$	$<10^{-6}$	$<10^{-6}$
3	0,001270	0,000075	0,000018	0,000004	0,000001	$<10^{-6}$	$<10^{-6}$
4	0,000988	0,000077	0,000022	0,000006	0,000002	$<10^{-6}$	$<10^{-6}$
5	0,000608	0,000059	0,000020	0,000007	0,000002	0,000001	$<10^{-6}$
6	0,000315	0,000037	0,000015	0,000006	0,000002	0,000001	$<10^{-6}$
7	0,000143	0,000020	0,000009	0,000004	0,000002	0,000001	$<10^{-6}$
8	0,000059	0,000009	0,000005	0,000003	0,000001	0,000001	$<10^{-6}$
9	0,000022	0,000004	0,000002	0,000001	0,000001	$<10^{-6}$	$<10^{-6}$
10	0,000008	0,000002	0,000001	0,000001	$<10^{-6}$	$<10^{-6}$	$<10^{-6}$
11	0,000003	0,000001	$<10^{-6}$	$<10^{-6}$	$<10^{-6}$	$<10^{-6}$	$<10^{-6}$
12	0,000001	$<10^{-6}$	$<10^{-6}$	$<10^{-6}$	$<10^{-6}$	$<10^{-6}$	$<10^{-6}$

Table 10. The probability of an attacker successfully conducting a double spend attack $N = 5$, $p = 0,7$, $q = 0,3$ for various j and k ($PI = 0,08$)

k	j						
	0	1	2	3	4	5	6
0	0,000408	–	–	–	–	–	–
1	0,003186	0,000090	0,000019	0,000004	0,000001	$<10^{-6}$	$<10^{-6}$
2	0,008155	0,000416	0,000107	0,000028	0,000007	0,000002	$<10^{-6}$
3	0,011987	0,000866	0,000273	0,000085	0,000026	0,000008	0,000002
4	0,012530	0,001163	0,000442	0,000165	0,000060	0,000021	0,000007
5	0,010381	0,001171	0,000531	0,000234	0,000099	0,000040	0,000015
6	0,007265	0,000960	0,000512	0,000264	0,000129	0,000059	0,000026
7	0,004473	0,000673	0,000418	0,000248	0,000139	0,000073	0,000036
8	0,002492	0,000417	0,000298	0,000203	0,000129	0,000076	0,000042
9	0,001282	0,000234	0,000191	0,000147	0,000105	0,000070	0,000043
10	0,000618	0,000121	0,000112	0,000097	0,000078	0,000057	0,000039

5 Results and Discussion

Based on the obtained expressions, we show what the required number of confirmations is in order to preserve the probability of an attacker's success below a predetermined "safe" value, for various hash rates. By "safe" value we mean a value P_S at which the upper limit of the probability of a double spending attack can be considered an acceptable risk.

Each user chooses a specific "safe" value for the probability of conducting a double-spending attack for himself, depending on the risks acceptable to him (transaction size, reputation risks, necessary speed of transaction, etc.). We will consider the meanings; $P_S = 0,1;\ 0,01; 0,001$ and a comparison of the results with the results presented in the work of Meni Rosenfeld [3].

Figures 1, 2 and 3 show the minimum number of confirmations necessary to maintain the likelihood of a successful double spend attack, depending on the attacker's hash, equal to or lower than the value. The hash rate of an honest network will still be considered $p = 1 - q$. When comparing the obtained results with the results calculated in accordance with the player's ruin model, given in the work of Meni Rosenfeld [3], we see that the probability of a successful double spend attack is lower than previously thought. In this connection, it may be limited to fewer necessary confirmations in the blockchain system with the same level of security. That in practice will significantly reduce the waiting time for a transaction and, as a result, significantly increase the speed of operations with blockchain systems.

So, if it is assumed that the attacker has 15% of the total network capacity at his disposal, then 2 confirmations are needed to keep the attacker's success rate not higher than 10%, while for the player's ruin model, 3 confirmations are expected. If the probability of the success of the attacker is not higher than 1% (under the same

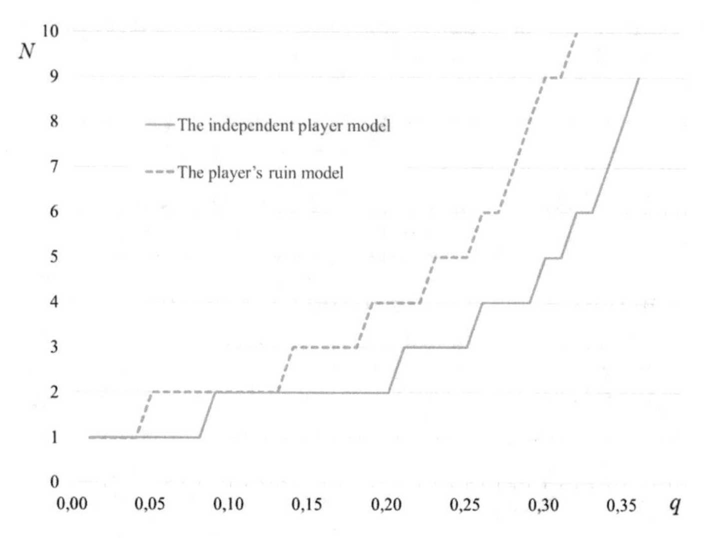

Fig. 1. The number of confirmations necessary to maintain the probability of success of an attacker at a level not exceeding $P_S = 0,1$. The solid line corresponds to the independent player model, the dashed line corresponds to the player's ruin model.

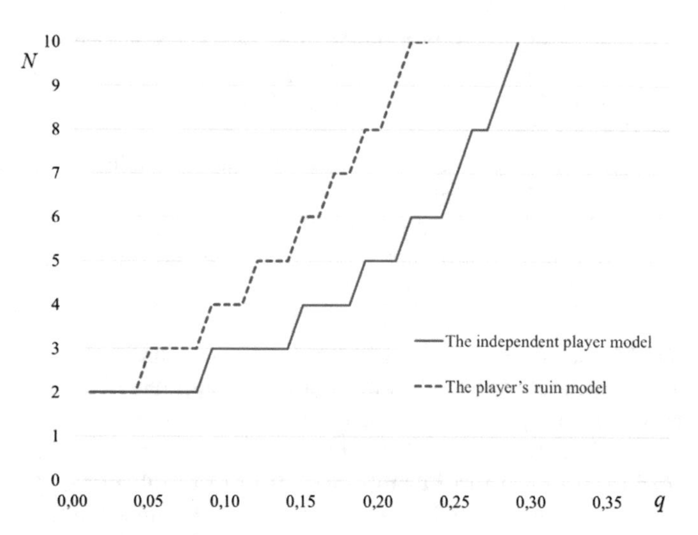

Fig. 2. The number of confirmations necessary to maintain the probability of success of an attacker at a level not exceeding $P_S = 0,01$. The solid line corresponds to the independent player model, the dashed line corresponds to the player's ruin model.

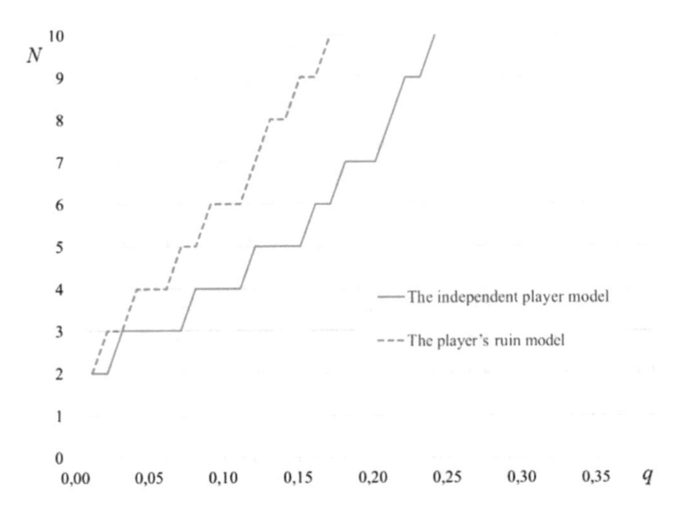

Fig. 3. The number of confirmations necessary to maintain the probability of success of an attacker at a level not exceeding. The $P_S = 0,001$ solid line corresponds to the independent player model; the dashed line corresponds to the player's ruin model.

conditions), 4 confirmations should be expected, and 6 confirmations for the player's ruin model. If we want to ensure the success rate of the attacker is not higher than 0.1%, we need to expect 5 confirmations, and for the player's ruin model - 9.

Thus, for the selected example, the required average statistical time spent waiting for confirmation of transactions can be reduced by $1/3$ (33%) for $P_S = 10\%$; by $2/6$ (33%) for $P_S = 1\%$ and by $4/9$ (44%) for $P_S = 0,1\%$. In other cases, from the above results, the average waiting time can be reduced by up to two times (for example with $P_S = 0,1$ and $q = 0,5 - 0,8;\ 0,19 - 2,0;\ 0,29$).

6 Conclusions

To analyze the security of one of the most popular consensus protocols, which are based on the proof of the work performed on the basis of hash functions, it is proposed to use a new model - the model of independent players. In the proposed model, events associated with the formation of blocks by an honest network and an attacker is considered independent events in general cases $p + q \neq 1$.

An expression was obtained analytically that characterizes the probability of an attacker conducting a successful double-spend attack on a blockchain system depending on the number of confirmations used, as well as the hashrate by an honest network and the attacker. Quantitative values of this probability are given.

Based on the results obtained, recommendations are given for determining the "safe" number of confirmations for successfully resisting a double spending attack on the blockchain system. A comparison is made with the results obtained by Meni Rosenfeld (using the player's ruin model).

It is shown that it is possible to limit itself to a smaller number of necessary confirmations in the blockchain system with the same level of security.

The practical result. The results obtained in practice will significantly (up to two times) reduce the waiting time for a transaction and, as a result, significantly increase the speed of operations with blockchain systems while maintaining a given level of security.

Academic result. Negate the results can be more cinnamon when the specific indicators are rounded up and the parameters are entered into the protocol by consensus based on the Proof of the Vicon robot, if the main mechanism is locked in by the consensus of the promising decentralized distribution of the technological systems and their implementation. This research might be useful for the improvement of various methods of information security, as well as other practical use [8–10].

References

1. Atlam, H.F., Alenezi, A., Alassafi, M.O., Wills, G.B.: Blockchain with internet of things: benefits, challenges, and future directions. Int. J. Intell. Syst. Appl. **10**(6), 40–48 (2018). https://doi.org/10.5815/ijisa.2018.06.05
2. Dey, S.: A proof of work: securing majority-attack in blockchain using machine learning and algorithmic game theory. Int. J. Wirel. Microwave Technol. **8**(5), 1–9 (2018). https://doi.org/10.5815/ijwmt.2018.05.01
3. Rosenfeld, M.: Analysis of hashrate-based double-spending, 13 p. (2014). https://arxiv.org/abs/1402.2009
4. Gervais, A., Ritzdorf, H., Karame, G.O., Capkun, S.: Tampering with the delivery of blocks and transactions in bitcoin. In: Proceedings of the 22nd ACM SIGSAC Conference on Computer and Communications Security - CCS 2015 (2015)
5. BitcoinWiki: Double-spending. https://en.bitcoinwiki.org/wiki/Double-spending
6. Shiryaev, A.N.: Probability. Graduate Texts in Mathematics (1996)
7. Nakamoto, S.: Bitcoin: A Peer-to-Peer Electronic Cash System, 9 p. (2009). https://bitcoin.org/bitcoin.pdf
8. Andrushkevych, A., Gorbenko, Y., Kuznetsov, O., Oliynykov, R., Rodinko, M.A.: A prospective lightweight block cipher for green IT engineering. In: Kharchenko, V., Kondratenko, Y., Kacprzyk, J. (eds.) Green IT Engineering: Social, Business and Industrial Applications. Studies in Systems, Decision and Control, vol. 171, pp. 95–112. Springer, Cham (2019). https://doi.org/10.1007/978-3-030-00253-4_5
9. Gorbenko, I., Kuznetsov, A., Gorbenko, Y., Vdovenko, S., Tymchenko, V., Lutsenko, M.: Studies on statistical analysis and performance evaluation for some stream ciphers. Int. J. Comput. **18**(1), 82–88 (2019)
10. Kovalchuk, L., Kaidalov, D., Nastenko, A., Rodinko, M., Shevtsov, O., Oliynykov, R.: Decreasing security threshold against double spend attack in networks with slow synchronization. In: IEEE INFOCOM 2019, Paris, France, pp. 216–221 (2019). https://doi.org/10.1109/INFCOMW.2019.8845301

Collaboration Based Simulation Model for Predicting Students' Performance in Blended Learning

Zhengbing Hu[1], Jun Su[2], and Yurii Koroliuk[3(✉)]

[1] School of Educational Information Technology, Central China Normal University, Wuhan, China
hzb@mail.ccnu.edu.cn

[2] School of Computer Science, Hubei University of Technology, Wuhan, China
sjhosix@gmail.com

[3] Chernivtsi Institute of Trade and Economics, Kyiv National University of Trade and Economics, Chernivtsi, Ukraine
yu_kor@ukr.net

Abstract. Due to the positive influence on students' learning outcomes, the interests of studying effective knowledge management has been risen recently. Developing and implementing effective strategies ensures to promote learning outcomes. By reviewing and examining various influence factors, this research study has predicted the major factors that may influence of learning outcomes in blended learning environment. A series of simulation experiments and factor analyses have been conducted in order to investigate collaboration during group learning process. The simulation model for blended learning environment employed in this research has drawn on the characteristics of the Structural Equation Model (SEM) of the blended learning process of 128 students. Both randomness of those student learning behaviors and the reaction to information overload have been considered during simulation modeling. The simulation model enables for greatly increasing statistical samples of student learning behavior analysis. Besides, this research has studied the impact of multiple factors to blended learning mode, these factors include: the size of learning group, the group composition according to previous performance, teaching material amount, and the teacher influence. Experimental results predict that the factors mentioned above can enhance collaborative interaction among students during writing and reading activity. The research results of the optimization restriction factors for blended learning environment achieved in this study can be useful reference for a teacher who are facing the similar challenges. The results of this paper can also be used to reveal and eliminate the problem of inefficient collaboration and poor student performance in blended learning environment. The model proposed in this paper can be integrated with most of decision support systems of universities.

Keywords: Collaborative learning · Interactive learning environments · Simulations · Teaching strategies · Predicting of academic performance

Z. Hu et al. (Eds.): ICCSEEA 2020, AISC 1247, pp. 653–667, 2021.
https://doi.org/10.1007/978-3-030-55506-1_58

1 Introduction

The informationization of the learning process has optimized the analysis results of student interaction learning, and it has implemented the iteration to the analysis results [1]. The widespread use of distance education technology and the use of logs to dynamically record learning processes allow many parameter data for different learning processes to be preserved and accumulated to the magnitude required for big data analysis. Recently, learning and analysis activities based on big data methods are increasing, and the introduction of big data analysis technology is one of the development trends of higher education in the future [2]. Successful problem solving of knowledge management supposes using accurate measurement of the large number of quantitative and qualitative data comprising adequate means of statistics and analysis.

The scope of data required for applying learning analysis is usually limited to parameters of individual students or a group of students [3]. The research results are often influenced by internal and external factors that influence individual or group learning. These factors include: different size of the study group, different length of the course, insufficiency of selected statistical parameters, the complexity of comparing results of test groups, different length of experimental time, different uniqueness of individual students, the presence of causal data, etc.

Thus, the main objective of this research is to predict the possible factors that influence of students' learning performance improvement by students' collaboration in a blended learning environment. By examining learning management parameters: the number of students in the group, the group composition based on learning performance in the past, the amount of training materials and teacher's influence. The data collected in this experiment ranged from statistical data of 128 students learning in blended learning mode which is further analyzed by using a combination of statistical and factor analysis methods, simulation modeling and table data visualization techniques.

The paper is structured as follows: Sect. 2 presents the background and related papers about effectiveness factor analysis methodology of blended learning environment. Section 3 describes data collection and analysis methods and models. Section 4 shows experimental results and related discussions. Section 5 is the conclusion of this paper. Finally, Sect. 6 presents the limitations and recommendations for future research.

2 Background and Related Work

In recent years, blended learning has become a trend, its principle concept [4] is a combination of learning modes of synchronization (face-to-face) and asynchronization (over the Internet), which achieved better outcomes by effective learning process [5].

Is blended learning efficient? This question can be answered by experience proof and factual results [5]. In general, the assessment performance of blended learning is like that of traditional face-to-face learning. In other words, blended learning is assessed by comparing with traditional face-to-face teaching [6]. Most other researches have realized that blended learning can improve students' learning outcome [7], but few of them gave reasons for the improvement. Some studies reveals that the main reason of such an improvement facilitates to establish collaborations [8] and cooperation [9] within a learning group. Besides, a teacher plays a role as a 'intermediary' who often raises questions. Also confirmed that asynchronous and synchronous e-learning mode has a positive impact on academic performance [10].

Effective teaching strategies are based on the improving of students' educational productivity. Forecasting such productivity under the influence of different factors is the current task when managing blended learning. The model description of the learning environment is laid in the basis of this forecast. Informatization of the learning process creates favorable preconditions of supplying forecasting models with qualitative and quantitative data of learning statistics. Thus, human-computer interaction in a blended learning environment enables a wide range of learning processes to be parameterized and statistically calculated. The collected parameters can be used for various learning content management goals. All the mentioned parameters are stored as log-files of learning management system (LMS), i.e., such as Moodle.

As follows, a large number of parameters and quantitative statistic data of student's educational behavior led to the emergence of some methods of predicting success. The most common is the use of regression forecasting models. However, most of them are based on linear dependencies, take into account a limited number of forecasting factors and describe only individual learning behavior of a student. The results of such forecast are not reliable for building effective collaborative blended learning strategies. To solve such issues A. Elbadrawy and others [11] suggest a class of collaborative multi-regression models. These models have the following parameters: student's past performance, engagement and course characteristics. But these models do not take into account the factor of tutor's active participation and the size of the training group, information overload. Another drawback of these models, together with other regressive ones, is that they give exact forecast only when large student groups are analyzed.

The example of application the another forecasting technology is the work of S. B. Kotsiantis [12]. The researcher suggests the use of machine learning techniques for forecasting students' grades. To construct the model there was used the data of virtual courses, e-learning log file, demographic and academic data of students, admissions/registration info, and so on. The analysis revealed Data Mining technologies that are the most suitable for forecasting. Though, the suggested forecasting model takes into account a limited number of factors, most of which are unmanageable. The main drawback of the model there remains an individual description of a student, not considering the group interaction peculiarities. A similar drawback is peculiar for the forecast model of S. Borkar and K. Rajeswari [13], based on the education Data Mining of the following parameters: graduation, attendance, assignment, unit test, university result. A. Mueen and others [14], C. Romero and others [15] introduce the parameter 'forum participation' to consider the group activity in the Data Mining of the forecasting model. But such model requires a large number of students to form a reliable data sampling and accurate forecast. Also, the factor of the tutor's active participation is not taken into account, etc.

U.R. Saxena and S.P Singh [16] have used Neuro-Fuzzy Systems for forecasting students' performance based on their CPA and GPA. Notwithstanding the fact that researchers confirm a high accuracy of the forecast, it is based only on two individual parameters of a student. Small sample size for building such a model is problematic and made about 20–26 cases of students' results.

K.D. Kolo and others [17] suggest a decision tree approach for predicting students' academic performance. To build the decision tree structure there was used the method of Chi-Square automatic interaction detection of factors: student's grade, student's status, student's gender, financial strength, attitude to learning. Nurafifah Mohammad Suhaimi and others [18] also used such classifier methods as Decision Tree compared with Support Vector Machine, Neural Network and Naïve Bayes for similar goals. Md Rifatul Islam Rifat and others [19] used six state-of-the-art classification algorithms for the prediction task of academic performance. However, such models do not describe the peculiarities of group interaction and cannot be widely applied when elaborating effective teaching strategies.

Therefore, the elaborating of adequate forecast models is complicated by that the process of blended learning is complex and dynamic. However, many of the measured data are stochastic, which requires a deep and representative statistical sample to reveal deterministic dependencies. Data sets in small scale will greatly affect the results of the analysis [20].

The reason for complexity increasing of the sample is objective. First of all, it is a small group scale. The number of groups ranges from 3 to 30, and the number is not enough for accurate factor analysis. Second, the process of collecting statistical information is long. Typically, information is collected during a learning cycle of six months or longer. During this time, the effects of factors may be unequal. There are possibilities occurring influence of factors that are not considered, i.e. the impact of the semester. Third, there are measurement errors associated with assessment for student learning effects.

Even adopting the most accurate model (with an accuracy of more than 86% [21]) in the above range (small scale 3–30 people), significant inconsistencies with the analysis results might be possibly occurred, which complicates learning management. On the one hand, assessment methods given by most hypothesis test methods (p-values, β-values, t-tests, confidence intervals, etc.) are difficult to assess students' performance in a small group. On the other hand, quantitative research can qualitatively explain the main learning mechanisms, student interaction, and information exchange. It is common practice to interpret these results in the form of structural equation modeling (SEM). However, thorough implementation of SEM in learning management requires further research especially for the aims of forecasting. Establishing a simulation model based on SEM is a helpful suggestion because it enables to verify models of the teaching strategy in the form of scenario collection of management parameters. Multiple simulation experiments can greatly improve the reliability of the predicting and therefore expand the research samples as endogenous and exogenous parameters of models.

3 Data, Methods and the Models

The data sample in this paper derived from 128 third-year students of economic cybernetics specialty, who had been studying for 2013–2018 academic years in Chernivtsi Institute of Trade and Economics of KNUTE (ChITE KNUTE). Blended learning mode was applied for those students in their learning process. The same subject in a semester is considered as an optional unit, all subjects were taught in Ukrainian language. Students participating in the study group with different performances. This study involved a total of 12 learning groups, there were 6–21 members in each group. Teaching work consists of two parts: classroom and online courses. Online courses were conducted by Moodle software over the Internet which can be found on ChITE KNUTE's website (http://www.dist.chtei-knteu.cv.ua:8080/).

The number of textbooks was determined by ECTS (European Credit Transfer System) and was divided into several grades ranging from 1.5, 3 to 4.5 points. The textbook was divided into 6–18 independent topics. The amount of work was the same for every week. Each topic was designed for students to learn theoretical knowledge to complete practical tasks. During the course learning, students attended lectures as required, students learned strategies for competing learning tasks, teachers evaluated students' achievements on previous learned subjects. Implementing learning tasks was through a forum, and an open asynchronous online discussion was conducted under teachers' guidance. Discussions were conducted in an asynchronous way, including posting, reading, and replying to posts written by group members or teachers. Learning outcome for each subjects is rated by 100 scores. The final scores for all subjects are calculated by the arithmetic mean value.

The measurement parameters of a blended mixed learning process include: student ID, group ID, achievement based on past course performance, final course achievement, subject number, ECTS academic credits, group size, theoretical material amount, amount of each student reading post, amount of each student witting post, amount of a teacher writing post. The log files generated by Moodle were analyzed in this study, the problems of accurately and comparatively evaluating parameters were occurred, it is not easy to identify some parameters including amount of theoretical textbooks, amount of readings posts for a student, amount of writing each post for a student, amount of writing each post for a teacher. Using the quantity indicator for writing or reading posts cannot clarify the actual amount of information exchange in a blended learning environment. In order to overcome this problem, it is necessary to calculate the number of characters written and read by each student in the learning process. The obtained characters can be converted to bytes using the method of [22]. Therefore, the information unit containing the characters of the 33-letter Ukrainian alphabet is shown as follows:

$$H_0 = \log_2 33 = 5.044\,\text{bits} \tag{1}$$

The goal of data analysis is to rely on other parameters valid in a blended learning environment and to maximize performance out of students. Reading and writing activity are treated as a direct and important factor, which represents degree of student's collaboration in a group as confirmed by K. Bielaczyc and A. Collins [23]. Proactive reading and posting are in turn related to other management and non-management factors in blended learning environment. The structure of the relationship model of these factors is shown in Fig. 1.

Learning process of 128 students in the statistical sample was analyzed in order to describe and confirm the importance of the relationship. In case of considering the number of students in a group, amount of reading and posting are accumulated. In order to effectively study other factors, at this stage, the total amount posted by the group reflects the influence of the group, group size and the total number of posts are not differentiated and are treated as the same factor. Post reading indicator is the proportion to all posts read by a student. The results of SEM analysis of the coefficients and the multiple linear regression indicator forms are given in Table 1.

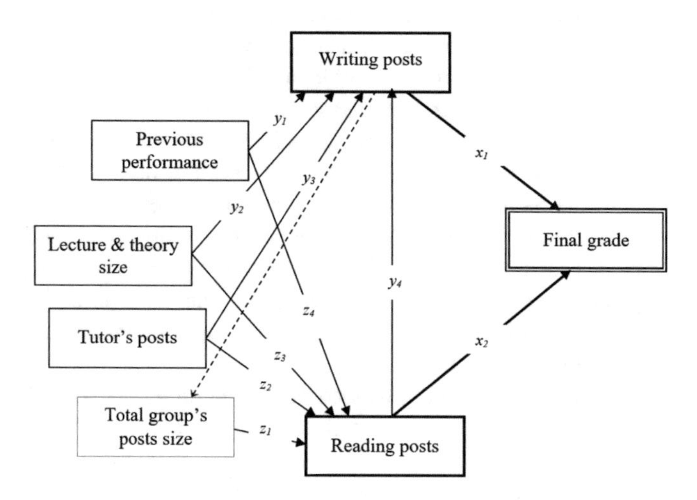

Fig. 1. SEM of student's performance

Table 1. SEM parameters of student's performance in a blended learning group

	Reading posts*					Writing posts**					Final grade***		
R^2	0.81					0.60					0.55		
Standard error	0.10					11.69					7.18		
F	0.007					0.02					0.013		
α	0.00					0.00					0.00		
Coefficients	Y-intercept	z_1	z_2	z_3	z_4	Y-intercept	y_1	y_2	y_3	y_4	Y-intercept	x_1	x_2
Value	−0.155	−0.001	−0.001	0.001	0.01	−83.19	1.28	0.091	−0.158	−15.44	61.67	0.295	19.96
t-value	−2.144	−10.35	−1.528	8.923	12.75	-9.810	10.13	5.47	−1.864	−2.011	34.19	8.13	6.70
p-value	0.034	0.00	0.129	0.00	0.00	0.00	0.00	0.00	0.065	0.046	0.00	0.00	0.00

Units of measurement: * from 0 to 1, ** Kbyte, *** from 0 to 100.

For the formation of SEM there were made the hypotheses about the existence of substantial connections between the studied input and output parameters. Hypothesis testing was performed by the means of MS Excel multiple regression of observed statistical parameters.

The analysis result of the significance of the SEM reveals that the value of index R^2 was higher than its critical value 0,349 (when $\alpha = 0.01$). The value of index F is always much smaller than the critical value, and the value of calculated α (significance of F) is close to zero. The t-value of absolute index exceeds 1.97 despite coefficients z_2 and y_3. The p-value is smaller than 0.05, despite the coefficients z_2 and y_3.

The results of the statistical analysis confirmed the relationship between input and output parameters of SEM. Factors affecting parameters of reading and postings include: the student's previous performance and the number of theoretical textbooks on the subject. Teacher guidance has a negative impact on students' reading and witting posts, but the impact is not big. The similar influence of teacher guidance on student learning is described in the work of Mazzolini and Maddison [24]. It has been confirmed that after completing the first two tasks, students use more suggestions from other group members than teacher's guidance [25]. In this study, a teacher can create conditions for conduct discussion and collaboration among students and supervise the process of ongoing discussion. In case of occurring passive discussions and off-topic situations, the teacher provides timely correction. The effects mentioned above can explain why multiple linear regression analysis shows a negative impact on the practice of writing discussion contents. Analysis confirms the existence of this impact. When discussing the actual online task, the output parameters of scores are generated by the positive influence of reading and writing posts.

Although dependency exists, the output SEM parameters contain an evident standard error index. In order to establish an effective knowledge management strategy, it is necessary to accurately predict the influence of available management factors. These calculations enable to generate a combination of certain factors in blended learning environment in order to identify the area of promoting students' learning performance. Obtained SEM parameters revealed the nature of output performance relying on input performance. To accurately describe a blended learning process, it needs to use proper methods to analyze students' behaviors upon collected parameters.

The nature of students' reaction to external stimuli is behaviors and a kind of clustering. This article has introduced some quantitative and qualitative methods to describe certain human behaviors. Therefore, in recent years, the methods of human-centered systems (HCS: Human Centered Systems) have often been used to describe the behavior of staff in industry, military, and stock markets, as well as in other environments [26]. When behavioral simulation model is used to study human behaviors, it is common to construct a single person's model (for example, a worker) at first that specific environmental model is established based on it, and the model is applied to a group of people in social networks afterwards [27].

In order to study students' learning behavior, a simulation model was established in this study, and the dependence of SEM parameters was calculated based on the model. A simulation model for each student was built using MATLAB's Simulink tool. The model uses some simulation programming elements such as 'Add', 'Subtraction', 'Division', 'Discrete Time Integrator', 'Uniform Random Number', 'Pulse Generator', 'Oscilloscope' and so on. The model contains a data input module (red ellipse 1 in Fig. 2) and a data output module (red ellipse 4 in Fig. 2). The input data is submitted for five parameters: Previous performance, Lecture & Theory size, Tutor's posts, Total student's write, Subjects. The input parameters are then multiplied by the SEM coefficients according to its structure (Gain blocks) and added (Subtract blocks). Multiple linear regression coefficients were described using SEM (indicated by Ellipse 2 in Fig. 2). Both the effects of standard error and the randomness of student behaviors are marked with Ellipse 3 in Fig. 2. Uniform Random Number generators are used to take into account random variables. The result of this processing is three discrete outputs indicated by Ellipse 4 in Fig. 2: Grade, Read, Write.

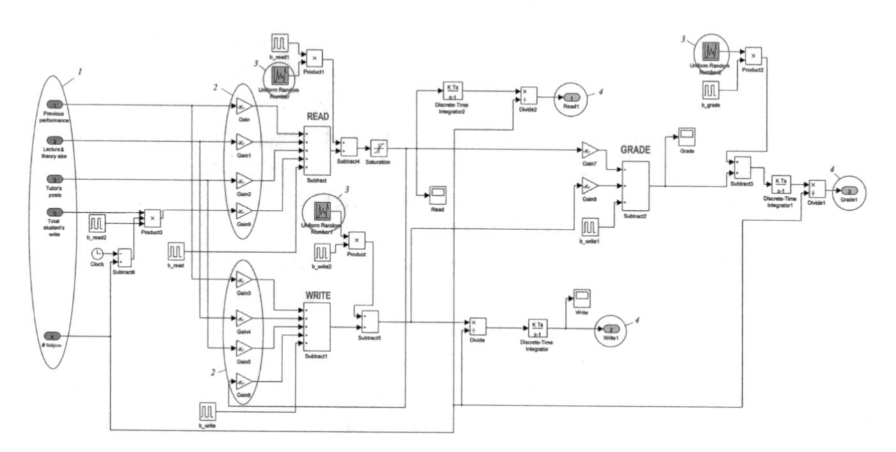

Fig. 2. Simulink/MATLAB model of a student's learning activity in a blended learning environment

A student's simulation model is in the status of discrete. The number of weeks is equivalent to the number of learning subjects. Each learning subject is an independent learning unit that highlights the characteristics of discreteness. Each learning subject contains a task, an online discussion module and a grade. This approach differs from a similar simulation model of continuous behaviors described by S. Elkosantini and D. Gien [27].

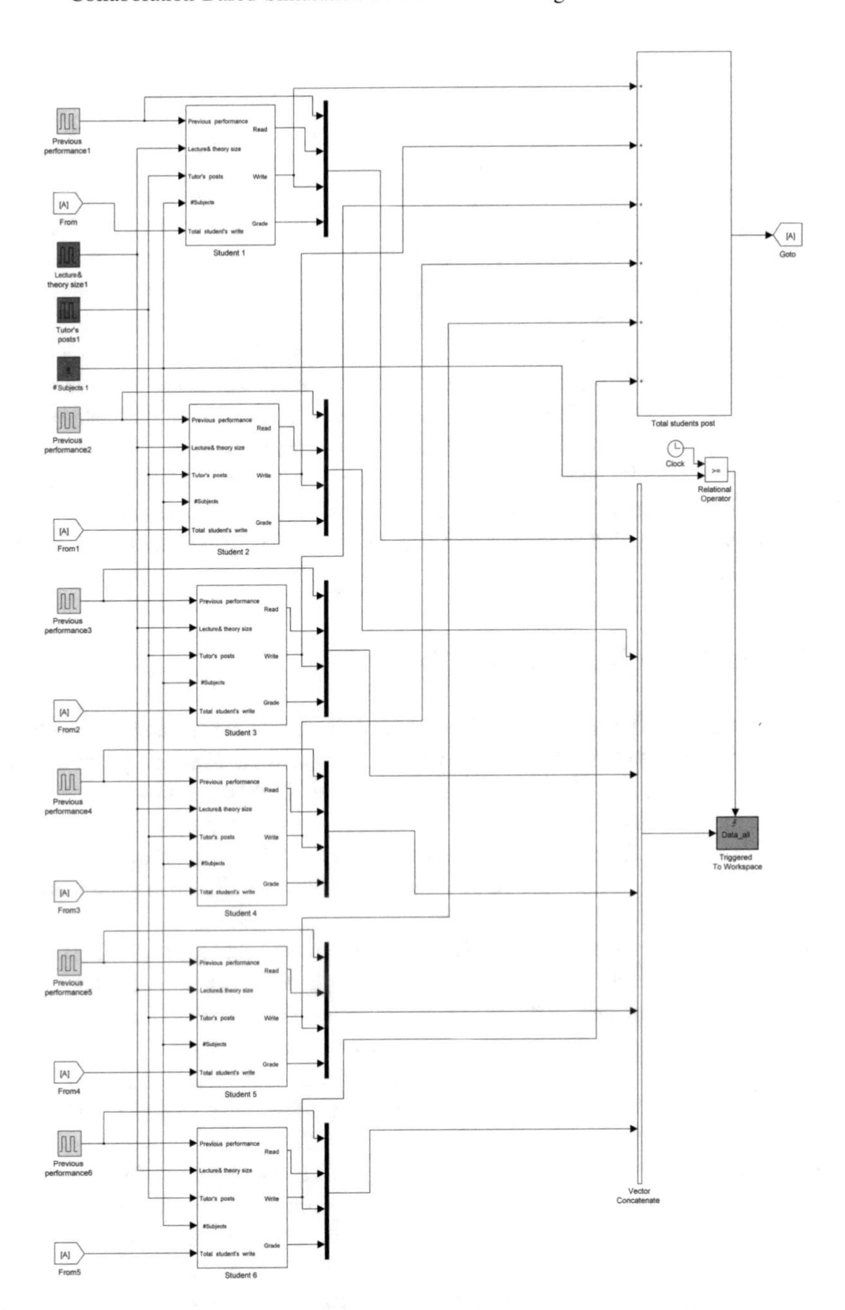

Fig. 3. Example of a learning group model (Simulink/MATLAB) with 6 students

Reading-writing practices and subject ratings are calculated in one cycle of the model. The final score is comprehensively calculated based on the results of each subject, and the students' scores are given by the real assessment system in a blended learning environment. The individual student models are combined into a learning collaboration network and model of a group is formed on this basis. Figure 3 presents a blended learning environment for a learning group with 6 students. The input parameters of the model include: the number of the textbook, the number of posts written by a teacher, the number of subjects (the course lasted for several weeks), and previous performance of each student. The input parameters were formed using a pulse generator with the period of signaling - one training week. The inputs for each student model (Student blocks) were formed individually, taking into account their training peculiarities, number of subjects, tutor's activity and others (Fig. 3, right side). Thus, there were formed the initial conditions of the experimental forecast. The outputs of each student sub-model were commuted to form a repository of student-produced information in the form of posts (Total students' posts block). The results of the model calculations were exported to an external file for later study using MATLAB software for further analysis.

4 Results Analysis and Discussion

By calculating the experimental results obtained from the experiment with different input parameter sets, seven groups of models were established, in which the numbers of students were 3, 6, 8, 10, 12, 15, 20 students. Students in these groups are completely different in their previous performance. The average scores of the group model ranges from 60 to 90. Some students got 60–95 scores before joining the group. In a blended learning environment, the performance indicator is dynamically changed accordingly as a student's actual performance changes.

There are three kinds of textbooks in the model: six subjects (capacity: 160 Kbyte), 12 subjects (capacity: 340 Kbyte), and 18 subjects (capacity: 480 Kbyte). The number of selected subjects depends on the ECTS credits required by the courses. The influence after a teacher posting 30–110 Kbytes was analyzed and examined.

Modeling 105 combinations of scenario input parameters received results in form of table are converted in form of matrix by the XYZ-Gridding tool of the OriginPro application, which is further converted to a visualization matrix by the tool – 'Color Map Surface (OriginPro)' (see Fig. 4 and 5). Due to the restriction of the number of input factors, it makes more difficult to accurately estimate achieved results, such an inferiority may result in misinterpreting other factors. However, the simulation model fully inherits the accuracy of the SEM indicators (Table 1: R^2 for reading activity is equal 0.81, R^2 for writing activity is equal 0.6, R^2 for grading score is equal 0.55) that gives a reason to consider the data of Fig. 4 and 5 reliable. In addition to the SEM accuracy indicators, there was held the comparing of academic performance for two academic groups of students who completed the course. The first group consisted of 6 students who studied 18 subjects. When doing the tasks, the tutor's activity was 35.6 Kbyte posts per student. The results of the course showed that the average performance has increased by 2.71 points. Simulation modeling predicts a 1.9 point performance increase (downward deviation 30%). The second group consisted of 7 students who studied 6 subjects. In this case, the tutor's activity was 10.4 Kbyte posts per student. After completing the course, the average performance has increased by 2.17 points. Simulation modeling predicts a 2.7 point performance increase (upward deviation 25%). The performance rates obtained correspond to the calculated R^2.

By the influences of a teacher's guidance and the factor of student's previous performance, the experiment in this research studied students' writing practice in blended learning environment (Fig. 4).

Figure 4 presents the absolute index of the writing practice (vertical axis) for three different course sizes: 6, 12, 18 subjects. The absolute index of writing practice was predicted in Kbyte of written posts by the entire modeling group. The horizontal axes show the initial conditions of prediction: tutor's posts per student (Kbyte) and previous performance (in a 100 point scale).

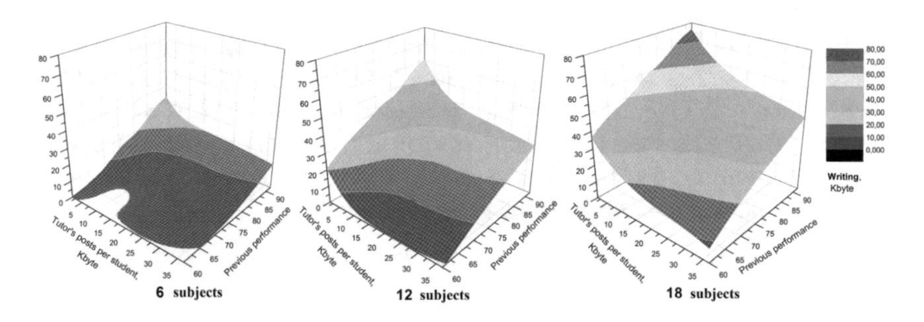

Fig. 4. Predicting results of writing practice in blended learning environment

As we can see from Fig. 4, for courses with more lectures and theoretical subjects, writing amount is larger. However, in the case of different number of subjects, the study did not find the differences in the form of exercises. With the decreasing of a teacher's guidance, the amount of exercises for students with different previous scores increases. For students with good previous scores, the growth speed is faster. As a result, students are tending to establish online collaboration more actively when solving specific tasks, finding out key points of the task, and solving difficult theoretical problems.

For students who had more than 75 scores previously, the study found that the role of a teacher is only assistance and the teacher's influence in promoting learning performance is not evident, the teacher can not optimize the performance greatly. For students with worse previous performance, more teacher's guidance can lead to reduce the amount of posts. Through verification, the proper explanation for the above phenomena is that students tend to read the posts by teachers and by the students who had better previous learning performance than to write their own posts, which suggests that these students become knowledge receiver according to Gillies [28].

A small peak appearing in Fig. 4 reflects that active posters are made up of a small number of students who usually have better performance. AbuSeileek [29] confirmed this characteristic. Although the results obtained by this experiment are different than

Qiu's study [30], there exists a direct proportion between the total number of posts posted by the students and the total number of posts posted by a teacher. Operational modeling shows that more teacher's guidance provides more information and might result in information overload. Students write less posts due to the influence of information overload, which was confirmed by Qiu [30].

To perform objective analysis, additional average performance of a group (the difference between the final grade of the course and the previous grade) was observed. The experimental group mixed the previous scores. The homogenous case was not considered because it is impossible to set up multiple groups with 6 students with the same level of previous performance.

Figure 5 represents the added average academic performance of the training group (vertical axis) for three different course sizes: 6, 12, 18 subjects. The added average academic performance was predicted as the difference between the received and previous course performance of the group. The horizontal axes show the initial prediction conditions: tutor's posts per student (Kbyte) and the size of the training group.

Figure 5 shows that the amount of theoretical material greatly affects critical performance indicator. In case of 6 subjects, even if teacher's guidance is reduced, the scores of groups with different numbers are promoted. For those groups with more than 10 student, with the increase in the number of posts posted by a teacher, the average scores of the groups are declined. As shown below, students in larger groups rely more on teacher's guidance than on collaborations of students.

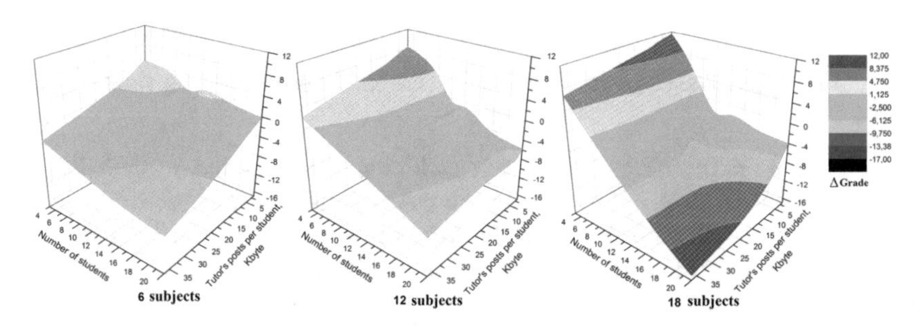

Fig. 5. Predicting results of the students' performance (course grade) in blended learning

In case of 18 subjects, teams with up to 8 students achieved very high additional scores at any level of teacher's guidance. In case of large groups, the bottleneck of optimizing learning performance still exists, but only when a group contains 10 students. A further increase in the number of a group can suddenly reduce the ratio of performance increasing. Students can not deal with the large number of posts sent by group members. More teacher's guidance may decline the average scores of these groups. This result indicates that a teacher should pay more attention on groups with lower initiative and less problems. However, teacher's guidance can not significantly affect the collaborations among students. The study results revealed the significant influence of group size in information overload that also was described by Hewitt and Brett [31] and Pfister and Oehl [32].

The proven accuracy of the predictions and the consistency of the results with the work of other scientists give reason to claim that the research objectives were achieved. The simulated forecast conditions have shown the utmost ability to optimize the management of LMS in ensuring active collaboration and increasing the success of training groups under the conditions of blended learning.

5 Conclusion

This paper uses the predicting results of the effects obtained in a blended learning environment as the outcome of implementing different knowledge management strategies. Input management parameters include learning group size, grouping based on previous learning performance, the amount of learning materials and the cycle of learning a subject, the teacher's guidance, appropriate learning strategies are chosen by considering the combination the parameters in blended learning environment. By analyzing the whole process of students learning in a blended environment, correlated simulation modeling experiment and the student collaborations within a group are evaluated objectively. The simulation model considers random dependence of the learning behavior set of 128 students in blended learning environment described by SEM.

Our study revealed that the success of achieving group's additional scores depends greatly on group size and the amount of learning materials. According to the maximum number of learning subjects and the maximum course cycle, small groups of up to 8 students can achieve 2–12 additional scores. This study suggests that the role of teachers is only 'assistance' in the case of different course scale and group size, teachers is not able to solve the bottleneck of restricting performance improvement in blended learning environment. Teacher's guidance is helpful to avoid the occurrence of information overload.

6 Limitations and Future Work

The limitations of this study are related to the fact that the data contained in constructing SEM and simulation model is only relevant to the exchange of textual information among students and the learning environment. The estimation of the theoretical textbook does not consider the included graphic images. This study did not consider the time delay between learning the textbook and starting the discussion. The quality index of the discussion posts was also not analyzed.

Due to the restriction of the number of input factors, it makes more difficult to accurately estimate achieved results, such an inferiority may result in misinterpreting other factors. However, the simulation model fully inherits the accuracy of the SEM indicators.

Therefore, future studies could explore the possibility of improving the SEM and the simulation model. In particular, it is assumed taking into account graphic data interchange in a blended learning environment. Specification of the SEM indicators will be held by considering new statistic data of the observation.

Acknowledgments. This scientific work was partially supported by RAMECS and self-determined research funds of CCNU from the colleges' primary research and operation of MOE (CCNU19TS022).

References

1. Dyckhoff, A.L., Zielke, D., Bültmann, M., Chatti, M.A., Schroeder, U.: Design and implementation of a learning analytics toolkit for teachers. Educ. Technol. Soc. **15**(3), 58–76 (2012)
2. Siemens, G., Long, P.: Penetrating the fog: analytics in learning and education. Educause **46** (5), 30–40 (2011)
3. Daniel, B.: Big data and analytics in higher education: opportunities and challenges. Br. J. Educ. Tech. **46**(5), 904–920 (2015). https://doi.org/10.1111/bjet.12230
4. Garrison, D.R., Kanuka, H.: Blended learning: uncovering its transformative potential in higher education. Internet High. Educ. **7**(2), 95–105 (2004). https://doi.org/10.1016/j.iheduc.2004.02.001
5. Means, B.: The effectiveness of online and blended learning: a meta-analysis of the empirical literature. Teach. Coll. Rec. **115**(3) (2013). http://www.tcrecord.org/library/content.asp?contentid=16882/. Accessed 23 May 2019
6. Means, B., Toyama, Y., Murphy, R., Bakia, M., Jones, K.: Evaluation of evidence-based practices in online learning: a meta-analysis and review of online learning studies. US Department of Education, Washington, DC (2009)
7. Zhao, Y., Lei, J., Yan, B., Lai, C., Tan, H.S.: What makes the difference? A practical analysis of research on the effectiveness of distance education. Teach. Coll. Rec. (2005). https://doi.org/10.1111/j.1467-9620.2005.00544.x
8. Kolloffel, B., Eysink, T.H.S., de Jong, T.: Comparing the effects of representational tools in collaborative and individual inquiry learning. Int. J. Comput.-Supported Collab. Learn. **6**(2), 223–251 (2011). https://doi.org/10.1007/s11412-011-9110-3
9. Johnson, D.W., Johnson, R.T.: Cooperation and the use of technology. In: Johanssen, D.H. (ed.) Handbook of Research on Educational Communications and Technology, 2nd edn, pp. 785–811. Lawrence Erlbaum Associates, Mahwah (2004)
10. Dada, E.G., Alkali, A.H., Oyewola, D.O.: An investigation into the effectiveness of asynchronous and synchronous e-learning mode on students' academic performance in National Open University (NOUN), Maiduguri Centre. Int. J. Mod. Educ. Comput. Sci. (IJMECS) **11**(5), 54–64 (2019). https://doi.org/10.5815/ijmecs.2019.05.06
11. Elbadrawy, A., Studham, R.S., Karypis, G.: Collaborative multi-regression models for predicting students' performance in course activities. In: Proceedings of the Fifth International Conference on Learning Analytics and Knowledge - LAK 2015 (2015). https://doi.org/10.1145/2723576.2723590
12. Kotsiantis, S.B.: Use of machine learning techniques for educational proposes: a decision support system for forecasting students' grades. Artif. Intell. Rev. **37**(4), 331–344 (2011). https://doi.org/10.1007/s10462-011-9234-x
13. Borkar, S., Rajeswari, K.: Predicting students academic performance using education data mining. Int. J. Comput. Sci. Mob. Comput. **2**(7), 273–279 (2013)
14. Mueen, A., Zafar, B., Manzoor, U.: Modeling and predicting students' academic performance using data mining techniques. Int. J. Mod. Educ. Comput. Sci. (IJMECS) **8** (11), 36–42 (2016). https://doi.org/10.5815/ijmecs.2016.11.05

15. Romero, C., López, M.-I., Luna, J.-M., Ventura, S.: Predicting students' final performance from participation in on-line discussion forums. Comput. Educ. **68**, 458–472 (2013). https://doi.org/10.1016/j.compedu.2013.06.009
16. Saxena, U.R., Singh, S.P.: Integration of neuro-fuzzy systems to develop intelligent planning systems for predicting students' performance. IEEE Technol. Eng. Educ. (ITEE) **7**(4), 1–7 (2012)
17. Kolo, K.D., Adepoju, S.A., Alhassan, J.K.: A decision tree approach for predicting students academic performance. Int. J. Educ. Manag. Eng. (IJEME) **5**, 12–19 (2015). https://doi.org/10.5815/ijeme.2015.05.02
18. Nurafifah, M.S., Abdul-Rahman, S., Mutalib, S., Hamid, N.H., Ab Malik, A.M.: Review on predicting students' graduation time using machine learning algorithms. Int. J. Mod. Educ. Comput. Sci. (IJMECS) **11**(7), 1–13 (2019). https://doi.org/10.5815/ijmecs.2019.07.01
19. Rifat, M.R.I., Al Imran, A., Badrudduza, A.S.M.: Educational performance analytics of undergraduate business students. Int. J. Mod. Educ. Comput. Sci. (IJMECS) **11**(7), 44–45 (2019). https://doi.org/10.5815/ijmecs.2019.07.05
20. Xie, K., Di Tosto, G., Lu, L., Cho, Y.S.: Detecting leadership in peer-moderated online collaborative learning through text mining and social network analysis. Internet High. Educ. **38**, 9–17 (2018). https://doi.org/10.1016/j.iheduc.2018.04.002
21. Marbouti, F., Diefes-Dux, H.A., Madhavan, K.: Models for early prediction of at-risk students in a course using standards-based grading. Comput. Educ. **103**, 1–15 (2016). https://doi.org/10.1016/j.compedu.2016.09.005
22. Shannon, C.E.: A mathematical theory of communication. Bell Syst. Tech. J. **27**(3), 379–423 (1948). https://doi.org/10.1002/j.1538-7305.1948.tb01338.x
23. Bielaczyc, K., Collins, A.: Fostering knowledge-creating communities. In: O'Donnell, A.M., Hmelo-Silver, C.E., Erkens, G. (eds.) Collaborative Learning, Reasoning, and Technology, pp. 61–98. Lawrence Erlbaum Associates, Mahwah (2006)
24. Mazzolini, M., Maddison, S.: Sage, guide or ghost? The effect of instructor intervention on student participation in online discussion forums. Comput. Educ. **40**(3), 237–253 (2003). https://doi.org/10.1016/S0360-1315(02)00129-X
25. Moallem, M.: An interactive online course: a collaborative design model. Educ. Tech. Res. Devel. **51**(4), 85–103 (2003). https://doi.org/10.1007/BF02504545
26. Elkosantini, S.: Toward a new generic behavior model for human centered system simulation. Simul. Model. Pract. Theory **52**, 108–122 (2015). https://doi.org/10.1016/j.simpat.2014.12.007
27. Elkosantini, S., Gien, D.: Integration of human behavioural aspects in a dynamic model for a manufacturing system. Int. J. Prod. Res. **47**(10), 2601–2623 (2009). https://doi.org/10.1080/00207540701663490
28. Gillies, R.M.: Teachers' and students' verbal behaviours during cooperative and small-group learning. Br. J. Educ. Psychol. **76**(2), 271–287 (2006). https://doi.org/10.1348/000709905X52337
29. AbuSeileek, A.F.: The effect of computer-assisted cooperative learning methods and group size on the EFL learners' achievement in communication skills. Comput. Educ. **58**(1), 231–239 (2012). https://doi.org/10.1016/j.compedu.2011.07.011
30. Qiu, M.: A mixed methods study of class size and group configuration in online graduate course discussions. ProQuest Dissertations and Theses (2010)
31. Hewitt, J., Brett, C.: The relationship between class size and online activity patterns in asynchronous computer conferencing environments. Comput. Educ. **49**(4), 1258–1271 (2007). https://doi.org/10.1016/j.compedu.2006.02.001
32. Pfister, H.R., Oehl, M.: The impact of goal focus, task type and group size on synchronous net-based collaborative learning discourses. J. Comput. Assist. Learn. **25**(2), 161–176 (2009)

Experimental Evaluation of Phishing Attack on High School Students

R. Marusenko[1], V. Sokolov[2](✉), and V. Buriachok[2]

[1] Taras Shevchenko National University of Kyiv, Kiev, Ukraine
marusenko.r@gmail.com
[2] Borys Grinchenko Kyiv University, Kiev, Ukraine
{v.sokolov,v.buriachok}@kubg.edu.ua

Abstract. The effectiveness of phishing attacks is being analyzed by many researchers. At the same time, researches often deal with the random sample of people suffered a phishing attack and are limited with analysis of consequences of unrelated cases without conducting an actual phishing experiment. Experiments typically involve a small number of respondents. The novelty of present study is to analyze the educational institution' susceptibility to phishing attack. Authors demonstrate a methodology of creating a group of targets homogeneous in age, place of study, level of knowledge and to conduct an experiment on a large group of respondents (3,661 people). The methodology of gathering and filtering of email addresses using open sources of information is explained. Emotionally neutral text of a phishing email to minimize the deceptive effect of the letter was formulated. The experiment showed the success rate of the attack on a large sample of students at 10.8%, and demonstrated the vulnerability of the educational institution's infrastructure to the hidden preparation and conduct of the attack. Novelty of methodology includes use of a phishing letter that includes a questionnaire to gather statistics on responders' awareness of phishing nature. It made possible to compare respondents' beliefs with the real susceptibility to phishing based on sensitive data they provided in return to the phishing letter. We show how the data collected by phishing can be personalized and conclude that respondents need further training to detect phishing attacks. We also argue necessary organizational, infrastructural measures, recommendations of necessary mail server configuration changes.

Keywords: Attack · Fishing · Social engineering · Sensitive information · Personal information

1 Introduction

Personal data protection is vital in the open data age. Users often rely on default protection and do not take an active position on non-disclosure and storage of sensitive information. In recent years, there has been active development of data-gathering technologies, which at first glance cannot be used to the detriment of the individual.

Nevertheless, accidentally leaked personal information can be associated with such a block of impersonal data and used to de-anonymize a person's profile with all the consequences for the individual.

Z. Hu et al. (Eds.): ICCSEEA 2020, AISC 1247, pp. 668–680, 2021.
https://doi.org/10.1007/978-3-030-55506-1_59

Besides, many users unaware that the dissemination of any information about themselves and their behavior using existent information technologies can lead to negative consequences elsewhere—information can be used for further attacks by social engineering methods.

One of these methods is phishing. A person himself here plays a key role, becoming a victim of cyberattacks because of his actions.

The current study reveals principles of collection and filtering of email addresses of desired targets, phishing letter design and mailing, evaluation of results and elaboration of phishing prevention recommendations.

2 Related Works

Scientists prove that social engineering methods in general and phishing in particular rely on internal characteristics of human nature and external circumstances that are based on human weaknesses [1]. We agree with the suggestion that among others, trust, susceptibility are the factors that affect the vulnerability to email phishing [2]. Therefore, it requires a thorough examination.

The present study enhances our former research on the attack in educational institutions using fake access points and creating a phishing page [3]. Nowadays, the main task of attackers is not to cause obvious harm to the victims but to quietly and inconspicuously parasitize on their hardware [4]. Previous research has shown that users' awareness of even technical specialties is insufficient. Therefore, attention to the development of methods for raising awareness of users and technical measures reducing the number of potential attacks on objects of information activity should be paid.

Some works devoted to phishing deal with the random sample of people suffered real phishing attacks [5–8]. This restricts an opportunity to gather relevant statistics. On the other hand, the controlled phishing experiments analyzed typically involve a small number of respondents and hardly can show the tendencies [2, 9] includes respondents informed that they are part of the phishing attack study [10]. Other studies are built on models that predict tendencies, but do not include real respondents [2, 11]. Some studies did not show significant differences in the perception of social engineering methods depending on the gender, previous training of the respondents [10]. In order to evaluate this, we suggest that the sample of respondents should be enlarged.

The significance of the present study lies in the analysis of the susceptibility to the phishing attack of a large group of targets homogeneous in age, place of study, level of knowledge and to conduct an experiment on a large group of respondents. The second aspect includes gathering of centralized statistics concerning victims' self-evaluation about their ability to detect phishing.

3 Problem Statement

Among the cyber-attack methods that are directly related to the human factor, social engineering has a significant role to play. About 31% of the 2018 attacks were committed using social engineering techniques [12].

Social engineering in the context of cybersecurity is a kind of psychological influence on a person to induce him to take certain actions to disclose certain information.

Techniques of social engineering can be used for:

- Direct access to sensitive, confidential information, personal data.
- Creation by the inaction of the victim the conditions for further receipt of such information.

Collecting data about other targets, i.e. the use of influence to further work with the information system, a third party, an organization, etc. In most cases, incidents involving the use of social engineering methods are caused by uncritical or inattentive behavior, as well as by lack of experience in detecting and recognizing malicious patterns of attacker's behavior.

In general, the features of social engineering attacks are low cost, no need for special knowledge, difficulty in tracking, and the ability to carry out for a long time.

According to EY Global Information Security Survey 2018–2019 [5] phishing heads the top-10 list of most common cyberattacks, counts remarkable 22%. Cisco's analysis of the past year also revealed that phishing remains a very effective way of stealing sensitive information [6] and was one of the most common causes of cyber incidents in Europe in 2018 [7].

SANS admits [8] that in last year phishing email rates increased 250% (more than one out of every 200 emails received by users).

Phishing is closely linked to the victim's behavior, as it is determined by the victim's trust. Hence, raising awareness of this vector of cyberattacks and being able to recognize them are the main methods for reducing the effectiveness of phishing attacks. Training on information security is one of the elements of the security measures of ISO/IEC 27001 [13]. Therefore, creating the design of the experiment, we planned to examine and link the respondents' awareness with the results of the impact of social engineering methods.

While the universal methods of prevention do not exist, it is stated, that awareness and training in the field of a particular type of attack are crucial [11]. On the other hand, user education intended to be the strongest link may at the same time be the weakest one [11]. Some studies show that people, who have less knowledge in IT security are harder to get interested in security issues when they most likely are the ones needing this knowledge more [9]. Awareness training requires not only delivering information to recipients, yet also their desire to learn how to identify confidential information and protect it [1]. The development of new automated information processing systems is also one of the possible solutions [14]. Therefore, we suggest recommendations on organizational measures as a result of our experiment.

A comparative statistical analysis of Positive Technologies shows that among the surveyed categories of victims of cyberattacks for 2018, educational institutions ranked fourth (7%) [3]. Moreover, for educational institutions, social engineering was the main vector of attacks (38%). This demonstrates the need for further analysis of the vulnerability of this type of institution, as well as the development of a methodology for preventing and minimizing the harm of such attacks by conducting training with potentially vulnerable recipients.

Our work aims are to investigate the possibility of an effective phishing attack within the educational institution, to evaluate its effectiveness for further development of methods of prevention and minimization of such attacks. The object of the study is the assessment of awareness of the possibility of malicious seizure of sensitive data by phishing for the potential further use by social engineering methods. The authors plan to involve the large sample for a better understanding of existing tendencies in the target group of respondents.

The scientific novelty of the work is to evaluate the level of vulnerability of the educational infrastructure for phishing attacks, to assess the level of preparation for their detection, evaluate the success rate of attack and to formulate recommendations on existing and potential vector-level attacks, as well as measures to counter such attacks.

4 Method of Research

The key factors for a successful phishing attack are users' trust in the source of the information (the initiator of the attack), which allows obtaining credible information voluntarily, as well as an incentive to provide information and not to leave the initiator's email unanswered. These factors determined the design of the experiment.

The work was carried out solely to study the level of awareness of the respondents. Any sensitive data that was sent with the consent of users, such as the aliases of such users, was not disclosed to protect them and were used for statistical purposes stated in the questionnaire and the consent of users to this expressed when sending the data.

The authors set the following tasks:

1. To create a pool of valid email addresses of the domain of the educational institution, elaborate the method of filtering them to get the addresses of students. Such a step was necessitated by the need to prepare for mailing that would be hidden from detection. Sending a large number of emails to non-existing addresses could potentially elicit a response from the administrator of the mail domain. On the other hand, sending out all valid addresses would result in letters being sent to the teaching staff, unit employees, who would also expose unauthorized mailing. To get the large sample of respondents authors needed to collect as many email addresses for mailing as possible.
2. To carry out a mailing without explicit deception (substitution of the sender's address), without creating a fake and visually identical site, without prompting the sending of information that the respondent would not have to send under other conditions (restoration of "compromising" password, etc.). Such a design is used to

avoid detection by the automated methods of phishing detection [15] and is aimed to evaluate the susceptibility to entry-level phishing.

3. To send letters containing a questionnaire whose stated purpose was in line with the true purpose of the research—to analyze respondents' awareness of information security threats and to evaluate the results.

Respondents who responded to the email and completed the questionnaire by providing sensitive information about themselves were given a link to Google's interactive material [16] designed to improve their ability to detect phishing emails and sites.

The questionnaire letter was sent from the student's address of the educational institution's mailing domain (without any modifications of sender credentials). It was expected, that going over the link provided by a third party who is not a teacher or employee of the teaching unit should indicate the lack of attention of the respondent to security issues. The more information such a respondent sends in return, the higher the risk to become a victim of real phishing for him is expected.

The design of the letter was maintained in the most neutral style (without emotional of charged language) which is used for sending teaching materials by the educational staff (Fig. 1). The letter requested the questionnaire to be completed by the given date added to prevent the letter to be shelved.

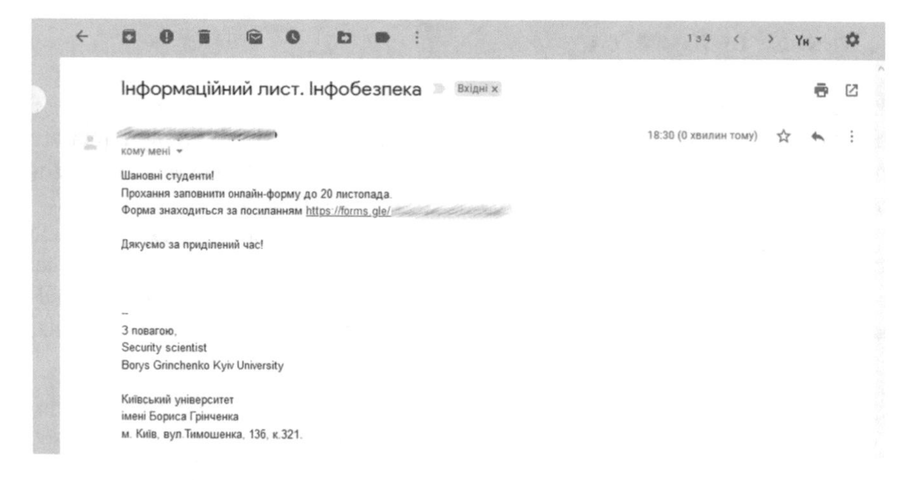

Інформаційний лист. Інфобезпека Вхідні x

18:30 (0 хвилин тому)

кому мені

Шановні студенти!
Прохання заповнити онлайн-форму до 20 листопада.
Форма знаходиться за посиланням https://forms.gle/

Дякуємо за приділений час!

--
З повагою,
Security scientist
Borys Grinchenko Kyiv University

Київський університет
імені Бориса Грінченка
м. Київ, вул.Тимошенка, 136, к.321.

Fig. 1. Sample of the letter with reference to questionnaire

The questionnaire was designed using a Google form template and contained nine questions about the essence of information security, some of which were open-ended, while others provided answer choices. Four questions were also added for further profiling of the respondents and for the possibility of validation the data entered. The questionnaire pursued two goals. First, to collect respondents' self-evaluation about their ability to detect phishing. Second, to give them the possibility to input sensitive data. The fact of sending a response to a phishing letter is considered as a success of the phishing, the amount of information provided by the respondent demonstrates the degree of success.

The list of mailing addresses was created using the technique real attacker may use. The Gmail-based corporate mail functionality enables us to get a list of addresses in the given domain containing the same combination of letters that we enter in the field of the recipient of the letter. By manual search of 676 two-letter combinations, 5,099 addresses were obtained, of which 3,933 corresponded to the student's email address template. The total number of students in the study domain is expected to be 8,783 students [17]. Thus, obtaining all student addresses would require automation of the process to run through at least 17,576 three-letter combinations that can be implemented by the interested intruder. Given facts, clearly show the necessity to rethink the policy of free access to these sensitive data.

We found that the assignment of email addresses to students and teaching staff is based on a single algorithm, which is described and placed in open access. Understanding such an algorithm allowed us to sort the emails with 100% accuracy without recourse to third parties. The algorithm also provides an ability to reveal the faculty and a year of entry of a particular student.

The examination of email addresses revealed 268 addresses of former students. The total number of sent letters is 3,665 letters, out of which there are four non-existent addresses. The emails collected were searched for duplicates. 246 addresses were identified as the assigned to students with the same surname and initials. It is very likely that students who study at two faculties at the same time have received our mails twice. The study of the specificity of their responses to received letters can also be conducted.

The response time limit for respondents was set up to 10 days. The answers collection form was open for 20 days.

It is found that the attack described in this study can be carried out from the outside of the mail domain from any address and even with the substitution of a nonexistent one. To do this, it is enough to select the names and faculty of the entrants of the desired educational institution from a national open database of entrants to educational institutions (can be accessed online) and to generate email addresses using the script and a template of email address, mentioned earlier. However, we predict less accuracy of this approach, as not all students recommended for enrollment begin their studies, also some individuals may leave the university.

The data collected is, by its nature, a collection of impersonal data. At the same time, using social engineering methods (mentioned national educational database), it is possible to uniquely identify the end respondents and use the resulting set of data to further attack using social engineering methods.

5 Research Results

The analysis of the qualitative component was carried out without further automation, whereas the quantitative indicators were calculated in Microsoft Office Excel.

In total, 400 responses were received (after which the experiment was stopped and the analysis started). By gender, women accounted for 67%, men—31%, and unknown —2%.

The total percentage of questionnaires received is 10.8% of the number of emails sent. Hence, it can be stated that more than 10% of the respondents (students) are ready to open a letter from an unknown person, go to the link from it, fill out the questionnaire providing data about themselves and send it back.

One of the questions of the questionnaire was to examine the respondents' conscious experience in dealing with certain types of cyber threats. Only 24 out of 400 respondents (7.1%) confirmed that they unintentionally sent their sensitive data to the stranger (Fig. 2).

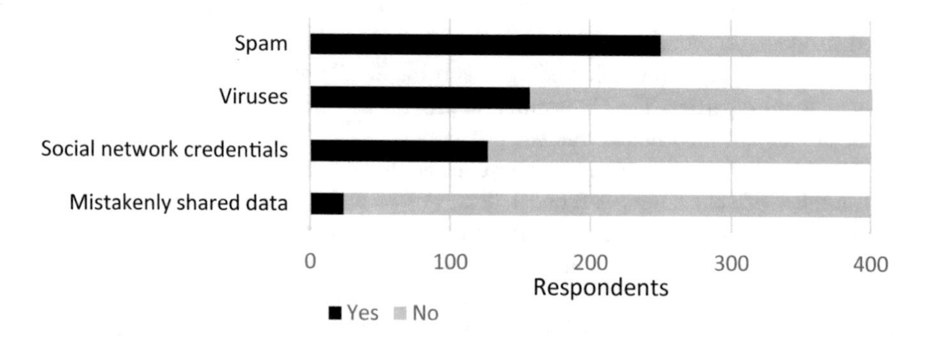

Fig. 2. Number of respondents who faced certain types of attacks.

We can assume that the majority of respondents (92.9%) who provided us their data were not aware of the fact that by filling out a questionnaire they also may send sensitive data to an unknown person.

The distribution of answers to the question of whether the respondent reads the provisions on consent to use personal data revealed the next distribution: 15% always read, 60% read from time to time, 25% never read.

Several questions of the questionnaire were based on the Likert scale. We used the Kendall W method to determine the possible concordance of respondents' answers to three questions (downloading unlicensed software, the trustworthiness of information on the Internet, user consent on personal data). Kendall's W for 391 respondents equals 0.0202 (p-value equals 0.00037). The significance level at the 0.05 shows that respondents' answers are not consistent. This may indicate that the respondents in the sample do not agree on the assessment of the block of given questions. Unfortunately, the use of open-ended questions restricts us in correlation search. On the other hand, composing a questionnaire without open questions would make phishing research impossible.

Nevertheless, we conducted a correlation analysis of the dependence of the answers to the questions already mentioned on the gender of the respondents. The Mann-Whitney U Test tool was used. For the dependence of downloading illegal software on the gender of the respondents p-value equals 0.4752 (255 females and 116 males) were obtained. This is the only significant correlation found. Other questions did not reveal statistically significant differences between the answers of the respondents of a different gender.

To conclude we assume that statistical analysis of the data obtained in our type of research requires further work. Complex instruments to obtain more data or additional surveys conducted to check the data obtained because of the initial "phishing attack" could be used.

A selective review of the answers provided showed that at least some students indicate real aliases by which their social media pages can be identified. Moreover, from these pages, it is possible to get photos, other facts about the respondents for further social engineering attacks.

Students who submitted replies to the questionnaire generally did not identify the sender as a potential attacker. In particular, 281 respondents provided their email address, 115 people provided the second email address, 326—their birthdays. We believe that the number of students who granted their sensitive data: a nickname from the social network, email address, including a second email other than corporate, date of birth is an indicator of the success of phishing.

The distribution of network students' use is shown in Fig. 3. In connection with the aliases, they provided in return to the phishing letter this data contains the probable vector of further social engineering attack.

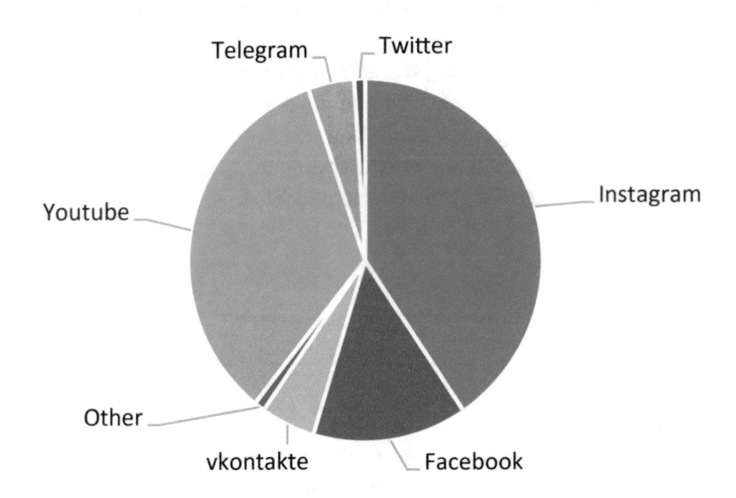

Fig. 3. Social networks usage.

At the same time, some respondents secure their sensitive data. One respondent sent a counter-notification asking whether or not to provide specific information that he did not intend to share. In addition, three students on email addresses, 11 students on a pseudonym, 23 on the date of birth answered in the form of "It doesn't matter," entered irrelevant data, etc. We include in this group those respondents who, although they sent the questionnaire but without answers to one or more questions. As shows questions about sensitive data (email, birthday versus social network respondent uses) caused more refusal to answer (Fig. 4). However, the percentage of obtained responses is still high.

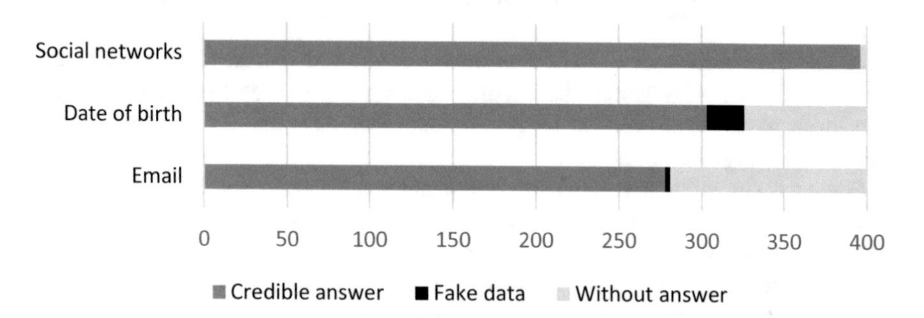

Fig. 4. Distribution of responses for sensitive and other data.

It should be noted that the person's details were still sent via an unknown link in a letter provided by an unknown person. According to the email address generation policy discussed above, even such responses can be personalized and exploited by an attacker.

Algorithms for behavioral analysis and processing behavior data have been explored and used for a long time for human behavior prediction, marketing, pattern recognition, product promoting and distribution of job responsibilities [18]. Patterns of user behavior are also important for social engineering. Therefore, we provided an analysis of the response time of a group of respondents to emails sent in a single packet (simultaneously).

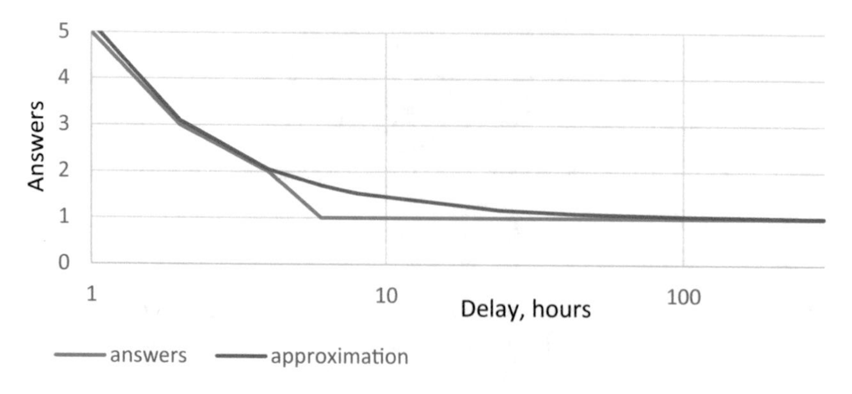

Fig. 5. Response time for emails sent simultaneously.

Questionnaires were sent in blocks of up to 200 emails at a time so that mailing is not detected as spam. The analysis of the time past from the moment of mailing to the receipt of the answers showed that the *mode* of the time of the answers lies within the first hour after the mailing. The latest replies were received within 10 days of mailing (Fig. 5). Most blocks of emails were sent between 9 AM and 10 AM on weekdays. We speculate that response time may be explained by the usage of mobile mail clients. The second factor to consider is an indication of the deadline before which the respondent's response is expected. However, this hypothesis needs further analysis.

The function of response time calculation for n respondents can be approximated to be

$$f(n) = 1 + \frac{4}{3}\pi\frac{1}{n}$$

where n is delay, hour. The observations on timings may be used either by an attacker to avoid detection and by the specialists who maintain automated spam (phishing) detection systems.

We experimentally proved that phishing to a large homogenous group of targets is a low-cost attack. The experiment did not require the involvement of third parties, special knowledge (no special software or hardware was used), could be maintained for a long period of time (the experiment lasted unnoticed for 20 days). We became aware of the single fact that the student questioned a cybersecurity department teacher regarding the nature of the mailing. This did not prevent the further continuation of the experiment and did not result in blocking of the attacker.

The current study allows us to evaluate the threat of phishing attacks to the educational institution as high enough. This confirms the conclusions drawn earlier that even after years of elaboration even an obviously suspicious email can be skipped by a spam filter and eventually opened by a computer user [4].

Our study can be summarized in the next recommendations on phishing attacks prevention in high school.

- *Infrastructural measures.* Establishing of the demilitarized zone (DMZ) to eliminate phishing outside of post domain, maintaining a firewall; mail server antivirus; introduction of anti-phishing authentication mechanisms for site access [19] as well as browser add-ons [20].
- *Organizational measures.* Increasing of students' awareness on possibilities of sensitive data extraction [21]; additional cybersecurity training; acquaintance with high school security policy and corporate network user rights; signing non-disclosure type agreement (NDA); nondisclosure of the student email generation algorithm.
- *Mail server configuring.* Limitation of address book visibility; application of role-based access control (RBAC) model to the different categories of users and as well as mail subdomain architecture; elaboration policy on user management; setting of bulk mailing limitations; automated traffic anomalies detection [22]; use of the digital signature for verifying the authenticity and for non-repudiation of mail origin.

- *Attack counteraction measures.* Spam, malware user notification; automatic blocking of suspicious users without notification; email delivery status notifications (DSNs) policy for the prevention of email scanning; real-time security administration; abnormal traffic detection [23, 24].

6 Conclusions and Further Research

The described experiment was conducted without creating artificial conditions, and therefore can be reproduced by anyone. The experiment demonstrated that the respondent might have the impression that he answers to some insignificant questions. At the same time, the array of such impersonal data can be easily linked to other data obtained from other open sources. In its aggregate, such information has some value for further attacks. The success rate of the attack on a large sample of students with the letter crafted without emotional of charged language (10.8%) demonstrated the vulnerability of the educational institution's infrastructure to the hidden preparation and conduct of the attack. The data collected by phishing can be personalized and therefore respondents need further training to detect phishing attacks.

Further development of this study may include sending letters from outside the institution, as well as from an unreliable address. Besides, the rate of the opened (read) letters can be calculated and their number correlated with the total number of responses. More detailed profiling of respondents by age, gender, a field of study can be conducted.

The general conclusion of our study is that even without stimulating responses by deception, the unknown sender can gain a large percentage of answers. Such a rate would be higher in case of masking the phishing nature of the letter under the imperative request of the administration, bank, e-services provider, etc. This, in turn, indicates the need to raise respondents' awareness of cyber-threats and establishing other infrastructural measures suggested.

Acknowledgments. This scientific work was partially supported by RAMECS and self-determined research funds of CCNU from the colleges' primary research and operation of MOE (CCNU19TS022). All experiments were conducted at Borys Grinchenko Kyiv University [25].

References

1. Fan, W., Lwakatare, K., Rong, R.: Social engineering: I-E based model of human weakness for attack and defense investigations. Int. J. Comput. Netw. Inf. Secur. (IJCNIS) **9**(1), 1–11 (2017). https://doi.org/10.5815/ijcnis.2017.01.01
2. Albladi, S.M., Weir, G.R.S.: User characteristics that influence judgment of social engineering attacks in social networks. Hum. Cent. Comput. Inf. Sci. **8**(1), 5 (2018). https://doi.org/10.1186/s13673-018-0128-7

3. Sokolov, V.Y., Kurbanmuradov, D.M.: Method of counteraction in social engineering on information activity objectives. Cybersecur. Educ. Sci. Tech. **1**, 6–16 (2018). https://doi.org/10.28925/2663-4023.2018.1.616
4. Sokolov, V.Y., Korzhenko, O.Y.: Analysis of recent attacks based on social engineering techniques. In: Dorenskyi, O.P. (ed.) All-Ukrainian Scientific and Practical Conference of Higher Education Applicants and Young Scientists "Computer Engineering and Cyber Security: Achievements and Innovations, pp. 361–363. Kropyvnytskyi (2018). https://doi.org/10.5281/zenodo.2575459
5. Kessel, P.: EY Global Information Security Survey 2018–19, London (2019)
6. Cisco: Annual Cybersecurity Report, San Jose (2018)
7. Campbell, N., Lautenbach, B.: Telstra Security Report 2018, Sydney (2018)
8. Pescatore, J.: SANS Top New Attacks and Threat Report, Swansea (2019)
9. Floderus, S., Rosenholm, L.: An Educational Experiment in Discovering Spear Phishing Attacks, Karlskrona (2019). urn:nbn:se:bth-18446
10. Broadhurst, R., et al.: Phishing and cybercrime risks in a university student community. Int. J. Cybersecur. Intell. Cybercrim. **2**(1), 4–23 (2019). https://doi.org/10.2139/ssrn.3176319
11. Chaudhry, J.A., Chaudhry, S.A., Rittenhouse, R.G.: Phishing attacks and defenses. Int. J. Secur. Appl. (IJSIA) **10**(1), 247–256 (2016). https://doi.org/10.14257/ijsia.2016.10.1.23
12. Positive Technologies: Cybersecurity Threatscape 2018. Trends and Forecasts, London (2019)
13. Funk, G., et al.: Implementation Guideline ISO/IEC 27001:2013. A practical guideline for implementing an ISMS in accordance with the international standard ISO/IEC 27001:2013, Berlin (2017)
14. Gururaj, H.L., Swathi, B.H., Ramesh, B.: Threats, consequences and issues of various attacks on online social networks. Int. J. Educ. Manag. Eng. (IJEME) **8**(4), 50–60 (2018). https://doi.org/10.5815/ijeme.2018.04.05
15. Zareapoor, M., Seeja, K.R.: Feature extraction or feature selection for text classification: a case study on phishing email detection. Int. J. Inf. Eng. Electron. Bus. (IJIEEB) **7**(2), 60–65 (2015). https://doi.org/10.5815/ijieeb.2015.02.08
16. Can you spot when you're being phished? (2018). https://phishingquiz.withgoogle.com. Accessed 28 Nov 2019
17. Statistical portrait (2019). http://kubg.edu.ua/prouniversitet/vizytivka/statistichni-portret.html. Accessed 28 Nov 2019. (Publication in Ukrainian)
18. Rahman, M.M.: Mining social data to extract intellectual knowledge. Int. J. Intell. Syst. Appl. (IJISA) **4**(10), 15–24 (2012). https://doi.org/10.5815/ijisa.2012.10.02
19. Wang, B., Wei, Y., Yang, Y., Han, J.: Design and implementation of anti-phishing authentication system. Int. J. Wirel. Microwave Technol. (IJWMT) **1**(6), 38–45 (2011). https://doi.org/10.5815/ijwmt.2011.06.06
20. Gupta, R., Shukla, P.K.: Experimental analysis of browser based novel anti-phishing system tool at educational level. Int. J. Inf. Technol. Comput. Sci. (IJITCS) **8**(2), 78–84 (2016). https://doi.org/10.5815/ijitcs.2016.02.10
21. Yin, M., Luo, J., Cao, D., et al.: User name alias extraction in emails. Int. J. Image Graph. Sig. Process. (IJIGSP) **3**, 1–9 (2011). https://doi.org/10.5815/ijigsp.2011.03.01
22. Nasr, A.A., Ezz, M.M., Abdulmaged, M.Z.: An intrusion detection and prevention system based on automatic learning of traffic anomalies. Int. J. Comput. Netw. Inf. Secur. (IJCNIS) **1**, 53–60 (2016). https://doi.org/10.5815/ijcnis.2016.01.07

23. Bernacki, J., Kołaczek, G.: Anomaly detection in network traffic using selected methods of time series analysis. Int. J. Comput. Netw. Inf. Secur. (IJCNIS) **9**, 10–18 (2015). https://doi.org/10.5815/ijcnis.2015.09.02
24. Abdulhamid, S.M., Shuaib, M., Osho, O., et al.: Comparative analysis of classification algorithms for email spam detection. Int. J. Comput. Netw. Inf. Secur. (IJCNIS) **1**, 60–67 (2018). https://doi.org/10.5815/ijcnis.2018.01.07
25. Buriachok, V., Sokolov, V.: Implementation of active learning in the master's program on cybersecurity. In: Hu, Z., Petoukhov, S., Dychka, I., He, M. (eds) Advances in Computer Science for Engineering and Education II. ICCSEEA 2019. Advances in Intelligent Systems and Computing, vol. 938, pp. 610–624. Springer, Cham (2020). https://doi.org/10.1007/978-3-030-16621-2_57

Author Index

Z. Hu et al. (Eds.): ICCSEEA 2020, AISC 1247, pp. 681–683, 2021.
https://doi.org/10.1007/978-3-030-55506-1

Printed in the United States
By Bookmasters